THE ROLE OF MICROTUBULES IN CELL BIOLOGY, NEUROBIOLOGY, AND ONCOLOGY

CANCER DRUG DISCOVERY AND DEVELOPMENT

BEVERLY A. TEICHER, PhD

SERIES EDITOR

THE ROLE OF MICROTUBULES IN CELL BIOLOGY, NEUROBIOLOGY, AND ONCOLOGY

Edited by

TITO FOJO, MD, PhD

Medical Oncology Branch, Center for Cancer Research,
National Cancer Institute, Bethesda, MD

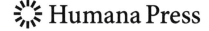 Humana Press

Editor
Tito Fojo, MD, PhD
Medical Oncology Branch
Center for Cancer Research
National Cancer Institute
Bethesda, MD

Series Editor
Beverly A. Teicher, PhD
Department of Oncology Research
Genzyme Corporation
Framingham, MA

ISBN: 978-1-58829-294-0   e-ISBN: 978-1-59745-336-3

Library of Congress Control Number: 2008920860

Cover Illustration: Figure 2B, Chapter 4, "Microtubule-Associated Proteins and Microtubule-Interacting Proteins: *Regulators of Microtubule Dynamics*," by Maria Kavallaris, Sima Don, and Nicole M. Verrills.

Printed on acid-free paper

9 8 7 6 5 4 3 2 1

springer.com

This book is dedicated to the memory of Dr. George A. Orr, a dear friend and colleague. George was a talented and creative scientist who set very high standards for himself and expected the same from his collaborators. He loved to share concepts and ideas, and many of us had the privilege of interacting and collaborating with him. He inspired and encouraged many young scientists and stimulated them to do their best work. About his own accomplishments, he was extremely humble.

George did his undergraduate and PhD studies at Queen's University, Belfast, N. Ireland. His postdoctoral work was done with Dr. Jeremy Knowles at the University of Oxford and then at Harvard University. George came to the Albert Einstein College of Medicine as an assistant professor in the Department of Molecular Pharmacology in 1978 and rose through the ranks to become a full professor in 1989.

George had many scientific interests but he was particularly dedicated to the development and application of new technologies to enhance our insights into the mode of action of drugs. His participation in the field of microtubule pharmacology and proteomics has opened new avenues of research to all of us. He is sorely missed.

v

PREFACE

I want to thank all who contributed to this first edition for their hard work and professionalism, and especially for their patience. I hope the readers will find this volume as helpful as I have found it.

There is no doubt that the family of proteins we call the tubulins and the microtubules that they form when they aggregate are extremely important in the cell and, as we are increasingly learning, important in diseases that afflict so many. This field of investigation is a testament to how important both basic and clinical sciences are in understanding disease mechanisms and making inroads into therapies. Without the basic science knowledge that has been accumulated, to which the authors of this work have contributed greatly, we would not be in the position we find ourselves of increasingly understanding disease and advancing therapies. As I read the chapters, I was humbled to think of the insights that so many have contributed to this field, and again became aware of how the collaborative effort of so many is needed to understand the complexities of nature. By working together, many have helped to advance this field. Because of their efforts, we find ourselves with the wealth of knowledge contained in this book. This knowledge gives us so much insight even as it challenges us to continue working. Thanks again to all of the wonderful collaborators for their excellence and their patience.

Tito Fojo, MD, PhD

CONTENTS

CONTRIBUTORS

SUSAN L. BANE, PhD • *Department of Chemistry, State University of New York at Binghamton, Binghamton, NY*

ASOK BANERJEE, PhD • *Department of Biochemistry, University of Texas Health Science Center, San Antonio, TX*

MURALIDHAR BEERAM, MD • *Division of Hematology Medical Oncology, Department of Medicine, University of Texas Health Science Center, San Antonio, TX*

KATIUSCIA BONEZZI, PhD • *Laboratory of Biology and Treatment of Metastasis, Department of Oncology, Mario Negri Institute for Pharmacological Research, Milan, Italy*

DIANE BRAGUER, PhD • *Formation de Recherche en Evolution-Centre National de la Recherche Scientifique 2737, Faculté de Pharmacie, Université de la Méditerranée, Marseille, France*

LUC BUÉE, PhD • *INSERM U815, Lille Cedex, France*

JANIS BUNKER • *Department of Neurology, University of California, San Francisco, CA*

FERNANDO CABRAL, PhD • *Department of Integrative Biology and Pharmacology, University of Texas-Houston Medical School, Houston, TX*

MANON CARRÉ • *UMR-Centre National de la Recherche Scientifique 2737, Faculté de Pharmacie, Université de la Méditerranée, Marseille, France*

A. DIMITRIOS COLEVAS, MD • *Stanford University School of Medicine, Stanford, CA*

JOHN J. CORREIA, PhD • *Department of Biochemistry, University of Mississippi Medical Center, Jackson, MS*

ANDRÉ DELACOURTE, PhD • *INSERM Research Center 815, INSERM U422, Lille Cedex, France*

SIMA DON • *Children's Cancer Institute Australia for Medical Research, Randwick, New South Wales, Australia*

KENNETH H. DOWNING, PhD • *Life Sciences Division, Lawrence Berkeley National Laboratory, Berkeley, CA*

STUART C. FEINSTEIN, PhD • *Neuroscience Research Institute and Department of Molecular, Cellular and Developmental Biology, University of California, Santa Barbara, CA*

TITO FOJO, MD, PhD • *Medical Oncology Branch, Center for Cancer Research, National Cancer Institute, Bethesda, MD*

PARASKEVI GIANNAKAKOU, PhD • *Department of Medicine, Division of Hematology and Oncology, Weill Medical College of Cornell University, New York, NY*

RAFFAELLA GIAVAZZI, PhD • *Laboratory of Biology and Treatment of Metastasis, Department of Oncology, Mario Negri Institute for Pharmacological Research, Milan, Italy*

LEE M. GREENBERGER, PhD • *Enzon Pharmaceuticals, Inc., Piscataway, NJ*

ERNEST HAMEL, MD • *Toxicology and Pharmacology Branch, Developmental Therapeutics Program, National Cancer Institute at Frederick, National Institutes of Health, NCI-Frederick, Frederick, MD*

RUTH HOGUE ANGELETTI, PhD • *Laboratory for Macromolecular Analysis and Proteomics, Albert Einstein College of Medicine, Bronx, NY*

SUSAN BAND HORWITZ, PhD • *Department of Molecular Pharmacology, Albert Einstein College of Medicine, Bronx, NY*

MARY ANN JORDAN, PhD • *Department of Molecular, Cellular, and Developmental Biology, University of California, Santa Barbara, Santa Barbara, CA*

MARIA KAVALLARIS, PhD • *Pharmacoproteomics Program, Children's Cancer Institute Australia for Medical Research, Randwick, New South Wales, Australia*

SHARON LOBERT, RN, MSN, PhD • *Department of Biochemistry and School of Nursing, University of Mississippi Medical Center, Jackson, MS*

FRANK LOGANZO, PhD • *Wyeth Pharmaceuticals, Oncology Research, Pearl River, NY*

RICHARD F. LUDUEÑA, PhD • *Department of Biochemistry, University of Texas Health Science Center, San Antonio, TX*

EVA NOGALES, PhD • *Howard Hughes Medical Institute, Molecular and Cell Biology Department, University of California, Berkeley, CA; and Life Sciences Division, Lawrence Berkeley National Laboratory, Berkeley, CA*

GEORGE A. ORR, PhD (deceased) • *Department of Molecular Pharmacology, Albert Einstein College of Medicine, Bronx, NY*

DAN L. SACKETT, PhD • *Laboratory of Integrative and Medical Biophysics, National Institute of Child Health and Human Development, National Institutes of Health, Bethesda, MD*

ORIT SCHARF, PhD • *PSI International, Inc., Fairfax, VA*

NICOLAS SERGEANT, PhD • *INSERM U422, Lille Cedex, France*

JAMES P. SNYDER, PhD • *Department of Chemistry, Emory University, Atlanta, GA*

CHRIS H. TAKIMOTO, MD, PhD, FACP • *South Texas Accelerated Research Therapeutics (START), San Antonio, TX*

GIULIA TARABOLETTI, PhD • *Laboratory of Biology and Treatment of Metastasis, Department of Oncology, Mario Negri Institute for Pharmacological Research, Milan, Italy*

PASCAL VERDIER-PINARD, PhD • *Department of Molecular Pharmacology, Albert Einstein College of Medicine, Bronx, NY*

LYUDMILA A. VERESHCHAGINA, PhD • *PSI International, Inc., Fairfax, VA*

NICOLE M. VERRILLS, PhD • *School of Biomedical Sciences, University of Newcastle, Callaghan, New South Wales, Australia*

FANG WANG, PhD • *Laboratory for Macromolecular Analysis and Proteomics, Albert Einstein College of Medicine, Bronx, NY*

LESLIE WILSON, PhD • *Department of Molecular, Cellular, and Developmental Biology, University of California, Santa Barbara, Santa Barbara, CA*

LIST OF COLOR PLATES

The images listed below appear in the color insert that follows page 478.

Companion CD-ROM

The CD-ROM that accompanies this book contains all of the images that appear in color in the book's insert.

1

An Overview of Compounds That Interact with Tubulin and Their Effects on Microtubule Assembly

Ernest Hamel

CONTENTS

SUMMARY

Over the last quarter of a century new classes of compounds that interfere with cell division as a result of binding to tubulin αβ-dimers, oligomers, or polymers have been described with seemingly ever-greater frequency. The cytological hallmark of tubulin interactive agents is the accumulation in drug-treated cell cultures of a high proportion of cells that appear to be arrested in mitosis. These cells have condensed chromosomes, no nuclear membrane, and a deformed or absent mitotic spindle. Flow cytometry for DNA content demonstrates large numbers of tetraploid (G2/M) or even octaploid cells.

Key Words: Tubulin; taxoid site; laulimalide site; tubulin alkylation; colchicine site; vincer domain.

1. INTRODUCTION

Antimitotic agents vary widely in molecular structure, even among compounds that appear to interact in the same binding region of tubulin, and represent natural products obtained from a wide variety of organisms, synthetic compounds, and synthetic analogs of the former. Antitubulin compounds range in complexity from the simple sulfhydryl reagent 2,4-dichlorobenzyl thiocyante (MW, 204) to the macrocyclic polyether halichondrins and spongistatins (MWs > 1100). The current interest in drugs that interact with tubulin is a result in large part of the clinical importance of the taxoids paclitaxel and docetaxel as anticancer agents *(1,2)*, but also to the possibility that additional agents might overcome resistance to antimitotic therapy, have reduced toxicity in patients,

From: *Cancer Drug Discovery and Development: The Role of Microtubules in Cell Biology, Neurobiology, and Oncology* Edited by: Tito Fojo © Humana Press, Totowa, NJ

and/or that antitubulin drugs used in combination might have synergistic clinical benefit, as has been repeatedly observed in cell culture studies (3–17). Moreover, drugs that interact with tubulin are useful in the treatment of parasitic diseases of humans and animals (18), and colchicine remains important as a treatment for familial Mediterranean fever and occasionally for other inflammatory conditions (19).

2. COMPOUNDS THAT INDUCE TUBULIN ASSEMBLY BY BINDING AT THE TAXOID OR LAULIMALIDE SITES

Currently, there are at least five well-described modes by which antimitotic agents interact with tubulin. In the first two modes, the compound is believed to bind with high affinity to polymerized tubulin vs the $\alpha\beta$-tubulin heterodimer. There are at least two polymer-based binding sites, to which drugs can bind, and this discussion will be restricted to the lead compounds, all of them natural products, with substantial cytotoxic activity (arbitrarily defined as yielding reported IC50s less than 50 nM). There are now eight such natural products.

The best described of the two binding sites is that to which paclitaxel (20) binds (21). Besides paclitaxel, natural products known to bind to the taxoid site are epothilones A and B (22,23), discodermolide (24,25), eleutherobin (26,27), cyclostreptin (also known as FR182877) (28–31), and dictyostatin 1 (32–34). The structures of these compounds are shown in Fig. 1. The first publication describing the assembly promoting properties of paclitaxel appeared in 1979 (20), and it was not until 1995 that a paper describing additional assembly inducing compounds, epothilones A and B, appeared (22).

The assembly promoting properties of laulimalide (structure in Fig. 2) were initially described in 1999 (35). Laulimalide has been shown to bind to a different site than the taxoid site on tubulin polymer (36). The evidence for a different site for laulimalide is its failure to inhibit the binding of taxoids to tubulin polymers, its stoichiometric incorporation with paclitaxel into polymer (36), and its ability to act synergistically with paclitaxel in promoting tubulin assembly (37). A second assembly promoting compound, peloruside A (38) (structure shown in Fig. 2), has been found to bind at the same site as laulimalide. Peloruside A does not inhibit taxoid binding to polymer but does inhibit laulimalide binding (39).

These eight natural products lead to the hyperassembly of microtubules of enhanced stability in cell free reaction mixtures. Typically, assembly reactions can occur under conditions where there is normally little or no tubulin polymerization: at reduced temperatures, in the absence of GTP, in the absence of microtubule-associated proteins or another assembly promoting reaction component (such as glycerol or glutamate), and at reduced tubulin concentrations. These properties indicate hypernucleation of assembly, and consistent with this conclusion is the routine observation that the microtubules formed with drug are shorter and more abundant than microtubules formed without drug. Microtubules formed in the presence of drug are also resistant to disassembly induced by cold temperatures, dilution, or Ca^{2+}. In all cases studied thus far, assembly promoting compounds bind to tubulin polymer stoichiometrically (i.e., in an amount equivalent to the tubulin content of the polymer; for e.g., see ref. 36). In cells, assembly promoting drugs cause formation of abnormal spindles in mitotic cells (those with condensed chromosomes), whereas in interphase cells the drugs cause increased numbers of misorganized microtubules that are typically shorter and bundled as compared with the microtubule cytoskeleton of control cells.

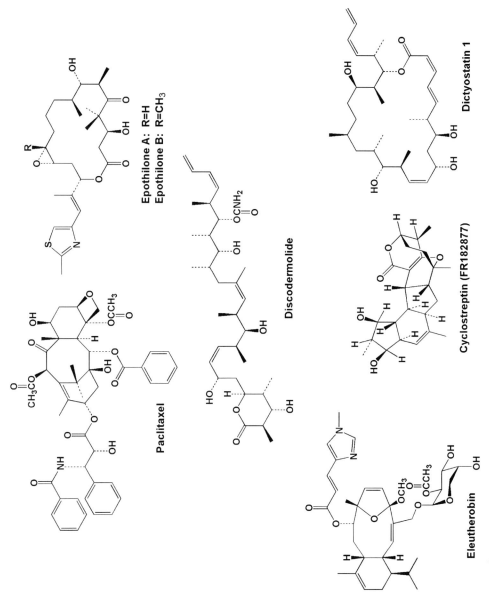

Fig. 1. Structures of paclitaxel, epothilones A and B, discodermolide, eleutherobin, cyclostreptin, and dictyostatin 1. The configurations of the chiral center, shown for dictyostati are as proposed by Paterson et al. (*33,104*).

3

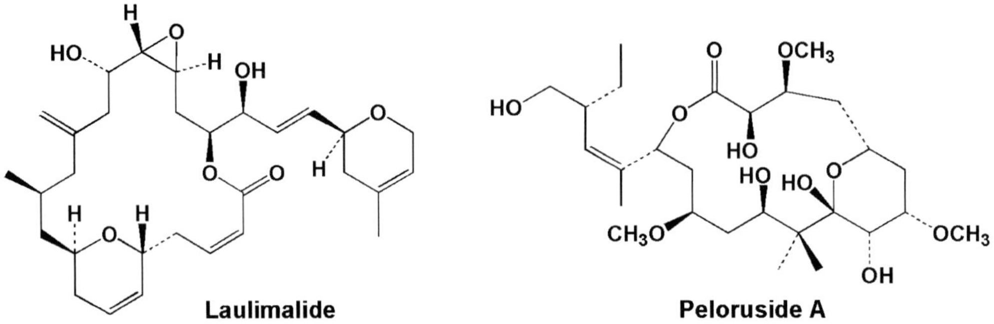

Fig. 2. Structures of laulimalide and peloruside A.

3. COMPOUNDS THAT INHIBIT TUBULIN ASSEMBLY BY ALKYLATING β-TUBULIN OR BY BINDING IN THE COLCHICINE SITE OR THE VINCA DOMAIN

Drugs interacting with tubulin by the other three well-described mechanisms all inhibit microtubule assembly. These agents either alkylate tubulin amino acid residues, bind in the colchicine site, or bind in the vinca domain. As a general rule, in cells treated with compounds that inhibit tubulin assembly, microtubules disappear. At lower drug concentrations, mitotic cells display abnormal spindles, whereas at higher concentrations there is no spindle at all. The microtubule network in interphase cells becomes progressively sparser until it completely vanishes as the drug concentration increases.

Highly specific covalent bond formation with tubulin amino acid residues, generally cysteines, occurs with at least four compounds that arrest cells in mitosis (structures in Fig. 3). Alkylation of Cys-239 of β-tubulin alone causes loss of tubulin's ability to polymerize. The covalent interaction of 2,4-dichlorobenzyl thiocyanate with tubulin occurs at multiple cysteine residues, but Cys-239 of β-tubulin is the most reactive *(40)*. The reaction with Cys-239 eliminates the ability of tubulin to assemble into polymer but has much less effect on the ability of tubulin to bind either colchicine or GTP. The IC_{50} of 2,4-dichlorobenzyl thiocyanate for inhibition of growth of two cell lines was in the 200–500 n*M* range *(41)*.

Subsequently, 2-fluoro-1-methoxy-4-pentafluorophenylsulfonamidobenzene (T138067) *(42)* was shown to react exclusively with Cys-239. T138067, which has structural analogy to colchicine site drugs, inhibits the binding of colchicine (structure shown in Fig. 4) to tubulin, and colchicine inhibits the covalent interaction of T138067 with β-tubulin. The IC_{50}s of T138067 with a number of cell lines were in the 10–50 n*M* range.

Originally, 4-*tert*-butyl-[3-(2-chloroethyl)ureido]benzene, an arylchloroethylurea, was also thought to react exclusively with Cys-239, and 4-*tert*-butyl-[3-(2-chloroethyl)ureido]benzene was cytotoxic toward several cell lines in the low micromolar range *(43)*. Newer analogs *(44)* have somewhat improved cytotoxicity and alkylate β-tubulin in cells more rapidly. However, subsequent work has shown that this class of drugs alkylates intracellular tubulin at Glu-198 of β-tubulin *(45)*.

Finally, the natural product ottelione A (RPR112378) *(46)* inhibits tubulin polymerization and reacts with a single tubulin cysteine residue, which has thus far not been identified. Ottelione A also inhibits colchicine binding to tubulin and is a structural analog

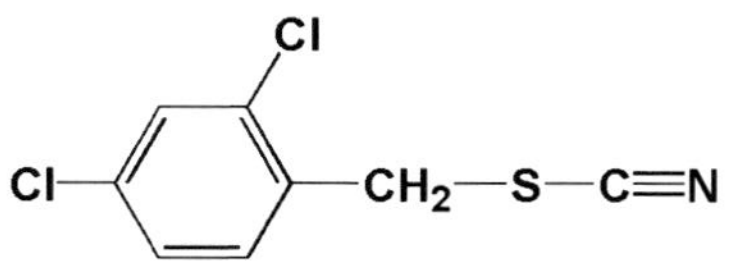

2-4-dichlorobenzyl thiocyanate

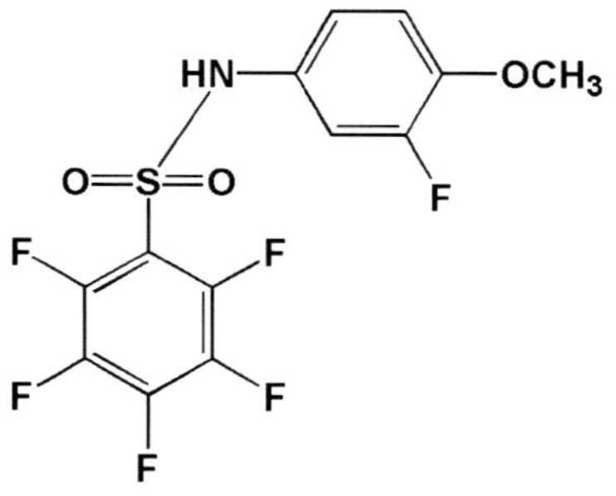

2-fluoro-1-methoxy-4-pentafluorophenylsulfonamidobenzene

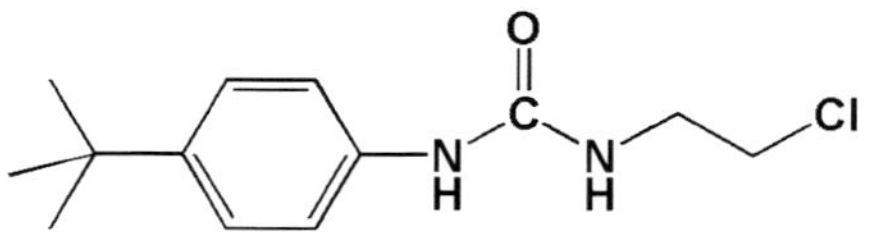

4-*tert*-butyl-[3-(2-chloroethyl)ureido]benzene

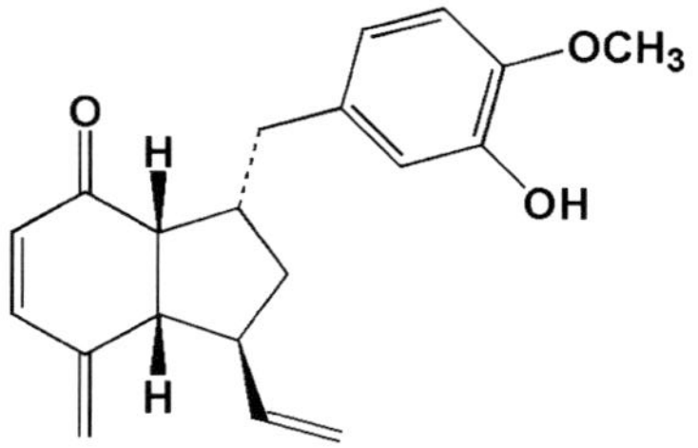

Ottelione A

Fig. 3. Structures of antimitotic compounds that alkylate tubulin sulfhydryl group(s).

5

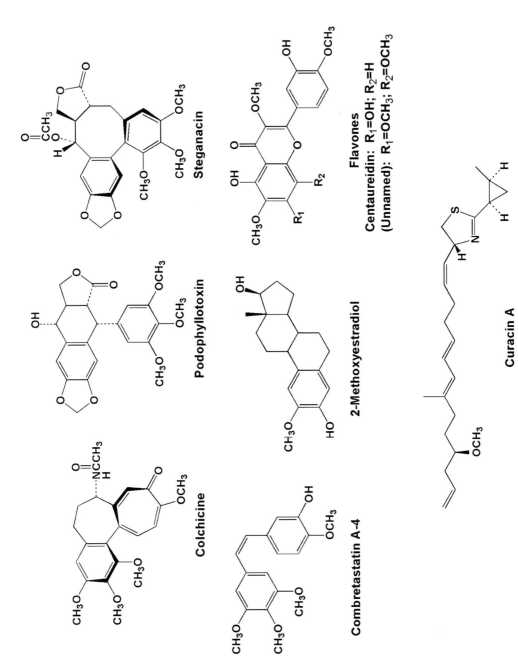

Fig. 4. Structures of natural products that bind in the colchicine site. The diagram of colchicine shows the (aS, 7S) configuration proposed by Brossi and his collaborators (105,106).

6

of colchicine site drugs. It is the most cytotoxic of the sulfhydryl reactive drugs, with an IC_{50} in human epithelial (KB) cells of 20 pM.

Unlike T138067 and ottelione A, most antimitotic drugs that inhibit colchicine binding to tubulin do not alkylate the protein. They seem to bind at a common site on β-tubulin close to the interface between the α- and β-subunits (47), and colchicine analogs with chloroacetyl groups attached to the A ring alkylate both Cys-239 and Cys-354 of β-tubulin (48,49). These compounds tend to be relatively simple structurally, compared with those inducing assembly or binding in the vinca domain, but are nevertheless structurally diverse. They include both natural products such as podophyllotoxin (50), steganacin (51), the combretastatins (52), 2-methoxyestradiol (53), flavones (54–56), and the curacins (57) (examples are shown in Fig. 4) and a wide variety of synthetic compounds, such as carbamates (58–60), heterocyclic ketones (61), and benzoylphenylureas (62). The examples of synthetic colchicine site compounds shown in Fig. 5 are only a small sample of the number of agents that have been reported. As inhibitors of cell growth, colchicine site drugs generally yield IC_{50}s in the low nanomolar to midmicromolar range.

Besides binding to unassembled αβ-tubulin dimer, colchicine, and assumedly other drugs that bind in the colchicine site, has a limited ability to enter microtubules. Small amounts of the tubulin–colchicine complex will copolymerize with tubulin free of drug, and the colchicine remains bound to the microtubules (63,64). This would indicate that the colchicine site on tubulin is not entirely masked or obliterated in the polymer.

The only colchicine site ligand readily available in a radiolabeled form is colchicine itself. The binding reaction of colchicine to tubulin has a number of unusual properties (65) that most other drugs that bind at the colchicine site do not share. The binding interaction does not readily occur on ice, and becomes progressively more rapid at temperatures in the 20–40°C range. Once bound, the dissociation of colchicine from tubulin is extremely slow, so that the binding is sometimes described, incorrectly, as irreversible. In contrast, most of the colchicine site drugs with which the author's laboratory has worked bind readily to tubulin in the cold and readily dissociate. Thus, the extent of inhibition observed can be highly variable, depending not only on reaction time and temperature, but more importantly on the relative binding and dissociation rates of colchicine and the potential inhibitor. In at least two cases (66,67), the author's laboratory initially concluded that a new class of synthetic molecules bound at a "new" site on tubulin, only to find that modification of reaction conditions (68) or synthesis of stronger analogs (unpublished data) revealed that the agents bound weakly to the colchicine site.

The last group of drugs with a relatively well-defined mechanism for interacting with tubulin bind in a region described as the "vinca domain." Almost all of these compounds inhibit the binding of vinblastine and vincristine to tubulin, but different inhibitory patterns are obtained when data are evaluated by the "classic" methods used in enzyme kinetics (Lineweaver-Burk and Hanes analyses). Both competitive and noncompetitive patterns occur. Thus, the author proposed that the vinca domain contains both the vinca site, where the competitive inhibitors bind, and nearby sites, where noncompetitive inhibitors bind. The strong inhibition observed with many noncompetitive inhibitors was postulated to derive from their interfering sterically with the vinca-binding site as a result of the close proximity of the different binding sites (69).

Drugs that bind in the vinca site itself are shown in Fig. 6. These include (1) the vinca alkaloids themselves, exemplified by vinblastine and vincristine, which competitively

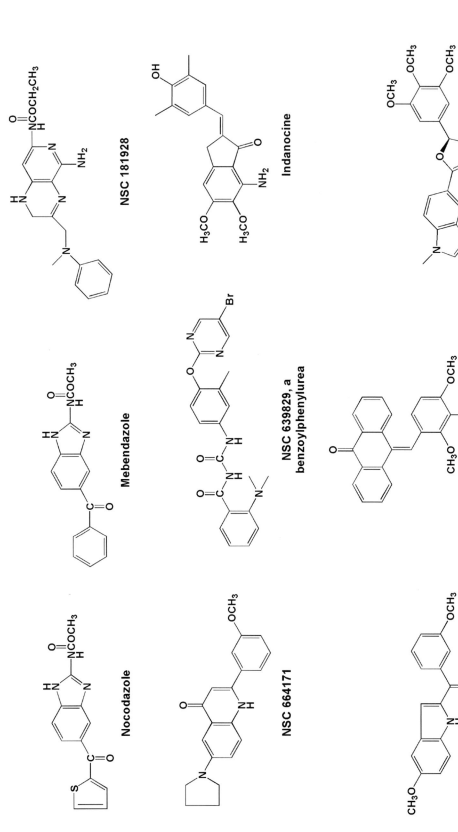

Fig. 5. Structures of synthetic compounds that inhibit the binding of colchicine to tubulin and thus probably bind in the colchicine site. Examples include the carbamates nocodazole, mebendazole, and NSC 181928 (first row); a "second generation" benzoylphenylurea NSC 639829 in clinical trials, and indanocine (*107*) (middle row); a benzylidene-9(10*H*)-anthracenone (*108*), a 2-aroylindole derivative (*107*) (middle row); a benzylidene-9(10*H*)-anthracenone derivative-...and a indolyloxazoline derivative (*110*) (bottom row).

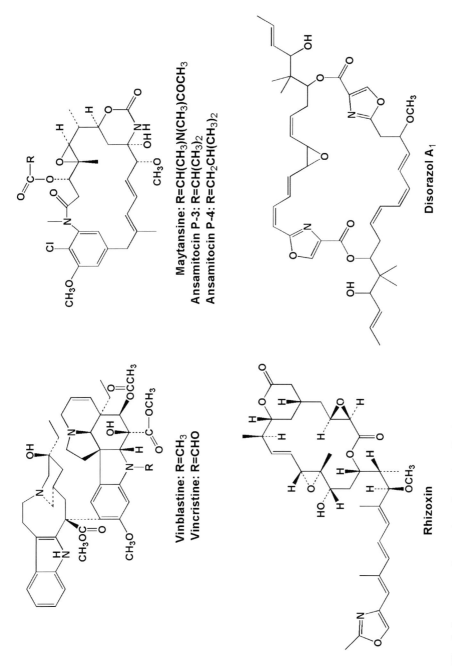

Vinblastine: R=CH₃
Vincristine: R=CHO

Maytansine: R=CH(CH₃)N(CH₃)COCH₃
Ansamitocin P-3: R=CH(CH₃)₂
Ansamitocin P-4: R=CH₂CH(CH₃)₂

Rhizoxin

Disorazol A₁

Fig. 6. Structures of natural products described as competitive inhibitors of the binding of vinblastine and/or vincristine to tubulin.

9

inhibit each other's binding to tubulin *(69,70)*; (2) the maytansinoids, exemplified by the plant product maytansine *(69,71,72)* and the fermentation products ansamitocins P-3 and P-4 *(73)*; (3) the fungal macrolide rhizoxin *(69)*; and (4) the myxobacterial macrocycle disorazol A$_1$ *(74–76)*. It should be noted that, in contrast to the competitive inhibition the author has observed with maytansine and rhizoxin vs vincristine or vinblastine *(69,72)*, Takahashi et al. *(77)* interpreted their binding data as indicating maytansinoids and rhizoxin bound at a common site on tubulin distinct from the vinca alkaloid-binding site. IC$_{50}$s of these compounds against a variety of cell lines range from 0.5 to 1 n*M* for maytansine to 20–40 n*M* for vinblastine. There is evidence that the vinca site, like the colchicine site, persists to a limited extent in microtubules, both at ends of microtubules and along their length. Binding of vinblastine to microtubules has been demonstrated to occur at a restricted number of αβ-tubulin dimers in the polymer *(78,79)*. The vinblastine site has recently been shown to be formed not by a single αβ-tubulin heterodimer but by the α-subunit of one heterodimer and the β-subunit of a second dimer *(80)*, consistent with the isodesmic tubulin assembly reaction induced by the drug *(81)* and with an early photoaffinity labeling study *(82)*.

Noncompetitive inhibitors of vinca alkaloid binding to tubulin can be viewed as falling into two structural classes, macrocyclic polyethers and peptides/depsiptides. The former group is the smaller and consists of two families of complex molecules, the halichondrins *(72,83)* and the spongistatins *(70,84)*, the latter also known as the altohytrins *(85)*. The most thoroughly studied member of each of these families in terms of interactions with tubulin are halichondrin B *(72)* and spongistatin 1 *(70,84)*. These compounds are shown in Fig. 7, together with a simpler analog of halichondrin B, NSC 707389 *(86)*, which is presently in clinical trials as an anticancer agent. Both halichondrin B and spongistatin 1 inhibit the binding of radiolabeled vinblastine to tubulin in a noncompetitive manner *(70,72)*. In addition, spongistatin 1 is a noncompetitive inhibitor of the binding of one of the peptide antimitotics, dolastatin 10, to tubulin *(70)*. As dolastatin 10 is itself a noncompetitive inhibitor of vinca alkaloid binding to tubulin *(69)*, this suggests that the vinca domain may consist of at least three distinct drug-binding regions. The macrocyclic polyethers are highly cytotoxic, with low picomolar IC$_{50}$s obtained with the most potent spongistatins *(70)*.

The larger structural class of compounds that probably bind in the vinca domain is a structurally varied group of antimitotic peptides and depsipeptides. These molecules are all natural products, and their common feature is that they all contain highly modified amino acid residues. They are derived from a wide variety of organisms, and they display a wide range of cytotoxic activity (low micromolar IC$_{50}$s obtained with the phomopsins and ustiloxins; low picomolar IC$_{50}$s obtained with the most active cryptophycins and tubulysins—*see* ref. 87 for additional details). The peptides and depsipeptides will be divided here into two groups, those that have been shown to inhibit vinca alkaloid binding (Fig. 8) and those that have not (Fig. 9).

Strong inhibition of vinblastine, vincristine, and/or rhizoxin binding to tubulin has been observed with the fungus-derived agents phomopsin A *(88)* and ustiloxin A *(89)*; with the marine agents dolastatin 10 *(69)*, hemiasterlin *(90)*, and vitilevuamide *(91)*; with the cyanobacterial agent cryptophycin 1 *(92–94)*; and with the myxobacterial agent tubulysin A *(95)*. The related compound, tubulysin D, is more cytotoxic than tubulysin A *(96)*, and the absolute configuration of the former compound was recently established *(97)*.

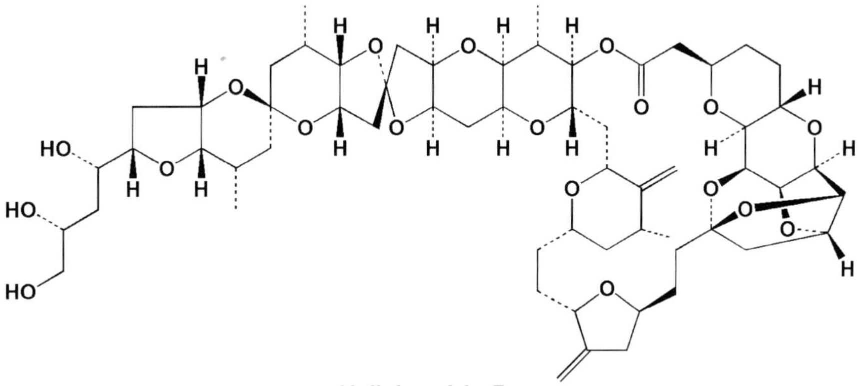

Halichondrin B

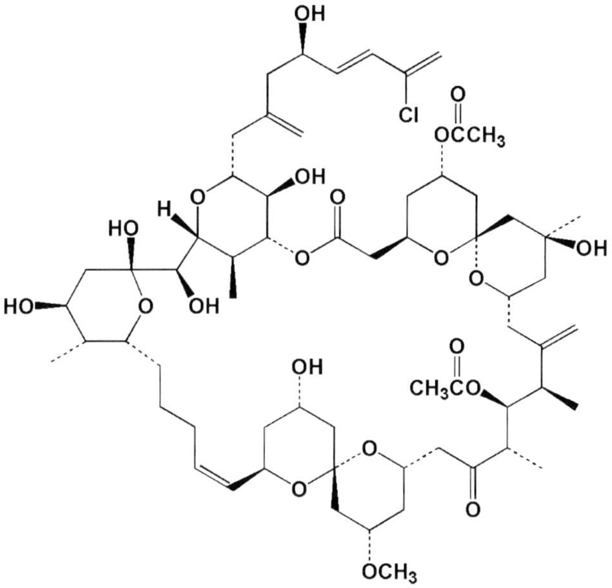

Spongistatin 1

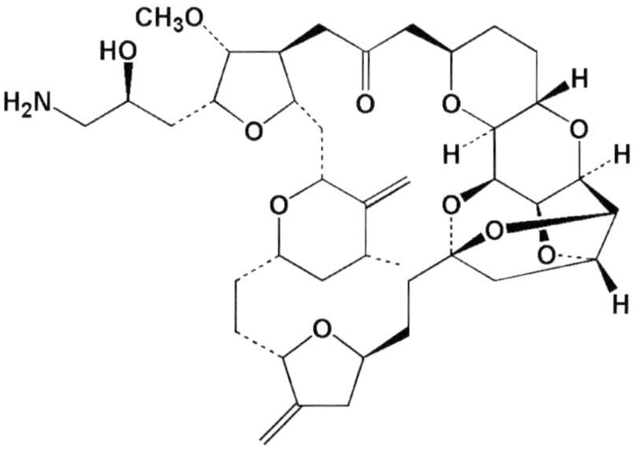

NSC 707389

Fig. 7. Structures of macrocyclic polyethers that inhibit the binding of vinblastine to tubulin.

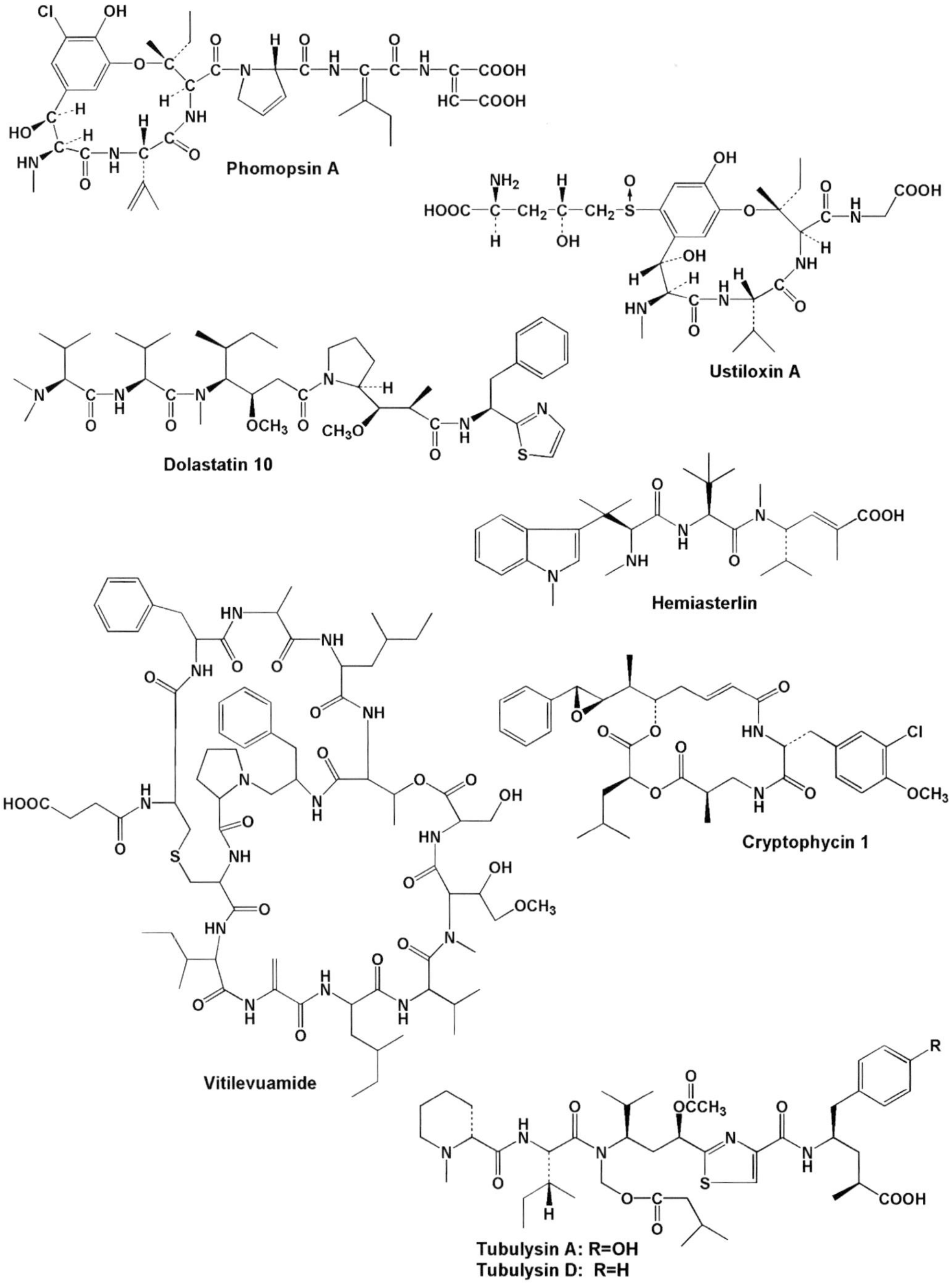

Fig. 8. Structures of peptides and depsipeptides that inhibit the binding of vinblastine and/or vin-cristine to tubulin. The configurations of the chiral centers for tubulysins A and D are as proposed by Höfle et al. *(97)* for tubulysin D.

For the structures shown in Fig. 8, the author has assumed that the chiral centers are identical in tubulysins A and D. Noncompetitive inhibition of vinblastine or vincristine binding by Lineweaver-Burk analysis or by the related Hanes analysis has been shown for phomopsin A *(69)*, dolastatin 10 *(69)* and a synthetic analog *(98)*, hemiasterlin *(90)*, vitilevuamide *(91)*, and cryptophycin 1 *(93)*.

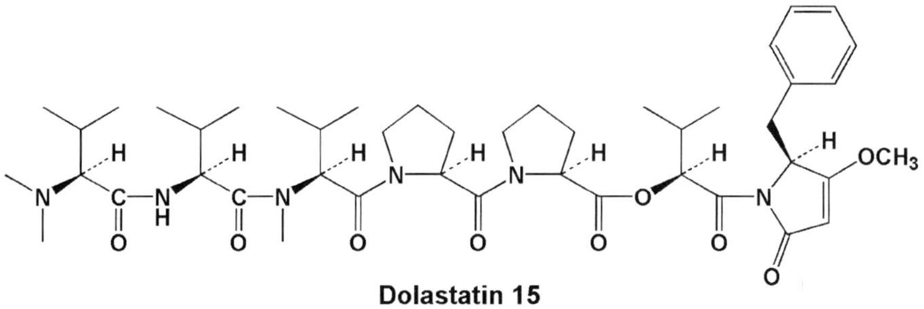

Dolastatin 15

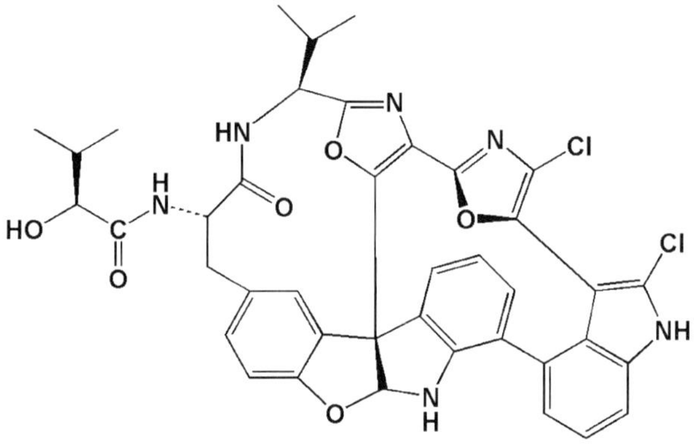

Diazonamide A

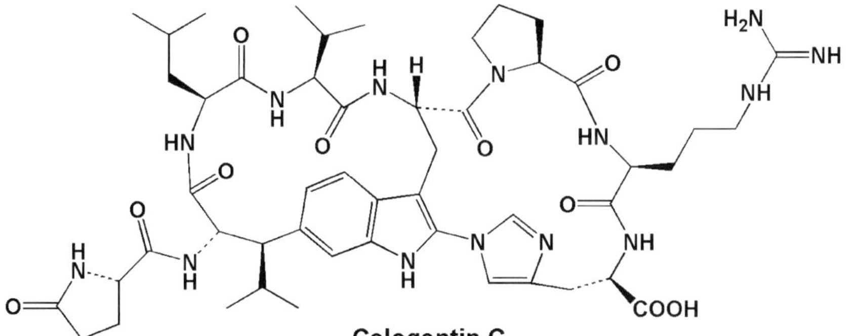

Celogentin C

Fig. 9. Structures of a depsipeptide and two peptides that have not been reported to inhibit vinca alkaloid binding to tubulin or have been shown not to affect vinca alkaloid binding (*see* text).

The synthesis of radiolabeled dolastatin 10 *(99)* permitted ready evaluation of the interaction of an additional antimicrotubule drug with tubulin. Kinetic analysis of inhibitory effects of a number of vinca domain drugs has indicated that spongistatin 1 is a noncompetitive inhibitor of dolastatin 10 binding to tubulin *(70)*, whereas an analog of dolastatin 10 *(70)*, cryptophycin 1 *(93)*, hemiasterlin *(90)*, and phomopsin A (unpublished data) are competitive inhibitors.

Figure 9 shows the structures of three additional peptides and depsipeptides. Dolastatin 15 *(100)* and diazonamide A *(101)*, both marine products, have no apparent inhibitory effect on the binding of vinblastine to tubulin. Recent experiments with radiolabeled dolastatin 15 demonstrated that the compound bound weakly to tubulin, with an apparent dissociation equilibrium constant of about 30 μM. The binding of dolastatin 15 to tubulin was inhibited strongly by dolastatin 10, phomopsin A, halichondrin B, and maytansine; moderately by vinblastine and vincristine; weakly by cryptophycin 1; and not at all by a potent analog of dizaonamide A. These results suggest dolastatin 15 does bind weakly in the vinca domain but leave the binding site of diazonamide A still unknown *(102)*.

The last peptide shown in Fig. 9 is the plant product celogentin C *(103)*. It inhibits microtubule assembly and is the most potent inhibitor among a growing number of structurally related peptides. Thus far, no data have been published describing the effects of any of this group of peptides either on vinca alkaloid binding to tubulin or on cell growth.

4. SUMMARY

In summary, tubulin is the primary target of a large and ever-growing number of small molecules. Their mechanisms of interaction with the protein are varied. As new agents are discovered, the complexity of interactions of drugs with tubulin only seems to increase. In particular, the simultaneous binding of laulimalide and paclitaxel to polymer revealed an unexpected new dimension to drugs that induce tubulin assembly; and it is becoming increasingly more difficult to rationalize the binding of the structurally diverse antimitotic peptides and depsipeptides to a single, well-defined site on the tubulin molecule. Finally, cyclostreptin has been shown very recently to react covalently with two amino acid residues of β-tubulin (III), Thr–218 and Asn–226, providing new insights into the mechanism of drug binding to the taxoid site.

REFERENCES

1. Rowinsky EK, Donehower RC. Paclitaxel (Taxol). New Engl J Med 1995;332:1004–1014.
2. Cortes JE, Pazdur R. Docetaxel. J Clin Oncol 1995;13:2643–2655.
3. Aoe K, Kiura K, Ueoka H, et al. Effect of docetaxel with cisplatin or vinorelbine on lung cancer cells. Anticancer Res 1999;19:291–299.
4. Batra S, Karlsson R, Witt L. Potentiation by estramustine of the cytotoxic effect of vinblastine and doxorubicin in prostatic tumor cells. Int J Cancer 1996;68:644–649.
5. Budman DR, Calabro A. In vitro search for synergy and antagonism: evaluation of docetaxel combinations in breast cancer cell lines. Breast Cancer Res Treat 2002;74:41–46.
6. Budman DR, Calabro A, Kreis W. Synergistic and antagonistic combinations of drugs in human prostate cancer cell lines in vitro. Anti-cancer Drugs 2002;13:1011–1016.
7. Budman DR, Calabro A, Wang LG, et al. Synergism of cytotoxic effects of vinorelbine and paclitaxel in vitro. Cancer Invest 2000;18:695–701.
8. Carles G, Braguer D, Sabeur G, Briand C. The effect of combining antitubulin agents on differentiated and undifferentiated human colon cancer cells. Anti-cancer Drugs 1998;9:209–221.
9. Culine S, Roch I, Pinguet F, Romieu G, Bressolle F. Combination paclitaxel and vinorelbine therapy: in vitro cytotoxic interactions and dose-escalation study in breast cancer patients previously exposed to anthracyclines. Int J Oncol 1999;14:999–1006.
10. Giannakakou P, Villalba L, Li H, Poruchynsky M, Fojo T. Combinations of paclitaxel and vinblastine and their effects on tubulin polymerization and cellular cytotoxicity: characterization of a synergistic schedule. Int J Cancer 1998;75:57–63.
11. Knick V, Eberwein D, Miller CG. Vinorelbine tartrate and paclitaxel combinations: enhanced activity against in vivo P388 murine leukemia cells. J Natl Cancer Inst 1995;87:1072–1077.

12. Kreis W, Budman DR, Calabro A. Unique synergism or antagonism of combinations of chemotherapeutic and hormonal agents in human prostate cancer cell lines. Br J Urol 1997;79:196–202.
13. Madari H, Panda D, Wilson L, Jacobs RS. Dicoumarol: a unique microtubule stabilizing natural product that is synergistic with taxol. Cancer Res 2003;63:1214–1220.
14. Martello LA, McDaid HM, Regl DL, et al. Taxol and discodermolide represent a synergistic drug combination in human carcinoma cell lines. Clin Cancer Res 2000;6:1978–1987.
15. Photiou A, Shah P, Leong JK, Moss J, Retsas S. In vitro synergy of paclitaxel (Taxol) and vinorelbine (Navelbine) against human melanoma cell lines. Eur J Cancer 1997;33:463–470.
16. Roch I, Bressolle F, Pinguet F. In vitro schedule-dependent interaction between paclitaxel and vinorelbine in A2780 parental and multidrug-resistant human ovarian cancer cell lines. Int J Oncol 1997;11:1379–1385.
17. Speicher LA, Barone L, Tew KD. Combined antimicrotubule activity of estramustine and taxol in human prostatic carcinoma cell lines. Cancer Res 1992;52:4433–4440.
18. McKellar QA, Scott EW. The benzimidazole anthelmintic agents—a review. J Vet Pharmacol Ther 1990;13:223–247.
19. Ben-Chetrit E, Levy M. Colchicine: 1998 update. Semin. Arthritis Rheu 1998;28:48–59.
20. Schiff PB, Fant J, Horwitz SB. Promotion of microtubule assembly in vitro by taxol. Nature (London) 1979;277:665–667.
21. Parness J, Horwitz SB. Taxol binds to polymerized tubulin in vitro. J Cell Biol 1981;91:479–487.
22. Bollag DM, McQueney PA, Zhu J, et al. Epothilones, a new class of microtubule-stabilizing agents with a taxol-like mechanism of action. Cancer Res 1995;55:2325–2333.
23. Kowalski RJ, Giannakakou P, Hamel E. Activities of the microtubule-stabilizing agents epothilones A and B with purified tubulin and in cells resistant to paclitaxel (Taxol®). J Biol Chem 1997;272: 2534–2541.
24. Ter Haar E, Kowalski RJ, Hamel E, et al. Discodermolide, a cytotoxic marine agent that stabilizes microtubules more potently than taxol. Biochemistry 1996;35:243–250.
25. Kowalski RJ, Giannakakou P, Gunasekera SP, Longley RE, Day BW, Hamel E. The microtubule-stabilizing agent disocodermolide competitively inhibits the binding of paclitaxel (Taxol) to tubulin polymers, enhances tubulin nucleation reactions more potently than paclitaxel, and inhibits the growth of paclitaxel-resistant cells. Mol Pharmacol 1997;52:613–622.
26. Long BH, Carboni JM, Wasserman AJ, et al. Eleutherobin, a novel cytotoxic agent that induces tubulin polymerization, is similar to paclitaxel (Taxol®). Cancer Res 1998;58:1111–1115.
27. Hamel E, Sackett DL, Vourloumis D, Nicolaou KC. The coral-derived natural products eleutherobin and sarcodictyins A and B: effects on the assembly of purified tubulin with and without microtubule-associated proteins and binding at the polymer taxoid site. Biochemistry 1999;38:5490–5498.
28. Sato B, Muramatsu H, Miyauchi M, et al. A new antimitotic substance, FR182877: I. Taxonomy, fermentation, isolation, physico-chemical properites and biological activities. J Antibiotics 2000;53: 123–130.
29. Sato B, Nakajima H, Hori Y, Hino M, Hashimoto S, Terano H. A new antimitotic substance, FR182877: II. The mechanism of action. J Antibiotics 2000;53:204–206.
30. Yoshimura S, Sato B, Kinoshita T, Takase S, Terano H. A new antimitotic substance, FR182877: III. Structure determination with errata published in vol 55, pp. C-1. J Antibiotics 2000;53:615–622.
31. Edler MC, Buey RM, Gussio R, et al. Cyclostreptin (FR182877), an antitumor tubulin-polymerizing agent deficient in enhancing tubulin assembly despite its high affinity for the taxoid site. Biochemistry 2004;44:11,525–11,538.
32. Isbrucker RA, Cummins J, Pomponi SA, Longley RE, Wright AE. Tubulin polymerizing activity of dictyostatin-1, a polyketide of marine sponge origin. Biochem Pharmacol 2003;66:75–82.
33. Paterson I, Britton R, Delgado O, Meyer A, Poullennec KG. Total synthesis and configurational assignment of (-)-dictyostatin, a microtubule-stabilizing macrolide of sponge origin. Agnew Chem Int Ed 2004;43:2–6.
34. Madiraju C, Edler MC, Hamel E, et al. Tubulin assembly, taxoid site binding, and cellular effects of the microtubule-stabilizing agent dictyostatin. Biochemistry 2005;44:15,053–15,063.
35. Mooberry SL, Tien G, Hernandez AH, Plubrukarn A, Davidson BS. Laulimalide and isolaulimalide, new paclitaxel-like microtubule-stabilizing agents. Cancer Res 1999;59:653–660.
36. Pryor DE, O'Brate A, Bilcer G, et al. The microtubule stabilizing agent laulimalide does not bind in the taxoid site, kills cells resistant to paclitaxel and epothilones, and may not require its epoxide moiety for activity. Biochemistry 2002;41:9109–9115.

37. Gapud EJ, Bai R, Ghosh AK, Hamel E. Laulimalide and paclitaxel: a comparison of their effects on tubulin assembly and their synergistic action when present simultaneously. Mol Pharmacol 2004;66: 113–121.

38. Hood KA, West LM, Rouwé B, et al. Peloruside A, a novel antimitotic agent with paclitaxel-like microtubule-stabilizing activity. Cancer Res 2002;62:3356–3360.

39. Gaitanos TN, Buey RM, Díaz JF, et al. Peloruside A does not bind to the taxoid site on β-tubulin and retains its activity in multidrug-resistant cell lines. Cancer Res 2004;64:5063–5067.

40. Bai R, Lin CM, Nguyen NY, Liu TY, Hamel, E. Identification of the cysteine residue of β-tubulin alkylated by the antimitotic agent 2,4-dichlorobenzyl thiocyanate, facilitated by separation of the protein subunits of tubulin by hydrophobic column chromatography. Biochemistry 1989;28: 5606–5612.

41. Abraham I, Dion RL, Duanmu C, Gottesman MM, Hamel E. 2,4-Dichlorobenzyl thiocyanate, an antimitotic agent that alters microtubule morphology. Proc Natl Acad Sci USA 1986;83:6839–6843.

42. Shan B, Medina JC, Santha E, et al. Selective, covalent modification of β-tubulin residue Cys-239 by T138067, an antitumor agent with in vivo efficacy against multidrug-resistant tumors. Proc Natl Acad Sci USA 1999;96:5686–5691.

43. Legault J, Gaulin JF, Mounetou E, et al. Microtubule disruption induced in vivo by alkylation of β-tubulin by 1-aryl-3-(2-chloroethyl)ureas, a novel class of soft alkylating agents. Cancer Res 2000;60:985–992.

44. Mounetou E, Legault J, Lacroix J, C-Gaudreault R. A new generation of N-aryl-N-(1-alkyl-2-chloroethyl)ureas as microtubule disrupters: synthesis, antiproliferative activity, and β-tubulin alkylation kinetics. J Med Chem 2003;46:5055–5063.

45. Bouchon B, Chambon C, Mounetou E, et al. Alkylation of β-tubulin on Glu 198 by a microtubule disrupter. Mol Pharmacol 2005;68:1415–1422.

46. Combeau C, Provost J, Lancelin F, et al. RPR112378 and RPR115781: two representatives of a new family of microtubule assembly inhibitors. Mol Pharmacol 2000;57:553–563.

47. Ravelli RBG, Gigant B, Curmi PA, et al. Insight into tubulin regulation from a complex with colchicine and a stathmin-like domain. Nature (London) 2004;428:198–202.

48. Bai R, Pei XF, Boyé O, et al. Identification of cysteine 354 of β-tubulin as part of the binding site for the A ring of colchicine. J Biol Chem 1996;271:12,639–12,645.

49. Bai R, Covell DG, Pei XF, et al. Mapping the binding site of colchicinoids on β-tubulin: 2-chloroacetyl-2-demethylthiocolchicine covalently reacts predominantly with cysteine 239 and secondarily with cysteine 354. J Biol Chem 2000;275:40,443–40,452.

50. Wilson L. Properties of colchicine binding protein from chick embryo brain. Interactions with vinca alkaloids and podophyllotoxin. Biochemistry 1970;9:4999–5007.

51. Schiff PB, Kende AS, Horwitz SB. Steganacin: an inhibitor of HeLa cell growth and microtubule assembly in vitro. Biochem Biophys Res Commun 1978;85:737–746.

52. Lin CM, Singh SB, Chu PS, et al. Interactions of tubulin with potent natural and synthetic analogs of the antimitotic agent combretastatin: a structure-activity study. Mol Pharmacol 1988;34:200–208.

53. D'Amato RJ, Lin CM, Flynn E, Folkman J, Hamel E. 2-Methoxyestradiol, an endogenous mammalian metabolite, inhibits tubulin polymerization by interacting at the colchicine site. Proc Natl Acad Sci USA 1994;91:3964–3968.

54. Beutler JA, Cardellina JH 2nd, Lin CM, Hamel E, Cragg GM, Boyd MR. Centaureidin, a cytotoxic flavone from *Polymnia fruticosa* inhibits tubulin polymerization. Bioorg Med Chem Lett 1993;3:581–584.

55. Lichius JJ, Thoison O, Montagnac A, Païs M, Guéritte-Voegelein F, Sévenet T. Antimitotic and cytotoxic flavonols from *Zieridium pseudobtusifolium* and *Acronychia porteri*. J Nat Prod 1994;57:1012–1016.

56. Shi Q, Chen K, Li L, et al. Cytotoxic and antimitotic flavonols from *Polanisia dodecandra*. J Nat Prod 1995;58:475–482.

57. Blokhin AV, Yoo HD, Geralds RS, Nagle DG, Gerwick WH, Hamel E. Characterization of the interaction of the marine cyanobacterial natural product curacin A with the colchicine site of tubulin and initial structure-activity studies with analogs. Mol Pharmacol 1995;48:523–531.

58. Hoebeke J, Van Nijen G, De Brabander M. Interaction of oncodazole (R 17934), a new antitumoral drug, with rat brain tubulin. Biochem Biophys Res Commun 1976;69:319–324.

59. Friedman PA, Platzer EG. Interaction of anthelmintic benzimidazoles and benzimidazole derivatives with bovine brain tubulin. Biochim Biophys Acta 1978;544:605–614.

60. Hamel E, Lin CM. Interactions of a new antimitotic agent, NSC-181928, with purified tubulin. Biochem Biophys Res Commun 1982;104:929–936.
61. Li L, Wang HK, Kuo SC, et al. Synthesis and biological evaluation of 3′,6,7-substituted 2-phenyl-4-quinolones as antimitotic antitumor agents. J Med Chem 1994;37:3400–3407.
62. Paull KD, Lin CM, Malspeis L, Hamel E. Identification of novel antimitotic agents acting at the tubulin level by computer-assisted evaluation of differential cytotoxicity data. Cancer Res 1992;52:3892–3900.
63. Sternlicht H, Ringel I. Colchicine inhibition of microtubule assembly via copolymer formation. J Biol Chem 1979;254:10,540–10,548.
64. Sternlicht H, Ringel I, Szasz J. The co-polymerization of tubulin and tubulin-colchicine complex in the absence and presence of associated proteins. J Biol Chem 1980;255:9138–9148.
65. Hastie SB. Interactions of colchicine with tubulin. Pharmacol Ther 1991;51:377–401.
66. Jiang JB, Hesson DP, Dusak BA, Dexter DL, Kang GJ, Hamel E. Synthesis and biological evaluation of 2-styrylquinazolin-4(3H)-ones, a new class of antimitotic anticancer agents which inhibit tubulin polymerization. J Med Chem 1990;33:1721–1728.
67. Batra JK, Powers LJ, Hess FD, Hamel E. Derivatives of 5,6-diphenylpyridazin-3-one: synthetic antimitotic agents which interact with plant and mammalian tubulin at a new drug-binding site. Cancer Res 1986;46:1889–1893.
68. Lin CM, Kang GJ, Roach MC, et al. Investigation of the mechanism of the interaction of tubulin with derivatives of 2- styrylquinazolin-4(3H)-one. Mol Pharmacol 1991;40:827–832.
69. Bai R, Pettit GR, Hamel E. Binding of dolastatin 10 to tubulin at a distinct site for peptide antimitotic agents near the exchangeable nucleotide and vinca alkaloid sites. J Biol Chem 1990;265: 17,141–17,149.
70. Bai R, Taylor GF, Cichacz ZA, et al. The spongistatins, potently cytotoxic inhibitors of tubulin polymerization, bind in a distinct region of the vinca domain. Biochemistry 1995;34:9714–9719.
71. Mandelbaum-Shavit F, Wolpert-DeFilippes MK, Johns DG. Binding of maytansine to rat brain tubulin. Biochem Biophys Res Commun 1976;72:47–54.
72. Bai R, Paull KD, Herald CL, Malspeis L, Pettit GR, Hamel E. Halichondrin B and homohalichondrin B, marine natural products binding in the vinca domain of tubulin: discovery of tubulin-based mechanism of action by analysis of differential cytotoxicity data. J Biol Chem 1991;266:15,882–15,889.
73. Higashide E, Asai M, Ootsu K., et al. Ansamitocin, a group of novel maytansinoid antibiotics with antitumour properties from Nocardia. Nature (London) 1977;270:721–722.
74. Irschik H, Jansen R, Gerth K, Höfle G, Reichenbach H. Disorazol-A, an efficient inhibitor of eukaryotic organisms isolated from myxobacteria. J Antibiotics 1995;48:31–35.
75. Elnakady YA, Sasse F, Lünsdorf H, Reichenbach H. Disorazol A$_1$, a highly effective antimitotic agent acting on tubulin polymerization and inducing apoptosis in mammalian cells. Biochem Pharmacol 2004;67:927–935.
76. Elknakady YA. Untersuchungen zum wirkungsmechanismus von disorasol, einem neuen antimitotischen stoff aus myxobakterien. Ph.D. Dissertation, Technischen Universität Carolo-Wilhelmina, Braunschweig, Germany. 2001; pp 112.
77. Takahashi M, Iwasaki S, Kobayashi H, Okuda S, Murai T, Sato Y. Rhizoxin binding to tubulin at the maytansine-binding site. Biochim Biophys Acta 1987:926:215–223.
78. Wilson L, Jordan MA, Morse A, Margolis RL. Interaction of vinblastine with steady-state microtubules in vitro. J Mol Biol 1982;159:125–149.
79. Jordan MA, Margolis RL, Himes RH, Wilson L. Identification of a distinct class of vinblastine binding sites on microtubules. J Mol Biol 1986;187:61–73.
80. Gigant B, Wang C, Ravelli RBG, et al. Structural basis for the regulation of tubulin by vinblastine. Nature (London) 2005;435:519–522.
81. Timasheff SN, Andreu JM, Na CC. Physical and spectroscopic methods for the evaluation of the interactions of antimitotic agents with tubulin. Pharmacol Ther 1991;52:191–210.
82. Safa AR, Hamel E, Felsted RL. Photoaffinity labeling of tubulin subunits with a photoactive analogue of vinblastine. Biochemistry 1987;26:97–102.
83. Hirata Y, Uemura D. Halichondrins—antitumor polyether macrolides from a marine sponge. Pure Appl Chem 1986;58:701–710.
84. Bai R, Cichacz ZA, Herald CL, Pettit GR, Hamel E. Spongistatin 1, a highly cytotoxic, sponge-derived, marine natural product that inhibits mitosis, microtubule assembly, and the binding of vinblastine to tubulin. Mol Pharmacol 1993;44:757–766.

85. Kobayashi M, Aoki S, Sakai H, et al. Altohytrin A, a potent anti-tumor macrolide from the Okinawan marine sponge *Hyrtios altum*. Tetrahedron Lett 1993;34:2795–2798.
86. Towle MJ, Salvato KA, Budrow J, et al. In vitro and in vivo anticancer activities of synthetic macrocyclic ketone analogs of halichondrin B. Cancer Res 2001;61:1013–1021.
87. Hamel E, Covell DG. Antimitotic peptides and depsipeptides. Curr Med Chem—Anti-Cancer Agents 2002;2:19–53.
88. Lacey E, Edgar JA, Culvenor CCJ. Interaction of phomopsin A and related compounds with purified sheep brain tubulin. Biochem Pharmacol 1987;36:2133–2138.
89. Li Y, Koiso Y, Kobayashi H, Hashimoto Y, Iwasaki S. Ustiloxins, new antimitotic cyclic peptides: interaction with porcine brain tubulin. Biochem Pharmacol 1995;49:1367–1372.
90. Bai R, Durso NA, Sackett DL, Hamel E. Interactions of the sponge-derived antimitotic tripeptide hemiasterlin with tubulin: comparison with dolastatin 10 and cryptophycin 1. Biochemistry 1999;43:14,302–14,310.
91. Edler MC, Fernandez AM, Lassota P, Ireland CM, Barrows LR. Inhibition of tubulin polymerization by vitilevuamide, a bicyclic marine peptide, at a site distinct from colchicine, the vinca alkaloids and dolastatin 10. Biochem Pharmacol 2002;63:707–715.
92. Kerksiek K, Mejillano MR, Schwartz RE, Georg GI, Himes RH. Interaction of cryptophycin 1 with tubulin and microtubules. FEBS Lett 1995;377:59–61.
93. Bai R, Schwartz RE, Kepler JA, Pettit GR, Hamel E. Characterization of the interaction of cryptophycin 1 with tubulin: binding in the *Vinca* domain, competitive inhibition of dolastatin 10 binding, and an unusual aggregation reaction. Cancer Res 1996;56:4398–4406.
94. Smith CD, Zhang X. Mechanism of action of cryptophycin: interaction with the *Vinca* alkaloid domain of tubulin. J Biol Chem 1996;271:6192–6198.
95. Khalil MWM. Tubulysin aus myxobakterien: untersuchungen zum wirkungsmechanismus. Ph.D. Dissertation, Technischen Universität Carolo-Wilhelmina, Braunschweig, Germany. 1999; pp 122.
96. Sasse F, Steinmetz H, Heil J, Höfle G, Reichenbach H. Tubulysins, new cytostatic peptides from myxobacteria acting on microtubuli: production, isolation, physico-chemical and biological properties. J Antibiotics 2000;53:879–885.
97. Höfle G, Glaser N, Leibold T, Karama U, Sasse F, Steinmetz H. Semisynthesis and degradation of the tubulin inhibitors epothilone and tubulysin. Pure Appl Chem 2003;75:167–178.
98. Natsume T, Watanabe J, Tamoki S, Fujio N, Miyasaka K, Kobayashi M. Characterization of the interaction of TZT-1027, a potent antitumor agent, with tubulin. Jpn J Cancer Res 2000;91:737–747.
99. Bai R, Taylor GF, Schmidt JM, et al. Interaction of dolastatin 10 with tubulin: induction of aggregation and binding and dissociation reactions. Mol Pharmacol 1995;47:965–976.
100. Bai R, Friedman SJ, Pettit GR, Hamel E. Dolastatin 15, a potent antimitotic depsipeptide derived from *Dolabella auricularia*: interaction with tubulin and effects on cellular microtubules. Biochem Pharmacol 1992;43:2637–2645.
101. Cruz-Monserrate Z, Vervoort HC, Bai R, et al. Diazonamide A and a synthetic structural analog: disruptive effects on mitosis and cellular microtubules and analysis of their interactions with tubulin. Mol Pharmacol 2003;63:1273–1280.
102. Cruz-Monserrate Z, Mullaney JT, Harran PG, Pettit GR, Hamel E. Dolastatin 15 binds in the vinca domain of tubulin as demonstrated by Hummel-Dreyer chromatography. Eur J Biochem 2003;270:3822–3828.
103. Kobayashi J, Suzuki H, Shimbo K, Takeya K, Morita H. Celogentins A-C, new antimitotic bicyclic peptides from the seeds of *Celosia argentea*. J Org Chem 2001;66:6626–6633.
104. Paterson I, Britton R, Delgado O, Wright AE. Stereochemical determination of dictyostatin, a novel microtubule-stabilising macrolide from the marine sponge *Corallistidae sp*. Chem Commun 2004;632–633.
105. Yeh HJC, Chrzanowska M, Brossi A. The importance of the phenyl-tropolone 'aS' configuration in colchicine's binding to tubulin. FEBS Lett 1988;229:82–86.
106. Brossi A, Boyé O, Muzaffar A, et al. aS,7S-absolute configuration of natural (-)-colchicine and allo-congeners. FEBS Lett 1990;262:5–7.
107. Leoni LM, Hamel E, Genini D, et al. Indanocine, a microtubule-binding indanone and a selective inducer of apoptosis in multidrug-resistant cancer cells. J Natl Cancer Inst 2000;92:217–224.
108. Beckers T, Reissmann T, Schmidt M, et al. 2-Aroylindoles, a novel class of potent, orally active small molecule tubulin inhibitors. Cancer Res 2002;62:3113–3119.

109. Prinz H, Ishii Y, Hirano T, et al. Novel benzylidene-9(10*H*)-anthracenones as highly active antimicrotubule agents: synthesis, antiproliferative activity, and inhibition of tubulin polymerization. J Med Chem 2003;46:3382–3394.
110. Li Q, Woods KW, Claiborne A, et al. Synthesis and biological evaluation of 2-indolyloxazolines as a new class of tubulin polymerization inhibitors: discovery of A-289099 as an orally active antitumor agent. Bioorg Med Chem Lett 2002;12:465–469.
111. Buey RM, Calvo E, Barasoain I, et al. Cyclostreptin binds covalently to microtubule pores and lumenal taxoid binding sites. Nature Chem Biol 2007;3:117–125.

2 Molecular Mechanisms of Microtubule Acting Cancer Drugs

John J. Correia and Sharon Lobert

SUMMARY

Here the molecular mechanism of antimitotic drugs, biological compounds that bind to tubulin and microtubules and suppress microtubule dynamics are reviewed. A common feature of tubulin-interacting compounds is that binding to tubulin is linked to assembly, either the stabilization of a microtubule lattice by compounds like the taxanes and epothilones, or the induction of alternate, nonmicrotubule polymer forms. The nonmicrotubule polymers arise from tubulin heterodimers or at microtubule ends with compounds like colchicine, vinca alkaloids, dolastatin, and cryptophycin-52. Their mechanism of action is strongly coupled to the mechanism of microtubule assembly, especially structural features that affect nucleotide binding, GTP hydrolysis, stabilization of longitudinal and lateral protofilament contacts, and endwise growth and disassembly dynamics. Quantitative analysis of drug binding and microtubule or nonmicrotubule polymer formation can be a useful tool in drug design, as in many cases the energetics are predictive of IC50 values and clinical doses. Furthermore, these drugs allosterically disrupt the regulation of microtubule dynamics, whereas the regulatory factors themselves may play an important role in drug resistance. Thus, the development of compounds that selectively target regulation of mitotic spindle dynamics and kinetochore capture, by chemical genetics for example, may result in useful and effective therapeutic tools.

Key Words: Allosteric regulation; antimitotics; kinetics; microtubules; polymers; thermodynamics; tubulin.

From: *Cancer Drug Discovery and Development: The Role of Microtubules in Cell Biology, Neurobiology, and Oncology* Edited by: Tito Fojo © Humana Press, Totowa, NJ

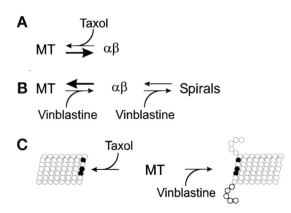

Fig. 1. Alterations in microtubule polymer levels can cause drug resistance. **(Panel A)** Taxol resistance occurs when microtubules are destabilized (indicated by the large dissociation arrow) thus opposing the formation of taxol-stabilized microtubules. Microtubule destabilization can reflect alterations in isotype levels, a decrease in activity or concentration of microtubule stabilizers, or an increase in activity or concentration of microtubule destabilizers *(7)*. **(Panel B)** Vinblastine resistance occurs when microtubules are stabilized (indicated by the large association arrow) thus opposing the formation of vinblastine-induced spirals. Microtubule stabilization can reflect alterations in isotype levels, an increase in activity or concentration of microtubule stabilizers, or a decrease in activity or concentration of microtubule destabilizers *(7)*. **(Panel C)** This mass action model can easily be extended to include the formation of + end microtubule structures during kinetochore attachment. This can represent a growing (blunt end) or a shrinking (curved end) dynamic conformation; or it could represent conformations that favor or disfavor kinetochore interaction and attachment; or it could represent a microtubule stabilized by Clip-170 or destabilized by MCAK (*see* Fig. 4 in Chapter 8).

1. INTRODUCTION

To understand how antimitotic drugs work one must understand the mechanism of microtubule assembly and regulation. This can be highlighted by the observation made by Cabral, Raff, and coworkers *(1–6)* that cells can develop resistance to drugs simply by varying the fraction of tubulin in the microtubule polymers (Fig. 1). Lower the microtubule polymer concentration and the cells become resistant to drugs like taxol (Fig. 1A); raise the microtubule polymer concentration and cells become resistant to drugs like vinblastine (Fig. 1B). Knowing that taxol binds preferentially to microtubules and not tubulin heterodimers, and that vinblastine destabilizes microtubules and favors the formation of tubulin spirals only begins to explain how this (mass action) mechanism of resistance might work. This model can also explain the cross-resistance between drugs like vincristine and vinblastine as well as cellular resistance to taxanes concurrent with sensitivity to vinca alkaloids. If a cell line or a tumor is resistant to both a taxane and a vinca alkaloid the primary mechanism of resistance cannot be a microtubule polymer-based mechanism, the emphasis of this review. Thus, the question can be asked, do the molecular mechanisms of antimitotic drugs explain these observations about drug resistance?

This review critically covers the essential features of tubulin and microtubule structure, microtubule assembly and dynamics, and the role of microtubule associated proteins, referred to as MAPs and dynamics regulators, in controlling microtubule assembly.

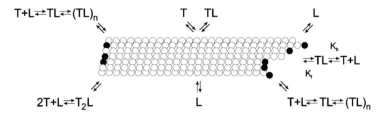

Fig. 2. Interactions between ligands or drugs and tubulin or microtubules can occur by numerous mechanisms that mimic the binding of regulatory factors. Depicted in this image are binding event to the microtubule walls (e.g., taxol) or the microtubule ends. The ends are not labeled and can represent events at either the – or + end, or both. The blunt and jagged end depicts a growing microtubule, possibly a sheet. End binding can be a direct event where the drug (L) binds to one end, both ends, or some specific state like GTP-Tb (shaded spheres), GDP-Tb (open spheres), or a specific structure like a curved protofilament or a sheet. Alternatively, a liganded-tubulin–drug complex (TL) can bind to the ends where the same qualifiers apply (one, both, and so on). These two end binding events (L or TL) represent classic stoichiometric inhibition or poisoning of microtubule assembly (e.g., colchicine). Sequestering occurs when dissociated tubulin binds to a ligand (T + L) and makes a stable complex [TL or $(TL)_n$]. The complex can be a ring (dolastatin) or a spiral (vinca alkaloids) and typically forms in a cooperative manner. There is evidence that these complexes can bind to the ends as well, and thus a spiral polymer [$(TL)_n$] can add directly to the ends or can grow from the ends, also in a cooperative manner. The T_2L complex represents a stathmin–tubulin complex that is reported to bind to microtubule ends, or the smallest typical drug-induced polymer that might bind. Drugs that bind at the interdimer interface (vinca alkaloids, dolastatin, and hemiasterlin) necessarily begin with a 1:2 stoichiometry (one drug per interface) *(49,60,61)*. The T or TL complex binding to the walls represents subunit rearrangements that must occur during lattice reorganization or katanin mediated microtubule breaking at lattice defects *(106)*. The binding of L to the microtubule wall typical refers to the stabilizing effect of taxol, but recent evidence suggests that destabilizers like vinblastine may also bind to the lattice and contribute to microtubule depolymerization. All of these reactions compete with one another and with the conformational states that occur within the lattice and at the ends.

Using a limited number of examples, three modes of drug interaction with tubulin and microtubules are discussed (*see* Fig. 2),

1. Substoichiometric poison;
2. Alternate polymer formation;
3. Microtubule stabilization.

   The information is integrated in an attempt to understand how drugs target microtubules and how cells and tumors resist their effects. Only minor consideration of pharmaco-dynamics and no discussion of MDR phenotypes are included here. Although clearly important therapeutic considerations, drug uptake, localization, and delivery to targets are considered to be independent of tubulin/microtubule based mechanisms, except where the mechanism directly impacts them. The focus is the interaction of antimitotic drugs with tubulin and microtubules and the disruption of microtubule dynamics and regulation. The point of view for this review is that assembly involves competition between numerous polymer forms (including different microtubule lattices) that are altered (stabilized or destabilized) by the interaction of regulatory factors and drugs. One can treat drugs as mimics of these regulatory factors because drugs induce tubulin polymer forms that are similar to tubulin assemblies in the normal cytoskeleton. This approach provides quantitative understanding of the tubulin cytoskeleton while also allowing a comparison and a distinction between the role of assembly in normal regulation and the role of drug-induced assembly in cytotoxicity.

There have been numerous reviews of microtubule assembly and regulation, many of these from a biophysical and thermodynamic perspective *(7–16)*. In 1991, one of the author's group wrote *(15)*: "Thus, to interpret results, the relative affinity of these drugs, their relative concentration, and the stability of alternate polymer forms are all involved in the competing equilibria and must be considered." An accurate "interpretation of data rests upon the framework of the model chosen, and thus strongly reinforces the idea that without a clear picture of the molecular events at the ends of microtubules, (an understanding of) the effects of drugs on microtubule dynamics will continue to be problematic." Fifteen years later this approach applies more than ever. The goal here is a quantitative dissection of tubulin and microtubule interactions that describe the effect of antimitotic drugs on the tubulin cytoskeleton. Selective examples that use this approach will be highlighted.

2. MICROTUBULE ASSEMBLY

Microtubules are polar cytoskeletal structures with a plus (+) and minus (–) end (*see* Chapter 10 in this book) comprised of $\alpha\beta$-tubulin heterodimers organized head-to-tail ($\alpha\beta\alpha\beta$) along (typically *12–15*) protofilaments that laterally contact to make a closed lattice. The molecular structure of a microtubule is derived from electron crystallography on (antiparallel and inverted) Zn-induced sheets *(17)* that were later docked into a 20 Å reconstruction of the microtubule lattice *(18)*. Refined structures have been described *(19–22)*. Assembly requires an unfavorable in vitro nucleation step (off γ-tubulin bound to centrosomes in vivo) and GTPMg bound to tubulin where GTP becomes nonexchangeable when buried in the lattice. GTP hydrolysis occurs at this buried longitudinal interface between dimers with the α-Glu254 catalytic group residing on the previous dimer in the protofilament *(23)* Thus, GTP hydrolysis is necessarily linked to assembly, but only in polymers with straight quaternary structures like microtubules. (The role of association in hydrolysis and the importance of the α-Glu254 catalytic group is not universally appreciated in the field. GTP hydrolysis requires tubulin association, but tubulin association does not necessarily cause GTP hydrolysis.) The polar structure of the microtubule requires that subunit addition at the + end involves formation of a $\alpha\beta_{GTP}\alpha\beta_{GTP}^{(+)}$ linkage ([+] indicates + end of the microtubule) where the buried GTP can be cleaved but the exposed β_{GTP} is a site for subunit addition. At the – end subunit addition involves formation of a $^{(-)}\alpha\beta_{GTP}\alpha\beta_{GDP}$ linkage ([–] indicates – end of the microtubule) where the newly buried GTP can be cleaved and an exposed α-chain is the site for subunit addition. The presence of GTP hydrolysis thus produces two distinct ends, a + end that is growing by producing an exposed nucleotide site *(24)*, and a – end that is growing by burying a β_{GTP} that can be immediately cleaved. This gives rise to microtubule ends with distinct stability (*see* discussion below and ref. *24*). The nucleotide at the + end is directly exchangeable with free nucleotide *(25)*, whereas the nucleotide at the – end is nonexchangeable and requires tubulin dissociation before exchange. These features are represented in the model of a microtubule protofilament shown in Scheme 1. In this mechanism the $\alpha\beta_{\mathbf{GTP}}$ added at the – end undergoes hydrolysis, albeit at some rate, while the $\alpha\beta_{\mathbf{GTP}}$ added at the + end is a site for further growth and can only undergo hydrolysis when buried by another tubulin heterodimer. Lateral interactions not depicted here are also known to strongly influence the cooperativity of assembly (*see* Fig. 2) and the rate of hydrolysis is dependent upon the quaternary structure of the MT end.

$$\alpha\beta_{\textbf{GTP}} \rightarrow {}^{(-)}\alpha\beta_{\textbf{GTP}}\alpha\beta_{\text{GDP}}\alpha\beta_{\text{GDP}}....\alpha\beta_{\text{GDP}}\alpha\beta_{\text{GDP}}\alpha\beta_{\textbf{GTP}}{}^{(+)} \leftarrow \alpha\beta_{\textbf{GTP}}$$

$$\downarrow$$

$${}^{(-)}\alpha\beta_{\textbf{GDP}}\alpha\beta_{\text{GDP}}\alpha\beta_{\text{GDP}}....\alpha\beta_{\text{GDP}}\alpha\beta_{\text{GDP}}\alpha\beta_{\textbf{GTP}}\alpha\beta_{\textbf{GTP}}{}^{(+)} \leftarrow \alpha\beta_{\textbf{GTP}}$$

$$\downarrow$$

$${}^{(-)}\alpha\beta_{\textbf{GDP}}\alpha\beta_{\text{GDP}}\alpha\beta_{\text{GDP}}....\alpha\beta_{\text{GDP}}\alpha\beta_{\text{GDP}}\alpha\beta_{\textbf{GDP}}\alpha\beta_{\textbf{GTP}}{}^{(+)}$$

Scheme 1

The involvement of irreversible GTP hydrolysis in the assembly process produces a nonequilibrium polymerization mechanism known as dynamic instability where both ends are able to undergo excursions of growth or disassembly with abrupt transition between them (*26, 27; see* also Chapter 10 in this book). The growth phase involves the addition of GTP containing subunits whereas the rapid disassembly phase involves the loss of GDP containing subunits. The transitions between growth and disassembly involve the frequency of catastrophe, pause, and rescue. Analysis of the dynamic properties of microtubules reveals that each end exhibits an overlapping distribution of rates and frequencies. The + end grows faster than the – end, whereas the catastrophe frequency is slightly higher at the + end, and the rescue frequency is slightly higher at the – end (*28,29*). This differential end effect is in part structural and in part because of heterogeneity in the microtubule lattice where protofilament number can vary within the same microtubule. Chretien and Fuller (*30*) have described 14 types of microtubule structures that define the probability of energetic configurations possible between subunits and protofilaments. To accommodate different protofilament numbers and helical repeats the protofilaments are skewed about their axis (radial and tangential bending) and longitudinally shifted to modify lateral interactions mediated mostly through the M loop. In addition, growth and shortening are cooperative processes involving lateral and longitudinal structural transitions. In general growing ends are blunt or sheet-like, whereas disassembling ends display curved or peeling protofilaments or ram's horns (*see* Fig. 1C, refs.*14,29,31*). It is thought that each structural configuration has an intrinsically different rate of growth and disassembly, and random conversion between configurations alters those rates, thus giving rise to stochastic dynamic instability behavior (*28,32–34*). Drug binding will disrupt both microtubule assembly and dynamics, depending upon where and how the drug incorporates into the lattice or competes with the molecular interactions. A number of examples are given in Fig. 2 where the emphasis is on different modes of disrupting the energetic and structural configurations primarily (but not exclusively) at microtubule ends.

Microtubule assembly is influenced by numerous solution variables including ionic strength, nucleotide content (GXP), pH and Mg^{+2} concentration. Each of these effectors plays a role in the stability of tubulin polymers. The exchangeable nucleotide sits in the interface between two dimers along a microtubule protofilament, explaining in part why it becomes nonexchangeable in the microtubule lattice (*17,18; see* also Fig. 3). The orientation of the dimer interface is sensitive to the nucleotide content preferring a

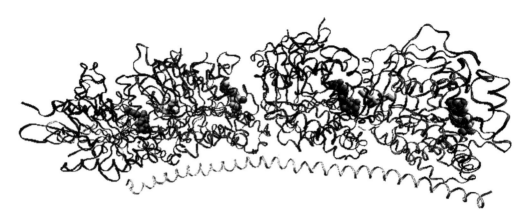

Fig. 3. A ribbon diagram representation of ISA0.pdb, the structure of stathmin–colchicines-tubulin *(48)*, was rendered with VMD. Two tubulin αβ-dimers are seen to interact with a long helix of the stathmin-like domain of RB3, plotted α1β1α2β2 going left to right. GTP on the α-chains is shown in orange, GDP on the β-chains in yellow. A Mg ion is shown in silver located near each nucleotide. Two colchicines are shown in green, one at each intradimer interface opposite the nonexchangeable GTP. To view this figure in color, see the insert and the companion CD-ROM.

straight quaternary structure with GTP, and a curved or helical structure with GDP. This explains the preference of GTP-tubulin to make microtubules, and GDP-tubulin to make small oligomers, rings or spirals *(15,35)*. The interaction is primarily a repulsive electrostatic effect caused by the presence of GTP in the curved oligomer or GDP in the straight oligomer *(35,36)*. In microtubules the presence of GDP has the effect of destabilizing the lattice by conformational strain *(37)*. Lateral interactions between protofilaments oppose this strain and stabilize the straight quaternary structure along the length of the microtubule, although ends are prone to splay apart into curved polymers. Recent structural data suggest that this strain causes the intradimer microtubule interface to curve *(22)*. Conformational strain is the driving force for splaying microtubule ends and rapid disassembly rates. The binding of MAPs (MAP2, tau, and so on) or taxanes stabilizes the microtubule lattice, whereas the binding of stathmin, XKCM1, vinblastine, or colchicine destabilizes the microtubule lattice, primarily by end-wise processes *(38–40;* *see* also Chapter 10 in this book). Destabilization of the lattice can mean formation of alternate polymer forms at the ends (Fig. 1), or dissociation into tubulin dimers, curved oligomers, (vinblastine) spirals or discrete (1:2 stathmin–tubulin) complexes (Fig. 2). Stability of each polymer is sensitive to the same effectors that stabilize microtubules (ionic strength, nucleotide content, pH, and Mg^{+2} concentration). This is because every polymer form appears to use the same αβ-interface and is thus GXP-dependent. In addition, all tubulin polymer forms necessarily bring the highly charged carboxyl tail of each subunit into close proximity, thus making polymer formation ionic strength and Mg^{+2}-dependent. Surprisingly, Mg^{+2}-induced rings (and GMPCPP-induced tubes) pack the carboxyl tail domain, usually on the outside of a microtubule, into the center of the ring (or tube; refs. *41,42)*, just like the images of curved protofilaments at microtubule ends visually suggest. This explains the enhanced requirement for divalent cation to suppress charge–charge repulsions and the altered thermodynamics of rings ($\Delta S \sim 0$) relative to microtubules and spirals ($\Delta S \gg 0$) *(43)*. Removing the carboxyl tail by subtilisin digestion enhances formation of all polymer forms while also promoting additional lateral packing, for example sheets of rings *(44,45)*. The effect of pH appears to be limited to polymers that involve lateral interactions. Microtubule assembly is favored by lower pH *(46)*, with the caveat that tubulin is only soluble and stable between pH 6 and 7.5. Vinca alkaloid induced spirals are pH independent *(35)* but lateral condensation of spirals into paracrystals is also favored by low pH *(47)*. A recent crystal structure of a stathmin-like domain with two tubulin dimers (Fig. 3) revealed interactions along the length of the stathmin helix that might account for the pH dependence for formation of this complex (*48*, and more recently *49*). Two histidines in the stathmin helix (H78, H129) interact with E417 on both β-subunits. It is not clear if identical interaction sites on tubulin cause the pH dependence of microtubules or vinca-induced paracrystals. It is worth noting that because both stathmin binding to dimers and microtubule formation are favored by low pH, both reactions are intrinsically weaker (in the absence of other factors) at cellular pH.

3. THE LINKAGE BETWEEN COMPETING EQUILIBRIA AT MICROTUBULE ENDS

A model of the tubulin–microtubule-antimitotic drug system is represented in Fig. 2. Microtubule growth proceeds by GTP-tubulin addition to the ends. At the resolution of negative stain electron microscopy *(31)* growing ends appear blunt although they probably display a jagged appearance at the molecular or subunit level representing protofilament and sheet elongation *(32)*. The addition of drugs (L or ligand) can disrupt microtubule assembly and dynamics by three primary modes:

1. Substoichometric poisoning;
2. Alternate polymer formation; or
3. Microtubule stabilization.

Details related to structure and molecular mechanism will be discussed in later sections. (Note the terms ligand and drug are interchangeably used here, but further, the authors mean to suggest that ligand can also refer to regulatory factors that control microtubule assembly or dynamics and compete with drugs.)

3.1. Substoichiometric Vs Stoichiometric Poisoning

Substoichiometric poisoning occurs if a drug can poison the end of a microtubule. This can occur by direct addition of drug (L) or by the addition of a tubulin–drug complex (TL formed with affinity K_b and binding to microtubule ends with affinity K_i; *see* Fig. 2). For this class of drugs it is considered unlikely that a drug can bind to a microtubule end and not also bind to a free tubulin heterodimer, and thus the pathway involving TL binding is preferred *(50)*. Alterations in longitudinal or lateral contacts induced by the drug binding to microtubule ends alter the affinity of the next subunit. Depending upon the cooperativity of the contacts at the end, microtubule growth is poisoned. Slow dissociation of the drug or the drug complex can also prevent subunits at the end from dissociating thus suppressing microtubule dynamics. This mechanism (first described for colchicine) is typically known as substoichiometric poisoning *(51)*. Alternatively, in stoichiometric poisoning the ligand or drug can bind to a tubulin heterodimer in solution making a complex that has reduced or no affinity for the end. By mass action this lowers the concentration of active tubulin heterodimers. This will reduce the rate of growth (rate $= k_g \times n \times [T]_{total} / (1 + K_b \times [L])$; where n is the number of sites of subunit addition, typically 13, T is the concentration of tubulin heterodimers, k_g is the rate of growth, and K_b is the affinity of the drug for tubulin heterodimers) and favor the frequency of catastrophe and disassembly. This mechanism is known as sequestration and exhibits a stoichiometric effect on microtubule assembly implying a stoichiometric or significant fraction of the tubulin must be sequestered before inhibition is observed. The colchicine family of drugs exhibits both substoichiometric and stoichiometric modes of microtubule inhibition (52,53). For colchicine and colchicine analogs K_b for binding to tubulin heterodimers is not linked thermodynamically to K_i for binding to microtubule ends (52,53). This means some colchicine analogs are effective at inhibiting 50% of microtubule growth when only 1–2% of the free tubulin heterodimer has bound drug, whereas other analogs only achieve 50% inhibition when >90% of the free tubulin heterodimer has bound drug. Distinct analog-dependent alterations in the structure of the free heterodimer-drug complex vs the ability of the complex to disrupt lateral contacts at the microtubule end appear to be critical for differential drug activity. Recent structural information about the colchicine-binding site in a tubulin–stathmin-like domain complex instead suggests differential radial and tangential bending or kinking at the intradimer interface occurs in a colchicine analog-dependent manner (48). This differentially affects the ability of the drug–tubulin complex to bind to microtubule ends and poison growth. There is currently no analysis of the differential consequences in vivo of sequestration vs substoichiometric mechanisms. These concepts may also apply to stathmin, other end binding proteins, and drugs that bind to other sites on tubulin that sequester dimers.

3.2. Alternate, Nonmicrotubule Polymer Formation

A second mode by which drugs can disrupt microtubule assembly and dynamics is by alternate or nonmicrotubule polymer formation. Tubulin has the intrinsic ability to form small curved oligomers (often called storage forms), favored by protein concentration, GDP, MAPs, and Mg^{+2} binding (35,54–56). Alternate polymer formation (rings and spirals) is especially common with the vinca alkaloid class of antimitotic drugs (including vinca alkaloids, dolastatin, and cryptophycin). The high affinity and cooperativity exhibited in ligand-linked polymer formation (see Fig. 4) has the effect of kinetically sequestering both the drug and heterodimers in these complexes (35,57,58). This has implications for drug accumulation and turnover (59). An additional observation is that the polymers themselves can act as inhibitors of microtubule dynamics, just as a colchicine–tubulin complex does. This is represented in Fig. 2 as the addition of $(TL)_n$ to a microtubule end although in principle any drug–tubulin n-mer has microtubule inhibition activity. As vinca alkaloids bind to the interdimer interface between tubulin heterodimers and near the exchangeable nucleotide (49,60,61), the smallest active ligand-linked polymer is a 1:2 drug:tubulin complex (Fig. 2). Alternatively, because formation of a 1:1 drug–tubulin complex is weak, direct drug addition to a microtubule end most likely occurs at a similar interdimer interface (49), which creates an alternate polymer attached to the end [αβVαβ], where V in this instance represents a vinca alkaloid. This alters both longitudinal and lateral interactions with a preference for suppressing dynamics at the + end (16; see Sections 3.6–3.8). Scheme 2 represents a vinca bound to a protofilament at each microtubule end during assembly. As in normal microtubule disassembly, curvature at the + end breaks different lateral contacts than curvature introduced at

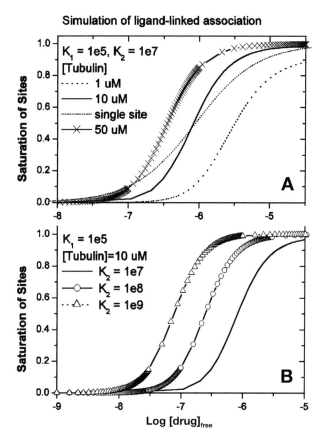

Fig. 4. Simulations of ligand-linked self-association of tubulin, where degree of association (and saturation of drug binding sites) are plotted vs log free drug concentration. K_1 corresponds to drug binding to the heterodimer, K_2 corresponds to association of the liganded complex TL to a spiral polymer *(35).* **Panel A** represents a family of assembly curves ($K_1 = 1 \times 10^5$, $K_1 = 1 \times 10^7$), as a function of tubulin concentration (1, 10, and 50 μM). These data correspond to vinblastine under typical in vitro conditions, pH 6.9. These curves highlight the importance of protein concentration in driving the reaction toward spiral assembly. To demonstrate the sharpness of the transition induced by this mechanism, a curve corresponding to noncooperative single site binding (10 μM) is also presented. **Panel B** presents a family of curves at a fixed tubulin concentration (10 μM), a fixed K_1 (1×10^5) and varying K_2 (1×10^7, 1×10^8, and 1×10^9). Note that 1% assembly or saturation of drug binding levels occur at 263, 76, and 24 nM, respectively. These data correspond to vinblastine, vincristine, and a drug with stronger assembly potential (e.g., dolastatin 10). The addition of MAPs and the crowded environment of the cytoplasm will shift these curves to lower drug concentrations. Similar polymer growth characteristics are assumed to occur at microtubule ends.

the – end. The GXP at the plus and minus end indicates vinca alkaloids induce interdimer curvature that disrupts GTP hydrolysis, and thus both GTP and slightly more so GDP *(16)* are stable in a vinca-induced interface. This is represented by placing GXP near the V site *(49).* This scheme is also sensitive to lateral contacts and is undoubtedly influenced by factors like MAPs or end binders that stabilize lateral contacts.

$$^{(-)}\ \alpha\beta_{GXP}V\alpha\beta_{GDP}\alpha\beta_{GDP}\alpha\beta_{GDP} \ldots\ldots \alpha\beta_{GDP}\alpha\beta_{GDP}\alpha\beta_{GXP}V\alpha\beta_{GTP}\ ^{(+)}$$

Scheme 2

3.3. Stabilization of Microtubules

A third and final mode by which drugs (L) can disrupt microtubule assembly and dynamics is by stabilization of microtubules. A prominent class of antimitotic drugs bind to and stabilize the microtubule lattice, represented by L binding to the microtubule wall (*see* Fig. 2). The assay to identify them involves enhancement of microtubule assembly as typically measured by turbidity or pelleting. One measures the critical concentration, C_c, the tubulin concentration below which nucleation of microtubule assembly will not occur as a function of drug concentration. C_c corresponds to the propagation constant for polymer growth, $C_c = 1/K_p$, which determines the overall free energy for ligand linked polymerization, $\Delta G = -RT \ \text{Ln} \ K_p$. Numerous thermodynamic studies (43,62–65) have investigated the role of GTP hydrolysis, pH, Mg^{+2}, temperature, and solution additives like glycerol. Dissecting the energetics of drug binding (L + M ↔ LM) from the energetics of assembly (T + M ↔ TM) complicate the interpretation of the data (*see* ref. 65 and discussion below). The overall conclusion is microtubule assembly is an entropically driven reaction ($\Delta S > 0$) involving the burying of a hydrophobic surface (the interdimer interface) and the release of water (43). The mechanism of stabilizers like taxanes involves stoichiometric binding to the lattice and stabilization of lateral protofilament contacts, with the maximum effect in vitro being at 1:1 binding ratios. Stabilizers inhibit both microtubule growth and shortening and thus suppress microtubule dynamics (66). An effect on microtubule ends is inferred.

3.4. Thermodynamic Approaches

The later discussion will involve quantitative evaluation of drug–tubulin interaction data. Based upon the introduction it can be now stated that the interaction will be the sum total of all interactions in the system, expressed and parsed as free energy as a function of solution and structure factors.

$$\Delta G_{(\text{drug–tubulin})} = \Delta G_{\text{binding}} + \Delta G_{\text{GXP}} + \Delta G_{\text{Mg}} + \Delta G_{\text{pH}} + \Delta G_{\text{polymer}}$$
$$+ \Delta G_{\text{lattice}} + \Delta G_{\text{MAP}} + \Delta G_{\text{coupling}}$$

The ΔG contributions for GXP, Mg^{+2}, and pH correspond to the discussion above for solution variables. The ΔG for binding and polymer formation reflect the linkage between the drug binding site and the drug-induced formation of polymer. The ΔG for lattice interactions reflects the plasticity of lateral interactions that can occur, especially in a microtubule lattice, and how they perturb the protofilament affinity term. The ΔG_{MAP} term reflects the contributions from other proteins or factors that regulate polymer stability and dynamics. Finally, the $\Delta G_{\text{coupling}}$ reflects the facts that every other term in the sum is dependent upon one another. For example, GDP binding to the E-site favors a curved protofilament, but adapts a straight conformation in the microtubule lattice; GTP binding to the E-site favors a straight protofilament, but can adapt a curved protofilament in rings, spirals or a stathmin–tubulin complex; these constrained states cost energy that gets tallied in the $\Delta G_{\text{coupling}}$ term. Structural data (22,48,49) suggest a major contributor to this $\Delta G_{\text{coupling}}$ term are rearrangements at the intradimer interfaces.

3.5. Polymer Mass Vs Dynamics

A normal microtubule cytoskeleton maintains a significant fraction of polymerized tubulin while also allowing dynamic instability to occur during morphogenesis and cell division *(67)*. The regulation of stability and dynamics is complex and involves tubulin concentrations, tubulin isotype distributions, GTP hydrolysis, and stabilizing and destabilizing regulatory factors *(2,6,7)*. As drugs affect both polymer stability and dynamics, the mechanism of cytotoxic action must (in a concentration dependent manner) also impact both thermodynamic and kinetic aspects of assembly. Two prominent camps exist concerning the cellular mode of action of drugs, one stressing the role of polymer levels *(68)* and a second stressing the role of microtubule dynamics *(69)*. It is clearly established that taxol resistance can be generated by reducing polymer levels, whereas colchicine or vinblastine resistance can be generated by increasing polymer levels *(1,2,4–6; see* Fig. 1). In cases where microtubule stability is the cause of resistance, these alterations also produce cells that are hypersensitive to drugs (i.e., reduced polymer levels induce colchicine and vinblastine sensitivity, whereas increased polymer levels induce taxol sensitivity). Extreme alterations in microtubule levels produce taxol- or colchicine-dependence *(6)*. Alternatively, Jordan and Wilson have documented the impact of a large numbers of effectors on microtubule dynamics, in vitro and in vivo *(38,39,50,66,69;* reviewed in ref. *35)*. All drugs tested thus far suppress dynamics, often with a + end preference. Drug-dependence suggests that drug binding stabilizes a required but unstable structure; a poison becomes therapeutic. To resolve the importance of polymer levels vs dynamics it would be helpful to know if every example of reduced polymer levels exhibits increased dynamics and vice versa? This seems unlikely as taxol suppression of dynamics rescues taxol-dependent cells *(69)*, whereas the same cells are hypersensitive to drugs like colchicine and vinblastine, which also suppress dynamics?

There is a tendency in the drug field to discount modes of action characterized by in vitro studies, claiming that clinical doses are so low that major rearrangements in the polymer levels or polymer structure are unlikely *(65,70; see* ref. *68* for opposing data and the later discussion on vinca alkaloids). It has been shown that taxol shifts the structure of kinetochore microtubule ends to a blunt assembly conformation (81% vs 32% in controls; B. McEwen, unpublished), that IC_{50} taxol doses can cause microtubule bundling *(71)*, and that the formation of functional kinetochore contacts and tension are disrupted *(72)*. These interactions undoubtedly reflect a distinct set of proteins and a distinct set of structural conformations at the + end *(7,8)*. How drugs mediate (in the case of drug-dependence) or disrupt (in the case of mitotic arrest) these interactions or conformational states appears to be the issue. That polymer levels predict these phenotypes may simply indicate the intimate connection between dynamics and polymer stability, especially in the extremes where polymer levels are toxically low or high. The hypothesis is that drugs restore a dynamic balance, especially in drug-dependent cases, between growing and shrinking + end kinetochore microtubule lattices. Taxol restores the balance to a growing conformation, at least until it again becomes toxic at higher concentrations, whereas colchicine or vinblastine restores the balance to a disassembling conformation. Thus, the suppression of dynamics may imply distinct and different (+) end conformations for each class of drugs. This would explain why taxol-resistant cells are vinca alkaloid-sensitive, the microtubule end conformation favors vinca alkaloid binding. It also implies the ability of other regulatory factors to interact with drug-stabilized ends. Thus, a resolution of the opposing views may be that dynamics and structure are inseparable, tightly coupled in the parlance of thermodynamics, especially at microtubule ends, and drug-interaction models that take this into account need to be developed and tested *(73)*.

3.6. The Colchicine Site

The colchicine family of drugs (including podophyllotoxin, combretastatin, and staganacin) bind at the intradimer interface *(48)*, near the N-site GTP, poison microtubule assembly substoichiometrically *(50,51)*, suppress microtubule dynamics *(72,74)* and induce sheet-like alternate polymer forms *(62)*. Colchicine itself (and a limited number of related compounds) induces GTPase activity, which is assumed to require formation of a straight polymer. This is based upon the concentration-dependence of GTPase activity *(75)*, the identification of α-Glu254 as the catalytic group on another dimer *(23)*, and the observation that GTPase activity only resides in straight protofilament structures like microtubules and Zn-sheets. Ravelli et al. *(48)* recently published a crystal structure of GDP-tubulin bound to RB3, a stathmin-like domain, and colchicine (Fig. 3). There are two tubulin dimers and two intradimer colchicine molecules in the structure. They conclude that colchicine sterically inhibits formation of a straight protofilament (*see* their Fig. 3C) arguing this is the mechanism of poisoning microtubule ends. This is consistent with the important role of K_i, the binding of a colchicine complex to microtubule ends, in defining the activity of colchicine analogs *(52,53)*. The problem is how can colchicine induce GTPase activity without forming a straight protofilament at some point? It has been observed by quantitative analytical ultracentrifugation (AUC) analysis that colchicine binding inhibits the formation of a human stathmin–GDP-tubulin complex by 1–2 kcal/mol *(76)*. This suggests that the structure reported by Ravelli et al. *(48)* is actually a constrained structure where the stathmin-like domain and the E-site GDP dominate the energetics, whereas colchicine binding to this complex costs energy. Thus, the Ravelli structure *(48)* is not the only possible RB3 structure and colchicine binding alone must also be able to induce a straight polymer with the necessary tertiary rearrangements required to hydrolyze GTP. Upon forming the GDP product the polymer interface curves, but at a cost ($\Delta G_{coupling}$) and the Ravelli structure reflects the cost.

Two additional factors contribute to the cytotoxicity of colchicine. Colchicine binding to tubulin is essentially irreversible *(77)*, thus making it impossible to clear from the cell without tubulin degradation. This should dramatically increase the lifetime and toxicity of colchicine in cells and tissues. Second, the GTPase activity should be a constant drain on the energy source of microtubule assembly in vitro and metabolic processes in vivo. Nonetheless, there is continuing interest in colchicine-like analogs, especially for the development of endothelial specific activity *(78)*. A novel Indanone compound, Indanocine was found to kill mutidrug resistant cancer cells during G1, a surprising and potentially new approach to tubulin-based therapy *(78)*. Indanocine resistant cells exhibited the equally surprising trait of being cross resistant to colchicine and vinblastine *(79)*. The mutation (β-Lys350Asn) turned out to involve N-site GTP binding and not the colchicine binding site. This is consistent with the Ravelli et al. *(48)* conclusions that colchicine binding is sensitive to bending at the intradimer or N-site interface just as vinblastine binding appears to be, and that mutations that alter the ability of the intradimer interface to adapt radial and tangential bending will allosterically inhibit the effects of multiple drugs. In general, this suggests that tubulin utilizes both the intradimer and the interdimer nucleotide binding sites as allosteric transducers of conformational changes, thus conveying information that selects bending modes that are then useful for construction of alternate polymer forms. It is worth emphasizing that this new view of cooperative tubulin assembly is based upon structural and mutation data *(22,48,49,79)* and supports a major role for rearrangements at both the intra- and interdimer interfaces.

The central theme for colchicine's effectiveness as a drug appears to be the affinity of the drug-complex for microtubule ends *(52,53)*. It is presumed that a single site occupied per end will suppress growth and dynamics *(51)*. Andreu and collaborators conclude that binding to microtubule ends (K_i) is not thermodynamically linked to binding to free heterodimers (K_b). This is reminiscent of microtubule assembly where irreversible GTP hydrolysis produces a polymer that exhibits nonequilibrium behavior, dynamic instability, where the off rate for GDP-subunits is not thermodynamically linked to the on rate of GTP-subunits *(67)*. Colchicine binding is essentially irreversible because of a conformational change in tubulin *(77)*. This strongly suggests irreversible colchicine binding induces straight polymer growth and GTP hydrolysis, followed by a colchicine-induced conformational change to GDP-tubulin curved polymers. It is this complex (TL) that can act as a substoichiometric inhibitor of microtubule ends. Different analogs demonstrate intrinsically different ability to induce straight or curved polymers *(52)*, consistent with their selective ability to inhibit microtubule assembly. The exact form of the inhibitory site at microtubule ends is undoubtedly a curved quaternary structure. If microtubule assembly can be 50% inhibited by 1% drug binding to the free tubulin, either the affinity of these drugs for ends is 50-fold higher than free tubulin or there is a significant cooperativity between binding of the TL complex and a conformational transition at the microtubule end that favors a paused or poisoned state. There have been two studies on the effects of colchicine on microtubule dynamics *(74,80)*. Both studies are consistent with suppression of growth and stabilization of microtubule ends. Neither of these studies compared colchicine analogs. It would be of predictive value to compare dynamics of a family of colchicine analogs where the energetics could be correlated with polymer formation, structural transitions at microtubule ends, and IC_{50} values.

3.7. The Vinca Alkaloid Site

The vinca alkaloid class of antimitotics (vinblastine, dolastatin, cryptophycin-52) bind to a hydrophobic pocket at the interdimer interface *(49)* and induce tubulin to form curved polymers, typically spirals or rings. There have been numerous quantitative studies of the energetics of vinca alkaloid induced spiral formation, primarily by Timasheff and coworkers *(54,55)*, and more recently by Correia and Lobert *(35,36,56,57)*. The simplest mechanism to describe the polymerization process is a ligand-linked mechanism where drug binds to a tubulin heterodimer with affinity K_1, and this drug–tubulin complex associates with another heterodimer or spiral polymer with affinity K_2. The data indicate that K_1 is weak, whereas the magnitude of K_2 strongly depends upon the drug moiety. Based upon the new structure of vinblastine bound to a RB3-tubulin–colchicine complex *(49)*, the magnitude of K_2 reflects interactions between a buried, and presumably nonexchangeable, vinblastine, and both α- and β-tubulin at the interdimer interface. To demonstrate the characteristics of this mechanism a simulation has been performed as a function of protein concentration (1, 10, and 50 µM; *see* Fig. 4A). The values of K_1 ($1 \times 10^5/M$) and K_2 ($1 \times 10^7/M$) correspond to in vitro data for vinblastine with pure tubulin *(35)*. Increasing protein concentration shifts the curve to the left, cooperatively (*see* noncooperative single site

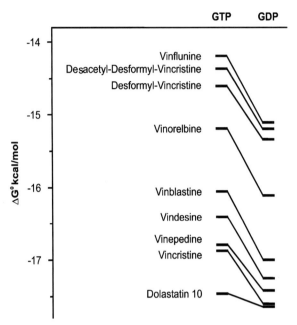

Fig. 5. The energetics of spiral formation for a family of vinca alkaloids as a function of GXP content. The free energy correspond to $\Delta G = -RT\ Ln\ K_1K_2$ in a Pipes, Mg, EGTA, 50 µM GXP, pH 6.9 solution *(36,37)*. The enhancement caused by GDP reflects the preferential formation of curved spiral polymers. GDP enhancement is expected to be true for all drugs in this class of compounds that induce polymer formation. Alternatively, GTP inhibition reflects $\Delta G_{coupling}$ or the strain induced by placing GTP at the E-site of a drug-induced curved spiral or ring polymer.

curve) driving assembly into spiral polymers by mass action. Note the degree of self-association coincides with saturation of drug binding because of the linkage of the tight self-association step (K_2) to the weaker drug binding step (K_1). This phcnomena reveals why one must use a thermodynamic linked-model for rigorous analysis of the binding data, properly taking protein and free drug concentration into account in the analysis. Otherwise the K_{app} inferred from the midpoint of the curve will vary with protein concentration *(54,55)*.

4. THE PREDICTIVE VALUE OF BINDING DATA—I

A large number of vinca analogs have been studied by these quantitative methods. A representative set of data is presented in Fig. 5, where the ΔG for the overall reaction K_1K_2 is plotted for eight vinca alkaloid analogs. Note that the formation of curved spiral polymers is enhanced in the presence of GDP by approx –0.8 kcal/mol (a factor of 4 in K) for each drug. The ability to form a spiral increases by more than 3 kcal/mol, going from the weakest (vinflunine) to the strongest (vincristine) drug. This difference primarily resides in K_2, the polymerization step, whereas K_1, drug binding to tubulin, is nearly constant ($\sim 1 \times 10^5/M$) *(35)*. This variation in K_2 corresponds to a 500-fold difference in spiraling potential, and must reflect the influence of each chemical modification on the stability of the drug–tubulin interdimer (and possibly intradimer) interface. These data allowed us to speculate that spiraling potential is a critical factor in the mechanism of action of these drugs. This was in part supported by the

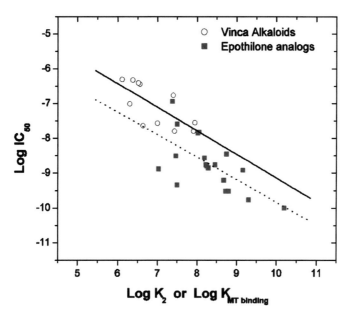

Fig. 6. Log IC_{50} vs log K for two classes of antimitotic drugs. For the vinca alkaloids the data are from AUC measurements of spiral formation *(37)* where K_2 corresponds to the addition of a drug–tubulin-GDP complex to a growing spiral polymer. (The slope of the line is –0.678 with r^2 = 0.50 and p = 0.016; for GTP-tubulin the slope = –0.704, r^2 = 0.417, p = 0.032, data are not shown.) The Epothilone data are for microtubule binding constants *(65)* and correspond to the affinity of the drug to a GDP-lattice that has been stabilized by crosslinking, done to uncouple drug binding from microtubule assembly. (The slope of the line is –0.647 with r^2 = 0.412 and p = 0.003). To view this figure in color, see the insert and the companion CD-ROM.

observation that vincristine with a large spiraling potential is cytotoxic at very low doses, both in cell culture and in patients. Alternatively, vinflunine, a new drug in clinical trials, with a very weak spiraling potential, is cytotoxic at a relatively high concentration. To investigate this further a study was initiated to measure spiraling potential by quantitative analysis of in vitro polymer formation and IC_{50} values in a leukemia cell line *(81)*. The data are presented in Fig. 6 as a log IC_{50} vs Log K_2 plot. As presented the data strongly suggest that spiraling potential inversely correlates with IC_{50} values, implying the ability to make a spiral polymer is the mechanism of action of this class of drugs. What does this mean in the context of suppression of microtubule dynamics in vivo?

The work by Andreu and coworkers *(52,53)* on colchicine analogs suggested the relevant parameter is K_1, the affinity of the drug complex for microtubule ends. Likewise, a number of studies have investigated the effect of low vinblastine concentrations on microtubule dynamics *(39,82)* and concluded there is a steep and cooperative suppression of both growth and shortening rates at the + end. At higher concentrations both ends are affected. Consistent with the model, if vinblastine or any vinca alkaloid complex binds to the end of a microtubule, longitudinal and lateral contacts will be disrupted. The minimum complex required corresponds to a 2:1 (T_2L; *see* Fig. 2) tubulin–drug complex. This necessarily introduces curvature at the inter- and intradimer interfaces *(49)*, and cooperatively poisons microtubule growth. To investigate this further the formation of spiral polymers has been simulated as a function of K_2 (10^7, 10^8, $10^9/M$) or spiraling potential (Fig. 4B). (As noted earlier, as the vinca alkaloid-binding site is at the interdimer interface *(49,60,61)*, polymer assembly and drug binding are structurally and energetically coupled such that ligand saturation and spiral polymerization are coincident.) These data correspond to in vitro conditions for vinblastine, vincristine, and possibly dolastatin 10 (*see* Fig. 5). At 10 nM drug only a few percent of tubulin forms spiral polymers at the highest K_2 value. This might lead one to believe this mechanism cannot account for suppression of microtubule assembly. However, recall the ability of colchicine analogs to suppress 50% of microtubule growth when only 1% of the free tubulin is liganded *(52)*. In this simulation 1% assembly levels occur at 263, 76, and 24 nM, respectively. Second, these polymerization constants correspond to in vitro conditions; both macromolecular crowding *(83)* and the presence of MAPs should shift these curves to the left or to lower drug concentrations. The extent of this effect is not currently known, but 1–2 orders of magnitude increase in K_2 would provide sufficient energy to make typical IC_{50} values easily achievable by this mechanism (*see* the IC_{50} scale in Fig. 6). Thus, it seems reasonable to propose that the in vivo mechanism of action of vinca alkaloids is the formation of spiral polymers at the end of microtubules. Spiral polymers growing off the ends of microtubules have been visualized at high drug concentrations *(40)*. Analogous to the mode of action of colchicine, just one T_2L complex would be required (hence substoichiometric) to suppress microtubule dynamics. Additional drug-induced growth off this (nucleation) site would constitute cooperative association, enhancement of drug potency, and kinetic suppression of dynamics.

Thus, rather than stressing the low concentration of free drug, one must stress the activity of drug-tubulin polymers and their ability to bind to or nucleate upon microtubule ends. The distinction made in the Perez-Ramirez et al. *(53)* colchicine study between binding to tubulin K_b and inhibiting microtubule assembly K_i undoubtedly applies here, with the addition that a ligand-linked mechanism *(54)* implicitly adds cooperativity to the mode of action of this class of drugs. As the vinca site is in the interdimer interface, all alternate polymers formed by the vinca class must necessarily disrupt the straight protofilament conformation, although one can predict drugs might bind with higher affinity to structures at microtubule ends than to free spiral (or ring) polymers. For example, double spirals are known to form *(84)* and these can be preferentially nucleated at microtubule ends simply because of the side-by-side arrangement of protofilaments *(40)*. Cooperative interactions with MAPs or factors that favor curved ends may be important, although small differences in polymer structure may contribute positive or negative effects on inhibition *(7,8)*. It has been shown that stathmin binding to tubulin competes with spiral formation *(76)*, especially at low pH. At higher pH there is evidence for stathmin binding directly to spirals, analogous to the reported ability to bind to microtubule ends *(85)*. Furthermore, it is now known vinblastine can bind to a stathmin–tubulin complex *(49)*. The implication of these observations for drug resistance is not clear, but may represent an explanation for the reported upregulation of stathmin in cancer cells *(86,87)*.

There have been a large number of recent studies on other members of this class of drugs, including Dolastatin 10, Cryptophycin 1, Cryptophycin 52, and Hemiasterlin *(60,61,88–93)*. These compounds also bind at the inter-dimer vinca site, induce single tubulin rings, and suppress microtubule dynamics by binding to protofilament ends. In general, Cryptophycin rings are smaller (24 nm diameter; 8–9 tubulin heterodimers) than Dolastatin 10 or Hemiasterlin rings (44.6 nm; *see* ref. *60,61*) reflecting differences in the radial and tangential bending induced by drug binding. As a group, they are extremely potent inhibitors of cell proliferation, presumably because of their tight formation of complexes at microtubule ends *(86,89)*. Unfortunately, there are no quantitative data on their affinity for tubulin rings; either the authors do not report them, or they form complexes in the picomolar concentration range and the affinity has not been estimated *(93)*. The estimates for Dolastatin 10 (Fig. 5) come from studies on tubulin in the micromolar concentration range and must be taken as a lower limit. Nonetheless, Boukaria et al. *(93)* do report an inverse correlation of cytotoxicity with apparent stability of the drug-induced rings, i.e., small IC_{50} vs large ring formation affinity. The cooperative polymerization induced by these drugs should act as a sink for drug binding with ring or spiral polymer assembly soaking up any available drug and shifting passive transport toward assembly-linked drug uptake. This has been described for Dolastatin 10 and Cryptophycin 1 and 52 *(60,61,89)* and implies that drugs that bind tightly and cooperatively to polymers will exhibit a narrow therapeutic window because all cells will accumulate toxic concentrations of drug *(81,89)*. This may explain why Dolastatin 10 failed phase 2 drug studies, and suggests a similar fate for other drugs of this class that induce high affinity, highly cooperative polymer assembly.

There are additional members of the vinca alkaloid class of drugs that bind competitively to the same site as vinblastine but appear to inhibit all forms of tubulin assembly, including rings and spiral. Maytansine is the best example of this subclass of drugs *(94)*. It is currently not understood how this binding mode should be compared with alternate polymer inducers. A sequestering mechanism seems to apply and thus a comparison of binding affinity vs spiral formation seems appropriate. But colchicine drugs distinguish themselves by their ability to poison microtubule ends *(52,53)*, and thus microtubule end binding by either the drug or a drug–tubulin complex may be more relevant. A microtubule dynamics study of a halichondrin B analog (E7389) demonstrated that this compound suppresses microtubule growth but not disassembly, rescue, or catastrophe *(95)*. Thus, binding to a single site on microtubule ends might prove to be the correct mode of action to rank drugs like maytansine. These compounds should also suppress binding to tubulin by sequestering proteins like stathmin with consequences that are not well understood. It is conceivable that new functions for tubulin complexes will be revealed by studies with these drugs.

4.1. Microtubule Stabilizers (Taxol)

This class of natural compounds (taxanes, epothilones A and B, eleutherobin, and discodermolide) bind to microtubules at the M loop, located between B7 and H9 on the β-subunit, and stabilize lateral contacts between protofilaments, primarily by electrostatic interactions *(16,94)*. The structure of taxol bound to Zn-induced sheets revealed taxol makes contact with the S9-S10 loop and interacts with H7, the core helix *(16,96)*, and the loop between S7 and H9. Assembly in the presence of stoichiometric amounts of taxol increases subunit spacing by 0.15 nm (4–4.15 nm), thus counteracting the 3–6% shortening induced by GTP hydrolysis *(97,98)*. Taxol binding reduces the number of protofilaments from 13 to 12 *(99)* although other workers suggest addition after assembly has only a minor effect *(97)*. Taxol also increases surface lattice defects two to sixfold *(97; see* discussion of katanin below). These interaction sites are consistent with missense mutations that confer drug resistance *(100,101)* although assignment of a mutation site as a drug binding site need not be absolute *(see* below; ref. *102)* and is not expected to be frequent in diploid cells (two copies per cell) expressing multiple tubu-lin genes *(6)*. Expression of the nonmutated genes should dilute out the effect of the mutated gene product unless the mutated gene product is the major species. Surprisingly, taxol differentially affects dynamics of microtubule ends, suppressing the rate and extent of shortening at the + end at low drug concentrations, whereas inhibit-ing growth and shortening at both ends at higher drug concentrations *(66)*. Recent stud-ies in vivo with fluorescent taxoids demonstrate a preferential interaction with newly polymerized sites *(103)*. The polarity of the microtubule and the structural basis of GTP-hydrolysis requires the + end, but not the − end, of a growing microtubule will have GTP-tubulin in the cap. If taxol binding increases the subunit spacing, and GTP-hydrolysis decreases subunit spacing, then by thermodynamic linkage taxol prefers binding to a GTP lattice structure, and thus the GTP-cap will have a higher affinity for taxol than the GDP-microtubule core. A comparison of microtubule assembly (*Cc* measurements) in GTP vs the nonhydrolyzable analog GMPCPP *(43)* suggests at least a 10-fold (−1.3 kcal/mol) stability of a GTP-lattice over a GDP-lattice, and implies a similar energetic preference for taxol binding to GTP-sites. Other structural constraints ($\Delta G_{coupling}$) may apply, but this prediction has not been tested by cryo-EM methods because the addition of GMPCPP to Zn-induced sheets cause their dissolution. Note that as the number of taxol binding sites at the + end GTP-cap are small relative to the number of total sites in a microtubule (~13 vs ~1250/μM), the effects of taxol on micro-tubule dynamics are expected to be gradual *(66)*. An equilibrium assumption allows us to calculate:

$$\text{Preferential binding to + end} = [\text{GTP-Cap sites}] \times K_{end} / [\text{GDP-sites}] \times K_{walls}$$

During rapid growth microtubules extend as sheet projections *(32,104)*. If the same lateral cap model applies to sheets, only the extreme end of the sheet will display high affinity taxol sites. By this same structural argument, taxol must also inhibit the interac-tion of microtubules with dynamics regulators that prefer peeling protofilaments (MCAK, stathmin), whereas activating the binding of + end binders that prefer a GTP-cap (MAPs, CLIP-170; refs.*7,8,105*). Given the plasticity of microtubule structure, this may be too simplistic and many similar structures *(30)* may exist that favor binding of different classes of proteins *(106,107)*. This has yet to be extensively tested.

5. THE PREDICTIVE VALUE OF BINDING DATA—II

Buey et al. *(65)* performed a thermodynamic analysis of epothilone analogs and demonstrated a relationship between binding affinity, microtubule stabilization, and cytotoxicty. Because drug binding is tight and linked to microtubule assembly they used crosslinked microtubules and a competitive binding assay (displacement of Flutax-2) to extract the energetics. The binding affinity is an excellent predictor of microtubule sta-bilization (or critical concentration *Cc*, plotted as log–log; $r^2 = 0.62$) and an even better predictor of IC$_{50}$ ($r^2 = 0.76$ for a limited data set; *see* Fig. 6). As with the vinca alkaloid correlation data described above, tight binding affinity predicts low IC$_{50}$ whereas weak binding affinity predicts a high IC$_{50}$. The agreement with the vinca alkaloid data seems to strengthen the correlation (although this agreement corresponds to measurements in two different cell lines and should be treated with caution). The authors argue that bind-ing affinity is a better tool for drug design than *Cc* because clinically the in vivo drug concentration is too low to significantly affect microtubule stability. This is directly related to questions about the role of microtubule structure, polymer levels, and dynamics play in cytotoxicity discussed earlier *(64,65,81)*. The correlation with IC_{50} suggests that binding affinity reflects events that have a direct impact upon microtubule + end structure, possibly suppression of microtubule flux or treadmilling, and passage through mitotic check points. The simplest interpretation of this correlation data is that the binding sites implicated in the IC_{50} measurements directly reflect the same interactions probed by the binding experiments and follow the same rank order, strongest to weakest, as the binding data. Other factors like MAP binding or kinetochore attachment may alter the affinity, but these effects are likely to be the same or similar for each drug of that class. Otherwise one must hypothesize different binding sites in vitro and in vivo. A preferred role for microtubule ends or the GTP-cap are highly likely.

5.1. Mutations That Affect Microtubule Stability

As stability of alternate polymer forms contributes to effectiveness of drugs, it is not surprising that mutations that affect polymer stability are mechanisms for drug resistance *(102,108,109)*. Thus, if taxol stabilizes microtubules, then mutations that destabilize microtubules will cause taxol-resistance. The mutations can occur in either α- or β-tubulin *(4)*, and they can affect lateral or longitudinal contacts *(100)*. Cabral *(6)* has argued convincingly that destabilizing mutations should be more prevalent than taxol binding site mutations except in cases where a single isotype predominates in the cell line or tissue. In general, there is a paucity of structural information about the role of the tubulin sequence on function, especially microtubule formation. This is because of the inability to do overexpression and site directed mutagenesis studies with mammalian tubulins. Consequently, the best data come from mutational work in yeast *(110–112)* and comparative sequence analysis between Antarctic fish tubulin and mammalian tubulins *(113)*. The Antarctic fish tubulins assemble extremely well at the body temperature of the fish (near 0°C) displaying the same critical concentration of comparative mammalian species at their body temperatures (for example, human at 37°C) with significantly less dynamic behavior. This is accomplished by sets of unique residue substitutions that map to lateral, interprotofilament contacts and to the hydrophobic cores of both α- and β-tubulin (Fig. 7). A single comparable mutation in yeast, F200Y, does not significantly affect temperature stability, although it does cause drug resistance and may influence microtubule dynamics through + end interactions with other regulatory complexes. It is apparent that drug-induced mutations affecting microtubule stability have become a major source of information about mammalian and human tubulin structure-function. Similar work needs to be pursued with tubulin isotypes. For example, what exactly is it about the sequence and structure of rat β-tubulin class 3 and 5 that destabilize microtubules, poisons cells and makes them taxol dependent *(114,115)*?

5.2. The Severing Activity of Katanin

The interaction of tubulin (T) and liganded-tubulin (TL) with the microtubules walls (Fig. 2) represents subunit rearrangements that must occur during lattice reorganization. This primarily refers to katanin-mediated microtubule breaking at lattice defects *(116)*. It can reflect subunit exchange induced by taxol or taxotere binding, which causes a change in protofilament number *(117)*. Or it reflects normal transitions between different protofilament numbers observed in typical microtubule distributions *(118,119)*. Katanin hexameric oligomers target these lattice defects (or protofilament transition

Fig. 7. A ribbon diagram representation of 1JFF.pdb, the structure of tubulin derived from Zn-induced sheets *(17)*, where the mutations that cause stabilization in cold adapted fish are highlighted in red *(111)*, was rendered with VMD. The N-site GTP and E-site GDP are marked in yellow with β-tubulin at the top of the figure. A Mg ion is labeled black. Note the mutations involve lateral contacts at the M loop and interior residues in both α- and β-tubulin. How these interior mutations effect polymer stability and dynamics are not fully understood, but may allosterically interact through both lateral interfaces and longitudinal intra- and interdimer interfaces. These mutations, if present in mammalian cells, are anticipated to cause vinca alkaloid resistance, as described in the text *(see* Fig. 1B). To view this figure in color, see the insert and the companion CD-ROM.

sites) and facilitate subunit loss and microtubule severing *(120)*. Microtubules exhibit a very low frequency of spontaneous severing or tubulin dissociation from the walls (k_{off} = 10^{-8} s^{-1}; ref. *121*), and thus katanin acts as an ATPase-dependent cooperative enzyme *(122)*. This activity causes microtubule depolymerization at the minus-end during metaphase poleward flux *(123)* and is presumably required for chromosome segregation. Although katanin will sever taxol stabilized microtubules there is no quantitative data on the effect of drugs or the GTP analog GMPCPP. One can certainly presume that factors that stabilize the microtubule lattice (taxol, GMPCPP) will resist katanin activity or the frequency of katinin-induced severing. Taxol suppresses microtubule flux in mitotic spindles *(124)* although other factors may contribute to this decrease in dynamics (e.g., Eg5) *(125)*. It is not known if other ligands (TL, e.g., colchicine) can participate in this process, or what their impact is on katanin activity.

6. SUMMARY

This review has focused on the molecular mechanism of action of tubulin interacting compounds that exhibit antimitotic activity. The emphasis has been on the quantitative linkage between drug binding, polymorphic tubulin structures, and the solution and regulatory factors that influence preferred binding modes and the polymer forms stabilized. Numerous other factors eventually come into play upon mitotic arrest, including apoptotic response and activation or avoidance of cell death pathways. Although critical to drug efficacy, these have not been the focus of this review. The intent of this overview is to introduce the idea that drugs and regulatory factors share numerous common features, and thus their overlapping activities can be allosterically coupled, enhancing or diminishing drug effects, and leading to unexplored mechanisms or features of drug resistance. Only limited examples were discussed and further quantitative work needs to be done to investigate the importance of these factors as predictive variables. The role of regulatory factors may be especially pertinent to the clinical need to develop drugs that exhibit cell type and tumor specificity.

ACKNOWLEDGMENTS

The author and his group thank Fernando Cabral, Susan Bane, Bruce McEwen, Eva Nogales, and David Odde for helpful discussions during the writing of this chapter. They especially thank Tito Fojo for his critical and helpful assistance during this process. The authors take full responsibility for the content and opinions expressed therein.

REFERENCES

1. Cabral F. The isolation of Chinese hamster ovary cell mutants requiring the continuous presence of taxol for cell division. J Cell Biol 1983;97:22–29.
2. Cabral F, Barlow SB. Resistance to antimitotic agents as genetic probes of microtubule structure and function. Pharmac Ther 1991:52:159–171.
3. Rudolph JE, Kimble M, Hoylt HD, Subler MA, Raff EC. *Drosophilia* β-tubulin sequences: a developmentally regulated isoform β3, the testes-specific β2, and an assembly-defective mutation of the testes specific isoform reveal both an ancient divergence in metazoan isotypes and structural constraints of the β-tubulin function. Mol Cell Biol 1987:7: 2231–2242.
4. Schibler MJ, Cabral F. Maytansine mutants of Chinese hamster ovary cells with an alteraction in α-tubulin. Cancer J Bioche 1985;63:503–510.
5. Schibler MJ, Huang B. The col$^r_4$ and col$^r_{15}$ β-tubulin mutations in *Chlamydomonas reinhardtii* confer altered sensitivity to microtubule inhibitors and herbicides by enhancing microtubule stability. J Cell Biol 1991;113:605–614.
6. Cabral F. Factors determining cellular mechanism of resistance to antimitotic drugs. Drug Res Updates 2001;4:3–8.
7. Cassimeris L. Accessory protein regulation of microtubule dynamics throughout the cell cycle. Curr Opin Cell Biol 1999;11:134–141.
8. Howard J, Hyman AA. Dynamics and mechanics of the microtubule plus end. Nature 2003;422:753–758.
9. Timasheff SN, Grisham LM. In vitro assembly of cytoplasmic microtubules. Ann Rev Biochem 1980;49:565.
10. Correia JJ, Williams RC Jr. Mechanisms of assembly and disassembly of microtubules. Ann Rev Biophys Bioeng 1983;12:211–235.
11. Engelborghs Y. Physiochemical aspects of tubulin-interacting antimitotic drugs. In:Avila J, ed. Microtubule Proteins. CRC Press, Inc., Boca Raton, 1990:1–35.
12. Erickson HP, O'Brien ET. Microtubule dynamic instability and GTP hydrolysis. Ann Rev Biophys Biomol Struct 1992;21:145–166.

13. Purich DL, Kristofferson D. Microtubule assembly: a review of progress, principles, and perspectives Adv Protein Chem 1984;36:133–212.

14. Desai A, Mitchison TJ. Microtubule polymerization dynamics. Annu Rev Cell Dev Biol 1997;13:83–117.

15. Correia JJ. Effects of antimitotic agents on tubulin-nucleotide interactions Pharmacol Ther 1991;52:127–147.

16. Correia JJ, Lobert S. Physiochemical aspects of tubulin-interacting antimitotic drugs. Curr Pharm Design 2001;7:1213–1228.

17. Nogales E, Wolff SG, Downing KH. Structure of the αβ tubulin dimer by electron crystallography. Nature 1998;391:199–203.

18. Nogales E, Whittaler M, Milligan RA, Downing KH. High resolution structure of the microtubule. Cell 1999;96:79–88.

19. Lowe J, Li H, Downing KH, Nogales E. Refined structure of αβ–tubulin at 3.5 Å resolution. J Mol Biol 2001;313:1045–1057.

20. Meuer-Grob P, Kasparian J, Wade RH. Microtubule structure at improved resolution. Biochem 2001;40:8000–8008.

21. Li H, DeRosier DJ, Nicholson WV, Nogales E, Downing KH. Microtubule structure at 8 Å resolution. Structure 2002;10:1317–1328.

22. Krebs A, Goldie KN, Hoenger A. Structural rearrangements in tubulin following microtubule formation. EMBO Rep 2005;6:227–232.

23. Nogales E, Downing KH, Amos LA, Lowe J. Tubulin and FtsZ form a distinct family of GTPases Nat Struct Biol 1998;5:451–458.

24. Nogales E. Structural Insights into microtubule function. Annu Rev Biochem 2000;69:277–302.

25. Mitchison TJ. Localization of an exchangeable GTP binding site at the plus end of microtubules. Science 1993;261:1044–1047.

26. Mitchison TJ, Kirschner M. Dynamic instability of microtubule growth. Nature 1984;312:237–242.

27. Horio T, Hotani H. Visualization of the dynamic instability of individual microtubules by dark-field microsciopy. Nature 1986;321:605–607.

28. Kowalski RJ, Williams RC Jr. Unambiguous classification of microtubule-ends in vitro: dynamic properties of the plus- and minus-ends. Cell Motil Cytoskel 1993;26:282–290.

29. Tran PT, Joshi P, Salmon ED. How tubulin subunits are lost from the shortening ends of microtubules. J Struc Biol 1997;118:107–118.

30 Chretien D, Fuller SD. Microtubules switch occasionally into unfavorable configurations during elongation. J Mol Biol 2000;298:663–676.

31. Mandelkow EM, Mandelkow E, Milligan RA. Microtubule dynamics and microtubule caps: A time-resolved cryo-electron microscopy study. J Cell Biol 1991;114:977–991.

32. Chretien D, Fuller SD, Karsenti EJ. Structure of growing microtubule ends: two-dimensional sheets close into tubes at variable rates. Cell Biol 1995;129:1311–1328.

33. Gildersleeve RF, Cross AR, Cullen KE, Fagen AP, Williams RC Jr. Microtubules grow and shorten at intrinsically variable rates. J Biol Chem 1992;267:7995–8006.

34. Odde DJ, Cassimeris L, Buettner HM. Kinetics of microtubule catastrophe assessed by probabilistic analysis. Biophys J 1995;69:796–802.

35. Lobert S, Correia JJ. Energetics of vinca alkaloid interactions with tubulin. Methods Enzymol Energ Macromol Part C (2000) 323:77–103.

36. Vulevic B, Lobert S, Correia JJ. Role of guanine nucleotides in the vinca alkaloid-induced self association of tubulin: effects of GMPCPP and GMPCP. Biochemistry 1997;36:12,828–12,835.

37. Caplow M, Ruhlen RL, Shanks J. The free energy of hydrolysis of a microtubule-bound nucleotide triphosphate is near zero: all of the free energy from hydrolysis is stored in the microtubule. J Cell Biol 1994;127:779–788.

38. Toso RJ, Jordan MA, Farrell KW, Matsumoto B, Wilson L. Kinetic stabilization of microtubule dynamic instability in vitro by vinblastine. Biochem 1993;32:1285–1293.

39. Wilson L, Jordan MA, Morse A, Margolis RL. Interaction of vinblastine with steady-state microtubules in vitro. J Mol Biol 1983;159:125–149.

40. Desai A, Verma S, Mitchison TJ, Walczak CE. Kin I kinesins are microtubule-destabilizing enzymes. Cell 1999;96:69–78.

41. Nogales E, Wang H-W, Niederstrasser H. Tubulin rings: which way do they curve. Curr Opin Struct Biol 2003;13:256–261.

42. Wang H-W, Nogales Eva. The nucleotide-dependent bending flexibility of tubulin regulates micro-tubule assembly. Nature 2005, 435:911–915.

43. Vulevic B, Correia JJ. Thermodynamic and Structural Analysis of Microtubule Assembly: The Role of GTP Hydrolysis. Biophys J 1997;72:1357–1375.

44. Lobert S, Hennington BS, Correia JJ. Multiple Sites for Subtilisin Cleavage of Tubulin: Effects of Divalent Cations. Cell Motil Cytoskel 1993;25:282–297.

45. Peyrot V, Briand C, Andreu JM. C-terminal cleavage of tubulin by subtilisin enhances ring formation. ABB 1990;279:328–337.

46. Lee JC, Timasheff SN. In vitro reconstitution of calf brain microtubules:effects of solution variables. Biochemistry 1977;16:1754–1764.

47. Rai SS, Wolff J. The C terminus of β-tubulin regulates vinblastine-induced tubulin polymerization. PNAS 1998;95:4253–4257.

48. Ravelli RBG, Gigant B, Curmi PA, et al. Insight into tubulin regulation from a complex with colchicines and a stathmin-like domain. Nature 2004;428:198–202.

49. Gigant B, Wang C, Ravelli RBG, et al. Structural basis for the regulation of tubulin by vinblastine. Nature 2005;435:519–527.

50. Skoufias DA, Wilson L. Mechanism of inhibition of microtubule polymerization by colchicine: Inhibitory potencies of unliganded colchicine and tubulin-colchicine complexes. Biochemistry 1992;31:738–746.

51. Margolis RL, Wilson L. Addition of colchicine-tubulin complex to microtubule ends: The mechanism of substoichiometric colchicine poisoning. PNAS 1977;74:3466–3470.

52. Barbier P, Peyrot V, Leynadier D, Andreu JM. The active GTP- and GDP-liganded states of tubulin are distinguished by the binding of chiral isomers of ethyl 5-amino-2-methyl-1,2-dihydro-3-phenylpyrido [3,4-b]pyrazin-7-yl carbamate. Biochemistry 1998;37:758–768.

53. Perez-Ramirez B, Andreu JM, Gorbuoff MJ, Timasheff SN. Stoichiometric and substoichiometric inhibition of tubulin self-assembly by colchicine analogues. Biochemistry 1996;35:3277–3285.

54. Na GC, Timasheff SN. Velocity sedimentation study of ligand-induced proteins self-association. Methods Enzymol 1985;117:459.

55. Na GC, Timasheff SN. Measurement and analysis of ligand-binding isotherms linked to protein self-association. Methods Enzymol 1985;117:496.

56. Sontag CA, Stafford WF, Correia JJ. A comparison of weight average and direct boundary fitting of sedimentation velocity data for indefinite polymerizing systems. Biophys Chem 2004;108:215–230.

57. Lobert S, Vulevic B, Correia JJ. Interaction of vinca alkaloids with tubulin: a comparison of vinblastine, vincristine and vinorelbine. Biochemistry 1996;35:6806–6814.

58. Theusius D, Dessen P, Jallon JM. Mechanism of bovine liver glutamate dehydrogenase self-assembly. I. Kinetic evidence of for a random association of polymer units. J Mol Biol 1975;92:413–432.

59. Vierdier-Pinard P, Kelper JA, Petit GR, Hamel E. Sustained intracellular retention of Dolaststain 10 causes its potent antimitotic activity. Mol Pharm 2000;57:180–187.

60. Lo M-C, Aulabaugh A, Krishnamurthy G, et al. Probing the interaction of HTI-286 with tubulin using a Stilbene analogue. JACS 2004;126:9898–9899.

61. Krishnamurthy G, Cheng W, Lo MC, et al. Biophysical characterization of the interactions of HTI-286 with tubulin heterodimer and microtubules. Biochemistry 2003;42:13,484–13,495.

62. Andreu JM, Wagenknecht T, Timasheff SN. Polymerization of the tubulin-colchicine complex: Relationship to microtubule assembly. Biochemistry 1983;22:1556–1566.

63. Diaz JF, Menendez M, Andreu JM. Thermodynamics of ligand-induced assembly of tubulin. Biochemistry 1993;32:10,067–10,077.

64. Diaz JF, Andreu JM. Assembly of purified GDP-tubulin into microtubules induced by taxol and tax-otere: Reversibility, ligand stoichiometry, and competition. Biochemistry 1993;32:2747–2755.

65. Buey RM, Diaz JF, Andreu JM, et al. Interaction of Epothilone analogs with the Paclitaxel binding site: Relationship between binding affinity, microtubule stabilization, and cytotoxicity. Chem Biol 2004;11:225–236.

66. Derry WB, Wilson L, Jordan MA. Substoichiometric binding of taxol suppresses microtubule dynamics. Biochemistry 1995;34:2203–2211.

67. Kirschner M, Mitchison T. Beyond self-assembly: from microtubules to morphogenesis. Cell 1986;45: 329–342.

68. Barlow SB, Gonzalez-Garay ML, Cabral F. Paclitaxel-dependent mutants have severely reduced microtubule assembly and reduced tubulin synthesis. J Cell Sci 2002;115:3469–3478.

69. Goncalves A, Braguer D, Kamath K, et al. Resistance to Taxol in lung cancer cells associated with increased microtubule dynamics. PNAS 2001;98:11,737–11,741.

70. Boukari H, Nossal R, Sackett DC. Stability of drug-induced tubulin rings by fluorescence correlation spectroscopy. Biochemistry 2003;42:1291–1300.

71. Rowinsky EK, Donehower RC, Jones RJ, Tucker RW. Microtubule changes and cytotoxicity in leukemic cell lines treated with taxol. Cancer Res 1988;48:4093–4100.

72. Skoufias DA, Andreassen PR, Lacroix FB, Wilson L, Margolis RL. Mammalian mad2 and bub1/bubR1 recognize distinct spindle-attachment and kinetochore-tension checkpoints. PNAS 2001;98:4492–4497.

73. Schlistra MJ, Martin SR, Bayley PM. The effect of podophyllotoxin on microtubule dynamics. J Biol Chem 1989;264:8827–8834.

74. Vandecanedelaere A, Martin SR, Schlistra MJ, Bayley PM. Effects of the tubulin-colchicine complex on microtubule instability. Biochemistry 1994;33:2792–2801.

75. Heusele C, Carlier M-F. GTPase activity of the tubulin-colchicine in relation with tubulin-tubulin interactions. Biochem Biophys Res Comm 1981;103:332–338.

76. Sontag C, Stafford W, Lobert S, Alday H, Correia JJ. OP18/stathmin competes with vinca alkaloid-induced tubulin spiral formation. 2007, submitted.

77. Pyles E, Bane Hastie S. Effect of the B ring and the C-7 substituent on the kinetics of colchicinoid-tubulin associations. Biochemistry 1993;32:2329–2336.

78. Leoni LM, Hamel E, Genini D, et al. Indanocine, a microtubule-binding indanone and a selective inducer of apoptosis in multidrug-resistant cancer cells. J Nat Cancer Insti 2000;92:217–224.

79. Hua XH, Genini D, Gussio R, et al. Biochemical genetic analysis of indanocine resistance in human leukemia1. Cancer Res 2001;61:7248–7254.

80. Panda D, Daijo JE, Jordan MA, Wilson L. Kinetic stabilization of microtubule dynamics at steady state in vitro by substoichiometric concentrations of tubulin-colchicine complex. Biochemistry 1995;34:9921–9929.

81. Lobert S, Fahy J, Hill BT, Duflos A, Entievant C, Correia JJ. Vinca Alkaloid-Induced Tubulin Spiral Formation Correlates with Cytotoxicity in the Leukemic L1210 Cell Line. Biochemistry 2000;39: 12,053–12,062.

82. Panda D, Jordan MA, Chu KC, Wilson L. Differential effects of vinblastine on polymerization and dynamics at opposite microtubule ends. J Biol Chem 1996;271:29,807–29,812.

83. Hall D, Minton AP. Effects of inert volume-excluding macromolecules on protein fiber formation. II. Kinetic models for nucleated fiber growth. Biophy Chem 2004;107:299–316.

84. Nogales E, Medrano FJ, Diakun GP, Mant GR, Towns-Andrews E, Bordas J. The effect of tempera-ture on the structure of vinblastine-induced polymers of purified tubulin: detection of a reversible conformational change. J Mol Biol 1995;254:416–430.

85. Belmont CD, Mitchison TJ. Identification of a protein that interacts with tubulin dimers and increases the catastrophe rate of microtubules. Cell 1996;84:623–631.

86. Alli E, Bash-Babula J, Yang J-M, Hait WN. Effect of Stathmin on the Sensitivity to Antimicrotubule Drugs in Human Breast Cancer. Cancer Res 2002;62:6864–6869.

87. Iancu C, Mistry SJ, Arkin S, Wallenstein S, Atweh GF. Effects of stathmin inhibition on the mitotic spindle. J Cell Sci 2001;114:909–916.

88. Panda D, Himes RH, Moore RE, Wilson L, Jordan MA. Mechanism of action of the unusually potent microtubule inhibitor Cryptophycin 1. Biochemistry 1997;36:12,948.

89. Panda D, DeLuca K, Williams D, Jordan MA, Wilson L. Antiproliferative mechanism of action of cryptophycin-52: Kinetic stabilization of microtubule dynamics by high-affinity binding to micro-tubule ends. Proc Natl Acad Sci USA 1998;95:9313.

90. Bai R, Durso NA, Sackett DL, Hamel E. Interactions of the sponge-derived antimitotic tripeptide Hemiasterlin with tubulin: Comparison with Dolastatin 10 and Cryptophycin 1. Biochemistry 1999; 38:14,302–14,310.

91. Barbier P, Gregoire C, Devred F, Sarrazin M, Peyrot V. In vitro effect of Cryptophycin 52 on micro-tubule assembly and tubulin: Molecular modeling of the mechanism of action of a new antimitotic drug. Biochemistry 2001;40:13,510–13,519.

92. Watts NR, Cheng N, West W, Steven A, Sackett DL. The Cryptophycin-tubulin ring structure indi-cates two points of curvature in the tubulin dimer. Biochemistry 2002;41:12,662–12,669.

93. Boukari H, Nossal R, Sackett DC. Stability of drug-induced tubulin rings by fluorescence correlation spectroscopy. Biochemistry 2003;42:1291–1300.

94. Huang AB, Lin CM, Hamel E. Maytansine inhibits nucleotide binding at the exchangeable site of tubulin. Biochem Biophys Res Commun 1985;128:1239–1246.

95. Kamath K, Wilson L, Cabral F, Jordan MA. βIII-tubulin induces paclitaxel resistance in association with reduced effects on microtubule dynamic instability. J Biol Chem 2005;280:12,902–12,907.

96. Amos LA, Lowe J. How taxol stabalises microtubule structure. Chem Biol 1999;6:R65–R69.

97. Arnal I, Wade RH. How does taxol stabilize microtubules? Curr Biol 1995;5:900–908.

98. Hyman AA, Chrétien D, Arnal I, Wade RH. Structural changes accompanying GTP hydrolysis in microtubules: information from a slowly hydrolyzable analogue guanylyl-(a,b)-methylenediphosphonate. J Cell Biol 1995;128:117–125.

99. Diaz JF, Valpuesta JM, Chacon P, Diakun G, Andreu JM. Changes in microtubule protofilament number induced by taxol binding to an easily accessible site—internal microtubule dynamics. J Biol Chem 1998;273:33,803–33,810.

100. Giannakakou P, Gusso R, Nogales E, et al. A common pharmacophore for epothilone and taxanes: Molecular basis for drug resistance conferred by tubulun mutations in human cancer cells. Proc Natl Acad Sci USA 2000;97:2904–2909.

101. Giannakakou P, Sackett D, Kang YK, Butters JT, Fojo T. Paclitaxel-resistant human ovarian cancer cells have mutant β-tubulins that exhibit impaired paclitaxel-driven polymerization J Biol Chem 1997;272:17,118–17,125.

102. Gonzalez-Garay ML, Chang L, Blade K, Menick DR, Cabral F. A β-Tubulin leucine cluster involved in microtubule assembly and paclitaxel resistance. J Biol Chem 1999;274:23,875–23,882.

103. Evangelio JA, Abal M, Barasoain I, et al. Fluorescent taxoids as probes of the microtubule cytoskeleton. Cell Motil Cytoskel 1998;39:73–90.

104. Arnal I, Karsenti E, Hyman AA. Structural transitions at microtubule ends correlate with their dynamic properties in Xenopus egg extracts. J Cell Biol 2000;149:767–774.

105. Diamantopoulos GS, Perez F, Goodson HV, et al. Dynamic localization of CLIP-170 to microtubule plus ends is coupled to microtubule assembly. J Cell Biol 1999;144:99–112.

106. Tirnauer JS, Bierer BE. EB1 proteins regulate microtubule dynamics, cell polarity, and chromosome stability. J Cell Biol 2000;149:761–766.

107. Niederstrasser H, Salehi-Had H, Gan EC, Walczak C, Nogales E, XKCM1 acts on a single protofilament and requires the c terminus of tubulin. J Mol Biol 2002;316:817–828.

108. Poruchynsky MS, Giannakakou P, Ward Y, et al. Accompanying protein alterations in malignant cells with a microtubule-polymerizing drug-resistance phenotype and a primary resistance mechanism. Biochem Pharm 2001;62:1469–1480.

109. Poruchynsky MS, Kim JH, Nogales E, et al. Tumor cells resistant to a microtubule-depolymerizing hemiasterlin analogue HTI-286 have mutations in α- or β-tubulin and increased microtubule stability. Biochemistry 2004;43:13,944–13,954.

110. Richards KL, Anders KR, Nogales E, Schwartz K, Downing K, Botstein D. Structure–Function Relationships in Yeast Tubulins. Mol Biol Cell 2000;11:1887–1903.

111. Paluh JL, Killilea AN, Detrich HW 3rd, Downing KH. Meiosis-specific failure of cell cycle progression in fission yeast by mutation of a conserved β-tubulin residue. Mol Biol Cell 2004;15:1160–1171.

112. Bode CJ, Gupta ML Jr, Suprenant KA, Himes RH. The two α-tubulin isotypes in budding yeast have opposing effects on microtubule dynamics in vitro. EMBO Rep 2003;4:94–99.

113. Detrich WD III, Parker SK, Williams RC Jr, Nogales E, Downing KH. Cold adaptation of microtubule assembly and dynamics. J Biol Chem 2000;275:37,038–37,047.

114. Hari M, Yang H, Zeng C, Canizales M, Cabral F. Expression of class III β-tubulin reduces microtubule assembly and confers resistance to paclitaxel. Cell Motil Cytoskel 2003;56:45–56.

115. Bhattacharya R, Cabral F. A ubiquitous β-tubulin disrupts microtubule assembly and inhibits cell proliferation. Mol Biol Cell 2004;15:3123–3131.

116. Davis LJ, Odde D, Block SM, Gross SP. The importance of latticed efects in Katanin-mediated microtubule severing in vitro. Biophys J 2002;82:2916–2917.

117. Diaz JF, Valpuesta JM, Chacon P, Diakun G, Andreu JM. Changes in microtubule protofilament number induced by taxol binding to an easily accessibly site. J Biol Chem 1998;273:33,803–33,810.

118. Chretien D, Fuller SD. Microtubules switch occasionally into unfavorable configurations during elongation. J Mol Biol 2000;298:663–676.

119. Chretien D, Metoz F, Verde F, Karsenti E, Wade RH Lattice defects in microtubules: protofilament numbers vary within individual microtubules. J Cell Biol 1992;117:1031–1040.

120. Hartman JJ, Vale RD. Microtubule disassembly by ATP-dependent oligomerization of the AAA enzyme katanin. Science 1999;286:782–785.

121. Dye RB, Flicker PF, Lien DY, Williams RC Jr. End-stabilizer microtubules observed in vitro: stability, subunit interchange, and breakage. Cell Motil Cytoskel 1992;21:171–196.

122. Vale RD. AAA proteins: lords of the ring. J Cell Biol 2000;139:F13–F19.
123. McNally FJ, Okawa K, Iwanatsu A, Vale RD Katanin, the microtubule-severing ATPase, is concentrated at centrosomes. J Cell Sci 1996;109:561–567.
124. Desai A, Maddox PS, Mitchison TJ, Salmon ED. Anaphase a chromosome movement and poleward spindle microtubule flux occur at similar rates in *Xenopus* extract spindles. J Cell Biol 1998;141: 703–713.
125. Miyamoto DT, Perlman ZE, Burbank KS, Groen AC, Mitchison TJ. The kinesin Eg5 drives poleward microtubule flux in *Xenopus laevis* egg extract spindles. J Cell Biol 2004;167:813–818.

3

Microtubule Dynamics

*Mechanisms and Regulation
by Microtubule-Associated Proteins
and Drugs In Vitro and in Cells*

Mary Ann Jordan and Leslie Wilson

CONTENTS

SUMMARY

Microtubules are dynamic cellular polymers. Their dynamics are exquisitely regulated and are essential to many cellular activities including mitosis, cell division, signaling, adhesion, directed migration, polarization, vesicle and protein delivery to and retrieval from the plasma membrane, and remodeling of cell shape and organization. The cytoskeleton, which includes microtubules, actin filaments, and intermediate filaments (such as vimentin, lamin, and keratin) is not a stable structure, but is a continuously evolving machine. The machine remakes or reorganizes itself in response to outside signals regulating cell activities (*1*). In Subheading 1, the authors describe the dynamic behaviors of microtubules in vitro with purified microtubules, their mechanisms, and some of the many ways that dynamics are modulated by microtubule-targeted drugs and endogenous cellular regulators. In Subheading 2, the authors describe microtubule dynamics and their regulation in living cells.

Key Words: Dynamic instability; treadmilling; polymerization; mitosis; migration; MAPs; taxol; Vinca.

1. MICROTUBULE POLYMERIZATION AND DYNAMICS IN VITRO

1.1. Dynamic Instability and Treadmilling: The Unusual Nonequilibrium Dynamic Behaviors of Microtubules

Microtubule polymerization in vitro occurs by a nucleation-elongation mechanism. The slow formation of a poorly-characterized microtubule nucleus is followed by rapid elongation of the microtubule at both ends by the reversible, noncovalent addition of

From: *Cancer Drug Discovery and Development: The Role of Microtubules in Cell Biology,
Neurobiology, and Oncology* Edited by: Tito Fojo © Humana Press, Totowa, NJ

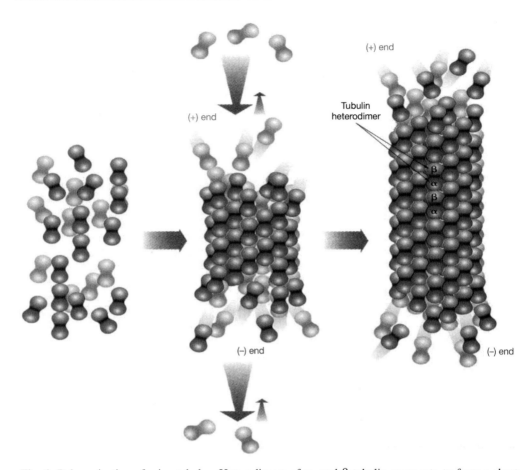

Fig. 1. Polymerization of microtubules. Heterodimers of α- and β-tubulin aggregate to form a short microtubule nucleus. Nucleation is followed by elongation of the microtubule at both ends by the reversible, noncovalent addition of tubulin dimers. Both ends can also shorten. The plus (+) end of the microtubule is kinetically more dynamic than the opposite or minus (–) end, growing and shortening over longer distances than the (–) end. Adapted from ref. *61*. Reprinted by permission of Chemistry and Biology. To view this figure in color, see the insert and the companion CD-ROM.

αβ-tubulin heterodimers, as shown in Fig. 1 (reviewed in ref. *2*). Tubulin dimers (Table 1), which are $46 \times 80 \times 65$ Å in size (width, length, and depth, respectively) are organized in the microtubule in "head-to-tail" fashion in the form of protofilaments with the long axis of the dimer parallel to the long axis of the microtubule (Fig. 1) *(3)*. As a rule in cells microtubules are made up of 13 protofilaments that form a closed tube, but microtubules assembled in vitro can have variable numbers of protofilaments, with a mean of 14 *(4,5)*. The protofilament number also can vary along the length of a single microtubule *(4)*. The two ends of microtubules differ, both structurally and kinetically. One end, termed the plus end, has β-tubulin exposed to the solvent, and α-tubulin faces the solvent at the opposite or minus end *(3,6)*.

The orientation of the αβ-tubulin dimers in the microtubule together with the hydrolysis of guanosine 5′-triphosphate (GTP) that occurs during polymerization confers unusual and distinct dynamic properties to the opposite microtubule ends *(7,8)*. Microtubules are not simple equilibrium polymers, but because GTP is hydrolyzed during normal assembly, they are capable of unusual nonequilibrium dynamic behaviors. One such behavior, called dynamic instability, is characterized by switching between relatively slow growth and rapid shortening at the two microtubule ends *(8)*. The second behavior, termed treadmilling, is characterized by net assembly at the plus ends and disassembly at the minus ends, which creates an intrinsic flow or flux of tubulin subunits from the plus to the minus ends *(7)*. Both behaviors require the hydrolysis of GTP to guanosine 5'-diphosphate (GDP) and inorganic phosphate (P_i) as, or shortly after, tubulin adds to microtubule ends (discussed in Subheading 1.1.2.2). In order to understand their various dynamic behaviors, how they are produced, and how the behaviors are regulated in cells, the authors will first summarize the current state of knowledge about the underlying mechanisms giving rise to these dynamics.

1.1.1. DYNAMIC INSTABILITY

Dynamic instability *(8,9)* is a process in which the individual microtubule ends switch between phases of relatively slow sustained growth and phases of relatively rapid shortening (Fig. 2A). Under certain conditions in vitro, some microtubules can shorten until they totally disappear, creating a net increase in the mean polymer length within the population that is accompanied by a decrease in the total number of microtubules *(8)*. The rates and extents of growth and shortening in vitro are considerably more extensive at the plus ends than at the minus ends *(10–12)*. Dynamic instability in vitro can be analyzed by computer-enhanced video microscopy using differential interference contrast microscopy or dark field microscopy *(10–13)*, and it can be analyzed under a variety of conditions. These conditions include shortly after polymerization is initiated and the microtubule polymer mass is increasing *(10–13)*, when microtubules are undergoing net disassembly, and when the microtubules are at or close to polymer mass steady state and the polymer mass and soluble tubulin concentrations are not changing (which probably best mimics most intracellular conditions) *(14–16)*.

Dynamic instability is characterized by a number of well-defined and quantifiable parameters *(10–14)*. As an example of the differences in parameter values at opposite microtubule ends, the dynamic instability parameters at plus and minus ends of microtubule-associated protein (MAP)-free bovine brain microtubules at steady state in vitro are shown in Table 2. As indicated previously, at these conditions the polymer mass has reached a plateau and the polymer and subunit concentrations are constant. The commonly-analyzed dynamic instability parameters include the rate of growth, the rate of shortening (frequently termed "shrinking"), the transition frequency from growth or an attenuated (or paused) state to shortening (called the "catastrophe" frequency) and the transition frequency from shortening to growth or to an attenuated state (called the "rescue" frequency). Periods of pause or attenuation are defined operationally as times when any changes in microtubule length that may be occurring are below the resolution of the light microscope (usually <0.2 μm). Also of value in characterizing dynamic instability is determination of the percent of time the microtubules are growing, shortening, or remain in an attenuated (paused) state. An additional calculated parameter termed "dynamicity" is highly useful to describe in a single parameter the overall rate of visually detectable exchange of tubulin dimers at microtubule ends *(14)*.

Dynamic instability is an intrinsic property of the tubulin backbone of the microtubule. It occurs with microtubules made up of highly purified tubulin and thus does not require any nontubulin proteins *(14,17,18)*. It is noteworthy that the shortening rates can

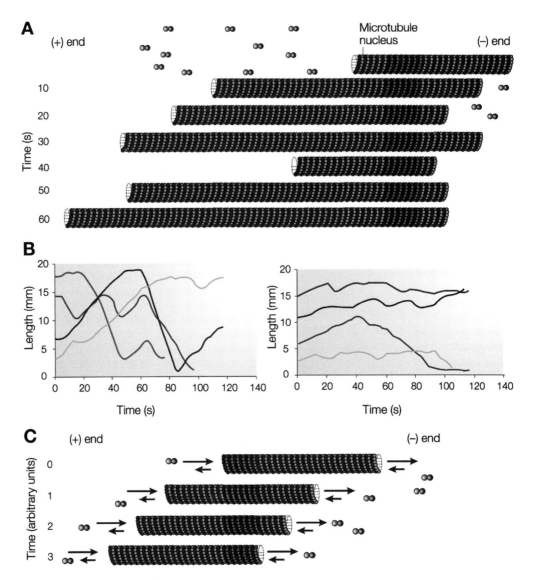

Fig. 2. Microtubules undergo dynamic instability and treadmilling. (**A**) Changes in length of a single microtubule over time. Microtubules grow and shorten stochastically over time by addition and loss of tubulin subunits from their ends. Changes in length at the plus ends are greater than at the minus ends. Microtubules also undergo phases of pause or attenuated dynamics. (**B**) Life history traces of the lengths of four individual microtubules in the absence of drug (left) and in the presence of a microtubule-targeted drug (right). The microtubules were assembled from purified bovine brain tubulin and the changes in length were traced by differential interference-contrast time-lapse microscopy. In the absence of drugs, dynamics are fast, with many length changes. In the presence of a drug such as paclitaxel or vinblastine, dynamics are suppressed. (**C**) Treadmilling microtubule. Tubulin subunits are added at the plus end of the microtubule at time 0, treadmill through the microtubule and are lost from the minus end of the microtubule at time 3. The length of the microtubule is unchanged. Treadmilling is brought about by the different tubulin critical concentrations at the opposite ends. From ref. *63* by permission of Nat. Rev. Cancer (http://www.nature.com). To view this figure in color, see the insert and the companion CD-ROM.

Table 1
Definitions

Tubulin heterodimer (dimer)	The building block of the microtubule. The microtubule subunit is a heterodimer of α- and β-tubulin
Protofilament	Most microtubules are made up of 13 protofilaments that run the length of the microtubule. Tubulin dimers are linked head to tail to form the protofilaments
Critical concentration (C_c)	At tubulin dimer concentrations above the C_c, the dimers polymerize into microtubules, whereas at concentrations below the C_c, microtubules depolymerize. Critical concentrations can differ at opposite microtubule ends
Dynamic instability	Microtubule ends transition stochastically between episodes of slow growing and rapid shortening. The transitions at opposite microtubule ends occur independently of each other
Treadmilling (flux)	Net addition of tubulin at one end of a microtubule and the balanced net loss at the other end, owing to differences in the tubulin critical concentrations at opposite microtubule ends

be extremely rapid (as rapid as 10,000 subunits per second *(19,20)* and that shortening appears to occur by the loss of large segments of peeling protofilaments *(21,22)*. It is also noteworthy that the rates of growth and shortening can be highly variable within a microtubule population and even within an individual microtubule *(17,20)*. The reasons for the high degree of variability are not known, but may be because of the differences in the microtubule lattice from one microtubule to another or within a single microtubule. Brain microtubules made up of different purified β-tubulin isotypes *(23)* display intrinsically different dynamics parameters in vitro. Especially interesting is the fact that microtubules made up of highly purified tubulin from a rapidly dividing human tumor cell line (HeLa), have intrinsically slow dynamics as compared with microtubules made from highly purified brain tubulin *(18)*. The reasons for these differences are not understood, but may be because of the intrinsic differences in the dynamic instability behaviors conferred by the different individual isotypes or because of one or several of the many post-translational modifications that tubulin is known to undergo *(24)*. Finally, the authors note that the dynamic instability behavior of microtubules is tightly controlled in cells by regulatory molecules that interact with microtubule surfaces and ends (discussed in Subheading 2.4).

1.1.2. TREADMILLING

Protein polymer treadmilling was first described theoretically for actin filaments by Albrecht Wegner *(25)*, and was based on the idea that differential use of the energy released during ATP hydrolysis at opposite ends of actin filaments could create different critical subunit concentrations (for definition *see* Table 1) at the opposite ends of the filaments that would give rise to a flow or flux of subunits from one end to the other at steady state *(see* Fig. 2C). There is a built-in bias in observation of microtubules by microscopic methods that favors observation of dynamic instability. Dynamics generated

Table 2
Dynamic Instability Parameters at Opposite Microtubule, Ends at Steady State In Vitro

Parameter	Plus end	Minus end
Rate (μm/min)		
Growing	1 ± 0.1	0.46 ± 0.07
Shortening	14.2 ± 1.2	13.2 ± 4.7
Length change (μm/event)		
Growing	2.3 ± 0.2	1 ± 0.1
Shortening	5.8 ± 0.7	2.8 ± 0.7
Percent time in phase		
Growing	80.4	76.7
Shortening	11.9	6.9
Attenuated	7.7	16.4
Transition frequencies (per minute)		
Catastrophe	0.31 ± 0.06	0.19 ± 0.05
Rescue	0.78 ± 0.28	20 ± 0.6
Dynamicity (μm/min)	2.3	0.78

Adapted from ref. 12.

by end length fluctuations more than micrometers in length can be readily visualized by direct observation, but dynamics owing to treadmilling are more subtle as they involve the flow of subunits through what may appear to be stationary polymers. Treadmilling of microtubules was initially demonstrated in vitro through a biochemical assay with brain microtubule proteins rich in stabilizing MAPs, including tau and MAP2, by observing the incorporation of [^{3}H]GTP into the polymer population at steady state, conditions at which there was no change in the concentration of soluble tubulin or microtubule polymer (7). The radiolabeled GTP, which exchanged rapidly with soluble tubulin subunits and was hydrolyzed during assembly, was linearly incorporated into the polymer and retained in the polymers as [^{3}H]GDP. Direct proof that thc tubulin was fluxing through the polymer included experiments using a pulse of [^{3}H]GTP, which was retained in the polymers until sufficient time had passed for the pulse to have fluxed through the polymers and be lost from the opposite ends (7).

The treadmilling rate of brain microtubules rich in MAPs in vitro is slow (only 0.01–0.02 μm/min) (7,26). These slow rates suggested that treadmilling was too slow to be of functional use in cells. However, it is now clear that the slow treadmilling rate of the MAP-rich microtubules used in the early experiments was a result of the stabilizing MAPs (MAP2, tau), which slow the treadmilling rate as well as suppress the dynamic instability of the microtubules (14,27,28). Specifically, analysis of the treadmilling rates in vitro of microtubules completely depleted of MAPs in glycerol-containing buffer that suppresses dynamic instability, demonstrated that the treadmilling rates in vitro can be quite rapid (~0.2 μm/min [28]). These experiments also demonstrate that, like dynamic instability, treadmilling is an intrinsic property of microtubules made up only of tubulin.

1.1.2.1. How is a Treadmilling Microtubule Created? In the authors' view, the hypothesis, initially advanced by Wegner (25) for actin treadmilling, that nucleotide hydrolysis creates different critical subunit concentrations at the opposite polymer ends, provides the strongest mechanistic explanation for intrinsic treadmilling. With different critical concentrations at the opposite microtubule ends, and when the microtubules are at or near steady state, the higher critical subunit concentration at the minus end gives rise to shortening at this end, whereas the lower critical concentration at the plus end of the microtubule results in net growth *(28,29)*. The overall concentration of soluble tubulin will be maintained in between the two different critical concentrations at the opposite ends and a continuous flow of subunits through the microtubule will occur. Using this model, it was calculated that very small increases in the critical concentration at the minus ends owing to slight increases in the minus end dissociation rate constant, will cause very large increases in the treadmilling rate. For example, a 50% increase in the off rate constant at the minus end of a microtubule would increase the treadmilling rate approx 18-fold. Thus relatively modest increases of the minus end off rate constant would increase the treadmilling rate of microtubules observed in vitro (0.2 μm/min) to levels similar to those observed in cells (0.5–5 μm/min) *(30–32)*. A mechanism in which a regulatory protein binds to the end of a microtubule and modestly modulates the off rate constant is clearly one that the cell could use to very efficiently change treadmilling rates *(28)*.

In contrast, Walker et al. *(11)* and Grego et al. *(33)* have suggested that biased polar dynamic instability (growing and shortening at both microtubule ends, with net growing occurring at one end and net shortening occurring at the opposite end) creates treadmilling. There is no question that biased polar dynamic instability at opposite microtubule ends can give rise to net growth at one end of a microtubule and net shortening at the opposite end as shown using fluorescence speckle microscopy (described further in Subheading 2.3.2) *(33)*. In fact, when fluorescence speckle microtubules were attached to glass slides by kinesin motor proteins and the dynamic instability excursions at the opposite microtubule ends analyzed, many of the microtubules achieved treadmilling, both in a plus-to-minus direction and in a minus-to-plus direction. However, such a biased polar dynamic instability mechanism involving visibly observable episodes of growth and shortening is not a necessary requirement for treadmilling. The strongest evidence that intrinsic treadmilling does not depend on dynamic instability behavior comes from the work of Panda et al. *(28)*, who showed clearly that under conditions in which dynamic instability was strongly suppressed, robust treadmilling of MAP-free microtubules occurred in the absence of appreciable dynamic instability behavior. Considerable evidence in cells also strongly supports the idea that treadmilling can occur in the absence of detectable dynamic instability behavior *(31,32)* and thus far, all treadmilling observed in cells is from the plus-to-minus ends; no minus-to-plus end treadmilling has been observed. Although treadmilling and dynamic instability may be mechanistically distinct, it seems clear that they can occur simultaneously *(29)*. One or the other behavior could predominate depending on the conditions, and some microtubule treadmilling in cells clearly includes a component of dynamic instability. Specifically, the robust treadmilling that occurs in microtubule cortical arrays of plant root epidermal cells appears to have a component of dynamic instability behavior at the microtubule plus ends, but not at the minus ends *(34)*.

For a microtubule to treadmill, both ends of the microtubule must be available for subunit exchange. If, for example, the minus end of a microtubule were anchored at a centrosome in a manner that blocked tubulin exchange at the end, that microtubule would not be able to treadmill. However, it is important to emphasize that tethering a microtubule near its end does not necessarily block the end *(26)*. One of the major sites of microtubule treadmilling is in mitotic spindles, where the plus and minus ends of the kinetochore microtubules are tethered at the kinetochores and spindle poles, respectively *(32,35,36)*. In this situation, the ends, although clearly tethered, remain free for subunit exchange and rapid flow of tubulin owing to treadmilling from the plus ends tethered at the kinetochores and the minus ends tethered to the spindle poles *(32)*. The optimal conditions for treadmilling also require that the microtubules be at or near polymer mass-steady state. For example, if the subunit concentration is below the critical subunit concentration at both the plus and minus microtubule ends, tread-milling cannot occur, as both ends will shorten. The flow of tubulin from plus-to-minus ends of spindle microtubules has been called flux, rather than treadmilling *(35,36)*, because the mechanistic basis of the flow could be created by motor mole-cules or other regulatory molecules acting to cause microtubule shortening at the minus ends. However, recent evidence indicates that the flow of tubulin dimers through microtubules in mitotic spindles, where the microtubules are tethered at their plus and minus ends and remain at a fairly constant length, is indeed owing, perhaps in large part, to the intrinsic ability of microtubules to treadmill *(32)*. Treadmilling clearly must be modulated in cells, and such modulation is likely to be affected by molecules acting at the ends or along the surfaces of the microtubules to increase or decrease subunit flow.

1.1.2.2. The Stabilizing "Cap." The mechanism responsible for dynamic instability is postulated to involve a reversible gain and loss of a so-called "stabilizing cap" at the microtubule ends *(37–48)* (Fig. 3). Tubulin dimers require GTP or GDP bound in the exchangeable nucleotide-binding site of β-tubulin (the E-site) in order to polymerize and depolymerize efficiently. Hydrolysis of the GTP to GDP and Pi is closely coupled to tubulin addition to the microtubule ends, so most of the microtubule length is made up of tubulin with GDP bound in the E-sites (tubulin-GDP) rather than GTP (tubulin-GTP). With respect to the dynamic instability behavior of microtubules, the hypothesis is that when the cap is present at a microtubule end, the end is stabilized and the micro-tubule can grow. But when the cap is lost, the unstable (strained) tubulin-GDP micro-tubule core is exposed and the microtubule shortens rapidly.

What is the "stabilizing cap"? The simple answer is that we do not really know, but a number of good models and supporting data for them have been described. Considerable evidence indicates that hydrolysis of GTP to GDP in tubulin is associated with a change in tubulin conformation *(21,22,49–51)*. For example, electron micro-scopic evidence indicates that tubulin-GDP has the propensity to adopt a "curved" confor-mation, which can be readily visualized at microtubule ends *(21,22,49,50)*. Specifically during a rapid shortening event, protofilaments made up of tubulin-GDP are seen to "peel away" from the microtubule in an outward curving manner, revealing the curved nature of the tubulin-GDP. Although evidence is scant, tubulin-GTP is thought to have a straight conformation, or a straighter conformation than tubulin-GDP (*see* ref. *47*). The tubulin-GDP, which makes up the core of the microtubule appears to prefer a true curved conformation as occurs in peeling protofilaments, but is held in a straight but strained form by the structural constraints imposed by lattice packing in the microtubule core (nicely described in ref. *47*). The hypothesis is that as long as the stabilizing cap is present at the microtubule end, the tubulin-GDP in the core is held in a strained "straight" form, adopting a true "curved" form only as it dissociates from the microtubule end.

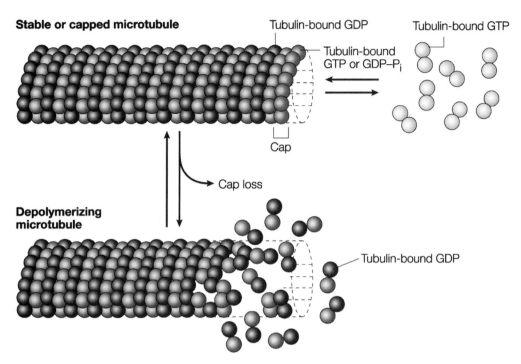

Fig. 3. Dynamic instability and the GTP cap. Tubulin-bound GTP is hydrolyzed to tubulin-GDP and inorganic phosphate (P_i) at the time that tubulin adds to the microtubule ends, or shortly thereafter. Ultimately, the P_i dissociates from the microtubule, leaving a microtubule core consisting of tubulin with stoichiometrically-bound GDP. A microtubule end containing tubulin-bound GTP or GDP-P_i is stable or "capped," against depolymerization. Hydrolysis of tubulin-bound GTP and the subsequent release of P_i induce conformational changes in the tubulin molecules that destabilizes the microtubule polymer, resulting in catastrophe and shortening of the microtubule. Adapted from ref. *63* by permission of Nat. Rev. Cancer (http://www.nature.com). To view this figure in color, see the insert and the companion CD-ROM.

Interestingly, the free energy change that occurs when tubulin-GTP is hydrolyzed to tubulin-GDP in the microtubule core is very small because the majority of the free energy is stored in the microtubule, possibly as a repulsive force between subunits *(52)*. One attractive hypothesis is that a slight bending of short protofilament regions near the end of a capped microtubule relaxes the microtubule near its end, resulting in a structure that serves as the stabilizing cap *(47)*.

The size and chemical nature of the stabilizing cap, also poorly understood, is of considerable interest, especially because molecules that regulate plus-end microtubule growth and track the plus ends (discussed further in Subheading 2.4) must be able to distinguish the capped end of a microtubule from the core. Considerable evidence supports the hypothesis that the stabilizing cap at microtubule ends is extremely small and that it probably consists of as little as a single layer of tubulin dimers bound either to GTP or GDP-P_i (tubulin with both GDP and P_i bound in the E-site) (Fig. 3) *(41,42,46,48)*. For example, Vandecandelaere et al. *(46)* monitored microtubule assembly and P_i release simultaneously using a fluorescence assay for P_i release and found that nucleotide hydrolysis kept pace with tubulin addition at rates as high as 200 molecules per second per microtubule. Similarly, Stewart et al. *(41)* were unable to detect the GTP cap under conditions in which a cap greater in size than 200 subunits could have been detected.

These and other studies provide compelling evidence that when microtubules are either growing rapidly or are growing relatively slowly at steady state, the rate of GTP hydrolysis is very closely coupled to the rate of tubulin addition and thus the cap is very small.

The chemical nature of the cap, that is, whether it consists of tubulin-GTP, tubulin-GDP-P_i, or a mixture of the two has remained controversial *(37,38,40,48,53–56)*. The experimental problems are several fold. First is that the cap is exceedingly difficult to detect experimentally. Second, determining the chemical nature of the cap (i.e., whether it is tubulin-GTP or tubulin-GDP-P_i) is complicated by any hydrolysis of GTP or dissociation of P_i that might continue to occur during the "dead time" (or processing time) required to separate microtubules from soluble tubulin. The work of several authors *(53–55)* supports the view that the cap is made up of tubulin-GTP. In contrast, evidence from the Carlier lab *(37,38,40)* and from the authors laboratory *(48,57)* indicates that the chemical form of the cap may be tubulin-GDP-P_i rather than tubulin-GTP. Interestingly, the structural cap model recently advanced by Janosi et al. *(47)* can accommodate a cap made up of tubulin-GTP or tubulin-GDP-P_i.

The authors work supports the idea that the stabilizing cap may consist of a single layer of tubulin-GDP-Pi at each microtubule end. Specifically it was found that MAP-rich bovine-brain microtubules pulsed with [γ^{32}P]GTP and centrifuged through stabilizing sucrose-cushions containing nonradioactive orthophosphate (P_i), in turn, contain small amounts of GDP-$^{32}P_i$ in a stably bound form *(48,57)*. The microtubules contain an average of about 26 molecules of tubulin-GDP-Pi and do not contain any detectable tubulin-GTP. This value is approximately twice the number of protofilaments in the microtubules (a mean of 14), suggesting that there are approx 13 molecules of tubulin-GDP-P_i in a highly stable nonexchangeable form at each microtubule end. Strong evidence indicates that the stably bound P_i is at the extreme microtubule end. For example, the number of stably bound GDP-P_i molecules per microtubule remains constant regardless of the microtubule length between mean lengths of 2.5 µm and 11 µm *(57)*. The limitation of such an approach to measure the size and chemical nature of the cap directly is that the [^{32}P]GTP can hydrolyze or be lost during the time required to collect the microtubules in the stabilizing cushions (2 h). Thus, the data cannot prove that the stabilizing cap is made up of tubulin-GDP-P_i. But because the GDP-P_i is stably bound and nonexchangeable, it can be concluded that the 26 molecules of tubulin-GDP-P_i at the extreme microtubule ends is either the cap, or is a stable remnant of the cap that distinguishes the microtubule ends from the rest of the polymer.

1.2. Modulation of Microtubule Dynamics by Regulatory MAPs and Drugs: In Vitro Mechanisms

The dynamics of microtubules, not just their presence, are critical to the cellular functions of the microtubules. The dynamics must be under very tight control (*see* Subheading 2). Although investigators are just beginning to learn about the many ways microtubule polymerization and dynamics are regulated in cells, it can already be seen that the number of molecules regulating the dynamics of microtubules is large, and that regulatory molecules, often functioning in concerted fashions *(58,59)*, can act by many distinct mechanisms. However, still very little is known about how most regulatory molecules work.

A number of well-known drugs potently suppress the dynamic instability and treadmilling dynamics of microtubules, and thereby perturb cellular processes dependent on the dynamics *(60–63)*. The mechanisms by which these drugs act to perturb microtubule dynamics are becoming understood, and understanding how these drugs work can help us understand how microtubule regulatory molecules modulate microtubule dynamics *(64)*. The important points the authors want to emphasize here in terms of the regulation of dynamics in cells, based on some of the known actions of microtubule-targeted drugs, are that (1) tubulin and microtubules have a large number of sites that can be targeted by cellular regulatory molecules and (2) remarkably small numbers of regulatory molecules binding to any individual microtubule are sufficient to exert powerful effects on the dynamic behavior of that microtubule. In this section, the understanding of the mechanisms of action in vitro of three of the best understood microtubule-targeted drugs (vinblastine, colchicine, and paclitaxel), of an important neuronal regulatory MAP (tau), and of a "catastrophe"-promoting factor (MCAK [mitotic centromere-associated kinesin]) is summarized.

Microtubule-targeted drugs have been studied and used as tools in cell biology for many years, beginning well before the dynamic behaviors of microtubules were discovered (reviewed in refs. *60,62,65*). Early experiments with these drugs involved use of high drug concentrations and made use of their abilities to inhibit or promote microtubule polymerization. For example, with the antimitotic actions of colchicine or nocodazole, high concentrations of the drugs depolymerized most of the spindle microtubules, resulting in a mitotic block *(66)*. Of course, when the spindle microtubules are completely destroyed, mitosis cannot occur. Similarly, any cellular process dependent on either the presence of microtubules or their dynamics will be blocked if the microtubules are destroyed. However, in many instances, the microtubules must not only be present, they must be dynamic to carry out their functions. This is certainly the case in mitosis, where extremely rapid microtubule dynamics are required for spindles to function properly (*see* in Subheading 2.5.1). It is now known that most of the drugs that act on tubulin and microtubules can powerfully modulate microtubule dynamics at 10- to 100-fold lower concentrations than those required to increase or decrease the microtubule polymer mass. It is these actions of the drugs that have allowed us to appreciate the importance of microtubule dynamics in cellular processes and, to think about how regulatory molecules might act to control dynamics. The actions of three major microtubule-targeted drugs on dynamics are described next.

1.2.1. Vinblastine

The interactions of vinblastine and other Vinca alkaloids with tubulin and microtubules have been extensively studied (reviewed in refs. *60,62,65,67*). Vinblastine, a drug that inhibits microtubule polymerization at high drug concentrations, binds both to soluble tubulin and directly to tubulin in the microtubule (Fig. 4). Its binding to the microtubule ends with high affinity is responsible for its effects on microtubule dynamics. Vinblastine binds to the β-tubulin subunit at a distinct region known as the Vinca-binding domain *(68,69)*. The binding of vinblastine to soluble tubulin is rapid and reversible, but relatively weak ($K_A \sim 2 \times 10^4/M$) *(67,70,71)*. The binding of vinblastine to soluble tubulin induces a conformational change in the tubulin, which not only causes tubulin to self-associate but also results in an increase in the affinity of vinblastine for the tubulin *(60,67,72)*. The ability of vinblastine to increase the affinity of tubulin for itself and the increase in the affinity of vinblastine for self-associated tubulin, very likely play important roles in the ability of vinblastine to bind to the microtubule and to stabilize microtubules kinetically.

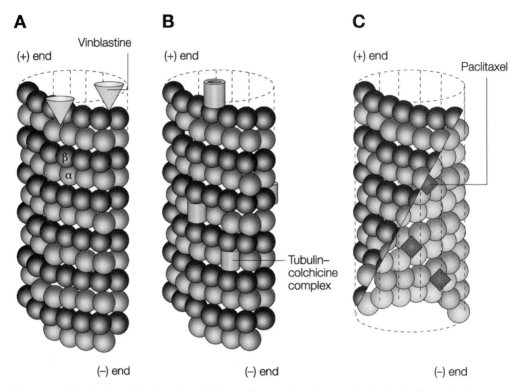

Fig. 4. Antimitotic drugs bind to microtubules at diverse sites that can mimic the binding of endogenous regulators. (**A**) A few molecules of vinblastine bound to high-affinity sites at the microtubule plus end suffice to suppress microtubule dynamics. (**B**) Colchicine forms complexes with tubulin dimers and copolymerizes into the microtubule lattice, suppressing microtubule dynamics. (**C**) A microtubule cut away to show its interior surface. Taxol binds along the interior surface of the microtubule, suppressing its dynamics. Adapted from ref. *63* by permission of Nat. Review Cancer (http://www.nature.com). To view this figure in color, see the insert and the companion CD-ROM.

Although it is not settled whether free vinblastine or vinblastine–tubulin aggregates bind to microtubule ends, it is clear that vinblastine indeed binds to the extreme microtubule ends with high affinity (1 μ*M*) *(70)*. Scatchard analysis with radiolabeled vinblastine reveals that there are about 16–17 high-affinity binding sites (k_d, 1–2 μ*M*) at the ends of an individual microtubule. Interestingly, at high concentrations, vinblastine also binds stoichiometrically with markedly reduced affinity (k_d, 0.3 m*M*) to tubulin along the length of the microtubule *(73,74)*. The inhibition of microtubule polymerization by high concentrations of vinblastine appears to be brought about by several mechanisms. First, inhibition of polymerization, which occurs at vinblastine levels that are far substoichiometric to the tubulin concentration, in part is caused by the binding of the drug to soluble tubulin and induction of tubulin aggregation, which prevents the tubulin from self-assembling into *bona fide* microtubule polymers. In addition, the binding of vinblastine at high concentrations to the low-affinity sites along the microtubule surface causes depolymerization by the "peeling away" of protofilament strands at both microtubule ends *(73)*. Perhaps most importantly, inhibition of polymerization is brought about by the binding of vinblastine or vinblastine–tubulin aggregates to microtubule plus ends, which prevents further tubulin addition.

However, the effects of low-vinblastine concentrations on microtubules are different from those exerted at high-vinblastine concentrations. At low concentrations, vinblastine does not appreciably decrease the polymer mass. Instead, it powerfully suppresses dynamic instability and treadmilling by binding in small numbers to the microtubule plus ends. The high affinity of vinblastine for plus ends appears uniquely responsible for its potent effects on dynamics. Specifically, 0.14 μM vinblastine suppresses the rate of microtubule treadmilling by 50%. More impressive is the fact that this powerful suppression of treadmilling occurs when an average of only 1–2 molecules of vinblastine are bound to the microtubule (70). With respect to its ability to suppress dynamic instability, at 0.1 μM in vitro, it strongly and selectively reduces the rate and extent of microtubule growth and shortening at plus ends and increases the percent of time the microtubules spend in an attenuated or paused state, neither growing nor shortening detectably. These effects of vinblastine on dynamic instability are produced exclusively at the plus ends (12). Because vinblastine reduces the rate and extent of shortening of a steady-state microtubule, it must be concluded that the binding of a few molecules of vinblastine to a microtubule plus end induces a conformational change in the tubulin at the end that strengthens the affinity of the tubulin to which it is bound for adjacent tubulin subunits near the end; i.e., it "strengthens" the stabilizing cap. Interestingly, vinblastine reduces the quantity of stably-bound GDP-P_i at microtubule ends (48). Thus, the increase in cap stability may be brought about in association with a change in the chemical nature of the cap.

1.2.2. COLCHICINE

The interaction of colchicine with tubulin and microtubules presents another variation in the mechanisms by which microtubule-targeted drugs inhibit microtubule function. Considerable evidence has indicated that the binding of colchicine to tubulin is an unusually slow process involving a conformational change in the tubulin that locks the colchicine into a site in the form of a final-state tubulin–colchicine complex from which the colchicine very poorly dissociates (reviewed in refs. 60,75). Importantly an X-ray crystal structure of a stathmin-domain–tubulin-complex containing bound colchicine has revealed the location of the colchicine-binding site in tubulin (76). Analysis of the structure reveals that when bound to tubulin, colchicine is indeed buried in the intermediate domain of the β-tubulin subunit, boxed in by strands S8 and S9, loop T7, and helices H7 and H8 of β-tubulin. Colchicine also interacts with loop T5 of the α-tubulin subunit. As occurs with the Vinca alkaloids, at high concentrations, colchicine depolymerizes microtubules, and at low concentrations it powerfully suppresses microtubule dynamics. Colchicine inhibits microtubule polymerization substoichiometrically (at concentrations well below the concentration of tubulin free in solution, reviewed in ref. 60), indicating that it inhibits microtubule polymerization by binding to microtubule ends rather than to the soluble tubulin pool. Unlike vinblastine, which acts selectively at the plus ends, colchicine can act at both microtubule ends e.g., (77). However, free colchicine itself does not appear to bind directly to the microtubule ends. Instead, colchicine first forms the poorly-reversible final-state tubulin–colchicine complex, which then copolymerizes along with free tubulin into the microtubule at both ends (77–80), reviewed in ref. 60. Unlike vinblastine, free colchicine and tubulin–colchicine complexes do not bind along the length of the microtubule (81). It is noteworthy that when the final-state tubulin–colchicine complexes are incorporated at the microtubule ends, the ends remain competent to grow and shorten. However, as with vinblastine, the ability of the ends to both grow and shorten is seriously compromised, resulting in powerful effects on microtubule dynamics.

Like vinblastine, colchicine inhibits the rate of treadmilling at steady state without appreciably affecting the mass of assembled microtubules (80,82). Relatively few colchicine–tubulin complexes incorporated at the microtubule plus end are sufficient to potently suppress the treadmilling rate (80). Also like vinblastine, the binding of colchicine to microtubule plus ends suppresses the rate and extent of microtubule growth and shortening, and increases the fraction of time that the microtubules remain in an attenuated state, neither growing nor shortening detectably (15,16). Tubulin–colchicine complexes may adopt a conformation in the microtubule lattice that disrupts the lattice in a way that slows but does not prevent new tubulin addition and loss. Importantly, the incorporated tubulin–colchicine complex must bind more tightly to its tubulin neighbors than tubulin itself does so that the normal rate of tubulin dissociation is reduced. Finally, because tubulin–colchicine complexes strongly reduce the catastrophe frequency and increase the rescue frequency, the tubulin–colchicine complex may modulate the mechanism responsible for gain and loss of the stabilizing GTP or GDP-Pi cap (48). Colchicine can reduce the quantity of stably-bound GDP-Pi at microtubule ends (48). Thus, ends containing incorporated tubulin–colchicine complexes are functionally "capped" but the cap itself must be structurally or chemically altered.

1.2.3. PACLITAXEL (TAXOL® BRISTOL–MYERS SQUIBB)

Taxol, which binds to tubulin along the length of the microtubule (3,83,84), acts quite differently than the drugs that act at the microtubule ends. Taxol was discovered to stabilize microtubules in 1979, when Schiff and Horwitz (85) made the surprising discovery that unlike the Vinca alkaloids, high concentrations of taxol stimulated microtubule polymerization. The taxanes bind poorly to tubulin in solution, but instead bind with high affinity to tubulin incorporated into the microtubules (86,87). The binding site for taxol is in the β-tubulin subunit, and its location, which is on the inside surface of the microtubule (Fig. 4C), is known with precision because the electron crystal structure of tubulin was carried out with the tubulin complexed with taxol (83). Although the binding site for taxol is on the inside surface of the microtubule, taxol gains access its binding site very rapidly from bulk solution (87). The mechanism by which this occurs is not clear, but seems to involve diffusion in some manner through small openings in the microtubule lattice or fluctuations of the microtubule lattice (87). The binding of taxol to its site on the inside microtubule surface stabilizes the microtubule, and increases microtubule polymerization presumably by inducing a conformational change in the tubulin that in an unknown manner increases its affinity for neighboring tubulin molecules (88). There is one taxol-binding site on every molecule of tubulin in a microtubule, and the ability of taxol to increase microtubule polymerization is associated with nearly 1:1 stoichiometric binding of taxol to tubulin in microtubules. Thus, if a typical microtubule consists of approx 10,000 tubulin molecules, then the ability of taxol to increase microtubule polymerization requires the binding of approx 5000 taxol molecules/microtubule.

Whereas large numbers of taxane molecules are required to increase microtubule polymerization, the binding of small numbers of taxol molecules stabilizes the dynamics of the microtubules without increasing microtubule polymerization. Taxol suppresses both the rate of treadmilling and the dynamic instability behavior of purified microtubules in vitro (89–91). For example, only one taxol molecule bound per about

200–600 tubulin molecules in a microtubule can powerfully reduce the rate and extent of microtubule shortening. Although taxol inhibits the rate and extent of growth at plus ends of steady-state microtubules, this is not a direct action of taxol on the growth rate but rather an indirect one, caused by the reduction in the rate and extent of shortening (at steady state, if subunits are prevented from coming off the microtubule by the drug, the concentration of subunits available to go back on is also reduced). Unlike the actions of colchicine and vinblastine (described earlier), taxol does not affect the transition frequencies between growth and shortening at the microtubule ends. In contrast to the strong stabilizing effects of taxol on microtubule plus ends, substoichiometric ratios of taxol bound to tubulin in microtubules do not affect minus end dynamic instability *(92)*. Thus, in mitotically-blocked cells, taxol can potently suppress dynamic at plus ends of spindle microtubules, whereas its impotence at minus ends permits continued microtubule depolymerization at the spindle poles *(93)*.

1.3. Mechanisms of Cellular Regulators of Microtubule Dynamics: Mimicking the Actions of Microtubule-Targeted Drugs

The mechanisms by which regulatory proteins modulate microtubules in vitro are not well understood except in a very few cases. In view of the ability of small numbers of drug molecules to effectively modulate microtubule dynamics, it should not have been surprising that cellular proteins that regulate microtubule dynamics might mimic the actions of the drugs. Such appears to be the case with tau, one of the best-studied neuronal stabilizing MAPs *(15,16,94)*, whose effects on microtubule dynamics are described in detail elsewhere in this volume (*see* Chapter 21). Briefly, the tau proteins are important neuronal MAPs that can bind to tubulin at the microtubule surface with very high stoichiometries. At high concentrations in vitro tau powerfully increases the rate and extent of microtubule polymerization and stabilizes the microtubules against disassembly (reviewed in ref. *95*). Whereas the specific location of the tau-binding sites on the microtubule surface is not known, it is clear that tau exerts its effects on microtubule polymerization by binding along the microtubule surface in large numbers. However, similar to the effects of low taxol concentrations on microtubule dynamics, low concentrations of tau strongly inhibit treadmilling and dynamic instability without appreciably modifying the microtubule polymer mass *(15,16,28,95)*. For example, at a ratio of four-repeat tau (a form of tau containing four microtubule-binding regions) to brain tubulin of 1:16, tau reduced the mean microtubule shortening rates and the lengths by 54% and 44% *(95)*. Similar but weaker effects were observed at tau: tubulin molar ratios as low as 1:332. Tau also potently inhibits the rate of microtubule treadmilling in vitro *(28)*. Interestingly, data have been obtained indicating that taxol and tau may bind in a similar region of tubulin *(84)*.

MCAK is another microtubule regulatory protein. It appears to modulate microtubule dynamics by binding to microtubule ends with very low stoichiometry *(58,96–99)*. MCAK is a member of the Kin I (internal eatalytic domain kinesin) subfamily of kinesin-related proteins. It does not translocate along microtubules, but rather, destabilizes microtubules both in cells and in vitro by acting as a "catastrophe-promoting factor." Data obtained in vitro with steady-state microtubules made up of highly purified HeLa cell microtubules and full-length bacterially-expressed MCAK dimers indicates that only a single molecule of MCAK dimers per 263 molecules of tubulin can increase the steady-state catastrophe frequency at microtubule plus ends approx 6.7-fold *(99)*. Thus, like the actions of microtubule-targeted drugs, microtubule regulatory proteins can modulate microtubule dynamics by binding to microtubule surfaces and ends with remarkably low stoichiometries.

2. MICROTUBULE DYNAMICS IN CELLS

Microtubule dynamics are exquisitely regulated in cells. Dynamically unstable microtubules probe the intracellular space, interacting with other cellular components including the chromosomal kinetochores, the cell cortex, and the dynamic actin cytoskeleton *(100–102)*. Treadmilling (fluxing) microtubules act as inducers of mechanical tension in the mitotic spindle *(103,104)*, and in addition, treadmilling microtubules can reorganize themselves spatially by translocating through the cytoplasm *(34)*. In this section, the following is described:

1. The organization of microtubules in cells.
2. The methods used to measure microtubule dynamics in cells.
3. The regulation of microtubule dynamics in cells. As examples, the regulation of microtubule dynamics in migrating cells and in mitosis is described.
4. The use of cancer chemotherapeutic drugs to kill cancer cells by misregulation of microtubule dynamics.

2.1. Organization of Microtubules in Cells

In interphase in many cell types, microtubules are nucleated at the centrosome or microtubule-organizing center located near the cell nucleus *(105,106)*, and they extend outward toward the cell periphery (Fig. 5). The centrosome contains large complexes called γ-tubulin ring complexes (γ-TuRCs), consisting of nine or more proteins and including a species of tubulin called γ-tubulin. Microtubule nucleation can occur at other cellular sites, but microtubules generally nucleate at the centrosome. Following nucleation at the centrosome, the plus ends of microtubules grow out toward the cell periphery, whereas the minus ends generally remain at the centrosome. In some cells, such as in epithelial cells and neuronal cells, microtubules are released from the centrosome and are then transported long distances outward from the centrosome *(105,106)*.

2.2. Measurement of Microtubule Dynamics in Living Cells

2.2.1. FLUORESCENCE-RECOVERY-AFTER-PHOTOBLEACHING (FRAPPING)

Microtubule dynamics in cells can be measured in a number of ways. Early studies measured the dynamics of large populations of microtubules using the technique of fluorescence-recovery-after-photobleaching in which all the cell's microtubules were made fluorescent by microinjection of fluorescent tubulin into the cell and subsequent incorporation of the marked tubulin into microtubules. A region of the cell was then exposed to laser light to bleach the fluorescent microtubules locally in that region, and the rate of recovery of the fluorescent tubulin into the bleached microtubules was measured. Using this method it became clear that microtubule dynamics were variable and were highly regulated by the cell. For example, it was found that mitotic microtubules had 18-fold greater dynamics than interphase microtubules *(107)*, that cells in sparse cultures had twofold faster dynamics than cells in confluent cultures, that fibroblasts had faster dynamics than epithelial cells, and that populations of microtubules with

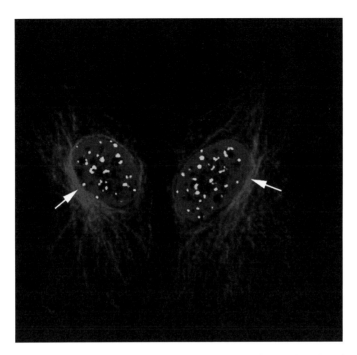

Fig. 5. Microtubules in two human osteosarcoma cells in interphase of the cell cycle. Microtubules extend outward from the microtubule-organizing centers (arrows) toward the periphery of each cell. Microtubules are in red, chromatin is in blue, and centromeres are in green. Image reproduced with permission from ref. *168*. To view this figure in color, see the insert and the companion CD-ROM.

different dynamic properties coexisted in the same cell *(108–111)*. Thus, the dynamics of microtubules in cells vary with the cell type, with their intracellular locations, with the cell's activities and throughout the cell cycle.

2.3. Dynamics at Plus Ends of Individual Uniformly Labeled Fluorescent Microtubules

More recently, using sensitive cooled charge-coupled–device cameras, it has become possible to measure the dynamics of *individual* microtubules in living cells microinjected with a small amount of fluorescent tubulin (generally rhodamine tubulin, and generally amounting to approx 10% of the total cellular tubulin) or expressing fluorescent tubulin (such as GFP-tubulin). The growing and shortening dynamics of individual microtubules are prominent and easily measured in the thin peripheral lamellar regions of interphase cells. Figure 6 shows a 16-s time-lapse sequence of fluorescently labeled microtubules in MCF7 human breast cancer cells. The microtubule marked by an arrowhead shortens, whereas the two microtubules marked by arrows lengthen during the time of recording. To determine how microtubule lengths change with time, the ends of the individual growing and shortening microtubules are traced by a cursor on succeeding recorded time-lapse frames, and their rates, lengths, and durations of growing and shortening are calculated. This technique has proved extremely valuable for determining the effects of microinjected proteins like MAP2 and tau and drugs like taxol, vinblastine, epothilone, and discodermolide *(112–117)* on the dynamics

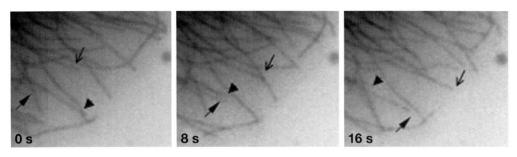

Fig. 6. Time-lapse images of dynamic microtubules in the peripheral region of a living MCF7 human breast cancer cell. Three dynamic microtubules that changed length over the course of 16 s are indicated by thick and thin arrows and an arrowhead, respectively. From Kamath and Jordan *(116)*.

of microtubules in cells. Even though the dynamics of only the peripherally located microtubule plus ends in cells are available for measurement by this technique, the results obtained have proved relevant to the effects of the experimental agents in other cell types and in mitotic cells.

2.3.1. CAGED FLUORESCENCE

Translational movements of individual microtubules can be distinguished from tread-milling of individual microtubules in cells by using microinjection of "caged" fluorescent tubulin. After incorporation of the caged fluorescent tubulin into microtubules, a localized region of the fluorescence in the microtubule lattice is activated (or uncaged) by laser illumination and the movements of the microtubules are monitored by the movement of the photoactivated fluorescent spots *(118,119)*. For example, a recent study of microtubule movement in living cells using photoactivation of caged fluorescent tubulin showed that peripheral, noncentrosome-associated microtubules are moved into the forming spindle region by transport or sliding interactions, not by treadmilling. Images obtained by this technique are shown in Fig. 7 *(119)*. In this image, the uncaged, fluorescent regions of the microtubules were translocated into the spindle from a distant part of the cell during the period of observation. In another classic study, Mitchison generated a fluorescent bar across the mitotic spindle of metaphase cells by microbeam photoactivation of fluorescence in the microtubules. The fluorescent zone moved pole-wards at 0.3–0.7 µm/min providing strong evidence for polewards flux or treadmilling in kinetochore microtubules *(35)*.

2.3.2. FLUORESCENCE SPECKLE MICROSCOPY

An even newer technique, fluorescence speckle microscopy (FSM), has recently been developed for use with either sensitive cooled charge-coupled–device cameras or epifluorescent microscopes or with confocal microscopy (Fig. 8). It employs similar methods to those described earlier for producing fluorescent microtubules (microinjection or expression of fluorescent tubulin), but takes advantage of variations in fluorescence intensity along the microtubule lattice that are induced by using intentionally low levels of fluorescent tubulin (0.1–0.5% of the total cellular tubulin may be labeled). When such low levels of fluorescent tubulin are coassembled with unlabeled tubulin subunits into the microtubules, a fluorescent speckle pattern is produced throughout the microtubule.

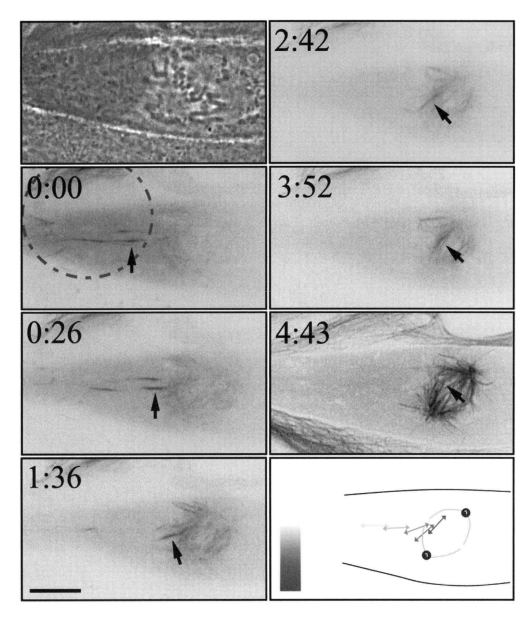

Fig. 7. Caged fluorescence. Photoactivation of fluorescence in an LLCPK1 cell expressing PA-GFP-α-tubulin. At the upper left is a phase contrast image of the cell at the time of photoactivation. The region of activation is shown as a pink overlay. Contrast has been reversed and photoactivated regions are in dark contrast; time in minutes; arrow marks the position of a photoactivated bundle of microtubules. Peripheral microtubules are transported into the spindle, not treadmilling during the time-course of observation. The whole cell was photoactivated at time 4:43 to show the entire microtubule array. A schematic diagram of the position of the activated bundle over time is represented by the color scale. By permission of Cell Press. Courtesy of Pat Wadsworth *(119)*. To view this figure in color, see the insert and the companion CD-ROM.

The movements of these speckles stand out to the eye, and their movements can be recorded by time-lapse microscopy and measured. The advantage of speckle microscopy is that it delivers simultaneous kinetic information over large areas of a cell, and gives

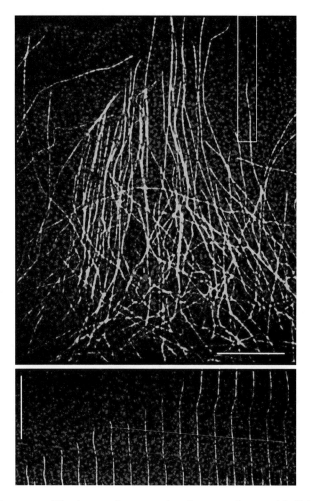

Fig. 8. Speckle microscopy. The image shows a migrating newt lung epithelial cell that has been comicroinjected with green fluorescent tubulin and red fluorescent actin and imaged by dual-wave length FSM. The movements of microtubules and F-actin are coupled in the lamellum. The top panel shows a single FSM image, the leading edge is at the top. The bottom panel shows a time montage of the boxed region near the top right of the top panel. This microtubule is transported rearward while simultaneously growing toward the leading edge. The white horizontal line tracks the retrograde movement of a speckle on the microtubule, which is moving at the same velocity as immediately adjacent speckles in the lamellar actin meshwork. Frames are at 10-s intervals. Scale bar represents 10 μm in both panels. Courtesy of C. Waterman-Storer *(102)*. To view this figure in color, see the insert and the companion CD-ROM.

higher spatial and temporal resolution than is attainable when a microtubule is uniformly fluorescent *(120,121)*. FSM has been used, for example, to distinguish between the simultaneous growth and transport of microtubules in the growth cones of neurons *(122)* and to measure rates of treadmilling or flux in spindle microtubules *(103)*.

In a variation on speckle microscopy, the extensive role of treadmilling in the formation of highly structured microtubule arrays at the cell cortex in the plant *Arabidopsis* was elucidated by photobleaching marks on fluorescent microtubules in a cortical microtubule array. The bleached spots remained stationary, but each microtubule as a

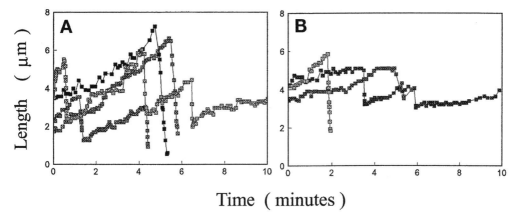

Fig. 9. Life history plots of typical length changes at opposite microtubule ends at steady state. Plus ends of bovine brain microtubules undergo greater length changes at more rapid rates than minus ends. Adapted from Panda et al. *(12)*.

whole appeared to move as a result of the balanced polymerization at one end and depolymerization at the other end. The entire microtubule polymer was not translocated, but changed location as a result of treadmilling *(34)*.

2.4. Regulation of Cellular Microtubule Dynamics and the Question of Minus End Dynamics in Cells

Microtubule dynamics are exquisitely controlled in cells. Several classes of proteins associate with microtubules and play critical roles in the spatial and temporal regulation of microtubule dynamics. These include the microtubule plus-end binding proteins (+TIPs), minus-end binding proteins, proteins that bind along the lengths of microtubules, GTPases and guanine nucleotide exchange factors. Specifically, (TIPs include APC (adenomatous polyposis coli) tumor suppressor protein, cytoplasmic linker proteins (CLIPs), CLIP-associating proteins (CLASPs), cytoplasmic dynein/dynactin, p150[Glued,] and end binding protein1 (EB1), an APC-interacting protein *(123–126)*. Minus-end binding proteins include the γ-TuRCs *(106)*, stathmin *(127)*, and possibly ninein *(128)*; for recent information on stathmin *see* also refs. *76,129*. MAP2, MAP4, and tau are the best-known examples of proteins that bind along the lengths of microtubules (*see* refs. *97,130*). The GTPase Ran, and the chromatin-bound guanine nucleotide exchange factor RCC1 are additional recently discovered important microtubule regulatory proteins *(131)*.

As mentioned in Subheading 1.1.1, minus ends of microtubules are dynamic in vitro but their dynamics differ from those of plus ends. As shown in Table 2 and Fig. 9, with respect to their dynamic instability behavior, minus ends are more stable than plus ends, exhibiting slower growth rates, fewer catastrophes, and more rescues than plus ends *(11,12,92)*. Interestingly, however, *in cells* the minus ends of microtubules have not been observed to polymerize or grow. Rather, they remain stable or undergo only depolymerization *(132)*. It is not clear whether the stability of minus ends in cells results from minus-end capping factors (such as γ-TuRCs or ninein) or post-translational modifications of tubulin that prevent assembly onto minus ends. What is clear is that minus ends in cells do not undergo even transient growth upon severing or breakage of the microtubule.

Drugs and MAPs clearly affect dynamics at minus ends differently than their effects on plus ends. One of the best examples in cells is that low concentrations of the drug taxol prevent microtubule assembly at kinetochore plus-ends in newt lung cell spindles without having any effect on minus-end disassembly at the poles *(93)*. This preferential effect of taxol on microtubule plus-end dynamics is consistent with a preferential plus-end taxol effect in vitro *(92)*. As mentioned in Subheading 1.2.1, vinblastine also affects dynamics only at microtubule plus ends *(12)*. In another example of the differences in effects at opposite microtubule ends among various microtubule regulatory molecules, the protein stathmin preferentially enhances the catastrophe rates at minus-ends in vitro *(127)*.

2.5. Microtubule Dynamics in Migrating Cells

In migrating cells, microtubules extend outward from the centrosome toward the leading edge of the cell where they exhibit dynamic instability. There, dynamic microtubules and dynamic actin filaments interact, in concert with the small GTP-binding signaling proteins Rac, Rho, and Cdc42, to regulate each others' behavior and to bring about the directed movement of the cell. As a migrating cell extends forward under the influence of actin polymerization at its leading edge, microtubules also grow outward at their plus ends in a region just subjacent to the actin meshwork. As shown in Fig. 7, simultaneous with the net forward growth of the microtubule plus ends, the main body of each microtubule is carried rearward (in the opposite direction) along with the myosin-powered rearward movement of the recently polymerized actin. Thus the microtubules grow forward at their ends in apparent compensation for their overall rearward flow. In addition, as the microtubules grow, they also promote Rac1 activation, which mediates actin polymerization and lamellipodial protrusion *(102)*. Thus, the growth of dynamic microtubules is necessary to signal actin polymerization and, in a tightly regulated feedback system the actin in turn influences microtubule transport.

At the rear or trailing edge of the migrating cell, the microtubules behave differently. To enable forward movement, a migrating cell must continuously create new sites of substrate adhesion toward its leading edge, while it disassembles older adhesions near its trailing edge. Dynamic microtubules toward the retracting rear of the cell participate in a behavior called "targeting," during which a microtubule repeatedly grows toward and comes into contact with an adhesion site, and then retracts or shortens. This dynamic instability behavior may be repeated several times per minute. The dynamic microtubule targeting behavior is associated with dissolution of the older focal adhesions. It has been hypothesized that the microtubule plus-end tracking protein APC might be delivered to the focal adhesions through these pulsing dynamic microtubules. APC is found in the loosening adhesions where it binds a Rac-GTP exchange factor that participates in adhesion disassembly. Thus delivery of APC through dynamic microtubules is likely to be an important determinant of focal adhesion degradation *(133)*.

2.5.1. Mitosis

Mitosis is another striking example of tight intracellular regional and temporal control of microtubule dynamics *(101)*. At the onset of mitosis, the interphase microtubule network disassembles and is replaced by a new population of spindle microtubules that is many fold more dynamic than the microtubules in the interphase cytoskeleton. MAP4, a MAP which binds along the surface of microtubules and appears to stabilize and slow their dynamics, becomes phosphorylated by cyclin B-cdc2 kinase at the onset of mitosis, leading to a reduced association of MAP4 with microtubules *(134,135)*, and likely contributing to increased dynamics. The change in microtubule dynamics also appears to be regulated by Ran, a Ras-like GTPase and its chromatin-bound nucleotide-exchange factor, RCC1, that are released from the nucleus following the nuclear envelope breakdown at the beginning of prometaphase. RCC1 may mediate a high local concentration of Ran-GTP in the region of the chromosomes, thus producing a new state of increased dynamic behavior of spindle microtubules involving an increase in the catastrophe frequency and a three- to eightfold increase in the rescue frequency. The result is that a large number of very dynamic microtubules accumulate in the region of the chromosomes and induce reorganization of the astral microtubule arrays into a bipolar spindle *(131,136)*.

The highly dynamic spindle microtubules produced by these changes are required for the timely and correct attachment of chromosomes at their kinetochores during prometaphase after nuclear envelope breakdown (Fig. 10A), for the complex movements of the chromosomes bringing them to their properly aligned positions at the metaphase plate (called congression) (Fig. 10B), and for the synchronous separation of the chromosomes in anaphase and telophase after the metaphase-to-anaphase checkpoint is satisfied (Fig. 10C,D). For example, during prometaphase, microtubules emanating from each of the two spindle poles make vast growing and shortening excursions, essentially probing the cytoplasm until they "find" and become attached to chromosomes at their kinetochores *(100)*. Such microtubules must be able to grow for long distances (typically 5–10 µm), then shorten almost completely, then regrow again until they successfully become attached. After attachment to microtubules emanating from one spindle pole, microtubules from the opposite pole must be sufficiently dynamic to successfully attach to the sister kinetochore, thereby setting up a bipolar spindle. By means of the synchronous growing and shortening of the microtubules in bundles attached to sister kinetochores, the chromosomes gradually congress to the metaphase plate. The mechanism of regulation of the synchrony within a bundle is unknown.

During metaphase (Fig. 10B), the duplicated chromosomes that have congressed to the spindle equatorial region are under high tension. The tension is produced, at least in part, by dynamic microtubules attached at their plus ends to the kinetochores, with their minus ends extending toward the two spindle poles. The tension arises from microtubule treadmilling (or flux) as well as from dynamic instability of the kinetochore microtubules *(103)*. During metaphase, tubulin is continuously added to microtubule plus ends (at the kinetochores) and lost at their minus ends (at the poles) in balanced fashion (i.e., the microtubules treadmill) *(32,35,137)*. If tension is absent, cell-cycle progress from metaphase to anaphase is blocked *(138–140)*. The relative importance of flux and dynamic instability varies among cell types *(103)*. In cells such as tissue culture cells, where the growing and shortening events of dynamic instability play a major role in tension production, the chromosome pairs oscillate back and forth across the spindle midline as the microtubules attached to sister kinetochores alternately grow and shorten. In addition, the intercentromere distance periodically increases and decreases as the tug-of-war occurs between microtubules attached to opposite spindle poles *(141,142)* . In contrast, treadmilling or flux is the dominant determinant of spindle tension in *Xenopus* and *Drosophila* spindles, and oscillations arising from dynamic instability do not occur *(103,143)*.

A number of microtubule-binding proteins participate in microtubule interactions with the kinetochore, and some of these appear to regulate microtubule dynamics during mitosis. Among the potential regulators are MCAK/XKCM1 *(144–147)*, EB1 *(148)*, and CLASP-1 *(149)*; also *see* review in ref. *101*. Transitions from microtubule

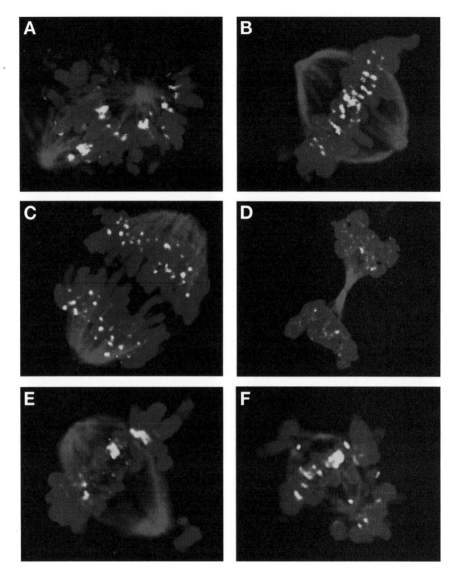

Fig. 10. Organization of the mitotic spindle in human osteosarcoma cells throughout mitosis and in the presence of antimitotic drugs. Microtubules are in red, chromosomes in blue, and centromeres/kineto-chores in green. (**A**) At prometaphase, the nuclear envelope has broken down, chromosomes are condensed and dynamic microtubules probe the cytoplasm until they contact a chromosome. (**B**) In early metaphase, most chromosomes have congressed to the equator to form the metaphase plate. (**C**) In anaphase the duplicated chromosomes have separated and are moving toward the spindle poles to form the two daughter cells. (**D**) In telophase, the separated chromosomes have reached the spindle poles and the cell is dividing to form two daughter cells. (**E**) In the presence of 10 n*M* taxol, some chromosomes remain at the spindle poles and have not congressed to the metaphase plate. (**F**) Similarly, in the presence of 50 n*M* vinflunine, some chromosomes remain at the spindle poles. In the presence of antimitotic drugs the diminished dynamic movements of chromosomes reduces the tension on the kinetochores, centromeres, and the conjoined chromosomes. These changes are associated with the arrest of mitosis at the metaphase–anaphase transition. Adapted from Okouneva et al. *(163)* and Kelling et al. *(164)*. To view this figure in color, see the insert and the companion CD-ROM.

growth to shortening at the kinetochore appear to be regulated, at least in part, by MCAK/XKCM1 and the Kin I kinesins, whereas CLASP-1 appears to have a less well understood effect on microtubule dynamics that might involve coassembly with tubulin heterodimers at the kinetochore *(149)*. The regulation of treadmilling at the kinetochore is not understood.

2.6. Dynamic Microtubules are Crucial to Passage Through the Mitotic Checkpoint

The spindle checkpoint ensures accurate chromosome segregation in mitosis and meiosis (reviewed in ref. *150*). If microtubule dynamics are perturbed during mitosis, the spindle checkpoint blocks mitosis. Thus, if chromosomes do not attain a bipolar spindle attachment because microtubules are not sufficiently dynamic to complete bipolar attachment or if there is insufficient tension on kinetochores *(151)*, the checkpoint blocks the activity of the anaphase-promoting complex, a large ubiquitin ligase required for chromosome segregation. On checkpoint activation, a checkpoint protein complex containing BubR1, Bub3, Mad2, and Cdc20 binds to the anaphase promoting complex, inhibits its ubiquitin ligase activity, and thus prevents destruction of sister chromatid cohesion and anaphase onset.

In anaphase, in order for chromosome segregation to occur (Fig. 11C), kinetochore microtubules must undergo net depolymerization resulting in chromosome movement to the spindle poles at the same time that another subpopulation of spindle microtubules (the interpolar microtubules) lengthens, resulting in lengthening of the entire spindle. In some cells, kinetochore microtubules continue to treadmill during anaphase. In elegant experiments Chen and Zhang *(32)* pulled a chromosome bivalent out of the spindle region by micromanipulation, severed its kinetochore microtubules, and then microinjected fluorescent tubulin into the cell. The gain or loss of tubulin at the newly formed free minus microtubule ends in the midregion of a half spindle was followed. The severed microtubules maintained a constant length as fluorescent tubulin added to the microtubules at their kinetochore (+) ends and treadmilled through the microtubules toward the severed (–) ends. These experiments showed that in grasshopper spermatocytes, kinetochore microtubules constantly treadmill during anaphase. In an unperturbed spindle, where the kinetochore microtubules remain attached to the spindle pole, the microtubules shorten at their minus ends by means of an unknown microtubule depolymerizing factor, thus reeling in the chromosomes.

Many cells eventually undergo apoptosis (programmed cell death) if they are blocked in mitosis by checkpoint activation *(152,153)*. It has been found that suppression of microtubule dynamics by drugs such as taxol and Vinca alkaloids appears to be a common mechanism by which these drugs block mitosis and kill tumor cells. Figure 11E,F shows human osteosarcoma cells after incubation with 10 n*M* taxol and 50 n*M* vinflunine, respectively. Many chromosomes are stuck at the spindle poles, unable to congress to the metaphase plate. At least one reason that cancer cells are relatively sensitive to these drugs as compared with normal cells is that many cancer cells divide more frequently than normal cells and thus frequently pass through a stage of vulnerability to mitotic poisons. However, the mechanisms by which normal cells escape the actions of antimitotic drugs are not well known.

2.7. The Importance of Microtubule Dynamics to Mitosis is Central to the Chemotherapeutic Mechanism of a Large Class of Cancer Drugs

The success of many microtubule-active agents as anticancer drugs is owing to their ability to affect microtubule dynamics and block mitosis and cell division. A very large number of chemically diverse substances originating from natural sources can bind to soluble tubulin or directly to tubulin in microtubules. Many of these compounds have been found to suppress microtubule dynamics. Dynamic microtubules are essential for completion of mitosis. Thus, most of these compounds are antimitotic agents and inhibit cell proliferation by suppression of microtubule dynamics during the particularly vulnerable stage of the cell cycle, namely, mitosis.

The microtubule-targeted antimitotic drugs are usually classified into two major groups. One group, known as microtubule destabilizing agents, inhibits microtubule polymerization at high concentrations and includes a number of compounds such as the Vinca alkaloids (vinblastine, vincristine, vinorelbine, vindesine, and vinflunine), cryptophycins, halichondrins, dolastatins, estramustine, 2-methoxyestradiol, colchicine, and combretastatins *(62,63,154)* that are used clinically or are under clinical investigation for treatment of cancer. In addition, this group includes a very large number of natural compounds that have not undergone clinical development for cancer therapy including the antitussive noscapine *(155)*, maytansine, rhizoxin, spongistatins, podophyllotoxin, steganacins, phenylahistins, and curacins *(154)*; herbicides that inhibit microtubule polymerization *(156)*; antifungal and antihelmintic agents *(157)*; and some psychoactive drugs *(158–160)*. The second major group is known as microtubule stabilizing agents. These agents stimulate microtubule polymerization and include paclitaxel (Taxol; the first identified in this class), docetaxel (Taxotere® Sanofi-Aventi), the epothilones, discodermolide, the eleutherobins, sarcodictyins, laulimalide, rhazinalam, and certain steroids and polyisoprenyl benzophenones *(62,161)*. It is likely that, throughout evolution, plants and animals have "taken advantage" of the importance of microtubule dynamics in cell function to independently evolve a vast number compounds that mimic endogenous regulators of microtubule dynamics for their own self protection from predation.

2.8. Stabilizers vs Destabilizers? Do Microtubule-Targeted Chemotherapeutic Drugs Work by Altering the Mass of Microtubules or by Suppressing Their Dynamics or Both?

The classification of drugs as microtubule "stabilizers" or "destabilizers" is overly simplified and often leads to confusion. The reason is that both drugs that increase and drugs that decrease microtubule polymerization at high concentrations often potently suppress microtubule dynamics at 10- to 100-fold lower concentrations, as described in detail in Subheading 1. The steady state balance between tubulin addition and loss at both microtubule ends is an amazingly useful phenomenon in cells. It allows microtubule dynamics to be speeded up or slowed down in cells in such a way that the dynamic behavior of microtubule ends can be modulated locally or temporally, whereas still maintaining a necessary mass of polymerized microtubules in cells. Thus, the sensitivity of microtubule dynamics to regulation means that both kinds of microtubule-regulating drugs, the taxanes and Vinca alkaloids, can kinetically stabilize the microtubules without changing the microtubule polymer mass. Before the importance of microtubule dynamics in cell function was appreciated, it was thought that the effects of the drugs on microtubule polymer mass were the most important actions responsible for their chemotherapeutic actions. Now that the role of microtubule dynamics in cell function is appreciated, it appears that the most important action of taxanes, Vinca alkaloids, and similar drugs may be the suppression of spindle microtubule dynamics, which results in the slowing or blocking of mitosis at the metaphase/anaphase transition and induction of apoptotic cell death. Thus, at a very basic mechanistic level, these two classes of drugs act similarly. However, at higher concentrations, there are clear differences in their cellular effects on microtubule mass and spindle organization. At high drug concentrations in human patients, they may work, in addition, by increasing or decreasing microtubule polymer levels. However, attainment and maintenance of high polymer-mass-changing drug concentrations may not be necessary for their therapeutic efficacy. The role of these differences in their tumor specificity, secondary antitumor mechanisms, and side effects is not known.

2.9. A Major Mechanism of Many Antimitotic Anticancer Drugs is Suppression of Mitotic Spindle Microtubule Dynamics

2.9.1. THE VINCA ALKALOIDS

As described in Subheading 2, vinblastine inhibits treadmilling of microtubules in vitro with the binding of only 1–2 molecules of vinblastine per microtubule and strongly inhibits dynamic instability at microtubule plus ends. In cancer cells, at low but clinically relevant concentrations (e.g., 0.5–10 nM in HeLa cells [162]), slowing of microtubule growth and shortening and/or treadmilling blocks mitotic progression. Specifically, using the technique of time-lapse microscopy of fluorescent microtubules in living BS-C-1 cells, it was found that concentrations of vinblastine that block cells in mitosis (3–64 nM) without inducing microtubule depolymerization, significantly slow microtubule dynamic instability in interphase cells. For example, the dynamicity was reduced by 75% at 32 nM vinblastine (112).

To directly examine the effects of Vinca alkaloids in mitotic, rather than interphase cells, the centromeres of human osteosarcoma cells were made fluorescent by transfection with green fluorescent protein-labeled centromere-binding protein B (GFP-CENP-B). After treatment with low concentrations of Vinca alkaloids, centromere movement (which depends on microtubule dynamics) was suppressed at concentrations that blocked mitosis, thus indicating that suppression of spindle microtubule dynamics by the Vinca alkaloids is strongly associated with mitotic block (163). Suppression of dynamics has at least two downstream effects on the spindle: it prevents the mitotic spindle from assembling normally, and it reduces the tension at the kinetochores of the chromosomes. Mitotic progress is delayed in a metaphase-like state with condensed chromosomes often stuck at the spindle poles, unable to congress to the spindle equator (Fig. 11E,F). The cell-cycle signal to the anaphase-promoting complex to pass from metaphase into anaphase is blocked and the cells eventually die by apoptosis. Thus, at low concentrations, mitotic block is owing to suppression of microtubule dynamics rather than to microtubule depolymerization (62,112,163).

2.9.2. TAXOL AND RELATED DRUGS

Taxol and its semisynthetic analog, docetaxel (Taxotere) were among the most important new additions to the chemotherapeutic arsenal in the late 20th century. As described in Subheading 1, taxol can bind to a large number of sites on the inside surface of a microtubule. Nearly stoichiometric binding is associated with increased microtubule polymer mass through stabilization of the assembled microtubules. However, lower concentrations of taxol suppress microtubule dynamics by binding to only a fraction of the tubulin dimers in a microtubule. In CaOv3 human ovarian cancer cells and A498 human kidney carcinoma cells, suppression of microtubule dynamics by low concentrations of taxol is associated with mitotic arrest in the absence of any detectable increase in microtubule polymer mass or significant microtubule bundling *(114)*. As with the Vinca alkaloids, the suppression of spindle microtubule dynamics prevents the dividing cancer cells from progressing from metaphase into anaphase, and the cells eventually die by apoptosis *(114,153,164)*.

Recently, the authors compared the effects of a number of antimitotic drugs at their IC$_{50}$s for inhibition by 50% of proliferation of a single cell type (MCF7 cells). With taxol, epothilone B, E7389 (a synthetic halichondrin analog), 2-methoxyestradiol, vinflunine, and vinblastine, suppression of microtubule dynamics strongly correlated with mitotic block confirming that suppression of microtubule dynamics is an important aspect of each of their anticancer mechanisms. Although there were interesting differences between the microtubule polymerizing and the microtubule depolymerizing drugs, all the antimitotic drugs stabilized microtubule dynamics at low concentrations. Mitotic arrest was correlated with nearly complete inhibition of most parameters of microtubule dynamic instability by the microtubule "stabilizers" taxol and epothilone B (*see* video clip). In contrast, mitotic arrest correlated primarily with inhibition of the microtubule growing rates and lengths by the microtubule "destabilizers" vinflunine, vinblastine, and E7389 *(165)*.

2.10. Studies Into the Mechanisms of Drug Resistance Highlight the Importance of Microtubule Dynamics to Cell Function

As microtubule dynamics are important to mitosis and their suppression leads to mitotic block and subsequent tumor cell death, it was hypothesized that cells might overcome the effects of microtubule-targeted drugs by altering their inherent microtubule dynamics. To test this hypothesis, Goncalves et al. *(166)* used A549 human lung tumor cells made resistant to taxol by selection in increasing concentrations of taxol. In the resistant A549-T12 and A549-T24 cell lines (9- and 17-fold resistant to taxol, respectively), when taxol was removed from the medium, it was found that microtubule dynamics speeded up so that their dynamicity was increased from 57% to 167%, as compared with the parental cells. When grown in taxol, their microtubule dynamics were suppressed back down to a level similar to that of parental cells in the absence of taxol. Interestingly, in the absence of taxol (when dynamics were speeded up), the resistant cells became blocked at the metaphase/anaphase transition of mitosis and displayed abnormal mitotic spindles containing uncongressed lagging chromosomes. These results indicate that for successful mitosis, microtubule dynamics must fall in a narrow range, and dynamics that are either too fast or too slow prevent the completion of mitosis.

2.10.1. FINAL NOTE

We are beginning to appreciate the many ways that microtubule dynamics are regulated and function in cells. That microtubule ends are dynamic and can function independent of changes in overall microtubule mass, gives microtubules signaling functions as well as structural functions. Thus microtubules *en masse* influence cell shape and serve as railroad tracks for translocation of cell components. At the same time, their dynamics function importantly in polarization, migration, adhesion, and mitosis. What a marvelous invention!

ACKNOWLEDGMENTS

We thank Dr. Jamie Bishop for critical reading of the manuscript. This work was supported by NIH grants CA57291 and NS13560.

REFERENCES

1. Machesky L, Bornens M. Cell structure and dynamics. Curr Opin Cell Biol 2003;15:2–5.
2. Hyams JS, Lloyd CW, eds. Microtubules. New York, NY, Wiley-Liss, Inc., 1994.
3. Nogales E, Whittaker M, Milligan R, Downing K. High-resolution model of the microtubule. Cell 1999;96:79–88.
4. McEwen B, Edelstein SJ. Evidence for a mixed lattice in microtubules reassembled in vitro. J Mol Biol 1977;139:123–143.
5. Pierson GB, Burton PR, Himes RH. Alterations in number of protofilaments in microtubules assembled in vitro. J Cell Biol 1977;76:223–228.
6. Meurer-Grob P, Kasparian J, Wade RH. Microtubule structure at improved resolution. Biochemistry 2001;40(27):8000–8008.
7. Margolis RL, Wilson L. Opposite end assembly and disassembly of microtubules at steady state *in vitro*. Cell 1978;13:1–8.
8. Mitchison TJ, Kirschner M. Dynamic instability of microtubule growth. Nature 1984;312:237–242.
9. Desai A, Mitchison T. Microtubule polymerization dynamics. Annu Rev Cell Dev Biol 1997;13: 83–117.
10. Horio T, Hotani H. Visualization of the dynamic instability of individual microtubules by dark-field microscopy. Nature 1986;321:605–607.
11. Walker RA, O'Brien ET, Pryer NK, et al. Dynamic instability of individual microtubules analyzed by video light microscopy: Rate constants and transition frequencies. J Cell Biol 1988;107:1437–1448.
12. Panda D, Jordan MA, Chin K, Wilson L. Differential effects of vinblastine on polymerization and dynamics at opposite microtubule ends. J Biol Chem 1996;271:29,807–29,812.
13. Hotani H, Horio T. Dynamics of microtubules visualized by darkfield microscopy: Treadmilling and dynamic instability. Cell Motil Cytoskeleton 1988;10:229–236.
14. Toso RJ, Jordan MA, Farrell KW, Matsumoto B, Wilson L. Kinetic stabilization of microtubule dynamic instability in vitro by vinblastine. Biochemistry 1993;32(5):1285–1293.
15. Panda D, Daijo JE, Jordan MA, Wilson L. Kinetic stabilization of microtubule dynamics at steady state in vitro by substoichiometric concentrations of tubulin-colchicine complex. Biochemistry 1995;34:9921–9929.
16. Panda D, Goode BL, Feinstein SC, Wilson L. Kinetic stabilization of microtubule dynamics at steady state by tau and microtubule-binding domains of tau. Biochemistry 1995;34:11,117–11,127.
17. Billger MA, Bhatacharjee G, Williams RC Jr. Dynamic instability of microtubules assembled from microtubule-associated protein-free tubulin: Neither variability of growth and shortening rates nor rescue requires microtubule-associated proteins. Biochemistry 1996;35:13,656–13,663.
18. Newton C, DeLuca J, Himes RH, Miller HP, Jordan MA, Wilson L. Intrinsically slow dynamic instability of HeLa cell microtubules in vitro. J Biol Chem 2002;277:42,456–42,462.
19. O'Brien ET, Salmon ED, Walker RA, Erickson HP. Effects of magnesium on the dynamic instability of individual microtubules. Biochemistry 1990;29:6648–6656.
20 Gildersleeve RF, Cross AR, Cullen KE, Fagan LP, Williams JRC. Microtubules grow and shorten at intrinsically variable rates. J Biol Chem 1992;267:7995–8006.
21. Simon JR, Salmon ED. The structure of microtubule ends during the elongation and shortening phases of dynamic instability examined by negative-stain electron microscopy. J Cell Sci 1990;96: 571–582.
22. Mandelkow E, Mandelkow E, Milligan, R. Microtubule dynamics and microtubule caps: a time-resolved cryo-electron microscopy study. J Cell Biol 1991;114:977–991.

23. Panda D, Miller HP, Banerjee A, Luduena RF, Wilson L. Microtubule dynamics in vitro are regulated by the tubulin isotype composition. Proc Natl Acad Sci USA 1994;91:11,358–11,362.

24. Luduena RF. Multiple forms of tubulin: different gene products and covalent modifications. Int Rev Cytology 1998;178:207–275.

25. Wegner A. Head to tail polymerization of actin. J Mol Biol 1976;108:139–150.

26. Margolis RL, Wilson L. Microtubule treadmills-possible molecular machinery. Nature (London) 1981;293:705–711.

27. Pryer NK, Walker RA, et al. Brain microtubule-associated proteins modulate microtubule dynamic instability *in vitro*. J Cell Sci 1992;103:965–976.

28. Panda D, Miller HP, Wilson L. Rapid treadmilling of MAP-free brain microtubules in vitro and its suppression by tau. Proc Natl Acad Sci USA 1999;96:12,459–12,464.

29. Margolis RL, Wilson L. Microtubule treadmilling: What goes around comes around. Bioessays 1998;20:830–836.

30. Keating TJ, Peloquin JG, Rodionov VI, Momcilovic D, Borisy GG. Microtubule release from the centrosome. Proc Natl Acad Sci USA 1997;94:5078–5083.

31. Rodionov VI, Borisy GG. Microtubule treadmilling in vivo. Science 1997;275:215–218.

32. Chen W, Zhang, D. Kinetochore fibre dynamics outside the context of the spindle during anaphase. Nature Cell Biol 2004;6:227–231.

33. Grego S, Cantillana V, Salmon ED. Microtubule treadmilling in vitro investigated by fluorescence speckle and confocal microscopy. Biophys J 2001;81:66–78.

34. Shaw SL, Kamyar R, Ehrhardt DW. Sustained microtubule treadmilling in Arabidopsis cortical arrays. Science 2003;300:1715–1718.

35. Mitchison TJ. Poleward microtubule flux in the mitotic spindle; evidence from photoactivation of fluorescence. J Cell Biol 1989;109:637–652.

36. Mitchison TJ, Salmon ED. Poleward kinetochore fiber movement occurs during both metaphase and anaphase-A in newt lung cell mitosis. J Cell Biol 1992;119:569–582.

37. Carlier MF, Didry D, Simon C, Pantaloni, D. Mechanism of GTP hydrolysis in tubulin polymerization: characterization of the kinetic intermediate microtubule GDP-Pi using phosphate analogs. Biochemistry 1989;28:1783–1791.

38. Carlier MF. Role of nucleotide hydrolysis in the dynamics of actin filaments and microtubule. Int Rev Cytol 1989;115:139–170.

39. Bayley PM, Schilstra MJ, Martin SM. Microtubule dynamic instability: numerical simulation of microtubule transition properties using a lateral cap model. J Cell Sci 1990;X:33–48.

40. Melki R, Carlier MF, Pantaloni D. Direct evidence for GTP and GDP-Pi intermediates in microtubule assembly. Biochemistry 1990;28:8921–8932.

41. Stewart RJ, Farrell KW, Wilson L. Role of GTP hydrolysis in microtubule polymerization: evidence for a coupled hydrolysis mechanism. Biochemistry 1990;29:6489–6498.

42. Walker RA, Pryer NK, Salmon ED. Dilution of individual microtubules observed in real time in vitro: evidence that the cap size is small and independent of elongation rate. J Cell Biol 1991;114:73–81.

43. Erickson HP, O'Brien ET. Microtubule dynamic instability and GTP hydrolysis. Annu Rev Biophys Biomol Struct 1992;21:145–166.

44. Drechsel DN, Kirschner MW. The minimum cap size required to stabilize microtubules. Curr Biol 1994;4:1053–1061.

45. Caplow M, Shanks J. Evidence that a single monolayer tubulin-GTP cap is both necessary and sufficient to stabilize microtubules. Mol Biol Cell 1996;7:663–675.

46. Vandecandelaere A, Brune M, Webb MR, Martin SR, Bayley PM. Phosphate release during microtubule assembly. What stabilizes growing microtubules? Biochemistry 1999;20:1918–1924.

47. Janosi IM, Chretien D, Flybjerg H. Structural microtubule cap: stability, catastrophe, rescue, and third state. Biophys J 2002;83:1317–1330.

48. Panda D, Miller H, Wilson L. Determination of the size and chemical nature of the stabilizing cap at microtubule ends using modulators of polymerization dynamics. Biochemistry 2002;41:1609–1617.

49. Hyman A, Chretien D, Arnal I, Wade R. Structural changes accompanying GTP hydrolysis in microtubules: information from a slowly hydrolyzable analogue guanylyl-(alpha,beta)-methylenediphosphonate. J Cell Biol 1995;128:117–125.

50. Muller-Reichert T, Chretien D, Hyman AA. Structural changes at microtubule ends accompanying GTP hydrolysis: information from a slowly hydrolysable analogue of GTP, guanylyl (a,b) methylenediphosphonate. Proc Natl Acad Sci USA 1998;95:3661–3666.
51. Wang HW, Nogales E. Nucleotide-dependent bending flexibility of tubulin controls microtubule assembly. Nature 2005;435:911–915.
52. Caplow M, Ruhlen RL, Shanks J. The free energy for hydrolysis of a microtubule-bound nucleotide triphosphate is near zero; all of the free energy for hydrolysis is stored in the microtubule lattice. J Cell Biol 1994;127:779–788.
53. Caplow M, Ruhlen R, Shanks J, Walker RA, Salmon ED. Stabilization of microtubules by tubulin-GDP-Pi subunits. Biochemistry 1989;28:8136–8141.
54. Trinczek B, Marx A, Mandelkow EM, Murphy DB, Mandelkow E. Dynamics of microtubules from erythrocyte marginal bands. Mol Biol Cell 1993;4:323–335.
55. Caplow M, Shanks J. Microtubule dynamic instability does not result from stabilization of microtubules by tubulin-GDP-Pi subunits. Biochemistry 1998;37:12,994–13,002.
56. Caplow M, Fee L. Concerning the chemical nature of tubulin subunits that cap and stabilize microtubules. Biochemistry 2003;42:2122–2126.
57. L. Wilson, H.P. Miller, and D. Panda, to be published.
58. Cassimeris L. Accessory protein regulation of microtubule dynamics throughout the cell cycle. Curr Opin Cell Biol 1999;11:134–141.
59. Kinoshita K, Arnal I, Desai A, Drechsel DN, Hyman AA. Reconstitution of physiological microtubule dynamics using purified components. Science 2001;294(5545):1340–1343.
60. Wilson L, Jordan MA. Pharmacological probes of microtubule function. Microtubules.Hyams J, Lloyd C. New York: John Wiley and Sons, Inc.; 1994;59–84.
61. Wilson L, Jordan MA. Microtubule dynamics: taking aim at a moving target. Chem Biol 1995;2: 569–573.
62. Jordan MA. Mechanism of action of antitumor drugs that interact with microtubules and tubulin. Curr Med Chem Anticancer Agents 2002;2:1–17.
63. Jordan MA, Wilson L. Microtubules as a target for anticancer drugs. Nat Rev Cancer 2004;4: 253–265.
64. Wilson L, Panda D, Jordan MA. Modulation of microtubule dynamics by drugs: a paradigm for the actions of cellular regulators. Cell Struct Function 1999;24:329–335.
65. Jordan MA, Wilson L, Vallee R, ed. Academic press. The use of drugs to study the role of microtubule assembly dynamics in living cells. Molecular Motors and the Cytoskeleton, Meth Enzymol 1998;298:252–276.
66. Dustin P. Microtubules. Berlin, Springer-Verlag, 1984 2nd edition pp 1–482.
67. Lobert S, Correia J. Energetics of Vinca alkaloid interactions with tubulin. Meth Enzymol 2000; 323:77–103.
68. Bai RB, Pettit GR, Hamel E. Dolastatin 10. a powerful cytostatic peptide derived from a marine animal. Inhibition of tubulin polymerization mediated through the vinca alkaloid binding domain. Biochem Pharmacol 1990;39:1941–1949.
69. Bai RB, Pettit GR, Hamel E. Binding of dolastatin 10 to tubulin at a distinct site for peptide antimitotic agents near the exchangeable nucleotide and vinca alkaloid sites. J Biol Chem 1990;265: 17,141–17,149.
70. Wilson L, Jordan MA, Morse A, Margolis RL. Interaction of vinblastine with steady-state microtubules in vitro. J Mol Biol 1982;159:129–149.
71. Correia JJ, Lobert S. Physiochemical aspects of tubulin-interacting antimitotic drugs. Curr Pharm Des 2001;7(13):1213–1228.
72. Na GC, Timasheff SN. Thermodynamic linkage between tubulin self-association and the binding of vinblastine. Biochemistry 1980;19(7):1347–1354.
73. Jordan MA, Margolis RL, Himes RH, Wilson L. Identification of a distinct class of vinblastine binding sites on microtubules. J Mol Biol 1986;187:61–73.
74. Singer WD, Jordan MA, Wilson L, Himes RH. Binding of vinblastine to stabilized microtubules. Mol Pharmacol 1989;36(3):366–370.
75. Hastie SB. Interactions of colchicine with tubulin. Pharmacol Ther 1991;512:377–401.
76. Ravelli RB, Gigant B, Curmi PA, et al. Insight into tubulin regulation from a complex with colchicine and a stathmin-like domain. Nature 2004;428:198–202.

77. Farrell KW, Wilson L. The differential kinetic stabilization of opposite microtubule ends by tubulin-colchicine complexes. Biochemistry 1984;23:3741–3748.
78. Sternlicht H, Ringel I. Colchicine inhibition of microtubule assembly via copolymer formations. J Biol Chem 1979;254:10,540–10,550.
79. Sternlicht H, Ringel I, Szasz J. Theory for modelling the copolymerization of tubulin and tubulin-colchicine complex. Biophys J 1983;42:255–267.
80. Skoufias D, Wilson L. Mechanism of inhibition of microtubule polymerization by colchicine: Inhibitory potencies of unliganded colchicine and tubulin-colchicine complexes. Biochemistry 1992;31:738–746.
81. Margolis RL, Rauch CT, Wilson L. Mechanism of colchicine dimer addition to microtubule ends: Implications for the microtubule polymerization mechanism. Biochemistry 1980;19:5550–5557.
82. Wilson L, Farrell KW. Kinetics and steady-state dynamics of tubulin addition and loss at opposite microtubule ends: the mechanism of action of colchicine. Ann NY Acad Sci 1986;466:690–708.
83. Nogales E, Wolf SG, Khan IA, Luduena RF, Downing KA. Structure of tubulin at 6.5A and location of the taxol-binding site. Nature 1995;375:424–427.
84. Kar S, Fan J, Smith MJ, Goedert MJ, Amos LA. Repeat motifs of tau bind to the insides of microtubules in the absence of taxol. EMBO J 2003;22:70–77.
85. Schiff PB, Horwitz SB. Taxol assembles tubulin in the absence of exogenous guanosine 5′-triphosphate or microtubule-associated proteins. Biochemistry 1981;20:3247–3252.
86. Diaz JF, Andreu JM. Assembly of purified GDP-tubulin into microtubules induced by taxol and taxotere: reversibility, ligand stoichiometry, and competition. Biochemistry 1993;32:2747–2755.
87. Diaz JF, Barasoain I, Andreu JM. Fast kinetics of Taxol binding to microtubules. Effects of solution variables and microtubule-associated proteins. J Biol Chem 2003;278:8407–8419.
88. Nogales E. Structural insights into microtubule function. Annu Rev Biophys Biomol Struct 2001; 30:397–420.
89. Wilson L, Miller HP, Farrell KW, Snyder KB, Thompson WC, Purich DL. Taxol stabilization of microtubules in vitro: dynamics of tubulin addition and loss at opposite microtubule ends. Biochemistry 1985;24:5254–5262.
90. Jordan MA, Toso RJ, Thrower D, Wilson L. Mechanism of mitotic block and inhibition of cell proliferation by taxol at low concentrations. Proc Natl Acad Sci USA 1993;90:9552–9556.
91. Derry WB, Wilson L, Jordan MA. Substoichiometric binding of taxol suppresses microtubule dynamics. Biochemistry 1995;34:2203–2211.
92. Derry WB, Wilson L, Jordan MA. Low potency of taxol at microtubule minus ends: implication for its anti-mitotic and therapeutic mechanism. Cancer Res 1998;58:1177–1184.
93. Waters JC, Mitchison TJ, Rieder CL, Salmon ED. The kinetochore microtubule minus-end disassembly associated with poleward flux produces a force that can do work. Mol Biol Cell 1996;7: 1547–1558.
94. Panda D, Samuel S, Massie M, Feinsntein S, Wilson L. Differential regulation of microtubule dynamics by 3-repeat and 4-repeat tau: Implications for the onset of neurodegenerative disease. Proc Natl Acad Sci USA 2003;100:9548–9553.
95. Feinstein SC, Wilson L. Inability of tau to properly regulate neuronal microtubule dynamics: a loss-of-function mechanism by which tau might mediate neuronal cell death. Biochimica et Biphysica Acta 2005;1739:268–279.
96. Curmi PA, Anderson SSL, Lachkar S, et al. The stathmin/tubulin interaction in vitro. J Biol Chem 1997;272:25,029–25,036.
97. Cassimeris L, Spittle C. Regulation of microtubule-associated proteins. Int Rev Cytol 2001;210:163–226.
98. Ovechkina Y, Wordeman L. Unconventional motoring: An overview of the kin C and Kin I kinesins. Traffic 2003;4:367–375.
99. Newton CN, Wagenbach M, Ovechkina Y, Wordeman L, Wilson L. MCAK, a kin I kinesin, increases the catastrophe frequency of steady-state HeLa cell microtubules in an ATP-dependent manner in vitro. FEBS Lett 2004;572:80–84.
100. Hayden JJ, Bowser SS, Rieder C. Kinetochores capture astral microtubules during chromosome attachment to the mitotic spindle: direct visualization in live newt cells. J Cell Biol 1990;111: 1039–1045.
101. McIntosh J, Grishchuk EL, West RR. Chromosome-microtubule interactions during mitosis. Annu Rev Cell Dev Biol 2002;18:193–219.

102. Rodriguez O, Schaefer A, Mandato CA, Forscher P, Bement WM, Waterman-Storer CM. Conserved microtubule-actin interactions in cell movement and morphogenesis. Nat Cell Biol 2003;5:599–609.

103. Maddox P, Desai A, Oegama K, Mitchison TJ, Salmon ED. Poleward microtubule flux is a major component of spindle dynamics and anaphase a in mitotic Drosophila embryos. Curr Biol 2002; 12:R651–R653.

104. Nedelac F, Surrey T, Karsenti E. Self-organisation and forces in the microtubule cytoskeleton. Curr opin Cell Biol 2003;15:118–124.

105. Abal M, Piel M, Bouckson-Castaing V, Mogensen M, Sibarita JB, Bornens M. Microtubule release from the centrosome in migrating cells. J Cell Biol 2002;159(5):731–737.

106. Job D, Valiron O, Oakley B. Microtubule nucleation. Curr Opin Cell Biol 2003;15:111–117.

107. Saxton WM, Stemple DL, Leslie RJ, Salmon ED, Zavortink M, McIntosh JR. Tubulin dynamics in cultured mammalian cells. J Cell Biol 1984;99:2175–2186.

108. Pepperkok R, Bre MH, Davoust J, Kreis TE. Microtubules are stabilized in confluent epithelial cells but not in fibroblasts. J Cell Biol 1990;111:3003–3012.

109. Wadsworth P, McGrail M. Interphase microtubule dynamics are cell type-specific. J Cell Sci 1990; 95:23–32.

110. Shelden E, Wadsworth P. Observation and quantification of individual microtubule behavior in vivo: microtubule dynamics are cell-type specific. J Cell Biol 1993;120:935–945.

111. Wadsworth P, Bottaro DP. Microtubule dynamic turnover is suppressed during polarization and stimulated in hepatocyte growth factor scattered Madin-Darby canine kidney epithelial cells. Cell Motil Cytoskeleton 1996;35:225–236.

112. Dhamodharan RI, Jordan MA, Thrower D, Wilson L, Wadsworth P. Vinblastine suppresses dynamics of individual microtubules in living cells. Mol Biol Cell 1995;6:1215–1229.

113. Dhamodharan RI, Wadsworth P. Modulation of microtubule dynamic instability in vivo by brain microtubule associated proteins. J Cell Sci 1995;108:1679–1689.

114. Yvon AM, Wadsworth P, Jordan MA. Taxol suppresses dynamics of individual microtubules in living human tumor cells. Mol Biol Cell 1999;10:947–949.

115. Honore S, Kamath K, Braquer D, Wilson L, Briand C, Jordan MA. Suppression of microtubule dynamics by discodermolide by a novel mechanism is associated with mitotic arrest and inhibition of tumor cell proliferation. Mol Cancer Therapeutics 2003;2:1303–1311.

116. Kamath K, Jordan MA. Suppression of microtubule dynamics by epothilone B in living MCF7 cells. Cancer Res 2003;63:6026–6031.

117. Bunker JM, Wilson L, Jordan MA, Feinstein SC. Modulation of microtubule dynamics by tau in living cells: Implications for development and neurodegeneration. Mol Biol Cell 2004;15:2720–2728.

118. Yvon AM, Gross DJ, Wadsworth P. Antagonistic forces generated by myosin II and cytoplasmic cynein regulate microtubule turnover, movement, and organization in interphase cells. Proc Natl Acac Sci USA 2001;98:8656–8661.

119. Tulu US, Rusan NM, Wadsworth P. Peripheral, non-centrosome-associated microtubules contribute to spindle formation in centrosome-containing cells. Curr Biol 2003;13:1844–1899.

120. Waterman-Storer CM, Salmon ED. Fluorescent speckle microscopy of microtubules: how low can you go? FASEB J 1999;13(suppl 2):S225–S230.

121. Danuser G, Waterman-Storer CM. Quantitative fluorescent speckle microscopy: where it came from and where it is going. J Microscopy 2003;211:191–207.

122. Zhou FQ, Waterman-Storer CM, Cohan CS. Focal loss of actin bundles causes microtubule redistribution and growth cone turning. J Cell Biol 2002;157:839–849.

123. Schuyler SC, Pellman D. Microtubule "plus-end tracking proteins": the end is just the beginning. Cell 2001;105:421–424.

124. Vaughan K. Surfing, regulating and capturing: are all microtubule-tip-tracking proteins created equal? Trends Cell Biol 2004;14:491–6.

125. Gundersen G, Gomes E, Wen Y. Cortical control of microtubule stability and polarization. Curr Opin Cell Biol 2004;16:106–12.

126. Galjart N. CLIPs and CLASPs and cellular dynamics. Nature Reviews Molecular Cell Biology 2005;6:487–98.

127. Manna T, Thrower D, Miller HP, Curmi P, Wilson L. Stathmin strongly increases the minus end catastrophe frequency and induces rapid treadmilling of bovine brain microtubules at steady state in vitro. J Biol Chem 2006;281:2071–2078.

128. Wittmann T, Bokoch GM, Waterman-Storer CM. Regulation of microtubule destabilizing activity of Op18.stathmin downstream of Rac1. J Biol Chem 2004;279:6196–6203.

129. Mogensen MM, Malik A, Piel M, Bouckson-Castaing V, Bornens M. Microtubule minus-end anchorage at centrosomal and non-centrosomal sites: the role of ninein. J Cell Sci 2000;113:3013–3023.

130. Karsenti E, Vernos I. The mitotic spindle: a self-made machine. Science 2001;294:543–547.

131. Wilde A, Lizarraga SB, Zhang L, et al. Ran stimulates spindle assembly by altering microtubule dynamics and the balance of motor activities. Nat Cell Biol 2001;3:221–227.

132. Dammermann A, Desai A, Oegama K. The minus end in sight. Curr Biol 2003;13:R614–R624.

133. Small JV, Kaverina I. Microtubules meet substrate adhesions to arrange cell polarity. Curr Opin Cell Biol 2003;15:40–47.

134. Ookata K, Hisanaga S, Bulinski JC, et al. Cyclin B interaction with microtubule-associated protein 4 (MAP4) targets p34cdc2 kinase to microtubules and is a potential regulator of M-phase microtubule dynamics. J Cell Biol 1995;128:849–862.

135. Chang W, Gruber D, Chari S, et al. Phosphorylation of MAP4 affects microtubule properties and cell cycle progression. J Cell Sci 2001;114(pt 15):2879–2887.

136. Rusan NM, Fagerstrom CJ, Yvon AMC, Wadsworth P. Cell cycle-dependent changes in microtubule dynamics in living cells expressing green fluorescent protein-alpha tubulin. Mol Biol Cell 2001;12:971–980.

137. Wilson PJ, Forer A. Effects of nanomolar taxol on crane-fly spermatocyte spindles indicate that acetylation of kinetochore microtubules can be used as a marker of poleward tubulin flux. Cell Motil Cytoskeleton 1997;37:20–32.

138. Li X, Nicklas RB. Mitotic forces control a cell-cycle checkpoint. Nature 1995;373:630–632.

139. Nicklas RB, Ward SC, Gorbsky GJ. Kinetochore chemistry is sensitive to tension and may link mitotic forces to a cell cycle checkpoint. J Cell Biol 1995;130:929–939.

140. Gorbsky GJ. Cell cycle checkpoints: arresting progress in mitosis. BioEssays 1997;19:193–197.

141. Rieder C, Schultz A, Cole R, Sluder G. Anaphase onset in vertebrate somatic cells is controlled by a checkpoint that monitors sister kinetochore attachment to the spindle. J Cell Biol 1994;127:1301–1310.

142. Shelby RD, Hahn KM, Sullivan KF. Dynamic elastic behavior of alpha-satellite DNA domains visualized in situ in living human cells. J Cell Biol 1996;135:545–557.

143. Desai A, Maddox PS, Mitchison TJ, Salmon ED. Anaphase A chromosome movement and poleward spindle microtubule flux occur at similar rates in Xenopus extract spindles. J Cell Biol 1998;141:703–713.

144. Walczak CE, Mitchison TJ, Desai A. XKCM1: a Xenopus kinesin-related protein that regulates microtubule dynamics during mitotic spindle assembly. Cell 1996;84:37–47.

145. Desai A, Verma S, Mitchison TJ, Walczak CE. Kin I kinesins are microtubule-destabilizing enzymes. Cell 1999;96(1):69–78.

146. Maney T, Wagenbach M, Wordeman L. Molecular dissection of the microtubule depolymerizing activity of mitotic centromere-associated kinesin. J Biol Chem 2001;276(37):34,753–34,758.

147. Walczak CE. Ran hits the ground running. Nat Cell Biol 2001;3:E69–E70.

148. Tirnauer JS, Canman JC, Salmon ED, Mitchison TJ. EB1 targets to kinetochores with attached polymerizing microtubules. Mol Biol Cell 2002;13:4308–4316.

149. Maiato H, Fairley E, Reider CL, Swedlow JR, Sunkel CE, Earnshaw WC. Human CLASP1 is an outer kinetochore component that regulates spindle microtubule dynamics. Cell 2003;113:891–904.

150. Yu H. Regulation of APC–Cdc20 by the spindle checkpoint. Curr Opin Cell Biol 2002;14:706–714.

151. Skoufias DA, Andreassen P, Lacroix F, Wilson L, Margolis RL. Mammalian mad2 and bub1/bubR1 recognize distinct spindle-attachment and kinetochore-tension checkpoints. Proc Natl Acad Sci USA 2001;10:4492–4497.

152. Jordan MA, Wilson L. Microtubule polymerization dynamics, mitotic block, and cell death by paclitaxel at low concentrations. Taxane Anticancer Agents. Georg GI, Chen TT, Ojima I, Vyas DM. Washington DC, American Chemical Society, 1995;138–153.

153. Jordan MA, Wendell KL, Gardiner S, Derry WB, Copp H, Wilson L. Mitotic block induced in HeLa cells by low concentrations of paclitaxel (Taxol) results in abnormal mitotic exit and apoptotic cell death. Cancer Res 1996;56:816–825.

154. Hamel E, Covell DG. Antimitotic peptides and depsipeptides. Curr Med chem Anticancer Agents 2002;2:19–53.

155. Zhou J, Gupta K, Aggarwal S, et al. Brominated derivatives of noscapine are potent microtubule-interfering agents that perturb mitosis and inhibit cell proliferation. Mol Pharmacol 2003;63(4):799–807.
156. Hoffman JC, Vaughn KC. Mitotic disrupter herbicides act by a single mechanism but vary in efficacy. Protoplasma 1994;179:16–25.
157. Lacey E, Gill JH. Biochemistry of benzimidazole resistance. Acta Trop 1994;56:245–262.
158. Cann JR, Hinman ND. Interaction of chlorpromazine with brain microtubule subunit protein. Mol Pharmacol 1975;11:256–267.
159. Boder GB, Paul DC, Williams DC. Chlorpromazine inhibits mitosis of mammalian cells. Eur J Cell Biol 1983;31:349–353.
160. Lobert S, Ingram J, Correia J. Additivity of dilantin and vinblastine inhibitory effects on microtubule assembly. Cancer Res 1999;59:4816–4822.
161. Jimenez-Barbero J, Amat-Guerri F, Snyder JP. The solid state, solution and tubulin-bound conformations of agents that promote microtubule stabilization. Curr Med Chem Anticancer Agents 2002;2(1):91–122.
162. Jordan MA, Thrower D, Wilson L. Mechanism of inhibition of cell proliferation by Vinca alkaloids. Cancer Res 1991;51(8):2212–2222.
163. Okouneva T, Hill BT, Wilson L, Jordan MA. The effects of vinflunine, vinorelbine, and vinblastine on centromere dynamics. Mol Cancer Therapeutics 2003;2:427–436.
164. Kelling J, Sullivan K, Wilson L, Jordan MA. Suppression of centromere dynamics by taxol in living osteosarcoma cells. Cancer Res 2003;63.
165. Jordan MA, Kamath K, Manna T, et al. The primary antimitotic mechanism of action of the synthetic halichondrin E7389 is suppression of microtubule growth. Mol. Cancer Ther 2005;4:1086–1095.
166. Goncalves A, Braguer D, Kamath K, et al. Resistance to taxol in lung cancer cells associated with increased microtubule dynamics. Proc Natl Acad Sci USA 2001;98:11,737–11,741.
167. J. Kelling. Anti-Cancer Agents. Curr Med Chem 2:1–17.

4

Microtubule-Associated Proteins and Microtubule-Interacting Proteins

Regulators of Microtubule Dynamics

Maria Kavallaris, Sima Don, and Nicole M. Verrills

CONTENTS

SUMMARY

Microtubules are regulated by a range of proteins that interact with tubulin and regulate their stability. A large number and variety of microtubule-associated proteins (MAPs) and microtubule-interacting proteins have been indentified and they exhibit cell and tissue specific expression. MAPs and microtubule-interacting proteins carry out a wide range of functions including regulation of microtubule stability, cross-linking microtubules and mediate interactions of microtubules with other proteins in the cell. The dynamic nature of microtubules and their range of cellular functions is dependent of the interaction and regulation of MAPs and microtubule-interacting proteins.

Key Words: Microtubules; tubulin; microtubule associated proteins; stathmin; microtubule stability.

1. INTRODUCTION

Microtubules perform a spectrum of functions and are a major component of the cytoskeleton. They are dynamic structures involved in many cellular processes including cell division, intracellular transport, and certain forms of cellular movement. The diversity of microtubule functions are regulated and influenced by microtubule-associated proteins (MAPs) and a range of tubulin and microtubule-interacting proteins. Indeed, the complex process that leads to the defined temporal and spatial organization of

From: *Cancer Drug Discovery and Development: The Role of Microtubules in Cell Biology, Neurobiology, and Oncology* Edited by: Tito Fojo © Humana Press, Totowa, NJ

microtubules involves both structural and regulatory coordination of MAPs and associated proteins. Microtubules are polymers that consist of αβ-tubulin heterodimers that are present in all eukaryotes. They are highly dynamic structures that are constantly growing and shortening, behavior referred to as dynamic instability. Multiple α- and β-tubulin isotypes have been described in mammalian cells and they display both developmental and tissue-specific expression (1). The various isotypes share a high degree of amino acid homology although they vary in their carboxy terminal region. It has been suggested that the role of different β-tubulin isotypes may be to influence microtubule dynamics by binding distinct MAPs in this region (2). Multiple factors can regulate the assembly and stability of microtubules including differential expression of different tubulin isotypes, regulation of monomer folding through tubulin-folding cofactors, microtubule nucleation, and post-translational modification of tubulin proteins. Additionally, various functions of microtubules involve their interaction with a large number of microtubule interacting proteins, which influence the regulation and distribution of microtubules in the cell through stabilizing or destabilizing effects. There is a wide variety of MAPs that provide the functional diversity of microtubules. This chapter will focus on MAPs that are primarily involved in regulating microtubule stability and dynamics.

2. STRUCTURAL AND FUNCTIONAL FEATURES OF MAPS

Structural MAPs were initially identified as proteins that copurified with tubulin in vitro, promoted microtubule assembly and stabilized microtubules. A variety of structural MAPs can interact on the surface of microtubules and reduce their dynamic behavior by stimulating tubulin polymerization. The precise nature of MAP binding to microtubules is not known. MAPs have a positive charge and it has been generally believed that they interact with the negatively charged C-terminal end of tubulin, facilitating the tubulin subunits to form the microtubule polymer. The exact sequences on tubulin that interact with MAPs are not well defined. This is partially because of the fact that little is known about the tertiary structure of the C-terminal region of tubulin.

Structural MAPs such as Tau, MAP2, and MAP4, are heat-stable and share a conserved molecular structure consisting of a projection domain and a carboxy-terminal microtubule-binding domain (3–8) (Fig. 1). This conserved domain contains three or four pseudorepeats. The N-terminal projection domain protrudes from the microtubule wall but does not bind microtubules, whereas the C-terminal microtubule binding domain (MTB) binds to the microtubule (9,10). The projection domain, projects away from the microtubule wall and is able to crosslink the microtubule with membranes, other microtubules or intermediate filaments (11).

MAPs colocalize with microtubules and MAP staining follows a similar structural pattern to microtubule staining (Fig. 2). As the predicted function of structural MAPs is to stabilize microtubules, there must be a mechanism that regulates these interactions in order for the dynamic feature of microtubules to exist. Phosphorylation of MAPs causes them to dissociate from microtubules, hence, allowing the microtubules to depolymerize leading to increased dynamic instability. The phosphorylation event occurs at the MTB of MAPs by several kinases such as the MAP kinases (12,13) and cdc2 kinase (14). Microtubule-affinity-regulating kinases phosphorylate MAPs at their MTB domain, causing their disengagement from the microtubule and leading to increased dynamic instability (15).

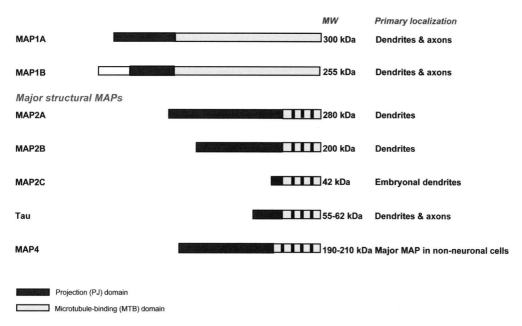

Fig. 1. Schematic diagram of the major MAPs. Protein structure, Protein size, and primary cellular localization. To view this figure in color, see the insert and the companion CD-ROM.

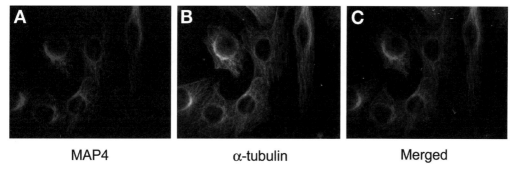

Fig. 2. MAP binding to microtubules. Human neuroblastoma, SHEP, cells costained with antibodies against MAP4 and α-tubulin and visualized using secondary antibodies tagged with Cy3 (red) and Cy2 (green), respectively. (**A**) MAP4 staining demonstrates that the MAP follows a microtubule-like distribution pattern. (**B**) α-Tubulin staining reveals the intricate microtubule network in these cells. (**C**) Merged image showing the costaining and localization of MAP4 and microtubules (yellow/orange). To view this figure in color, see the insert and the companion CD-ROM.

The organization of microtubules in different cell types is likely to be dependent on the types of MAPs expressed and the nature of the microtubule-organizing center. The basis for this is that the organization of microtubules varies in different cell types. As an example, interphase microtubules in cultured cells tend to radiate from the centrosome to the cell periphery *(16)* while microtubules in neurites run in parallel structures *(17)*. The major structural MAPs in neurons are MAP2 and tau *(12)*, whereas the most abundant and ubiquitous MAP in nonneuronal cells is MAP4. Many of the functional differences in the neuronal and nonneuronal MAPs have been identified in cells transfected with cDNA encoding MAP2, tau or MAP4. Distinct differences in the

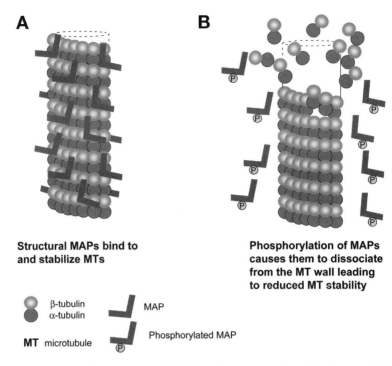

Fig. 3. Schematic model of interactions of MAPs with microtubules. **(A)** Structural MAPs such as MAP2, tau, and MAP4 can bind through their MBD to the microtubule wall and suppress microtubule dynamics by stabilizing microtubules. **(B)** The interaction of MAPs with microtubules is regulated by phosphorylation. Phosphorylation causes MAPs to dissociate from the microtubule wall, which in turn increases microtubule dynamics. NB: The size of the projection domain varies between the various MAPs (*see* Fig. 1) and therefore this figure is a generalized representation of MAP-microtubule interactions. To view this figure in color, see the insert and the companion CD-ROM.

organization of microtubules have been noted in cells transfected with cDNA encoding these three MAPs. Dense microtubule bundling is induced with MAP2 or Tau *(18–20)* whereas this does not occur with MAP4 *(21,22)*. MAPs are differentially expressed in different tissues and are critical regulators of microtubule assembly and stability (Fig. 3).

2.1. MAP1

The microtubule-associated proteins 1A (MAP1A) and 1B (MAP1B) are high molecular weight protein complexes that are distantly related and are encoded by distinct genes that reside on chromosomes 15q15.3 and 5q13, respectively. Both MAP1A and MAP1B are synthesized as a polyprotein (referred to as the heavy chain) coupled to a light chain (LC), i.e., MAP1A-LC2, 28 kDa, and MAP1B-LC1, 34 kDa *(23,24)*. MAP1 proteins are thought to play a role in regulating the neuronal cytoskeleton. MAP1A is abundant in the adult brain, whereas MAP1B is a neuritogenesis-associated MAP *(25)*. MAP1A binds to microtubules through a novel acidic binding motif *(26)*, whereas MAP1B binds to microtubules through a series of basic motifs *(27)*. A third light chain, LC3 (18 kDa), is a subunit of MAP1A and MAP1B and has been proposed that expression of LC3 can regulate the microtubule binding activity of MAP1A and MAP1B *(28)*.

MAP1B is the first MAP to be expressed during embryonal development of the nervous system. The product of the MAP1B gene is a precursor polypeptide that most likely undergoes proteolytic processing to generate the final MAP1B heavy chain and LC1 light chain (reviewed in ref. *29*). In contrast to other microtubule-stabilizing proteins such as MAP2 and tau that differ in their molecular structure (Fig. 1), transfection of MAP1B into COS cells did not result in the formation of microtubule bundles *(30)*. However, microtubules in MAP1B-transfected cells were stabilized against nocodazole-induced depolymerization and were enriched in acetylated α-tubulin (indicative of stable microtubules) *(30)*. Therefore, MAP1B can stabilize microtubules in cells, albeit less efficiently than MAP2 and tau, but it cannot induce microtubule bundling.

In mice, MAP1B is expressed at high levels during embryogenesis and reaches maximum levels at 2–3 wk of age followed by a decrease in MAP1B and increased expression of other MAPs such as MAP1A. The highest levels of MAP1B are found in regions that show extensive growth of axons and motor neurons, retinal ganglion cells, olfactory epithelium, and nerve layer of olfactory bulb *(31–33)*. The distribution and function of MAP1B can be modified by phosphorylation. Studies on neuronal growth cones have revealed that MAP1B can interact with both microtubules and actin microfilaments *(34,35)*. Gene knockout studies of the murine MAP1B gene suggested an important role in development and function of the nervous system *(36,37)*.

2.2. Tau

Brain tissue is an abundant source of microtubules and many studies have traditionally focused on brain MAPs. The most studied MAPs are the heat stable proteins MAP2 and Tau. Both these neuronal MAPs share homologous repeats in the carboxy-terminal region that contribute to microtubule binding and stability *(38)*. Tau has been shown by covalent crosslinking experiments to bind to two distinct sites on the C-terminal third of both α- and β-tubulin *(39)*.

The Tau gene is located on chromosome 17q21. Tau tissue expression is predominantly neuronal, although very low levels are detected in oligodendrocytes and astrocytes. Tau mRNA and proteins have also been detected in diverse tissues of nonneuronal origin such as heart, kidney, lung, muscle, pancreas, and testis *(40–42)*. The 5′ UTR to the end of the 3′ UTR of Tau, spans 133.9kb and contains 16 exons *(41,43)*. Although exons 1 and 14 are transcribed, they are not translated. Tau exons are alternatively spliced in a tissue-specific and a developmentally regulated manner. Certain isoforms of Tau mRNA such as 4A, 6, and 8 are never found in brain. However, 4A is found in peripheral nervous tissue and results in a tau protein isoform originally named big tau for its higher molecular weight *(44)*.

Transcription of Tau starts at a unique site at the start of exon 1 and stops at one of two alternate polyadenylation sites downstream of exon 14. This leads to two transcripts of 2 and 6 kb encoding tau, with the 6 kb being the major isoform in brain. Interestingly the two transcripts differ in their subcellular localization, with the 2 kb isoform found in the cell body and the 6 kb isoform found in axons *(45)*. Tau gene regulation is poorly understood. It is thought that the 3′ UTR contains *cis*-elements that localize tau mRNA to specific cell compartments. The half-life of Tau mRNA is relatively long in neuronal cells (~10 h).

There are three adult-brain-specific alternatively spliced coding exons, exons 2, 3, and 10. Incorporation of exon 10 adds one microtubule-binding repeat domain to the three other exons that alter the tau protein binding capacity on microtubules. Tau exists as two classes of protein because of alternate splicing of exon 10, 3 repeat (3R)- and 4 repeat (4R)-tau isoforms *(46)*. Further, alternate splicing of exons 2 and 3 allows for a total of six isoforms, three each of 3R and 4R isoforms. In the foetal brain only the shortest isoform is expressed *(46)*.

Tau is localized mainly in axons, which has several protein isoforms ranging in size from 55 to 62 kDa *(47)*. Tau proteins were shown to be potent promoters of tubulin polymerization *(48)*, and their overexpression by microinjection and transfection led to increased microtubule mass accumulation, increased microtubule stability, and extensive microtubule bundling *(18,49)*. Inhibition of Tau isoforms by antisense oligonucleotides prevented the establishment of axonal polarity and outgrowth in primary cerebellar neurons *(50)*. Furthermore, sense and antisense transfection analysis of Tau function in PC 12 cells demonstrated the role of tau in promoting net microtubule assembly, the rate of neurite elongation and stability of neurites *(51)*. In contrast to the in vitro studies outlined above, mice with targeted disruption of the Tau gene exhibited subtle changes in microtubule organization in small caliber axons and despite being chronically deficient in tau, were relatively normal *(52)*. Interestingly, the total absence of this usually predominant MAP did not alter microtubule stability or neurite growth properties in primary neuronal cultures from mice. A possible explanation for the stark differences between the in vitro and in vivo studies is that there is sufficient MAP protein redundancy or adequate plasticity to mask the requirement for tau during development in the mouse. Gene targeting of another predominant axonal MAP, MAP1B also produced a viable and relatively normal mouse *(37)*. However, mating *tau* and *MAP1B* null animals demonstrated that loss of both axonal MAPs is lethal by 4 wk of age *(53)*. It appears that both *tau* and MAP1B are required for the genesis of axon tracts and that these two MAPs function synergistically.

There has been a strong association between the accumulation of filamentous tau and a number of neurodegenerative diseases, referred to as tauopathies. These include, Pick's disease, frontal temporal dementia, and Parkinsonism linked to chromosome 17 (FTDP-17), and Alzheimer's disease *(54)*. All these conditions lead to major filamentous deposits of hyperphosphorylated tau with associated neuronal loss in the affected regions. Alzheimer's disease is the most common human cognitive neurodegenerative disease. The principle pathology in the brain neurons of affected individuals are insoluble tangled filaments, which are predominantly made up of tau aggregations *(55)*. The identification of tau mutations in a set of dominantly inherited cases of FTDP-17 was a major breakthrough in unraveling neuronal degradation and death. A direct link between errors in tau and abnormal accumulation of tau-containing filaments in neurons, have been directly associated with neuronal failure (reviewed in ref. *54*). Two classes of tau mutations have been identified that lead to morphologically distinct filaments. One class of mutations can alter the binding of tau to microtubules and leads to the production of abnormal twisted filamentous structures that are made up of all six isoforms that occur primarily in neurons. The other class is splicing mutations that change the ratio of tau with four microtubule-binding domains (4R) to types with only three microtubule-binding domains. These mutations cause the appearance of filamentous structures in both neurons and glia. Despite the differential effects of the two classes of mutations, they both lead to FTDP-17, possibly because they cause a net increase in the unbound pool of tau. Increased free tau in neurons could lead to sequestration of free tubulin as was observed in lamprey anterior bulbar cells overexpressing human tau *(56)*. These cells contained tau filaments entangled with tubulin.

2.3. MAP2

MAP2 is a neuronal-specific high molecular weight protein (M_r ~280 kDa) *(57)*, which is able to bind both tubulin and actin through its tubulin-binding domain and modify the stability of microtubules and microfilaments *(58,59)*. Both MAP2-tubulin and MAP2-actin interactions are regulated by phosphorylation *(59)*, however, the phosphorylation events that reduce the ability of MAP2 to bind to the microtubules enhance its ability to interact with the actin cytoskeleton *(60)*. The MAP2 gene is located on chromosome 2q34-q35. The expression of MAP2 is predominantly neuronal and more specifically, in dendrites. Multiple isoforms of MAP2 exist in human cells, with high molecular weight isoforms MAP2a and MAP2b, and low molecular weight isoforms MAP2c and MAP2d being expressed at different stages of development *(61)* and differentiation *(62)*. MAP2a and MAP2b are concentrated in dendrites of neuronal cells, with MAP2a being expressed at the late stage of development and MAP2b being present at both embryonic and adult stages. MAP2c is also expressed in dendrites, but can be found in axons and glial cells *(25)*. Different isoforms of MAP2 stabilize microtubules to different extents. Although MAP2c is usually expressed at the early stages of development and then replaced by high molecular weight MAP2 isoforms *(61)*, it has been shown to be expressed continuously in the adult mammalian olfactory system and in the retina *(33)*. MAP2c is expressed at high levels in early brain development. MAP2a and MAP2b localize exclusively to dendrites, whereas MAP2c is found in all cell compartments *(63,64)*, suggesting that the various MAP2 isoforms have distinct cellular functions. MAP2d is produced by developmentally regulated alternate splicing of MAP2 *(65,66)*.

Apart from regulation of cytoskeletal dynamics, other functions of MAP2 include neuronal morphogenesis, where it has been tentatively implicated in neuronal outgrowth and polarity, microtubule crosslinking in dendrites as well as organelle trafficking in axons and dendrites *(25,67)*. Inhibition of MAP2 expression by antisense transfection led to inhibition of processes associated with neuronal differentiation such as formation and outgrowth of neurites *(68)*. Moreover, in MAP2-deficient mice, a reduction in microtubule density in dendrites and a reduction of dendritic length were observed, demonstrating that this protein is essential for dendrite formation and growth *(69)*. MAP2c is also involved in the neurite initiation process, both by stabilizing microtubules and re-organizing filamentous actin structures in forming neurites *(70)*.

MAPs such as MAP2, have been associated with changes in microtubule stability in antimicrotubule drug resistant cells. Vincristine-resistant neuroblastoma cells, BE/VCR10 that display increased levels of polymerized tubulin, were investigated for changes in MAP expression *(71)*. A marked decrease in MAP2c was identified in the vincristine-selected cells compared with the drug-sensitive parental, BE(2)-C cells. In contrast, no major changes were observed in MAP2a, MAP2b, or MAP4 expression in the resistant cells. Several studies have suggested that MAP2c promotes microtubule polymerization to a much lesser extent than high molecular weight MAPs *(72,73)*. Moreover, the significant decrease in expression of MAP2c in BE/VCR10 cells is likely to be contributing to hyperstable microtubules by shifting the equilibrium toward MAP2 isoforms such as MAP2a and MAP2b that are stronger promoters of microtubule stability than MAP2c. This in turn would counteract the microtubule destabilizing effects of vincristine, thus leading to a selective growth advantage of these cells in the presence of the drug.

2.4. MAP4

The majority of studies on structural MAPs have focused on those originally identified in neuronal tissues and cells. The major MAP found in cells of non-neuronal origin is MAP4. The Map4 gene resides on chromosome 3p21, and encodes for a heterogeneous thermostable polypeptide with a molecular weight of 190–210 kDa (Fig. 1). The heterogenous nature of the MAP4 protein may be as a result of alternative splicing during transcription *(74)* and/or post-translation phosphorylation *(75)*. It was first isolated from human HeLa cells as a 210 kDa protein *(76)* the same size as that found in mouse tissues *(77)*. However, in rat and bovine species, the molecular weights are 200 kDa and 190 kDa, respectively as identified on sodium dodecyl sulfate-polyacrylamide gel electrophoresis *(78)*. Homologs of MAP4 have been identified in all vertebrates and even nematodes *(79)*. Microtubules with wide ranging functions are associated with MAP4 including those involved in organelle and vesicular transport, cell shape during differentiation and spindle microtubules during mitosis *(16,80)*.

Overexpression of either full-length MAP4 or the microtubule-binding domain of MAP4 has been shown to stabilize microtubules in stably transfected mouse L^{tk} cells that either contained the full-length MAP4 (L-MAP4) or its L-MTB *(22)*. More stable microtubules were observed in both transfected cell lines, as determined by exposure to nocodazole, an agent that depolymerizes microtubules *(22)*. In addition to microtubule stability, cell growth was affected. Of interest was a study that aimed to see if this effect on cell growth influenced the in vivo function of microtubules. Overexpressing MAP4 was found to inhibit vesicular transport such as recycling of the transferrin receptor, low-density lipoprotein (LDL), and some Golgi elements in addition to stabilizing microtubules *(81)*. This finding added support for the importance of MAP4 in regulating microtubule function.

Similarly, biochemical studies have demonstrated the in vitro ability of MAP4 to stabilize microtubules and dampen dynamic instability *(76,82)*, and these findings are supported in vivo by transfection *(21,22)* and microinjection studies *(83)*. Antisense RNA directed toward MAP4 caused a decrease in total tubulin, a decrease in polymerized tubulin, slow recovery from drug-induced microtubule depolymerization, and a less polar and flattened cell shape *(84)*. This suggests that MAP4 regulates the assembly level of microtubules and through this mechanism may be involved in controlling cell spreading and shape. In addition, post-translational modifications of MAPs have been shown to regulate microtubule dynamics. For example, phosphorylation of MAP4 by cyclin B-cdc2 kinase renders microtubules more dynamic in vitro *(85)* and reduces MAP4's capacity to stimulate in vitro polymerization of microtubules *(86)*. A nonphosphorylatable mutant of MAP4 was shown to bind microtubules more avidly than wild-type MAP4 in vitro, and microtubules in cells expressing nonphosphorylatable MAP4 were more resistant to nocodazole depolymerization *(87)*.

MAP4 can localize both to interphase (Fig. 2) and mitotic microtubules *(88)*, which suggests that it may contribute to the role of microtubules in proliferation and differentiation of cells. MAP4 has the same structural organization as the other members of the MAP family of proteins, i.e., it has an acidic domain, which projects from the surface of the microtubule and a basic domain that interacts with the microtubule wall *(21)*. The assembly promoting portion of the microtubule-binding domain has a sequence that is responsible for binding to the microtubules *(89)*. This indicates that MAP4 interacts directly with microtubules and that altering this interaction can influence microtubule dynamics and assembly.

In the mouse, expression of MAP4 was found in most tissues. MAP4 was predominant in specific cell types indicating that it may play a functional role in particular tissues or cells *(80)*. Multiple isoforms of MAP4 appear to be tissue and developmental stage specific *(88)*. A detailed analysis of MAP4 microtubule-binding domains has revealed that the isoforms are distinct from each other *(88)*. Together, these data support the view that the different MAP4 isoforms may have differing roles in specific tissues.

Polymerized microtubules exist in equilibrium with free tubulin dimers. The main function of MAP4 is to stabilize microtubules, thus increased expression of MAP4 may shift this equilibrium toward more stable microtubules leading to decreased levels of free tubulin dimers. An antisense study demonstrated that depleting MAP4 expression reduced the level of cellular tubulin and also decreased the stability of microtubules *(84)*. These contradicted previous results, which showed that the absence of MAP4 had no effect on microtubule stability *(90)*. It should be noted that these studies differed in their approach. The initial MAP4 functional study involved microinjection of antibody that blocked MAP4 binding to microtubules into single cells and studied the effect on microtubule dynamics *(90)*. This approach did limit the studies that could be performed because of the low cell number. In contrast, the other group stably expressed antisense MAP4 in cells, which allowed a more detailed analysis of microtubule dynamics, morphology and cell cycle to be performed *(84)*.

Changes in MAP4 expression have been associated with resistance to anitimicrotubule agents. In leukemia cells selected for resistance to the microtubule destabilizing agents vincristine, a significant increase in MAP4 protein expression was detected *(91)*. The vincristine-resistant cells had an acquired mutation in β-tubulin and increased polymerized tubulin levels. The increased MAP4 protein was found to be associated with the microtubule fraction, suggesting it was not phosphorylated, and therefore contributing to the increased microtubule stability in these cells.

In murine cells, induction of the tumor suppressor gene, p53 transcriptionally represses MAP4 *(92)*. Increased MAP4 expression, which occurs when p53 is mutated, increases microtubule polymerization and paclitaxel binding, resulting in heightened sensitivity to the microtubule stabilizing drug paclitaxel and reduced sensitivity to the microtubule destabilizing drug, vinblastine *(93)*. Thus, controlling this p53-dependent regulation of MAP4 is a potential approach to enhance the action of antimicrotubule drugs. DNA damage increased wild-type p53 and decreased MAP4 expression, resulting in decreased sensitivity to paclitaxel and increased sensitivity to vinblastine in cell lines *(94)*. This regulation is now being tested in cancer clinical trials where the DNA damaging agent, doxorubicin is used to induce p53, and hence repress MAP4, followed by sequential treatment with vinorelbine *(95)*.

3. OTHER REGULATORS OF MICROTUBULE STABILITY

The focus in the preceding section was on MAPs that bind to and stabilize microtubules. There are however, other regulators of microtubule function that affect the dynamic properties of microtubules by affecting the stability of microtubules. Microtubule-destabilizing proteins include a group of proteins that upon binding to the microtubules or the tubulin heterodimers, decrease the total microtubule polymer mass. Examples of such proteins include stathmin, mitotic centromere-associated kinesin (MCAK), and katanin and examples of microtubule-stabilizing proteins include stable tubulin only protein (STOP), End-binding (EB) 1, CLIP-170, tumor overexpressed gene (TOG), and surviving. Many of these microtubule dynamic regulators play a pivotal role in cell division.

3.1. Other Microtubule Stabilizing Proteins

3.1.1. STABLE TUBULIN ONLY PROTEIN (STOP)

Another group of MAPs that promote microtubule stabilization are STOP proteins. These calmodulin-regulated and calmodulin-binding proteins are encoded by a single gene that displays highly variable tissue specificity because of mRNA splicing and alternate promotor use. The STOP gene is localized to chromosome 13q3. An interesting feature of these proteins is that, unlike other MAPs such as MAP2 and tau, they stabilize microtubule polymers against cold induced depolymerization *(96)*. STOPs are expressed in both neuronal and nonneuronal cell types, and a lack of these proteins is thought to be associated with defects in synapse function in mice *(97)*. STOP proteins are also involved in neuronal differentiation and are required for normal neurite formation *(98)*.

3.1.2. END-BINDING PROTEIN 1 (EB1) AND OTHER PLUS END TRACKING PROTEINS

EB1 belongs to an evolutionary conserved family that regulates microtubule assembly and stability. EB1 was originally identified by its physical association with the carboxy-terminal portion of the adenomatous polyposis coli (APC) tumor suppressor protein, an APC domain commonly mutated in familial and sporadic forms of colorectal neoplasia *(99)*. EB1 associates with APC and p150^{glued} (also known as dynactin 1), a component of the dynactin complex. EB1 colocalizes with microtubules, preferentially at their plus ends, throughout the cell cycle *(100)*. Another term for the behavior of proteins that associate on microtubule ends is "plus-end tracking' proteins (+TIP) *(101,102)*.

Overexpression of EB1 induces microtubule bundles that are more resistant to nocodazole and more acetylated than regular microtubules *(100)*. Analysis of the functional domains of EB1 protein family members revealed overlapping but distinct regions that contain the C-terminal signature that associates with APC and p150^{glued} *(103)*. Interestingly, the APC or p150^{glued} binding domains are not required for EB1- or the EBF3-induced microtubule bundling. EB1 is a potent inducer of microtubule polymerization and is a specific marker of growing microtubule tips *(104–109)*. Apart from the interaction of EBI with microtubule tips, both EB1 and APC localize to centrosomes and are functional components of this system *(110)*. EB1 plays an important role in anchoring the minus end of microtubules to the centriole. EB1 binding also reflects kinetochore directionality during cell division suggesting that EB1 may have a functional role in the polymerization of kinetochore microtubules and/or attachment *(111)*.

In recent years the number of identified +TIPs and their proposed mechanisms of action have been increasing (reviewed in ref. *112*). The first evidence for the existence of +TIP was observed with live cell imaging between a fusion protein of a MAP called cytoplasmic linker protein, CLIP-170, and green fluorescent protein *(113)*. CLIP-170 is a nucleotide-sensitive MAP that links microtubules to endocytic vesicles in vitro *(114,115)*. The plus end tracking of CLIP-170 is mediated by microtubule treadmilling

(116). CLIP-170 proteins add to the plus ends of growing microtubules, but then soon these molecules dissociate behind a region of new microtubule growth giving the illusion that they are moving along with the microtubule although individual CLIP-170 molecules are stationary. All CLIP-170 family members have the ability to plus-end track and these proteins contain one or more conserved MTBs, referred to as CAP-Gly at their N-terminal end. This domain is also found in the p150^{glued} subunit of dynactin (a complex that regulates cytoplasmic dynein, a motor protein), certain tubulin folding factors and a member of the kinesin motor superfamily *(117).*

CLIP-170 and CLIP-115 associate specifically with the ends of growing microtubules as plus-end tracking proteins and may act as microtubule stabilizing factors. Two CLIP-associated proteins (CLASPs) have been identified *(118).* CLASPs have a number of functional features such as:

1. Ability to bind CLIPs and microtubules;
2. Colocalize with the CLIPs at microtubule plus ends; and
3. Exhibit microtubule-stabilizing effects in transfected cells.

CLASP2 appears to be involved in orientating stabilized microtubules toward the cells leading edge *(118).* Evidence suggests that CLIP-170 may mediate the association of dynein/dynactin to microtubule plus ends, and it binds to kinetochores in a dynein/dynactin-dependent fashion, in both cases, through its C-terminal domain. The C-terminal domain contains two zinc finger motifs that are thought to mediate protein–protein interactions. A protein implicated in normal brain development, LIS1 was originally identified in a rare brain malformation called Lissencephaly I, and it interacts with dynein and other proteins. LIS1 has been shown to colocalize and directly interact with CLIP-170. The recruitment of LIS1 to kinetochores is a dynein/dynactin dependent process, and CLIP-170 is dependent on its binding to LIS1 *(119).* Overexpression of CLIP-170 leads to localization of phospho-LIS1 and dynactin to stabilized microtubule bundles. LIS1 appears to play a role as a regulated adapter between CLIP-170 and cytoplasmic dynein at sites involved in microtubule transport, and/or in the control of microtubule dynamics.

3.1.3. Dis1/TOG

In recent years, a new family of MAPs, the human analog of TOG (Dis1/TOG) family has been identified as regulators of microtubule function *(120).* The Dis1/TOG family is highly conserved in evolution. Unlike the structural MAPs, the localization and function of Dis1/TOG proteins are not dependent on their microtubule-binding activity and they tend to perform their diverse roles by interacting with other regulatory molecules such as microtubule motors and centrosomal proteins. A 6,449 bp cDNA, termed colonic, hepatic tumor over-expressed gene (ch-TOG) that is highly expressed in human tumors and brain has been described and it encodes for a 218 kDa TOG polypeptide. The distribution of TOG is cell cycle-dependent and is associated with centrosomes and spindles in mitotic cells *(121).* In contrast, during interphase it concentrates in the perinuclear cytoplasm. TOG cosediments with paclitaxel-stabilized microtubules and TOG can promote microtubule assembly in vitro *(121).* TOG displays high homology to XMAP215, a previously described MAP from Xenopus eggs, and appears to be important for microtubule rearrangements and spindle assembly in dividing cells *(120).*

3.1.4. Survivin

Survivin is a member of the inhibitor of apoptosis gene family and functions both as a suppressor of apoptotic cell death and as a regulator of cell division. It is localized to both kinetochores and mitotic spindle microtubules, and increases microtubule stability by either directly regulating microtubule dynamics or by recruitment of MAPs and motor proteins *(122,123)*. Wild-type p53 negatively regulates survivin expression at the gene and protein levels, by repressing the survivin promoter *(124)*. The role of survivin in the inhibition of apoptosis and in the regulation of microtubule dynamics may potentially facilitate evasion from checkpoint mechanisms of cell cycle arrest and promote resistance to antimicrotubule agents *(122)*.

3.2. Microtubule Destabilizing Proteins

3.2.1. Stathmin

A major tubulin-regulatory protein is stathmin (also referred to as oncoprotein 18, OP18; metablastin; p19), which was first identified as a highly overexpressed protein in leukemia *(125,126)* and as a protein that underwent phosphorylation in response to extracellular stimuli *(127)*. Stathmin is a well-conserved, ubiquitous, cytosolic phosphoprotein that destabilizes microtubules and forms specific complexes with tubulin dimers *(128,129)*.

The stathmin gene resides on chromosome 1q36, and expression has been detected in all tissues, with the highest levels found in brain, neurons, testis, and leukemic lymphocytes *(130)*. Stathmin belongs to a family of structurally related proteins that are also expressed in the nervous system and include superior cervical ganglion (SCG)10, SCG10-like protein (SCLIP), RB3 and two splice variants RB3' and RB3'' that displays microtubule destabilizing activity *(131,132)* (Fig. 4). Stathmin interacts directly with microtubules *(129)*, through a complex of one stathmin molecule to two tubulin heterodimers *(133,134)*. Stathmin binds to the region around helix 10 of α-tubulin dimer, a region involved in longitudinal interactions in the microtubule, sequestering the dimer and linking two α/β-tubulin heterodimers *(135)*. Insight into the interaction of stathmin with tubulin has been gained with the aid of a 3.5 Å model of tubulin in complex with colchicine and with the stathmin-like domain of RB3 *(136,137)*. The RB3-stathmin-like domain resembles a hook-like appearance that holds two bound tubulin heterodimers (Fig. 5).

The interaction of stathmin with tubulin is directly dependent on the degree of stathmin phosphorylation, whereby increasing phosphorylation inhibits binding *(138,139)*. The sequential cell cycle-dependent phosphorylation on four serine residues, Ser16, 25, 38, and 63 abolishes the microtubule-destabilizing activity of stathmin, with phosphorylation of Ser63 contributing substantially to the inactivation (reviewed in ref. *140*). Two possible mechanisms of microtubule destabilization by stathmin have been proposed:

1. Sequestration of tubulin heterodimers resulting in depletion of the soluble tubulin available for polymerization (reviewed in ref. *141*);
2. Stimulation of microtubule catastrophe *(129)*.

It appears that both mechanisms can occur depending on pH, with tubulin-sequestering and catastrophe-enhancing activity observed at pH 6.8, and only catastrophe-enhancing at pH 7.5 *(142)*. Decreased stathmin expression favors increased microtubule polymerization *(143)* whereas increased expression reduces microtubule polymer mass *(138,144)*.

Stathmin/Op18

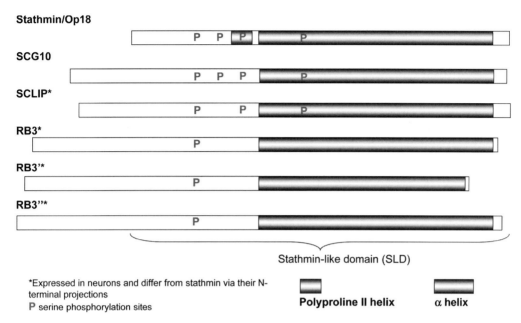

Fig. 4. Schematic diagram of the stathmin-family of proteins. The stathmin-like domain is highly conserved among the family members. Stathmin is unique among the family members as it contains a polyproline II helix in the N-terminal region. Apart from the remaining N-terminal region, which is not structured, the remaining protein has a α-helical structure. Adapted from ref. *168*. To view this figure in color, see the insert and the companion CD-ROM.

Interestingly, increased stathmin levels have been identified in acute leukemia *(125,145)*, lymphoma *(145,146)*, neuroblastoma *(147)*, and breast cancer *(148)* compared with normal tissue. Levels of stathmin phosphorylation significantly correlated with a high percentage of leukemic blast cells in S-phase and white blood count at diagnosis, suggesting that this could be a valuable target to inhibit the proliferation of leukemia cells *(149)*. A somatic mutation in stathmin was recently identified in an esophageal adenocarcinoma *(150)*. Transfection studies of the mutant stathmin into NIH3T3 cells resulted in foci formation and tumor growth when transplanted in immunodeficient mice. Cells expressing the mutant stathmin had altered tubulin ultra structure, and cell cycle analysis revealed a doubling in the percentage of cells in G2/M. The mutant stathmin had decreased specific phosphorylation, suggesting that mutations in stathmin affecting the phosphorylation capacity have profound effects on cell homeostasis that may lead to tumorigenicity *(150)*. These studies suggest that regulation of microtubule dynamics may have a causal relationship to cancer development.

Cell lines displaying overexpression of stathmin show decreased microtubule polymerization, decreased paclitaxel binding, and decreased paclitaxel sensitivity *(151)*. In paclitaxel-resistant A549 lung cancer cells with an α-tubulin mutation *(152)* and increased microtubule dynamic instability *(153)*, the active nonphosphorylated form of stathmin was increased about twofold, whereas the inactive phosphorylated forms were barely detected *(152)*. The two-dimensional polyacrylamide gel electrophoresis analysis of the childhood tumor neuroblastoma found that more aggressive neuroblastomas that coincidently are less responsive to therapy, have reduced phosphorylation of stathmin *(147)*.

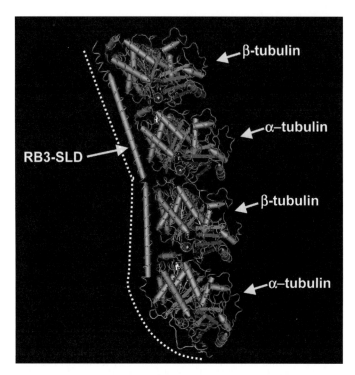

Fig. 5. Structure of tubulin-colchicine:RB3-tubulin complex. Insight into the interaction of stathmin with tubulin has been gained with the aid of a 3.5 Å model of tubulin in complex with colchicine and with the stathmin-like domain of RB3. The RB3-stathmin-like domain resembles a hook that holds two bound tubulin heterodimers (shadowed by the white dotted line for clarity) (Protein Data Bank entry 1SA1) *(137)*. To view this figure in color, see the insert and the companion CD-ROM.

Stathmin, like MAP4 is negatively regulated by wild-type p53 *(154,155)*. Antisense inhibition of stathmin was found to be synergistic with paclitaxel in inhibiting growth and clonogenic potential of K562 leukaemia cells *(156)*. In combination with stathmin inhibition, paclitaxel induced more severe mitotic abnormalities and increased apoptotic effects. Inhibition of stathmin expression also resulted in increased sensitivity to paclitaxel and decreased sensitivity to vinblastine *(157)*. In support, overexpression of stathmin in human lung carcinoma cells led to the increased sensitivity to *vinca* alkaloids vindesine and vincristine, but no changes in sensitivity to taxanes were observed *(158)*. In human breast cancer cell lines that harbored mutations in p53 overexpression of stathmin was associated with decreased levels of microtubule polymer, and consequently, with decreased binding of paclitaxel and increased binding of vinblastine. Interestingly, stathmin-overexpressing cells were less sensitive to both of these antimicrotubule agents *(151)*. Although the resistance of stathmin-overexpressing cells to paclitaxel was predicted according to altered microtubule dynamics and drug binding, resistance to vinblastine was unexpected, as the increased binding of the drug to the cells was observed. However, by measuring the amount of mitotic-specific phosphoproteins in controls and stathmin-overexpressing cells treated with vinblastine, it was found that stathmin decreases the number of cells entering mitosis. This impediment in cell progression from G_2 to mitosis was thought to be responsible for diminishing the cytotoxic effects of vinblastine *(151)*.

3.2.2. KATANINS

Other microtubule-destabilizing proteins include Xenopus Kinesin Catastrophe Modulator 1 (XKCM1), Xenopus Kinesin Superfamily protein (XKIF)2 and katanin, XKCM1 and XKIF2 are Kin I kinesins that unlike most kinesins that use ATP to move along a microtubule, depolymerize microtubules. Katanin is a heterodimeric microtubule-severing protein that localizes to centrosomes *(159)* and is responsible for the majority of M-phase severing activity in *Xenopus* eggs *(160)*, and is essential for releasing microtubules from neuronal chromosomes. *(161)*. The kinesin-like proteins XKCM1 and XKIF2 are required for the formation and maintenance of the mitotic spindle. These proteins cause rapid microtubule disassembly in vitro, binding preferentially to microtubule ends, both plus and minus, and cause destabilization of the microtubule lattice *(162,163)*. However, unlike stathmin, XKCM1 does not seem to possess tubulin-sequestering activity (reviewed in ref. *164*).

3.2.3. MITOTIC CENTROMERE-ASSOCIATED KINESIN

In the mitotic spindle, kinetochore microtubules facilitate the separation and segregation of chromosomes *(123)*. The functions of kinetochores include the attachment of the sister chromatids to the opposing poles of the spindle, generating force for the pole ward chromosome movement, positioning chromosomes on the spindle plane, and control of anaphase/metaphase transition. Several motor and accessory proteins such as kinesins and dyneins, centromere proteins (CENP-B, -C, -E and -F), inner centromere protein (INCENP), and MCAK are involved in performing these functions *(123,165)*. MCAK belongs to a unique group of motor proteins that are not motile but can destabilize microtubules. It is a member of the Kin I subfamily of kinesin-related proteins and it is an ATPase that catalytically depolymerizes microtubules by increasing the rate of dissociation of tubulin from the ends of microtubules *(166)*.

The correct attachment of kinetochore microtubules is essential for the correct separation and segregation of chromosomes. The serine/threonine kinase Aurora B belongs to a family of proteins that constitutes key regulators in the execution of mitotic events. Aurora B is thought to regulate kinetochore-microtubule attachments and promote correct chromosome biorientation. MCAK can depolymerize microtubules. Both Aurora B and MCAK proteins have comparable functions where they localize to mitotic centromeres. They regulate microtubule dynamics, correct chromosome congression, and correct erroneous kinetochore-microtubule attachments. Aurora B has been shown to phosphorylate and regulate MCAK both in vitro and in vivo *(167)*. Aurora B activity was essential for the localization of MCAK to centromeres, but not to spindle poles. Upon Aurora B-induced phosphorylation of serine 196 in the neck region of MCAK, microtubule depolymerization activity was inhibited *(167)*. Demonstrating a direct link between the microtubule depolymerase MCAK and Aurora B kinase.

SUMMARY

The organization of interphase microtubules and mitotic spindles are dramatically altered during the cell cycle and development and this organization is highly dependent on interactions between microtubules and accessory proteins. The highly coordinated events that lead to separation and segregation of chromosomes are highly dependent of microtubule dynamics. The molecular mechanisms underlying this dynamic behavior are multifaceted and not yet fully understood. What is clear however is that microtubule regulating factors such as MAPs and other microtubule interacting proteins that either reduce or enhance microtubule dynamics during different phases of the cell cycle, are essential for many cellular functions.

ACKNOWLEDGMENTS

Immunofluorescence images in Fig. 2 were used with the kind permission of Sela Pouha, Children's Cancer Institute Australia for Medical Research. M. Kavallaris is supported by a National Health and Medical Research Council RD Wright Career Development Award No. 300580.

REFERENCES

1. Luduena RF. Multiple forms of tubulin: different gene products and covalent modifications. Int Rev Cytol 1998;178:207–275.
2. Nogales E. Structural insight into microtubule function. Annu Rev Biophys Biomol Struct 2001; 30;397–420.
3. Lewis SA, Cowan NJ. Complex regulation and functional versatility of mammalian alpha- and beta-tubulin isotypes during the differentiation of testis and muscle cells. J Cell Biol 1988;106: 2023–2033.
4. Lee G, Cowan N, Kirschner M. The primary structure and heterogeneity of tau protein from mouse brain. Science 1988;239:285–288.
5. Goedert M, Wischik CM, Crowther RA, Walker JE, KlugA. Cloning and sequencing of the cDNA encoding a core protein of the paired helical filament of Alzheimer disease: identification as the microtubule-associated protein tau. Proc Natl Acad Sci USA 1988;85:4051–4055.
6. Kindler S, Schulz B, Goedert M, Garner CC. Molecular structure of microtubule-associated protein 2b and 2c from rat brain. J Biol Chem 1990;265:19,679–19,684.
7. Aizawa H, Emori Y, Murofushi H, Kawasaki H, Sakai H, Suzuki K. Molecular cloning of a ubiquitously distributed microtubule-associated protein with Mr 190,000. J Biol Chem 1990;265: 13,849–13,855.
8. West RR, Tenbarge KM, Olmsted JB. A model for microtubule-associated protein 4 structure. Domains defined by comparisons of human, mouse, and bovine sequences. J Biol Chem 1991;266:21,886–21,896.
9. Aizawa H, Murofushi H, Kotani S, Hisanaga S, Hirokawa N, Sakai H. Limited chymotryptic digestion of bovine adrenal 190,000-Mr microtubule-associated protein and preparation of a 27,000-Mr fragment which stimulates microtubule assembly. J Biol Chem 1987;262:3782–3787.
10. Brandt R, Lee G. Functional organization of microtubule-associated protein tau. Identification of regions which affect microtubule growth, nucleation, and bundle formation in vitro. J Biol Chem 1993;268:3414–3419.
11. Chapin SJ, Bulinski JC. Microtubule-stabilisation by assembly-promoting microtubule-associated proteins: a repeat performance. Cell Motil Cytoskeleton 1992;23:236–243.
12. Illenberger S, Drewes G, Trinczek B, et al. Phosphorylation of microtubule-associated proteins MAP2 and MAP4 by the protein kinase p110mark. Phosphorylation sites and regulation of microtubule dynamics. J Biol Chem 1996;271:10,834–10,843.
13. Ebneth A, Drewes G, Mandelkow E. Phosphorylation of MAP2c and MAP4 by MARK kinases leads to the destabilization of microtubules in cells. Cell Motil Cytoskel 1999;44:209–224.
14. Ookata K, Hisanaga S, Sugita M, et al. MAP4 is the in vivo substrate for CDC2 kinase in HeLa cells: identification of an M-phase specific and a cell cycle-independent phosphorylation site in MAP4. Biochemistry 1997;36:15,873–15,883.
15. Drewes G, Ebneth A, Mandelkow EM. MAPs, MARKs and microtubule dynamics. Trends Biochem Sci 1998;23:307–311.
16. Bulinski JC, Borisy GG. Immunofluorescence localization of HeLa cell microtubule-associated proteins on microtubules in vitro and in vivo. J Cell Biol 1980;87:792–801.
17. Balint E, Cheng M, Rupp B, Grimley PM, Aszalos A. Cytoskeletal modulation of plasma membrane events induced by interferon-alpha. J Interferon Res 1992;12:249–255.

18. Kanai Y, Takemura R, Takeshi O, et al. Expression of multiple tau isoforms and microtubule bundle formation in fibroblasts transfected with a single tau cDNA. J Cell Biol 1989;109:1173–1184.
19. Kanai Y, Chin J, Hirokawa N. Microtubule bundling by tau proteins in vivo: analysis of functional domains. EMBO J 1992;11:3953–3961.
20. Burgin KE, Ludin B, Ferralli J, Matus A. Bundling of microtubules in transfected cells does not involve an autonomous dimerization site on the MAP2 molecule. Mol Biol Cell 1994;5:511–517.
21. Olson KR, McIntosh JR, Olmsted JB. Analysis of MAP 4 function in living cells using green fluorescent protein (GFP) chimeras. J Cell Biol 1995;130:639–650.
22. Nguyen HL, Chari S, Gruber D, Lue CM, Chapin SJ, Bulinski JC. Overexpression of full- or partial-length MAP4 stabilizes microtubules and alters cell growth. J Cell Sci 1997;110:281–294.
23. Hammarback JA, Obar RA, Hughes SM, Vallee RB. MAP1B is encoded as a polyprotein that is processed to form a complex N-terminal microtubule-binding domain. Neuron 1991;7:129–139.
24. Langkopf A, Hammarback JA, MullerR, Vallee RB, Garner CC. Microtubule-associated proteins 1A and LC2. Two proteins encoded in one messenger RNA. J Biol Chem 1992;267:16,561–16,566.
25. Tucker RP. The roles of microtubule-associated proteins in brain morphogenesis: a review. Brain Res Brain Res Rev 1990;15:101–120.
26. Cravchik A, Reddy D, Matus A. Identification of a novel microtubule-binding domain in microtubule-associated protein 1A (MAP1A). J Cell Sci 1994;107:661–672.
27. Noble M, Lewis SA, Cowan NJ. The microtubule binding domain of microtubule-associated protein MAP1B contains a repeated sequence motif unrelated to that of MAP2 and tau. J Cell Biol 1989;109:3367–3376.
28. Mann SS, Hammarback JA. Molecular characterization of light chain 3. A microtubule binding subunit of MAP1A and MAP1B. J Biol Chem 1994;269:11,492–11,497.
29. Gonzalez-Billault C, Jimenez-Mateos EM, Caceres A, Diaz-Nido J, Wandosell F, Avila J. Microtubule-associated protein 1B function during normal development, regeneration, and pathological conditions in the nervous system. J Neurobiol 2004;58:48–59.
30. Takemura R, Okabe S, Umeyama T, Kanai Y, Cowan NJ, Hirokawa N. Increased microtubule stability and alpha tubulin acetylation in cells transfected with microtubule-associated proteins MAP1B, MAP2 or tau. J Cell Sci 1992;103:953–964.
31. Tucker RP, Binder LI, Matus AI. Neuronal microtubule-associated proteins in the embryonic avian spinal cord. J Comp Neurol 1988;271:44–55.
32. Tucker RP, Matus AI. Developmental regulation of two microtubule-associated proteins (MAP2 and MAP5) in the embryonic avian retina. Development 1987;101:535–546.
33. Viereck C, Tucker RP, Matus A. The adult rat olfactory system expresses microtubule-associated proteins found in the developing brain. J Neurosci 1989;9:3547–3557.
34. Mansfield SG, Diaz-Nido J, Gordon-Weeks PR, Avila J. The distribution and phosphorylation of the microtubule-associated protein MAP 1B in growth cones. J Neurocytol 1991;20:1007–1022.
35. Garcia Rocha M, Avila J. Characterization of microtubule-associated protein phosphoisoforms present in isolated growth cones. Brain Res Dev Brain Res 1995;89:47–55.
36. Edelmann W, Zervas M, Costello P, et al. Neuronal abnormalities in microtubule-associated protein 1B mutant mice. Proc Natl Acad Sci USA 1996;93:1270–1275.
37. Takei Y, Kondo S, Harada A, Inomata S, Noda T, Hirokawa N. Delayed development of nervous system in mice homozygous for disrupted microtubule-associated protein 1B (MAP1B) gene. J Cell Biol 1997;137:1615–1626.
38. Schoenfeld TA, Obar RA. Diverse distribution and function of fibrous microtubule-associated proteins in the nervous system. Int Rev Cytol 1994;151:67–137.
39. Chau MF, Radeke MJ, de Ines C, Barasoain I., Kohlstaedt LA, Feinstein SC. The microtubule-associated protein tau cross-links to two distinct sites on each alpha and beta tubulin monomer via separate domains. Biochemistry 1998;37:17,692–17,703.
40. Neve RL, Harris P, Kosik KS, Kurnit DM, Donlon TA. Identification of cDNA clones for the human microtubule-associated protein tau and chromosomal localization of the genes for tau and microtubule-associated protein 2. Brain Res 1986;387:271–280.
41. Andreadis A, Brown WM, Kosik KS. Structure and novel exons of the human tau gene. Biochemistry 1992;31:10,626–10,633.
42. Gu Y, Oyama F, Ihara Y. Tau is widely expressed in rat tissues. J Neurochem 1996;67:1235–1244.
43. Poorkaj P, Kas A, D'Souza I, et al. A genomic sequence analysis of the mouse and human microtubule-associated protein tau. Mamm Genome 2001;12:700–712.

44. Goedert M, Spillantini MG, Crowther RA. Cloning of a big tau microtubule-associated protein characteristic of the peripheral nervous system. Proc Natl Acad Sci USA 1992;89:1983–1987.
45. Wang Y, Loomis PA, Zinkowski RP, Binder LI. A novel tau transcript in cultured human neuroblastoma cells expressing nuclear tau. J Cell Biol 1993;121:257–267.
46. Goedert M, Spillantini MG, Jakes R, Rutherford D, Crowther RA. Multiple isoforms of human microtubule-associated protein tau: sequences and localization in neurofibrillary tangles of Alzheimer's disease. Neuron 1989;3:519–526.
47. Henriquez JP, Cross D, Vial C, Maccioni RB. Subpopulations of tau interact with microtubules and actin filaments in various cell types. Cell Biochem Funct 1995;13:239–250.
48. Cleveland DW, Hwo S-Y, Kirschner MW. Purification of tau, a microtubule-associated protein that induces assembly of microtubules from purified tubulin. J Mol Biol 1977;116:207–225.
49. Drubin DG, Kirschner MW. Tau protein function in living cells. J Cell Biol 1986;103:2739–2746.
50. Caceres A, Kosik KS. Inhibition of neurite polarity by tau antisense oligonucleotides in primary cerebellar neurons. Nature 1990;343:461–463.
51. Esmaeli-Asad B, McCarty JH, Feinstein SC. Sense and antisense transfection analysis of tau function: tau influences net microtubule assembly, neurite outgrowth and neuritic stability. J Cell Sci 1994;107:869–879.
52. Harada A, Oguchi K, Okabe S, et al. Altered microtubule organisation in small-calibre axons of mice lacking tau protein. Nature 1994;369:488–491.
53. Takei Y, Teng J, Harada A, Hirokawa N. Defects in axonal elongation and neuronal migration in mice with disrupted tau and map1b genes. J Cell Biol 2000;150:989–1000.
54. Garcia ML, Cleveland DW. Going new places using an old MAP: tau, microtubules and human neurodegenerative disease. Curr Opin Cell Biol 2001;13:41–48.
55. Price DL, Sisodia SS. Mutant genes in familial Alzheimer's disease and transgenic models. Annu Rev Neurosci 1998;21:479–505.
56. Hall GF, Chu B, Lee S, Liu Y, Yao J. The single neurofilament subunit of the lamprey forms filaments and regulates axonal caliber and neuronal size in vivo. Cell Motil Cytoskel 2000;46:166–182.
57. Murphy DB, Borisy GG. Association of high-molecular-weight proteins with microtubules and their role in microtubule assembly in vitro. Proc Natl Acad Sci USA 1975;72:2696–2700.
58. Serrano L, Avila J, Maccioni RB. Controlled proteolysis of tubulin by subtilisin: localisation of the site for MAP2 interaction. Biochemistry 1984;23:4675–4681.
59. Selden SC, Pollard TD. Phosphorylation of microtubule-associated proteins regulates their interaction with actin filaments. J Biol Chem 1983;258:7064–7071.
60. Ozer RS, Halpain S. Phosphorylation-dependent localization of microtubule-associated protein MAP2c to the actin cytoskeleton. Mol Biol Cell 2000;11:3573–3587.
61. Tucker RP, Binder LI, Viereck C, Hemmings BA, Matus AI. The sequential appearance of low- and high-molecular weight forms of MAP2 in the developing cerebellum. J Neurosci 1988;8:4503–4512.
62. Falconer MM, Vaillant A, Reuhl K, Laferriere N, Brown DL. The molecular basis of microtubule stability in neurons. Neurotoxicology 1994;15:109–122.
63. Tucker RP, Matus AI. Microtubule-associated proteins characteristic of embryonic brain are found in the adult mammalian retina. Dev Biol 1988;130:423–434.
64. Meichsner M, Doll T, Reddy D, Weisshaar B, Matus A. The low molecular weight form of microtubule-associated protein 2 is transported into both axons and dendrites. Neuroscience 1993;54:873–880.
65. Doll T, Meichsner M, Riederer BM, Honegger P, Matus A. An isoform of microtubule-associated protein 2 (MAP2) containing four repeats of the tubulin-binding motif. J Cell Sci 1993;106:633–639.
66. Ferhat L, Ben-Ari Y, Khrestchatisky M. Complete sequence of rat MAP2d, a novel MAP2 isoform. Comptes Rendus de l Academie des Sciences - Serie Iii, Sciences de la Vie CR Acad Sci III 1994;317:304–309.
67. Sanchez C, Diaz-Nido J, Avila J. Phosphorylation of microtubule-associated protein 2 (MAP2) and its relevance for the regulation of the neuronal cytoskeleton function. Prog Neurobiol 2000;61:133–168.
68. Dinsmore JH, Solomon F. Inhibition of MAP2 expression affects both morphological and cell division phenotypes of neuronal differentiation. Cell 1991;64:817–826.
69. Harada A, Teng J, Takei Y, Oguchi K, Hirokawa N. MAP2 is required for dendrite elongation, PKA anchoring in dendrites, and proper PKA signal transduction. J Cell Biol 2002;158:541–549.

70. Dehmelt L, Smart FM, Ozer RS, Halpain S. The role of microtubule-associated protein 2c in the reorganization of microtubules and lamellipodia during neurite initiation. J Neurosci 2003;23: 9479–9490.

71. Don S, Verrills NM, Liaw TYE, et al. Neuronal-associated microtubule proteins class III_β-tubulin and MAP2c in neuroblastoma: Role in resistance to microtubule-targeted drugs. Mol Cancer Ther 2004;3:1137–1146.

72. Viereck C, Tucker RP, Matus A. The adult rat olfactory system expresses microtubule-associated proteins found in the developing brain. J Neurosci 1989;9:3547–3557.

73. Tucker RP. The roles of microtubule-associated proteins in brain morphogenesis: a review. Brain Res Rev 1990;15:101–120.

74. Chapin SJ, Lue CM, Yu MT, Bulinski JC. Differential xpression of alternatively spliced forms of MAP4: a repertoire of structurally different microtubule-binding domains. Biochemistry 1995;34:2289–2301.

75. Chapin SJ, Bulinski JC. Non-neuronal 210 x 10(3) Mr microtubule-associated protein (MAP4) contains a domain homologous to the microtubule-binding domains of neuronal MAP2 and tau. J Cell Sci 1991;98(Pt 1):27–36.

76. Bulinski JC, Borisy GG. Microtubule-associated proteins from cultured HeLa cells. Analysis of molecular properties and effects on microtubule polymerization. J Biol Chem 1980;255:11, 570–11,576.

77. Parysek LM, Asnes CF, Olmsted JB. MAP4: Occurrence in mouse tissues. J Cell Biol 1984;99: 1309–1315.

78. Olmsted JB. Non-motor microtubule-associated proteins. Curr Biol 1991;3:52–58.

79. Goedert M. Tau protein and the neurofibrillary pathology of Alzheimer's disease. Ann N Y Acad Sci 1996;777:121–131.

80. Parysek LM, Wolosewick JJ, Olmsted JB. MAP 4: a microtubule-associated protein specific for a subset of tissue microtubules. J Cell Biol 1984;99:2287–2296.

81. Bulinski JC, McGraw TE, Gruber D, Nguyen HL, Sheetz MP. Overexpression of MAP4 inhibits organelle motility and trafficking in vivo. J Cell Sci 1997;110:3055–3064.

82. Murofushi H, Kotani S, Aizawa H, Hisanaga S, Hirokawa N, Sakai H. (1986) Purification and characterization of a 190-kD microtubule-associated protein from bovine adrenal cortex. J Cell Biol 1997;103:1911–1919.

83. Yoshida T, Imanaka-Yoshida K, Murofushi H, Tanaka J, Ito H, Inagaki M. Microinjection of intact MAP-4 and fragments induces changes of the cytoskeleton in PtK2 cells. Cell Motil Cytoskeleton 1996;33:252–262.

84. Nguyen HL, Gruber D, Bulinski JC. Microtubule-associated protein 4 (MAP4) regulates assembly, protomer-polymer partitioning and synthesis of tubulin in cultured cells. J Cell Sci 1999;112:1813–1824.

85. Ookata K, Hisanaga S, Bulinski JC, et al. Cyclin B interaction with microtubule-associated protein 4 (MAP4) targets p34cdc2 kinase to microtubules and is a potential regulator of M-phase microtubule dynamics. J Cell Biol 1995;128:849–886.

86. Kitazawa H, Iida J, Uchida A, et al. Ser787 in the proline-rich region of human MAP4 is a critical phosphorylation site that reduces its activity to promote tubulin polymerization. Cell Struct Funct 2000;25:33–39.

87. Chang W, Gruber D, Chari S, et al. Phosphorylation of MAP4 affects microtubule properties and cell cycle progression. J Cell Sci 2001;114:2879–2887.

88. Chapin SJ, Bulinski JC. Cellular microtubules heterogeneous in their content of microtubule-associated protein 4 (MAP4). Cell Motil Cytoskel 1994;27:133–149.

89. Tokuraku K, Katsuki M, Nakagawa H, Kotani S. A new model for microtubule-associated protein (MAP)-induced microtubule assembly. The Pro-rich region of MAP4 promotes nucleation of microtubule assembly in vitro. Eur J Biochem 1999;259:158–166.

90. Wang XM, Peloquin JG, Zhai Y, Bulinski JC, Borisy GG. Removal of MAP4 from microtubules in vivo produces no observable phenotype at the cellular level. J Cell Biol 1996;132:345–357.

91. Kavallaris M, Tait AS, Walsh BJ, et al. Multiple microtubule alterations are associated with Vinca alkaloid resistance in human leukemia cells. Cancer Res 2001;61:5803–5809.

92. Murphy M, Hinman A, Levine AJ. Wild-type p53 negatively regulates the expression of a microtubule-associated protein. Genes Dev 1996;10:2971–2980.

93. Zhang CC, Yang JM, White E, Murphy M, Levine A, Hait WN. The role of MAP4 expression in the sensitivity to paclitaxel and resistance to vinca alkaloids in p53 mutant cells. Oncogene 1998;16: 1617–1624.
94. Zhang CC, Yang J-M, Bash-Babula J, et al. DNA damage increases sensitivity to Vinca alkaloids and decreases sensitivity to taxanes through p53-dependent repression of microtubule-associated protein 4. Cancer Res 1999;59:3663–3670.
95. Bash-Babula J, Toppmeyer D, Labassi M, et al. A Phase I/pilot study of sequential doxorubicin/ vinorelbine: effects on p53 and microtubule-associated protein 4. Clin Cancer Res 2002;8: 1057–1064.
96. Bosc C, Andrieux A, Job D. STOP proteins. Biochemistry 2003;42:12,125–12,132.
97. Andrieux A, Salin PA, Vernet M, et al. The suppression of brain cold-stable microtubules in mice induces synaptic defects associated with neuroleptic-sensitive behavioral disorders. Genes Dev 2002;16:2350–2364.
98. Guillaud L, Bosc C, Fourest-Lieuvin A, et al. STOP proteins are responsible for the high degree of microtubule stabilization observed in neuronal cells. J Cell Biol 1998;142:167–179.
99. Su LK, Burrell M, Hill DE, et al. APC binds to the novel protein EB1. Cancer Res 1995;55: 2972–2977.
100. Bu W, Su LK. Regulation of microtubule assembly by human EB1 family proteins. Oncogene 2001; 20:3185–3192.
101. Schuyler SC, Pellman D. Microtubule plus-end-tracking proteins: The end is just the beginning. Cell 2001;105:421–424.
102. Schroer TA. Microtubules don and doff their caps: dynamic attachments at plus and minus ends. Curr Opin Cell Biol 2001;13:92–96.
103. Bu W, Su LK. Characterization of functional domains of human EB1 family proteins. J Biol Chem 2003;278:49,721–49,731.
104. Morrison EE, Askham JM. EB 1 immunofluorescence reveals an increase in growing astral microtubule length and number during anaphase in NRK-52E cells. Eur J Cell Biol 2001;80:749–753.
105. Mimori-Kiyosue Y, Shiina N, Tsukita S. The dynamic behavior of the APC-binding protein EB1 on the distal ends of microtubules. Curr Biol 2000;10:865–868.
106. Minami Y, Sakai H. Effects of microtubule-associated proteins on network formation by neurofilament-induced polymerization of tubulin. FEBS Lett 1986;195:68–72.
107. Morrison EE, Wardleworth BN, Askham JM, Markham AF, Meredith DM. EB1, a protein which interacts with the APC tumour suppressor, is associated with the microtubule cytoskeleton throughout the cell cycle. Oncogene 1998;17:3471–3477.
108. Nakamura M, Zhou XZ, Lu KP. Critical role for the EB1 and APC interaction in the regulation of microtubule polymerization. Curr Biol 2001;11:1062–1067.
109. Tirnauer JS, Bierer BE. EB1 proteins regulate microtubule dynamics, cell polarity, and chromosome stability. J Cell Biol 2000;149:761–766.
110. Louie RK, Bahmanyar S, Siemers KA, et al. Adenomatous polyposis coli and EB1 localize in close proximity of the mother centriole and EB1 is a functional component of centrosomes. J Cell Sci 2004;117:1117–1128.
111. Tirnauer JS, Canman JC, Salmon ED, Mitchison TJ. EB1 targets to kinetochores with attached, polymerizing microtubules. Mol Biol Cell 2002;13:4308–4316.
112. Carvalho P, Tirnauer JS, Pellman D. Surfing on microtubule ends. Trends Cell Biol 2003;13:229–237.
113. Perez F, Diamantopoulos GS, Stalder R, Kreis TE. CLIP-170 highlights growing microtubule ends in vivo. Cell 1999;96:517–527.
114. Pierre P, Scheel J, Rickard JE, Kreis TE. CLIP-170 links endocytic vesicles to microtubules. Cell 1992;70:887–900.
115. Rickard JE, Kreis TE. Identification of a novel nucleotide-sensitive microtubule-binding protein in HeLa cells. J Cell Biol 1990;110:1623–1633.
116. Waterman-Storer CM, Desai A, Bulinski JC, Salmon ED. Fluorescent speckle microscopy, a method to visualize the dynamics of protein assemblies in living cells. Curr Biol 1998;8:1227–1230.
117. Li S, Finley J, Liu ZJ, et al. Crystal structure of the cytoskeleton-associated protein glycine-rich (CAP-Gly) domain. J Biol Chem 2002;277:48,596–48,601.
118. Akhmanova A, Hoogenraad CC, Drabek K, et al. Clasps are CLIP-115 and -170 associating proteins involved in the regional regulation of microtubule dynamics in motile fibroblasts. Cell 2001;104: 923–935.

119. Coquelle FM, Caspi M, Cordelieres FP, et al. LIS1, CLIP-170's key to the dynein/dynactin pathway. Mol Cell Biol 2002;22:3089–3102.
120. Ohkura H, Garcia MA, Toda T. Dis1/TOG universal microtubule adaptors - one MAP for all? J Cell Sci 2001;114:3805–3812.
121. Charrasse S, Schroeder M, Gauthier-Rouviere C, et al. The TOGp protein is a new human microtubule-associated protein homologous to the Xenopus XMAP215. J Cell Sci 1998;111(Pt 10):1371–1383.
122. Giodini A, Kallio MJ, Wall NR, et al. Regulation of microtubule stability and mitotic progression by survivin. Cancer Res 2002;62:2462–2467.
123. Mollinedo F, Gajate C. Microtubules, microtubule-interfering agents and apoptosis. Apoptosis 2003;8:413–450.
124. Mirza A, McGuirk M, Hockenberry TN, et al. Human survivin is negatively regulated by wild-type p53 and participates in p53-dependent apoptotic pathway. Oncogene 2002;21:2613–2622.
125. Hanash SM, Baier LJ, McCurry L, Schwartz SA. Lineage-related polypeptide markers in acute lymphoblastic leukemia detected by two-dimensional gel electrophoresis. Proc Natl Acad Sci USA 1986;83:807–811.
126. Melhem RF, Zhu XX, Hailat N, Strahler JR, Hanash SM. Characterization of the gene for a proliferation-related phosphoprotein (oncoprotein 18) expressed in high amounts in acute leukemia. J Biol Chem 1991;266:17,747–17,753.
127. Sobel A, Tashjian AH Jr. Distinct patterns of cytoplasmic protein phosphorylation related to regulation of synthesis and release of prolactin by GH cells. J Biol Chem 1983;258:10,312–10,324.
128. Sobel A. Stathmin: a relay phosphoprotein for multiple signal transduction? Trends Biochem Sci 1991;16:301–305.
129. Belmont L, Mitchison T. Identification of a protein that interacts with tubulin dimers and increases the catastrophe rate of microtubules. Cell 1996;84:623–631.
130. Bieche I, Maucuer A, Laurendeau I, et al. Expression of stathmin family genes in human tissues: non-neural-restricted expression for SCLIP. Genomics 2003;81:400–410.
131. Ozon S, Maucuer A, Sobel A. The stathmin family — molecular and biological characterization of novel mammalian proteins expressed in the nervous system. Eur J Biochem 1997;248:794–806.
132. Charbaut E, Curmi PA, Ozon S, Lachkar S, Redeker V, Sobel A. Stathmin family proteins display specific molecular and tubulin binding properties. J Biol Chem 2001;276:16,146–16,154.
133. Curmi PA, Andersen SS, Lachkar S, et al. The stathmin/.tubulin interaction in vitro. J Biol Chem 1997;272:25,029–25,036.
134. Jourdain L, Curmi P, Sobel A, Pantaloni D, Carlier MF. Stathmin:a tubulin-sequestering protein which forms a ternary T2S complex with two tubulin molecules. Biochemistry 1997;36:10,817–10,821.
135. Wallon G, Rappsilber J, Mann M, Serrano L. Model for stathmin/Op18 binding to tubulin. EMBO J 2000;19:213–222.
136. Gigant B, Curmi PA, Martin-Barbey C, et al. The 4 A X-ray structure of a tubulin:stathmin-like domain complex. Cell 2000;102:809–816.
137. Ravelli RBG, Gigant B, Curmi PA, et al. Insight into tubulin regulation from a complex with colchicine and a stathmin-like domain. Nature 2004;428:198–202.
138. Larsson N, Marklund U, Gradin HM, Wandzioch E, Cassimeris L, Gullberg M. Control of microtubule dynamics by oncoprotein 18: dissection of the regulatory role of multisite phosphorylation during mitosis. Mol Cell Biol 1997;17:5530–5539.
139. Horwitz SB, Shen H-J, He L, et al. The microtubule-destabilising activity of metablastin (p19) is controlled by phosphorylation. J Biol Chem 1997;13:8129–8132.
140. Lawler S. Microtubule dynamics: if you need a shrink try stathmin/Op18. Curr Biol 1998;8:R212–R214.
141. Andersen SS. Balanced regulation of microtubule dynamics during the cell cycle: a contemporary view [published erratum appears in Bioessays 1999 Apr;21(4):363]. Bioessays 1999;21:53–60.
142. Howell B, Larsson N, Gullberg M, Cassimeris L. Dissociation of the tubulin-sequestering and microtubule catastrophe-promoting activities of oncoprotein 18/stathmin. Mol Biol Cell 1999;10:105–118.
143. Howell B, Deacon H, Cassimeris L. Decreasing oncoprotein 18/stathmin levels reduces microtubule catastrophes and increases microtubule polymer in vivo. J Cell Sci 1999;112(Pt 21):3713–3722.

144. Marklund U, Larsson N, Gradin HM, Brattsand G, Gullberg M. Oncoprotein 18 is a phosphorylation-responsive regulator of microtubule dynamics. EMBO J 1996;15:5290–5298.

145. Roos G, Brattsand G, Landberg G, Marklund U, Gullberg M. Expression of oncoprotein 18 in human leukemias and lymphomas. Leukemia 1993;7:1538–1546.

146. Brattsand G, Roos G, Marklund U, et al. Quantitative analysis of the expression and regulation of an activation-regulated phosphoprotein (oncoprotein 18) in normal and neoplastic cells. Leukemia 1993;7:569–579.

147. Hailat N, Strahler J, Melhem R, et al. N-myc gene amplification in neuroblastoma is associated with altered phosphorylation of a proliferation related polypeptide (Op18). Oncogene 1990;5:1615–1618.

148. Curmi PA, Nogues C, Lachkar S, et al. Overexpression of stathmin in breast carcinomas points out to highly proliferative tumours. Br J Cancer 2000;82:142–150.

149. Melhem R, Hailat N, Kuick R, Hanash SM. Quantitative analysis of Op18 phosphorylation in childhood acute leukemia. Leukemia 1997;11:1690–1695.

150. Misek DE, Chang CL, Kuick R, et al. Transforming properties of a Q18-E mutation of the microtubule regulator Op18. Cancer Cell 2002;2:217–228.

151. Alli E, Bash-Babula J, Yang J-M, Hait WN. Effect of stathmin on the sensitivity to antimicrotubule drugs in human breast cancer. Cancer Res 2002;62:6864–6869.

152. Martello LA, Verdier-Pinard P, Shen H-J, He L, Orr GA, Horwitz SB. Elevated levels of microtubule destabilizing factors in a Taxol-resistant/dependent A549 cell line with an α-tubulin mutation. Cancer Res 2003;63:1207–1213.

153. Goncalves A, Braguer D, Kamath K, et al. Resistance to Taxol in lung cancer cells associated with increased microtubule dynamics. Proc Natl Acad Sci USA 2001;98:11,737–11,742.

154. Murphy M, Ahn J, Walker KK, et al. Transcriptional repression by wild-type p53 utilizes histone deacetylases, mediated by interaction with mSin3a. Genes Dev 1999;13:2490–2501.

155. Johnsen JI, Aurelio ON, Kwaja Z, et al. p53-mediated negative regulation of stathmin/Op18 expression is associated with G_2/M cell-cycle arrest. Int J Cancer 2000;88:685–691.

156. Iancu C, Mistry SJ, Arkin S,Atweh GF. Taxol and anti-stathmin therapy: a synergistic combination that targets the mitotic spindle. Cancer Res 2000;60:3537–3541.

157. Iancu C, Mistry SJ, Arkin S, Wallenstein S, Atweh GF. Effects of stathmin inhibition on the mitotic spindle. J Cell Sci 2001;114:909–916.

158. Nishio K, Nakamura T, Koh Y, Kanzawa F, Tamura T, Saijo N. Oncoprotein 18 overexpression increases the sensitivity to vindesine in the human lung carcinoma cells. Cancer 2001;91:1494–1499.

159. McNally FJ. Modulation of microtubule dynamics during the cell cycle. Curr Opin Cell Biol 1996;8:23–29.

160. McNally FJ, Thomas S. Katanin is responsible for the M-phase microtubule-severing activity in Xenopus eggs. Mol Biol Cell 1998;9:1847–1861.

161. Ahmad FJ, Yu W, McNally FJ, Baas PW. An essential role for katanin in severing microtubules in the neuron. J Cell Biol 1999;145:305–315.

162. Desai A, Hyman A. Microtubule cytoskeleton: No longer an also Ran. Curr Biol 1999;9:R704–R707.

163. Walczak CE, Mitchison TJ, Desai A. XKCM1: a Xenopus kinesin-related protein that regulates microtubule dynamics during mitotic spindle assembly. Cell 1996;84:37–47.

164. Walczak CE. Microtubule dynamics and tubulin interacting proteins. Curr Opin Cell Biol 2000;12:52–56.

165. Rieder CL, Salmon ED. The vertebrate cell kinetochore and its roles during mitosis. Trends Cell Biol 1998;8:310–318.

166. Hunter AW, Caplow M, Coy DL, et al. The kinesin-related protein MCAK is a microtubule depolymerase that forms an ATP-hydrolyzing complex at microtubule ends. Mol Cell 2003;11:445–457.

167. Lan W, Zhang X, Kline-Smith SL, et al. Aurora B phosphorylates centromeric MCAK and regulates its localization and microtubule depolymerization activity. Curr Biol 2004;14:273–286.

5

The Post-Translational Modifications of Tubulin

*Richard F. Ludueña
and Asok Banerjee*

Contents

Summary

The tubulin molecule is unusual because of the number and nature of post-translational modifications that it undergoes. These modifications may be involved in regulating microtubule stability and interactions with microtubule-associated proteins, but they also may have as yet undiscovered functions. Certain of these modifications are found in many other proteins; these include phosphorylation of a serine residue in β-tubulin and acetylation of a lysine residue in α-tubulin. Other modifications occur exclusively, or almost exclusively in tubulin. Among these are the removal and addition of a tyrosine at the C-terminus of α and the addition of several glutamate or glycine residues to the γ-carboxyl group of glutamate residues in the C-terminal regions of both α and β. The identification of the mechanisms by which these modifications occur and of their roles in microtubule assembly and function are currently very active topics of research; they will be addressed in this chapter.

Key Words: Tubulin; α-tubulin; β-tubulin; post-translational modification; phosphorylation; acetylation; tyrosinolation; polyglutamation; polyglycylation; deglutamylation.

From: *Cancer Drug Discovery and Development: The Role of Microtubules in Cell Biology, Neurobiology, and Oncology* Edited by: Tito Fojo © Humana Press, Totowa, NJ

1. INTRODUCTION

The tubulin molecule is subject to a large number of different post-translational modifications (Table 1). With the possible exception of collagen, no protein is modified in as many different ways. Some of these modifications, such as phosphorylation and acetylation are common to many proteins. Others, such as polyglutamylation, are very rare, whereas still others, such as tyrosinolation/detyrosinolation and polyglycylation, have so far been reported only in tubulin. Many of these modifications are widespread among eukaryotic tubulins. In this review, the various post-translational modifications will be defined, and described or speculated on their functional significance. Wherever the known enzymes are involved in these modifications shall be described. The post-translational modifications of tubulin have been reviewed before *(1–5)*. Here, the authors shall concentrate on the more recent findings.

2. TYROSINOLATION/DETYROSINOLATION (α)

The genes for most of the vertebrate α-tubulin isotypes encode a tyrosine at the C-terminus, preceded by a glutamate (*see* Table 7 in Chapter 6). One exception is α4, whose C-terminal residue is glutamate. The C-terminal tyrosine is removed by a tubulin carboxypeptidase and can be added back by the enzyme tubulin-tyrosine ligase in a reaction requiring ATP but not involving any kind of ribosome or RNA *(6)*. The ligase can also add a tyrosine to the α4 isotype *(7)*. The ligase has been purified and characterized *(8,9)*. Purification of tubulin carboxypeptidase is still incomplete *(10–12)*. The favored substrate of the ligase is the free tubulin dimer, whereas the carboxypeptidase prefers the microtubule *(13,14)*.

Tubulin monomers that contain the C-terminal tyrosine are often referred to as "Tyr-tubulin," whereas the ones lacking the C-terminal tyrosine are called "Glu-tubulin." Thus, Tyr-tubulin can be detyrosinolated to form Glu-tubulin, which can then be retyrosinolated to form Tyr-tubulin. Immunofluorescent staining of cultured cells revealed that these two tubulin classes are able to form two different subsets of microtubules. Microtubules containing Tyr-tubulin are present in the interphase network as well as the metaphase spindle, whereas those containing Glu-tubulin are absent from the mitotic spindle *(15)*. In general, Glu-tubulin is enriched in stable microtubules, whereas microtubules containing Tyr-tubulin are labile *(16,17)*. Glu-tubulin is often seen in axonemes, sperm manchettes, basal bodies, centrioles, centrosomes and the primary cilium as well as in the perinuclear region *(18–22)*. In addition, the pattern of tyrosinolation in specific tissues can change during development *(23,24)*. Microtubules made of Glu-tubulin appear to interact with the intermediate filament protein vimentin through a kinesin-dependent mechanism *(25–27)*, suggesting a role for this modification in microtubule-intermediate filament "crosstalk."

The tyrosinolation/detyrosinolation cycle may play a role in axoneme motility. In sea urchin sperm flagella, outer doublets 1, 5, and 6 are preferentially tyrosinolated, whereas doublets 3 and 8 are largely detyrosinolated *(28)*. The former group corresponds to the place where the flagellum bends. Perhaps tyrosinolation makes the microtubule more flexible in this region. The tyrosinolation/detyrosinolation cycle has been seen not only in mammals, but also in a variety of invertebrates, plants, and protists *(19,20,29,30)* (Table 1). In corn, tyrosinolation/detyrosinolation is highly tissue-specific. For example,

Table 1
Phylogenetic Distribution of Tubulin Post-Translational Modifications[a]

Tubulin	Modification	Animals	Plants	Fungi	Protists
α	Acetylation	Yes	Yes	No	Yes
	Detyrosinolation	Yes	Yes	No	Yes
	Deglutamylation	Yes	No	No	No
	Polyglutamylation	Yes	Yes	No	Yes
	Polyglycylation	Yes	No	No	Yes
	Phosphorylation	Yes	No	No	Yes
	Palmitoylation	Yes	No	Yes	No
β	Phosphorylation	Yes	No	No	No
	Polyglutamylation	Yes	No	No	Yes
	Polyglycylation	Yes	No	No	Yes

[a]Source: Refs. 6,19,20,29–31,47–60,75–79,102–104,114,121,122,135,137.

the α6 isotype is tyrosinolated in pollen and detyrosinolated in leaves and anthers *(30)*. Similarly, in pine and onion, all the microtubules in the root tip contain Tyr-tubulin but not Glu-tubulin *(31)*.

How does the tyrosinolation/detyrosinolation cycle function at the molecular level? There are two models that address this question. First, the presence or absence of the C-terminal tyrosine can act as a signal to another protein, rather than changing any of the intrinsic properties of the protein itself. For example, microtubules made of Glu-tubulin interact with a complex at the growing end, thereby stabilizing the microtubule *(32,33)*. The Glu-tubulin by itself does not stabilize the microtubule *(34)*. The fact that in rat seminiferous epithelium detyrosinolation is associated with microtubule depolymerization is consistent with the hypothesis that the C-terminal residue is only a signal, and that the meaning of the signal may vary with different tissues *(35)*. In the second model, removal of the tyrosine can alter the conformation of the C-terminal end and conceivably change the intrinsic properties of the tubulin molecule itself *(36)*. The presence of tyrosine may affect the likelihood that the C-terminus would project out from the tubulin molecule rather than lie down along the tubulin surface; thus, the C-terminal tyrosine could affect the overall conformation of the tubulin molecule. Both models could be true. The second model is likely to occur less frequently because the C-terminal end is already highly negatively charged and the tyrosine is an uncharged residue. On the other hand, the tyrosine could engage in a hydrophobic interaction with another hydrophobic residue, possible even another tubulin dimer in a microtubule.

One additional possibility is that tyrosine can be part of the nitric oxide signaling system. The tubulin-tyrosine ligase can incorporate nitrotyrosine, itself generated by the interaction of nitric oxide with tyrosine, onto α-tubulin *(37)*. This can accompany neuronal differentiation induced by nerve growth factor *(38)*. Tubulin with a C-terminal nitrotyrosine inhibits differentiation of muscle cells *(39)*. However, incorporation of nitrotyrosine into tubulin is irreversible and can cause disorganization of the cell, perhaps by altering the relative proportions of stable and unstable microtubules and or by affecting the binding of other proteins to the C-terminal region *(40)*.

3. DEGLUTAMYLATION (α)

This is a modification that probably happens to Glu-tubulin: the C-terminal glutamate is removed *(41)*. The resulting α-tubulin, lacking both the C-terminal tyrosine and the penultimate glutamate, is referred to as Δ2-tubulin. The enzyme that removes the glutamate has not been identified. In contrast to the story with the C-terminal tyrosine, once the penultimate glutamate is removed, it is never added back. In other words, Δ2-tubulin is unable to participate in the tyrosinolation/detyrosinolation cycle. Δ2-tubulin represents about 35% of mammalian brain α-tubulin; it is also found in sea urchin sperm flagella and cilia *(42)*. The levels of Glu-tubulin and Δ2-tubulin have been found to increase significantly in rat cardiac muscle just before heart failure, indicating a possible role of these tubulins in the stabilization of microtubules during stress conditions *(43)*. When Banerjee and Kasmala *(44)* examined taxol-induced microtubule assembly of Tyr-tubulin and Δ2-tubulin, they found that the former polymerized much faster and to a greater extent than did the latter. On the other hand, when assembly was studied in the presence of magnesium and glycerol, Tyr-tubulin assembled much more slowly than did Δ2-tubulin. There is some evidence that cold-resistant microtubules in the gerbil cochlea consist of Δ2-tubulin that is acetylated and polyglycylated *(45)*. This is consistent with the finding that Tyr-tubulin from bovine brain had 1–4 glutamates added, whereas Δ2-tubulin had 1–3 glycines added instead *(46)*. In other words, there is some coordination among the post-translational modifications of tubulin.

4. ACETYLATION (α)

α-Tubulin undergoes acetylation at the ε-amino group of lysine 40 *(47,48)*. The monoclonal antibody specific for this modification has been widely used for studying the distribution of acetylated tubulin in various cells and tissues *(49)*. Tubulin acetylation has been observed in both vertebrates and invertebrates and in plants and protists as well *(29,31,47–60)* (*see* Table 9 in Chapter 6). Acetylation of tubulin even occurs in the primitive protist *Giardia (61)*. Acetylated tubulin is enriched in the neuronal growth cones and also in the leading edges of fibroblasts *(62,63)*. In chondrocytes, acetylation has been seen in centrioles, centrosomes, primary cilia, and the perinuclear region *(22)*. Mature cochlear cells contain acetylated tubulin *(23)*. The microtubules of the spermatid manchette are also acetylated *(18)*. An α-tubulin acetyltransferase and a tubulin deacetylase have been identified in *Chlamydomonas (64,65)*. In addition, indirect studies with the histone deacetylase 6 (HDAC6) demonstrated that acetylation of tubulin may regulate microtubule-dependent cell motility *(66)*. HDAC6 is able to deacetylate tubulin in vivo *(67)*. Overexpression of HDAC6 increased tubulin deacetylation and induced chemotactic cell movement, indicating that acetylation of α-tubulin may perform important roles in regulating cell signaling and homeostasis.

In general, acetylation is most common in microtubules that are very stable, such as axonemes and axostyles *(68)*. However, it appears that a cold-resistant subpopulation of microtubules in other cells contains acetylated tubulin *(45)*. As tubulin is generally purified by cycles of assembly and disassembly in a process that is based on the cold-lability of microtubules, the possibility that a substantial fraction of the acetylated tubulin whose properties would be well worth studying is discarded should always be considered. Acetylated microtubules appear to be more resistant to colchicines, although not to thiobendazole *(69)*. Correlation should not imply causation, however, Palazzo et al. *(70)* found that acetylation does not make microtubules more stable to nocodazole, and they suggested that rather than acetylation conferring stability, stable microtubules become acetylated. On the other hand, Matsuyama et al. *(71)* observed that inhibiting HDAC6 increased microtubule stability in vivo. Recently the drug tubacin was identified *(72)*; this compound specifically inhibits the class II HDAC6 and blocks deacetylation of α-tubulin in cultured mammalian cells without affecting that of the histones. Interestingly, the microtubules of these cells do not become more stable as a result *(73)*. These somewhat contradictory results suggest that the acetylation/stabilization connection is not a simple one. Also, Casale et al. *(74)* reported that acetylated tubulin inhibited brain plasma membrane Na$^+$, K$^+$-ATPase. How that relates to microtubule stability is not clear.

5. POLYG108LUTAMYLATION (α,β)

This is an unusual modification in which a glutamate is added to the γ-carboxyl group of specific glutamate residues in the C-terminal region of tubulin through an isopeptide (α/γ) linkage. To this added glutamate are then added several others through α/α linkages, resulting in as many as 17 glutamates in a single chain *(19)*. Immunohistochemistry and mass spectrometry have been used to study the distribution of polyglutamylation both phylogenetically and at the subcellular level. Polyglutamylation of tubulin occurs in animals (vertebrates, echinoderms, insects, and nematodes), plants, and protists *(19,20,30,59,75–79)* (Table 1). The fact that *Giardia* tubulin is polyglutamylated suggests that this modification is very ancient *(78)*.

As is the case with acetylation, polyglutamylation is common in stable microtubules, such as those of centrioles, basal bodies, axonemes, and axostyles *(19,20,68,76,79–82)*. However, glutamylation also occurs on microtubules that are less stable, such as neuronal microtubules *(76,81)* and cochlear microtubules (where it rises and falls during development) *(23,24)*. However, the pattern of glutamylation is different in very stable microtubules. In axonemes, for example, both α and β are polyglutamylated; in neuronal tubulin, mainly α is thus modified *(76,82,83)*. Nevertheless, during neuronal differentiation, β becomes polyglutamylated as well *(83,84)*. Interestingly, the gall-midge *Asphondylia ruebsaameni*, an insect that has an aberrant sperm flagellum, containing hundreds of doublet microtubules, has the reverse pattern in its flagellum, where α is only slightly glutamylated and β has high glutamylation *(79)*.

Very stable microtubules are usually polyglutamylated, whereas less stable microtubules are sometimes monoglutamylated *(83–87)*. The microtubules of centrioles, basal bodies, and axonemes are polyglutamylated *(19,83)*. Polyglutamylation also occurs in brain microtubules *(46,87)*. Mitotic and meiotic spindles in germ cells are generally made of monoglutamylated tubulin, although mitotic spindles in HeLa cells are polyglutamylated *(83)*.

The distribution of glutamylation in axonemes is very complex. Generally, the proximal parts of the axoneme are likely to be polyglutamylated, the more distal parts monoglutamylated *(79,80,83)*. In regular axonemes, doublets 1, 5, and 6 (where the axoneme bends during wave propagation) are less glutamylated than the rest of the outer doublets *(80)*. Interestingly, these doublets 1, 5, and 6 are also enriched in tyrosinolated tubulin *(28)*. In contrast to these, acetylation of α is spread evenly throughout the axoneme *(80)*. Insects have additional accessory microtubules along the periphery of the sperm flagellum. These accessory microtubules are less glutamylated *(74)*.

The mechanism by which this modification occurs is probably quite complex. One enzyme must scan the C-terminal end of tubulin to find an appropriate glutamate residue to which to add another glutamate through an α/γ linkage. After that, a second enzyme, a tubulin polyglutamylase, adds a series of glutamates to the first one through α/α linkages (88). A tubulin polyglutamylase has been purified from the trypanosome *Crithidia* (89). It forms a complex with the tubulin dimer and can add glutamates to both α- and β-tubulins from brain (89). In this respect, the *Crithidia* enzyme is different from the putative mammalian brain equivalent, which preferentially glutamylates α. A similar enzyme has been observed in HeLa cells, perhaps existing as two isozymes (90). The fact that polyglutamylation was found in brain Tyr-tubulin and Glu-tubulin, but not in Δ2-tubulin, suggests that the penultimate glutamate residue must be essential for the tubulin polyglutamylase (46).

What is the function of polyglutamylation and how does it serve it? Polyglutamylation influences binding of microtubule-associated proteins (MAPs) to tubulin in an unusual bell-shaped pattern, with least binding seen with tubulin with no glutamates or with six glutamates added, the highest binding is seen in tubulin with 3–4 glutamates added (91–93). It is thus possible that polyglutamylation influences binding of MAPs and kinesin and may thus play an important regulatory role in the cell. A specific role for tubulin polyglutamylation in pigment granule transport in the Atlantic cod has been proposed (94). Polyglutamylation adds a number of negative charges to a region of the molecule that is already strongly negatively charged. It is therefore highly likely that the polyglutamylated C-terminus will project from the microtubule surface. Such a projection could interact with MAPs and kinesin. Despite being common in stable microtubules, polyglutamylation of tubulin does not increase cold-resistance (45). However, Antarctic fish tubulin is only slightly polyglutamylated, this probably allows dynamic behavior at cold-temperatures (95).

6. POLYGLYCYLATION (α,β)

In addition to polyglutamylation, both α- and β-tubulin also undergo polyglycylation by the addition of multiple glycine residues to the γ-carboxyl groups of specific glutamic acid residues near the C-terminus (96–99). As with the former modification, the first glycine is connected to the glutamate residue through an α/γ linkage. The remaining glycines are added to the first one by α/α linkages. This is a large-scale modification: in *Paramecium*, up to 34 glycines can be added to a single tubulin molecule (96,100), also, about 60% of β-tubulin in bull sperm is polyglycylated (97). Unlike polyglutamylation where one chain of glutamates is added to a single specific residue, in polyglycylation, the glycines are added to more than one glutamate residue in the C-terminal region. In *Paramecium* β-tubulin, for example, the glycines are added to glutamates in or near the axonemal signal sequence (101). In *Paramecium*, the glycylated residues are glu437, glu438, glu439, and glu441 (bold-faced in the following sequence: EGEF**EEEGE**Q). The most abundant polyglycylated β isoform has two glycines added to glu437, two to glu438, one to glu439, and one to glu441.

Polyglycylation is a widely distributed modification, occurring in animals (vertebrates, echinoderms, mollusks, and insects) and protists, but apparently not in plants or fungi (30,102–104) (Table 1). In *Giardia*, thought to be among the simplest eukaryotes, both α and β are polyglycylated (61). Polyglycylation is particularly common in microtubules forming stable organelles, such as axonemes and basal bodies (105). Interestingly, polyglycylation has not been observed in centrioles (82). If this last observation continues to hold up, the absence of polyglycylation might constitute one of the few biochemical differences between basal bodies and centrioles. In the unusual sperm flagellum of the gall-midge *A. ruebsaameni* both α and β are polyglycylated to a small extent *(79)*. Small amounts of polyglycylation are seen in the accessory microtubules of bee sperm *(79)*. Polyglycylation is also observed in less stable microtubules, such as those of the cochlea and neurons *(24,45,46)*. In these cases, however, the extent of polyglycylation appears to be much less; up to three attached glycines have been observed in neuronal tubulin *(46,106)*.

The mechanism by which polyglycylation occurs is unclear. The polyglycylating activity has never been isolated although a deglycylating activity has been observed in *Paramecium (106)*. This finding suggests that there is a polyglycylation cycle. The first step is likely to be the addition of a single glycine to a glutamate through an α/γ linkage. Such monoglycylated tubulin has been seen in newly formed cilia and basal bodies *(105)*. The next step would then be the addition of more glycines through α/α linkages. The third step would be the removal of some or all of these glycines by a deglycylase.

The role of polyglycylation in axonemes is fairly clear. When the glycylated gluta-mates of *Tetrahymena* β-tubulin are replaced by other amino acids, the result is highly deleterious *(107)*. Mutants lack central pair microtubules and the B-tubule of the outer doublets *(108)* (the structure of the axoneme is described in Chapter 1). In contrast, similar inhibition of polyglycylation of α had no effect *(108)*. These results imply that polyglycylation of β is essential for the proper architecture of the axoneme. As polygly-cylation occurs in the signal sequence of β, one could argue that the role of this particular sequence in isotypes such as mammalian βIVb is to provide a site for polyglycylation, which in turn is necessary for proper construction of the axoneme, specifically the cen-tral pair and the B-tubules. Such a model is consistent with two other findings. First, when purified axonemal outer doublet microtubules are incubated with mammalian brain tubulin dimers, the latter polymerizes onto the A-tubule much more readily than onto the B-tubule *(109)*. Brain tubulin, lacking the extensive polyglycylation, would have difficulty forming a B-tubule. Second, the extent of polyglycylation appears to be uniform along the length of the axoneme, in contrast to polyglutamylation, which is highest at the proximal end and then decreases distally *(79)*. If polyglycylation specifies and maintains the unusual architecture of a microtubular organelle, then one would expect it to be uniform through-out that organelle. On the other hand, a modification such as polyglutamylation, which may form a gradient, would not be necessary for architecture but rather for function. In fact, a gradient of modifications has been proposed to act as a vernier to regulate the place-ment along the axonemal microtubules of radial spokes during movement *(110)*.

The role of polyglycylation in other microtubules is unclear. Perhaps it defines a sta-ble subset of microtubules. Polyglycylated microtubules in the gerbil cochlea are more resistant to cold; however, these microtubules also contain Δ2-tubulin and acetylated tubulin, and so one cannot uniquely attribute cold resistance to the presence of polyg-lycylation *(45)*. However, the observation of Banerjee *(46)* that polyglycylation occurs on Δ2-tubulin and not on Tyr-tubulin suggests a fine coordination among these post-translational modifications. This, in turn, raises the possibility that polyglycylation of neuronal α-tubulin may be functionally significant. In evaluating its function, it must be kept in mind that the cerebral cortex and other brain structures contain cilia *(111)*. These are nonmotile cilia lacking the central pair microtubules *(112)*. In at least one case, the ependymal cilia, known to contain the βIV isotype, which contains the axonemal signal sequence, is the prime site for polyglycylation *(113)*.

7. PHOSPHORYLATION (α,β)

Phosphorylation of mammalian brain β-tubulin was the first post-translational modification of tubulin to be discovered *(114)*. Later work showed that phosphorylation of mammalian β-tubulin was restricted to the βIII isotype; the phosphorylated residue is a serine near the C-terminal end *(115–117)*. βIII is phosphorylated in brain but not in the testis *(118)*. Avian βVI is also phosphorylated near the C-terminal end at ser441 *(119)*. Both α and β are phosphorylated in sea urchin sperm in both the A- and B-tubules of the outer doublets *(120)*. Only α is phosphorylated in *Chlamydomonas* axonemes, however, both serine and threonine are phosphorylated *(121)*. Carrot tubulin is also phosphorylated *(122)*. Tyrosine phosphorylation occurs in α-tubulin in activated human B-lymphocytes *(123)*. Both α and β are phosphorylated on tyrosine residues in nerve growth cones from fetal rats and in uterine smooth muscle of the rat *(124–126)*. Little is known about the phylogenetic distribution of tubulin phosphorylation at specific sites.

The precise mechanism of phosphorylation of tubulin is not yet clear. The tyrosine kinase Syk has been implicated in phosphorylation of lymphocyte α-tubulin *(123)*. In the brain, serine phosphorylation of βIII may involve a form of casein kinase but other kinases could be involved as well *(127–129)*.

The function of tubulin phosphorylation is also not clear. Khan and Ludueña *(130)* removed some of the phosphate on βIII using phosphatase 2A, the resulting tubulin assembled normally in the absence of MAPs but its assembly in the presence of MAP2 was significantly decreased. This implies that the phosphate may promote interaction with MAPs, a reasonable supposition as MAPs bind to the phosphorylated area *(131)*. Phosphorylation of βIII accompanies maturation and differentiation of neurons *(132–134)*. As speculated in Chapter 6, phosphorylation of βIII may be a way to enhance its assembly into microtubules, whereas also decreasing the dynamic behavior of the resulting microtubules by promoting their interaction with MAPs. The function of phosphorylation in axonemes and other microtubule organelles is not known.

8. PALMITOYLATION (α)

Tubulin, like many other proteins, can undergo palmitoylation *(135,136)*, this modification involves addition of the fatty acid palmitate to the sulfhydryl group of a cysteine residue. The presence of a covalently linked palmitate could allow anchoring of a protein to a membrane. The major palmitoylated residue is cys376 of α-tubulin *(136)*. The modification has been seen in mammals and yeast. Mutation of the palmitoylated residue in *Saccharomyces cerevisiae* has a striking effect during mitosis and changes the position of the nucleus *(137)*. Thus, it is possible that palmitoylation may play a role in proper orientation of the mitotic spindle.

9. OTHER MODIFICATIONS

Terashima et al. *(138)* used NAD-arginine ADP-ribosyltransferase to add an ADP-ribosyl group to bovine brain tubulin. The effect of the modification was to inhibit microtubule assembly. However, the fact that the enzyme used was from the chicken and that this modification has not been demonstrated in vivo means that its physiological significance is not clear.

Neuronal tubulin may last a long time and accumulate covalent modifications characteristic of aged proteins. These modifications could conceivably be catalyzed by reactive oxygen species. One of these is the isomerization of asparagine and aspartate residues to generate isoaspartate *(139)*. Another is the formation of lysinoalanine crosslinks between tubulin molecules *(140)*. It is possible that these modifications could eventually compromise neuronal function if the modified tubulin is not turned over fast enough. Other than that, the physiological significance of these modifications is not clear.

10. EVOLUTION OF THE POST-TRANSLATIONAL MODIFICATIONS OF TUBULIN

The post-translational modifications of tubulin are so numerous, and some, the most bizarre and complex, are very widespread among eukaryotic tubulin, that it is likely that they can tell us about the evolution of tubulin and its isotypes. They will be examined one at a time. So far, *phosphorylation* of tubulin appears randomly distributed. The best-studied examples occur near the C-terminus of certain vertebrate tubulin β-isotypes (βIII and βVI), which have a serine in this region. It is known that certain tyrosines are phosphorylated in mammalian tubulin, that both α- and β-tubulin are phosphorylated in sea urchin flagella and that α is phosphorylated in *Chlamydomonas* flagella, although the specific phosphorylated residues have not been identified *(120,121,123–126)*. Although phosphorylation is widespread, there is as yet no compelling evidence to suggest that there is a single tubulin phosphorylation site conserved among eukaryotic phyla. Certainly, there are many β-tubulins that do not have a serine near the C-terminus. Similarly, there is no evidence of a tubulin-specific kinase. Also, phosphorylation is such a common modification of proteins that it is easy to imagine a tubulin mutating to acquire a phosphorylation site when such a modification may have been adaptive. In short, tubulin phosphorylation does not seem ancient and there is no reason to imagine that the first α- and β-tubulins were phosphorylated. *Palmitoylation* of α-tubulin is probably more ancient, having been observed in both mammals and yeast *(135–137)*. Palmitoylation has not been demonstrated in plants or protists. However, the palmitoylation site, cys376, is conserved among the plants and protists so it is conceivable that this modification arose very early in evolution.

How old is the *tyrosinolation/detyrosinolation* cycle? It has been observed among animals, plants, and protists but not among the fungi *(6,19,20,29,30)*. One would expect that for an α-tubulin to participate in this cycle, its C-terminus should end with either EY or E. That is the case for most of the animal, plant, and protist α-tubulins. All of the vertebrate α's, except for α8 and αTT1, fit this model *(see* Table 7 in Chapter 6). Therefore, it is likely that this is a very old modification. However, the cycle is a readily reversible process that can regulate microtubule stability. The regulation appears to involve an indirect signaling mechanism rather than a direct structural change *(32,33)*. One might expect, therefore, that the cycle with its associated tubulin-specific ligase and carboxypeptidase and the proteins involved in signaling would have appeared somewhat later in evolution, when microtubules had evolved to serve a variety of complex functions beyond axonemal motility and centriole and basal body formation. On the other hand, the degree of tyrosinolation varies among the axonemal outer doublets of sea urchins *(28)* in a pattern that has been argued helps to regulate motility. It has been seen that the degree of polyglutamylation varies in a complementary way, both modifications working to ensure that one set of doublets is more flexible than the others. Having two regulatory systems working together may allow more precise control of motility but it is likely that one system would be enough to generate some degree of motility. As polyglutamylation, according to present knowledge, is more widespread and common (involving both α and β instead of just α) than the tyrosinolation/detyrosinolation cycle, it is likely that the latter evolved more recently. If future experiments reveal that the degree of tyrosinolation varies in the outer doublets of the axonemes of a variety of protists then there might be need to revise this conclusion and accept that the cycle is equally as ancient as polyglutamylation. An analogous argument could be made for *deglutamylation*, which also requires α-tubulins to end in either EY or E. Although this modification has so far only been observed in animals, so many α-tubulins end in either EY or E that it is possible that deglutamylation will be observed in other eukaryotes. However, for the moment it should be assumed that deglutamylation is considerably younger than most of the other modifications.

Acetylation of α at lys40 is undoubtedly a very old modification, occurring in animals, plants, and protists, but not fungi *(29,31,47–60)*. As one would expect, lys40 is a highly conserved residue, except in fungi. Acetylation is seen in α-tubulins in axonemes and centrioles, suggesting that it may have been a feature of the earliest microtubule organelles. In striking contrast to other α-tubulins, mammalian and avian α8 lacks lys40 and also cannot undergo deglutamylation nor participate in the tyrosinolation/detyrosinolation cycle *(141)*. This strongly suggests that the α8 isotype has a unique functional significance. It is interesting that, besides α-tubulin, the eukaryotic proteins that have been reported to have acetylated internal lysines consist almost entirely of proteins associated with the nucleus (histones, transcription factors, nuclear import factors, the nuclear receptor HNF-4, proliferating cell nuclear antigen, and nonhistone chromosomal proteins) *(142,143)*. Many ribosomal proteins (that, like nuclear proteins, interact with nucleic acids) are also acetylated *(144)*. Acetylation of lysines has also been seen in a DNA-binding archaeal protein *(145)*. As will be seen, a similar pattern is observed in polyglutamylation.

Polyglycylation has been observed in plants, animals, and protists and, indeed, in every organism that has been examined *(30,102–104)*. Although it is present in both α and β, it appears to be intimately associated with the signal sequence for axoneme formation. This signal sequence, or something very similar, is common to all nonfungal β-tubulins. Polyglycylation has been observed in both basal bodies and axonemes but not in centrioles. Should this difference hold up then it might be that polyglycylation arose after basal bodies and centrioles acquired different functions. If, as argued earlier, polyglycylation is required for formation of doublet microtubules, then it must be necessary for the addition of the B-tubule to the A-tubule of the axonemal doublets. This would account for its presence in basal bodies, where the B-tubule also adds to the A-tubule, after which the C-tubule is added to create a triplet. As the same process must occur in centrioles, which also have triplet microtubules, it is strange that polyglycylation would not be present in these organelles. *Polyglutamylation* may be one of the oldest, perhaps the oldest, of tubulin post-translational modifications *(19,20,30,59,75–79)*. It occurs in animals, plants, and protists and in both α- and β-tubulins. It is associated with centrioles, basal bodies, and axonemes. If polyglutamylation occurs in γ or ε (although the lack of glutamates in their C-terminal regions argues against this), then polyglutamylation may have appeared before the evolution of α- and β-tubulin and could be very ancient indeed.

11. SUMMARY

Tubulin is clearly subject to a wide variety of post-translational modifications. The large number of modifications is manifested in the multiple species of tubulin seen by isoelectric focusing *(146)*. Many of these, such as detyrosinolation, polyglutamylation, and polyglycylation are rare or absent in other proteins, but are widespread throughout the world of eukaryotic tubulins, suggesting that they are very ancient. It is difficult to prepare tubulin that has not been modified, which makes it hard to find a substrate for identifying and purifying the modification enzymes. Nonmodified tubulin would also be a good control in experiments to test hypotheses about the roles of these modifications. Recently, Shah et al. *(147)* have prepared nonmodified and viable recombinant mouse tubulin in *Escherichia coli*. The tubulin dimers consisted of $\alpha 2$ and βIVa. This major advance should allow some of these issues to be explored. However, the right questions need to be asked.

ACKNOWLEDGMENTS

Supported by grants to RFL (Welch Foundation AQ-0726, National Institutes of Health CA26376 and CA084986, US Army Breast Cancer Research Program W81XWH-05-1-0238 and DAMD17-01-1-0411, US Army Prostate Cancer Research Program DAMD17-02-1-0045 and W81XWH-04-1-0231) and to AB (National Institutes of Health CA59711 and US Army Breast Cancer Research Program DAMD17-98-1-8244). We thank our collaborators: H. William Detrich, Israr Khan, Melvyn Little, and Hans-Peter Zimmermann. We thank Lorraine Kasmala and Veena Prasad for skilled technical assistance.

REFERENCES

1. Ludueña RF. The multiple forms of tubulin: different gene products and covalent modifications. Int Rev Cytol 1998;178:207–275.
2. MacRae TH. Tubulin post-translational modifications—enzymes and their mechanisms of action. Eur J Biochem 1997;244:265–278.
3. Westermann S, Weber K. Post-translational modifications regulate microtubule function. Nat Rev Mol Cell Biol 2003;4:938–947.
4. Erck C, Frank R, Wehland J. Tubulin-tyrosine ligase, a long-lasting enigma. Neurochem Res 2000;25:5–10.
5. Lafenechère L, Job D. The third tubulin pool. Neurochem Res 2000;25:1–8.
6. Barra HS, Arce CA, Rodríguez JA, Caputto R. Some common properties of the protein that incorporates tyrosine as a single unit and the microtubule proteins. Biochem Biophys Res Commun 1974;60: 1384–1390.
7. Gu W, Lewis SA, Cowan NJ. Generation of antisera that discriminate among mammalian α-tubulins. Introduction of specialized isotypes into cultured cells results in their coassembly without disruption of normal microtubule function. J Cell Biol 1988;106:2011–2022.
8. Raybin D, Flavin M. Enzyme which specifically adds tyrosine to the alpha chain of tubulin. Biochemistry 1977;16:2189–2194.
9. Ersfeld K, Wehland J, Plessmann Dodermont H, Gerke V, Weber K. Characterization of the tubulin-tyrosine ligase. J Cell Biol 1993;120:725–732.
10. Argaraña CE, Barra HS, Caputto R. Tubulinyl-tyrosine carboxypeptidase from chicken brain: properties and partial purification. J Neurochem 1980;34:114–118.
11. Kumar N, Flavin M. Preferential action of a brain detyrosinolating carboxypeptidase on polymerized tubulin. J Biol Chem 1981;256:7678–7686.
12. Arce CA, Barra HS. Association of tubulinyl-carboxypeptidase with microtubules. FEBS Lett 1983; 157:75–78.
13. Wehland J, Weber K. Tubulin-tyrosine ligase has a binding site on β-tubulin: a two domain structure of the enzyme. J Cell Biol 1987;104:1059–1067.

14. Webster DR, Gundersen GG, Bulinski JC, Borisy GG. Assembly and turnover of detyrosinated tubulin *in vivo*. J Cell Biol 1987;105:265–276.

15. Gundersen GG, Kalnoski MH, Bulinski JC. Distinct populations of microtubules: tyrosinated and nontyrosinated α-tubulin are distributed differently *in vivo*. Cell 1984;38:779–789.

16. Kreis TE. Microtubules containing detyrosinated tubulin are less dynamic. EMBO J 1987;6: 2597–2606.

17. Khawaja S, Gundersen GG, Bulinski JC. Enhanced stability of microtubules enriched in detyrosinated tubulin is not a direct function of detyrosination level. J Cell Biol 1988;106:141–149.

18. Kierszenbaum AL. Intramanchette transport (IMT): managing the making of the spermatid head, centrosome, and tail. Mol Reprod Devel 2002;63:1–4.

19. Geimer S, Teltenkötter A, Plessmann U, Weber K, Lechtreck KF. Purification and characterization of basal apparatuses from a flagellate green alga. Cell Motil Cytoskeleton 1997;37:72–85.

20. Mansir A, Justine JL. The microtubular system and posttranslationally modified tubulin during spermatogenesis in a parasitic nematode with amoeboid and aflagellate spermatozoa. Mol Reprod Devel 1998;49:150–167.

21. Mencarelli C, Bré MH, Levilliers N, Dallai R. Accessory tubules and axonemal microtubules of *Apis mellifera* sperm flagellum differ in their tubulin isoform content. Cell Motil Cytoskeleton 2000;47:1–12.

22. Poole CA, Zhang ZJ, Ross JM. The differential distribution of acetylated and detyrosinated alpha-tubulin in the microtubular cytoskeleton and primary cilia of hyaline cartilage chondrocytes. J Anat 2001;199:393–405.

23. Tannenbaum J, Slepecky NB. Localization of microtubules containing posttranslationally modified tubulin in cochlear epithelial cells during development. Cell Motil Cytoskeleton 1997;38:146–162.

24. Saha S, Slepecky NB. Age-related changes in microtubules in the guinea pig organ of Corti. Tubulin isoform shifts with increasing age suggest changes in micromechanical properties of the sensory epithelium. Cell Tissue Res 2000;300:29–46.

25. Gurland G, Gundersen GG. Stable, detyrosinated microtubules function to localize vimentin intermediate filaments in fibroblasts. J Cell Biol 1995;131:1264–1290.

26. Liao G, Gundersen GG. Kinesin is a candidate for cross-bridging microtubules and intermediate filaments. Selective binding of kinesin to detyrosinated tubulin and vimentin. J Biol Chem 1998;273: 9797–9803.

27. Kreitzer FG, Liao G, Gundersen GG. Detyrosination of tubulin regulates the interaction of intermediate filaments with microtubules *in vivo* via a kinesin-dependent mechanism. Mol Biol Cell 1999;10:1105–1138.

28. Pechart I, Kann MKL, Levilliers N, Bré MH, Fouquet JP. Composition and organization of tubulin isoforms reveals a variety of axonemal models. Biol Cell 1999;91:685–697.

29. Day R, Criel GRJ, Walling MA, MacRae TH. Posttranslationally modified tubulins and microtubule organization in hemocytes of the brine shrimp, *Artemia franciscana*. J Morphol 2000;244:153–166.

30. Wang W, Vignani R, Scali M, Sensi E, Cresti M. Post-translational modification of α-tubulin in *Zea mays* L are highly tissue specific. Planta 2004;218:460–465.

31. Gilmer S, Clay P, MacRae TH, Fowke LC. Tyrosinated, but not detyrosinated, α- tubulin is present in root tip cells. Protoplasma 1999;210:92–98.

32. Idriss HT. Man to *Trypanosome*: the tubulin tyrosination/detyrosination cycle revisited. Cell Motil Cytoskeleton 2000;45:173–184.

33. Infante AS, Stein MS, Zhai Y, Borisy GG, Gundersen GG. Detyrosinated (Glu) microtubules are stabilized by an ATP-sensitive plus-end cap. J Cell Sci 2000;113:3907–3919.

34. Webster DR, Wehland J, Weber K, Borisy GG. Detyrosination of α tubulin does not stabilize microtubules *in vivo*. J Cell Biol 1990;141:175–185.

35. Correa LM, Miller MG. Microtubule depolymerization in rat seminiferous epithelium is associated with diminished tyrosination of α-tubulin. Biol Reprod 2001;64:1644–1652.

36. Ponstingl H, Little M, Krauhs E, Kempf T. Carboxy-terminal amino acid sequence of α-tubulin from porcine brain. Nature 1979;282:423–424.

37. Kalisz HM, Erck C, Plessmann U, Wehland J. Incorporation of nitrotyrosine into α-tubulin by recombinant mammalian tubulin-tyrosine ligase. Biochim Biophys Acta 2000;148:131–138.

38. Cappelletti G, Maggioni MG, Tedeschi G, Maci R. Protein tyrosine nitration is triggered by nerve growth factor during neuronal differentiation of PC12 cells. Exp Cell Res 2003;288:9–20.

39. Chang W, Webster DR, Salam AA, et al. Alteration of the C-terminal amino acid of tubulin specifically inhibits myogenic differentiation. J Biol Chem 2002;277:30,690–30,698.
40. Eiserich JP, Estévez AG, Bamberg TV, et al. Microtubule dysfunction by posttranslational nitrotyrosination of α-tubulin: a nitric oxide-dependent mechanism of cellular injury. Proc Nat Acad Sci USA 1999;96:6365–6370.
41. Paturle-Lafanechère L, Eddé B, Denoulet P, Van Dorsselaer A, Mazarguil H, Le Caer JP. Characterization of a major brain tubulin variant which cannot be tyrosinated. Biochemistry 1991;30:10,523–10,528.
42. Mary J, Redeker V, Le Caer JP, Promé JC, Rossier J. Class I and IVa β-tubulin isotypes expressed in adult mouse brain are glutamylated. FEBS Lett 1994;353:89–94.
43. Belmadani S, Poüs C, Ventura-Clapier R, Fischmeister R, Méry PF. Post-translational modifications of cardiac tubulin during chronic heart failure in the rat. Mol Cell Biochem 2002;237:39–46.
44. Banerjee A, Kasmala LT. Differential assembly kinetics of α-tubulin isoforms in the presence of paclitaxel. Biochem Biophys Res Commun 1998;245:349–351.
45. Bane BC, MacRae TH, Xiang H, Bateman, Slepecky NB. Microtubule cold stability in supporting cells of the gerbil auditory sensory epithelium: correlation with tubulin post-translational modifications. Cell Tissue Res 2002;307:57–67.
46. Banerjee A. Coordination of posttranslational modifications of bovine brain α-tubulin. Polyglycylation of Δ2-tubulin. J Biol Chem 2002;277:46,140–46,144.
47. L'Hernault SW, Rosenbaum JL. Chlamydomonas α-tubulin is posttranslationally modified by acetylation on the ε-amino group of a lysine. Biochemistry 1985;24:473–478.
48. LeDizet M, Piperno G. Identification of an acetylation site of Chlamydomonas α-tubulin. Proc Nat Acad Sci USA 1987;84:5720–5724.
49. Piperno G, Fuller MT. Monoclonal antibodies specific for an acetylated form of α-tubulin recognize the antigen in cilia and flagella from a variety of organisms. J Cell Biol 1985;101:2085–2094.
50. Gallo JM, Precigout E. Tubulin expression in trypanosomes. Biol Cell 1988;64:137–143.
51. Wolf KW, Regan CL, Fuller MT. Temporal and spatial pattern of differences in microtubule behaviour during Drosophila embryogenesis revealed by distribution of a tubulin isoform. Development 1988;102:311–324.
52. Siddiqui SS, Aamodt E, Rastinejad F, Culotti J. Anti-tubulin monoclonal antibodies that bind to specific neurons in Caenorhabditis elegans. J Neurosci 1989;9:2963–2972.
53. Wilson PJ, Forer A. Acetylated α-tubulin in spermatogenic cells of the crane fly Nephrotoma suturalis: Kinetochore microtubules are selectively acetylated. Cell Motil Cytoskeleton 1989; 14:237–250.
54. Souto-Padron T, Cunha e Silva NL, de Souza W. Acetylated alpha-tubulin in Trypanosoma cruzi: Immunocytochemical localization. Mem Inst Oswaldo Cruz 1993;88:517–528.
55. Delgado-Viscogliosi P, Brugerolle G, Viscogliosi E. Tubulin post-translational modifications in the primitive protist Trichomonas vaginalis. Cell Motil Cytoskeleton 1996;33:288–297.
56. Wolf KW. Cytology of Lepidoptera VIII. Acetylation of α-tubulin in mitotic and meiotic spindles of two Lepidoptera species, Ephesia kuehniella (Pyralidae) and Pieris brassicae (Pieridae). Protoplasma 1996;190:88–98.
57. Wolf KW. Acetylation of α-tubulin in male meiotic spindles of Pyrrhocoris apterus, an insect with holokinetic chromosomes. Protoplasma 1996;191:148–157.
58. Huang RF, Lloyd CW. Gibberellic acid stabilises microtubules in maize suspension cells to cold and stimulates acetylation of α-tubulin. FEBS Lett 1999;443:317–320.
59. Noël C, Gerbod D, Fast NM, et al. Tubulins in Trichomonas vaginalis: molecular characterization of α-tubulin genes, posttranslational modifications, and homology modeling of the tubulin dimer. J Eukaryot Microbiol 2001;48:647–654.
60. Lazareva EM, Polyakov VY, Chentsov YS, Smirnova EA. Time and cell cycle dependent formation of heterogeneous tubulin arrays induced by colchicine in Triticum aestivum root meristem. Cell Biol Int 2003;27:633–646.
61. Campanati L, Bré MH, Levilliers N, de Souza W. Expression of tubulin polyglycylation in Giardia lamblia. Biol Cell 1999;91:499–506.
62. Piperno G, LeDizet M, Chang XJ. Microtubules containing acetylated α-tubulin in mammalian cells in culture. J Cell Biol 1987;104:289–302.

63. Robson SJ, Burgoyne RD. Differential localisation of tyrosinated, detyrosinated, and acetylated α-tubulins in neurites and growth cones of dorsal root ganglion neurons. Cell Motil Cytoskeleton 1989;12:273–282.

64. Greer K, Maruta H, L'Hernault SW, Rosenbaum JL. α-Tubulin acetylase activity in isolated *Chlamydomonas* flagella. J Cell Biol 1985;101:2081–2084.

65. Maruta H, Greer K, Rosenbaum JL. The acetylation of alpha-tubulin and its relationship to the assembly and disassembly of microtubules. J Cell Biol 1986;103:571–579.

66. Hubbert C, Guardiola A, Shao R, et al. HDAC6 is a microtubule-associated deacetylase. Nature 2002;417:455–458.

67. Zhang Y, Caron C, Matthias G, Hess D, Khochbin S, Matthias P. HDAC-6 interacts with and deacetylates tubulin and microtubules in vivo. EMBO J 2003;22:1168–1179.

68. Boggild AK, Sundermann CA, Estridge BH. Localization of post-translationally modified α-tubulin and pseudocyst formation in tritrichomonads. Parasitol Res 2002;88:468–474.

69. Pisano C, Battistoni A, Antoccia A, Degrassi F, Tanzarella C. Changes in microtubule organization after exposure to a benzimidazole derivative in Chinese hamster cells. Mutagenesis 2000;15: 507–515.

70. Palazzo A, Ackerman B, Gundersen GG. Tubulin acetylation and cell motility. Nature 2003;421:230.

71. Matsuyama A, Shimazu T, Sumida Y, Saito A, Yoshimatsu Y, Seigneurin–Berny D. In vivo destabilization of dynamic microtubules by HDAC6-mediated deacetylation. EMBO J 2002;21:6820–6831.

72. Haggarty SJ, Koeller KM, Wong JC, Butcher RA, Schreiber SL. Multidimensional chemical genetic analysis of diversity-oriented synthesis-derived deacetylase inhibitors using cell-based assays. Chem Biol 2003;10:383–396.

73. Haggarty SJ, Koeller KM, WOng JC, Grozinger CM, Schreiber SL. Domain-selective small-molecule inhibitor of histone deacetylase 6 (HDA6)-mediated tubulin deacetylation. Proc Nat Acad Sci USA 2003;100:4389–4394.

74. Casale CH, Alonso AC, Barra HS. Brain plasma membrane Na$^+$,K$^+$-ATPase is inhibited by acetylated tubulin. Mol Cell Biochem 2001;216:85–92.

75. Pucciarelli S, Ballarini P, Miceli C. Cold-adapted microtubules: Characterization of tubulin posttranslational modifications in the Antarctic ciliate *Euplotes focardii*. Cell Motil Cytoskeleton 1997;38: 329–340.

76. Bobinnec Y, Marcaillou C, Debec A. Microtubule polyglutamylation in *Drosophila melanogaster* brain and testis. Eur J Cell Biol 1999;78:671–674.

77. Huitorel P, White D, Fouquet JP, Kann ML, Cosson J, Gagnon C. Differential distribution of glutamylated tubulin isoforms along the sea urchin sperm axoneme. Mol Reprod Devel 2002;62:139–148.

78. Boggild AK, Sundermann CA, Estridge BH. Post-translational glutamylation and tyrosination in tubulin of tritrichomonads and the diplomonad *Giardia intestinalis*. Parasitol Res 2002;88:58–62.

79. Mencarelli C, Caroti D, Bré MH, et al. Glutamylated and glycylated tubulin isoforms in the aberrant sperm axoneme of the gall-midge fly, *Asphondylia ruebsaameni*. Cell Motil Cytoskeleton 2004;58: 160–174.

80. Fouquet JP, Kann ML, Péchart I, Prigent Y. Expression of tubulin isoforms during the differentiation of mammalian spermatozoa. Tissue Cell 1997;29:573–583.

81. Bonnet C, Denarier E, Bosc C, Lazereg S, Denoulet P, Larcher JC. Interaction of STOP with neuronal tubulin is independent of polyglutamylation. Biochem Biophys Res Commun 2002;297:787–793.

82. Million K, Larcher JC, Laoukili J, Bourguignon D, Marano F, Tournier F. Polyglutamylation and polyglycylation of α- and β-tubulins during in vitro ciliated cell differentiation of human respiratory epithelial cells. J Cell Sci 1999;112:4357–4366.

83. Kann ML, Soues S, Levilliers N, Fouquet JP. Glutamylated tubulin: diversity of expression and distribution of isoforms. Cell Motil Cytoskeleton 2003;55:14–25.

84. Wolff A, de Néchaud B, Chillet D, et al. Distribution of glutamylated α- and β-tubulin in mouse tissues using a specific monoclonal antibody, GT335. Eur J Cell Biol 1992;59:425–432.

85. Verdier-Pinard P, Wang F, Martello L, Burd B, Orr GA, Horwitz SB. Analysis of tubulin isotypes and mutations from taxol-resistant cells by combined isoelectrofocusing and mass spectrometry. Biochemistry 2003;42:5349–5357.

86. Audebert S, Desbruyères E, Gruszczynski C, et al. Reversible polyglutamylation of α- and β-tubulin and microtubule dynamics in mouse brain neurons. Mol Biol Cell 1993;4:615–626.

87. Rao S, Åberg F, Nieves E, Horwitz SB, Orr GA. Identification by mass spectrometry of a new α-tubulin isotype expressed in human breast and lung carcinoma cell lines. Biochemistry 2001;40: 2096–2103.

88. Westermann S, Plessmann U, Weber K. Synthetic peptides identify the minimal substrate require-ments of tubulin polyglutamylase in side chain elongation. FEBS Lett 1999;459:90–94.

89. Westermann S, Schneider A, Horn EK, Weber K. Isolation of tubulin polyglutamylase from *Crithidia*; binding to microtubules and tubulin, and glutamylation of mammalian brain α- and β-tubulins. J Cell Sci 1999;112:2185–2193.

90. Regnard C, Desbruyères E, Denoulet P, Eddé B. Tubulin polyglutamylase: isozymic variants and reg-ulation during the cell cycle in HeLa cells. J Cell Sci 1999;112:4281–4289.

91. Boucher D, Larcher JC, Gros F, Denoulet P. Polyglutamylation of tubulin as a progressive regulator of *in vitro* interactions between the microtubule-associated protein tau and tubulin. Biochemistry 1994;33:12,471–12,477.

92. Larcher JC, Boucher D, Lazereg S, Gros F, Denoulet P. Interaction of kinesin motor domains with α- and β-tubulin subunits at a tau-independent binding site. Regulation of polyglutamylation. J Biol Chem 1996;271:22,117–22,124.

93. Bonnet C, Boucher D, Lazereg S, et al. Differential binding regulation of microtubule-associated proteins MAP1A, MAP1B, and MAP2 by tubulin polyglutamylation. J Biol Chem 2001;276:12,839–12,848.

94. Klotz A, Rutberg M, Denoulet P, Wallin M. Polyglutamylation of Atlantic cod tubulin: immunochem-ical localization and possible role in pigment granule transport. Cell Motil Cytoskeleton 1999; 44:263–273.

95. Redeker V, Frankfurter A, Parker SK, Rossier J, Detrich HW 3rd. Posttranslational modification of brain tubulins from the Antarctic fish *Notothenia coriiceps*: reduced C-terminal glutamylation corre-lates with efficient microtubule assembly at low temperature. Biochemistry 2004;43:12,265–12,274.

96. Redeker V, Levilliers N, Schmitter JM, et al. Polyglycylation of tubulin: A posttranslational modifi-cation in axonemal microtubules. Science 1992;266:1688–1691.

97. Rüdiger M, Plessman U, Rüdiger AH, Weber K. β tubulin of bull sperm is polyglycylated. FEBS Lett 1995;264:147–151.

98. Weber K, Schneider A, Müller N, Plessman U. Polyglycylation of tubulin in the diplomonad *Giardia lamblia*, one of the oldest eukaryotes. FEBS Lett 1996;393:27–30.

99. Mary J, Redeker V, Le Caer JP, Rossier J, Schmitter JM. Posttranslational modifications in the C-terminal tail of axonemal tubulin from sea urchin sperm. J Biol Chem 1994;271:9928–9933.

100. Vinh J, Langridge JI, Bré MH, et al. Structural characterization by tandem mass spectrometry of the posttranslational polyglycylation of tubulin. Biochemistry 1999;38:3133–3139.

101. Raff EC, Fackenthal JD, Hutchens JA, Hoyle HD, Turner FR. Microtubule architecture specified by a β-tubulin isoform. Science 1997;275:70–73.

102. Bressac C, Bré MH, Darmanaden-Delorme J, Laurent M, Levilliers N, Fleury A. A massive new post-translational modification occurs on axonemal tubulin at the final step of spermatogenesis in Drosophila. Eur J Cell Biol 1995;67:346–355.

103. Bré MH, Redeker V, Quibell M, et al. Axonemal tubulin polyglycylation probes with two monoclonal antibodies: Widespread evolutionary distribution, appearance during spermatozoan maturation and possible function in motility. J Cell Sci 1996;109:727–738.

104. Levilliers N, Fleury A, Hill AM. Monoclonal and polyclonal antibodies detect a new type of post-translational modification in axonemal tubulin. J Cell Sci 1995;108:3013–3028.

105. Ittode F, Clérot JC, Levilliers N, Bré MH. Tubulin polyglycylation: a morphogenetic marker in cili-ates. Biol Cell 2000;92:615–628.

106. Bré MH, Redeker V, Vinh J, Rossier J, Levilliers N. Tubulin polyglycylation: differential posttransla-tional modification of dynamic cytoplasmic and stable axonemal microtubules in *Paramecium*. Mol Biol Cell 1998;9:2655–2665.

107. Xia L, Hai B, Gao Y, et al. Polyglycylation of tubulin is essential and affects cell motility and divi-sion in *Tetrahymena thermophila*. J Cell Biol 2000;149:1097–1106.

108. Thazhath R, Liu C, Gaertig J. Polyglycylation domain of β-tubulin maintains axonemal architecture and affects cytokinesis in *Tetrahymena*. Nat Cell Biol 2002;4:256–259.

109. Bergen LG, Borisy GG. Head-to-tail polymerization of microtubules in vitro. Electron microscope analysis of seeded assembly. J Cell Biol 1980;84:141–150.

110. Cibert C. Entropy and information in flagellar axoneme cybernetics: a radial spokes integrative func-tion. Cell Motil Cytoskeleton 2003;54:296–316.

111. Dahl HA. Fine structure of cilia in rat cerebral cortex. Z Zellforsch Mikr Anat 1963;60:369–386.

112. Händel M, Schulz S, Stanarius A, et al. Selective targeting of somatostatin receptor 3 to neuronal cilia. Neurosci 1999;89:909–926.

113. Jensen-Smith HC, Ludueña RF, Hallworth R. Requirement for the βI and βIV tubulin isotypes in mammalian cilia. Cell Motil Cytoskeleton 2003;55:213–220.
114. Eipper BA. Rat brain microtubule protein. Purification and determination of covalently bound phosphate and carbohydrate. Proc Nat Acad Sci USA 1972;69:2283–2287.
115. Alexander JE, Hunt DF, Lee MK, et al. Characterization of posttranslational modifications in neuron-specific class III β-tubulin by mass spectrometry. Proc Nat Acad Sci USA 1991;88:4685–4689.
116. Ludueña RF, Zimmermann HP, Little M. Identification of the phosphorylated β-tubulin isotype in differentiated neuroblastoma cells. FEBS Lett 1988;230:142–146.
117. Díaz-Nido J, Serrano L, López-Otin C, Vandekerckhove J, Ávila J. Phosphorylation of a neuronal-specific β-tubulin isotype. J Biol Chem 1990;265:13,949–13,954.
118. Lee MK, Tuttle JB, Rebhun LI, Cleveland DW, Frankfurter A. The expression and posttranslational modification of neuron-specific β-tubulin isotype during chick embryogenesis. Cell Motil Cytoskeleton 1990;17:118–132.
119. Rüdiger M, Weber K. Characterization of the post-translational modifications in tubulin from the marginal band of avian erythrocytes. Eur J Biochem 1993;218:107–116.
120. Stephens RE. Structural chemistry of the axoneme: Evidence for chemically and functionally unique tubulin dimers in outer fibers. Soc Gen Physiol Ser 1975;30:181–204.
121. Piperno G, Luck DJ. Phosphorylation of axonemal proteins in *Chlamydomonas reinhardtii*. J Biol Chem 1976;251:2161–2167.
122. Koontz DA, Choi JH. Evidence of phosphorylation of tubulin in carrot suspension cells. Physiol Plant 1993;87:576–583.
123. Peters JD, Furlong MT, Asai DJ, Harrison ML Geahlen RL, Syk, activated by cross-linking the B-cell antigen receptor, localizes to the cytosol where it interacts with and phosphorylates α-tubulin on tyrosine. J Biol Chem 1996;271:4755–4762.
124. Matten WJ, Aubry M, West J, Maness PF. Tubulin is phosphorylated at tyrosine by pp60^{c-src} in nerve growth cone membranes. J Cell Biol 1990;111:1959–1970.
125. Atashi JR, Klinz SG, Ingraham CA, Matten WT, Schacher M, Maness PF. Neural cell adhesion molecules modulate tyrosine phosphorylation of tubulin in nerve growth cone membranes. Neuron 1992;8:831–842.
126. Joseph MK, Fernström MA, Soloff MS. Switching of β- to α-tubulin phosphorylation in uterine smooth muscle of parturient rats. J Biol Chem 1982;257:11,728–11,733.
127. Serrano L, Díaz-Nido J, Wandosell F, Ávila J. Tubulin phosphorylation by casein kinase II is similar to that found in vivo. J Cell Biol 1987;105:1731–1739.
128. Crute BE, Van Buskirk RG. A casein kinase-like kinase phosphorylates β-tubulin and may be a microtubule-associated protein. J Neurochem 1992;59:2017–2023.
129. Takahashi M, Tomizawa K, Sato K, Ohtake A, Omori A. A novel tau-tubulin kinase from bovine brain. FEBS Lett 1995;372:59–64.
130. Khan IA, Ludueña RF. Phosphorylation of $β_{III}$-tubulin. Biochemistry 1996;35:3704–3711.
131. Littauer UZ, Giveon D, Thierauf M, Ginzburg I, Ponstingl H. Common and distinct tubulin binding sites for microtubule-associated proteins. Proc Nat Acad Sci USA 1986;83:7162–7166.
132. Fanarraga ML, Avila J, Zabala JC. Expression of unphosphorylated class III β-tubulin isotype in neuroepithelial cells demonstrates neuroblast commitment and differentiation. Eur J Neurosci 1999;11:517–527.
133. Gard DL, Kirschner MW. A polymer-dependent increase in phosphorylation of β-tubulin accompanies differentiation of a mouse neuroblastoma cell line. J Cell Biol 1985;100:764–774.
134. Aletta JM. Phosphorylation of type III β-tubulin in PC12 cell neurites, during NGF-induced process outgrowth. J Neurobiol 1996;31:461–475.
135. Caron JM. Posttranslational modification of tubulin by palmitoylation. I. *In vivo* and cell-free studies. Mol Biol Cell 1997;8:621–636.
136. Ozols J, Caron JM. Posttranslational modification of tubulin by palmitoylation. II. Identification of sites of palmitoylation. Mol Biol Cell 1997;8:637–645.
137. Caron JM, Vega LR, Fleming J, Bishop R, Solomon F. Single site α-tubulin mutation affects astral microtubules and nuclear positioning during anaphase in *Saccharomyces cerevisiae*: possible role of palmitoylation of α-tubulin. Mol Biol Cell 2001;12:2672–2687.
138. Terashima M, Yamamori C, Tsuchiya M, Shimoyama M. ADP-Ribosylation of tubulin by chicken NAD-arginine ADP-ribosyltransferase suppresses microtubule formation. J Nutr Sci Vitaminol 1999;45:393–400.

139. Najbauer J, Orpiszerski J, Aswad DW. Molecular aging of tubulin: Accumulation of isoaspartyl sites in vitro and in vivo. Biochemistry 1996;35:5183–5190.
140. Correia JJ, Lipscomb LD, Lobert S. Nondisulfide crosslinking and chemical cleavage of tubulin subunits: pH and temperature dependence. Arch Biochem Biophys 1993;300:105–114.
141. Stanchi F, Corso V, Scannapieco P, et al. TUBA8: a new tissue-specific isoform of α-tubulin that is highly conserved in human and mouse. Biochem Biophys Res Commun 2000;270:1111–1118.
142. Polevoda B, Sherman F. The diversity of acetylated proteins. Genome Biol 2002;**3**/5/reviews/0006.1 (on-line).
143. Naryzhny SN, Lee H. The post-translational modifications of proliferating cell nuclear antigen: acetylation, not phosphorylation, plays an important role in the regulation of its function. J Biol Chem 2004;279:20,194–20,199.
144. Odintsova TI, Muller EC, Ivanov AV, et al. Characterization and analysis of posttranslational modifications of the human large cytoplasmic ribosomal subunit proteins by mass spectrometry and Edman sequencing. J Protein Chem 2003;22:249–258.
145. Wardleworth BN, Russell RJ, Bell SD, Taylor GL, White MF. Structure of Alba: an archaeal protein modulated by acetylation. EMBO J 2002;21:4654–4662.
146. Williams RC, Shah C, Sackett D. Separation of tubulin isoforms by isoelectric focusing in immobilized pH gradient gels. Anal Biochem 1999;275:265–267.
147. Shah C, Xu CZQ, Vickers J, Williams R. Properties of microtubules assembled from mammalian tubulin synthesized in *Escherichia coli*. Biochemistry 2001;40:4844–4852.

6 The Isotypes of Tubulin

Distribution and Functional Significance

Richard F. Ludueña
and Asok Banerjee

CONTENTS

SUMMARY

The tubulin molecule is an α/β heterodimer. In most eukaryotes both α- and β-tubulin consist of isotypes encoded by different genes and differing in amino acid sequence. Differences among isotypes are often highly conserved in evolution, suggesting that they have functional significance. The complex isotype families in mammals, *Drosophila* and higher plants have been particularly well studied. Different isotypes often have different cellular and tissue distributions. In addition, purified isotypes display different properties including assembly, GTPase, conformation, dynamics, and ability to interact with anti-tumor drugs. The different cellular, tissue, and species distribution, as well as their primary structures and their *in vitro* properties give clues as to the possible functions of the different isotypes, which will be discussed in this chapter.

Key Words: Tubulin; α-tubulin; β-tubulin; βI; βII; βIII; βIV; βV; βVI; isotypes; anti-tumor drugs; evolution; axonemes; cilia; flagella.

1. INTRODUCTION

Tubulin, the subunit protein of microtubules is an α/β heterodimer *(1,2)*. The full amino acid sequences of α and β were first determined in 1981 and found to be 41% identical *(3,4)*. The existence of tubulin isotypes was confirmed in this same work. The amino acid sequences of the peptides, obtained from pig brain tubulin, showed heterogeneity at various positions, indicating that at least four forms of α and two forms of β were expressed in pig brain, presumably encoded by different genes *(3,4)*. Since that

From: *Cancer Drug Discovery and Development: The Role of Microtubules in Cell Biology,
Neurobiology, and Oncology* Edited by: Tito Fojo © Humana Press, Totowa, NJ

time genes for α- and β-tubulin have been sequenced from a large number of eukaryotes. Many of these organisms contain multiple genes for α or β, or both, generally encoding proteins of different amino acid sequence. These different proteins will be referred to as *isotypes* of α or β, meaning proteins encoded by different genes with different amino acid sequences. More recently, other very different forms of tubulin have been discovered, designated as γ, δ, ε, ζ, η, θ, ι, and κ. Still others may be waiting in the wings. These tubulins, together with α and β, are generally grouped together as the tubulin superfamily. Some related proteins have been observed in prokaryotes as well. The tubulin superfamily and the related prokaryotic proteins will be discussed in Chapter 7. In addition to the genetically encoded forms of tubulin, multiple forms of α and β exist, differing in their post-translational modifications. These will be discussed in Chapter 5.

The existence of tubulin isotypes had been predicted long before 1981. In 1967, Behnke and Forer *(5)* had suggested that in view of the different stability of microtubules performing different functions, there must be different forms of tubulin. This proposal was later elaborated into the multitubulin hypothesis, which proposed the existence of such forms, each one responsible for a specific function *(6)*. As will be seen here, the multitubulin hypothesis is fundamentally correct, although not all isotypes can be explained this way, and, in those cases where the hypothesis applies, the functional differences are often far more subtle and complex than originally envisioned. The area of tubulin isotypes has been reviewed before *(7–9)*. Here the concentration will be on discoveries made since 1998.

2. DISTRIBUTION OF TUBULIN ISOTYPES

2.1. Phylogenetic Distribution

The existence of tubulin isotypes has been demonstrated in many organisms (Tables 1–3). It is clear that organisms in every eukaryotic phylum exhibit multiple isotypes of both α- and β-tubulin. This is particularly true for the higher eukaryotes. Among the animals, in every case where multiple isotypes of α and β have been searched for, they have been found. One possible exception is the sea urchin *Lytechinus*, where a single α-tubulin gene was reported *(10)*. However, since this was published, multiple isotypes of α have been found in the sea urchins *Paracentrotus* and *Strongylocentrotus (11,12)*; hence, it is very likely that further investigation will reveal multiple isotypes of α in *Lytechinus* as well. Plants have a similar story. Multiple isotypes of both α- and β-tubulin have been found in every plant that has been investigated. In short, there are no plants or animals that have been found to express either a single α or a single β isotype. Every plant and animal that has been studied expresses multiple isotypes of both α and β.

Protists and fungi, however, are a more complex story. Among the fungi there have been organisms, such as *Candida* or *Histoplasma*, which express only a single α and a single β *(8)*. Others have a single β with multiple α. Interestingly, the converse pattern of a single α with multiple β has not been seen in fungi. Within the different phyla of fungi, all appear to contain species that express multiple α or multiple β isotypes, or both. Expression of a single α or single β is restricted to the ascomycetes and microsporidia (Table 2). In view of the pattern observed with plants and animals, one is tempted to conclude that multicellularity favors the existence of multiple isotypes of α- and β-tubulin.

Table 1
Isotypes of α-, β-, and γ-Tubulin: Animals[a]

Genus	Phylum/ division	Number of isotypes			Differences in expression?	References
		α	β	γ		
Homo (human)	Chordate	7	8	2	Yes	*8,76,181,185*
Macaca (Rhesus monkey)	Chordate	ND	≥6	ND	–	*69*
Sus (pig)	Chordate	≥4	≥2	ND	Yes	*3,4*
Bos (cow)	Chordate	ND	≥4	ND	Yes	*64*
Odocoileus (deer)	Chordate	ND	≥2	ND	–	*96*
Canis (dog)	Chordate	ND	≥4	≥1	–	*302,303*
Mus (mouse)	Chordate	6	7	2	–	*8,42,45,304*
Rattus (rat)	Chordate	≥3	≥4	≥1	–	*183,305–311*
Gallus (chicken)	Chordate	5	7	≥1	Yes	*8,63,173,312*
Xenopus (clawed frog)	Chordate	≥2	≥2	≥1	Yes	*8,43,313*
Notothenia (rockcod)	Chordate	≥8	≥4	ND	Yes	*8,288,314*
Chionodraco (icefish)	Chordate	≥4	≥2	ND	–	*288*
Gadus (Atlantic cod)	Chordate	ND	≥4	ND	Yes	*98,293*
Oncorhynchus (salmon)	Chordate	4	ND	ND	–	*8*
Salmo (trout)	Chordate	≥2	ND	ND	–	*8*
Torpedo (electric eel)	Chordate	≥2	ND	ND	Yes	*8*
Danio (zebrafish)	Chordate	2	ND	ND	–	*315*
Ictalurus (catfish)	Chordate	ND	≥2	ND	–	*97*
Mustelus (dogfish shark)	Chordate	ND	≥2	ND	–	*97*
Myxine (hagfish)	Chordate	2	ND	ND	–	*316*
Branchiostoma (lancelet)	Chordate	2	ND	ND	–	*316*
Halocynthia	Tunicata	ND	>2	ND	Yes	*17*
Ciona	Tunicata	3	ND	ND	–	*316*
Oikopleura	Tunicata	10	ND	ND	–	*317*
Paracentrotus (sea urchin)	Echinodermata	4	3	ND	–	*8,11*

(Continued)

Table 1 *(Continued)*

Genus	Phylum/ division	Number of isotypes			Differences in expression?	References
		α	β	γ		
Lytechinus (sea urchin)	Echinodermata	1	2	ND	–	*10*
Strongylocentrotus (sea urchin)	Echinodermata	3	≥1	2	–	*12,318*
Gecarcinus (land crab)	Arthropoda	>4	ND	ND	Yes	*319*
Homarus (lobster)	Arthropoda	2	≥1	ND	–	*316*
Heliothis (moth)	Arthropoda	ND	≥2	ND	Yes	*8*
Bombyx (moth)	Arthropoda	≥3	≥4	ND	Yes	*19*
Drosophila (fruit fly)	Arthropoda	4	3	2	Yes	*8,49,320*
Octopus	Mollusca	ND	≥2	ND	Yes	*8*
Aplysia (sea hare)	Mollusca	2	ND	ND	–	*321*
Hirudo (leech)	Annelida	2	ND	ND	–	*322*
Trichostrongylus	Nematoda	ND	≥2	ND	Yes	*209*
Caenorhabditis	Nematoda	4	3	≥1	Yes	*8,21,22, 208,323*
Cyathostomum	Nematoda	ND	≥3	ND	–	*324,325*
Cylicocyclus	Nematoda	ND	≥3	ND	–	*325,326*
Haemonchus	Nematoda	≥1	4	ND	–	*8,207*
Cooperia	Nematoda	ND	≥2	ND	–	*327*
Brugia	Nematoda	ND	≥2	ND	–	*8*
Gyrodactylus	Platyhelminthes	ND	3	ND	–	*328*
Echinococcus (tapeworm)	Platyhelminthes	ND	≥3	ND	–	*329*
Schmidtea	Platyhelminthes	>1	ND	ND	Yes	*20*
Schistosoma	Platyhelminthes	2	ND	ND	–	*8*

[a]The table gives either the actual number of isotypes or else states that there are at least that number. The symbol "≥" as in "≥4" means that there are 4 known isotypes but that there is a reasonable probability of more, based on information from closely related organisms. For more information, *see* ref. *8*, Table 1. For purposes of comparison, the isotypes of γ-tubulin, when known, are included in this table, although γ-tubulin will be discussed further in Chapter 7.

Various patterns of tubulin isotype expression are observed among the protists. Several, such as *Physarum* or *Trichomonas* express multiple isotypes of both α and β; some, such as *Euplotes* express a single α and multiple β; others, such as *Chlamydomonas* and *Plasmodium* have the reverse pattern. *Dictyostelium* expresses only one α and only one β. The widespread occurrence of multiple isotypes among the protists may reflect the complex cellular architecture of some of these organisms.

The knowledge of isotypes of γ-tubulin is still in its infancy, but it is clear that these occur. Multiple γ isotypes have been observed among the animals, plants, and protists, but not among the fungi. γ-tubulin, which is thought to nucleate microtubules, is found

Table 2
Isotypes of α-, β- and γ-Tubulin: Plants and Fungi[a]

Genus	Phylum/ division	Number of isotypes			Differences in expression?	References
		α	β	γ		
Plants						
Daucus (carrot)	Angiosperm	≥1	>4	ND	Yes	24
Pisum (pea)	Angiosperm	ND	3	ND	–	8
Glycine (soybean)	Angiosperm	ND	3	ND	Yes	8,47
Solanum (potato)	Angiosperm	ND	≥2	ND	–	8
Eucalyptus	Angiosperm	>1	ND	ND	–	8
Zinnia	Angiosperm	ND	≥3	ND	Yes	8
Prunus (plum)	Angiosperm	>1	ND	ND	Yes	8
Oryza (rice)	Angiosperm	3	3	≥1	Yes	8,330,331
Triticum (wheat)	Angiosperm	>1 6	ND	Yes	–	25,332
Arabidopsis (cress)	Angiosperm	4	8	2	Yes	8,333
Nicotiana (tobacco)	Angiosperm	2	5	≥1	Yes	334–338
Hordeum (barley)	Angiosperm	5	≥3	≥1	Yes	28,339
Gossypium (cotton)	Angiosperm	≥5	≥6	ND	–	26,340
Lupinus (lupine)	Angiosperm	ND	≥2	1	Yes	8,341
Populus (aspen)	Angiosperm	3	ND	ND	–	342,343
Cosmos (sunflower)	Angiosperm	≥2	ND	ND	Yes	44
Zea (corn)	Angiosperm	≥6	8	2	Yes	8,344,345
Eleusine (goosegrass)	Angiosperm	≥3	≥4	ND	–	346,347
Miscanthus	Angiosperm	8	ND	ND	–	348
Anemia (fern)	Angiosperm	2	2	1	–	349,350
Physcomitrella (moss)	Bryophyta	2	5	1	–	351–353
Fungi						
Histoplasma	Ascomycota	1	1	ND	–	8
Aspergillus	Ascomycota	2	2	1	Yes	8
Colletotrichum	Ascomycota	2	2	ND	Yes	8,33,354
Candida	Ascomycota	1	1	1	–	8,355,356
Neurospora	Ascomycota	2	1	1	Yes	8,35,357
Trichoderma	Ascomycota	ND	≥3	ND	–	8,358
Hypocrea	Ascomycota	ND	2	ND	–	8
Paracoccidioides	Ascomycota	2	ND	ND	Yes	36
Botryotinia	Ascomycota	ND	1	ND	–	359
Erysiphe (grass mildew)	Ascomycota	ND	1	ND	–	8
Epichloe	Ascomycota	ND	1	ND	–	8
Saccharomyces	Ascomycota	2	1	1	No	8,213
Schizosaccha-romyces	Ascomycota	2	1	1	No	8,313
Pneumocystis	Ascomycota	1	1	ND	–	8,360
Geotrichum	Ascomycota	ND	2	ND	–	8
Conidiobolus	Zygomycota	2	≥1	ND	–	361,362
Rhizopus	Zygomycota	3	3	ND	–	361,362
Basidiobolus	Zygomycota	2	2	ND	–	362

(Continued)

Table 2 (Continued)

Genus	Phylum/ division	Number of isotypes			Differences in expression?	References
		α	β	γ		
Spiromyces	Zygomycota	ND	2	ND	–	362
Mortierella	Zygomycota	ND	2	ND	–	362
Powellomyces	Chytridiomycota	3	ND	ND	–	361,362
Allomyces	Chytridiomycota	ND	2	ND	–	363
Spizellomyces	Chytridiomycota	ND	2	ND	–	361
Harpochytrium	Chytridiomycota	ND	2	ND	–	361
Glomus	Glomeromycota	ND	2	ND	–	364
Suillus	Basidiomycota	ND	2	ND	–	365
Cryptococcus	Basidiomycota	ND	2	ND	No	366
Encephalitozoon	Microsporidia	1	1	1	–	8,367,368

[a]See explanation under Table 1.
For more information, *see* ref. *8*, Table 2.

Table 3
Isotypes of α-, β-, and γ-Tubulin: Protists[a]

Genus	Phylum/ division	Number of isotypes			Differences in expression?	References
		α	β	γ		
Cryptosporidium	Apicomplexa	ND	1	1	–	8,369,370
Toxoplasma	Apicomplexa	3	3	ND	–	8,38
Babesia	Apicomplexa	ND	1	ND	–	8
Plasmodium	Apicomplexa	2	1	1	–	8,371
Eimeria	Apicomplexa	ND	1	ND	–	8
Physarum (slime mold)	Mycetozoa	3	4	2	Yes	8,372
Chloromonas (snow alga)	Chlorophyta	2	ND	ND	–	373
Chlamydomonas	Chlorophyta	2	1	1	–	8,374
Polytomella	Chlorophyta	ND	2	ND	–	8
Volvox	Chlorophyta	ND	2	ND	–	8,375
Paramecium	Ciliophora	2	1	1	–	8,376
Tetrahymena	Ciliophora	1	2	1	–	8,377
Stylonichia	Ciliophora	2	1	ND	–	8
Euplotes	Ciliophora	1	4	2	–	378,379
Moneuplotes	Ciliophora	3	≥1	2	–	8,380,381
Histriculus	Ciliophora	1	ND	ND	–	382
Moneuplotes	Ciliophora	5	ND	ND	–	381
Tintinnopsis	Ciliophora	2	ND	ND	–	383
Strobilidium	Ciliophora	3	ND	ND	–	383
Metacylis	Ciliophora	2	ND	ND	–	383
Laboea	Ciliophora	3	ND	ND	–	383
Strombidinopsis	Ciliophora	6	ND	ND	–	383
Favella	Ciliophora	2	ND	ND	–	383
Opisthonecta	Ciliophora	2	ND	ND	–	384

(Continued)

Table 3 *(Continued)*

Genus	Phylum/ division	Number of isotypes			Differences in expression?	References
		α	β	γ		
Halteria	Ciliophora	6	ND	ND	–	*381*
Metopus	Ciliophora	3	ND	ND	–	*381*
Heliophrya	Ciliophora	3	ND	ND	–	*381*
Nyctotherus	Ciliophora	2	ND	ND	–	*381*
Dictyostelium	Acrasiomycota	1	1	1	–	*8,38*
Ectocarpus	Phaeophyta	ND	2	ND	–	*8*
Chondrus	Rhodophyta	ND	2	ND	–	*8*
Achlya	Oomycota	ND	1	ND	–	*8*
Amphidinium (dinoflagellate)	Dynophyceae	2	ND	ND	–	*386*
Reticulomyxa	Rhizopoda	2	2	1	–	*8,387*
Naegleria	Heterolobosea	≥4	>1	ND	Yes	*8,37*
Leishmania	Euglenozoa	≥1	2	1	Yes	*8,388,389*
Trypanosoma	Euglenozoa	1	1	1	–	*390,391*
Trichomonas	Parabasalidea	2	≥3	ND	Yes	*8,392*
Tritrichomonas	Parabasalidea	≥2	ND	ND	–	*393*
Trichonympha	Parabasalidea	2	≥1	ND	–	*394*
Hypotrichomonas	Parabasalidea	2	3	ND	–	*395*
Monocercomonas	Parabasalidea	ND	2	ND	–	*396*
Pelvetia (brown alga)	Phaeophyceae	≥2	ND	ND	–	*397*
Bigelowiella	Cercozoa	3	3	ND	–	*398*
Pyrsonympha	Oxymonadida	2	ND	ND	–	*394*
Streblomastix	Oxymonadida	5	1	ND	–	*399*

^aSee explanation under Table 1.
For more information, *see* ref. *8*, Table 3.

in centrosomes as well as other microtubule organelles. It is conceivable that fungi, which lack centrosomes, may not require more than a single isotype of γ-tubulin.

2.2. Tissue, Cellular, and Subcellular Distribution

What functions do isotypes serve? Why have the differences among isotypes in groups such as the vertebrates been so widely conserved? The fact of this conservation argues that the differences must matter, but it does not prove it. One could argue that all isotypes are completely interchangeable functionally and that there is a space of certain amino acid sequences that are compatible with function. Evolution has randomly filled at least part of this space. In other words, conceivably, mammalian βI, for example, could accept certain mutations and still assemble into a microtubule that would perform all microtubule-mediated functions; mammalian βIII could do likewise. However, βI could not mutate into βIII, because the intermediate forms would not be viable. This would help to account for the preservation of isotype differences in evolution. What about the fact that isotypes often differ in their tissue distribution? One could further argue that when a certain tissue differentiates, a cassette of genes is expressed that happens to include one particular isotype and not another. The fact that certain isotypes

Table 4
Tubulin Isotypes in the Inner Ear of the Gerbil[a]

Cell	βI	βII	βIII	βIV
Cochlea (adult)				
Outer hair cell	+	–	–	+
Inner hair cell	+	+	–	–
Outer pillar cell	–	+	–	+
Inner pillar cell	–	+	–	+
Deiters cell	+	+	–	+
Schwann cell	+	?	?	?
Neurons	+	+	+	?
Afferent dendrites	–	–	–	–
Cochlea (developing)				
Outer hair cell	+	+	–	+
Inner hair cell	+	+	–	+
Outer pillar cell	+	+	–	+
Inner pillar cell	+	+	–	+
Deiters cell	+	+	–	+
Afferent dendrites	+	+	+	–
Vestibular organ (adult)				
Type I hair cell	+	–	–	+
Type II hair cell	+	–	–	+
Supporting cell	+	+	–	+
Schwann cell	+	?	?	?
Neurons				
Axons, soma	+	+	+	–
Dendrites	+	+	+	–
Calyx	–	–	+	–

Source: Adapted from refs. 13–15.
[a]Absence of signal could indicate either that the isotype as not present in the tissue or that extensive post-translational modification made it undetectable to the antibody.

have their expression regulated by particular factors is consistent with this model, as will be discussed later. The result would be tissues expressing different isotypes. Each isotype would be participating in certain generic processes such as mitosis as well as tissue-specific processes such as secretion in the liver or axonemal motility in tracheal epithelia. However, by this model the isotypes would be interchangeable. In other words, if the liver isotype were to be expressed in the tracheal epithelia and not in the liver and conversely for the tracheal isotype, processes such as secretion and axonemal motility would not be compromised. The ideal way to prove that the structural differences among isotypes have functional significance is to demonstrate that the isotypes are not functionally interchangeable. As will be seen later, this has been done in a few cases. In addition, there is a great deal of other evidence that supports the hypothesis that isotype differences are functionally significant.

As just discussed, the differences in tissue distribution among tubulin isotypes do not constitute definitive evidence that the isotype differences are functionally significant. Nevertheless, in many cases, the distribution of isotypes among tissues and even among different cell types in the same tissue is extremely complex. Table 4 shows the distribution

Table 5
Distribution of Tubulin Isotypes in Maize[a]

Cell/tissue	β1	β2	α1	α3	α5
Seedling root tip cells	+	–	+	+	–
Seedling leaf epidermis	+?	–	+	+	ND
Male meiocytes	+	+	+	+	ND
Pollen tubes					
Axial microtubules	–	–	+	–	–
Microtubules associated with either vegetative nuclei or sperm cells	–	–	+	+?	+?

[a]*Source:* Ref. 400.

of isotypes in different cells of the inner ear. It is striking that adjacent cells can have different isotype compositions, even in cells that perform similar although not identical functions such as the inner and outer hair cells of the cochlea *(13)*. In addition, the pattern of isotypes changes during development. For example, the inner and outer hair cells express the same set of β isotypes (βI, βII, and βIV) in early development and then the outer hair cells stop making βII, whereas the inner hair cells stop expressing βIV *(14)*. It is difficult to ascribe this complexity to different cassettes of genes.

In certain cases, isotype distributions appear to differ within the same cell. For example, some of the neurons of the gerbil vestibular organ have a portion, called the calyx, which is like a cup enveloping the adjacent hair cell. Although the rest of the neuron contains βI, βII, and βIII, only βIII occurs in the calyx (Table 4) *(15)*. If the cell is unable to discriminate among the isotypes, how is it able to arrange that βI and βII, but not βIII, be restricted from entering the calyx? If the cell is able to distinguish the isotypes from each other, then it is easy to imagine that the different isotypes can perform different functions. Nevertheless, there is still one way out of the dilemma posed by the vestibular neurons, a way that would still allow us to maintain the functional interchangeability of isotypes. A highly elaborate series of temporally regulated cassettes of genes can be posited, such that only βIII is expressed when the calyx is forming, whereas all three isotypes are expressed before that time.

Isotype distributions are also complex in other animals. This complexity has been seen in the frog *Rana (16)*, the tunicate *Halocynthia (17)*, sea urchins *(18)*, the fruit fly *Drosophila*, the moths *Heliothis* and *Bombyx (19)*, the mollusc *Patella*, the nematode *Brugia*, and even the platyhelminth *Schmidtea (20)* (reviewed in refs. *8,9*).

An intriguing example of isotype distributions has been observed in the nematode *Caenorhabditis elegans*. These organisms contain touch receptor neurons whose microtubules are made up of 15 protofilaments, instead of the more usual 11 protofilaments as are the other microtubules of *C. elegans*. The tubulin dimers that constitute these "giant" microtubules consist of a unique α and a unique β isotype *(21,22)*. As will be argued later for the mammalian βVI isotype, it is possible that microtubules of unique morphology require unique tubulin isotypes. Of the other isotypes in *C. elegans*, some interchangeability has been observed, but one β isotype is required for centrosomes to be stable *(23)*.

Isotype distributions are complex in plants as well as shown for maize in Table 5. Similar complex tissue distributions of plant tubulin isotypes have been found in *Arabidopsis*, soybean, carrot *(24)*, wheat *(25)*, tobacco, and plum (reviewed in refs. *8,9*).

More recently, complex distributions of isotypes have been reported in cotton *(26)*, rice *(27)*, and barley *(28)*. Interestingly, one of the rice β-tubulin genes encodes three different mRNA species, thereby creating even more isotypes *(29)*, a rare example of tubulin isotypes arising by alternative splicing.

The relative levels of plant tubulin isotypes appear to be controlled by hormones such as gibberellin *(27)* as well as by factors that selectively degrade the mRNAs for particular isotypes *(30)*. The story of barley is probably typical. Schröder et al. *(28)* did not attempt to study the entire set of tubulin isotypes, but only the five α isotypes in the leaf. They found that α3 was probably constitutive, being expressed at each stage of leaf development. α2 and α4 were found largely in meristematic cells, declining during later stages of differentiation (α2 declined more rapidly than α4). α1 and α5, however, appeared very transiently only in the rapidly growing cells; these cells contain microtubule bundles that determine the later morphology of the leaf cells *(31)*. This work teaches a valuable lesson in indicating that an important tubulin isotype can appear, do its job, and then quickly disappear, and hence may escape detection in experiments. Mutants of two specific isotypes of *Arabidopsis* α-tubulin altered the growth pattern of the hypocotyls and the pattern of microtubules in the root *(32)*, as one would expect given the different tissue distributions of the *Arabidposis* isotypes.

Fungi are simpler than plants and animals. Nevertheless, differences in tubulin isotype expression have been observed in *Colletotrichum (33)*, *Aspergillus*, *Fusarium (34)*, *Neurospora (35)*, and *Paracoccidioides (36)* (reviewed in refs. *8,9*). In some cases, these organisms have one isotype that is high during the vegetative phase and one during conidiation *(33,34)*.

Differences in expression are occasionally seen in protists, even though they are single-celled organisms. This is sometimes the case in different stages of the life cycle and has been observed in *Plasmodium*, *Leishmania*, *Physarum*, *Naegleria (37)*, and *Toxoplasma (38)* (reviewed in ref. *8*). In the case of *Physarum*, for example, of the three stages of its life cycle—amoeba, plasmodium, and flagellate—a different β isotype predominates at each stage, whereas the α isotypes differ as well, but not so strikingly *(39,40)*.

One oddity of tubulin distribution is that when an organism with multiple isotypes has an α or a β isotype of unusual sequence that isotype is often associated with the reproductive system. In *Drosophila*, the α4 isotype is only 67% identical to the other three α isotypes; it is uniquely expressed in the oocyte and the early embryo *(41)*. The mouse αTT1 isotype, sharing about 70% identity to the other α, is expressed only in the testis *(42)*. *Xenopus* has an unusual α expressed in the ovary *(43)*. The platyhelminth *Schmidtea* has a highly divergent α expressed only in the testis *(20)*. Sunflower pollen has a unique α-tubulin, much more basic than other α. It is even thought to have a different tertiary structure with an altered H1/B2 loop facing into the interior of the microtubule *(44)*. The fungus *Colletotrichum* has a divergent β expressed only in its conidia *(33)*. The protist *Naegleria* expresses three α isotypes, one of which is only 61.9% identical to the other two α; the unusual α is not expressed in the flagellate, but only in the dividing amoeba, where it is found in the spindle *(37)*. It is hard to account for these divergent isotypes being restricted to the reproductive tissues. If the divergent isotype in one organism had a striking resemblance to the corresponding isotype in another, one could argue that the isotypes shared a particular structural feature that is necessary to perform a certain function related to reproduction; formation of the meiotic spindle would be a tempting candidate. However, the divergent isotypes not only do not resemble each other, they can occur in either male or female reproductive organs. As most of these divergent isotypes are α, it may be that there is a particular function carried out by the β-subunit in, say, meiosis and that this function does not involve α at all. Perhaps the α- and β-subunits in reproductive cells are expressed in the same cassette of genes. If only β is performing a stringent function, one could then argue that this situation leaves α free to diverge significantly in the course of evolution.

Not all of the highly divergent isotypes occur in reproductive tissues, however. Mammals and birds have a very divergent β isotype whose expression is restricted to haematopoietic tissues, including erythrocytes and platelets (45,46). This will be discussed later in more detail. The soybean produces a divergent β isotype; low levels of this isotype are expressed in the cotyledon, and high levels in the hypocotyl, when the soybean is grown in the absence of light (47). It may be that an isotype that performs only a single function is more likely to diverge in the course of evolution than one that is involved in a large number of processes.

3. FUNCTIONS OF TUBULIN ISOTYPES

3.1. Tubulin Isotypes in **Drosophila**

The most unambiguous demonstration of isotype-specific functions comes from a series of experiments done in the fruit fly *Drosophila*. Early experiments showed that mutation of the testis-specific β2 isotype caused inability to form axonemes of normal morphology and function (48–51). Similar results were obtained when β2 was replaced by the divergent β3 isotype; meiosis was blocked as well (52). It is interesting that loss of β2 blocks meiosis but not mitosis. This observation may be connected with the fact that in *Drosophila*, the meiotic spindle is surrounded by a membranous structure (53). Conceivably, the β2 isotype may play a role in interactions of the meiotic spindle with that membrane.

Alterations in the β3 isotype also result in specific changes. This isotype appears for a short time during embryogenesis. Mutants of β3 have poor sensory perception. Microtubules in the chordotonal sensory organ are more highly crosslinked in the mutant than in the wild-type. The authors suggest that increased crosslinking may inhibit flexibility during development leading to impaired function later on (54). Perhaps β3 has a smaller propensity to form crosslinks. In addition, β3 expression correlates with muscle development whereas β1 expression is induced by attachment to the epidermis (55).

The α-tubulin isotypes of *Drosophila* also appear to have specific functions. Komma and Endow (56) showed that the α67C isotype binds to the motor protein Ncd whereas the α84B isotype does not. Mutations in α67C alter meiosis I and decrease the accuracy of chromosome segregation (57). Hutchens et al. (58) found that replacement of the α84B with the very similar (98% identical) α85E led to synthesis of abnormal axonemes, often lacking the central pair microtubules as well as the outer singlet, or accessory, microtubules, characteristic of insect sperm flagella.

3.2. Mammalian Tubulin Isotypes: the β-Isotypes

In addition to *Drosophila*, a good deal is now known, or at least hypothesized, about the functional assignments of the β-tubulin isotypes in mammals. This will be reviewed later. As will be seen, the functional significance of some of the isotypes is fairly certain, others are speculative, and some are completely unknown.

Table 6
Vertebrate β-Tubulin Isotypes[a]

Designation	Species	C-terminal sequence	Distribution
Class Ia	Human	YQDATAEEEEDFGEEAEEEA	Widespread
	Mouse	YQDATAEEEEDFGEEAEEEA	
	Rat	YQDATAEEEEDFGEEAEEEA	
	Chicken	YQDATAEEEEDFGEEAEEEA	
Class Ib	Human	YQDATAEEEEDFGEEAEEEA	Retina
Class II	Human	YQDATADEQGEFEEEEGEDEA	Brain, muscle, and so on
	Mouse	YQDATADEQGEFEEEEGEDEA	
	Rat	YQDATADEQGEFEEEEGEDEA	
	Chicken	YQDATADEQGEFEEEGEDEA	
	Gadus	YQDATADEEGEFDEEAEEDG	
	Notothenia	YQDATAEEEGEFEEEGEYEDGA	
Class III	Human	YQDATAEEEGEMYEDDEEESEAQGPK	Neurons, Sertoli, and so on
	Rat	YQDATAEEEGEMYEDDDEESERQGPK	
	Chicken	YQDATAEEEGEMYEDDEEESEQGAK	
	Xenopus	YQDATAEEEGEMYEDDEEESEGQGK	
	Gadus	YQDATAEEEENFDEEADEEIA	
Class IVa	Human	YQDATAEQGEFEEEAEEEVA	Brain
	Mouse	YQDATAEEGEFEEEAEEEVA	
Class IVb	Human	YQDATAEEEGEFEEEAEEEVA	Widespread, esp. in ciliated tissues, sperm
	Mouse	YQDATAEEGEFEEEAEEEVA	
	Rat	YQDATAEEGEFEEEAEEEVA	
	Chicken	YQDATAEEEGEFEEEAEEEAE	
	Gadus	YQDATAEEEGEFEEEGEEELA	
	Notothenia	YQDATAEEEGEFEEEGEEDLA	
Class V	Human	YQDATANDGEEAFEDEEEEIDG	Unknown
	Mouse	YQDATVNDGEEAFEDEDEEEINE	
	Chicken	YQEATANDGFEAFEDDEEEINE	
	Xenopus	YQEATANDEEEAFEEDEEEVNE	
Class VI	Human	FQDAKAVLEEDEEVTEEAE MEPEDKGH	Platelets, bone marrow
	Mouse	FQDVRAGLEDSEEDAEEAEV EAEDKDH	
	Chicken	YQDATADVEEAEASPEKET	
Class VII	Human	YQDATAEGEGV	Unknown
Unclassified	*Notothenia*	YQDATADEMGEYEEDEIEDE EEVRHDVRH	

[a]*Source:* From refs. *45,63,67,69,181,182,217,307–310,401–403.*
The chicken has two forms of βII, differing from each other at 2 out of 445 positions *(63).*

3.2.1. βI

The βI isotype appears to be the most widespread among mammalian tissues (Table 6). It has been seen in almost every tissue that has been examined *(8,59)*. It is also found in many avian tissues *(60)*. It is highly conserved in evolution: although the avian and mammalian lines diverged 310 million years ago (mya) *(61)*, chicken and mouse βI are identical in all 444 residues *(45,62,63)*. The relative amounts of βI in different tissues are very variable. In cow brains βI constitutes about 3–4% of the total β-tubulin *(64)*; by contrast, in the thymus βI appears to be the major β isotype *(60)*. In fact, thymus tubulin was used as the positive control in the selection of the monoclonal antibody to βI *(59)*. However, βI is probably not a constitutive tubulin. In the gerbil cochlea, for example, βI is expressed in hair cells but not in pillar cells *(13)*. Also, follicle-stimulating hormone induces expression of βI in rat granulosa cells *(65)*, suggesting that it may be performing a specific function, although one could argue that the hormone is merely stimulating cell proliferation, which would in turn require microtubule assembly. βI is clearly not constitutive in zebrafish, where its expression is limited to the nervous system throughout development, and in the adult brain is restricted to the regions where proliferation is occurring *(66)*. Higher primates appear to have two very similar forms of βI, but they are unlikely to differ in function *(67–69)*. Mice and chickens have only a single βI.

What might be the role of βI? Narishige et al. *(70)* found that cardiac hypertrophy is accompanied by increased βI and βII. They speculated that βI may play a role in increasing microtubule stability. There is some evidence indicating specific roles for βI. First, it is found in a variety of mammalian cilia, including those of nasal epithelia, tracheal epithelia, vestibular epithelia, and oviduct epithelia *(15,71,72)*. Traces of βI have been observed in mouse sperm as well *(9)*. As will be discussed further later, the major constituent of ciliary and flagellar axonemes is βIV, which has the signal sequence (EGE-FEEE) proposed by Raff et al. *(73)* to be a requirement for a β-tubulin to be incorporated into axonemes. However, although βI lacks that signal sequence, it is conceivable that the signal sequence requirement does not apply to all of the microtubules in the axoneme. Certainly, the structure of axonemal microtubules is sufficiently complicated that it is easy to visualize that there are more than enough functions to be distributed among two isotypes. For example, one could speculate that βI could form one or both of the central pair microtubules or the B-tubules of the outer doublets.

The clearest evidence for a specific function for βI was obtained by Lezama et al. *(74)* MDCK cells, βI was relatively depleted in the cortical regions of MDCK cells, an area that is rich in actin filaments. They also observed that overexpression of βI tubulin in MDCK cells and incorporation of exogenous βI tubulin into microtubules interferes with adhesion and spreading. They suggest that βI may interfere with the actin–tubulin interaction. Very recently another possible function for βI was suggested. Yanagida et al *(75)* found that human fibrillarin forms a complex with the α3 and βI isotypes of tubulin. Fibrillarin is involved in ribosome assembly and processing of rRNA *(75)*. The specific role of tubulin in this process is unknown.

3.2.2. βII

The brain is the source of the tubulin used in the vast majority of experimentation in vitro. As βII constitutes 58% of the total β-tubulin in bovine brain *(64)*, one could say that βII is the best studied of the tubulin isotypes. For this reason, it is highly ironic that so little is known about βII specific function. However, as βII is highly conserved in evolution, it probably has a particular role to play. βII has a considerably more restricted distribution than does βI. βII is prominent in the brain, where it is expressed in both neurons and glia. βII is also found in skeletal and smooth muscle and in connective tissue *(76)*. It is found in the breast, adrenal, and testis as well *(77,78)*. In other tissues where βII occurs, it is more likely to be restricted to a single cell type than is βI. For example, in the skin, where βI is expressed in each of the three layers of the stratum malpighii, βII is concentrated in only one of these layers, the stratum granulosum *(59)*.

βII is more widespread in early development. In fetal rats, not only does βII occur in muscles, nerves, and connective tissue but also in the retina, chondrocytes, and endothelial cells *(77)*. Not surprisingly, βII also is found in neural stem cells *(79)*. Unlike βI and βIV, βII is generally not associated with axonemal microtubules except for those of the cilia of olfactory epithelia *(71)*. The significance of this finding is uncertain. An immunogold electron microscopic study of axonemes in retinal and tracheal cilia showed that βII was present near the axonemes but did not form part of their microtubules, unlike βIV, which was clearly incorporated into the axonemal microtubules *(80)*. βII, thus, is probably not adapted to function in axonemal microtubules. One study in HeLa interphase cells found that βII was concentrated in the perinuclear region and the periphery of these cells. Cold treatment (which causes microtubules to break up) resulted in βII being associated with the centrosome and the cell periphery; nocodazole treatment had the same effect *(81)*. This finding raises the possibility that βII may play a role in anchoring microtubules to the centrosome and the cell periphery.

A highly unusual property of βII has recently been discovered. Ranganathan et al. *(82)* observed that βII, but not βI, βIII, or βIV, occurred in the cell nuclei of prostate tumors and benign prostate hyperplasia. A later study showed that βII was present in the nuclei of cultured rat kidney mesangial cells in interphase in nonmicrotubule form *(83)*. This will be discussed later on. The possibility will be raised that βII may play a role in organizing the nuclear membrane during mitosis. Even if βII has a function involving the cell nucleus and mitosis, however, this does not seem sufficient to explain its very high concentration in neurons, which undergo little or no cell division and, which appear to have a very high ratio of cytoplasm to nucleus. A similar argument would apply to muscles, which are also rich in βII *(76)*, although it is perhaps relevant that in muscle, microtubules are nucleated by the nuclear membrane rather than by the centrosome *(84)*. In nerves and muscles, βII probably has other functions, totally unrelated to mitosis, but what these functions may be is a complete mystery.

3.2.3. βIII

3.2.3.1. Unusual Characteristics of βIII. The βIII isotype has six distinguishing characteristics, each of which is probably relevant to developing an understanding of its functional significance.

1. *βIII is highly conserved in evolution.* As is the case with βI, there are only two differences in the amino acid sequences of chicken and human βIII *(85,86)*.
2. *βIII has a highly unusual distribution of cysteines.* All the vertebrate β isotypes have cysteines at positions 12, 127, 129, 201, 211, 303, and 354. The more widely distributed β isotypes—βI, βII, and βIV—also have a cysteine at position 239. βIII lacks this cysteine but has a cysteine at position 124 instead, where βI, βII, and βIV have a serine. The significance of these cysteines will be discussed later.
3. *βIII has an extremely narrow distribution in normal adult tissues.* It is the most abundant in the brain, where it is found only in neurons and not in glial cells (by contrast, βII is found in both) *(87)*. Its absence from glial cells has made βIII a useful marker for neuronal differentiation *(88,89)*. βIII synthesis can be induced by factors such as androgens *(90)*, STEF *(91)*, and nerve growth factor *(92)*. The latter, when combined with retinoic acid, can cause human umbilical cord blood cells to synthesize βIII as well as

other neuronal proteins *(92)*. βIII also occurs in Sertoli cells and, in small amounts, in the vestibular organ, the nasal epithelia, and the colon *(93)*. In other adult tissues that have been examined, βIII appears to be absent. However, βIII is found in a large number of cancers and is also widespread in some developing tissues.

4. *When tubulin is reduced and carboxymethylated, βIII has a unique electrophoretic mobility on polyacrylamide gels in the system of Laemmli (94,95)*. This feature has made it easy to measure its levels in the brains of different vertebrates. βIII accounts for 25% of the total β-tubulin in the brains of cows and 20% in deer brains *(96)*. The fact that, unlike the more abundant βII, βIII occurs only in neurons and not in glial cells, however, suggests that the relative amount of βIII in neurons must be very high indeed. This is consistent with the observation that βIII accounts for $25.7 \pm 0.7\%$ of the total β in bovine cerebral gray matter and only $20.7 \pm 0.5\%$ in white matter, the latter being enriched in glial cells *(135)*. In the brains of chickens, dogfish shark, and catfish, βIII accounts, respectively, for 14%, 8–17% and 10%, of the total β *(96,97)*. Interestingly, in a cold-adapted fish, the Atlantic cod *Gadus morhua*, βIII accounts for 30% of the total β tubulin *(98)*. However, in other cold-adapted fishes, the Antarctic cod *Notothenia* and the Antarctic icefish *Chaenocephalus*, βIII accounts for 8–12% and 4%, respectively *(97)*.

5. *βIII is phosphorylated at a serine near the C-terminus (99)*. Except for βVI, the other vertebrate β isotypes have no serines in this region and thus cannot be phosphorylated here.

6. *The dynamic behavior* in vitro *of microtubules made of the αβIII dimer is higher than that of microtubules made of either the αβII or αβIV dimers (100)*.

3.2.3.2. βIII is likely to be less sensitive to reactive oxygen species (ROS) and free radicals. Let these observations be put together to see if they point to a specific functional role for βIII. The unusual cysteine distribution is a good place to begin. It has long been known that microtubule assembly in vitro and in vivo is exquisitely sensitive to sulfhydryl-oxidizing agents *(101)*. Cys239 in β is very reactive and its oxidation inhibits assembly *(102,103)*. In other words, a tubulin molecule oxidized at cys239 cannot assemble onto a microtubule *(104)*. βIII lacks cys239 and has ser239 instead; βV and βVI also have ser239. It has been shown that the αβIII and αβVI dimers are significantly less reactive with alkylating agents than are the other isotypes and that the polymerization of αβVI is less inhibited by alkylation *(96,105)*. It must be emphasized that the presence of a serine at position 239 is highly unusual among tubulins. Outside of βIII, βV, and βVI, every other animal β-tubulin contains cys239 *(8)*. Also, almost every plant and protist β-tubulin has a cysteine at either position 239 or 238 or both. Fungal β-tubulins are virtually the only ones without a cysteine in this area. βIII also contains a cysteine at position 124 and this is even more unusual. Except for βV and avian βVI, there is no β-tubulin in any eukaryote with a cysteine at position 124. The fact that cys124 and ser239 are both highly conserved in the evolution of βIII and highly unusual in the universe of β-tubulins strongly indicates that these particular residues must play a major role in the function of βIII.

It is probably not a coincidence that cys124 is very close to the highly conserved cys127 and cys129. Most β-tubulins have cysteines at these positions. These three cysteines (124, 127, and 129) constitute a cysteine cluster. A cysteine cluster of identical topography occurs in Von Willebrand's protein, a giant serum protein that promotes blood coagulation. In Von Willebrand's protein, the cysteine cluster is the site of interchain disulfide bonds *(106)*. Von Willebrand's protein also contains sets of vicinal cysteines (with two residues between the cysteines)—as with cys124 and cys127 in βIII—that appear to undergo sulfhydryl-disulfide interchanges during polymerization *(107)*. It is conceivable that such an interchange occurs when the αβIII dimer polymerizes. On the other hand, it is possible that an intrachain disulfide forms in response to oxidation; This possibility will be pursued later. Although it is generally thought that disulfide bridges in proteins cannot form in the cytosol, recent evidence indicates that the SV40 protein Vp1 forms transitory intrachain and interchain disulfides whereas folding and oligomerizing in the cytoplasm. The mature virus has no disulfides of any kind *(108)*.

The absence of a cysteine at position 239 in βIII is probably very telling, especially when one considers the effects of nitric oxide (NO) and ROS on microtubules. ROS are generated by mitochondria and can also be found in the diet. These species can react with sulfhydryl groups. In view of the overall high reactivity of the sulfhydryl group of cys239, it is easy to imagine it reacting with ROS. In addition, certain tissues synthesize NO, which is itself a free radical capable of reacting with sulfhydryl groups. One ROS, O_2^- (superoxide anion), reacts with NO to make peroxynitrite (ONOO$^-$) *(109)*. ONOO$^-$ in turn reacts with tubulin to form disulfide bridges between the α- and β-subunits *(109,110)*. Cys239, which is close to the α/β interface *(111)*, probably participates in this disulfide, which inhibits microtubule assembly *(112)*. Lacking this cysteine, the assembly of βIII would not be inhibited. Thus, βIII is likely to be less sensitive to free radicals.

3.2.3.3. βIII is most likely to occur in tissues and tumors with elevated levels of ROS and free radicals.

Is a protective role of βIII consistent with its observed distribution? As mentioned earlier, βIII is highly concentrated in neurons. The neuronal isozyme of nitric oxide synthase (neuronal NOS [nNOS]) is elevated in the brain *(113)*. NO is produced by neurons, particularly at the synapses *(114–116)*. Although NO has not been shown to react directly with tubulin sulfhydryls, it does react with the microtubule-associated protein (MAPs) tau and it has been proposed that NO could thus play a regulatory role in neuronal differentiation *(117)*. Although there is no reason to imagine that ROS are especially high in neurons, it must be remembered that adult neurons rarely, if ever, reproduce. The turnover time of neuronal tubulin is unknown, but it is probably very long, perhaps in the range of weeks or months. This long turnover time of tubulin would give sufficient time for even a low concentration of ROS to react with tubulin and damage the microtubules. Thus, there is a clear advantage for neurons to form their microtubules from a tubulin isotype less likely to react with ROS, NO, or ONOO$^-$. Incidentally, nNOS is also elevated in muscles, which lack βIII. However, muscles appear to have little need for microtubules so the high NO may not be a problem there.

βIII is elevated in Sertoli cells *(118)*. These cells produce NO and are very rich in the inducible isotype of NOS (inducible NOS [iNOS]) *(119,120)*. The rest of the testis has very little iNOS *(120)*. Sertoli cells also have high levels of the enzyme superoxide dismutase, which they also secrete. This indicates that Sertoli cells operate in an environment rich in free radicals *(121)*. The model would predict, therefore, that Sertoli cells would be rich in βIII. βIII has also been seen in the vestibular organ of the gerbil inner ear. In this organ, which is responsible for balance, the βIII is concentrated in the calyx, a neuronal extension that cups the bottom ends of the hair cells; the dendrites, soma and axons of these particular neurons contain βI and βII in addition to βIII, but the calyx has only βIII *(15)*. It would seem, therefore, that βIII has some particular function in this region. It is probably not coincidental that vestibular neurons as well as the hair cells produce both NO and ROS

(122–125). Small amounts of βIII are present in the colon *(59)* and the nasal epithelium, where βIII is even found in the cilia *(71)*. Although these tissues are not known to produce ROS, it is possible that the colon and the nasal epithelia would be exposed to free radicals in the food we eat and the air we breathe. Interestingly, however, βIII, although present in fetal lung, is absent in adult lung *(126)*. Nevertheless, it would appear that the normal distribution of βIII in adult mammalian tissues is generally consistent with its playing a role in protecting microtubules from oxidation by NO, ROS, or ONOO⁻.

The presence of βIII in tumors is consistent with this model as well. Many tumors express βIII, including some of nonneuronal origin such as lymphomas *(127–135)*. This has been reviewed by Katsetos et al. *(136,137)*. Tubulin is the target for some of the most successful antitumor drugs such as the taxanes and *Vinca* alkaloids *(138,139)*. Hence, one could argue that cancer cells rely heavily on microtubules. Cancers generally function under oxidative stress, in which the ratio of ROS to antioxidants is abnormally high *(140–144)*. Cancer cells therefore need protection from the same ROS that may have helped to create the cancer in the first place. There is thus a selective advantage for cancer cells to make their microtubules from βIII. One might expect that the more aggressive cancers would have more oxidative stress and hence more ROS and more need for βIII. In fact, it has been observed that tumors of higher malignancy express higher levels of βIII *(134,137,145)*. A study of patients with nonsmall cell lung cancer showed that those whose tumors had elevated βIII responded less well to drugs and had a poorer prognosis *(146)*. When we compared the MCF-7 and BT-549 breast cancer cell lines, it was found that the latter, which has much more βIII than the former, and is resistant to taxol, vinblastine, and cryptophycin 1 *(147)*, also has a much higher level of free radicals (Chaudhuri and Ludueña, unpublished results).

A very interesting finding that may be relevant at this point is that of Carré et al. *(148)*. They found that tubulin occurs in mitochondrial membranes and that the membrane tubulin is enriched in βIII compared with the rest of the cellular tubulin. Mitochondrial membrane tubulin represents about 2% of total cellular tubulin *(148)*. Mitochondria are the cells' major producers of ROS. Perhaps the function of βIII in the mitochondrial membrane is to protect the cell from ROS; conceivably this could be the role of the unusual cys124 of βIII. The ROS could be neutralized by forming a disulfide bridge involving cys124 and either cys127 or cys129. That disulfide could then be reduced by subsequent reaction with the thioredoxin system, a set of proteins that cells use to protect themselves from free radicals. The thioredoxin system has been shown to reduce disulfide bridges in tubulin *(110)*. This is obviously highly speculative, but the possible role of mitochondrial membrane βIII in protecting cells from ROS does parallel the postulated role of βIII in protecting microtubule assembly from ROS.

3.2.3.4. The unusual dynamic behavior of βIII may be highly regulated. If βIII helps a cell cope with oxidative stress, why don't all cells use βIII for their microtubules and not bother with the other isotypes? Does βIII have a countervailing disadvantage? There is evidence that it does. When the αβII, αβIII, and αβIV dimers are allowed to assemble in vitro in the absence of MAPs, αβII, and αβIV begin to assemble immediately, but αβIII only assembles after a long lag-time *(149)*. Whether this is a phenomenon involving nucleation or elongation of microtubules is not clear. In the presence of either tau or MAP2, however, αβIII assembles without a lag-time, exactly as does αβII *(150)*. In another experiment, when βIII was transfected into Chinese hamster ovary (CHO) cells, microtubule assembly in the cells actually decreased *(151)*. It thus appears that βIII may have an intrinsically lesser ability to polymerize into microtubules. Cells such as neurons, which require large amounts of βIII, may compensate for its poorer polymerization by synthesizing tau, MAP2, or other MAPs. It is interesting that βIII and MAP2 are often synthesized concurrently, not only in neurons *(91,152)* but also in nonneuronal tumors *(90,135)*.

Another unique property of βIII is that microtubules formed of αβIII are considerably more dynamic in vitro than those formed of either αβII or αβIV *(100)*. This property may be very important in development. βIII is expressed in the embryonic nervous system in neurons as well as in cells that later stop expressing it *(136)*. βIII is also expressed in differentiating neuroblastoma cells *(153)* and in regenerating neurons *(154,155)*. It is possible that neurons undergoing rapid growth and differentiation require very dynamic microtubules *(136)*. A small amount of MAP2 is probably expressed at this stage *(156)* and may be sufficient to allow the αβIII dimer to form microtubules. At this stage βIII is not phosphorylated; once the neurons have matured, βIII becomes phosphorylated *(157)*.

3.2.3.5. A Model for βIII Function. All these observations and speculations could be put together into a model for the functional role of βIII, based in part on the ideas of Katsetos et al. *(136)*. In the embryonic nervous system, and perhaps in other cells as well *(158)*, the high dynamicity of microtubules made of αβIII helps the cells to grow and differentiate. At some point, the cells begin to express other isotypes such as βII and βIV that are less dynamic, possibly as a way to regulate the overall dynamic behavior of the microtubules. As the cells differentiate, glia, and other nonneuronal cells stop expressing βIII. In neurons, however, which are faced with problems caused by NO and ROS, βIII expression is retained in order to protect microtubules from oxidation. However, the high dynamicity conveyed by βIII is curtailed, perhaps by increased synthesis of other tubulin isotypes or MAPs, but also by phosphorylation, which is known to promote interaction of βIII with MAP2 *(159)*.

Although this is an attractive model, there are some potential problems with it. First, the high dynamicity of microtubules formed of αβIII was obtained using fully phosphorylated βIII from bovine brain *(100)*. The model assumes that phosphorylation would decrease dynamicity, and therefore predicts that nonphosphorylated βIII would have even higher dynamicity. This prediction, however, has yet to be tested. Second, one could argue that, according to the model, all embryonic tissues should express βIII, as they are all undergoing rapid growth and differentiation. However, this does not seem to be the case *(158)*. It may be that the βIII gene is one of a set of genes that is activated in embryogenesis only in the nervous system and a discrete set of other tissues. The remaining tissues in the embryo may need to find another way to create dynamic microtubules, perhaps using a different tubulin isotype.

3.2.4. βIV

Mammals have two forms of βIV, designated as βIVa and βIVb. The former is expressed only in the brain, whereas the latter is expressed in many tissues including the brain *(62)*. The sequence differences between the two are minor, always involving very conservative amino acid substitutions. Although it is conceivable that there is a functional difference between βIVa and βIVb the fact that birds only have a single βIV suggests that this is not likely to be the case. At any rate, the monoclonal antibody used to localize and purify βIV does not distinguish between βIVa and βIVb (150). Therefore, in this review the two isotypes will be referred collectively as βIV.

βIV has one very clear-cut function: it occurs in axonemes, the microtubule-based apparatus that powers cilia and flagella. In mammals, βIV has been localized in sperm flagella (9), and in cilia of the tracheal epithelium, brain ependyma, oviduct, efferent duct of the testis, vestibular hair cells, retinal rod cells, olfactory neurons, and esophageal progenitor cells (15,71,72,80,160). In fact, βIV has been found in every mammalian axoneme that has been tested (72). This finding is totally consistent with the prediction of Raff et al. (73) who proposed that any β-tubulin that forms part of an axonemal microtubule must contain, very close to the C-terminus, the sequence EGEFXXX (where X is D or E). Of the mammalian β-tubulin isotypes, βIVa, and βIVb are the only ones with this sequence. Therefore, it can be concluded that one function of βIV is to form the axonemal microtubules.

Exactly what role does βIV play in the axoneme? The axoneme is a highly specialized structure, consisting of at least 125 different polypeptides (161). In the middle are two singlet microtubules, known as the central pair. Along the periphery are nine doublet microtubules, known as the outer doublets. Each of these doublets consists of a complete microtubule called the A-tubule and an incomplete microtubule called the B-tubule (162). During motility the motor protein dynein that is connected to the A-tubule of each outer doublet interacts with and slides along the B-tubule of the adjacent outer doublet in a pattern that appears to be regulated by the central pair microtubules (163,164). When one considers the structure of the axoneme, one would imagine that the outer doublet is the specialized microtubule that requires a specific type of tubulin. In contrast, the central pair microtubules seem uncomplicated. In addition, the outer doublet cannot be formed in vitro. Recently, however, it has become clear that the central pair microtubules are special and that they have to rotate around each other to determine which outer doublet pairs slide past one another (163,164). Like the distributor of a car, the rotating central pair microtubules serially make contacts, through some bridging proteins, with specific outer doublets. It should not be surprising if this highly complex microtubule machinery requires a particular tubulin isotype. In fact, there is room in this scenario for more than one isotype. It has been mentioned earlier that many axonemes also contain βI (72).

It is possible that βIV is involved in determining axonemal microtubule structure rather than being directly required for motility. This is based on the observation that two of the cilia types in which βIV occurs are nonmotile: the retinal rod and the kinocilia of the vestibular hair cell (15,80). Whether βIV plays a role in intraflagellar transport is not clear (165).

In *Drosophila*, the β-tubulin isotype, β2, is the only one that contains the EGEFEEE motif and is the only one found in the sperm flagellar axoneme. If β2 loses this motif, or if β2 is replaced by β1, then the outer doublet microtubules are present but not the central pair. Clearly, the EGEFEEE motif is very important. Interestingly, however, if the axoneme motif is inserted into β1, then the outer doublets and central pair are all present, but the distal end of the axoneme is abnormal. This implies that the EGEFEEE motif is not enough to specify a proper axoneme; the other parts of the β2 isotype must also be important (166). Extending this finding to mammals, one could argue that βIV is required for proper formation of both the central pair and outer doublet microtubules.

What is the role of the EGEFEEE sequence? It appears to be the site for polyglycylation, a post-translational modification in which a series of glycines are attached to the γ-carboxyl group of glutamate residues. This modification is very common in axonemal tubulin. The polyglycyl side chain is thought to be necessary for the assembly of the central pair microtubules as well as the B-tubule (167). This modification will be discussed further in Chapter 5. From these observations it is probably safe to hypothesize that βIV is a major constituent of axonemal microtubules because it has a sequence that can be polyglycylated and that this polyglycylation is necessary to form the central pair and outer doublet microtubules.

However, the experiment on *Drosophila* described above implies that other parts of the βIV molecule are necessary for proper formation of the axonemal microtubules. Therein lies a problem for the mammalian sperm cell. The conformation of αβIV is significantly less rigid than that of either αβII or αβIII (168). The high levels of ROS in the testis we have already commented on. Although the precise susceptibility of the αβIV dimer to oxidation has never been tested, it is likely to be higher than that of the other isotypes. The need to protect βIV from oxidation may account for the presence of the protein thioredoxin-like 2 in sperm cells and tracheal cilia. This protein binds well to microtubules and is presumably capable of reducing any disulfide bridges that form in βIV (169).

Another possible function for βIV was observed in cultured rat kidney mesangial cells. The microtubules of these cells contain largely βI and βIV (170). When the cells are extracted, the βI microtubules disappear completely, but the βIV becomes associated with actin filaments (171). Interactions between microtubules and actin filaments are becoming well known (172). The results described here raise the possibility that βIV may be involved in these interactions. It may be that βI and βIV have opposite effects on microtubule–actin crosstalk.

3.2.5. βV

βV is the most intriguing of the β isotypes. It is highly conserved in evolution, suggesting that it may have a specific function. However, not only is that function unknown, even the normal distribution of βV is not known. Using mRNA measurements, Sullivan et al. (173) showed that in chickens βV is found in every tissue outside of the brain. Preliminary results with a monoclonal antibody to βV, however, suggest that it is found in mammalian brain but in relatively few other tissues (174). Further work will be necessary to resolve this. Perhaps the only clue to the function of βV is that it has the same distribution of cysteine residues as βIII. In other words, it has cys124 but lacks cys239. If, as was speculated earlier, cys124 allows βIII to react harmlessly with ROS and if the lack of cys239 allows βIII to form microtubules resistant to oxidation, then perhaps the same is true for βV. In fact, βV could conceivably do the job of βIII in tissues that lack that isotype.

3.2.6. βVI

βVI is the least conserved of the β isotypes. In fact, avian and mammalian βVI are so different from each other that it is not clear that they belong to the same isotype class. They have been grouped together because they are clearly associated with the hematopoietic system. In chickens, βVI forms the microtubules of the erythrocyte; in mammals, whose erythrocytes lack microtubules, βVI is found in platelets and in hematopoietic tissues such as bone marrow and spleen (45,46). βVI has a unique arrangement of cysteines.

Table 7
Vertebrate α-Tubulin Isotypes[a]

Designation	Species	C-terminal sequence	Distribution
Class I	Human α1	VDSVEGEEEGEEY	Mostly brain
	Human α3	VDSVEGEEEGEEY	Widespread
	Mouse α1	VDSVEGEGEEEGEEY	
	Mouse α2	VDSVEGEGEEEGEEY	
	Rat α	VDSVEGEGEEEGEEY	
	Chicken α1	VDSVEGEGEEEGEEY	
	Xenopus α1	TDSVEGEGEEEGEEY	
	Torpedo α	VDSVEGEGEEEGEEY	
	Notothenia α	VDSIEGDEEEEGEEY	
	Notothenia α	VDSIEGDGEEEGEEF	
	Salmon α	GDSIEGEGEEEGEEY	
Class II	Human α2	VDSVEAEAEEGEEY	Testis
	Mouse α3/7	VDSVEAEAEEGEEY	
	Rat α3/7	VDSVERKGEEGEEY	
	Trout α	VDSVEGEAEEGEEY	
Class III	Human α4	IDSYEDEDEGEE	Brain, muscle
	Mouse α4 IDSYEDEDEGEE		
	Rat α4	IDSYEDEDEGEE	
	Chicken α5	LDSYEDEEEGEE	
Class IV	Human α6	ADSADGEDEGEEY	Blood
	Mouse α6	ADSAEGDDEGEEY	
	Xenopus	ADSADAEDEGEEY	
	Notothenia α	ADSLGGEDEEGEEY	
	Notothenia α	ADSLGDEEDEEGEEY	
Class V	Human α8	TDSFEEENEGEEF	Heart, skeletal muscle, and testis
	Mouse α8	TDSFEEENEGEEF	
	Chicken α8	TDLFEDENEAGDS	
Class VI	Mouse αTT1	MGSVEAEGEEEDRDTSC CIMFSSSIGNRHPC	Testis
Class VII	Xenopus	TESGDGGEDEEDEY	Ovary
Unclassified	Danio	ADSTDDCGEDEEEY	

Source: From refs. *8,43,185,305,306,678,401–410.* The classification adopted here is based on that of Lewis and Cowan *(410).*

[a]Mouse α1 and mouse α2 differ from each other at 1 position. The mouse α3 and α7 genes have different nucleotide sequences but encode identical proteins, referred to as α3/7.

As is true for βIII and βV, βVI has ser239 instead of the assembly-critical cys239. Keeping to this same pattern, chicken βVI has cys124, although mammalian βVI has ser124 *(45).* Mouse and chicken βVI also have two extra cysteines: at positions 37 and 315 *(45,63).* One of these may be involved in the disulfide bridge that has been observed in mammalian platelet tubulin *(175).*

Platelet tubulin has been extensively studied. βVI constitutes about 90% of the total platelet β-tubulin *(176).* The transcription factor NF-E2 induces βVI synthesis in the megakaryocytes of the bone marrow *(177).* Platelets are formed by budding off from megakaryocytes *(178)*. Platelets have a marginal band at the periphery of the cell. The marginal band consists of a single microtubule about 100 μm long, wound around itself 7–12 times *(176,178,179)*. Inhibition of βVI synthesis results in platelets with a marginal band consisting of a single microtubule with only 2–3 coilings; platelets are spherical instead of discoid and, in some experiments, blood coagulation is compromised *(176,179,180)*.

Based on these results one could hypothesize that the peculiar structure of βVI lends itself to forming the marginal band microtubule. This hypothesis appears to be correct. Platelets lacking βVI contain a marginal band formed of the βI and βII isotypes. In the normal platelet, 95% of βVI is in the marginal band, whereas about 45% of βII and 58% of βI are incorporated *(179)*. It would thus seem that βVI is better adapted to forming this unusual microtubule organelle than are the other β isotypes.

3.2.7. βVII

Very little is known about βVII. Its sequence lacks most of the C-terminus and it is expressed in the brain *(181,182)*. Its function, distribution, and properties are completely unknown.

3.3. Mammalian Tubulin Isotypes: the α-Isotypes

There is not much to say about the specific functions, if any, of the α isotypes in mammals. Their tissue distributions seem much less complex, as far as is known, than the distributions of the β isotypes (Table 7). α1 is found mostly in brain but also in a variety of other tissues *(8)*. α2 is similar *(183,184)*. α3/7 is found only in the testis, where it is the major α isotype. α4 is widespread, especially in muscle and heart; α6 is also widespread, but less common than the others. α8 is considerably divergent in sequence, being only 89% identical to the other α (except for the even more divergent αTT1); it is found in heart, testis, and skeletal muscle, and at very low levels in the brain and pancreas *(185)*. The unusual isotype αTT1 is found only in the testis, where it is a minor component of the α population *(42)*.

The α1, α2, α3/7, α4, and α6 isotypes are at least 94% identical in amino acid sequence. In addition, the differences tend to be conservative such as ser/thr or ilu/val. When viewed in conjunction with the fact that the tissue distributions of several of these are quite similar, it is hard to imagine that the differences among these isotypes are functionally significant. However, α8, with its more divergent sequence and its highly restricted distribution, may be an exception. This isotype has a unique sequence at positions 35–45, which is TFDAQASKIND and TFGTQASKIND, respectively, in human and mouse α8. The equivalent sequence in α1, α2, α3/7, α4, and α6 is the completely different QMPSDKTIGGG. This region corresponds to a loop located on the inner microtubule wall that may play a role in contacts between adjacent protofilaments *(185)*. Conceivably, microtubules with α8 could have very different dynamics than those containing the other α isotypes. The fact that the unique features of α8 are highly conserved in evolution suggests that these are functionally significant.

Even though some functional differences among mammalian α isotypes are plausible, none have ever been demonstrated in vitro or in vivo. One approach to this question would be to develop more antibodies specific for α isotypes and then use them for detailed immunohistochemistry as well as to purify tubulin dimers homogeneous for their α-subunit. If the α isotypes exhibit complex cellular distributions, as is the case, for example, with the β isotypes of the cochlea that would be highly instructive. Similarly, comparison of the behavior in vitro of different α isotypes may reveal functional differences. However, such a comparison has to account for differences that may arise because of different extents of tyrosinolation/detyrosinolation and deglutamylation. In addition, it would be good to know which combinations of specific α and β isotypes occur in different tissues. At the moment, much remains to be done to allow us to understand the significance of the mammalian α isotypes.

3.4. Tubulin Isotypes May Have Functional Roles That Do Not Involve Microtubules

As the mammalian β isotypes have been surveyed, occasional indications of tubulins that may have functional properties of at least potential physiological relevance that do not involve being part of a microtubule have been seen. For example, βII, occurring in the nuclei of various cell types, has been seen as a reticulum rather than as a microtubule; it is in the form of an αβII dimer with apparently normal drug-binding properties *(83)*. Although most of the cells in which this occurs are abnormal, soluble αβII has been shown to compete with heterochromatin protein 1 for binding to the nuclear envelope *(186)*. Similarly, βIII occurs in mitochondrial membranes where it may act to protect the cell from ROS *(148)*. Cells treated to destroy the microtubules show βIV bound to actin. Whether an interaction between actin and soluble αβIV dimers occurs in intact cells is not clear. Nevertheless, the idea of tubulin acting in a nonmicrotubule context is not a new one. A spiral polymer of tubulin constitutes the conoid structure of the protist *T. gondii (187)*. A possibly analogous situation is provided by the enzyme glutathione peroxidase. Usually, the role of this selenium-containing protein is to protect cells from ROS *(188)*. The sperm isozyme of glutathione peroxidase has an additional function, however. After the sperm cell has matured, this isozyme polymerizes to form a sheath around the mitochondrion, losing its enzymatic activity in the process *(189)*. In a sense this is the converse of the situation in microtubules. On the one hand, there has been glutathione peroxidase that normally functions as a monomer, one isoform of which polymerizes, losing its original function. On the other hand, there has been tubulin that normally forms polymers, but that has at least one isoform that can abandon its role in forming that polymer and assume another function. Regardless of the applicability of this particular analogy, the idea that certain tubulin isotypes may have functions that do not involve forming microtubules may be worth pursuing.

An unusual finding that may speak to a possible nonmicrotubule role of βII is its occurrence in the nuclei of a wide variety of cells. This was first discovered in rat kidney mesangial cells *(83)*. In these cells, an antibody to βII strongly stained the nuclei but not the cytoplasm. The staining occurred throughout the nuclei, but was concentrated in the nucleoli. When the mesangial cells enter mitosis, the βII leaves the nuclei and helps to form the mitotic spindle. During telophase, βII enters the reforming nucleus. In contrast, βI and βIV, that constitute the interphase microtubule network, enter the spindle during mitosis, at the end of mitosis returning to the interphase network. These two isotypes never enter the nuclei.

The nuclear βII was in the form, not of a microtubule, but of a reticulum *(83)*. Western blot analysis of the purified nuclei indicated a band reactive with the antibody to βII that comigrated on gels with bovine brain βII. Cells from which the cytosol had been extracted showed α-tubulin in the nuclei as well. Treatment of the cells with fluorescent colchicine showed accumulation of fluorescence in the nuclei in a pattern indistinguishable from that of βII, suggesting that the nuclear tubulin was in the form of an αβII dimer, capable of binding to colchicine. Disruption of the nuclear βII staining with nocodazole, taxol, and vinblastine, corroborated this interpretation *(190,191)*. Microinjection of fluorescently labeled αβII into the cytosol of rat kidney mesangial cells resulted in accumulation of fluorescence in the nuclei. In contrast, microinjected fluorescent αβIII and αβIV did not enter the nuclei *(170)*. It thus appeared that there was a process that, in these cells, resulted in an αβII dimer entering the nuclei. The fact that micro-injected αβII only entered the nuclei after a cycle of cell division had been completed suggests that nuclear transport may not be involved in the process but rather that the nucleus assembles around the αβII dimer *(170)*.

Other studies revealed that only certain cultured nontransformed cells contained nuclear βII, whereas nuclear βII occurred in almost every cultured cancer cell *(192)*. A survey of about 200 tumors excised from patients showed nuclear βII in 74% of them *(76)*. In general, nuclear βII staining was very variable, depending on the tumor type. In tumors of the prostate, stomach, and colon, nuclear βII was seen in every sample studied. In contrast, only a few hepatic and brain tumors showed nuclear βII. In some excisions, nuclear βII occurred in almost every tumor cell, but sometimes in only a fraction. The intensity of nuclear staining also varied. The pattern of intranuclear staining was variable as well. In some cases, βII was concentrated in the nucleoli; in others it appeared to stain the entire nucleoplasm except the nucleoli. Cytoplasmic staining of βII was also highly variable. Many samples appeared to have βII only in their nuclei and not in the cytoplasm.

The study with human tumors revealed two unusual patterns. First, nuclear βII occurred in tumors of tissues such as the prostate, in which the normal tissue does not express βII. This would suggest that transformation leads cells first to express βII and then to localize it to the nuclei. Second, otherwise normal cells near the tumor would also contain nuclear βII. This was particularly striking in cases of breast cancers that had metastasized to the lymph nodes. Lymphocytes normally do not stain for βII. However, lymphocytes adjacent to the metastatic cancer cells contained nuclear βII. These results suggest that a cancer cell can influence adjacent normal cells to make βII and put it in the nuclei *(76)*. Analysis of a number of normal tissues indicated that most of them did not contain nuclear βII *(76)*. The exceptions were bone marrow, placenta, and pancreatic acinar cells.

What conclusions can be drawn from the story of nuclear βII? It is clearly not a normally widespread phenomenon, being found mostly in cancers and cultured cells. The presence of nuclear βII in cultured cells, tumors, placenta, and bone marrow may indicate an association with proliferation, but this does not explain its presence in the pancreas. A recent finding may cast some light on nuclear βII. Kourmouli et al. *(186)*, working with human endometrial carcinoma cells, examined heterochromatin protein 1, which binds to proteins associated with chromatin such as transcriptional regulators. It also binds to the nuclear envelope. The binding of heterochromatin protein 1 to the nuclear envelope is strongly inhibited by a soluble protein that was found to be a mixture of the α2βII and α6βII dimers. The specific α isotypes involved in this are probably incidental, but it is striking that the only β isotype in these dimers is βII. The authors report that the αβII dimer binds very tightly to the nuclear envelope, thus preventing heterochromatin protein 1 from binding there. These findings raise the possibility that a role of βII may be to control the interaction of chromatin with the nuclear membrane and perhaps also control the distribution of nuclear membrane fragments during mitosis. Such a possibility may explain the higher concentration of βII observed in the perinuclear region *(81)*. In addition, one could imagine that a relatively minor alteration in certain cell types, especially cancer cells, would result in βII remaining in the nuclei after mitosis is complete. Such alterations may involve a modification of βII or of heterochromatin protein 1 or of the nuclear envelope itself. For example, if heterochromatin protein 1 is altered so as to decrease its affinity for the nuclear envelope, the αβII dimer may stay bound to that membrane at the end of mitosis and may remain in the nucleus during interphase. Of course, the connection of heterochromatin protein 1 and βII may be only a coincidence. However, breaking up the nuclear envelope during prophase and putting it together again during telophase are functions mediated by microtubules *(193–196)*, so a specific connection between the nucleus and one tubulin isotype should not seem too outlandish. On the other hand, many cells appear to lack βII; how do these cells regulate nuclear envelope breakdown and reassembly if βII is important for this process? Are there subtle differences in the processing of the nuclear envelope in these cells? This may be worth examining.

3.5. Not All Isotype Differences are Functionally Significant

The data presented above argue strongly that certain tubulin isotypes have specific functions. This does not necessarily apply to all cases of isotypes, however. Many organisms have isotypes that differ from each other at only a few positions, and only with conservative amino acid substitutions. It is hard to imagine that these small differences are physiologically significant. The key evidence bearing on this point has to do with interchangeability of isotypes. For example, one of the two β isotypes of the fungus *Aspergillus* appears largely during conidiation. However, replacing it with the other one does not alter this process *(197,198)*. Similarly, *Aspergillus* has one α isotype involved in vegetative growth and another in sexual development. Using appropriate manipulations, the expressions of the two α isotypes were reversed. No effect on the viability of *Aspergillus* was observed, provided that particular levels of expression of each isotype were chosen. In fact, it took three copies of the vegetative isotype to replace the sexual isotype without altering the phenotype *(199)*. This experiment has an important implication, namely that when performing genetic manipulation of isotype expression, one must be careful to maintain the same level of total tubulin isotype expression. For example, it may be that the vegetative α isotype of *Aspergillus* is expressed at a lower level than is the sexual α isotype. In that case, replacing the latter with only one copy of the former would mean that the total amount of tubulin expressed would be lower than normal and that could have a deleterious effect. Alternatively, the extra β-tubulin, lacking its α partner, may be toxic to the cell. These factors have to be considered when weighing the results of this type of experiment.

Several early experiments using cultured cells suggested that the vertebrate isotypes were interchangeable. For example, in cultured cells, most of the β isotypes are able to form the mitotic spindle as well as the interphase microtubule network *(118,200–202)*. A similar result was obtained with the α isotypes *(203)*. These experiments, however, do not necessarily prove that the tubulin isotypes are interchangeable. As most cells, no matter what isotypes they express, have both a mitotic spindle and an interphase network, it is not surprising that each isotype could participate in forming these structures. However, cultured cells are less complex than cells *in situ*. The latter may have a specialized need for a particular tubulin isotype that would not arise in a cultured cell.

The above experiments, although very carefully performed, were not set up to address the kinds of subtle and varied possibilities that have been being reviewed for the mammalian β isotypes such as the ability to form axonemes, protect cellular microtubules from oxidation, interact with actin, or form marginal band microtubules.

3.6. Isotype Differences May be Generally Adaptive Without the Isotypes Having Specific Functions

It is possible that certain tubulin isotypes may not have specific functions but that the presence of different isotypes may be adaptive in that they may increase the repertoire of responses to environmental challenges. This is likely to be the case in some plant isotypes. For example, expression of certain β isotypes in *Arabidopsis* decreases in the cold whereas that of another β isotype increases *(204)*. A similar result was obtained in wheat, where lowering the ambient temperature to 4°C increased the expression of one α isotype and decreased that of another *(205)*. Under these conditions, the microtubules become more dynamic. The authors propose that microtubules act as a kind of temperature-sensor and that the cold-induced change in their behavior triggers specific cellular responses to the cold *(205)*. Such a model would not be possible without having different isotypes and yet a specific function cannot be assigned to each isotype.

Multiple isotypes may also play a role in resistance to toxins. For example, there is evidence that having multiple isotypes may make nematodes more resistant to benzimidazoles *(206–210)*. Warm-blooded mammals are more protected from the environment than are plants or nematodes. However, a great deal of evidence indicates that tumors expressing certain isotypes are more resistant to drugs, or that drug treatment may lead to increased expression of particular isotypes (reviewed in refs. *211,212*). Do these results—which will be discussed in more detail later on—speak to the hypothesis of multiple isotypes being generically adaptive in mammals? Certainly, as will be seen, the specific interpretations of these results are highly complex. It is hard to imagine that we evolved for a half-billion years in order to develop mechanisms of resistance to antitumor drugs. On the other hand, one must recall that many of these drugs are, or are derived from, natural products. Thus, it is not inconceivable that the relative amounts of tubulin isotypes may be adjusted in order to help cope with environmental toxins. The fact that several of these toxins are intended to heal, is an unfortunate complication.

Another mechanism by which tubulin isotypes can be generally, rather than specifically, adaptive, is to have them differ in functionally relevant properties such as their dynamic behavior. For example, the yeast *Saccharomyces cerevisiae* has two α-tubulin isotypes (Tub1 and Tub3) *(213)*. Microtubules made of Tub3 are less dynamic in vitro than are the wild-type microtubules. Conversely, microtubules made from Tub1 are more dynamic. The shrinkage rate and the catastrophe frequency for Tub1 are, respectively, four- and threefold more than the corresponding parameters for Tub3, resulting in Tub 1 microtubules having twice the dynamicity of Tub3 microtubules *(214)*. Perhaps, the cell can alter the relative proportions of the two isotypes in order to adapt its microtubule dynamicity to different conditions.

3.7. Altered Expression of Tubulin Isotypes in Drug-Resistant Cells and Tumors

One of the most interesting observations was reported by the Horwitz laboratory, which found that the levels of βI and βII rose 1.9-fold and 21-fold, respectively, in taxol-resistant murine cell lines *(215)*. Subsequently, Ranganathan et al. *(216,217)* observed that the levels of βIII and βIVa rose four to ninefold and three to fivefold, respectively, in estramustine-resistant DU-145 human prostate cancer cells. Taxol-resistant MCF-7 human breast cancer cells were found to express increased levels of βIII, βIVa and the tyrosinated form of α-tubulin *(218)*. Surveying the many studies done in this area, one of the most frequent results is that tumors expressing increased levels of βIII are more resistant to taxanes and estramustine *(146,212,218–221)*. Almost as frequently observed is that increases in βIV expression also accompany resistance to taxanes and vincristine *(222–224)*. In many fewer cases, taxane resistance involves increased expression of βI *(225)* or βII *(212,215,226)*. Increased βI expression also correlates with resistance to vincristine and E7010 *(222,227)*. Thus, there is ample evidence to suggest that cells alter the synthesis of certain tubulin isotypes in order to survive drug exposure.

How can one make sense of these complex findings? One can speculate that tumor cells elevate the synthesis of the isotype that has the weakest affinity for the drug in question. For example, Derry et al. *(228)* showed that the dynamics of microtubules made from the αβII dimer are significantly more sensitive to inhibition by taxol than are the dynamics of microtubules made from αβIII or αβIV. This is certainly consistent with the majority of the taxane studies, which showed resistance accompanied by increases in βIII or βIVa. These results are also corroborated by the observation that inhibition of the synthesis of βIII by the antisense phosphorothioate oligodeoxynucleotide increases the sensitivity of A549 lung cancer cells to taxol *(229)*. Similarly, transfection of βIII into CHO cells caused a slight increase in taxol resistance *(151)*. On the other hand, overexpression of βIII in human prostate cancer cells failed to affect the sensitivity to taxol *(230)*. Furthermore, studies with human ovarian tumor xenografts failed to detect any significant role of a specific tubulin isotype level on taxol sensitivity *(231)*. These investigators used patient samples (before or after chemotherapy with taxol) to establish a subset of 12 xenografts, and found no correlation between the tubulin isotype expression and the taxol sensitivity. Similarly, overexpression of βIVb in CHO cells did not create resistance to taxol *(232)*. Resistance to *Vinca* alkaloids is reported to be associated with decreased βIII expression *(233)*. This is not consistent with the finding of Khan and Ludueña *(159)* who showed that microtubule assembly of αβIII in the presence of tau was more sensitive to vinblastine inhibition than was assembly of either αβII or αβIV. In short, the hypothesis that tubulin isotypes that are elevated in drug-resistant tumor cells are those isotypes that interact less well with that drug in vitro is consistent with some studies but not others.

What other factors could account for these contradictions? First, there are certain experimental aspects to be considered. If Reverse Transcriptase Polymerase Chain Reaction (RT-PCR) indicates that the mRNA of one isotype increases much more than that of another isotype in response to a drug that does not necessarily mean that the protein levels of these isotypes increase in the same ratio. One isotype may be more sensitive to proteolysis, for example. Similarly, isotype-specific antibodies may detect a many-fold increase in one isotype and only a small percentage increase in another. However, if the latter is much more abundant in the cell than the former, then the small percentage increase in the latter may be much more significant physiologically than the large percentage increase in the former. In such cases, it is important to know the actual isotype levels rather than only the percentage increase or decrease. Also, it is possible that changes in post-translational modification in the antibody epitope (which is generally the C-terminus and the site of most modifications) may highly alter the detectability of the isotype even if its actual level remains the same. Second, other processes could be supervening that make the change in level of an isotype irrelevant. For example, taxol resistance is sometimes accompanied by mutations of isotypes such as βI *(221)*. A cell with a taxol-resistant βI may actually increase its resistance to taxol by making less of the other isotypes, including βIII. Third, resistance could reflect the assembly properties of the isotypes rather than their drug-binding ability. For example, microtubules containing βIII are less stable in vitro *(100)*. As taxol can increase microtubule assembly, one could argue that increased βIII would cause more resistance to taxol by making less stable microtubules *(233)*. Similarly, as vinblastine inhibits microtubule assembly, one would expect that decreased βIII would increase the overall stability of cellular microtubules and thus increase vinblastine resistance; this is exactly what has been observed *(233)*. Finally, it is possible that cells that more readily mutate to a drug resistant phenotype have higher concentrations of ROS and, hence, that they may have an increased requirement for βIII in order to protect their microtubules from the ROS. However, if, as it has been speculated, βV has the same protective function as does βIII, then, as βV is rarely measured in tumor cells, a scenario could be hypothesized where βV increases and βIII decreases, keeping the total tubulin concentration the same, but only the βIII decrease is detected. In that connection, it is interesting that a preliminary survey of 12 NIH cancer cell lines found that βV was expressed in 11 of them, generally at higher levels than βIII *(234)*.

The mechanism by which the expression of specific tubulin isotypes is altered in drug-resistant cancer cells is still obscure. Overexpression of the oncogenic epidermal growth factor receptor family of kinases has been reported to induce taxol resistance and also increase the expression of βIVa and βIVb *(235)*. Involvement of p53 has been implicated in modulating the expression of tubulin isotypes and drug resistance in human breast cancer cells *(224)*. Extensive analysis with isogenic stable cell lines overexpressing a specific tubulin isotype may shed light on these mechanisms.

3.8. Properties of Purified Mammalian Tubulin Isotypes In Vitro

If the differences among the tubulin isotypes are functionally significant, then it could be expected that the purified isotypes would behave differently from each other in vitro. To address this issue, monoclonal antibodies have been constructed specific for the mammalian βI, βII, βIII, and βIV isotypes *(59,64,150,236–239)*. These have been used to purify the αβII, αβIII, and αβIV dimers from bovine brain by immunoaffinity chromatography. A large number of parameters have been assayed in vitro. The dimers differ from each other in virtually every parameter that has been assayed. Assembly into microtubules is an obvious first parameter to examine. In the presence of either tau or MAP2, αβII and αβIII assemble more rapidly and to a higher extent than does αβIV *(150)*. In the absence of MAPs, but in the presence of 4 M glycerol, αβII and αβIV assemble rapidly with no lag time, whereas αβIII assembles only after a considerable lag-time *(149)*. This raises the possibility that αβIII has a harder time nucleating in vitro in the absence of nucleating factors such as γ-tubulin. Microtubules formed from αβIII are considerably more dynamic than those formed from either αβII or αβIV *(100)*. Possibly consistent with these findings is that the intrinsic GTPase activity of tubulin is the highest for αβIII than for either αβII or αβIV *(240)*. However, during microtubule assembly in the absence of MAPs, αβIII hydrolyzes GTP more slowly than do the other two dimers *(9)*. One must be cautious about extrapolating these results to the situation

Table 8
Tubulin Isotypes: Intrinsic GTPase Activity and Interactions With Antitumor Drugs[a]

Ligand	$\alpha\beta_{II}$	$\alpha\beta_{III}$	$\alpha\beta_{IV}$
Intrinsic GTPase			
Induced by colchicine (nmole/h/mL)	4.5	9.6	3
Induced by MTPT[b] (nmole/h/mL)	5.8	11.5	7.3
Interactions with antitumor drugs			
K_d for colchicine (M)	4.2	8.3	0.3
$k_{on,app}$ for colchicine binding (M/s)	132 ± 5	30 ± 2	236 ± 7
K_d for DAAC[c] (M)	0.4	0.7	0.3
k_2 for DAAC (s^{-1})	0.67	0.05	0.59
K_d for MTPT (M)	3	6.4	1.8
k_2 for MTPT (s^{-1})	4.22	2.07	5.28
K_d for thiocolchicine THC18 (M)	0.5	17	ND[d]
K_d for nocodazole (M)	0.52	1.54	0.29
K_d for IKP104 (M)	0.01	0.11	1.4–1.8
Suppressivity of dynamics to taxol[e]	3626	765	784
IC_{50} for vinblastine[f] (M)	0.6	2.1	0.6
IC_{50} for vinblastine[g] (M)	0.5	1.8	2

[a]*Source:* From refs. *159,228,240,260–262,411,412.*

[b]MTPT, 5-(2′,3′,4′-trimethoxyphenyl)-1-methoxytropone.

[c]DAAC, desacetamidocolchicine.

[d]ND, not determined.

[e]This is a parameter that indicates the sensitivity of the shortening rate to taxol *(228).*

[f]Microtubule assembly was measured in the presence of tau and a series of vinblastine concentrations.

[g]Microtubule assembly, as above, measured in presence of MAP2.

in vivo. Buffer conditions used in vitro may not be physiological and different cell types may have different MAPs that could create major differences in the relative assembly and dynamic properties of the isotypes.

Structural differences among the isotypes are also evident. For example, the mammalian βIII isotype is phosphorylated, whereas the others are not *(241).* Using differential scanning calorimetry, Schwarz et al. *(242)* found that αβIII is considerably more resistant to decay than is αβII. The half-times for decay at 37°C of colchicine-binding activity for αβII and αβIII were, respectively, 17 h and 50 h *(242).* Conformation was also probed using a series of sulfhydryl-reactive crosslinkers of the structure: $ICH_2–CONH–(CH2)_x–NHCO–CH_2I$, where x (the number of methylene groups) is either 2, 3, 4, 5, 6, 7, or 10 *(268).* The reagent with $x = 2$ forms two intrachain crosslinks in β-tubulin *(243,244).* One, designated β*, is between cys239 and cys354 and the other, designated β^s, connects cys12 to either cys201 or cys211 *(102,245).* When the series of crosslinkers were reacted with the different isotypes, the β* crosslink formed, as expected, in αβII and αβIV, but not in αβIII, which lacks cys239. However, the β^s crosslink did not form at all in αβIII, even though βIII has the cysteines involved. Also, in αβII, the β^s crosslink formed at high yield with the $x = 2$ crosslinker, and with the crosslinkers where $x = 4, 5, 6,$ and 7, but very little with the $x = 3$ and $x = 10$ compounds. In contrast, in αβIV, the β^s crosslink formed well with each crosslinker *(168).* These results suggest that at least one of the cysteines involved in the β^s crosslink is probably unavailable in αβIII and that it is available in αβII and αβIV, but even more so in the latter. These results are consistent with αβIII having a more rigid conformation than either αβII or αβIV, but also suggest that the conformation of αβIV is the least rigid of the three dimers.

Not surprisingly, the isotypes also differ in their ligand-binding properties (Table 8). This has been studied in more detail with colchicine and its analogs. Colchicine binds to tubulin in a slow, irreversible, and temperature-dependent manner *(246–252)*. The binding of drug to tubulin results in a promotion of drug fluorescence that has been used to characterize this interaction *(253,254)*. The binding of colchicine is a two-step process in which initial complex formation is followed by a slow conformational change resulting in the formation of a stable complex *(255,256)*. When the association kinetics are studied under pseudo-first-order conditions, the kinetics exhibit a biphasic pattern *(255–257)*. Biphasic kinetics are also observed for the faster-binding analogs of colchicine such as desacetamidocolchicine (DAAC) and the bicyclic analog 5-(2′,3′,4′-trimethoxyphenyl)-1-methoxytropone (MTPT), which binds to tubulin almost instantaneously *(257,258)*.

The origin of the biphasic kinetics in the colchicines–tubulin interaction was not clear until it was demonstrated that immunoaffinity depletion of the tubulin dimers to remove the αβIII dimer eliminated the slow phase, resulting in monophasic kinetics *(259,260)*. Furthermore, addition of αβIII to the αβIII-depleted tubulin restored the biphasic kinetics. Subsequent kinetic studies with the isotypically pure tubulin dimers demonstrated that the isotypes differ significantly in their on-rate constants for binding colchicine. The apparent on-rate constants $(k_{on,app})$ for αβII, αβIII, and αβIV are shown in Table 8. Scatchard analysis revealed that the isotypes also differ in their affinity constants for colchicine and its B-ring analogs *(261,262)*. Analysis of the binding kinetics of colchicine and its analogs indicated that not only does αβIII have the lowest affinity for colchicine, but that the rate (k_2) of the conformational change in tubulin that is part of the drug binding reaction is the slowest for αβIII (Table 8) *(261,262)*. The slow rate of this conformational change may reflect the higher rigidity of αβIII. If this is the case, then this may explain its lessened ability to interact with nocodazole and taxol, although the binding kinetics of these drugs with tubulin isotypes have not been studied in any detail.

The interaction of *Vinca* alkaloids with purified tubulin isotypes is more complicated. One study compared the effects of vinblastine on αβII, αβIII, and αβIV and measured vinblastine's ability to inhibit microtubule assembly and induce spiral aggregate formation *(159)*. The results were clear: microtubule assembly of αβIII was least sensitive to inhibition by vinblastine. Similarly, αβIII was the least susceptible to vinblastine-induced aggregation. Interestingly, although vinblastine induced αβIV to form spiral aggregates, αβIII generally formed amorphous aggregates instead *(159)*. A second study carefully and rigorously examined the effects of three *Vinca* alkaloids (vincristine, vinblastine, and vinorelbine) on self-aggregation of αβII and αβIII. Few significant differences between αβII and αβIII were noted *(239)*. Although the two studies appear to give contradictory results, this is not necessarily the case. No MAPs were present in the latter study, whereas they are present in the former. The study of Lobert et al. *(239)* suggests that the isotypes do not differ in terms of the specific tubulin–tubulin interactions or conformational changes involved in self-aggregation. The study of Khan and Ludueña *(159)* suggests that the isotypes differ either in their interactions with MAPs or else in the ability of vinblastine to interfere with the MAP-induced change in tubulin conformation that permits assembly. The work of Banerjee et al. *(64)* suggests that αβII and αβIII interact equally well with both MAP2 and tau, so the former model is unlikely. As αβIII has the most rigid conformation of the three dimers, it is not surprising that vinblastine's ability to interfere with the conformational change induced by the MAPs is the weakest in αβIII. Similarly, as a conformational change induced by vinblastine is likely to favor aggregation that change is likely to be least marked in αβIII. This is consistent with the observation that αβIII does not aggregate into spirals; perhaps the conformation of αβIII does not permit it to form spirals. A startling difference in vinblastine-induced aggregation was seen when vinblastine (20 *M*) was added to preparations of erythrocyte tubulin and brain tubulin from chickens *(105)*. The former consists largely of αβVI, whereas the latter is likely to be a mixture of αβI, αβII, αβIII, and αβIV *(64)*. About 42% of the brain tubulin aggregated into spirals whereas 74% of the erythrocyte tubulin formed spirals. Aggregation of the latter was so dramatic that the resulting flocculent precipitate was readily visible to the naked eye *(105)*. Clearly, αβVI has a unique ability to interact with vinblastine. Conceivably, the ability of βVI to form microtubules in which the protofilaments bend so as to form a circular microtubule may translate into a higher ability for the protofilaments to bend to form the vinblastine-induced spiral.

The most consistent finding, one obtained by a wide variety of experimental approaches, is that αβIII has a more rigid conformation than either αβII or αβIV. Could this have any bearing on the differences that have been discussed in vivo? First, a more rigid αβIII would hydrolyze GTP more slowly during microtubule assembly, as has been observed *(9)*. This would increase the growth rate as that depends on the presence of unhydrolyzed GTP at the microtubule end *(263)*. Second, a more rigid dimer is less likely to bind tightly to an adjacent dimer in the microtubule and thus the longitudinal dimer-dimer interactions will be weaker. Hence, the rate of shrinkage might be faster. In short, the increased dynamic behavior of αβIII microtubules may be a function of the rigidity of αβIII.

The basic limitation of the experiments in which purified tubulin isotypes are studied in vitro is that one only gets answers to the questions one asks. Assembly, GTPase, and drug-binding activities are fairly obvious and easy areas to investigate. The fact is, however that the number of proteins or other factors known to interact with tubulin is rising very quickly. To name but a few, in addition to the well-known MAPs, there have been various chaperones *(264,265)*, collapsin-response mediator protein 2 *(266)*, stable-tubulin-only polypeptide *(267)*, the importin/Ran-GTP system *(268)*, XMAP215 *(269)*, Fhit *(270)*, katanin *(271)*, aurora kinase *(272)*, stathmin *(273)*, clathrin-coated vesicles *(274)*, aggregosomes *(275)*, and the proteins of the axoneme, centrosome, and basal body *(276,277)*. In addition to mitosis and the other classical microtubule functions, microtubules are thought to be involved in processes such as determination of neuronal polarity and intramanchette transport *(278,279)*. Katanin, incidentally, has been shown to interact differently with two different β isotypes in *C. elegans (271)*. Recent work suggests that $G_s\alpha$ binds to the β-subunit of tubulin close to the GTP binding site *(280)*. As it has been discussed earlier, both the intrinsic and assembly-mediated GTPase activity of tubulin differ among the isotypes *(9,240)*, it is not unreasonable to expect that the binding and effects of $G_s\alpha$ may be isotype-specific as well. Someday these systems will be constructed and tested in vitro with purified tubulin isotypes. Dramatic differences among the isotypes in such experiments would strongly support the hypothesis that certain functions are mediated by different isotypes.

3.9. Structure–Function Correlations in Tubulin Isotypes

The differences in amino acid sequence among the isotypes of a given organism are generally clustered at the C-terminal ends. The fact that the sequences of the C-termini are usually highly conserved in evolution, even to minor differences, indicates that the C-termini are important. In addition, the C-termini contain the sites of most of the post-translational modifications, including phosphorylation, tyrosinolation/detyrosinolation, deglutamylation, polyglutamylation, and polyglycylation. The C-termini are highly negatively charged. As negative charges repel, the C-termini are likely to be projecting outward from the tubulin dimer and the microtubule. With such a model, it is very easy to imagine that the C-terminus serves as a signal for other proteins that help to determine the function of that isotype. Since a β-tubulin with the sequence EGEFEEE near its C-terminus is likely to form an axoneme *(73)*. Fackenthal et al. *(49)* found that removal of the C-terminus from the axonemal β2 isotype in *Drosophila* did not prevent that isotype from forming the axonemal microtubules, but those axonemes were not functional. Clearly, the signal sequence is necessary for successful function in the case of this isotype. The C-termini of α- and β-tubulin are also the sites where a variety of proteins bind; these include MAP2, tau, calponin, and the motor protein Ncd *(281–283)*. Interestingly, Burns and Surridge *(284)* noticed a correlation between the nature of the aromatic amino acid near the C-terminus of β isotypes and the amino acid at position 217/218. If the former is a tyrosine then the latter two are both threonines, whereas if the former is a phenylalanine, then the latter are other residues. This suggests that the C-terminus may occasionally lie down along the microtubule and interact with the residues at position 217/218. Thus, the "visibility" of the signal sequence may vary depending on circumstances.

The C-terminal sequence is not the whole story, however. Tubulin isotypes differ from each other at other places besides their C-termini. The lack of the assembly-critical cys239 in mammalian βIII is a case in point. Hoyle et al. *(285)* prepared a chimera of *Drosophila* β2 in which positions 1–344 were replaced by the corresponding sequence of β3. The remainder of the β2 contained the C-terminal sequence. β2 is the axonemal and meiotic isotype. If the C-terminal sequence were all that mattered then the chimeric tubulin should function equally well. In reality, the chimeric protein did not form outer doublet microtubules very well and was not able to carry out meiosis successfully. Thus, parts of the protein other than the C-termini must play a role in determining isotype function.

Other evidence supports this hypothesis. For example, a difference has been observed in the conformational rigidity among the αβII, αβIII, and αβIV dimers in the region in which a crosslink can be artificially formed between cys12 and either cys201 or cys211 *(168)*. Modeling studies indicate that this region is the binding pocket for the exchangeable GTP and that GTP binding is influenced by conformational changes in this region *(286)*. The kinetics of hydrolysis of this GTP, which determine the dynamic properties of the microtubule, will certainly be influenced by the conformational rigidity in this area, which in turn depends on the nature of the isotype. Similarly, the lateral and longitudinal bond energies in the microtubule have been estimated and could easily vary among the isotypes *(287)*. Specific amino acid substitutions at positions involved in lateral tubulin/tubulin interactions have been shown to promote cold stability *(288,289)*.

The simplest hypothesis about the structure/function correlations in tubulin isotypes is that the C-terminal sequence serves as a signal to other cellular proteins to determine at which cellular location, or in which population of microtubules, the isotype will perform its function. The rest of the protein is necessary for that function to be performed properly.

4. THE EVOLUTION OF TUBULIN ISOTYPES

4.1. Evolution of the Vertebrate β-Isotypes

Enough β tubulins from vertebrates have been sequenced to enable one to construct a rough family tree. As a first step, it can be asked, which β isotypes do *not* appear in both mammals and birds. Thus, birds have only a single βI and a single βIV. Mammals (mice, humans) have two βIV. Thus, the divergence of βIVa and βIVb must be dated after 310 mya, the date at which the ancestral lines of mammals and birds diverged *(61)* but before the divergence of the rodents and primates at 84 mya *(290)*. Although βIVa and βIVb differ in their tissue distributions—the former occurring in brain only and the latter in all tissues—there is as yet no evidence of a functional difference between them. Humans and rhesus monkeys have two βI but mice have only one *(69)*. Thus, the βI isotype diverged into two species sometime after 84 mya. As with βIVa and βIVb, the functional significance, if any, of the differences between βIa and βIb is as yet unknown.

As a second step, the vertebrate β isotypes should be grouped, based on their sequences, as follows:

1. Group 1: βI and βIV;
2. Group 2: βII;
3. Group 3: βIII and βV;
4. Group 4: βVII;
5. Group 5: βVI.

Groups 1–3 have been identified in the amphibian *Xenopus* (Table 6). Thus, these isotypes were probably present when vertebrates took their first step on land about 360 mya *(291)*. There is no distinction between βI and βIV in *Xenopus*. The separation of βI and βIV probably occurred at the time of the appearance of reptiles over 310 mya *(61)*. In contrast, βVII has been seen only in humans, so it may have appeared very recently. The mammalian and avian βVI are so different from each other that it is possible that each one may have appeared, separately, after 310 mya. βV occurs in birds, mammals, and amphibians, but not in fish. Thus, it probably diverged from βIII at least as early as 360 mya.

Studies of β isotypes in fish are illuminating. There are various β isotypes present in fish that do not have precise equivalents among other vertebrates. In addition to these, however, Groups 1, 2, and 3 can be recognized. Thus, these groups probably diverged from each other at or sometime after the appearance of the chordates about 590 mya *(292)*. A very intriguing experiment by Modig et al. *(293)* may cast light on the early evolution of vertebrate isotypes. The Atlantic cod, *Gadus morhua* has cold-stable microtubules. Transfection of fish βIV into human cells caused the microtubules of these cells to become cold-stable. The same result was observed upon transfection of βII. However, transfection of fish βIII did not confer cold stability. It is logical to assume that if βII and βIV are major structural components of fish microtubules, then they must be able to provide cold stability. In contrast, fish βIII is incapable of performing this function. As *Gadus* lives its entire life cycle at cold temperatures, it is unlikely that it could have microtubules made entirely of βIII, as such microtubules would be cold-labile. It is conceivable, of course, that certain MAPs could make βIII-microtubules cold-stable. This is unlikely, however, as microtubule cold-stability in Antarctic fish has been shown to reside in tubulin and not in MAPs *(294)*.

Thus βIII cannot be the major component of any microtubule population in fish and the invitation is given to speculate upon its function. In mammals, the earlier hypothesizing suggested that βIII had two major functions: (1) to form highly dynamic microtubules that may be particularly important in development, especially in the nervous system; and (2) to make microtubules resistant to ROS. Both of these functions could reasonably occur in fish. Vertebrate evolution has been suggested to be an example of neoteny, in which a larva attains sexual maturity without metamorphosing into an adult (295). The ancestors of the vertebrates may have had clearly differentiated larval and adult stages, the former motile, and the latter sessile. At one point in evolution, the larva acquired sexual maturity and the adult stage disappeared. Thus, vertebrates were able to grow in size and retain motility. Tunicates, which are nonvertebrate chordates with very simple nervous systems (296), have neither βII nor βIII. It is possible that βIII appeared at the time when the vertebrates diverged from the other chordates and that its high dynamicity made it useful in the rapid growth of the complex nervous system of vertebrates (297). In this connection, it is worth recalling that βIII is common and widespread in embryos, in which growth and development take place very rapidly (136,158). βIII is unlikely ever to have been the sole component of a microtubule, but it could confer dynamicity to microtubules in which it occurred. Microtubules made of mixtures of αβIII and αβII dimers are significantly more dynamic than those made of αβII alone, provided that αβIII predominates (100). It is perhaps not a coincidence that chordates appeared soon after the concentration of O_2 in the Earth's atmosphere reached 10% of present levels, the level required to form collagen and hence cartilage and bone (298). The higher O_2 level would have led to increased production of ROS. As vertebrates developed an advanced nervous system, βIII may have acquired the additional function of protecting the long-lived neuronal microtubules from ROS. If these arguments were correct, one would predict that cephalopod mollusks that are as ancient as the vertebrates and that are long-lived and have a complex nervous system (299), would also have a tubulin isotype capable of protecting the neuronal microtubules from ROS.

The development of the nervous system would also have entailed the appearance of βII, which presumably has a major, but as yet unknown, function in the nervous system. If microtubules play a role in reorganizing the nuclear envelope, and if this is an ancient function, βII may have retained this function, and the other isotypes may have lost it.

About 360 mya, the vertebrates emerged onto the land (300), thereby exposing themselves directly to the higher levels of O_2 present in the atmosphere as well as to the strong solar ultraviolet radiation that is capable of creating free radicals. There may have been a premium on protection of microtubules from ROS, not only in the brain, but in other tissues as well. If βV shares this function with βIII, as has been hypothesized, it is possible that βV appeared about this time to protect the microtubules of other tissues from ROS, whereas βIII performed that same function in neurons.

Full sequencing and analysis of reptile, amphibian, and fish genomes as well as those of the nonvertebrate chordates may flesh out, corroborate, or disprove some of these speculations. Further experiments on the evolution of nitric oxide synthase as well as careful studies of the tissue distribution of nitric oxide synthase and tubulin isotypes in fish would be very useful as well.

4.2. Evolution of the Vertebrate α-Isotypes

The evolution of the α isotypes in vertebrates is not as well understood as that of the β isotypes. It is clear that the class V α (called α8) are present in mammals and birds, but not, so far as it can be told, in amphibians or fish (185). Therefore, α8 probably appeared between 360 mya and 310 mya. The same argument could be made for class III. However, it appears that classes I, II, and IV were present in fish probably after the chordates appeared around 590 mya *(292)*. *Xenopus* ovarian α and mouse testis αTT1 are too unique to draw conclusions regarding their evolution.

4.3. Evolution of Tubulin Isotypes in the Other Eukaryotes

The knowledge of tubulin isotype evolution in other eukaryotes is quite limited. There are only a few phyla where multiple tubulin isotypes have been sequenced in more than one organism. Nevertheless, there are a few generalizations that are probably safe to make. With one possible exception, which will be discussed shortly, it is clear that, although all α-tubulins resemble each other and the β-tubulins do likewise, there is no specific resemblance between one particular isotype in one phylum and another particular isotype in another phylum. In other words, there is no close structural resemblance between, say, βIII in vertebrates and any β isotype in any other phylum. In brief, families of isotypes are phylum-specific. This has been seen to be true for the α- and β-tubulins in vertebrates. There are discernible families of α and β isotypes in arthropods, nematodes, and angiosperms *(8)*.

The possible exception to this pattern is βIV, the isotype with the signal sequence EGEFEEE, which is required for a β-tubulin to form part of an axoneme *(73)*. This sequence, or one very similar, occurs in at least one isotype in virtually every eukaryotic organism except fungi (which lack axonemes, basal bodies, and centrioles). In a sense, therefore, a tubulin containing this sequence has to be thought of as ancestral to all β-tubulins. It is highly unlikely that the signal sequence would spontaneously arise *de novo* three separate times, during the evolution of animals, plants, and protists. Therefore, a β-tubulin containing this sequence must have been present in the ancestral eukaryote. Fungi, presumably, would have lost this β-tubulin when they lost the complex microtubule apparatuses in which this tubulin is required. That said, however, beyond the signal sequence there is no overall specific quantifiably demonstrable similarity between, say, vertebrate βIV and the corresponding β2 isotype in *Drosophila*. Thus, if assigning to βIV the additional function of being involved in actin–microtubule crosstalk is correct, this function may have arisen secondarily in βIV. If a β-isotype in *Drosophila* also has this function, then that isotype need not be β2.

Various fish (*Notothenia* and *Danio*) have at least one α or β isotype that have no specific equivalent in amphibians, birds, or mammals. The specific functions of these isotypes are unknown. Conceivably, these may represent the survivors of a large pool of tubulin isotypes that arose when the vertebrates appeared. Speculating further, an intriguing correlation could be postulated. Most of today's animal phyla—at least those where fossil evidence is available—arose during the so-called Cambrian explosion, about 530 mya, when the ancestors of today's phyla shared their world with animals with unusual body plans who left no descendants *(301)*. Whatever geological, climatic, or ecological factors promoted the appearance of multiple body plans could also have impelled the diversification of tubulin isotypes. If the multiple isotypes had different functions, then in view of tubulin's important role in development, it is not difficult to imagine that different combinations of isotypes correlated with the appearance of specific phyla. If it is assumed that the earliest eukaryote had a single α and a single β isotype, then this tubulin would perhaps have been involved in different functions: not only mitosis and axonemal motility but perhaps nuclear envelope organization and actin–microtubule crosstalk as well. The appearance of multiple isotypes meant that these functions could have been distributed among different isotypes. Subsequent evolution of each phylum would involve essentially random selection from this pool of isotypes of particular ones performing whatever functions were adaptive for organisms in that phylum. For example, an early animal having several isotypes with the appropriate signal sequence for axonemal motility, could be imagined. In vertebrates, the ancestral βIV could have been randomly selected for subsequent evolution and the others lost; in arthropods the ancestral β2 would have been similarly selected. In such fashion, phyla would arise with unique families of isotypes, and there would be no specific similarity—other than the signal sequence—among the axonemal isotypes of the various phyla.

Two fish β isotypes may fit this hypothesis. *Gadus* has a β isotype (classified as βIII) that has the same cysteines as does βIII (i.e., it lacks cys239 and has cys124). Nevertheless, the C-terminus of the isotype lacks the basic residues seen in mammalian βIII. Conceivably, this isotype could have the putative antioxidizing property of mammalian βIII but lack the dynamic properties of βIII. In contrast, *Notothenia* has a β isotype (unclassified) that has the C-terminus with basic residues similar to those of βIII or βVI, but does not have the same cysteines as βIII. Perhaps, this isotype exhibits the dynamic behavior of βIII but lacks any antioxidant activity.

In pursuing the evolution of the isotypes of α- and β-tubulin a trail has been followed that fades out sometime before the beginning of the Paleozoic. Further insights may be provided when the story of α and β is compared with that of the other members of the tubulin superfamily, as will be discussed in Chapter 7.

ACKNOWLEDGMENTS

Supported by grants to RFL (Welch Foundation AQ-0726, National Institutes of Health CA26376 and CA084986, US Army Breast Cancer Research Program W81XWH-05-1-0238 and DAMD17-01-1-0411, US Army Prostate Cancer Research Program DAMD17-02-1-0045 and W81XWH-04-1-0231) and to AB (National Institutes of Health CA59711 and US Army Breast Cancer Research Program DAMD17-98-1-8244). The tremendous contributions of the late Mary Carmen Roach in creating several of the isotype-specific monoclonal antibodies is gratefully recalled. The collaborators: Drs. Larry Barnes, Kirk Beisel, Don Cleveland, W. Brent Derry, H. William Detrich, Charles DuMontet, Yves Engelborghs, Arlette Fellous, Richard Hallworth, Paul Hoffman, Elzbieta Izbicka, Mary Ann Jordan, Israr Khan, Jeffrey Kreisberg, George Langford, Ruth Lezama, John Liggins, Tom MacRae, Isaura Meza, Melvyn Little, Qing Lu, Carina Modig, Grace Moore, Doug Murphy, Dulal Panda, Brian Perry, Peter Ravdin, Robert Renthal, Barbara Schneider, Jyotsna Sharma, Julia Vent, Katherine Wall, Margareta Wallin, Consuelo Walss-Bass, Alex Weis, Leslie Wilson, Karen Woo, Keliang Xu, I-Tien Yeh, and Hans-Peter Zimmermann as well as students: Patrick Joe, Gerardo Elguezabal, Jonquille Eley, and Heather Jensen-Smith are thanked. Mohua Banerjee, Lorraine Kasmala, Veena Prasad, Patricia Schwarz, and Phyllis Trcka Smith, Virginia Boucher, Sebastien David, Herb Miller, and Margaret Miller are thanked for skilled technical assistance. Helpful suggestions, materials, or information from Asish Chaudhuri, Kenneth Downing, Charles DuMontet, Anna Lazzell, Linda Luduena, Susan Mooberry, Jack Tuszynski, Isao Tomita, Consuelo Walss-Bass, and Leslie Wilson are gratefully acknowledged.

REFERENCES

1. Bryan J, Wilson L. Are cytoplasmic microtubules heteropolymers? Proc Nat Acad Sci USA 1971;68:1762–1766.
2. Ludueña RF, Shooter EM, Wilson L. Structure of the tubulin dimer. J Biol Chem 1977;252:7006–7014.
3. Ponstingl H, Krauhs E, Little M, Kempf T. Complete amino acid sequence of α-tubulin from porcine brain. Proc Nat Acad Sci USA 1981;78:2757–2761.
4. Krauhs E, Little M, Kempf T, Hofer-Warbinek R, Ade W, Ponstingl H. Complete amino acid sequence of β-tubulin from porcine brain. Proc Nat Acad Sci USA 1981;78:4156–4160.
5. Behnke O, Forer A. Evidence for four classes of microtubules in individual cells. J Cell Sci 1967;2:169–192.
6. Fulton C, Simpson PA. Selective synthesis and utilization of flagellar tubulin. The multi-tubulin hypothesis. In: Goldman R, Pollard T, Rosenbaum J, eds. Cell Motility Vol. 3. Cold Spring Harbor, NY: Cold Spring Harbor Lab. Press 1976; 987–1005.
7. Ludueña RF. Are tubulin isotypes functionally significant. Mol Biol Cell 1993;4:445–457.
8. Ludueña RF. The multiple forms of tubulin: different gene products and covalent modifications. Int Rev Cytol 1998;178:207–275.
9. Lu Q, Moore GD, Walss C, Ludueña RF. Structural and functional properties of tubulin isotypes. Adv Struct Biol 1998;5:203–227.
10. Alexandraki D, Ruderman JV. Evolution of αq- and β-tubulin genes as inferred by the nucleotide sequences of sea urchin cDNA clones. J Mol Evol 1983;19:397–410.
11. Gianguzza F, Di Bernardo MG, Sollazzo M, et al. DNA sequence and pattern of expression of the sea urchin (*Paracentrotus lividus*) α-tubulin genes. Mol Reprod Dev 1989;1:170–181.
12. Edvardsen RB, Flaat M, Tewari R, et al. Most intron positions in *Oikopleura dioica* α-tubulin genes are unique: did new introns help to preserve and expand gene families? NCBI Accession no. AAM73988, AAM73989, AAM73990, 2002.
13. Hallworth R, Ludueña RF. Differential expression of β tubulin isotypes in the adult gerbil organ of Corti Hearing Res 2000;148:161–172
14. Jensen-Smith HC, Eley J, Steyger PS, Ludueña RF, Hallworth R. Cell type-specific reduction of β tubulin isotypes synthesized in the developing gerbil organ of Corti. J Neurocytol 2003;32:185–197.
15. Perry B, Jensen-Smith HC, Ludueña RF, Hallworth R. Differential expression of β tubulin isotypes in gerbil vestibular end organs. J Assoc Res Otorhinolaryngol (JARO) 2003;4:329–338 (on-line).
16. Wang T, Lessman CA. Isoforms of soluble α-tubulin in oocytes and brain of the frog (genus *Rana*): changes during oocyte maturation. Cell Mol Life Sci 2002;59:2216–2223.
17. Miya T, Satoh N. Isolation and characterization of cDNA clones for β-tubulin genes as a molecular marker for neural cell differentiation in the ascidian embryo. Int J Dev Biol 1997;41:551–557.
18. Costa S, Ragusa MA, Drago G, et al. Sea urchin neural α2 tubulin gene: isolation and promoter analysis. Biochem Biophys Res Commun 2004;316:446–453.
19. Kawasaki H, Sugaya K, Quan GX, Nohata J, Mita K. Analysis of α- and β-tubulin genes of *Bombyx mori* using an EST database. Insect Biochem Mol Biol 2003;33:131–137.
20. Simoncelli F, Sorbolini S, Fagotti A, Di Rosa I, Porceddu A, Pascolini R. Molecular characterization and expression of a divergent α-tubulin in planarian *Schmidtea polychroa*. Biochim Biophys Acta 2003;1629:26–33.
21. Savage C, Hamelin M, Culotti JG, Coulson A, Albertson DG, Chalfie M. mec-7 is a β-tubulin gene required for the production of 15-protofilament microtubules in *Caenorhabditis elegans*. Genes Dev 1989;3:870–881.
22. Fukushige T, Siddiqui ZK, Chou M, et al. MEC-12, an α-tubulin required for touch sensitivity in *C. elegans*. J Cell Sci 1999;112:395–403.
23. Wright AJ, Hunter CP. Mutations in a β-tubulin disrupt spindle orientation and microtubule dynamics in the early *Caenorhabditis elegans* embryo. Mol Biol Cell 2003;14:4512–4525.
24. Okamura S, Naito K, Sonehara K, et al. Characterization of the carrot β-tubulin gene coding a divergent isotype, β-2. Cell Struct Funct 1997;22:291–298.
25. Matzk F, Meyer H-M, Horstmann C, Balzer HJ, Bäumlein H, Schubert I. A specific α-tubulin is associated with the initiation of parthenogenesis in Salmon wheat lines. Hereditas 1997;126:219–224.
26. Whittaker DJ, Triplett BA. Gene-specific changes in α-tubulin transcript accumulation in developing cotton fibers. Plant Physiol 1999;121:181–188.

27. Yoshikawa M, Yang G, Kawaguchi K, Komatsu S. Expression analyses of β-tubulin isotype genes in rice. Plant Cell Physiol 2003;44:1202–1207.

28. Schröder J, Stenger H, Wernicke W. α-Tubulin genes are differentially expressed during leaf cell development in barley (*Hordeum vulgare* L.). Plant Mol Biol 2001;45:723–730.

29. Morello L, Bardini M, Sala F, Breviario D. A long leader intron of the *Ostub 16* rice β-tubulin gene is required for high-level gene expression and can autonomously promote transcription both in vivo and in vitro. Plant J 2002;29:33–44.

30. Ebel C, Gómez Gómez L, Schmit AC, Neuhaur-Url G, Boller T. Differential mRNA degradation of two β-tubulin isoforms correlates with cytosolic Ca^{2+} changes in glucan-elicited soybean cells. Plant Physiol 2001;126:87–96.

31. Hellmann A, Wernicke W. Changes in tubulin protein expression accompany reorganization of microtubular arrays during cell shaping in barley leaves. Planta 1998;204:220–225.

32. Abe T, Thitamadee S, Hashimoto T. Microtubule defects and cell morphogenesis in the lefty1 lefty2 tubulin mutant of *Arabidopsis thaliana*. Plant Cell Physiol 2004;45:211–220.

33. Buhr TL, Dickman MB. Isolation, characterization, and expression of a second β-tubulin-encoding gene from *Colletotrichum gloeosporiodes* f. Sp. *aeschynomene*. Appl Environ Microbiol 1994;60: 4155–4159.

34. Yan K, Dickman MB. Isolation of a β-tubulin gene from *Fusarium moniliforme* that confers cold-sensitive benomyl resistance. Appl Environ Microbiol 1996;62:3053–3056.

35. Monnat J, Ortega Perez R, Turian G. Molecular cloning and expression studies of two divergent α-tubulin genes in *Neurospora crassa*. FEMS Microbiol Lett 1997;150:33–41.

36. Silva WP, Soares RBA, Jesuino RSA, Izacc SMS, Felipe MSS, Soares CMA. Expression of α tubulin during the dimorphic transition of *Paracoccidioides brasiliensis*. Med Mycol 2001;39:457–462.

37. Chung S, Cho J, Cheon H, Paik S, Lee J. Cloning and characterization of a divergent α-tubulin that is expressed specifically in dividing amebae of *Naegleria gruberi*. Gene 2002;293:77–86.

38. Hu K, Suravajjala S, DiLullo C, Roos D, Murray J. Functional specialization of tubulin isoforms in Toxoplasma isoforms in *Toxoplasma gondii*. Am Soc Cell Biol Ann Meeting Abstr p. 424a.

39. Paul ECA, Buchschacher GL, Cunningham DB, Dove WF, Burland TG. Preferential expression of one β-tubulin gene during flagellate development in Physarum. J Gen Microbiol 1992;138:229–238.

40. Cunningham DB, Buchschacher GL, Burland TG, Dove WF, Kessler D, Paul ECA. Cloning and characterization of the altA α-tubulin gene of *Physarum*. J Gen Microbiol 1993;139:137–151.

41. Matthews KA, Rees D, Kaufman TC. A functionally specialized α-tubulin is required for oocyte meiosis and cleavage mitoses in *Drosophila*. Development 1993;117:977–991.

42. Hecht NB, Distel RJ, Yelick PC, et al. Localization of a highly divergent mammalian testicular α tubulin that is not detectable in brain. Mol Cell Biol 1988;8:996–1000.

43. Wu W-L, Morgan GT. Ovary-specific expression of a gene encoding a divergent α-tubulin isotype in *Xenopus*. Differentiation 1994;58:9–18.

44. Evrard J-L, Nguyen I, Bergdoll M, Mutterer J, Steinmetz A, Lambert A-M. A novel pollen-specific α-tubulin in sunflower: structure and characterization. Plant Mol Biol 2002;49:611–620.

45. Wang D, Villasante A, Lewis SA, Cowan NJ. The mammalian β-tubulin repertoire:hematopoietic expression of a novel heterologous β-tubulin isotype. J Cell Biol 1986;103:1903–1910.

46. Murphy DB, Wallis KT, Machlin PS, Ratrie H, Cleveland DW. The sequence and expression of the divergent β-tubulin in chicken erythrocytes. J Biol Chem 1987;262:14,305–14,312.

47. Guiltinan MJ, Ma D-P, Barker RF, et al. The isolation, characterization and sequence of two divergent β-tubulin genes from soybean (*Glycine max* L.). Plant Mol Biol 1987;10:171–184.

48. Fackenthal JD, Hutchens JA, Turner FR, Raff EC. Structural analysis of mutations in the *Drosophila* β2-tubulin isoform reveals regions in the β-tubulin molecule required for general and for tissue-specific microtubule functions. Genetics 1995;139:267–286.

49. Fackenthal JD, Turner FR, Raff EC. Tissue-specific microtubule functions in *Drosophila* spermatogenesis require the β-tubulin isotype-specific carboxy-terminus. Dev Biol 1993;158:213–227.

50. Fuller MT, Caulton JH, Hutchens JA, Kaufman TC, Raff EC. Genetic analysis of microtubule structure. A β-tubulin mutation causes the formation of aberrant microtubules in vivo and in vitro. J Cell Biol 1993;104:385–394.

51. Rudolph JE, Kimble M, Hoyle HD, Suber MA, Raff EC. Three *Drosophila* β-tubulin sequences: a developmentally regulated isoform (β3), the testis-specific isoform (β2), and an assembly-defective mutation of the testis-specific isoform (β2t[8]) reveal both an ancient divergence in metazoan isotypes and structural constraints for β-tubulin function. Mol Cell Biol 1987;7:2231–2242.

52. Hoyle HD, Raff EC. Two *Drosophila* β tubulin isoforms are not functionally equivalent. J Cell Biol 1990;11:1009–1026.
53. Kramer J, Hawley RS. The spindle-associated transmembrane protein Axs identifies a membranous structure ensheathing the meiotic spindle. Nature Cell Biol 2003;5:261–267.
54. Dettman RW, Turner FR, Hoyle HD, Raff EC. Embryonic expression of the divergent *Drosophila* β3-tubulin isoform is required for larval behavior. Genetics 2001;158:253–263.
55. Buttgereit D, Paululat A, Renkawitz-Pohl R. Muscle development and attachment to the epidermis is accompanied by expression of β3 and β1 tubulin isotypes, respectively. Int J Dev Biol 1996;40:189–196.
56. Komma DJ, Endow SA. Enhancement of the ncd microtubule motor mutant by mutants of αTub67C. J Cell Sci 1997;110:576–583.
57. Matthies HJG, Messina LG, Namba R, Greer KJ, Walker MY, Hawley RS. Mutations in the α-Tubulin 67C gene specifically impair achiasmate segregation in *Drosophila melanogaster*. J Cell Biol 1999;147:1137–1144.
58. Hutchens JA, Hoyle HD, Turner FR, Raff EC. Structurally similar *Drosophila* α-tubulins are functionally distinct in vivo. Mol Biol Cell 1997;8:481–500.
59. Roach MC, Boucher VL, Walss C, Ravdin PM, Ludueña RF. Preparation of a monoclonal antibody specific for the class I isotype of β-tubulin. The β isotypes of tubulin differ in their cellular distributions within human tissues. Cell Motil Cytoskeleton 1998;39:273–285.
60. Havercroft JC, Cleveland DW. Programmed expression of β-tubulin genes during development and differentiation of the chicken. J Cell Biol 1984;99:1927–1935.
61. Kumar S, Hedges, SB. A molecular timescale for vertebrate evolution. Nature 1998;392:917–920.
62. Lewis SA, Lee MGS, Cowan NJ. Five mouse tubulin isotypes and their regulated expression during development. J Cell Biol 1985;101:852–861.
63. Monteiro MJ, Cleveland DW. Sequence of chicken cβ7 tubulin. Analysis of a complete set of vertebrate β-tubulin isotypes. J Mol Biol 1988;199:439–446.
64. Banerjee A, Roach MC, Wall KA, Lopata MA, Cleveland DW, Ludueña RF. A monoclonal antibody against the type II isotype of β-tubulin. Preparation of isotypically altered tubulin. J Biol Chem 1988;263:3029–3034.
65. Grieshaber NA, Ko C, Grieshaber SS, Ji I, Ji TH. Follicle-stimulating hormone-responsive cytoskeletal genes in rat granulosa cells: class I β-tubulin, tropomyosin-4, and kinesin heavy chain. Endocrinology 2003;144:29–39.
66. Oehlmann VD, Berger S, Sterner C, Korsching SI. Zebrafish β tubulin 1 expression is limited to the nervous system throughout development, and in the adult brain is restricted to a subset of proliferative regions. Gene Expr Patterns 2004;4:191–198.
67. Lee MG, Lewis SA, Wilde CD, Cowan NJ. Evolutionary history of a multigene family: an expressed human β-tubulin gene and three processed pseudogenes. Cell 1983;33:477–487.
68. Hirakawa M, Yamaguchi H, Imai K, Shimada J. NCBI Accession no. BAB63321, 1999.
69. Crabtree DV, Ojima I, Geng X, Adler AJ. Tubulins in the primate retina: evidence that xanthophylls may be endogenous ligands for the paclitaxel-binding site. Bioorg Med Chem 2001;9:1967–1976.
70. Narishige T, Blade KL, Ishibashi Y, et al. Cardiac hypertrophy and developmental regulation of the β-tubulin multigene family. J Biol Chem 1999;274:9692–9697.
71. Woo K, Jensen-Smith HC, Ludueña RF, Hallworth R. Differential expression of β tubulin isotypes in gerbil nasal epithelia. Cell Tissue Res 2002;309:331–335.
72. Jensen-Smith HC, Ludueña RF, Hallworth R. Requirement for the βI and βIV tubulin isotypes in mammalian cilia. Cell Motil Cytoskeleton 2003;55:213–220.
73. Raff EC, Fackenthal JD, Hutchens JA, Hoyle HD, Turner FR. Microtubule architecture specified by a β-tubulin isoform. Science 1997;275:70–73.
74. Lezama R, Castillo A, Ludueña RF, Meza I. Over-expression of βI tubulin in MDCK cells and incorporation of exogenous βI tubulin into microtubules interferes with adhesion and spreading. Cell Motil Cytoskeleton 2001;50:147–160.
75. Yanagida M, Hayano T, Yamauchi Y, et al. Human fibrillarin forms a sub-complex with splicing factor 2-associated p32, protein arginine methyltransferases, and tubulins α3 and β1 that is independent of its association with preribosomal ribonucleoprotein complexes. J Biol Chem 2004;279:1607–1614.
76. Yeh I-T, Ludueña RF. The β_{II} isotype of tubulin is present in the cell nuclei of a variety of cancers. Cell Motil Cytoskeleton 2004;57:96–106.
77. Arai K, Shibutani M, Matsuda H. Distribution of the class II β-tubulin in developmental and adult rat tissues. Cell Motil Cytoskeleton 2002;52:174–182.

78. Dozier JH, Hiser L, Davis JA, et al. β class II tubulin predominates in normal and tumor breast tissues. Breast Cancer Res 2003;5:R157–R169.
79. Nakamura Y, Yamamoto M, Oda E, et al. Expression of tubulin βII in neural stem/progenitor cells and radial fibers during human fetal brain development. Lab Invest 2003;83:479–489.
80. Renthal R, Schneider BG, Miller MM, Ludueña RF. β_{IV} is the major β-tubulin isotype in bovine cilia. Cell Motil Cytoskeleton 1993;25:19–29.
81. Armas-Portela R, Parrales MA, Albar JP, Martinez C, Avila J. Distribution and characteristics of βII tubulin-enriched microtubules in interphase cells. Exp Cell Res 1999;248:372–380.
82. Ranganathan S, Salazar H, Benetatos CA, Hudes GR. Immunohistochemical analysis of β-tubulin isotypes in human prostate carcinoma and benign prostatic hypertrophy. Prostate 1997;30:263–268.
83. Walss C, Kreisberg JI, Ludueña RF. Presence of the β_{II}-isotype of tubulin in the nuclei of cultured rat kidney mesangial cells. Cell Motil Cytoskeleton 1999;42:274–284.
84. Bugnard E, Zaal KJM, Ralston E. Reorganization of microtubule nucleation during muscle differentiation. Cell Motil Cytoskeleton 2005;60:1–13.
85. Sullivan KF, Cleveland DW. Sequence of a highly divergent β tubulin gene reveals regional heterogeneity in the β tubulin polypeptide. J Cell Biol 1984;99:1754–1760.
86. Banerjee A. NCBI Accession no. AAL28094, 2001.
87. Burgoyne RD, Cambray-Deakin MA, Lewis SA, Sarkar S, Cowan NJ. Differential distribution of β-tubulin isotypes in cerebellum. EMBO J 1988;7:2311–2319.
88. Chen SS, Revoltella RP, Papini S, et al. Multilineage differentiation of rhesus monkey embryonic stem cells in three-dimensional culture systems. Stem Cells 2003;21:281–295.
89. Katsetos CD, Legido A, Perentes E, Mörk SJ. Class III β-tubulin isotype: a key cytoskeletal protein at the crossroads of developmental neurobiology and tumor neuropathology. J Child Neurol 2003;18:851–866.
90. Butler R, Leigh PN, Gallo JM. Androgen-induced up-regulation of tubulin isoforms in neuroblastoma cells. J Neurochem 2001;78:854–861.
91. Matsuo N, Hoshino M, Yoshizawa M, Nabeshima Y. Characterization of STEF, a guanine nucleotide exchange factor for Rac1, required for neurite growth. J Biol Chem 2002;277:2860–2868.
92. Sanchez-Ramos JR, Song S, Kamath SG, et al. Expression of neural markers in human umbilical cord blood. Exp Neurol 2001;171:109–115.
93. Correa LM, Miller MG. Microtubule depolymerization in rat seminiferous epithelium is associated with diminished tyrosination of α-tubulin. Biol Reprod 2001;64:1644–1652.
94. Laemmli UK. Cleavage of structural proteins during the assembly of the head of bacteriophage T_4. Nature 1970;227:680–685.
95. Little M. Identification of a second β chain in pig brain tubulin. FEBS Lett 1979;108:283–286.
96. Ludueña RF, Roach MC, Trcka PP, et al. β_2-Tubulin, a form of chordate brain tubulin with lesser reactivity toward an assembly-inhibiting sulfhydryl-directed cross-linking reagent. Biochemistry 1979; 21:4787–4794.
97. Detrich HW, Prasad V, Ludueña RF. Cold-stable microtubules from Antarctic fishes contain unique α tubulins. J Biol Chem 1987;262:8360–8366.
98. Modig C, Olsson P-E, Barasoain I, et al. Identification of β_{III}- and β_{IV}-tubulin isotypes in cold-adapted microtubules from Atlantic cod (Gadus morhua): antibody mapping and cDNA sequencing. Cell Motil Cytoskeleton 1999;42:315–330.
99. Alexander JE, Hunt DF, Lee MK, et al. Characterization of posttranslational modifications in neuron-specific class III β-tubulin by mass spectrometry. Proc Nat Acad Sci USA 1991;88:4685–4689.
100. Panda D, Miller HP, Banerjee A, Ludueña RF, Wilson L. Microtubule dynamics in vitro are regulated by the tubulin isotype composition. Proc Nat Acad Sci USA 1994;91:11358–11362.
101. Mellon MG, Rebhun LI. Sulfhydryls and the in vitro polymerization of tubulin. J Cell Biol 1976; 70:226–238.
102. Little M, Ludueña, RF. Structural differences between brain β1- and β2-tubulins: implications for microtubule assembly and colchicine binding. EMBO J 1985;4:51–56.
103. Bai RL, Lin CM, Nguyen NY, Liu TY, Hamel E. Identification of the cysteine residue of β-tubulin affected by the antimitotic agent 2,4-dichlorobenzyl thiocyanate, facilitated by separation of the protein subunits of tubulin by hydrophobic column chromatography. Biochemistry 1989;28:5606–5612.
104. Palanivelu P, Ludueña RF. Interactions of the τ-tubulin-vinblastine complex with colchicine, podophyllotoxin, and N,N′-ethylenebis(iodoacetamide). J Biol Chem 1982;257:6311–6315.

105. Ludueña RF, Roach MC, Jordan MA, Murphy, DB. Different activities of brain and erythrocyte tubulins toward a sulfhydryl group-directed reagent that inhibits microtubule assembly. J Biol Chem 1985;260:1257–1264.

106. Dong Z, Thoma RS, Crimmins DL, McCourt DW, Tuley EA, Sadler JE. Disulfide bonds required to assemble functional von Willebrand factor multimers. J Biol Chem 1994;260:6753–6758.

107. Mayadas TN, Wagner DD. Vicinal cysteines in the prosequence play a role in von Willebrand multimer assembly. Proc Nat Acad Sci USA 1992;89:3531–3535.

108. Li PP, Nakanishi A, Clark SW, Kasamatsu H. Formation of transitory intrachain and interchain disulfide bonds accompanies the folding and oligomerization of simian virus 40 Vp1 in the cytoplasm. Proc Nat Acad Sci USA 2002;99:1353–1358.

109. Beckman JS, Chen J, Ischiropoulos H, Crow JP. Oxidative chemistry of peroxynitrite. Methods Enzymol 1994;233:229–240.

110. Landino LM, Iwig JS, Kennett KL, Moynihan KL. Repair of peroxynitrite damage to tubulin by the thioredoxin reductase system. Free Radic Biol Med 2004;36:497–506.

111. Nogales E, Wolf SG, Downing KH. Structure of the αβ tubulin dimer by electron crystallography. Nature 1998;391:199–203.

112. Landino LM, Hasan R, McGaw A, et al. Peroxynitrite oxidation of tubulin sulfhydryls inhibits microtubule polymerization. Arch Biochem Biophys 2002;398:213–220.

113. Mungrue IN, Bredt DS. nNOS at a glance: implications for brain and brawn. J Cell Sci 2004;117:2627–2629.

114. Gally JA, Montague PR, Reeke GN, Edelman GM. The NO hypothesis: possible effects of short-lived rapidly diffusible signal in the development and function of the nervous system. Proc Nat Acad Sci USA 1990;87:3547–3551.

115. Bredt DS, Snyder SH. Nitric oxide, a novel neuronal messenger. Neuron 1992;8:3–11.

116. Blottner D, Luck G. Just in time and place: NOS/NO system assembly in neuromuscular junction formation. Microsc Res Tech 2001;55:171–180.

117. Cappelletti G, Tedeschi G, Maggioni MG, Negri A, Nonnis S, Maci R. The nitration of τ protein in neurone-like PC12 cells. FEBS Lett 2004;562:35–39.

118. Lewis SA, Cowan NJ. Complex regulation and functional versatility of mammalian α- and β-tubulin isotypes during the differentiation of testis and muscle cells. J Cell Biol 1988;106:2023–2033.

119. Lee NPY, Cheng CY. Regulation of Sertoli cell tight junction dynamics in the rat testis via the nitric oxide synthase/soluble guanylate cyclase/3′,5′-cyclic guanosine monophosphate/protein kinase G signaling pathway: an in vitro study. Endocrinology 2003;144:3114–3129.

120. Kon Y, Namiki Y, Endoh D. Expression and distribution of inducible nitric oxide synthase in the testis. Jpn J Vet Res 2002;50:115–123.

121. Mruk DD, Cheng CY. In vitro regulation of extracellular superoxide dismutase in sertoli cells. Life Sci 2000;67:133–145.

122. Holstein GR, Friedrick VI, Martinelli GP, Holstein GR. Monoclonal L-citrulline immunostaining reveal NO-producing vestibular neurons. Ann NY Acad Sci 2001;942:65–78.

123. Nie G, Wang J. Localization of nitric oxide synthase in the chicken vestibular system. J Clin Otorhinolaryngol 2002;16:426–427 (article in Chinese, abstract in English).

124. Takumida M, Anniko M. Simultaneous detection of both nitric oxide and reactive oxygen species in guinea pig vestibular sensory cells. ORL 2002;64:143–147.

125. Takumida M, Anniko M. Direct evidence of nitric oxide production in guinea pig vestibular sensory cells. Acta Otolaryngol 2000;120:134–138.

126. Katsetos CD, Kontogeorgos G, Geddes JF, et al. Differential distribution of the neuron-associated class III β-tubulin in neuroendocrine lung tumors. Arch Pathol Lab Med 2000;124:535–544.

127. Matsuzaki F, Harada F, Nabeshima Y, Fujii-Kuriyama Y, Yahara I. Cloning of cDNAs for two β-tubulin isotypes expressed in murine T cell lymphoma L5178Y and analysis of their translation products. Cell Struct Funct 1987;12:317–325.

128. Asai DJ, Remolona NM. Tubulin usage in vivo: A unique spatial distribution of the minor neuronal-specific β-tubulin isotype in pheochromocytoma cells. Dev Biol 1989;132:398–409.

129. Scott CA, Walker CC, Neal DA, et al. β-Tubulin epitope expression in normal and malignant epithelial cells. Arch Otolaryngol Head Neck Surg 1990;116:583–589.

130. Katsetos CD, Herman MM, Frankfurter A, Uffer S, Perentes E, Rubinstein LJ. Neuron-associated class III β-tubulin isotype, microtubule associated protein 2 and synaptophysin in human retinoblastomas in situ. Lab Invest 1991;64:45–64.

131. Maraziotis T, Perentes E, Karamitopoulou E, et al. Neuron-associated class III β-tubulin isotype, retinal S-antigen, synaptophysin, and glial fibrillary acidic protein in human medulloblastomas: a clinicopathological analysis of 36 cases. Acta Neuropathol 1992;84:355–363.
132. Furuhata S, Kameya T, Toya S, Frankfurter A. Immunohistochemical analysis of 61 pituitary adenomas with a monoclonal antibody to the neuron-specific β-tubulin isotype. Acta Neuropathol 1993;86:518–520.
133. Woulfe J. Class III β-tubulin immunoreactive intranuclear inclusions in human ependymomas and gangliogliomas. Acta Neuropathol 2000;100:427–434.
134. Katsetos CD, Del Valle L, Geddes JF, et al. Aberrant localization of the neuronal class III β-tubulin in astrocytomas. A marker for anaplastic potential. Arch Pathol Lab Med 2001;125:613–624.
135. Hisaoka M, Okamoto S, Koyama S, et al. Microtubule-associated protein-2 and class III β-tubulin are expressed in extraskeletal myxoid chondrosarcoma. Mod Pathol 2003;16:453–459.
136. Katsetos CD, Legido A, Perentes E, Mörk SJ. Class III β-tubulin isotype: a key cytoskeletal protein at the crossroads of developmental neurobiology and tumor neuropathology. J Child Neurol 2003;18:851–866.
137. Katsetos CD, Herman MM, Mörk SJ. Class III β-tubulin in human development and cancer. Cell Motil Cytoskeleton 2003;55:77–96.
138. Hardman JG, Limbird LE. Goodman and Gilman's The Pharmacological Basis of Therapeutics, 9th ed. McGraw-Hill, New York, 1996:1228, 1257–1261, 1603.
139. Mekhail TM, Markman M. Paclitaxel in cancer therapy. Expert Opin Pharmacother 2002;3:755–766.
140. Schiff R, Reddy P, Ahotupa M, et al. Oxidative stress and AP-1 activity in tamoxifen-resistant breast tumors in vivo. J Nat Cancer Inst 2000;92:1926–1934.
141. Brown NS, Bicknell R. Hypoxia and oxidative stress in breast cancer. Oxidative stress: its effects on the growth, metastatic potential and response to therapy of breast cancer. Breast Cancer Res 2001; 3:323–327.
142. Portakal O, Ozkaya O, Erden Inal M, Bozan B, Kosan M, Sayek I. Coenzyme Q10 concentrations and antioxidant status in tissues of breast cancer patients. Clin Biochem 2000;33:279–284.
143. Ray G, Batra S, Shukla NK, et al. Lipid peroxidation, free radical production and antioxidant status in breast cancer. Breast Cancer Res Treat 2000;59:163–170.
144. Punnonen K, Ahotupa M, Asaishi K, Hyoty M, Kudo R, Punnonen R. Antioxidant activities and oxidative stress in human breast cancer. J Cancer Res. Clin Oncol 1994;120:374–377.
145. Katsetos CD, Del Valle L, Geddes JF, et al. Localization of the neuronal class III β-tubulin in oligodendrogliomas: comparison with Ki-67 proliferative index and 1p/19q status. J Neuropathol Exp Neurol 2002;61:307–320.
146. Dumontet C, Isaac S, Souquet PJ, et al. Expression of class III β tubulin in non-small cell lung cancer is correlated with resistance to taxane chemotherapy. Electr J Oncol 2002;1:58–64.
147. Colmenares SU, DeLuca K, Jordan MA, Mooberry SL. Native overexpression of βIII isotype of tubulin in the BT-549 breast carcinoma line is associated with resistance to paclitaxel, vinblastine and cryptophycin 1. Proc Am Assn Cancer Res 1998;39:163.
148. Carré M, André N, Carles G, et al. Tubulin is an inherent component of mitochondrial membranes that interacts with the voltage-dependent anion channel. J Biol Chem 2002;277:33,644–33,669.
149. Lu Q, Ludueña RF. In vitro analysis of microtubule assembly of isotypically pure tubulin dimers. Intrinsic differences in the assembly properties of $\alpha\beta_{II}$, $\alpha\beta_{III}$, and $\alpha\beta_{IV}$ tubulin dimers in the absence of microtubule-associated proteins. J Biol Chem 1994;269:2041–2047.
150. Banerjee A, Roach MC, Trcka P, Ludueña RF. Preparation of a monoclonal antibody specific for the class IV isotype of β-tubulin. Purification and assembly of $\alpha\beta_{II}$, $\alpha\beta_{III}$, and $\alpha\beta_{IV}$ tubulin dimers from bovine brain. J Biol Chem 1992;267:5625–5630.
151. Hari M, Yang H, Zeng C, Canizales M, Cabral F. Expression of class III β-tubulin reduces microtubule assembly and confers resistance to paclitaxel. Cell Motil Cytoskeleton 2003;56:45–56.
152. Evans J, Sumners C, Moore J, et al. Characterization of mitotic neurons derived from adult rat hypothalamus and brain stem. J Neurophysiol 2001;87:1076–1085.
153. Ohuchi T, Maruoka S, Sakudo A, Arai T. Assay-based quantitative analysis of PC12 differentiation. J Neurosci Methods 2002;118:1–8.
154. Braun H, Schäfer K, Höllt V. βIII tubulin-expressing neurons reveal enhanced neurogenesis in hippocampal and cortical structures after a contusion trauma in rats. J Neurotrauma 2002;19:975–983.
155. Xu G, Pierson CR, Murakawa Y, Sima AAF. Altered tubulin and neurofilament expression and impaired axonal growth in diabetic nerve regeneration. J Neuropathol Exp Neurol 2002;61:164–175.

156. Harada A, Teng J, Takei Y, Oguchi K, Hirokawa N. MAP2 is required for dendrite elongation, PKA anchoring in dendrites, and proper PKA signal transduction. J Cell Biol 2002;158:541–549.

157. Fanarraga ML, Avila J, Zabala JC. Expression of unphosphorylated class III β-tubulin isotype in neuroepithelial cells demonstrates neuroblast commitment and differentiation. Eur J Neurosci 1999;11:517–527.

158. Molea D, Stone JC, Rubel EW. Class III β-tubulin expression in sensory and nonsensory regions of the developing avian inner ear. J Comp Neurol 1999;406:183–198.

159. Khan IA, Ludueña RF. Different effects of vinblastine on the polymerization of isotypically purified tubulins from bovine brain. Invest New Drugs 2003;21:3–13.

160. Daniely Y, Liao G, Dixon D, et al. Critical role of p63 in the development of a normal esophageal and tracheobronchial epithelium. Am J Physiol Cell Physiol 2004;287:C171–C181.

161. Pazour GJ, Agrin NS, Leszyk JD, Witman GB. Proteomic characterization of a eukaryotic cilium. American Society for Cell Biol Ann Meeting Abstracts. p. 55a. 2004.

162. Dustin P. Microtubules. Springer-Verlag, Berlin, 1984:149.

163. Smith EF. Regulation of flagellar dynein by the axonemal central apparatus. Cell Motil Cytoskeleton 2002;52:33–42.

164. Mitchell DR, Nakatsugawa M. Bend propagation drives central pair rotation in *Chlamydomonas reinhardtii* flagella. J Cell Biol 2004;166:709–715.

165. Sloboda RD. A healthy understanding of intraflagellar transport. Cell Motil Cytoskeleton 2002; 52:1–8.

166. Nielsen MG, Turner FR, Hutchens JA, Raff EC. Axoneme-specific β-tubulin specialization: a conserved C-terminal motif specifies the central pair. Curr Biol 2001;11:529–533.

167. Thazhath R, Liu C, Gaertig J. Polyglycylation domain of β-tubulin maintains axonemal architecture and affects cytokinesis in Tetrahymena. Nat Cell Biol 2002;4:256–259.

168. Sharma J, Ludueña RF. Use of N,N′-polymethylenebis(iodoacetamide) derivatives as probes for the detection of conformational differences in tubulin isotypes. J Prot Chem 1994;13:165–176.

169. Sadek CM, Jiménez A, Damdimopoulous AE, et al. Characterization of human thioredoxin-like 2. A novel microtubule-binding thioredoxin expressed predominantly in the cilia of lung airway epithelium and spermatid manchette and axoneme. J Biol Chem 2003;278:13,133–13,142.

170. Walss-Bass C, Kreisberg JI, Ludueña RF. Mechanism of localization of β_{II}-tubulin in the nuclei of cultured rat kidney mesangial cells. Cell Motil Cytoskeleton 2001;49:208–217.

171. Walss-Bass C, Prasad V, Kreisberg JI, Ludueña RF. Interaction of the β_{IV}-tubulin isotype with actin stress fibers in cultured rat kidney mesangial cells. Cell Motil Cytoskeleton 2001;49:200–207.

172. Kodoma A, Lechler T, Fuchs E. Coordinating cytoskeletal tracks to polarize cellular movements. J Cell Biol 2004;167:203–207.

173. Sullivan KF, Havercroft JC, Machlin PS, Cleveland DW. Sequence and expression of the chicken β5- and β4-tubulin genes define a pair of divergent β-tubulins with complementary patterns of expression. Mol Cell Biol 1986;6:4409–4418.

174. Banerjee A, Elguezabal G, Joe P, Lazzell A, Prasad V, Luduena RF. Distribution and characterization of the β_V isotype of tubulin in mammalian cells. Mol Biol Cell 2003;14:182A.

175. Ikeda Y, Steiner M. Sulfhydryls of platelet tubulin: Their role in polymerization and colchicine binding. Biochemistry 1978;17:3454–3459.

176. Italiano JE, Bergmeier W, Tiwari S, et al. Mechanisms and implications of platelet discoid shape. Blood 2003;101:4789–4796.

177. Lecine P, Italiano JE, Kim S-W, Villeval J-L. Shivdasaani RA. Hematopoietic-specific β1 tubulin participates in a pathway of platelet biogenesis dependent on the transcription factor NF-E2. Blood 2000;96:1366–1373.

178. Hartwig J, Italiano J. The birth of the platelet. J Thromb Haemostasis 2003;1:1580–1586.

179. Schwer HD, Lecine P, Tiwari S, Italiano JE, Hartwig JH. Shivdasani RA. A lineage-restricted and divergent β-tubulin isoform is essential for the biogenesis, structure and function of blood platelets. Curr Biol 2001;11:579–586.

180. White JG, de Alarcon PA. Platelet spherocytosis: a new bleeding disorder. Am J Hematol 2002;70:158–166.

181. DuMontet C, Viormery AV. Expression of a new β tubulin isotype in brain. Mol Biol Cell 1999;10:141a.

182. Van Geel M, van Deutekom JC, van Staalduinen A, et al. Identification of a novel β-tubulin subfamily with one member (TUBB4Q) located near the telomere of chromosome region 4q35. Cytogenet Cell Genet 2002;88:316–321.

183. Miller FD, Naus CC, Durand M, Bloom FE, Milner RJ. Isotypes of α-tubulin are differentially regulated during neuronal maturation. J Cell Biol 1987;105:3065–3073.
184. Przyborski SA, Cambray-Deakin MA. Developmental regulation of α-tubulin mRNAs during the differentiation of cultured cerebellar granule cells. Mol Brain Res 1996;36:179–183.
185. Stanchi F, Corso V, Scannapieco P, et al. TUBA8: a new tissue-specific isoform of α-tubulin that is highly conserved in human and mouse. Biochem Biophys Res Commun 2000;270:1111–1118.
186. Kourmouli N, Dialynas G, Petraki C, et al. Binding of heterochromatin protein 1 to the nuclear envelope is regulated by a soluble form of tubulin. J Biol Chem 2001;276:13,007–13,014.
187. Hu K, Roos DS, Murray JM. A novel polymer of tubulin forms the conoid of *Toxoplasma gondii*. J Cell Biol 2002;158:1039–1050.
188. Imai H, Nakagawa Y. Biological significance of phospholipid hydroperoxide glutathione peroxidase (PHGPx, GPx4) in mammalian cells. Free Radic Biol Med 2003;34:145–169.
189. Ursini F, Heim S, Kiess M, et al. Dual function of the selenoprotein PHGPx during sperm maturation. Science 1999;285:1393–1396.
190. Xu K, Ludueña RF. Characterization of nuclear β_{II}-tubulin in tumor cells: a possible novel target for taxol. Cell Motil Cytoskeleton 2002;53:39–52.
191. Walss-Bass C, Kreisberg JI, Ludueña RF. Effect of the anti-tumor drug vinblastine on nuclear β_{II}-tubulin in cultured rat kidney mesangial cells. Invest New Drugs 2003;21:15–20.
192. Walss-Bass C, Xu K, David S, Fellous A, Ludueña RF. Occurrence of nuclear β_{II}-tubulin in cultured cells. Cell Tissue Res 2002;308:215–223.
193. Ewald A, Zünkler C, Lourim D, Dabauvalle M-C. Microtubule-dependent assembly of the nuclear envelope in *Xenopus laevis* egg extract. Eur J Cell Biol 2001;80:678–691.
194. Salina D, Bodoor K, Enarson P, Raharjo WH, Burke B. Nuclear Envelope Dyn 2001;79:533–542.
195. Burke B, Ellenberg J. Remodelling the walls of the nucleus. Nat Rev Mol Cell Biol 2002;3:487–497.
196. Beaudouin J, Gerlich D, Daigle N, Eils R, Ellenberg J. Nuclear envelope breakdown proceeds by microtubule-induced tearing of the lamina. Cell 2002;108:83–96.
197. Weatherbee JA, May GS, Gambino J, Morris NR. Involvement of a particular species of β-tubulin (β3) in conidial development in *Aspergillus nidulans*. J Cell Biol 1985;101:706–711.
198. Oakley BR. Tubulins in *Aspergillus nidulans*. Fungal Genet Biol 2004;41:420–427.
199. Kirk KE, Morris NR. Either β-tubulin isogene product is sufficient for microtubule function during all stages of growth and differentiation in *Aspergillus nidulans*. Mol Cell Biol 1993;13:4465–4476.
200. Joshi HC, Yen TJ, Cleveland DW. In vivo coassembly of a divergent β-tubulin subunit (cβ6) into microtubules of different function. J Cell Biol 1987;105:2179–2190.
201. Lopata MA, Cleveland DW. In vivo microtubules are copolymers of available β-tubulin isotypes. Localization of each of six vertebrate β-tubulin isotypes using polyclonal antibodies elicited by synthetic peptide antigens. J Cell Biol 1987;105:1707–2730.
202. Lewis SA, Gu W, Cowan NJ. Free intermingling of mammalian β-tubulin isotypes among functionally distinct microtubules. Cell 1987;49:539–548.
203. Gu W, Lewis SA, Cowan NJ. Generation of antisera that discriminate among mammalian α-tubulins. Introduction of specialized isotypes into cultured cells results in their coassembly without disruption of normal microtubule function. J Cell Biol 1988;106:2011–2022.
204. Chu B, Snustad DP, Carter JV. Alteration of β-tubulin gene expression during low-temperature exposure in leaves of *Arabidopsis thaliana*. Plant Physiol 1993;103:371–377.
205. Abdrakhamanova A, Wang QY, Khokhlova L, Nick P. Is microtubule disassembly a trigger for cold acclimation? Plant Cell Physiol 2003;44:676–686.
206. Roos MH, Boersema JH, Borgsteede FHM, Cornelissen J, Taylor M, Ruitenberg EJ. Molecular analysis of selection for benzimidazole resistance in the sheep parasite *Haemonchus contortus* Mol Biochem Parasitol 1990;43:77–88.
207. Kwa MSG, Veenstra JG, Roos MH. Molecular characterisation of β-tubulin genes present in benzimidazole-resistant populations of *Haemonchus contortus*. Mol Biochem Parasitol 1993;60:133–144.
208. Driscoll M, Dean E, Reilly E, Bergholz E, Chalfie M. Genetic and molecular analysis of a *Caenorhabditis elegans* β-tubulin that conveys benzimidazole sensitivity. J Cell Biol 1989;109:2993–3003.
209. Grant WN, Mascord LJ. β-tubulin gene polymorphism and benzimidazole resistance in *Trichostrongylus colubriformes*. Int J Parasitol 1996;26:71–77.
210. Silvestre A, Cabaret J. Mutation in position 167 of isotype 1 β-tubulin gene of *Trichostrongylid* nematodes: role in benzimidazole resistance? Mol Biochem Parasitol 2002;120:297–300.

211. Burkhart CA, Kavallaris M, Horwitz SB. The role of β-tubulin isotypes in resistance to antimitotic drugs. Biochim Biophys Acta 2001;1471:O1–O9.
212. Orr GA, Verdier-Pinard P, McDaid H, Horwitz SB. Mechanisms of taxol resistance related to microtubules. Oncogene 2003;22:7280–7295.
213. Schatz PJ, Pillus L, Grisafi P, Solomon F, Botstein D. Two functional α-tubulin genes of the yeast *Saccharomyces cerevisiae* encode divergent proteins. Mol Cell Biol 1986;6:3711–3721.
214. Bode CJ, Gupta ML, Suprenant KA, Himes RH. The two α-tubulin isotypes in budding yeast have opposing effects on microtubule dynamics in vitro. EMBO Rep 2003;4:94–99.
215. Haber M, Burkhart CA, Regl DL, Madafiglio J, Norris MD, Horwitz SB. Altered expression of Mβ2, the class II β-tubulin isotype, in a murine J774.2 cell line with a high level of taxol resistance. J Biol Chem 1995;270:31,269–31,275.
216. Ranganathan S, Dexter DW, Benetatos CA, Chapman AE, Tew KD, Hudes GR. Increase of β_{III}- and β_{IVa}-tubulin isotypes in human prostate carcinoma cells as a result of estramustine resistance. Cancer Res 1996;56:2584–2589.
217. Ranganathan S, Dexter DW, Benetatos CA, Hudes GR. Cloning and sequencing of human βIII-tubulin cDNA: induction of βIII isotype in human prostate carcinoma cells by acute exposure to antimicrotubule agents. Biochim Biophys Acta 1998;1395:237–245.
218. Banerjee A. Increased levels of tyrosinated α-, β_{III}-, and β_{IV}-tubulin isotypes in paclitaxel-resistant MCF-7 breast cancer cells. Biochem Biophys Res Commun 2002;293:598–601.
219. Kavallaris M, Kuo DYS, Burkhart CA, et al. Taxol-resistant epithelial ovarian tumors are associated with altered expression of specific β-tubulin isotypes. J Clin Invest 1997;100:1282–1293.
220. Sangrajang S, Denoulet P, Laing NM, et al. Association of estramustine resistance in human prostatic carcinoma cells with modified patterns of tubulin expression. Biochem Pharmacol 1998;55:325–331.
221. Verdier-Pinard P, Wang F, Martello L, Burd B, Orr GA, Horwitz SB. Analysis of tubulin isotypes and mutations from taxol-resistant cells by combined isoelectrofocusing and mass spectrometry. Biochemistry 2003;42:5349–5357.
222. Sirotnak FM, Danenberg KD, Chen J, Fritz F, Danenberg PV. Markedly decreased binding of vincristine to tubulin in Vinca alkaloid-resistant Chinese hamster cells is associated with selective overexpression of α and β tubulin isoforms. Biochem Biophys Res Commun 2000;269:21–24.
223. Makarovsky AN, Siryaporn E, Hixson DC, Akerley W. Survival of docetaxel-resistant prostate cancer cells in vitro depends on phenotype alterations and continuity of drug exposure. Cell Mol Life Sci 2002;59:1198–1211.
224. Galmarini CM, Kamath K, Vanier-Viornery A, et al. Drug resistance associated with loss of p53 involves extensive alterations in microtubule composition and dynamics. Br J Cancer 2003;88: 1793–1799.
225. Giannakakou P, Sackett DF, Kang YK, et al. Paclitaxel-resistant human ovarian cancer cells have mutant β-tubulins that exhibit impaired paclitaxel-driven polymerization. J Biol Chem 1997;272:17,118–17,125.
226. Bernard-Marty C, Treilleux I, Dumontet C, et al. Microtubule-associated parameters as predictive markers of docetaxel activity in advanced breast cancer patients: results of a pilot study. Clin Breast Cancer 2002;3:341–345.
227. Iwamoto Y, Nishio K, Fukumoto H, Yoshimatsu K, Yamakido M, Saijo N. Preferential binding of E7010 to murine β3-tubulin and decreased β3-tubulin in E7010-resistant cell lines. Jpn J Cancer Res 1998;89:954–962.
228. Derry WB, Wilson L, Khan IA, Ludueña RF, Jordan MA. Taxol differentially modulates the dynamics of microtubules assembled from unfractionated and purified β-tubulin isotypes. Biochemistry 1997; 36:3554–3562.
229. Kavallaris M, Burkhart CA, Horwitz SB. Antisense oligonucleotides to class III β-tubulin sensitize drug-resistant cells to taxol. Br J Cancer 1999;80:1020–1025.
230. Ranganathan S, McCauley RA, Dexter DW, Hudes GR. Modulation of endogenous β-tubulin isotype expression as a result of human β_{III} cDNA transfection into prostate carcinoma cells. Br J Cancer 2001;85:735–740.
231. Nicoletti MI, Valoti G, Giannakakou P, et al. Expression of β-tubulin isotypes in human ovarian carcinoma xenografts and in a sub-panel of human cancer cell lines from the NCI-anticancer drug screen: correlation with sensitivity to microtubule active agents. Clin Can Res 2001;7:2912–2922.
232. Blade K, Menick DR, Cabral F. Overexpression of class I, II or IVb β-tubulin isotypes in CHO cells is insufficient to confer resistance to paclitaxel. J Cell Sci 1999;112:2213–2221.

233. Kavallaris M, Tait AS, Walsh BJ, et al. Multiple microtubule alterations are associated with Vinca alkaloid resistance in human leukemia cells. Cancer Res 2001;61:5803–5809.
234. Hiser L, Aggarwal A, Young R, et al. Comparison of β-tubulin mRNA and protein levels in 12 human cancer cell lines. Cell Motil Cytoskeleton 2006;63:41–52.
235. Montgomery RB, Guzman J, O'Rourke DM, Stahl WL. Expression of oncogenic epidermal growth factor receptor family kinases induces paclitaxel resistance and alters β-tubulin isotype expression. J Biol Chem 2000;275:17,358–17,363.
236. Banerjee A, Roach MC, Trcka P, Ludueña RF. Increased microtubule assembly in bovine brain tubulin lacking the type III isotype of β-tubulin. J Biol Chem 1990;265:1794–1799.
237. Lee MK, Tuttle JB, Rebhun LI, Cleveland DW, Frankfurter A. The expression and posttranslational modification of neuron-specific β-tubulin isotype during chick embryogenesis. Cell Motil Cytoskeleton 1990;17:118–132.
238. Lobert S, Frankfurter A, Correia JJ. Binding of vinblastine to phosphocellulose-purified and αβ-class III tubulin: the role of nucleotides and β-tubulin isotypes. Biochemistry 1995;34:8050–8060.
239. Lobert S, Frankfurter A, Correia JJ. Energetics of Vinca alkaloid interactions with tubulin isotypes: implications for drug efficacy and toxicity. Cell Motil Cytoskeleton 1998;39:107–121.
240. Banerjee A. Differential effects of colchicine and its B-ring modified analog MTPT on the assembly-independent GTPase activity of purified β-tubulin isoforms from bovine brain. Biochem Biophys Res Commun 1997;231:698–700.
241. Khan IA, Ludueña RF. Phosphorylation of β_{III}-tubulin. Biochemistry 1996;35:3704–3711.
242. Schwarz PM, Liggins JR, Ludueña RF. β-Tubulin isotypes purified from bovine brain have different relative stabilities. Biochemistry 1998;37:4687–4692.
243. Ludueña RF, Roach MC. Interaction of tubulin with drugs and alkylating agents. 1. Alkylation of tubulin by iodo[^{14}C]acetamide and N,N′-ethylenebis(iodoacetamide). Biochemistry 1981;20:4437–4444.
244. Roach MC, Ludueña RF. Different effects of tubulin ligands on the intrachain cross-linking of β_1-tubulin. J Biol Chem 1984;259:12,063–12,071.
245. Little M, Ludueña RF. Location of two cysteines in brain β_1-tubulin that can be cross-linked after removal of exchangeable GTP. Biochim Biophys Acta 1987;912:28–33.
246. Taylor EW. The mechanism of colchicine binding inhibition of mitosis I. Kinetics of inhibition and the binding of H^3-colchicine. J Cell Biol 1965;25:145–160.
247. Wilson L, Friedkin M. The biochemical events of mitosis. I. Synthesis and properties of colchicine labeled with tritium in its acetyl moiety. Biochemistry 1966;5:2463–2468.
248. Borisy GG, Taylor EW. The mechanism of action of colchicine: Binding of colchicine-^{3}H to cellular protein. J Cell Biol 1967;34:525–533.
249. Weisenberg RC, Borisy GG, Taylor EW. The colchicine-binding protein of mammalian brain and its relation to microtubules. Biochemistry 1968;7:4466–4479.
250. Wilson L. Properties of colchicine-binding protein from chick embryo brain. Interactions with Vinca alkaloids and podophyllotoxin. Biochemistry 1970;9:4999–5007.
251. Wilson L, Meza I. The mechanism of action of colchicine: colchicine-binding properties of sea urchin sperm tail outer doublet tubulin. J Cell Biol 1973;58:709–714.
252. Wilson L, Bryan J. Biochemical and pharmacological properties of microtubules. Adv Cell Mol Biol 1974;3:21–72.
253. Bhattacharyya B, Wolff J. Promotion of fluorescence upon binding of colchicine to tubulin. Proc Nat Acad Sci USA 1974;71:2627–2631.
254. Arai T, Okuyama T. Fluorometric assay of tubulin-colchicine complex. Anal Biochem 1975;69:443–448.
255. Garland D. Kinetics and mechanism of colchicine binding to tubulin: evidence for ligand-induced conformational change. Biochemistry 1978;17:4266–4272.
256. Lambeir A, Engelborghs Y. A fluorescence stopped flow study of colchicine binding to tubulin. J Biol Chem 1981;256:3279–3282.
257. Banerjee A, Ludueña RF. Kinetics of association and dissociation of colchicine-tubulin complex from brain and renal tubulin. Evidence for the existence of multiple isotypes of tubulin in brain with differential affinity to tubulin. FEBS Lett 1987;219:103–107.
258. Banerjee A, Barnes LD, Ludueña RF. The role of the B-ring of colchicine in the stability of the colchicine-tubulin complex. Biochim Biophys Acta 1987;913:138–144.
259. Banerjee A, Ludueña RF. Distinct colchicine binding kinetics of bovine brain tubulin lacking the type III isotype of β-tubulin. J Biol Chem 1991;266:1689–1691.

260. Banerjee A, Ludueña RF. Kinetics of colchicine binding to purified β-tubulin isotypes from bovine brain. J Biol Chem 1992;267:13,335–13,339.

261. Banerjee A, D'Hoore A, Engelborghs Y. Interaction of desacetamidocolchicine, a fast-binding analogue of colchicine with isotypically pure tubulin dimers $\alpha\beta_{II}$, $\alpha\beta_{III}$, and $\alpha\beta_{IV}$. J Biol Chem 1994;269:10,324–10,329.

262. Banerjee A, Engelborghs Y, D'Hoore A, Fitzgerald TJ. Interaction of a bicyclic analogue of colchicine with purified β-tubulin isoforms from bovine brain. Eur J Biochem 1997;246:420–424.

263. Carlier M-F. Role of nucleotide hydrolysis in the dynamics of actin filaments and microtubules. Int Rev Cytol 1989;115:139–170.

264. Guasch A, Aloria K, Pérez R, Avila J, Zabala JC, Coll M. Three-dimensional structure of human tubulin chaperone cofactor Am J Mol Biol 2002;318:1139–1149.

265. Saito Y, Yamagishi N, Ishihara K, Hatayama T. Identification of α-tubulin as an hsp105α-binding protein by the yeast two-hybrid system. Exp Cell Res 2003;286:233–240.

266. Fukata Y, Itoh TJ, Kimura T, et al. CRMP-2 binds to tubulin heterodimers to promote microtubule assembly. Nat Cell Biol 2002;4:583–591.

267. Bonnet C, Denarier E, Bosc C, Lazereg S, Denoulet P, Larcher JC. Interaction of STOP with neuronal tubulin is independent of polyglutamylation. Biochem Biophys Res Commun 2002;297:787–793.

268. Ems-McClung SC, Zheng Y, Walczak CE. Importin α/β and Ran-GTP regulate XCTK2 microtubule binding through a bipartite nuclear localization signal. Mol Biol Cell 2004;15:46–57.

269. Kinoshita K, Habermann B, Hyman AA. XMAP215: a key component of the dynamic microtubule cytoskeleton. Trends Cell Biol 2002;12:267–273.

270. Chaudhuri AR, Khan IA, Prasad V, Robinson AK, Ludueña RF, Barnes LD. The tumor suppressor protein Fhit. A novel interaction with tubulin. J Biol Chem 1999;274:24,738–24,382.

271. Lu C, Srayko M, Mains PE. The *Caenorhabditis elegans* microtubule-severing complex MEI-1/MEI-2 katanin interacts differently with two superficially redundant β-tubulin isotypes. Mol Biol Cell 2004;15:142–150.

272. Murata-Hori M, Tatsuka M, Wang YL. Probing the dynamics and functions of aurora B kinase in living cells during mitosis and cytokinesis. Mol Biol Cell 2002;13:1099–1108.

273. Curmi P, Andersen SSL, Lachkar S, et al. The stathmin/tubulin interaction in vitro. J Biol Chem 1997;272:25,029–25,036.

274. Rappoport JZ, Taha BW, Simon SM. Movement of plasma-membrane-associated clathrin spots along the microtubule cytoskeleton. Traffic 2003;4:460–467.

275. Garcia-Mata R, Gao Y-S, Sztul E. Hassles with taking out the garbage: aggravating aggresomes. Traffic 2002;3:388–396.

276. Geimer S, Melkonian M. The ultrastructure of the *Chlamydomonas reinhardtii* basal apparatus: identification of an early marker of radial asymmetry inherent in the basal body. J Cell Sci 2004;117:2663–2674.

277. Matsuura K, Lefebvre PA, Kamiya R, Hirono M. Bld10p, a novel protein essential for basal body assembly in Chlamydomonas: localization to the cartwheel, the first ninefold symmetrical structure appearing during assembly. J Cell Biol 2004;165:663–671.

278. Baas PW. Neuronal polarity: microtubules strike back. Nat Cell Biol 2002;4:E194–E195.

279. Kierszenbaum AL. Intramanchette transport (IMT): managing the making of the spermatid head, centrosome, and tail. Mol Reprod Dev 2002;63:1–4.

280. Layden Donati RJ, Oh J, Yang S, Johnson ME, Rasenick MM. Structural model of Gα-tubulin interaction. Am Soc Cell Biol Ann Meeting Abstracts 2004;p. 425A.

281. Littauer UZ, Giveon D, Thierauf M, Ginzburg I, Ponstingl H. Common and distinct tubulin binding sites for microtubule-associated proteins. Proc Nat Acad Sci USA 1986;83:7162–7166.

282. Fujii T, Koizumi Y. Identification of the binding region of basic calponin on α- and β-tubulins. J Biochem 1999;125:869–875.

283. Karabay A, Walker RA. Identification of Ncd tail domain-binding sites on the tubulin dimer. Biochem Biophys Res Commun 2003;305:523–528.

284. Burns RG, Surridge C. Analysis of β-tubulin sequences reveals highly conserved, coordinated amino acid substitutions. Evidence that these hot spots are directly involved in the conformational change required for dynamic instability. FEBS Lett 1990;271:1–8.

285. Hoyle HD, Hutchens JA, Turner FR, Raff EC. Regulation of β-tubulin β3 function and expression in *Drosophila* spermatogenesis. Dev Genet 1995;16:148–170.

286. Keskin O, Durell SR, Baahar I, Jernigan RL, Covell DG. Relating molecular flexibility to function: a case study of tubulin. Biophys J 2002;83:663–680.

287. Van Buren V, Odde DJ, Cassimeris L. Estimates of lateral and longitudinal bond energies within the microtubule lattice. Proc Nat Acad Sci USA 2002;99:6035–6040.
288. Detrich HW, Parker SK, Williams RC, Nogales E, Downing KH. Cold adaptation of microtubule assembly and dynamics. Structural interpretation of primary sequence changes present in the α- and β-tubulins of Antarctic fishes. J Biol Chem 2000;275:37,038–37,047.
289. Pucciarelli S, Miceli C. Characterization of the cold-adapted α-tubulin from the psychrophilic ciliate *Euplotes focardii*. Extremophiles 2002;6:385–389.
290. Murphy WJ, Elzirik E, Johnson WE, Zhang YP, Ryder OA, O'Brien SJ. Molecular phylogenetics and the origins of placental mammals. Nature 2001;409:614–618.
291. Colbert EH. Evolution of the Vertebrates, 3rd ed. New York: John Wiley and Sons; 1980.
292. Benton MJ, Ayala FJ. Dating the tree of life. Science 2003;300:1698–1700.
293. Modig C, Wallin M, Olsson P-E. Expression of cold-adapted β-tubulins confer cold-tolerance to human cellular microtubules. Biochem Biophys Res Commun 2000;269:787–791.
294. Detrich HW, Neighbors BW, Sloboda RD, Williams RC. Microtubule-associated proteins from Antarctic fishes. Cell Motil Cytoskeleton 1990;17:174–186.
295. Willmer P. Invertebrate Relationships. Patterns in Animal Evolution. Cambridge: Cambridge University Press; 1990.
296. Barrington EJW. Essential features of lower types. In: Wake MH, ed. Hyman's Comparative Vertebrate Anatomy, 3rd ed. Chicago: University of Chicago Press; 1979;57–86.
297. Northcutt RG. The comparative anatomy of the nervous system and sense organs. In: Wake MH, ed. Hyman's Comparative Vertebrate Anatomy, 3rd ed. Chicago: University of Chicago Press; 615–769.
298. Cloud P. Oasis in Space: Earth History from the Beginning. W.W. Norton and Co., New York.
299. Buchsbaum R. Animals Without Backbones. Chicago: University of Chicago Press.
300. Palmer D. Prehistoric Past Revealed: the Four Billion Year History of Life on Earth. University of California Press, Berkeley, 2003.
301. Gould SJ. Wonderful Life: the Burgess Shale and the Nature of History, W.W. Norton and Co., New York, 1989.
302. Arai K. Molecular cloning of isotype-specific regions of five classes of canine β-tubulin and their tissue distribution. NCBI Accession no. BAA96409, BAA96410, BAA96411, BAA96412, 1999.
303. Sidjanin DJ, Zangerl B, Johnson JL, et al. Cloning of the canine δ-tubulin cDNA (TUBD) and mapping to CFA9. Anim Genet 2002;33:161–162.
304. Kubo A, Hata M, Kubo A, Tsukita S. Gene-knockout analysis of two γ-tubulin isoforms in mice. NCBI Accession no. BAD27264, BAD27265, 2004.
305. Lemischka IR, Farmer S, Racaniello VR, Sharp PA. Nucleotide sequence and evolution of a mammalian α-tubulin messenger RNA. J Mol Biol 1981;151:101–120.
306. (no reference) NCBI Accession no. XP_232565, XP_237302.
307. Usui H, Miyazaki Y, Xin D, Ichikawa T, Kumanishi T. Cloning and sequencing of the rat cDNAs encoding class I β-tubulin. DNA Seq 1998;9:365–368.
308. Ginzburg I, Teichman A, Dodemont HJ, Behar L, Littauer UZ. Regulation of three β-tubulin mRNAs during rat brain development. EMBO J 1985;4:3667–3673.
309. Dennis KE, Spano A, Frankfurter A, Moody SA. *Rattus norvegicus* neuron-specific class III β-tubulin mRNA. NCBI Accession no. NP_640347, 2001.
310. Arai K. Preparation and characterization of a monoclonal antibody to class II β-tubulin isotype. NCBI Accession no. BAB72260, 2001.
311. Nakadai T, Okada N, Makino Y, Tamura T. Structure of rat γ-tubulin and its binding to HP33. DNA Res 1999;6:207–209.
312. Linhartová I, Novotná B, Sulimenko V, Dráberová E, Dráber P. γ-tubulin in chicken erythrocytes: changes in localization during cell differentiation and characterization of cytoplasmic complexes. Dev Dyn 2002;223:229–240.
313. Stearns T, Evans L, Kirschner M. γ-Tubulin is a highly conserved component of the centrosome. Cell 1991;65:825–836.
314. Parker SK, Detrich HW. Evolution, organization and expression of α-tubulin genes in the Antarctic fish *Notothenia coriiceps*. Adaptive expansion of a gene family by recent gene duplication, inversion, and divergence. J Biol Chem 1998;273:34,358–34,369.
315. Bormann P, Zumsteg VM, Roth LWA, Reinhard E. Target contact regulates GAP-43 and α-tubulin mRNA levels in regenerating retinal ganglion cells. J Neurosci Res 1998;52:405–419.

316. Edvardsen RB, Flaat M, Tewari R, et al. Most intron positions in *Oikopleura dioica* α-tubulin genes are unique: did new introns help to preserve and expand gene families? NCBI Accession no. AAM73981, AAM73982, AAM73986, AAM73987, AAM73991, AAM73992, AAM73993, AAM73995, AAM73996, AAM73997, 2002b.

317. Edvardsen,RB, Lerat E, Flaat M, et al. Hypervariable intron/exon organizations in the chordate *Oikopleura* and the nematode *Caenorhabditis*, two species with a very short life cycle. NCBI Accession no. AAO00725, AAP80593, AAP80594, AAP80595, AAP80596, AAP80597, AAP80598, AAP80599, AAP80600, AAP80601, AAP80602, AAP80603, 2002d.

318. Rogers GC, Chui KK, Lee EW, et al. A kinesin-related protein, KRP(180), positions prometaphase spindle poles during early sea urchin embryonic cell division. J Cell Biol 2000;150:499–512.

319. Varadaraj V, Kumari SS, Skinner DM. Molecular characterization of four members of the α-tubulin gene family of the Bermuda land crab *Gecarcinus lateralis*. J Exp Zool 1997;278:63–77.

320. Llamazares S, Tavosanis G, Gonzalez C. Cytological characterisation of the mutant phenotypes produced during early embryogenesis by null and loss-of-function alleles of the γTub37C gene in *Drosophila*. J Cell Sci 1999;112:659–667.

321. Moccia R, Chen D, Lyles V, et al. An unbiased cDNA library prepared from isolated *Aplysia* sensory neuron processes is enriched for cytoskeletal and translational mRNAs. J Neurosci 2002;23:9409–9417.

322. Fedorov A, Johnston H, Korneev S, Blackshaw S, Davies J. Cloning, characterisation and expression of the α-tubulin genes of the leech, *Hirudo medicinalis*. Gene 1999;227:11–19.

323. Bobinnec Y, Fukuda M, Nishida E. Identification and characterization of *Caenorhabditis elegans* γ-tubulin in dividing cells and differentiated tissues. J Cell Sci 2000;113:3747–3759.

324. Pape M, Schnieder T, von Samson-Himmelstjerna G. Investigation of diversity and isotypes of the β-tubulin cDNA in several small strongyle (*Cyathostominae*) species. J Parasitol 2002;88:673–677.

325. Von Samson-Himmelstjerna G, Harder A, Pape M, Schneider T. Novel small strongyle (*Cyathostominae*) β-tubulin sequences. Parasitol Res 2001;87:122–125.

326. Pape M, von Samson-Himmelstjerna G, Schnieder T. Characterisation of the β-tubulin gene of *Cylicocyclus nassatus*. Int J Parasitol 1999;29:1941–1947.

327. Njue AI, Prichard RK. Cloning two full-length β-tubulin isotype cDNAs from *Cooperia oncophora*, and screening for benzimidazole resistance-associated mutations in two isolates. Parasitology 2003;127:579–588.

328. Collins CM, Miller KA, Cunningham CO. Characterisation of a β-tubulin gene from the monogenean parasite, *Gyrodactylus salaris* Malmberg, 1957. Parasitol Res 2004;92:390–399.

329. Brehm K, Kronthaler K, Jura H, Frosch M. Cloning and characterization of β-tubulin genes from *Echinococcus multilocularis*. Mol Biochem Parasitol 2000;107:297–302.

330. Qin X, Gianì S, Breviario D. Molecular cloning of three rice α-tubulin isotypes: differential expression in tissues and during flower development. Biochim Biophys Acta 1997;1354:19–23.

331. Kim Y-K, Cha Y-K, Jun H-Y, Kim J-D, Choi J-S, Kim HR. Nucleotide sequence of a cDNA (OstubG2) encoding a γ-tubulin in the rice plant (*Oryza sativa*). NCBI Accession no. O49068, 2001.

332. Segal G, Feldman M. NCBI Accession no. AAD10487, AAD10488, AAD10489, AAD10490, AAD10492, AAD10493, 1996.

333. Liu B, Joshi HC, Wilson TJ, Silflow CD, Palevitz BA, Snustad DP. γ-Tubulin in *Arabidopsis*: gene sequence, immunoblot, and immunofluorescence studies. Plant Cell 1994;6:303–314.

334. Okamura S, Okahara K, Iida T, et al. Isotype-specific changes in the amount of β-tubulin RNA in synchronized tobacco BY2 cells. Cell Struct Funct 1999;24:117–122.

335. Okamura S, Hara M, Yamaguchi A. NCBI Accession no. AAB50565, 2000.

336. Okamura S, Yamaguchi A, Narita K, Morita M, Imanaka T. β-Tubulin isotypes in the tobacco BY2 cell cycle. Cell Biol Int 2003;27:245–246.

337. Breviario D, Linss M, Nick P. α-Tubulins from tobacco: gene cloning and expression studies. NCBI Accession no. CAD13176, CAD13177, CAD13178, 2001.

338. Kautz K, Schroeder J, Wernicke W. Characterization of γ-tubulin from tobacco. NCBI Accession no. CAC00547, 2000.

339. Schröder J. NCBI Accession no. CAA10664CAA70891, CAB76916, CAB76380, 2000.

340. Ji S, Liang X, Shi Y, Weu G, Lu Y, Zhu Y. Expression profile study and functional analysis of α- and β-tubulin isotypes during cotton fiber development. NCBI Accession no. AAL92026, AAN32988, AAN32989, AAN32991, AAN32995, AAQ92665, AAQ92664, AAQ92666, AAQ92667, AAQ92668, 2002.

341. Saibo NJM, Van Der Straeten D, Rodriges-Pousada C. *Lupinus albus* γ-tubulin: mRNA and protein accumulation during development and in response to darkness. Planta 2004;219:201–211.
342. Kalluri UC, Joshi CP. Molecular cloning of tubulin cDNA from aspen xylem. NCBI Accession no. AAO23139, 2002.
343. Wang Y-S, Tsai C-J. Isolation and characterization of cDNAs involved in vascular development of quaking aspen. NCBI Accession no. AAO63773, AAO63781, 2003.
344. Canaday J, Stoppin V, Endle MC, Lambert AM. NCBI Accession no. Q41808, 1994.
345. Canaday J, Stoppin V, Endle MC, Lambert AM. Identification of two maize cDNAs encoding γ-tubulin. NCBI Accession no. CAA58670, 1995.
346. Yamamoto E, Zeng L, Baird WV. α-Tubulin missense mutations correlate with antimicrotubule drug resistance in *Eleusine indica*. Plant Cell 1998;10:297–308.
347. Yamamoto E, Baird WV. Molecular characterization of four β-tubulin genes from dinitroaniline susceptible and resistant biotypes of *Eleusine indica*. Plant Mol Biol 1999;39:45–61.
348. Wu W, Schaal BA, Hwang CY, Chiang YC, Chiang TY. Molecular cloning and evolutionary analysis of *Miscanthus* α-tubulin genes. Am J Bot 2003;90:1513–1521.
349. Fuchs U, Moepps B, Maucher HP, Schraudolf H. Isolation, characterization and sequence of a cDNA encoding γ-tubulin protein from the fern *Anemia phyllitidis* L Sw Plant Mol Biol 1993;23: 595–603.
350. Moepps B, Maucher HP, Bogenberger JM, Schraudolf H. Characterization of the α- and β-tubulin gene families from *Anemia phyllitidis*. NCBI Accession no. CAA48929, CAA48930, 1993.
351. Fujita T, Nishiyama T, Hasebe M. Isolation of α tubulin cDNAs in *Physcomitrella patens*. NCBI Accession no. BAC24799, BAC24800, 2002.
352. Baur A, Gorr G, Jost W. Six β-tubulin genes from *Physcomitrella patens*. NCBI Accession no. AAQ88113, AAQ88114, AAQ88115, AAQ88116, AAQ88117, AAQ88118, 2003.
353. Wagner TA, Sack FD, Oakley BR, Oakley CE, Schwuchow J. Characterization of γ-tubulin from *Physcomitrella patens*. NCBI Accession no. AAD33883, 1999.
354. Takano Y, Oshiro E, Okuno T. Microtubule dynamics during infection-related morphogenesis of *Colletotrichum lagenarium*. Fungal Genet Biol 2001;34:107–121.
355. Daly S, Yacoub A, Dundon WE, Mastromei G, Islam K, Lorenzetti R. Isolation and characterization of a gene encoding α-tubulin from *Candida albicans*. Gene 1997;187:151–158.
356. Dujon B, Sherman D, Fischer G, et al. Genome evolution in yeasts. Nature 2004;430:35–44.
357. Heckmann S, Schliwa M, Kube-Granderath E. Primary structure of *Neurospora crassa* γ-tubulin. Gene 1997;199:303–309.
358. Mukherjee M, Hadar R, Mukherjee PK, Horwitz BA. Homologous expression of a mutated β-tubulin gene does not confer benomyl resistance on *Trichoderma virens*. J Appl Microbiol 2003;95:861–867.
359. Park S-Y, Jung O-J, Chung Y-R, Lee C-W. Isolation and characterization of a benomyl-resistant form of β-tubulin-encoding gene from the phytopathogenic fungus *Botryotinia fuckeliana*. Molecules Cells 1997;7:104–109.
360. Zhang J, Stringer JR. Cloning and characterization of an α-tubulin-encoding gene from rat-derived *Pneumocystis carinii*. Gene 1993;123:137–141.
361. Keeling PJ, Luker MA, Palmer JD. Evidence from β-tubulin phylogeny that microsporidia evolved from within the fungi. Mol Biol Evol 2000;17:23–31.
362. Keeling PJ. Congruent evidence from α-tubulin and β-tubulin gene phylogenies for a zygomycete origin of microsporidia. Fungal Genet Biol 2003;38:298–309.
363. Voigt K, Einax E. Oligonucleotide primers for the universal amplification of β-tubulin genes facilitate phylogenetic analyses in the regnum Fungi. Org Divers Evol 2003;3:185–194.
364. Corradi N, Kuhn G, Sanders IR. Monophyly of β-tubulin and H⁺-ATPase gene variants in *Glomus intraradices*: consequences for molecular evolutionary studies of AM fungal genes. Fungal Genet Biol 2004;41:262–273.
365. Juuti JT, Jokela S, Tarkka M, Paulin L, Lahdensalo J. Two phylogenetically highly distinct β-tubulin genes of the basidiomycete *Suillus bovinus*. NCBI Accession no. CAG27308, CAG27309, 2004.
366. Cruz MC, Edlind T. β-Tubulin genes and the basis for benzimidazole sensitivity of the opportunistic fungus *Cryptococcus neoformans*. Microbiology 1997;143:2003–2008.
367. MacDonald LM, Armson A, Thompson A, Reynoldson JA. Characterization of factors favoring the expression of soluble protozoan tubulin proteins in *Escherichia coli*. Protein Expr Purif 2003;29:117–122.
368. Katinka MD, Duprat S, Cornillot E, et al. Genome sequence and gene compaction of the eukaryote parasite *Encephalitozoon cuniculi*. Nature 2001;414:450–453.

369. Cacciò S, La Rosa G, Pozio E. The β-tubulin gene of *Cryptosporidium parvum*. Mol Biochem Parasitol 1997;89:4155–4159.
370. Abrahamsen MS, Templeton TJ, Enomoto S, et al. Complete genome sequence of the apicomplexan, *Cryptosporidium parvum*. Science 2004;304:441–445.
371. Maessen S, Wesseling JG, Smits MA, Konings RN, Schoenmakers JG. Theγ-tubulin gene of the malaria parasite *Plasmodium falciparum*. Mol Biochem Parasitol 1993;60:27–35.
372. Lajoie-Mazenc I, Détraves C, Rotaru V, et al. A single γ-tubulin gene and mRNA, but two γ-tubulin polypeptides differing by their binding to the spindle pole organizing centres. J Cell Sci 1996;109:2483–2492.
373. Willem S, Srahna M, Loppes R, Matagne RF. NCBI Accession no. AAB86648, AAB86649, 1997.
374. Silflow CD, Liu B, LaVoie M, Richardson EA, Palevitz BA. γ-Tubulin in *Chlamydomonas*: characterization of the gene and localization of the gene product in cells. Cell Motil Cytoskeleton 1999;42:285–297.
375. Mages W, Salbaum JM, Harper JF, Schmitt R. Organization and structure of *Volvox* α-tubulin genes. Mol Gen Genet 1988;213:449–458.
376. Dupuis-Williams P. γ-tubulin is necessary for basal body duplication in *Paramecium*. NCBI Accession no. CAA09992, 1998.
377. Joachimiak E, Miceli C, Kaczanowska J. Cloning and expression analysis of γ-tubulin gene in *Tetrahymena pyriformis*. NCBI Accession no. AAG44954, 2000.
378. Pucciarelli S, Ballarini P, Miceli C. Cold-adapted microtubules: Characterization of tubulin posttranslational modifications in the Antarctic ciliate *Euplotes focardii*. Cell Motil Cytoskeleton 1997;38:329–340.
379. Tan M, Liang A, Heckmann K. The two γ tubulin genes of *Euplotes octocarinatus* code for a slightly different protein. NCBI Accession no. P34786, CAA70745, 1996.
380. Tan M, Heckmann K. The two γ-tubulin-encoding genes of the ciliate *Euplotes crassus* differ in their sequences, codon usage, transcription initiation sites and poly(A) addition sites. Gene 1998;210: 53–60.
381. Katz LA, Israel RL. The fate of duplicated α-tubulin genes in ciliates. NCBI Accession no. AAL33680, AAL33681, AAL33682, AAL33683, AAL33684, AAL33685, AAL33686, AAL33691, AAL33692, AAL33694, AAL33695, AAL33697, AAL33698, AAL33699, AAL33700, AAL33713, AAL33714. AAL33715, AAL33718, AAL33719, AAL33724, AAL33725, 2001.
382. Pérez-Romero P, Villalobo E, Díaz-Ramos C, Calvo P, Santos-Rosa F, Torres A. α-Tubulin of *Histriculus cavicola*. Microbiología 1997;13:57–66.
383. Snoeyenbos-West OLO, Salcedo T, McManus GB, Katz LA. Insights into the diversity of choreotrich and oligotrich ciliates (Class: Spirotrichea) based on genealogical analyses of multiple loci. Int J Syst Evol Microbiol 2002;52:1901–1913.
384. Sanchez-Silva R, Torres A. α-Tubulin of peritrich ciliates. (Unpublished) NCBI Accession no. AAM50063, AAM50064, 2002.
385. Ueda M, Graf R, MacWilliams HK, Schliwa M, Euteneuer U. Centrosome positioning and directionality of cell movements. Proc Nat Acad Sci USA 1997;94:9674–9678.
386. Saldarriaga JF, McEwan ML, Fast NM, Taylor FJR, Keeling PJ. Multiple protein phylogenies show that *Oxyrrhus marina* and *Perkinsus marinus* are early branches of the dinoflagellate lineage. Int J Syst Evol Microbiol 2003;53:355–365.
387. Kube-Granderath E, Schliwa M. Unusual distribution of γ-tubulin in the giant fresh water amoeba *Reticulomyxa filosa*. Eur J Cell Biol 1997;72:287–296.
388. Libusová L, Sulimenko T, Sulimenko V, Hozák P, Dráber P. γ-Tubulin in *Leishmania*: cell cycle-dependent changes in subcellular localization and heterogeneity of its isoforms. Exp Cell Res 2004;295: 375–386.
389. Ivens AC, Lewis SM, Bagherzadeh A, Zhang L, Chan HM, Smith DF. A physical map of the *Leishmania* major Friedlin genome. Genome Res 1998;8:135–145.
390. Ersfeld K, Gull K. Partitioning of large and minichromosomes in *Trypanosoma brucei*. Science 1997;276:611–614.
391. Scott V. NCBI Accession no. CAA68866, 1996.
392. Noël C, Gerbod D, Fast NM, et al. Tubulins in *Trichomonas vaginalis*: molecular characterization of α-tubulin genes, posttranslational modifications, and homology modeling of the tubulin dimer. J Eukaryot Microbiol 2001;48:647–654.

393. Schneider A, Plessmann U, Felleisen R, Weber K. α-Tubulins of *Tritrichomonas mobilensis* are encoded by multiple genes and are not posttranslationally tyrosinated. Parasitol Res 1999;85: 246–248.

394. Moriya S, Tanaka K, Ohkuma M, Sugano S, Kudo T. Diversification of the microtubule system in the early stage of eukaryote evolution: elongation factor 1α and α-tubulin protein phylogeny of termite symbiotic oxymonad and hypermastigote protists. J Mol Evol 2001;52:6–16.

395. Moriya S, Gerbod D, Viscogliosi E. NCBI Accession no. BAC98828, 2003.

396. Gerbod D, Sanders E, Moriya S, et al. Molecular phylogenies of Parabasalia inferred from various protein coding gene sequences and comparison with small subunit rRNA-based trees. NCBI Accession no. AAQ19197, AAQ19198, AAQ19199, AAQ19200, AAQ19201, 2003.

397. Coffman HR, Kropf DL. The brown alga, *Pelvetia fastigiata*, expresses two α-tubulin sequences. NCBI Accession no. Q40831, Q40832, 1999.

398. Keeling PJ, Deane JA, McFadden GI. The phylogenetic position of α- and β- tubulins from the *Chlorarachnion* host and *Cercomonas* (Cercozoa). J Eukaryot Microbiol 1998;45:561–570.

399. Keeling PJ, Leander BS. Characterisation of a non-canonical genetic code in the oxymonad *Streblomastix strix*. J Mol Biol 2002;326:1337–1349.

400. Eun S-O, Wick SM. Tubulin isoform usage in maize microtubules. Protoplasma 1998;204:235–244.

401. Lee MG, Loomis C, Cowan NJ. Sequence of an expressed human β-tubulin gene containing ten Alu family members. Nucleic Acids Res 1984;12:5823–5836.

402. Strausberg RL, Feingold EA, Grouse LH, et al. Generation and initial analysis of more than 15,000 full-length human and mouse cDNA sequences. Proc Nat Acad Sci USA 2002;99:16,899–16,903. Swissprot Accession no. Q9BUF5, Q8AVU1.

403. Adachi J, Aizawa K, Akahira S, et al. Swissprot Accession no. Q9CUN8, 2000.

404. Smith DJ. The complete sequence of a frog α-tubulin and its regulated expression in mouse L-cells. Biochem J 1988;249:465–472.

405. Cowan NJ, Dobner PR, Fuchs EV, Cleveland DW. Expression of human α-tubulin genes: interspecies conservation of 3′- untranslated regions. Mol Cell Biol 1983;3:1738–1745.

406. Dode C, Weil D, Levilliers J, et al. Sequence characterization of a newly identified human α-tubulin gene (TUBA2). Genomics 1998;47:125–130.

407. Dobner PR, Kislauskis E, Wentworth BM, Villa-Komaroff L. Alternative 5′ exons either provide or deny an initiator methionine codon to the same α-tubulin coding region. Nucleic Acids Res 1987;15:199–218.

408. Klein SL, Strausberg RL, Wagner L, Pontius J, Clifton SW, Richardson P. Genetic and genomic tools for *Xenopus* research: The NIH *Xenopus* initiative. Dev Dyn 2002;225:384–391.

409. Song HD, Wu XY, Sun XJ, et al. Gene expression profiling in the zebrafish kidney marrow tissue. NCBI Accession no. AAQ097807, 2003.

410. Lewis SA, Cowan NJ. Tubulin genes: structure, expression and regulation. In: Avila J, ed. Microtubule Proteins. Boca Raton, Florida: CRC Press; 37–66.

411. Khan IA, Tomita I, Mizuhashi F, Ludueña RF. Differential interaction of tubulin isotypes with the antimitotic compound IKP-104. Biochemistry 2000;39:9001–9009.

412. Banerjee A, Kasmala LT, Hamel E, Sun L, Lee KH. Interaction of novel thiocolchicine analogs with the tubulin isoforms from bovine brain. Biochem Biophys Res Commun 1999;254:334–337.

APPENDIX

Nomenclature of Avian and Mammalian β-Tubulin Isotypes

Class	Human	Mouse	Rat	Chicken
Ia	HM40	Mβ5	rbt. 5	cβ7
II	Hβ9	Mβ2	rbt.1	cβ1/cβ2
III	Hβ4	Mβ6	rbt. 3	cβ4
IVa	H5β	Mβ4	rbt. 2	–
IVb	Hβ2	Mβ3	–	cβ3
V	–	–	–	cβ5
VI	Hβ1	Mβ1	–	cβ6
VII	Hβ4Q	–	–	–

Adapted from ref. *410.*

7

The Tubulin Superfamily

Richard F. Luduena
and Asok Banerjee

Contents

INTRODUCTION
γ-TUBULIN
δ-TUBULIN
ε-TUBULIN
η-TUBULIN
ζ-, θ-, ι-, AND κ-TUBULINS
THE EVOLUTION OF THE TUBULIN SUPERFAMILY
TUBULIN-LIKE PROTEINS IN PROKARYOTES
THE ORIGIN OF TUBULIN?
ACKNOWLEDGMENTS
REFERENCES

Summary

In addition to the well-known α- and β-tubulin, which constitute the tubulin dimer, there are other related forms of tubulin, including γ, δ, ε, η and others. γ- Tubulin plays a key role in the nucleation of microtubule assembly at the centrosome. The roles of the other members of the tubulin superfamily are still being explored. It is interesting that all of them are found either in the centrosome or the very similar basal body; there is evidence that some of these play significant roles in the assembly of these organelles. The proteins of the tubulin superfamily are also related to the prokaryotic protein FtsZ, which plays a key role in cell division. Comparison of the sequences of all of these proteins allows for speculation about their evolution. These proteins and their evolution will be discussed in this chapter.

Key Words: Tubulin; α-tubulin; β-tubulin; γ-tubulin; δ-tubulin; ε-tubulin; η-tubulin; evolution; centriole; centrosome; basal body.

1. INTRODUCTION

The tubulin superfamily includes α and β that have already been extensively discussed in Chapter 6, as well as the more distantly related γ, δ, ε, ζ, η, θ, ι, and κ (Table 1). Others may be awaiting discovery. Together with α and β, these tubulins constitute the tubulin superfamily. Several of these tubulins are widespread among eukaryotes and the

From: *Cancer Drug Discovery and Development: The Role of Microtubules in Cell Biology, Neurobiology, and Oncology* Edited by: Tito Fojo © Humana Press, Totowa, NJ

Table 1
The Tubulin Superfamily (γ-κ)

Tubulin	Genus	Phylum/Division	References
γ^a	Homo	Chordate	79
	Mus	Chordate	80
	Gallus	Chordate	25
	Xenopus	Chordate	81
	Strongylocentrotus	Echinodermata	82
	Drosophila	Arthropoda	31
	Caenorhabditis	Nematoda	28
	Oryza	Angiosperm	83
	Arabidopsis	Angiosperm	12
	Anemia	Angiosperm	10
	Physcomitrella	Bryophyta	84
	Aspergillus	Ascomycota	9
	Neurospora	Ascomycota	85
	Saccharomyces	Ascomycota	86
	Schizosaccharomyces	Ascomycota	81
	Ustilago	Basidiomycota	87
	Encephalitozoon	Microsporidia	88
	Plasmodium	Apicomplexa	89
	Physarum	Mycetozoa	29
	Chlamydomonas	Chlorophyta	21
	Tetrahymena	Ciliophora	24
	Euplotes	Ciliophora	90
	Dictyostelium	Acrasiomycota	91
	Reticulomyxa	Rhizopoda	92
	Leishmania	Euglenozoa	36
δ	Homo	Chordate	42
	Canis	Chordate	93
	Mus	Chordate	41
	Xenopus	Chordate	94
	Ciona	Urochordata	95
	Chlamydomonas	Chlorophyta	39
	Plasmodium	Apicomplexa	2
	Trypanosoma	Euglenozoa	50
ε	Mus	Chordate	96
	Xenopus	Chordate	97
	Chlamydomonas	Chlorophyta	46
	Paramecium	Ciliophora	44
	Giardia	Diplomonadida	98
	Trypanosoma	Euglenozoa	50
ζ	Trypanosoma	Euglenozoa	99
	Leishmania	Euglenozoa	100
	Xenopus	Chordata	4
η	Paramecium	Ciliophora	48,101
	Xenopus	Chordata	4
θ	Paramecium	Ciliophora	102
ι	Paramecium	Ciliophora	103
κ	Paramecium	Ciliophora	51

^{a}For γ-tubulin only some representative organisms are given, to illustrate that γ occurs in many phyla.

functions of some are starting to become clear. In addition, proteins similar to α and β have been found in bacteria, which also express a protein, FtsZ, which is more distantly related to tubulin. The recently discovered members of the tubulin superfamily, as well as their possible prokaryotic relatives, and speculation on the origin and evolution of these different forms of tubulin will be briefly reviewed. These proteins have been reviewed in more detail by others (1–8).

2. γ-TUBULIN

γ-Tubulin was first described in *Aspergillus* (9). It is 28–35% identical to α and β in amino acid sequence. γ-Tubulin appears to occur in all eukaryotes (Table 1) (1,2,10–14). In animals and protists, it is found in the centrosome, in the immediate vicinity of the centriole; in fungi, and plants that lack centrosomes; γ occurs, respectively, in the spindle pole body and the microtubule-organizing center (15–17). The fundamental function of γ is to nucleate microtubule assembly in vivo. The precise mechanism is controversial. Nucleation involves a complex of γ with at least two proteins in yeast and up to six in higher eukaryotes, the latter called a γ-tubulin ring complex (γ-TuRC) (16,18). In one model, the γTuRC has γ-tubulins associating laterally to bind longitudinally to α- and β-tubulins at the microtubule minus end. In another model, the γ-TuRC constitutes part of a microtubule protofilament at the minus end. Either way, the γ-TuRC acts to cap the minus end and prevent further addition of tubulin subunits there (3). As one would expect, γ-tubulin also nucleates microtubule assembly in vitro, binding to the minus end of the microtubule with a K_d of 0.1 nM (19).

In addition to being involved in the nucleation of microtubules of the mitotic spindle and interphase network, γ-tubulin appears to be present and involved in any system in which microtubule nucleation occurs. For example, it is a constituent of the basal body, which nucleates the axonemal microtubules (20). γ may play a specific role in the nucleation of the central pair microtubules. In *Chlamydomonas*, its location in the flagellar transition region is consistent with this hypothesis (21). In trypanosomes, inhibition of γ-tubulin synthesis results in paralyzed flagella that lack the central pair microtubules (22). A particular role for γ in synthesis of the central pair makes sense as these, unlike the outer doublet microtubules, are not rooted in the basal body microtubules, in addition, the central pair appear to have a precise point of origin (23).

In *Tetrahymena*, γ is located in the microtubule-organizing centers of the basal bodies, as well as those of the macronuclear and micronuclear envelopes and the pores of the contractile vacuole (24). In addition, γ occurs in chicken erythrocyte marginal bands, which do not appear to grow out of centrosomes (25). γ-Tubulin may serve other functions as well. It is present in the cytoplasm of various cells in a form not associated with microtubules; Zhou et al. (26) have shown that overexpression of γ increases the synthesis of α and β, raising the possibility that γ could regulate the synthesis of these other tubulins. Also, γ binds to dynamin, which is required for centrosome cohesion; thus γ could play a role in organizing the centrosome (27). In the nematode *Caenorhabditis*, γ-tubulin occurs close to the membrane at the apical end of intestinal cells (28).

Just as is the case with α- and β-tubulin, γ also exists in isotypes, but so far no organism has been found to have more than two γ-isotypes. Isotypes of γ have been seen in animals, plants, and protists (*see* Tables 1–3 in Chapter 1). In some cases, it is difficult to be sure about the presence of two isotypes. In *Physarum*, for example, there is only a single γ-tubulin gene; the two γ-isotypes, therefore, probably arise by alternative splicing *(29)*. The sunflower *Helianthus* has two γ-tubulins, with molecular weights of 52 kD and 58 kD, however, it is not clear if different genes encode these or if they are the result of post-translational modification *(30)*.

The functional significance of the individual γ-isotypes is as yet unknown. Generally, there is little difference in their distributions in vivo. However, in *Helianthus*, the 58 kD γ-tubulin is expressed in every tissue, whereas the 52 kD γ is found in undifferentiated cells; the former is associated with the nucleus and the latter is not *(30)*. *Drosophila* has one γ-isotype that is found in ovaries and embryos; removal of this isotype resulted in abnormal meiosis *(31)*. However, they could not rule out the possibility that the abnormal cells simply suffered from having too little γ. Although one study suggested that the two γ-isotypes in *Drosophila* have similar functions *(32)*, another study suggests the opposite *(33)*. One γ-isotype is always present in the centrosome and not in the mitotic spindle; the other is in soluble form and only occurs in the centrosome during mitosis. In humans, there are two γ-isotypes, whose sequences are 97.3% identical, but there is no evidence of a functional difference *(34)*. If the two γ-isotypes play specific functional roles, it may be that one is important in microtubule nucleation, whereas the other has another function, possibly regulating tubulin synthesis.

There is evidence suggesting that γ-tubulin is subject to post-translational modification. In yeast, phosphorylation of γ appears to regulate microtubule organization *(35)*. γ-Tubulins from both pigs and *Leishmania* exhibit multiple charge variants *(36,37)*. As there are only two genes for γ, one is tempted to conclude that these variants arise from post-translational modifications. Of the various modifications seen in α- and β-tubulin, there are only two that are likely to yield more than two charge variants: phosphorylation and polyglutamylation. In the former case, multiple sites will have to be adduced, which can be differentially phosphorylated. Polyglutamylation is a more promising candidate, as this modification is known to result in multiple charge variants in α and β *(38)*. However, γ-tubulin lacks the glutamates in the C-terminal region that are polyglutamylated *(20)*. On the other hand, there are some nearby glutamates that could conceivably be modified. At the moment, the question of post-translational modifications of γ-tubulin must remain open.

3. δ-TUBULIN

δ-Tubulin was first discovered in *Chlamydomonas*; when it is mutated there are fewer flagella and the basal bodies often consist of nine doublet microtubules instead of nine triplets *(39)*. δ-Tubulin has since been found in a variety of chordates and protists (Table 1). It is located in the basal bodies and centrosomes *(4)*. In the latter organelles, δ is particularly concentrated in the region between centrioles in the same centrosomes and between centrioles in adjacent centrosomes *(40)*. It often colocalizes with γ-tubulin in the centrosome *(41)*. However, in interphase mammalian cells, δ is often present largely in the cytoplasm, only concentrating at the centrosome during mitosis *(41)*. δ-Tubulin is also present in the manchette and flagellum of the sperm cell *(41)*. In addition, in the mouse testis δ is present in the intercellular bridge that forms following incomplete cytokinesis in spermatid development *(42)*.

The function of δ-tubulin is not clear. Deletion causes appearance of basal bodies with nine doublet microtubules in *Chlamydomonas (43)*. Its location in centrosomes suggests that it may play a structural role in organizing this organelle. It is also likely to be involved in basal body function. The fact that in some cells, it only becomes concentrated in centrosomes during mitosis, suggests that δ may have a particular mitotic function, perhaps in centrosome separation. Interestingly, δ in mice exists as two isoforms, differing in that one lacks a 31-amino acid segment that is present in the other segment *(41)*. This segment is encoded in a single exon, making it very probable that the two isoforms arise by alternative splicing. Interestingly, the large δ is expressed mostly in the testis and is present in other tissues at low concentrations. The small form is expressed in most cells. The fact that humans have only a single δ-isoform suggests that the two δ-isoforms in mice may have no functional significance. It is not known whether δ is subject to post-translational modification, however, its C-terminal region contains either one or no glutamates, so δ is not a good candidate for either polyglycylation or polyglutamylation.

4. ε-TUBULIN

ε-Tubulin was discovered by searching the human genome *(40)*. ε also occurs in various protists (Table 1). ε is located in both basal bodies and centrosomes. In *Paramecium* basal bodes, ε is concentrated at the two ends *(44)*. Centrioles occur in pairs, but they are not identical. The oldest of each pair has a subdistal appendage that projects out from the centriole *(45)*. These appendages are lacking in the basal body as well as in the younger centriole. ε-Tubulin is associated with the older centriole. It is also located in the basal body *(40)*. Purified ε-tubulin gives six spots on two-dimensional gel electrophoresis, suggesting the presence of post-translationally modified forms (Dupuis-Williams, personal communication). Polyglutamylation is an obvious candidate for creating such an electrophoretic pattern. Interestingly, however, there are no glutamates in the vicinity of the C-terminus of ε, making polyglutamylation somewhat unlikely.

Deletion of ε in *Paramecium* results in loss of the B- and C-tubules of the basal body triplets *(44)*. Similar results are obtained in *Chlamydomonas*, where deletion of ε also results in defective meiosis *(46)*. Depletion of ε in *Xenopus* inhibits centriole replication and disorganizes the mitotic spindle *(47)*. It appears likely that ε is required for formation of basal bodies and centrioles. It could conceivably form part of the B- and C-tubules but other models are also possible.

5. η-TUBULIN

η-Tubulin has been observed in only four organisms: the protists *Chlamydomonas* and *Paramecium* and the animals *Ciona* and *Xenopus*. There is no evidence for η in humans or mice. η is located in the basal bodies *(4)*. Studies in *Paramecium* have suggested that η is required for basal body duplication and may also interact with γ-tubulin *(48)*. It appears to interact with β-tubulin and may act as a minus-end capping protein *(49)*. η has only a single glutamate near its C-terminus; if either polyglycylation or polyglutamylation occur to η, this would be the likely site.

6. ζ-, θ-, ι-, AND κ-TUBULINS

The tubulin superfamily keeps growing. New tubulins are constantly being discovered whose function is as yet unknown. ζ-tubulin has been found in basal bodies of the related protists *Trypanosoma* and *Leishmania* as well as in the *Xenopus* genome; it appears to be absent from the genomes of yeast and *Arabidopsis (2,5,50)*. As plants and fungi lack basal bodies, it is reasonable to speculate that ζ may play a role in these organelles but what that role may be is unclear. ζ has two to three glutamates in its C-terminal region, raising the possibility of polyglutamylation or polyglycylation, but these have not yet been observed in ζ.

θ-, ι-, and κ-tubulins have been identified in the genome of *Paramecium (51)*. θ occurs in the basal body (Dupuis-Williams, personal communication). Other than that, the subcellular localization of ι and κ and the functional roles, if any, of θ, ι, and κ are unknown. θ is 56% identical in sequence to *Paramecium* β, ι is 29% identical to yeast β, and κ is 56% identical to *Paramecium* α, θ, and κ have three glutamates in their C-terminal regions, whereas ι has one, making them reasonable candidates for polyglutamylation or polyglycylation. It is certainly possible that more tubulins remain to be discovered as more genomes are sequenced. Fortunately, the Greek alphabet has quite a few more letters.

7. THE EVOLUTION OF THE TUBULIN SUPERFAMILY

It is not difficult to construct a family tree of the other tubulins, based on their amino acid sequences *(4)* (Fig. 1). The first major division in tubulin evolution was between the ζ/η/δ branch and the α/β/γ/ε/ι/κ/θ branch. As α and β are components of centriole and basal body microtubules, then there will be a division between a tubulin that constitutes the centriole and basal body microtubules and a tubulin (δ) that helps to organize these organelles. The next event would have been the division of the α/β/γ/ε/ι/κ/θ branch into ε and the α/β/γ/ι/κ/θ branch. Afterwards, γ diverged from the α/β/ι/κ/θ branch, thereby creating a tubulin that served to nucleate microtubules. Perhaps this corresponds to the creation of an interphase microtubule network, a mitotic spindle, and an axoneme with central pair microtubules. As far as known, γ is the only tubulin involved in nucleation of microtubules in vivo. Presumably the tubulin in the α/β/ι/κ/θ branch was a structural component of microtubules. The next event was the division of the α/β/ι/κ/θ branch into α/κ, β/θ, and ι-tubulins. κ and θ are so similar to α and β in *Paramecium* that it is conceivable that they diverged sometime in ciliate evolution, much later than the other tubulins. In fact, one might almost consider κ and θ to be isotypes of α and β, respectively. Although the functions of θ and ι-tubulin are unknown, the roles of α and β, the structural components of microtubules, are well known. At this time microtubules must have acquired their present morphology. The major role of β-tubulin in binding the exchangeable GTP raises the possibility that complex microtubule dynamic behavior may have appeared about this time. Also, about this time, the ζ/η/δ branch divided into ζ, η, and δ, probably giving the basal bodies and centrioles their final form.

It is interesting that all of the different tubulins (α, β, γ, δ, ε, η, θ, and ζ) whose subcellular locations are known occur in or are associated with basal bodies and/or the very similar centrioles. Several of them (α, β, γ, δ, ε, and η) appear to be necessary for the proper functioning of these organelles. Is it possible that these tubulins originated and diverged from each other, performing some noncentrosomal/basal body functions, while the original centrosome/basal body was evolving separately? Then it has to be postulated that these tubulins subsequently wandered into these organelles, all losing their original functions and acquiring their present roles. This is an unlikely scenario. Therefore, it can be concluded that these forms of tubulin must have evolved with the centrosome/basal body.

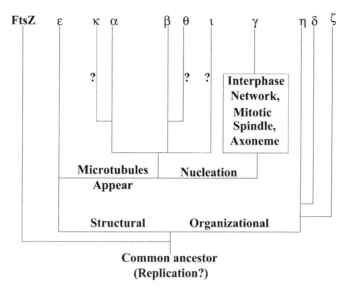

Fig. 1. Hypothesis for the evolution of the tubulin superfamily. The scheme is based on that of Dutcher *(4)*. A common ancestral protein, perhaps involved in replication, gave rise to FtsZ and the tubulin superfamily. The line from the first tubulin diverged into two, one constituting a structure, perhaps the ancestral centrosome/basal body, and the other helping to organize the structure. From the latter are derived η, δ, and ζ, whose roles may still be organizational. The structural tubulin gave rise to two branches: ε, and the common ancestor of α, β, and γ. At this point, it is likely that actual microtubules appeared for the first time and the α/β/γ line diverged into α/β and γ. The latter's major function appears to be to nucleate microtubules, which means that it was now possible to form an interphase microtubule network, a mitotic spindle and the central pair microtubules of the axoneme. The α/β ancestor then diverged into the α and β lines. Probably soon after this the *Prosthecobacter* α- and β-tubulin-like proteins *(58)* diverged, respectively, from α- and β-tubulin. Eventually, ι and θ diverged from β and κ branched off from α, but as ι, θ, and κ are thus far found only in *Paramecium*, it is not possible to determine whether these divergences were ancient or recent.

8. TUBULIN-LIKE PROTEINS IN PROKARYOTES

The authors analysis of the different forms of tubulin has brought us close to the origin of the eukaryotic cell. Can we go any further? Perhaps we can. Bacteria appear to contain a dynamic cytoskeleton *(6)*, a major component of which is the protein FtsZ, present in most of the eubacteria and archaea, whose primary structure shows some resemblance to that of tubulin *(52,53)*. Its three-dimensional structure is very similar to tubulin *(54,55)*. FtsZ is required for cell division in the eubacteria *Escherichia coli* and *Bacillus subtilis (7)*. In addition, FtsZ has GTPase activity, which is associated with polymerization, although the polymer is a ring rather than a microtubule *(8,56,57)*. These rings constrict during bacterial cell division and, in essence, pinch the cell in two *(8,57)*. FtsZ is present in, and participates in the replication of, chloroplasts and many mitochondria, indicating that it is very ancient indeed. FtsZ may have appeared before the origin of eukaryotes. It is intriguing that both tubulin and FtsZ are critically involved in cell division, although they operate by very different mechanisms. The sequence relationships between FtsZ and tubulin suggest that the two proteins may have had a common ancestor. Tubulin is involved in both cell motility and cell division, but as FtsZ appears to be involved only in cell division, it is perhaps more probable that the common ancestor of tubulin and FtsZ also played a role in cell division.

Recently, proteins similar to α- and β-tubulins have been found in the eubacterium *Prosthecobacter (58)*. The bacterial α and β have 31–35% and 34–37% sequence identity with eukaryotic α and β, respectively. The bacterial α and β are actually more similar to their eukaryotic homologs than they are to eukaryotic γ-, δ-, or ε-tubulins, hence, these latter probably branched off from the tubulin family tree before the appearance of the *Prosthecobacter* α and β. The *Prosthecobacter* β can self-assemble into protofilaments or rings provided that either GTP or GDP is present *(59)*. Also, α and β appear to form heterodimers. Either α or β by itself can hydrolyze GTP, but the two together hydrolyze GTP even faster *(59)*.

There are two possibilities to account for the *Prosthecobacter* proteins. One is that almost the entire evolution of the tubulins occurred among the prokaryotes, which would imply that the centrosome arose in prokaryotes. Development of eukaryotes would only have occurred after the appearance of α and β. Second, the bacterial α and β could be survivors of a very ancient gene transfer event from an early eukaryote into a prokaryote. Perhaps these bacterial tubulin analogs speak to a period in evolution when the primitive ancestors of α- and β-tubulin formed a structure other than a microtubule. Howerer, it should be noted that there is as yet no evidence that *Prosthecobacter* α and β are part of a cytoskeleton.

9. THE ORIGIN OF TUBULIN?

The authors now return to the basal body and centriole, two of the most architecturally complex organelles, consisting of nine triplet microtubules arranged in cylindrical array. The basal body consists of, or is associated with, most of the major classes of tubulin (α, β, γ, δ, ε, η, θ, and ζ) as well as several other structural and catalytic proteins, and exhibits at least three post-translational modifications characteristic of tubulin, namely acetylation, polyglutamylation, and polyglycylation, the last two generally acknowledged to be bizarre. The centriole, which otherwise closely resembles the basal body, may lack polyglycylation and is associated with fewer tubulin classes (α, β, γ, δ, and ε). In addition, a few centrioles, associated with development in *Drosophila* have been observed to consist of nine singlets or doublets instead of triplets. These are considered "immature" centrioles *(60,61)*. In general, however, this constellation of characteristics shown by the basal body/centriole appears to be irreducibly complex. With minor exceptions (such as the parasitic protist *Lecudina tuzetae*, whose gamete flagellar axoneme has six outer doublet with no central pair microtubules *[62]*), there are no simple basal body/centrioles, or any lacking the post-translational modifications, or made of any number of triplet microtubules other than nine. For all that Nature tells us, the basal body/centriole could have appeared suddenly like Athena, leaping fully formed out of the head of a prokaryotic Zeus. And yet, this cannot possibly be. There must once have been simpler organelles engaged perhaps in either motility or genome replication rather than both. These intermediates would eventually have disappeared, being replaced by the basal body and centriole known today. This would have happened at the time of the origin of eukaryotes, or even earlier. The existence of FtsZ and the *Prosthecobacter* α- and β-tubulin homologs indicates that much of the evolution of tubulin and perhaps of the centriole/basal body could have occurred in the prokaryotic stage of evolution.

At which stage of evolution did the common ancestor of FtsZ and tubulin live and the ancestral basal body/centriole arise? The post-translational modifications may provide a clue. Polyglutamylation has been seen, not only among the tubulins, but also in some nucleosome assembly proteins *(63)* called NAP1 and NAP2, which have, respectively, 9 and 10 glutamates added, distributed between two glutamates in their C-terminal regions. The polyglutamylation sites on these two proteins resemble those on tubulin. For NAP1 and NAP2 these are, respectively, (with the modified glutamates in boldface) YD**E**EGEEAD AND FE**E**GEEGEE. Polyglutamylation is an unusual modification not only because it may be rare outside of the tubulins, but also because its functional significance does not depend on the precise structure of the side chain. In other words, although the polyglutamylated residue is always a single glutamate in the C-terminal region that particular glutamate varies among the tubulin isotypes and it is not possible to state *a priori* which glutamate is modified. Also, the precise number of glutamates added is variable and seems to make relatively little difference. For example, for regulating MAP binding, the effect of three glutamates in the polyglutamyl side chain differs from that of six glutamates but not from that of four *(38)*. What is common to all forms of polyglutamylation, however, is the α/γ-linkage by which the polyglutamyl chain is connected to the glutamate residue in the C-terminal region.

Polyglycylation has a story that is similar but not identical to that of polyglutamylation. So far, polyglycylation has been observed only in tubulin and in no other protein. Also, in contrast to polyglutamylation, which modifies only a single glutamate residue near the C-terminus, polyglycylation modifies several. The actual number of glycines in the polyglycyl side chains is highly variable and the precise number does not seem to be important. Nor does it matter which of the possible polyglycylation sites is actually modified. Xia et al. *(64)* worked with *Tetrahymena* that has five polyglycylated glutamates in the β-chain and three in α; mutants lacking any one or two of the five polyglycylated glutamates in β were unaffected; mutations of any three of these glutamates were deleterious or lethal. Removal of the polyglycylated glutamates in α had no effect *(64)*. As has been seen, polyglycylation is critical for proper formation of the doublet microtubules of the axoneme and, presumably, of the triplet microtubules of the basal body as well, although this has not been directly examined. It would appear, however, that the one consistent feature is the presence of the α/γ-linkage joining the first glycine to a glutamate in the C-terminal region of tubulin.

On the face of it, polyglutamylation and polyglycylation seem strange. If all that is required is that a chain of glutamates or glycines be present, it is easier to envision a mutation in the tubulin gene that would simply add them to the C-terminus of tubulin in an α/α-linkage through ribosomal synthesis. Instead there is a cumbersome system whereby one enzyme adds glutamate or glycine to a glutamate through a α/γ-linkage and another enzyme adds further glutamates or glycines through α/α-linkages to the first one. As there are no mixed chains of glutamates and glycines, there has to be at least four enzymes, two for polyglutamylation and two for polyglycylation. Nature seems to have gone to a great deal of trouble to ensure that α/γ-linkages occur in the C-terminal region of tubulin.

If the rest of the tubulin molecule is ignored and only the C-termini as they actually are is focused on, what is seen is not a simple chain so much as a branching network (Fig. 2). The actual C-terminal peptides of many tubulin isotypes consist largely of glutamates and an occasional glycine, aspartate, alanine, or valine, mostly small amino acids. Attached to the C-terminal through α/γ-linkages are up to five side chains, one made of glutamates and the others of glycines, each as long as or even longer than the remainder of the C-terminal peptide. Nowhere else in the world of living organisms is

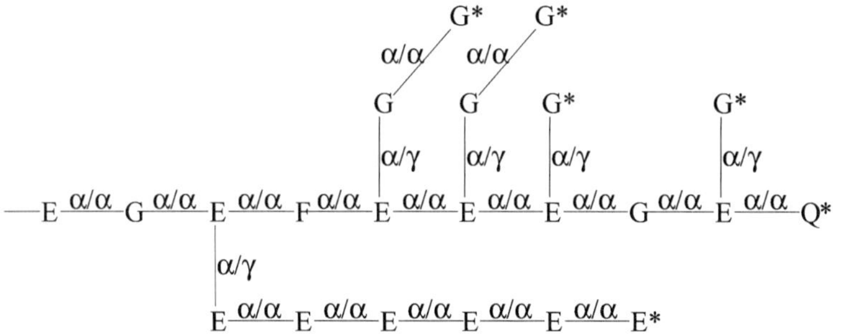

Fig. 2. Hypothetical C-terminus of β-tubulin of *Paramecium*, showing known polyglycylation sites *(104)* and assuming polyglutamylation at a site homologous to polyglutamylation sites in other organisms. α/γ- and α/α-linkages are shown. Bonds are not drawn to scale. For polyglycylation, the numbers of glycines found most commonly attached to the indicated residues are given. More glycines are often attached. Note that this protein has six C-termini (indicated with asterisks).

there anything like this. However, a similar structure has been observed in the nonliving world in prebiotic experiments. When glutamate and other amino acids are heated to 170°C under anhydrous conditions, they polymerize into the so-called proteinoids *(65)*. Proteinoid formation appears to require the presence of glutamate, which forms an imide that catalyzes polymerization *(66)*. Some branching as well as both α/α- and α/γ-linkages are thought to occur *(67)*. When placed in aqueous medium, proteinoids form cell-like structures called microspheres that exhibit enzymatic activities, including ATPase, and can induce polymerization of nucleotides and amino acids *(68,69)*. The microspheres can even divide *(70)* and stimulate the growth of neurites *(71)*. It is easy to envision a volcanic eruption or a meteorite impact—both very common on the early Earth *(72)*—creating temperatures that would allow proteinoids to form. Similarly, glycine is the most common amino acid formed in prebiotic experiments, followed closely by alanine, and more distantly by valine, aspartate, and glutamate *(73)*. Glycine-rich proteinoids were probably abundant at the time life originated and were likely to have contained alanine, valine, aspartate, and glutamate.

The first self-replicating systems could have arisen in the vicinity of proteinoids that could conceivably have formed a catalytic matrix to facilitate the earliest biological processes. What if a few polyglycyl chains joined to glutamates through α/γ-linkages helped a primitive microtubule become a doublet or even a triplet, thereby forming a stronger structure? What if a network of glutamates helped the ancestors of NAP1 and NAP2 stabilize a nucleic acid or helped a primitive tubulin form a microtubule? In this connection, it is interesting that glutamate, particularly at high concentrations, strongly stabilizes tubulin's conformation and also induces tubulin to polymerize *(74–77)*. Perhaps a function of polyglutamylation is to ensure that the tubulin molecule always has a supply of glutamate at hand to help in maintaining the molecular conformation. If glutamate- or glycine-rich proteinoids promoted the assembly of the first microtubules, there would have been a selective advantage to an organism that could mutate so as to synthesize a system of enzymes capable of adding chains of glutamates or glycines to proteins through α/γ-linkages. An organism with these capabilities could liberate itself from dependence on naturally occurring proteinoids and could explore new environments, new morphologies, and new biochemistries.

We have obviously wandered far into the realm of the speculative, piling guess on guess and hypothesis on hypothesis, pushing the story of tubulin to as far back as the origin of life itself. The only possible justification for dragging the reader over such a mountain of speculation is that such a discussion may lead to fruitful lines of experimentation. For example, it would be useful to search among prokaryote genomes for genes similar to those of the yet undiscovered eukaryotic enzymes that catalyze the polyglutamylation and polyglycylation reactions. These genes need not all be found in the same prokaryotic organism. The assemblage of genes could have arisen through lateral transfer. The recently advanced hypothesis of the "ring of life," suggesting that eukaryotes arose from a fusion of eubacteria and archaea *(78)* means that these genes could have arisen anywhere and not necessarily together. Purified γ, δ, ε, and other tubulins should be analyzed to see what kinds, if any, of post-translational modifications occur in these proteins. The line of speculation would imply that polyglutamylation and polyglycylation would be found in several of these. Otherwise, these modifications would not be as old as herein suggested. It should be seen if polyglutamylation and polyglycylation occur in any prokaryotic organisms. The ability of glutamate-rich or glycine-rich proteinoids to facilitate DNA replication or packaging or to regulate microtubule assembly could be tested. Modeling experiments *in silico* may reveal the functional significance of the post-translational modifications. Such experimentation would add to the knowledge of tubulin evolution and function and may shed light on the origin of eukaryotes and even on the first cells.

ACKNOWLEDGMENTS

Supported by grants to RFL (Welch Foundation AQ-0726, National Institutes of Health CA26376 and CA084986, US Army Breast Cancer Research Program W81XWH-05-1-0238 and DAMD17-01-1-0411, US Army Prostate Cancer Research Program DAMD17-02-1-0045 and W81XWH-04-1-0231) and to AB (National Institutes of Health CA59711 and US Army Breast Cancer Research Program DAMD17-98-1-8244). We thank Lorraine Kasmala and Veena Prasad for skilled technical assistance. We gratefully acknowledge helpful suggestions, materials, or information from Pascale Dupuis-Williams and Jack Tuszynski. Special thanks to Marc Kessler, Susan Osgood, Katy Kessler, Rev. Ted Hoskins, and Isle-au-Haut, Maine, for providing an environment conducive to thinking and writing.

REFERENCES

1. Oakley BR. γ-Tubulin. In: Microtubules, Hyams JS, Lloyd CW, eds. Wiley-Liss, New York, pp 33–45.
2. McKean PG, Vaughan S, Gull K. The extended tubulin superfamily. J Cell Sci 2001;114:2723–2733.
3. Moritz M, Agard DA. γ-Tubulin complexes and microtubule nucleation. Curr Op Struct Biol 2001;11:174–181.
4. Dutcher SK. Long-lost relatives reappear: identification of new members of the tubulin superfamily. Curr Op Microbiol 2003;6:634–640.
5. Dutcher SK. The tubulin fraternity: alpha to eta. Curr Op Cell Biol 2001;13:49–54.
6. Errington J. Dynamic proteins and a cytoskeleton in bacteria. Nature Cell Biol 2003;5:175–178.
7. Lutkenhaus J, Addinall SG. Bacterial cell division and the Z ring. Annu Rev Biochem 1997;66: 93–116.
8. Margolin W. Organelle: self-assembling GTPase caught in the middle. Curr Biol 2000;10:R328–R330.
9. Oakley CE, Oakley BR. Identification of γ-tubulin, a new member of the tubulin superfamily encoded by *mipA* gene of *Aspergillus nidulans*. Nature 1989;338:662–664.

10. Fuchs U, Moepps B, Maucher HP, Schraudolf H. Isolation, characterization and sequence of a cDNA encoding γ-tubulin protein from the fern *Anemia phyllitidis* L. Sw. Plant Mol Biol 1993;23:595–603.
11. Liang A, Heckmann K. The macronuclear γ-tubulin-encoding gene of *Euplotes octocarinatus* contains two introns and an in-frame TGA. Gene 1993;136:319–322.
12. Liu B, Joshi HC, Wilson TJ, Silflow CD, Palevitz BA, Snustad DP. γ-Tubulin in *Arabidopsis*: gene sequence, immunoblot, and immunofluorescence studies. Plant Cell 1994;6:303–314.
13. Lopez I, Khan S, Sevik M, Cande WZ, Hussey PJ. Isolation of a full length cDNA encoding *Zea mays* γ-tubulin. Plant Physiol 1995;107:309–310.
14. Scott V, Sherwin T, Gull K. γ-Tubulin in trypanosomes: Molecular characterization and localization to multiple and diverse microtubule organizing centres. J Cell Sci 1997;110:157–168.
15. Fuller SD, Gowen BE, Reinsch S, et al. The core of the mammalian centriole contains γ-tubulin. Curr Biol 1995;5:1384–1393.
16. Knop M, Schiebel E. Spc98p and Spc97p of the yeast γ-tubulin complex mediate binding to the spindle pole body via their interaction with Spc110p. EMBO J 1997;16:6985–6995.
17. Shimamura M, Brown RC, Lemmon BE, et al. γ-Tubulin in basal land plants: characterization, localization, and implication in the evolution of acentriolar microtubule organizing centers. Plant Cell 2004;16:45–59.
18. Inclán YF, Nogales E. Structural models for the self-assembly and microtubule interactions of γ-, δ- and ε-tubulin. J Cell Sci 2000;114:413–422.
19. Leguy R, Melki R, Pantaloni D, Carlier MF. Monomeric γ-tubulin nucleates microtubules. J Biol Chem 2000;275:21,975–21,980.
20. Kierszenbaum AL. Intramanchette transport (IMT): managing the making of the spermatid head, centrosome, and tail. Mol Reprod Dev 2002;63:1–4.
21. Silflow CD, Liu B, LaVoie M, Richardson EA, Palevitz BA. γ-Tubulin in *Chlamydomonas*: characterization of the gene and localization of the gene product in cells. Cell Motil Cytoskeleton 1999;42: 285–297.
22. McKean PG, Baines A, Vaughan S, Gull K. γ-Tubulin functions in the nucleation of a discrete subset of microtubules in the eukaryotic flagellum. Curr Biol 2003;13:598–602.
23. Geimer S, Melkonian M. The ultrastructure of the *Chlamydomonas reinhardtii* basal apparatus: identification of an early marker of radial asymmetry inherent in the basal body. J Cell Sci 2004;117: 2663–2674.
24. Shang Y, Li B, Gorovsky MA. *Tetrahymena thermophila* contains a conventional γ-tubulin that is differentially required for the maintenance of different microtubule-organizing centers. J Cell Biol 2002;158:1195–1206.
25. Linhartová I, Novotná B, Sulimenko V, Dráberová E, Dráber P. Gamma-tubulin in chicken erythrocytes: changes in localization during cell differentiation and characterization of cytoplasmic complexes. Dev Dynamics 2002;223:229–240.
26. Zhou J, Shu HB, Joshi HC. Regulation of tubulin synthesis and cell cycle progression in mammalian cells by γ-tubulin-mediated microtubule nucleation. J Cell Biochem 2002;84:472–483.
27. Thompson HM, Cao H, Chen J, Euteneuer U, McNiven MA. Dynamin 2 binds γ-tubulin and participates in centrosome cohesion. Nature Cell Biol 2004;6:335–342.
28. Bobinnec Y, Fukuda M, Nishida E. Identification and characterization of *Caenorhabditis elegans* γ-tubulin in dividing cells and differentiated tissues. J Cell Sci 2000;113:3747–3759.
29. Lajoie-Mazenc I, Détraves C, Rotaru V, et al. A single γ-tubulin gene and mRNA, but two γ-tubulin polypeptides differing by their binding to the spindle pole organizing centres. J Cell Sci 1996;109: 2483–2492.
30. Petitprez M, Caumont C, Barthou H, Wright M, Alibert G. Two γ-tubulin isoforms are differentially expressed during development in *Helianthus annuus*. Physiologia Plantarum 2001;111:102–107.
31. Llamazares S, Tavosanis G, Gonzalez C. Cytological characterisation of the mutant phenotypes produced during early embryogenesis by null and loss-of-function alleles of the γTub37C gene in *Drosophila*. J Cell Sci 1999;112:659–667.
32. Tavosanis G, Gonzalez C. γ-Tubulin function during female germ-cell development and oogenesis in *Drosophila*. Proc Nat Acad Sci USA 2003;100:10,263–10,268.
33. Raynaud-Messina B, Debec A, Tollon Y, Gares M, Wright M. Differential properties of two γ-tubulin isotypes. Eur J Cell Biol 2001;80:643–649.
34. Wise DO, Krahe R, Oakley BR. The γ-tubulin gene family in humans. Genomics 2000;67:164–170.

35. Vogel J, Drapkin B, Oomen J, Beach D, Bloom K, Snyder M. Phosphorylation of gamma-tubulin reg-ulates microtubule organization in budding yeast. Dev Cell 2001;1:621–631.
36. Dráber P, Sulimenko V. Association of γ-tubulin isoforms with tubulin dimers. Cell Biol Int 2003;27: 197–198.
37. Libusová L, Sulimenko T, Sulimenko V, Hozák Dráber P. γ-Tubulin in *Leishmania*: cell cycle-depend-ent changes in subcellular localization and heterogeneity of its isoforms. Exp Cell Res 2004; 295:375–386.
38. Boucher D, Larcher JC, Gros F, Denoulet P. Polyglutamylation of tubulin as a progressive regulator of *in vitro* interactions between the microtubule-associated protein tau and tubulin. Biochemistry 1994;33:12,471–12,477.
39. Dutcher SK, Trabuco EC. The *UNI3* gene is required for assembly of basal bodies of *Chlamydomonas* and encodes δ-tubulin, a new member of the tubulin superfamily. Mol Biol Cell 1998;9:1293–1308.
40. Chang P, Stearns T. δ-Tubulin and ε-tubulin: two new human centrosomal tubulins reveal new aspects of centrosome structure and function. Nature Cell Biol 2000;2:30–35.
41. Smrzka OW, Delgehyr N, Bornens M. Tissue-specific expression and subcellular localisation of mammalian δ-tubulin. Curr Biol 2000;10:413–416.
42. Kato A, Nagata Y, Todokoro K. δ-Tubulin is a component of intercellular bridges and both the early and perinuclear rings during spermatogenesis. Dev Biol 2004;269:196–205.
43. O'Toole ET, Giddings TH, McIntosh JR, Dutcher SK. Three-dimensional organization of basal bod-ies from wild-type and δ-tubulin deletion strains of *Chlamydomonas reinhardtii*. Mol Biol Cell 2003;14:2999–3012.
44. Dupuis-Williams P, Fleury-Aubusson A, Garreau de Loubresse N, et al. Functional role of ε-tubulin in the assembly of the centriolar microtubule scaffold. J Cell Biol 2002;158:1183–1193.
45. Nakagawa Y, Yamane Y, Okanoue T, Tsukita S, Tsukita S. Outer dense fiber 2 is a widespread centro-some scaffold component preferentially associated with mother centrioles: its identification from iso-lated centrosomes. Mol Biol Cell 2001;12:1687–1697.
46. Dutcher SK, Morrissette NS, Preble AM, Rackley C, Stanga J. ε-Tubulin is an essential component of the centriole. Mol Biol Cell 2002;13:3859–3869.
47. Chang P, Giddings TH, Winey M, Stearns T. ε-Tubulin is required for centriole duplication and micro-tubule organization. Nature Cell Biol 2003;5:71–76.
48. Ruiz F, Krzywicka A, Klotz C, et. al. The *SM19* gene, required for duplication of basal bodies in *Paramecium*, encodes a novel tubulin, η-tubulin. Curr Biol 2000;10:1451–1454.
49. Ruiz F, Dupuis-Williams P, Klotz C, et al. Genetic evidence for interaction between η- and β-tubulins. Eukaryotic Cell 2004;3:212–220.
50. Vaughan S, Attwood T, Navarro M, Scott V, McKean P, and Gull K. New tubulins in protozoal parasites. Curr Biol 2000;10:R258–R259.
51. Ruiz F. (Unpublished) NCBI Accession # CAE11219, 2003.
52. Vaughan S, Wickstead B, Gull K, Addinall AG. Molecular evolution of FtsZ protein sequences encoded within the genomes of Archaea, Bacteria, and Eukaryota. J Mol Evol 2004;58:19–39.
53. Wang X, Lutkenhaus J. FtsZ ring: the eubacterial division apparatus conserved in archaebacteria. Mol Microbiol 1996;21:313–319.
54. Lowe J, Amos LJ. Crystal structure of the bacterial cell-division protein FtsZ. Nature 1998; 391: 203–206.
55. Nogales E, Wolf SG, Downing KH. Structure of the αβ tubulin dimer by electron crystallography. Nature 1998;391:199–203.
56. Bi E, Lutkenhaus J. FtsZ ring structure associated with division in *Escherichia coli*. Nature 1991;354:161–164.
57. Addinall SG, Holland B. The tubulin ancestor, FtsZ, draughtsman, designer and driving force for bac-terial cytokinesis. J Mol Biol 2002;318:219–236.
58. Jenkins C, Samudrala R, Anderson I, et al. Genes for the cytoskeletal protein tubulin in the bacterial genus *Prosthecobacter*. Proc Nat Acad Sci USA 2002;99:17,049–17,054.
59. Sontag CA, Stricker J, Staley JT, Erickson HP. Self assembly of a new class of bacterial tubulins. American Society for Cell Biology Annual Meeting Abstracts. 2004; p 423a.
60. McDonald K, Morphew M. Improved preservation of ultrastructure in difficult-to-fix organisms by high pressure freezing and freeze substitution: 1. *Drosophila melanogaster* and *Stronglyocentrotus purpuratus* embryos. Microsc Res Tech 1993;24:465–473.

61. Callaini G, Whitfield WG, Riparbelli MG. Centriole and centrosome dynamics during the embryonic cell cycles that follow the formation of the cellular blastoderm in *Drosophila*. Exp Cell Res 1997;234:183–190.

62. Kuriyama R, Omoto CK, Gèze M, Bessse C, Schrével J. Changes in the distribution of microtubule/centrosome-containing structures in the protozoan, *Lecudina tuzetae*. American Society for Cell Biology Annual Meeting Abstracts. 2004; p 70a.

63. Regnard C, Desbruyères E, Huet JC, Beauvallet C, Pernollet JC, Eddé B. Polyglutamylation of nucleosome assembly proteins. J Biol Chem 2000;275:15,969–15,976.

64. Xia L, Hai B, Gao Y, et al. Polyglycylation of tubulin is essential and affects cell motility and division in *Tetrahymena thermophila*. J Cell Biol 2000;149:1097–1106.

65. Fox SW, Harada K. Thermal copolymerization of amino acids to a product resembling protein. Science 1958;128:1214.

66. Mark HF, Gaylord MG, Bikales NM. Encyclopedia of Polymer Science and Technology, Interscience Publishers, New York, vol 9, 1964; pp 284.

67. Fox SW, Dose K. Molecular Evolution and the Origin of Life. W.H. Freeman & Co., San Francisco, 1972.

68. Fox SW, Jungck JR, Nakashima T. From proteinoid microsphere to contemporary cell: formation of internucleotide and peptide bonds by proteinoid particles. Origins of Life 1974;5:227–237.

69. Nakashima T, Fox SW. Metabolism of proteinoid microspheres. Topics Curr Chem 1987;139:58–81.

70. Fox SW. Synthesis of life in the lab? Defining a protoliving system. Quart Rev Biol 1991;66: 181–185.

71. Fox SW, Hefti F, Hartikka J, et al. Pharmacological activities in thermal proteins: relationships in molecular evolution. Int J Quantum Chem Quantum Biol Symp 1987;14:347–349.

72. Cloud P. Oasis in Space: Earth History from the Beginning. W.W. Norton, and Co., New York, 1988.

73. Schlesinger G, Miller SL. Prebiotic synthesis in atmospheres containing CH_4, CO, and CO_2. J Mol Evol 1983;19:376–382.

74. Wilson L. Properties of colchicine-binding protein from chick embryo brain. Interactions with *Vinca* alkaloids and podophyllotoxin. Biochemistry 1970;9:4999–5007.

75. Hamel E, Lin CM. Stabilization of the colchicine-binding activity of tubulin by organic acids. Biochim Biophys. Acta 1981;675:226–231.

76. Hamel E, Lin CM. Glutamate-induced polymerization of tubulin: Characteristics of the reaction and application to the large-scale purification of tubulin. Arch Biochem Biophys 1981;209:29–40.

77. Hamel E, del Campo AA, Lowe MC, Waxman PG, Lin CM. Effects of organic acids on tubulin polymerization and associated guanosine 5'-triphosphate hydrolysis. Biochemistry 1982;21:503–509.

78. Rivera MC, Lake JA. The ring of life provides evidence for a genome fusion origin of eukaryotes. Nature 2004;431:152–155.

79. Zheng Y, Jung MK, Oakley BR. γ-Tubulin is present in *Drosophila melanogaster* and *Homo sapiens* and is associated with the membrane. Cell 1991;65:817–823.

80. Kubo A, Hata M, Kubo A, Tsukita S. Gene-knockout analysis of two gamma-tubulin isoforms in mice. NCBI Accession nos. BAD27264, BAD27265, 2004.

81. Stearns T, Evans L, Kirschner M. γ-Tubulin is a highly conserved component of the centrosome. Cell 1991;65:825–836.

82. Rogers GC, Chui KK, Lee EW, et al. A kinesin-related protein, KRP(180), positions prometaphase spindle poles during early sea urchin embryonic cell division. J Cell Biol 2000;150:499–512.

83. Kim YK, Cha YK, Jun HY, Kim JD, Choi JS, Kim HR. Nucleotide sequence of a cDNA (OstubG2) encoding a gamma-tubulin in the rice plant (*Oryza sativa*). NCBI Accession no. O49068, 2001.

84. Wagner TA, Sack FD, Oakley BR, Oakley CE, Schwuchow J. Characterization of gamma tubulin from *Physcomitrella patens*. NCBI Accession no. AAD33883, 1999.

85. Heckmann S, Schliwa M, Kube-Granderath E. Primary structure of *Neurospora crassa* γ-tubulin. Gene 1997;199:303–309.

86. Sobel SG, Snyder M. A highly divergent γ-tubulin gene is essential for cell growth and proper microtubule organization in *Saccharomyces cerevisiae*. J Cell Biol 1995;131:1775–1788.

87. Luo H, Perlin MH. The γ-tubulin-encoding gene from the basidiomycete fungus, *Ustilago violacea*, has a long 5'-untranslated region. Gene 1993;137:187–194.

88. Katinka MD, Duprat S, Cornillot E, et al. Genome sequence and gene compaction of the eukaryote parasite *Encephalitozoon cuniculi*. Nature 2001;414:450–453.

89. Maessen S, Wesseling JG, Smits MA, Konings RN, Schoenmakers JG. The γ-tubulin gene of the malaria parasite *Plasmodium falciparum*. Mol Biochem Parasitol 1993;60:27–35.

90. Tan M, Liang A, Heckmann K. The two gamma tubulin genes of *Euplotes octocarinatus* code for a slightly different protein. NCBI Accession nos. P34786, CAA70745, 1996.

91. Ueda M, Graf R, MacWilliams HK, Schliwa M, Euteneuer U. Centrosome positioning and directionality of cell movements. Proc Nat Acad Sci USA 1997;94:9674–9678.

92. Kube-Granderath E, Schliwa M. Unusual distribution of γ-tubulin in the giant fresh water amoeba *Reticulomyxa filosa*. Eur J Cell Biol 1997;72:287–296.

93. Sidjanin DJ, Zangerl B, Johnson JL, et al. Cloning of the canine delta tubulin cDNA (*TUBD*) and mapping to CFA9. Anim Genet 2002;33:161–162.

94. Piard-Ruster K, Stearns T. Characterization of *Xenopus* delta-tubulin. NCBI Accession no. AAL27450, 2001.

95. Inaba K, Satouh Y. Molecular cloning of ascidian delta-tubulin. NCBI Accession no. BAB85852, 2002.

96. (no reference) NCBI Accession no. XP 125543.

97. Chang P, Stearns T. *Xenopus* epsilon tubulin is a centrosomal protein. (NCBI Accession no.) AAN77278, 2001.

98. Morrison HG, McArthur AG, Adam RD, et al. Draft sequence of the *Giardia lamblia* genome. NCBI Accession no. EAA40536, 2003.

99. Berriman M, Hertz-Fowler CVA, hall N, et al. NCBI Accession no. CAB95398, 2002.

100. Ivens AC, Lewis SM, Bagherzadeh A, Zhang L, Chan HM, Smith DF. A physical map of the *Leishmania major* Friedlin genome. Genome Res 1998;8:135–145.

101. Ruiz F. NCBI Accession no. CAB99490, 2000.

102. Dupuis-Williams P. NCBI Accession no. CAD20607, 2000.

103. Dupuis-Williams P, Fleury-Aubusson A, Garreau de Loubresse N, Geoffroy H, Vayssie L, and Rossier J. Functional role of epsilon-tubulin in centriolar microtubule scaffold. NCBI Accession no. CAD20608, 2000.

104. Vinh J, Langridge JI, Bré MH, et al. Structural characterization by tandem mass spectrometry of the posttranslational polyglycylation of tubulin. Biochemistry 1999;38:3133–3139.

8 Tubulin Proteomics in Cancer

*Pascal Verdier-Pinard, Fang Wang,
Ruth Hogue Angeletti, Susan Band Horwitz,
and George A. Orr*

CONTENTS

SUMMARY

Human tubulin α/β heterodimers are encoded by six and seven genes for the α- and β-subunit, respectively. Each of these isotypes can undergo various posttranslational modifications. Most of the sequence specificity for each isotype and posttranslational modifications occur at the C-terminal part of tubulin. The biological significance of this so-called "tubulin code" and its regulation are unresolved questions in basic cytoskeleton research and in pathologies such as cancer. βIII-tubulin repeatedly appears as a potential marker of poor prognosis in different tumor types and of drug resistance. Nevertheless, because of limitations in present methods of analysis, it is still unclear if tubulin isotype expression profiles will provide useful biomarkers for the stratification of cancer patients prior to treatments including microtubulo-interacting drugs. Recent progresses in mass spectrometry-based analyses of tubulin isotype expression in cancer cells are presented in this chapter. Such approaches allow the validation of tools such as antibodies used in immunohistochemistry, identifies tubulin sequences expressed in human cells and posttranslational modifications, and offer new avenues for tubulin isotype quantitation in tumor and normal tissues.

From: *Cancer Drug Discovery and Development: The Role of Microtubules in Cell Biology,
Neurobiology, and Oncology* Edited by: Tito Fojo © Humana Press, Totowa, NJ

Key Words: Microtubules; tubulin; proteomics; MALDI-TOF MS (matrix assisted laser desorption ionization-time-of-flight mass spectrometer); ESI MS (electrospray ionization mass spectrometer); IEF (isoelectrofocusing); tubulin isotypes; drug sensitivity.

1. INTRODUCTION

Microtubules are important components of the cytoskeleton and play crucial roles in a diverse array of cellular processes including morphogenesis, motility, organelle, and vesicle trafficking, and chromosome segregation during mitosis. Microtubules also function as a scaffold for a large number of signaling proteins and may serve to localize and regulate their activities. Because of their pivotal role in cell function, particularly cell division, the microtubule cytoskeleton has emerged as an effective target for cancer chemotherapy. This chapter will focus on the use of proteomic technologies to define the tubulin-isotype composition of human cancer cell lines. Mass spectrometry (MS)-based analyses of tubulin are also useful for the determination of tubulin-ligand binding sites; but this aspect is beyond the scope of the present chapter.

2. COMPLEXITY OF TUBULIN-PROTEIN EXPRESSION

Tubulin, the basic building block of microtubules is a 100 kDa heterodimer consisting of single α- and β-polypeptide chains. In humans, six α-tubulin and seven β-tubulin genes are expressed and are differentially distributed in tissues (Table 1) (*see also* Chapter 10). Each isotype can also generate multiple isoforms because of extensive post-translational modifications, including polyglutamylation, polyglycylation, reversible tyrosination, phosphorylation, and acetylation (Table 1; Fig. 1) (*see* also Chapter 10). Consequently, numerous possible combinations of tubulin heterodimers, defined by isotype composition and the degree of post-translational modifications (PTM), can occur in cells and tissues. Whether the expression of multiple tubulin isotypes with their associated modifications is of functional significance, is an area of intense investigation. One exciting possibility is that the extensive tubulin structural diversity represents a complex code that can be deciphered and acted upon by interacting proteins *(1,2)*.

There is evidence from in vitro studies that the tubulin-isotype composition can influence both the dynamic and drug-binding properties of microtubules *(3–6)* and that the PTM may regulate interactions with associated structural and motors proteins *(7–9)*. When the amino acid sequences of the human α- or β-tubulin isotypes are aligned, most of the divergence is contained in the last 20 amino acids (Table 2) *(10)*. This domain has been termed the isotype-defining region and, significantly, all of the known PTM, with the exception of acetylation of Lys40, occur within this highly acidic region (Fig. 1) *(1)*. In the electron crystallographic model of the αβ-tubulin heterodimer, this C-terminal peptide is not resolved because of its unstructured nature. Both α and β C-terminal peptides lie as a flexible arm at the surface of the microtubule lattice *(11,12)*, where they are accessible to the enzymes responsible for the various PTM and for interactions with regulatory and motor proteins (Fig. 1).

Despite the importance of microtubules as a target for anticancer drugs, detailed information on tubulin composition in human non-neuronal cells is limited. The tubulin expression profiling in mammalian cell lines/tissues has been achieved at the mRNA level by reverse transcriptase-polymerase chain reaction (RT-PCR) and at the protein level by antibody-based approaches *(13)*. RT-PCR, although a convenient approach,

Table 1
Nomenclature, Tissue Distribution, C-Terminus Sequence and Post-Translational Modifications of Human Tubulin Isotypes

Isotype	Human gene	Tissue expression[a]	C-terminus sequence[b]	Post-translational modifications[c]
α				
Kα1	TUBA1/k-α1	Widely expressed	MAALEKDYEEVGVDSV EGEGEEEGEEY	glutamylation (E445)
bα1	TUBA3/b-α1	Mainly in brain	MAALEKDYEEVGVHS VEGEGEEEGEEY	glycylation (E445)
3	TUBA2	Testis specific	LAALEKDYEEVGVDS VEAEAE-EGEEY	detyrosination/ tyrosination
4	TUBA4	Brain muscle	MAALEKDYEEVGI DSYEDEDE—GEE-	removal of penultimate glutamate
6	TUBA6	Widely expressed	MAALEKDYEEVGA DSADGEDE—GEEY	acetylation (K40)
8	TUBA8	Heart, muscle testis	LAALEKDYEEVGTDS FEEENE—GEEF	
β				
I	HM40/TUBB	Constitutive	YQDATAEEEEDFGEEA EEEA	glutamylation (E435)
II	Hβ9/TUBB2	Major neuronal lung	YQDATADEQGEFEEE EGEDEA	glycylation (E445,E447, E448)
III	Hβ4/TUBB4	Minor neuronal testis	YQDATAEEEGEMYED DEEESEAQGPK	phosphorylation of βIII isotype (S444)
IVa	Hβ5/TUBB5	Brain specific	YQDATAEQGEFEEEAE EEVA[d] YQDATAEEGEFEEEAE EEVA	
IVb	Hβ2	Major testis	YQDATAEEEGEFEEEA EEEVA	
V	5β/βV	Uterus adeno-carcinoma	YQDATANDGEEAFEDEE EEIDG	
VI	Hβ1/TUBB1	Blood	FQDAKAVLEEDEEVTEE AEMEPEDKGH	

[a]Tubulin isotype distribution in human tissues has been performed mainly by RT-PCR, Western blotting, and immunohistochemistry.

[b]Amino acids differing from isotype 1 (k-α1) for α-tubulins or from isotype I (HM40/TUBB) for β-tubulins are highlighted in black.

[c]Position and amino acid modified by posttranslational modification are indicated in brackets and refer to studies performed on mammalian tubulin.

[d]Two βIVa-tubulin sequences with distinct C-termini were found in the NCBI protein database. The upper C-terminus sequence was found in human brain and lower sequence was found in a human oligodendroglioma and in mouse brain.

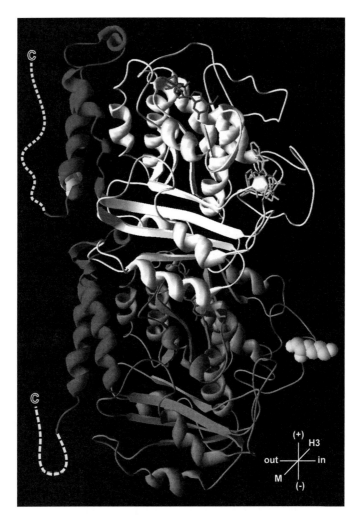

Fig. 1. Model of the tubulin α/β-heterodimer. β-tubulin is represented at the top in light gray and α-tubulin at the bottom in dark gray. The atoms of Lys40 on α-tubulin, where acetylation occurs as well as those of Glu435 on β-tubulin, where glutamylation occurs, are represented by yellow spheres. The extreme C-termini are represented in yellow as dashed random-shaped lines that were added to the structure. Their respective length is approximate, but shows that the average α-tubulin C-terminus is smaller than the average β-tubulin C-terminus. These C-termini are the locus of multiple posttranslational modifications (*see* Table 1). The domain forming the external surface of microtubules, which interact with MAPS and motors is colored in red. Taxol is represented in green whereas GTP and GDP are represented in red. The orientation diagram in the bottom right corner indicates that the tubulin is viewed sideway, the lumen of the microtubule (in) being on the right and the external surface (out) on the left. Two major lateral domains, the helix 3 (H3) and the M-poop (M) are in the back and in the front of each subunit, respectively. To view this figure in color, see the insert and the companion CD-ROM.

does not give direct information on the corresponding protein expression levels or to the type and extent of the various PTM. This is particularly relevant to β-tubulin mRNA because it appears to be autoregulated by the level of the free β-tubulin pool in cells *(14)*. Similarly, heterozygous transcriptional expression of a tubulin mutation in cancer cell lines selected for resistance to microtubule-interacting drugs does not reveal whether both wild-type and mutant tubulin proteins are expressed and if so, at

Table 2
MS Performance and Application in Proteomics Studies

Instrument	Resolution[a]	Mass accuracy[b] (ppm)	MS/MS capability	Application in proteomics	High throughput	LC capability[c]
Maldi-TOF	~10^3	100–1000	PSD	Peptide mass mapping	Yes	
Maldi-TOF-TOF	~10^3	100–1000	CID	Protein ID by MS/MS	Yes	
Maldi-QqTOF	~10^4	5–40	CID	Protein ID by MS/MS	Yes	
Maldi-ion trap	~10^3	300–1000	CID	Protein ID by MS/MS	Yes	
ESI-ion trap	~10^3	300–1000	CID	Protein ID and PTM by MS/MS		Yes
ESI-QqTOF	~10^4	5–40	CID	Protein ID and PTM by MS/MS		Yes
ESI-FT ICR MS	~10^5	<1–10	ECD	Protein ID and PTM by MS/MS and top–down MS/MS approach		Yes

[a]Resolution is defined as the full-width-at-half-maximum.
[b]Mass measurement accuracy in parts per million (ppm).
[c]The effluent of HPLC is delivered directly to the mass spectrometer. MALDI-TOF, matrix assisted laser desorption ionization time of flight mass spectrometer; MALDI-QqTOF, matrix assisted laser desorption ionization hybrid triple quadrupole time-of-flight mass spectrometer; MALDI-TOF-TOF, matrix assisted laser desorption ionization tandem time-of-flight mass spectrometer; MALDI-ion trap, matrix assisted laser desorption ionization ion trap mass spectrometer; ESI-ion trap, electrospray ionization ion trap mass spectrometer; ESI-QqTOF, electrospray ionization hybrid triple quadrupole time-of-flight mass spectrometer; ESI-FT ICR MS, electrospray ionization Fourier transform ion cyclotron resonance mass spectrometer; ECD, electron capture dissociation.

what ratio. The presence of a number of tubulin pseudogenes can also lead to erroneous expression data.

Antibodies directed against the C-terminal peptides of most of the β-tubulin, but not α-tubulin, isotypes are readily available and have been used to evaluate their respective isotype expression levels (see Chapter 10). However, such studies are semiquantitative at best, because of the differing binding affinities of the antibodies employed. Moreover, the isotype specificity of some of these antibodies has not been firmly established. Nevertheless, antibodies, which specificity has been validated by MS are essential tools for high throughput analysis of tenuous samples and for the determination of subcellular distribution of tubulin isotypes by immunofluorescence or electron microscopy. Antibodies recognizing glutamylated tubulin are also available, but they provide no information as to the length of the appended side chain. This is potentially important information as in vitro studies have indicated that the ability of certain structural and motor microtubule associated proteins (MAPs) to interact with tubulin varies with the length of the polyglutamyl side chain (7–9).

3. MS AS A TOOL IN MICROTUBULE RESEARCH

In an effort to expand the scope of tubulin isotype qualitative and quantitative analysis, it was decided to incorporate proteomic approaches into the analysis of tubulin-isotype expression in cancer cell lines and tissues. MS has become an essential component of biological experimentation because accurate mass measurement of peptides and proteins can define both their sequence and PTM. Mass spectrometers consist of six functional elements: sample inlet, ion source, mass analyzer, ion detector, vacuum system, and a dedicated computer. There are a variety of instruments available with different means of ionization, detection, sensitivity, and resolution. Different ionization methods (such as matrix-assisted laser desorption ionization (MALDI) and electrospray ionization [ESI]) can be combined with differing mass analyzers (such as time of flight, tandem time of flight, hybrid triple quadrupole time of flight, ion trap, and Fourier transform ion cyclotron resonance [FT ICR]) and differing ion detection systems, each with a different suitability for certain experiments (Table 2).

In MS, the mass-to-charge ratios of gas-phase peptide or protein ions are determined. Thus, the peptide or protein molecules must be ionized and the ions must be in the gas-phase. MALDI *(15)* and ESI *(16)* are two "soft ionization" methods. MALDI usually produces singly charged peptide ions. It is often coupled to a time-of-flight (TOF) mass analyzer. The isotopic peaks of a small peptide (<2000 Da) may be resolved in the linear mode of MALDI-TOF MS and the monoisotopic mass can be obtained with the mass measurement error within 100 ppm, which means that the mass measurement error will be less than 0.2 Da for a peptide with the mass of 2000 Da. The combination of two-dimensional electrophoresis, in-gel protease digestion and MALDI-TOF MS peptide mass mapping has become an important analytical technique for identification of proteins. In this method, the determined peptide molecular masses are compared with expected values computed from the database entries according to the enzymes' cleavage specificity. MALDI-TOF is very sensitive and requires small quantities of samples. It is the ideal technique for high-throughput application. Protein can also be identified using tandem mass spectrometry (MS/MS), in which the ionized peptide is selected and fragmented to collect the sequence information (MS/MS spectrum), followed by comparing the MS/MS spectrum to predicted tandem mass spectra computer generated from a sequence database. This method is more unambiguous than those achieved by peptide mass mapping as the MS/MS spectra provide peptide sequence information. Although postsource decay (PSD) spectra obtained with MALDI-TOF can provide amino acid sequence data of peptides, it is time-consuming and low-intensity precursor ions do not produce enough PSD fragmentation to allow derivation of even short sequence tags *(17)*. MALDI can also be coupled to hybrid triple quadrupole TOF (QqTOF), tandem time-of-flight (TOF/TOF) *(18)* or ion trap mass analyzer *(19)*. Each of these combinations can generate collision-induced dissociation (CID) MS/MS spectra of selected precursor ions (*see* Table 2).

ESI produces singly and/or multiply charged peptide ions. It is capable for on-line liquid chromatography–mass spectometry (LC-MS) analysis. It can be coupled with variety of mass analyzers such as ion trap, QqTOF, and FT ICR. The resolution of a mass spectrometer is defined as the ratio of the *m/z* value on the full-width-at-half-maximum of the mass peak or $\Delta m/z$. Therefore, mass spectrometers such as QqTOF and FT ICR with a resolution >10^4 and >10^5, respectively afford the detection of isotopic peaks of a large peptide ion with mass of ca. 5000 Da (full-width-at-half-maximum = 0.5 Da and 0.05 Da, respectively). The monoisotopic mass of a peptide can be easily measured with the mass accuracy of 5–40 ppm by QqTOF and less than 1–10 ppm by FT ICR, respectively, which means that the mass measurement error for a peptide with the mass of 2000 Da will be less than 0.08 Da and 0.02 Da by QqTOF and FT ICR, respectively. Low-flow rate ESI improves the sensitivity (20,21). Both nano- and micro-ESI are now widely used in biological MS (22). ESI produces multiply charged protein ions, allowing the m/z to fall within the range of most analyzers. In contrast to "bottom up" proteolysis characterization, "top down" FT ICR MS has been used for the detailed structural characterization of large proteins (23), in which ionized proteins can be trapped in the ICR cell and subjected to electron capture dissociation (24,25) to obtained the fragment ions of proteins.

As the carboxyl-terminal peptides are both characteristics of a specific tubulin isotype and are also the main site for PTM, generation and purification of these peptides from tubulin have been the basis for most MS-based characterization of tubulin across phyla (26–34). These carboxyl-terminal peptides can be generated from purified tubulin by chemical cleavage with cyanogen bromide (CNBr) or enzymatic cleavage with an endoprotease. In most studies, anion exchange enrichment of these acidic peptides was performed before reverse phase high-pressure liquid chromatography (HPLC) separation and MS. Antibodies against the isotype-defining domains have also been used to capture specific populations of carboxyl-terminal peptides on immunoaffinity columns. This approach is particularly useful when the analysis of discrete tubulin isoforms is sought. Frankfurter and colleagues (26) have combined CNBr cleavage and antibody capture of C-terminal peptides to demonstrate the glutamylation and phosphorylation of βIII-tubulin from bovine brain. It was sought to design methods for the isolation and analysis of tubulin from human cell lines and tissue that would be more direct from sample preparation to analysis.

4. ISOLATION OF TUBULIN FROM HUMAN CELL LINES

Tubulin can be isolated from tissues rich in microtubules through repeated cycles of polymerization at 37°C and depolymerization at 0°C (35). However, as some microtubules are cold-stable, this subpopulation would be lost during the purification process. Isolation of tubulin by anion ion exchange chromatography on diethylaminoethyl cellulose is well-established and can be used to purify soluble tubulin from a variety of sources (36). The process usually requires larger amounts of starting material with some variability in terms of yield and purity of tubulin, and is only worth pursuing when functional tubulin is required. To isolate tubulin from nonneuronal cell extracts, where the tubulin concentration is lower than the critical concentration needed for assembly at 37°C, Taxol is used for its ability to promote tubulin polymerization at these low concentrations. The use of Taxol as a tool to isolate microtubule-associated proteins was originally described by Vallee (37). It was demonstrated that Taxol-driven polymerization of tubulin from cytosolic extracts derived from a variety of cancer cell lines was quantitative and yielded highly enriched microtubule preparations (38). It should be noted that with some cell lines and almost every tissue, some F-actin often cosediments with the Taxol-stabilized microtubules. If necessary, F-actin can be eliminated by pretreating the cell lysates with a F-actin-stabilizing drug, for example, phalloidin, and preclearing by centrifugation. Conversely, Taxol-driven polymerization can be performed in the presence of a F-actin-depolymerizing drug, for example, latrunculin B that maintains actin in its monomeric form (G-actin).

5. DIRECT MALDI-TOF MS ANALYSIS OF C-TERMINAL TUBULIN PEPTIDES FROM TOTAL CELL EXTRACTS

In the initial analysis of human cell lines CNBr was used to release the C-terminal tubulin fragments *(39)*. The masses of the human α- and β-tubulin CNBr-derived C-terminal peptides are all unique and are in the 2000–4000 mass range thus facilitating direct MS analysis (Table 3). This was originally anticipated that it would be necessary to isolate tubulin from these cell lines before CNBr-release of the highly divergent C-terminal peptides. However, because of the acidic nature of these peptides, it was found that they could be analyzed directly by MALDI-TOF-MS in the negative mode without significant interference from other peptides released from this region of the gel. Selective detection in the negative mode is a characteristic feature of highly acidic peptides *(40)*. For MS analysis, total cell extracts are resolved by sodium dodecyl sulfate-polyacrylamide gel electrophoresis (SDS-PAGE), transferred to nitrocellulose, and the region of the blot corresponding to tubulin (~50 kDa) excised and digested with CNBr/formic acid. A rabbit polyclonal antibody prepared against a synthetic peptide corresponding to the final 12 amino acids of human Kα1 was used to follow the release of the CNBr C-terminal fragment. Under the digestion conditions *(39)*, i.e., CNBr (150 mg/mL) in 70% formic acid for 3.5 h at room temperature, no immunoreactivity remained on the nitrocellulose filter after the incubation period demonstrating that the CNBr cleavage/release of the C-terminal peptides was quantitative.

The MALDI-TOF mass spectra of the CNBr-released C-terminal tubulin peptides from the A549 lung-cancer cell line is shown in Fig. 1A. Three major, and several minor, molecular ions are clearly resolved. To confirm that these ions were derived from tubulin peptides, the three major ions were subjected to CID/PSD-MALDI-MS. The fragmentation patterns obtained confirmed that the molecular ions with *m/z* values of 2860.5 and 3367 corresponded to the C-terminal CNBr fragments of Kα1 and βI-tubulins, respectively. However, the molecular ion with a *m/z* value of 2590.4 did not correspond to the mass of any cloned α/β-tubulin C-terminal fragment in the databases at that time. Only a partial fragmentation pattern for this molecular ion was observed by CID/PSD-MALDI-MS. Nevertheless, fragmentation ions corresponding to the first several N-terminal amino acids of the peptide could be identified and they were clearly derived from an α-tubulin. The remaining sequence of this peptide was determined by a combination of tandem MS and MALDI-TOF MS analysis of partial carboxypeptidase Y digests. The amino acid sequence of this peptide shared similarities with the corresponding region of the mouse α6-isotype, but was clearly distinct *(39)*. More recently, the gene encoding the human α6-tubulin was identified *(41)* and the deduced sequence of its C-terminal peptide was identical to the sequence obtained by. The minor ion with a *m/z* value of 3478.9 in Fig. 2A was tentatively identified as the βIVb-isotype. This combined isoelectric focusing (IEF)-MS approach can be used to document the specificity of any putative tubulin isotype-specific antibody. There have been reports that the βII-tubulin isotype is a major component of the tubulin cytoskeleton in the MDA-MB 231 breast cell line *(42,43)*. In both cases, the identification was by immunological-based methods. The MS-based method, which measures isotype composition based on the CNBr-release of the highly divergent C-terminal peptides, found that the βII-isotype was lower than the level of detection in this cell line *(39)*. As βII-tubulin could be detected in bovine brain microtubule preparations, where it is the major isotype, it is not believed that the inability to detect the βII-isoform in this cell line was a problem inherent to the MS-based method. Similarly,

Table 3
Identity of Human Tubulin Isotypes: NCBI Accession Number, Isoelectric Point, and Mass

Tubulin isotype	Accession number NCBI	Protein mass (Da)	Isoelectric point	CNBr C-terminus mass (Da)[b]
α1/bα1	CAA25855	50,157.7	5.02	2882.9
α1/Kα1	AAC31959	50,151.6	4.94	2860.9
α3	Q13748	49,959.5	4.98	4152.2
α4	A25873	49,924.4	4.95	2633.6
α6	Q9BQE3	49,895.3	4.96	2590.6
α8	Q9NY65	50,093.5	4.94	4158.2
βI	AAD33873	49,670.8	4.78	3367.3
βII	AAH01352/ NP_001060[a]	49,953.1/ 49,907	4.78/4.78	3467.4
βIII	AAH007448	50,432.7	4.83	1624.6
βIVa	P04350/ NP_006078[a]	49,630.9/ 49,585.8	4.81/4.78	3350.4/3351.3
βIVb	P05217	49,831	4.79	3480.5
βV	NP_115914	49,857.1	4.77	3552.5
βVI	NP_110400	50,326.9	5.05	809.8

[a]Two different sequences were found in the NCBI protein database and since no source expressing nonmodified βII or βIVa has been identified, both should be considered.
[b]Calculated [M-H]− mass of the CNBr C-terminal peptides when analyzed by MALDI-TOF in the negative mode.

βIVa-mRNA was detected by RT-PCR in A549 cells, yet it was undetectable at the protein level in the same cell line.

Ion suppression effects, however, can be a significant problem in MALDI-TOF MS and the analysis of the C-terminal tubulin peptides derived from total cell extracts may under-represent specific isotypes (39). Ions for minor tubulin species could be suppressed by the presence of the more highly abundant tubulin C-terminal peptides and/or because of differential ionization efficiencies of the various peptides. This approach was re-evaluated by reducing the complexity of the sample before CNBr digestion (44). Tubulin was purified by the Taxol-based polymerization method from MDA-MB-231 cell extracts followed by the separation of α-tubulin and β-tubulin on SDS-PAGE gels before transfer to the nitrocellulose membrane. Tubulin can be resolved into two distinct bands by 1D SDS-PAGE using 95% pure SDS containing large aliphatic chain alkyl sulfate contaminants. With mammalian tubulin, the band with the slowest electrophoretic motility contains only α-tubulin and the fastest β-tubulin.

The region of the blot containing either α- or β-tubulin was CNBr-digested and MALDI-TOF MS analysis was performed (Fig. 2B,C). The m/z peaks intensity corresponding to the βIVb-peptide was increased significantly compared with that obtained from tubulin analyzed from unfractionated cell extracts (Fig. 2A) (39). Very low levels of the C-terminal peptide of tyrosinated α4-tubulin were also detected (Fig. 2B). However, the C-terminal peptide of the βII-isotype was still undetectable (44). It could be argued that the Taxol-dependent polymerization process selectively enriched for specific β-tubulin isotypes. For this reason, immunoblot analysis was used to follow the isotype composition during the tubulin isolation step. The data established that the isotype composition of the final Taxol-stabilized microtubule pellet was representative of the cellular tubulin composition (44).

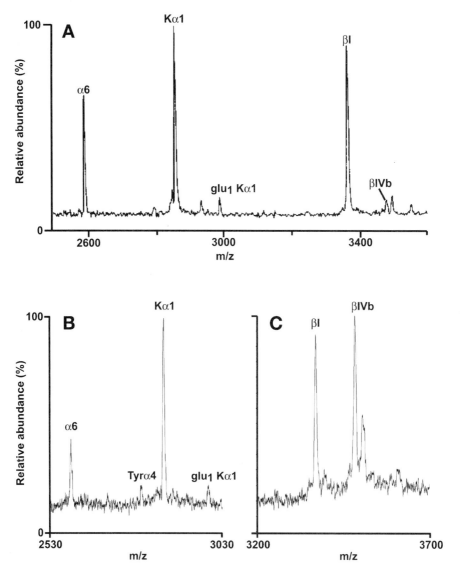

Fig. 2. Analysis of CNBr C-terminal peptides of tubulin isotypes by MALDI-TOF MS in the negative ion mode. CNBr digestion was performed either on mixed α- and β-tubulin **(Panel A)** or after electrophoretic separation of α-tubulin **(Panel B)** from β-tubulin **(Panel C)**. Panel A, α- and β-tubulin isotypes present in A549 total lysates; Panel B, α-tubulin isotypes present in Taxol-stabilized microtubules isolated from A549 lysates; and Panel C, β-tubulin isotypes present in Taxol-stabilized microtubules isolated from A549 lysates.

6. ANALYSIS OF TUBULIN-ISOTYPE EXPRESSION
BY COMBINED IEF AND MS

The isolation of tubulin from cell extracts using the Taxol-driven polymerization coupled to high-resolution isoelectrofocusing (IEF) was developed as a tool to visualize and analyze the various intact tubulin isoforms present in cancer cell lines *(44)*. The recent technological improvements in IEF involving the use of precast immobilized pH

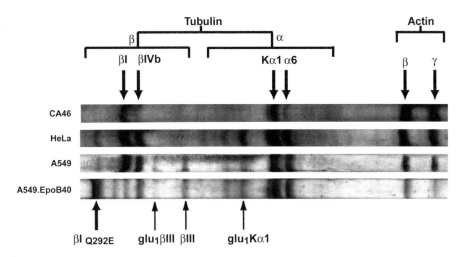

Fig. 3. Separation of tubulin isotypes by high-resolution isoelectrofocusing. Tubulin isotypes were resolved on IEF gel and Coomassie blue stained gels are presented. Cell lines are indicated on the left and each identified tubulin isotype is indicated by an arrow: CA46, Burkitt lymphoma; HeLa, cervical carcinoma; A549, nonsmall cell lung carcinoma; A549.EpoB40, an epothilone B-resistant A549 cell line. Actins that are present in the Taxol-stabilized microtuble preparations are found at higher pI values than tubulins.

gradient gel strips up to 24 cm in length and containing a one pH unit gradient from 4.5 to 5.5 were instrumental in separating tubulin isotypes that differed in pI by only 0.01 pH units (Table 3). An important additional advantage of this IEF approach is that charge altering mutations to tubulin in drug-resistant cell lines are readily detectable *(44)*.

After electrofocusing and protein staining, the location of each tubulin isotype on the immobilized pH gradient gel strip was definitely assigned by performing in-gel trypsin digestion followed by MALDI-TOF MS analysis. Mass analysis of each tryptic digest detected several isotype specific peptides that confirmed the assignment of each band to an individual tubulin isotype class *(44)*. The presence of the tubulin isotypes was confirmed in the MDA-MB 231 and A549 cell lines identified previously by MALDI-TOF MS analysis of C-terminal tubulin peptides *(39)*. Morever, a protein having the pI value predicted for recently cloned human α6-tubulin was detected and confirmed as being α6-tubulin by peptide mapping. The high-resolution IEF approach demonstrated that there was an excellent match between theoretical and measured pI of nonmodified βI-, βIVb-, βIII-, Kα1-, and α6-tubulin isotypes and their monoglutamylated isoforms (Fig. 3). This approach also provided a semiquantitative assessment of tubulin-isotype expression between drug-sensitive and -resistant cell lines based on Coomassie blue staining patterns. An increase in βIII-tubulin expression was detected in resistant cells *(44)*. A minor βIII-tubulin species that corresponded most likely to monoglutamylated or phosphorylated βIII-tubulin was also detected.

The ability to resolve and identify each tubulin isotype on high-resolution IEF gels also allowed to evaluate the specificity of a panel of tubulin isotype-specific antibodies *(44)*. The anti-β-tubulin isotype monoclonal antibodies produced and characterized by Dr. Lerdueña and coworkers *(10)* represent valuable tools that have helped to unravel the functional significance of the different tubulin isotypes. No cross-reactivity of the anti-βI-, βIII-, and βIV-tubulin antibodies was observed in the present study. However,

extensive cross-reactivity of the anti-βII-tubulin antibody with βI-tubulin was detected
(44). The mutation Q292E found in βI-tubulin in MDA-MB-231.K20T, a Taxol-resistant
cell line was very instrumental in revealing this cross-reactivity. This mutant βI-tubulin
is more acidic than βI and therefore focuses at a more acidic position on IEF strips than
wild-type βI-tubulin (Fig. 3). After transfer on nitrocellulose, the western signal patterns
generated with either the anti-βI- or the anti-βII-tubulin were undistinguishable *(44)*.

7. DIRECT ANALYSIS OF TUBULIN-ISOTYPE EXPRESSION BY ESI MS

Tubulin α- and β-subunits can also be separated by reverse-phase HPLC on C4, C8,
or C18-columns using linear aqueous acetonitrile gradients containing TFA and analyzed directly by ESI-MS *(38)*. If necessary, samples can be reduced and alkylated
before LC. Poorly soluble tubulins can be solubilized in 6 *M* guanidine-HCl with no
change in retention time. Figure 4 (inset) is a representative chromatogram showing the
total ion current of a typical human microtubule preparation. All the masses in the range
of human tubulins were found between 44 and 60 min of the gradient, which includes
the largest ion peak *(38)*. This was not an unanticipated result as tubulin is the major
protein component in the Taxol-stabilized microtubule pellets. After averaging the scans
between 44 and 60 min, overall profiles of the deconvoluted mass peaks for the different tubulin proteins were obtained in both cell lines (Fig. 4). The observed masses
matched closely those of βI-, βIVb-, α6-, tyrα4-, Kα1-, glu-Kα1-, and βIII-tubulin,
respectively (Table 3) *(38)*.

To confirm the identity of the A549-tubulin isotypes, tryptic mass mapping of the isotypes resolved on the C4 column was performed. In these experiments, the column effluent was split for fraction collection and for delivery to the ion trap mass spectrometer. The
total ion current was deconvoluted in 1 min segments across the gradient and fractions that
contained the same protein masses were pooled. After removal of solvents, tryptic digestion was performed and the resulting peptides were analyzed by MALDI-TOF MS and
tubulin isotype-specific tryptic peptides were identified *(38)*. Importantly, no βII- or βIVa-
isotype-specific tryptic peaks were observed during these mass mapping experiments.

In the field of drug resistance, mutations that occur to the primary drug target represent important mechanisms by which a cell can overcome the cytotoxic effects of a drug
(13). To date, the search for tubulin mutations in cell lines selected for resistance to
microtubule-stabilizing drugs such as Taxol and the epothilones has been largely
restricted to βI-tubulin. However, studies from this group have also detected mutations
to Kα1-tubulin in some drug-resistant cell lines *(45)*. Moreover, βIVb- and α6-tubulins
are present at significant levels in many of the cell lines examined *(38,44)*. Although
technically feasible, the complete sequencing of the multiple tubulin transcripts in human
cell lines would be a lengthy and difficult task partly because they are so highly conserved. In addition, whether mutations that occur to a single tubulin allele are expressed,
and to what levels, are important as the ratio of wild-type to mutant tubulin could have a
significant effect on the level of resistance. A major advantage of the LC/ESI-MS method
is that it can be used to analyze directly mutations that occur to the various tubulin isotypes in drug-resistant cell lines *(38)*. In a series of Taxol- and epothilone-resistant
human cell lines that harbor heterozygous mutations in either βI- or Kα1-tubulins, the
expression of the mutant tubulins was clearly observed at levels that appeared higher than
the wild-type isotype (Fig. 5) *(38)*. With the ion trap mass spectrometer that was used in

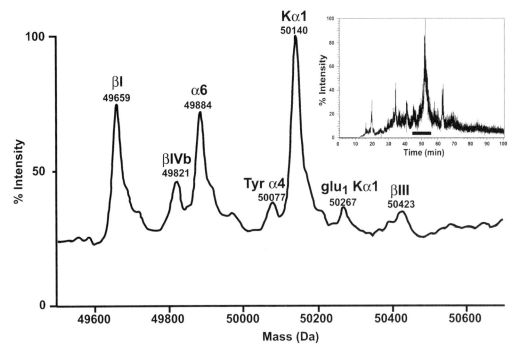

Fig. 4. LC-ESI-MS analysis of tubulin isotypes present in Taxol-stabilized microtubules from A549 cells. Tubulin isotypes were separated on a C4 HPLC column. The total ion chromatograph is shown in the inset. The deconvoluted mass spectrum was obtained by averaging scans in the retention time range (44–56 min) as indicated by the horizontal bar in the inset.

these studies, it was possible to discriminate wild-type and mutant βI-tubulins that differ in mass by 26 Da. By using a QqTOF or FT ICR mass spectrometers, it is expected to be able to detect expression of mutant tubulins that differ in mass by 15 Da or even lower.

8. OVERVIEW OF TUBULIN-ISOTYPE EXPRESSION IN DRUG-SENSITIVE AND RESISTANT CELL LINES

In the first tubulin proteomic manuscript evidence was presented for a new human α-tubulin isotype expressed in breast and lung cancer cell lines *(39)*. As discussed earlier this isotype was eventually shown to be α6-tubulin. In the most recent LC/ESI-MS manuscript, the first evidence was presented for the expression of the α4-tubulin isotype in human cell lines *(38)*. This α-tubulin isotype is the only isotype not to have a tyrosine or a phenylalanine residue as the last encoded residue. The MS analysis showed that the majority of α4-tubulin isotype was tyrosinated in the cell lines studied, indicating the presence of a functional tubulin tyrosine ligase. The most striking difference between the analysis of tubulin-isotype expression at the protein level *(38,39,44)* and previous studies at the mRNA level *(46)* is the absence of detectable βIVa- and βII-tubulin expression in both sensitive and resistant cell lines. As the data established that tubulin was quantitatively recovered from cells by Taxol-driven polymerization *(38)*, either there is an unknown bias in the RT-PCR results or there is a still unknown silencing mechanism for βII- and βIVa-tubulin mRNA translation. Increased βIII-tubulin expression levels

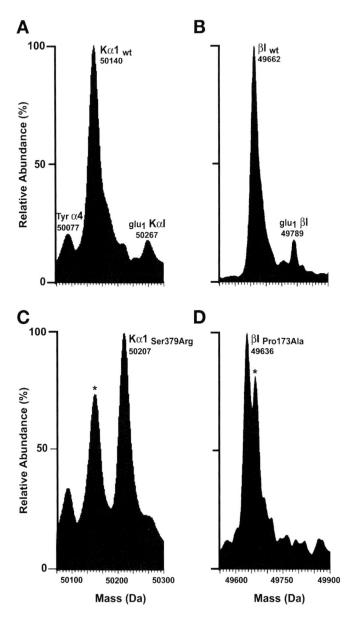

Fig. 5. LC-ESI-MS analysis of the Kα1- and βI-isotypes present in Taxol-stabilized microtubules from Taxol- and epothilone-resistant cell lines. The drug-sensitive parental cell lines express Kα1- and βI-tubulins (**Panel A** [A549] and **Panel B** [HeLa]) with the correct mass. A549-T12, a Taxol-resistant cell line expresses both the wild-type and mutant Kα1-tubulins (**Panel C**), whereas HeLa.EpoA9, an epothilone-resistant cell line, expresses both the wild-type and mutant βI-tubulins (**Panel D**). The asterix in panels C and D indicate the wild-type proteins.

have been repeatedly confirmed at the protein level using a variety of methods in cell lines resistant to Taxol *(13,47–49)*.

The analysis of tubulin-isotype expression in cancer cell lines has also allowed to determine which sequences of the main isotypes are expressed. This is important as multiple entries for some human-tubulin isotypes can be found in the databases *(38)*.

Because of the inability to detect expression of βIVa- and βII-tubulins at the protein level, it is still necessary to clarify which sequences are correct for these two isotypes (Table 3). An interesting observation is that the ratios of Kα1-tubulin: βI-tubulin and α6-tubulin: βIVb-tubulin are close to one in the cell lines examined so far. This trend suggests that there may be a preferential pairing of a specific β-tubulin isotype with a specific α-tubulin isotype.

In contrast to mammalian neuronal tubulins, PTM to tubulin in these human cancer cell lines are minimal. Glutamylation was limited to one residue and was present in a very small fraction of the tubulin. All of the α-tubulins were fully tyrosinated and Lys40 acetylation was lower than the level of detection. However, it is important to stress that the analyses were performed on the total tubulin population present in these cell lines. It is known that polyglutamylation is cell-cycle regulated and reaches a maximum during mitosis when spindle microtubules are extensively polyglutamylated *(50)*.

9. QUANTITATION IN TUBULIN PROTEOMICS

Differential expression of the several α- and β-tubulin isotypes has been proposed as a potential mechanism of resistance toward microtubule-stabilizing drugs. Unfortunately, the use of MS to quantify tubulin isotype levels between different biological samples is unreliable for several reasons, including different ionization potentials of peptides, ion suppression effects, and variation of responses by detectors. Nevertheless, in the past few years several labeling methods have been introduced that allow investigators to quantify relative proteins levels between biological samples. In some cases, absolute protein levels have been quantified. In the tubulin proteomics program, two quantitation methods are being integrated, one designed for cultured cell lines and the other for tissue biopsies.

9.1. Quantitation of Relative Tubulin Expression Levels in Human Cell Lines

Stable isotope labeling by amino acids in cell culture (SILAC) are being used to differentially label proteins in Taxol-sensitive and -resistant cell lines *(51–53)*. In the SILAC method, two cell lines are cultured for several generations in media containing either a "heavy" essential amino acid (e.g., ^{2}H n = 3-Leu or ^{13}C n = 6-Arg) or its normal "light" counterpart. After five doubling in the presence of the "heavy" medium, the majority of the "light" tubulin has been replaced by its "heavy" counterpart. For relative quantitation of protein expression levels, equal amounts of total cytosolic proteins from cell lines grown in "light" medium and "heavy" medium are mixed and tubulins are isolated by Taxol-induced polymerization and resolved by 1D SDS-PAGE. For relative quantitation of total α- and β-tubulin levels, the tubulin-containing region is excised from the gel and subjected to trypsin digestion followed by MALDI-TOF MS. In the case of the leucine-labeling protocol, for example, leucine-containing tryptic peptides will be represented b y two *m/z* peaks separated by a multiple of 3 Da in the mass spectra, depending on the number of leucines contained in the peptide. As both heavy and light tryptic peptides are chemically equivalent, the ratio of signal intensities for each peptide pair is a reflection of the relative protein levels in the two samples. For relative quantitation of individual tubulin isotype levels, samples will be separated by IEF and gel bands containing a particular isotype analyzed by tryptic mass mapping.

9.2. Quantitation of Absolute Tubulin Isotype Levels
in Tissue Biopsies

The SILAC quantitation method cannot be used for tissue-derived samples. In this situation, it is proposed to determine absolute tubulin isotype levels by the addition of isotopically labeled peptides, prepared by either synthetic (^{15}N) or enzymatic (^{18}O) procedures, as internal standards.

10. CONCLUSION

In the laboratories, proteomic methods are being developed that will allow the quantitative analysis of tubulin isotypes in human cell lines, tissues, and organs. Such information will provide the ability to resolve the important question regarding whether tubulin-isotype composition has a significant role in drug response and resistance.

ACKNOWLEDGMENTS

This work was supported in part by USPHS Grants CA 39821, CA 77263 and the National Foundation for Cancer Research (S.B.H.), AI49749 and Grant DAMD17-01-0123 from Department of Defense Breast Cancer Program (G.A.O).

REFERENCES

1. Westermann S, Weber K. Post-translational modifications regulate microtubule function. Nat Rev (Mol Cell Biol) 2003;4:938–947.
2. Bloom K. Microtubule composition:Cryptography of dynamic polymers. Proc Natl Acad Sci USA 2004;101:6839–6840.
3. Panda D, Miller HP, Banerjee A, Luduena RF, Wilson L. Microtubule dynamics in vitro are regulated by the tubulin isotype composition. Proc Natl Acad Sci USA 1994;91:11,358–11,362.
4. Lu Q, Luduena RF. In vitro analysis of microtubule assembly of isotypically pure tubulin dimers. Intrinsic differences in the assembly properties of alpha beta II, alpha beta III, and alpha beta IV tubulin dimers in the absence of microtubule-associated proteins. J Biol Chem 1994;269:2041–2047.
5. Derry WB, Wilson L, Khan IA, Luduena RF, Jordan MA. Taxol differentially modulates the dynamics of microtubules assembled from unfractionated and purified beta-tubulin isotypes. Biochemistry 1997;36:3554–3562.
6. Banerjee A, Roach MC, Trcka P, Luduena RF. Increased microtubule assembly in bovine brain tubulin lacking the type III isotype of beta-tubulin. J Biol Chem 1990;265:1794–1799.
7. Boucher D, Larcher JC, Gros F, Denoulet P. Polyglutamylation of tubulin as a progressive regulator of in vitro interactions between the microtubule-associated protein Tau and tubulin. Biochemistry 1994;33:12,471–12,477.
8. Larcher JC, Boucher D, Lazereg S, Gros F, Denoulet P. Interaction of kinesin motor domains with alpha- and beta-tubulin subunits at a tau-independent binding site. Regulation by polyglutamylation. J Biol Chem 1996;271:22,117–22,124.
9. Bonnet C, Boucher D, Lazereg S, et al. Differential binding regulation of microtubule-associated proteins MAP1A, MAP1B, and MAP2 by tubulin polyglutamylation. J Biol Chem 2001;276: 12,839–12,848.
10. Luduena RF. Multiple forms of tubulin: different gene products and covalent modifications. Int Rev Cytol 1998;178:207–275.
11. Nogales E, Wolf SG, Downing KH. Structure of the alpha beta tubulin dimer by electron crystallography. Nature 1998;391:199–203.
12. Nogales E. Structural insight into microtubule function. Annu Rev Biophys Biomol Struct 2001;30: 397–420.
13. Orr GA, Verdier-Pinard P, McDaid HM, Horwitz SB. Mechanisms of taxol resistance related to the microtubule. Oncogene 2003;22:7280–7295.

14. Cleveland DW, Sullivan KF. Molecular biology and genetics of tubulin. Annu Rev Biochem 1985;54:331–365.
15. Karas M, Hillenkamp F. Laser desorption ionization of proteins with molecular masses exceeding 10,000 daltons. Anal Chem 1988;60:2299–2301.
16. Fenn JB, Mann M, Meng CK, Wong SF, Whitehouse CM. Electrospray ionization for mass spectrometry of large biomolecules. Science 1989;246:64–71.
17. Kaufmann R, Haurand P, Kirsch D, Spengler B. Post-source decay and delayed extraction in matrix-assisted laser desorption/ionization-reflectron time-of-flight mass spectrometry. Are there trade-offs? Rapid Commun Mass Spectrom 1996;10:1199–1208.
18. Medzihradszky KF, Campbell JM, Baldwin MA, et al. The characteristics of peptide collision-induced dissociation using a high-performance MALDI-TOF/TOF tandem mass spectrometer. Anal Chem 2000;72:552–558.
19. Krutchinsky AN, Kalkum M, Chait BT. Automatic identification of proteins with a MALDI-quadrupole ion trap mass spectrometer. Anal Chem 2001;73:5066–5077.
20. Gale DC, Smith RD. Small volume and low-flow-rate electrospray ionization mass spectrometry of aqueous samples. Rapid Commun Mass Spectrom 1993;7:1017–1021.
21. Emmett MR, Caprioli RM. Micro-electrospray mass spectrometry:ultra-high sensitivity analysis of pepitdes and proteins. J Am Soc Mass Spectrom 1994;5:605–613.
22. Wilm M, Shevchenko A, Houthaeve T, et al. Femtomole sequencing of proteins from polyacrylamide gels by nano-electrospray mass spectrometry. Nature 1996;379:466–469.
23. Kelleher NL. Top-down proteomics. Anal Chem 2004;76:197A–203A.
24. Zubarev R, Horn D, Fridriksson E, et al. Electron capture dissociation for structural characterization of multiply charged protein cations. Anal Chem 2000;72:563–573.
25. Horn D, Ge Y, McLafferty F. Activated ion electron capture dissociation for mass spectral sequencing of larger (42 kDa) proteins. Anal Chem 2000;72:4778–4784.
26. Alexander JE, Hunt DF, Lee MK, et al. Characterization of posttranslational modifications in neuron-specific class III beta-tubulin by mass spectrometry. Proc Natl Acad Sci USA 1991;88:4685–4689.
27. Rudiger M, Weber K. Characterization of the post-translational modifications in tubulin from the marginal band of avian erythrocytes. Eur J Biochem 1993;218:107–116.
28. Redeker V, Levilliers N, Schmitter JM, et al. Polyglycylation of tubulin: a posttranslational modification in axonemal microtubules. Science 1994;266:1688–1691.
29. Rudiger A, Rudiger M, Weber K, Schomburg D. Characterization of post-translational modifications of brain tubulin by matrix-assisted laser desorption/ionization mass spectrometry: direct one-step analysis of a limited subtilisin digest. Anal Biochem 1995;224:532–537.
30. Rudiger M, Plessmann U, Rudiger AH, Weber K. Beta tubulin of bull sperm is polyglycylated. FEBS Lett 1995;364:147–151.
31. Mary J, Redeker V, Le Caer JP, Rossier J, Schmitter JM. Posttranslational modifications in the C-terminal tail of axonemal tubulin from sea urchin sperm. J Biol Chem 1996;271:9928–9933.
32. Bre MH, Redeker V, Vinh J, Rossier J, Levilliers N. Tubulin polyglycylation: differential posttranslational modification of dynamic cytoplasmic and stable axonemal microtubules in paramecium. Mol Biol Cell 1998;9:2655–2665.
33. Redeker V, Rossier J, Frankfurter A. Posttranslational modifications of the C-terminus of alpha-tubulin in adult rat brain: alpha 4 is glutamylated at two residues. Biochemistry 1998;37:14,838–14,844.
34. Vinh J, Langridge JI, Bre MH, et al. Structural characterization by tandem mass spectrometry of the posttranslational polyglycylation of tubulin. Biochemistry 1999;38:3133–3139.
35. Shelanski ML, Gaskin F, Cantor CR. Microtubule assembly in the absence of added nucleotides. Proc Natl Acad Sci USA 1973;70:765–768.
36. Sackett DL. Rapid purification of tubulin from tissue and tissue culture cells using solid-phase ion exchange. Anal Biochem 1995;228:343–348.
37. Vallee RB. Purification of brain microtubules and microtubule-associated protein 1 using taxol. Methods Enzymol 1986;134:104–115.
38. Verdier-Pinard P, Wang F, Burd B, Angeletti RH, Horwitz SB, Orr GA. Direct analysis of tubulin expression in cancer cell lines by electrospray mass spectrometry. Biochemistry 2003;42:12,019–12,027.
39. Rao S, Aberg F, Nieves E, Horwitz SB, Orr GA. Identification by mass spectrometry of a new alpha-tubulin isotype expressed in human breast and lung carcinoma cell lines. Biochemistry 2001;40:2096–2103.
40. Jai-nhuknan J, Cassady CJ. Negative ion postsource decay time-of-flight mass spectrometry of peptides containing acidic amino acid residues. Anal Chem 1998;70:5122–5128.

41. Strausberg RL, Feingold EA, Grouse LH, et al. Generation and initial analysis of more than 15,000 full-length human and mouse cDNA sequences. Proc Natl Acad Sci USA 2002;99:16,899–16,903.

42. Shalli K, Brown I, Heys SD, Schofield AC. Alterations of beta-tubulin isotypes in breast cancer cells resistant to docetaxel. FASEB J 2005;19:1299–1301.

43. Walss-Bass C, Xu K, David S, Fellous A, Ludueña RF. Occurrence of nuclear beta(II)-tubulin in cultured cells. Cell Tissue Res 2002;308:215–223.

44. Verdier-Pinard P, Wang F, Martello L, Burd B, Orr GA, Horwitz SB. Analysis of tubulin isotypes and mutations from taxol-resistant cells by combined isoelectrofocusing and mass spectrometry. Biochemistry 2003;42:5349–5357.

45. Martello LA, Verdier-Pinard P, Shen HJ, et al. Elevated Levels of Microtubule Destabilizing Factors in a Taxol-resistant/dependent A549 Cell Line with an alpha-Tubulin Mutation. Cancer Res 2003;63:1207–1213.

46. Kavallaris M, Kuo DY, Burkhart CA, et al. Taxol-resistant epithelial ovarian tumors are associated with altered expression of specific beta-tubulin isotypes. J Clin Invest 1997;100:1282–1293.

47. Ranganathan S, Benetatos CA, Colarusso PJ, Dexter DW, Hudes GR. Altered beta-tubulin isotype expression in paclitaxel-resistant human prostate carcinoma cells. Br J Cancer 1998;77:562–566.

48. Ranganathan S, Dexter DW, Benetatos CA, Hudes GR. Cloning and sequencing of human beta III-tubulin cDNA: induction of betaIII isotype in human prostate carcinoma cells by acute exposure to antimicrotubule agents. Biochim Biophys Acta 1998;1395:237–245.

49. Nicoletti MI, Valoti G, Giannakakou P, et al. Expression of beta-tubulin isotypes in human ovarian carcinoma xenografts and in a sub-panel of human cancer cell lines from the NCI-Anticancer Drug Screen: correlation with sensitivity to microtubule active agents. Clin Cancer Res 2001;7:2912–2922.

50. Regnard C, Desbruyeres E, Denoulet P, Edde B. Tubulin polyglutamylase: isozymic variants and regulation during the cell cycle in HeLa cells. J Cell Sci 1999;112(pt 23):4281–4289.

51. Ong SE, Blagoev B, Kratchmarova I, et al. Stable isotope labeling by amino acids in cell culture, SILAC, as a simple and accurate approach to expression proteomics. Mol Cell Proteomics 2002;1:376–386.

52. Ong SE, Kratchmarova I, Mann M. Properties of 13C-substituted arginine in stable isotope labeling by amino acids in cell culture (SILAC). J Proteome Res 2003;2:173–181.

53. Blagoev B, Kratchmarova I, Ong SE, Nielsen M, Foster LJ, Mann M. A proteomics strategy to elucidate functional protein-protein interactions applied to EGF signaling. Nat Biotechnol 2003;21:315–318.

9 Tubulin and Microtubule Structures

Eva Nogales and Kenneth H. Downing

SUMMARY

Microtubules are cytoskeletal polymers made of repeating αβ-tubulin heterodimers that play essential roles in all eukaryotic cells. The dynamic character of microtubules is key to their functions, as evidenced by the large number of natural compounds that bind tubulin, alter microtubule dynamics and result in mitotic arrest. The electron crystallographic structure of tubulin showed that the α- and β- tubulin monomers are very similar. They contain a nucleotide binding domain, an intermediate domain where Taxol binds in β-tubulin, and a C-terminal helical region that form the crest of microtubule protofilaments where motors and other associated proteins bind. The complex dynamic behavior of microtubules, while modulated by other cellular factors, is ultimately due to the tubulin subunit architecture and its intrinsic GTPase activity. Thus, knowledge of the structure of tubulin and the microtubule is essential for our understanding of vital microtubule properties with relevance to broadly used anticancer therapies.

Key Words: Tubulin; microtubule; dynamic instability; anti-mitotic agents; Taxol; epothilone; GTPase.

1. INTRODUCTION

Microtubules are made of repeating αβ-tubulin heterodimers that bind head to tail into protofilaments. Parallel association of protofilaments forms the microtubule wall and gives rise to a polar polymer. In the cell the "minus-end" of the microtubule, capped by α-monomers, is generally attached to the centrosome. The more dynamic "plus-end,"

From: *Cancer Drug Discovery and Development: The Role of Microtubules in Cell Biology, Neurobiology, and Oncology* Edited by: Tito Fojo © Humana Press, Totowa, NJ

capped by β-monomers, can actively explore the cell, and attaches to the kinetochore during mitosis. Microtubules associate with a number of other proteins, including motors and cellular regulators that either stabilize or destabilize microtubules. Each tubulin monomer binds one molecule of guanosine 5′-triphosphate (GTP), nonexchangeably in α-tubulin (N-site) and exchangeably in β-tubulin (E-site). GTP must be present at the E-site for tubulin to polymerize, but it is hydrolyzed and becomes nonexchangeable on polymerization. With guanosine 5′-diphosphate (GDP) at the E-site, the microtubule is in a metastable structure that is thought to be stabilized by a cap of remaining GTP-tubulin monomers at the ends (*see* Chapter 3). The loss of this cap results in rapid depolymerization *(1,2)*. These characteristics result in dynamic instability, an essential microtubule property that allows microtubules to search through the cell for targets such as kinetochores.

The amino acid sequence of tubulin is highly conserved across species, reflecting constraints imposed by microtubule structure and functions that rely on a large variety of interprotein interactions. On the other hand, both α- and β-monomers exist in numerous isotypic forms and undergo a variety of post-translational modifications that are likely to be important for specific interactions with the other proteins and in providing microtubules with different dynamic properties *(3)*. The stability and dynamics of microtubules are very actively regulated by a number of cellular factors *(4)*, as well as a variety of ligands, some with important anticancer properties *(5)*.

2. THE TUBULIN FOLD

The structure of the tubulin dimer was obtained by electron crystallography of zinc-induced tubulin sheets stabilized with the anticancer drug Taxol® (paclitaxel) *(6–8)*. These sheets, discovered in the 1970s *(9)*, are formed by the antiparallel association of protofilaments. The resulting polymer has no polarity and no overall curvature, so that it can grow in two-dimensions, making what can be considered a two-dimensional crystal. Addition of Taxol stabilizes the sheets against cold-induced depolymerization and aging, as it does microtubules. Using cryopreservation, low-dose imaging methods and image processing, a structural model of the tubulin dimer bound to Taxol was initially obtained at 3.7 Å resolution and subsequently refined to 3.5 Å (R factor of 23.2 and free R factor of 29.7). The current model includes residues α:2–34, α:61–439, β:2–437, one molecule of GTP, one of GDP, and one of Taxol. A ribbon diagram representation of the dimer structure is shown in Fig. 1.

The α- and β-tubulin monomers are very similar in structure and quite compact. The secondary and tertiary structures of the α- and β-monomers are essentially identical, as expected from their identify of >40% over the entire sequence of above 450 amino acids of the sequence *(10)*. Each monomer, as it is shown in Fig. 2, is formed by three sequential and functionally distinctive domains. The N-terminal region is a distinctive nucleotide-binding domain formed by six parallel β-strands (S1–S6) alternating with helices (H1–H6). Each of the loops (T1–T6) that joins the end of a strand with the beginning of the next helix is directly involved in interactions with the nucleotide. Within each monomer, nucleotide binding is completed by interaction with the N-terminal end of the core helix H7. The core helix connects the nucleotide-binding domain with the smaller, second domain, which is formed by three helices (H8–H10) and a mixed β-sheet (S7–S10). The C-terminal region is formed mainly by two antiparallel helices (H11–H12) that cross over the first

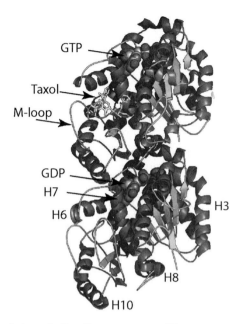

Fig. 1. Ribbon diagram of the tubulin dimer structure. The α-monomer is at the bottom, and β-monomer at the top. Atoms in the nucleotides are shown as solid spheres, Taxol is shown in yellow. The GTP in α is locked in place at the intradimer interface, whereas the E-site nucleotide in β is exposed at the surface of the dimer but becomes blocked when another dimer binds to the top of this dimer. To view this figure in color, see the insert and the companion CD-ROM.

two domains. The loop connecting strand S7 to helix H9 is termed the M-loop and is of particular importance in forming lateral contacts between tubulin molecules.

Figure 3 shows details of the nucleotide-binding region in α-tubulin. Residues involved directly in nucleotide binding and hydrolysis are highly conserved between α- and β-monomers, with some interesting exceptions. Residue α:Ala12 corresponds to β:Cys12; residue α:Val74 in H2, involved in phosphate binding, is substituted by β:Thr74; residue α:Phe141, in the glycine-rich T4 loop, is substituted by a β:Leu141. The sugar-binding T5 loop is very different for α- and β-monomers. Most of these residues, particularly those in T5, are involved in longitudinal contacts between monomers, and these differences are probably important for the relative strength and reversibility of the monomer–monomer and dimer–dimer contacts.

The specificity of tubulin for GTP is obtained by the hydrogen bonding of the 2-exocyclic amino group in GTP to the hydroxyl groups of Asn206 and Asn228, and by hydrogen bonding of the 6-oxo group to the amino group of Asn206. Otherwise, the base sits in a rather hydrophobic pocket defined by Ile16, Val131, Leu227, and Val231 on one side and Tyr224 on the other. The ribose group interacts with main chain and side chain groups in the T5 loop, whereas interaction with the phosphates is dominated by hydrogen bonds with the main chain amines in T1 (the equivalent of the P-loop in G proteins) and T4 (a glycine-rich segment that is considered the tubulin signature motif).

The only known structural homolog of tubulin is the bacterial FtsZ *(11)*. FtsZ is a ubiquitous protein in eubacteria and archaebacteria that is essential for bacterial cell division. FtsZ localizes at the site of septation during cell division and was identified

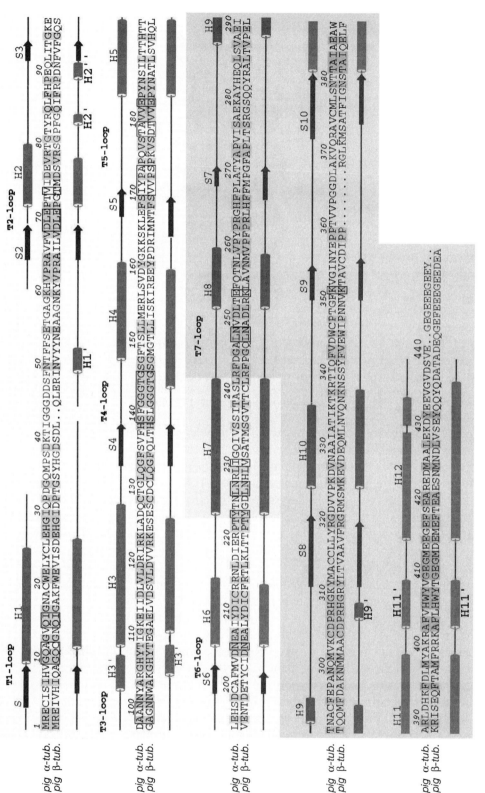

Fig. 2.

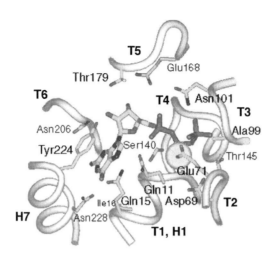

Fig. 3. Topological structures around the nonexchangeable nucleotide site. Residues of β-tubulin making principal contacts with the nucleotide are shown, along with the loop and helix segments in which they are found. The Mg^{++} ion is shown as the yellow sphere. To view this figure in color, see the insert and the companion CD-ROM.

from mutations that interfere with cytokinesis. FtsZ binds and hydrolyzes GTP, forms in vitro filaments reminiscent of tubulin protofilaments, and has the highest degree of sequence identity with tubulin. Although this degree of identity is not very high (around 10%), it is highly significant because it includes the glycine-rich, tubulin signature motif. These similarities led to the proposition that tubulin and FtsZ could be structurally related. This was emphatically demonstrated when the crystal structures of both proteins were obtained and later compared in detail *(6,7,11)*. Interestingly, it has been recently discovered that some bacteria also have tubulin genes, the function of which still remains a mystery *(12)*. It is not known yet if these bacteria also contain FtsZ, and if so, if these two proteins have related or distinctive functions.

In the refined structure of tubulin a magnesium ion was clearly seen at the N-site, but not at the E-site where the GTP had been hydrolyzed. The magnesium ion is bound by salt bridges with two highly conserved residues, Asp69 and Glu71 in the T2 loop. Its position agrees well with the position of the magnesium ion in the crystal structure of FtsZ bound to GTP *(13)*, and with its effect in stability of the αβ-tubulin dimer *(14)*.

3. STRUCTURE OF THE DIMER AND THE PROTOFILAMENT

Because the structure of tubulin was obtained from a polymerized form of the protein, the crystallographic model includes not only the structure of the dimer, as shown in Fig. 1, but also that of the whole protofilament. A segment of one protofilament is shown in

Fig. 2. *(Opposite page)* Secondary structure for α- and β-tubulin indicated on the pig sequences. Residues conserved between α- and β-tubulins are highlighted in cyan. Residues making direct contact with the nucleotide are boxed. Background colors indicate approximate domain boundaries: yellow—N-terminal, nucleotide-binding domain; green—core helix H7; blue—second or intermediate domain; purple—C-terminal domain. Regions not resolved in the crystal structure, including α35–60 and the C-termini of both monomers, are shown with lighter letters. Residue numbering used here is based on the alignment shown in this figure. Note that there are thus gaps in the β-sequence at residues 43–44 and 261–268. To view this figure in color, see the insert and the companion CD-ROM.

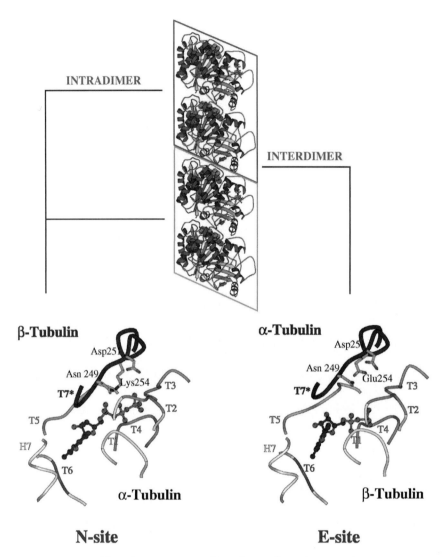

Fig. 4. Protofilament and interface structures. Two dimers from the crystal structure represent the protofilament and illustrate burial of the E-site nucleotide within the polymer. Details of nucleotide contacts from both monomers re shown for the intra- and interdimer interfaces. To view this figure in color, see the insert and the companion CD-ROM.

Fig. 4. The position of the two nucleotides in the α- and β-monomers is identical with respect to the monomer boundaries, but very different with respect to the dimer boundaries. The N-site nucleotide in α-tubulin (GTP) is buried at the monomer–monomer interface within the dimer, explaining the nonexchangeability at that site. On the other hand, the nucleotide at the E-site is partially exposed on the surface of the dimer, allowing its exchange with the solution. The longitudinal contacts are very similar between monomers within the dimer and between dimers. Thus the first consequence of the polymerization process is the burial of the E-site nucleotide at the newly formed interface, so that, in the microtubule, this nucleotide becomes nonexchangeable. In both the intra- and interdimer interfaces the nucleotide is directly involved in the contact between monomers. Thus, there is a region in the tubulin structure on the opposite side from the T1 to T6 loops that is involved in the interaction with the nucleotide in the next monomer along the protofilament. This region encompasses loop T7 and helix H8, in the second domain. The T7 loop includes residues that are highly conserved in both tubulin monomers as well as in the bacterial homolog FtsZ. The region of the intermonomer interaction around the nucleotides is shown in detail in Fig. 4.

The structural superposition of tubulin and FtsZ shows that the residues conserved between the two proteins are all localized to sites of interaction with the nucleotide, including loops T1-T6 and helix H7, as well as T7. This observation led to a model of FtsZ polymerization in which T7 would contribute to the interaction with the nucleotide in the adjacent molecule along the filament. Residues Asn249 and Asp251 in T7 are totally conserved in all α- and β-tubulins and all FtZ proteins (as well as in all γ-tubulins except that in *Saccharomyces cerevisiae*). These residues in tubulin are involved in the binding of the αβ-phosphates of the nucleotide of the next monomer. The interaction with the nucleotide across the longitudinal interface is completed by Lys254 in β-tubulin (within the H8 helix), which interacts with the γ-phosphate of the N-site nucleotide, and in α-tubulin by Glu254, which is in a position that would be close to the γ-phosphate of the E-site nucleotide (in the crystal structure, this nucleotide is GDP after hydrolysis during the formation of the sheets) (*see* Fig. 4). When the equivalent residue in FtsZ, aspartic acid, is mutated to alanine, the binding of GTP is not affected, but its hydrolysis is totally abolished *(15)*. Similarly, mutation of Glu254 (or Asp251) to alanine in the yeast α-tubulin TUB1 is a dominant lethal, and transient expression of the mutant α-tubulin results in hyperstable microtubules *(16)*. This effect is easily seen as a consequence of a loss of the ability to hydrolyze GTP. These results strongly support the idea that this residue is essential for the activation of hydrolysis concomitant to tubulin polymerization. The idea that tubulin is its own GTPase-activating protein, based on the linkage of hydrolysis with polymerization, is confirmed with the identification of an activating region in the molecule that includes mainly loop T7 and helix H8.

The tubulin structure thus explains how hydrolysis follows polymerization. The exchangeable nucleotide is bound to β-tubulin but exposed on the plus end of the microtubule. Addition of a new dimer to the end provides the catalytic mechanism to promote hydrolysis. The lack of hydrolysis at the N-site is related to differences between the residues that correspond to this catalytic group. Whereas Glu254 in α-tubulin is in an ideal position to be involved in the hydrolysis of the E-site nucleotide, the opposite charge of Lys254 in β-tubulin is likely to strengthen the monomer–monomer contacts through its interaction with the phosphate groups of the N-site nucleotide.

The longitudinal contact between monomers is very extensive. About 3000 $Å^2$ of surface area are buried with the formation of the dimer from the monomers, or in a contact between dimers. Apart from a very important van der Waals contribution, the character of the interface is mainly hydrophobic and polar, with minimal electrostatic interactions. The two surfaces involved in the interfaces are convoluted and highly complementary in shape, testifying to the age of the protein and the optimization of the contacts over millions of years. The high conservation of the αβ-tubulin sequences across species has been interpreted as a consequence of the restrictions imposed by tubulin self-assembly, as mutation of a residue in a surface involved in tubulin–tubulin contacts could dramatically affect polymerization unless a number of coordinated and complimentary mutations were to occur simultaneously across the interface. About 52% of the residues in the intradimer interface are totally conserved across species, whereas about 40% are conserved in the interdimer contact *(17)*.

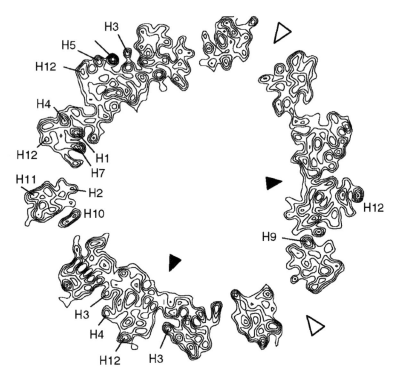

Fig. 5. Cross-section of the three-dimensional microtubule density map *(19)*. Because of the helical structure of the microtubule, each monomer is cut through at a different level. Some of the helices are marked for reference, and noise peaks in the background have been removed for clarity. Open arrowheads mark openings in the wall of the microtubule. Most interprotofilament interactions occur toward the inner part of the wall, as indicated by the filled arrowheads. To view this figure in color, see the insert and the companion CD-ROM.

The larger stretches of absolutely conserved residues cluster in a well-defined area across the longitudinal interface between tubulin monomers *(17)*. This region corresponds to the T3 loop and the loop and small helix H11′ between H11 and H12, on one side, and helix H8 on the other. In contrast, residues involved in lateral contacts between protofilaments cluster in regions of divergence among species and between α- and β-tubulins. This suggests that differences in dynamic instability observed in microtubules from different species may result from variations in residues involved in contacts between protofilaments.

4. MICROTUBULE STRUCTURE

Understanding the interactions between tubulin protofilaments as well as the dimer orientation within the microtubule was made possible by docking the crystal structure of the protofilament into a three-dimensional reconstruction of the microtubule obtained by cryo-electron microscopy *(18,19)*. The more recent reconstruction at 8 Å resolution, as shown in Figs. 5 and 6, allows for direct visualization of α-helices in the polymer and very precise positioning of the dimers with respect to each other. The resultant microtubule model, part of which is shown in Fig. 7A, confirms that the plus end of the microtubule is crowned by β-tubulin monomers exposing their nucleotide surface to the solution, whereas the minus end is crowned by α-monomers exposing their catalytic end.

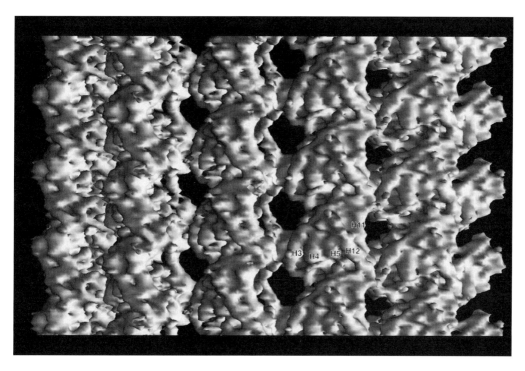

Fig. 6. Surface view of the microtubule density map, seen from the outside with the plus end up. Densities corresponding to several of the α-helices are marked for reference. To view this figure in color, see the insert and the companion CD-ROM.

This orientation has very important repercussions for the GTP-cap model of microtubule dynamics. As discussed later, a conformational change associated with the GTP-bound state appears to stabilize the polymerized state. Exposure of GTP at the plus end ensures that the rate of hydrolysis will be low until addition of a new dimer occurs. Thus, the end of the microtubule can be capped by molecules that contain GTP and maintain the integrity of the microtubule. Loss of this GTP cap, which can occur through GTP-GDP exchange within a dimer at the end, exchange of a GTP-containing dimer for a GDP-containing one, or hydrolysis from transient docking of a dimer onto the microtubule end, destabilizes the microtubule and can lead to depolymerization from the plus end. The ability to switch between stable and unstable microtubule ends leads to the dynamic behavior that is discussed more fully in Chapter 3.

The model also shows that the C-terminal helices H11 and H12 form the crest of the protofilaments on the outside surface of the microtubule. The bumpy inside surface of the microtubule is defined by loops H1-S2, H2-S3, and S9-S10. The latter loop is eight residues longer in α-tubulin, with the extra residues filling the pocket in α-tubulin that in β-tubulin forms part of the Taxol-binding site. The lateral contact between protofilaments is dominated by the interaction of the M-loop with loop H2-S3 and helix H3. As seen in Fig. 7A, the connection between protofilaments occurs toward the lumen of the microtubule, whereas there is a significant space between the protofilaments toward the outside. The M-loop is in a position where it could act as a hinge, allowing for variation in the amount of this space and accounting for the observed variability in the number of protofilament in a microtubule.

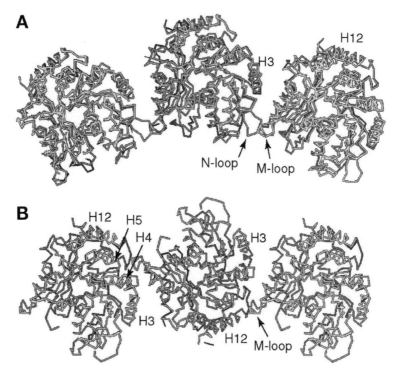

Fig. 7. Protofilament arrangements and contacts in (**A**) microtubules, (**B**) the Zn-sheets. In the sheets, adjacent protofilaments are antiparallel; the yellow protofilaments are seen from the plus end, the blue one from the minus end. All protofilaments in the microtubule are seen from the plus end. Whereas the M-loop makes the major contacts from one side of the molecule in both structures, it interacts mainly with helices H4, H5, and H12 in the sheets and with loop H1-S2 (the N-loop) and helix H3 in microtubules. To view this figure in color, see the insert and the companion CD-ROM.

There are small differences in the structure of tubulin in the microtubule with respect to the zinc sheets. The most prominent of these is the partial loss of structure in helix H6. This segment is stabilized by interprotofilament interactions in the sheets but may be mobile, or adopt different conformations in α- and β-tubulin, in the microtubule. Either way, it appears that the holes in the microtubule wall, seen in Fig. 6, are larger than they would be if H6 were in the same position as in the crystal structure. These holes could provide a path for entry of small ligands such as Taxol into the lumen of the microtubule, where their binding site is located.

5. LATERAL CONTACTS IN ZINC-SHEETS

Lateral contacts between protofilaments in the zinc sheets entail extensive interactions between homologous monomers (α–α, β–β). The tightest part of the interface involves the interaction of the M-loop in one monomer with helices H12 and H5 in the adjacent one. An end-on view of the sheet protofilaments is shown in Fig. 7B. Helix H12 has been identified as a major site for interaction of tubulin with motor proteins (18,20). The involvement of H12 in lateral interactions in the zinc sheets explains the poor binding of kinesin-like motors to these polymers, and their destabilizing effect at high concentrations (21).

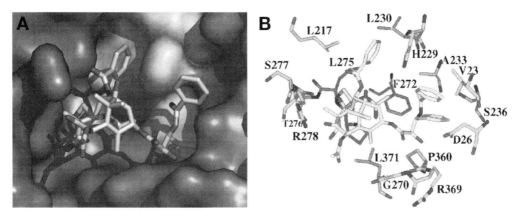

Fig. 8. The Taxol-binding site. (**A**) Taxol is shown over the tubulin surface, which is colored to show the nucleotide-binding domain in green and the intermediate domain in blue. (**B**) Detail of the taxol-binding pocket showing residues in closest contact with the drug. To view this figure in color, see the insert and the companion CD-ROM.

The crystallographic model includes a zinc ion at the lateral contact between α-tubulins. Its location and active, direct participation in lateral interactions strongly suggest that this zinc is at least partially responsible for the generation of the alternative, antiparallel protofilament–protofilament interaction in zinc-induced polymers. Other zinc ions could be required for the formation of the sheets, perhaps through their binding to lower affinity sites.

6. TAXOL-BINDING SITE

The crystal structure of tubulin bound to Taxol provides a good foundation for understanding both the binding and function of microtubule-stabilizing drugs. The Taxol molecule is well defined in the crystallographic density map, and the nature of its interactions with the protein can thus be well understood. The position of Taxol with respect to the protein surface is shown in Fig. 8A. This view is colored to show the binding site at the interface between the nucleotide-binding and intermediate domains. Interestingly, the densities for both the 2-phenyl side chain and the N′ phenyl group remained low through the crystallographic refinement, suggesting some degree of mobility of these groups. Whereas changes in the N′ group produce little effect on the binding and activity of Taxol derivatives, the 2-phenyl side chain is absolutely required for function (22). It is thus surprising that this part of the molecule is not more ordered. The crystallographically refined Taxol structure is in a conformation very similar to that determined independently by energy-based refinement (23), except for torsional rotations of the side-chain phenyl rings.

A considerable number of residues, mostly in the second domain of β-tubulin, are involved in direct contact with Taxol (Fig. 8B). In helix H1, Val23 makes hydrophobic contact with both the N′ and 3′ phenyl rings, whereas Asp26 is in hydrogen bonding distance from the nitrogen group in the side chain. In the H6-H7 loop, Leu217 and Leu219 make hydrophobic contact with the 2-phenyl ring, which is completed by His 229 and Leu230 in the core helix (H7). In the same helix Ala233 and Ser236 contact the 3′-phenyl group. The hydrophobic environment of the 3′-phenyl group is completed by Phe272 at the end of the B7 strand. The M-loop (the loop between S7 and H9) also plays a substantial

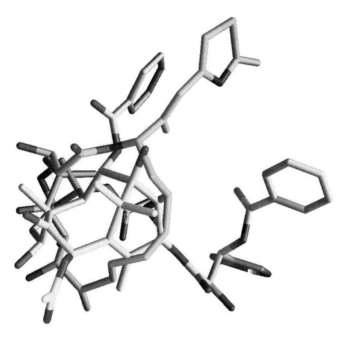

Fig. 9. Superposition of Taxol (yellow) and epothilone-A (blue) in the conformations and orientations in which they bind tubulin. Oxygen atoms are indicated in red, nitrogen in blue, and sulfur in green. To view this figure in color, see the insert and the companion CD-ROM.

role in the binding of the baccatin ring, in particular residues Pro274, Leu275, Thr276 (which contacts the essential oxetane ring), Ser277, and Arg278. The pocket is completed by residues in the B9-B10 loop (Pro360, Arg369, Gly370, and Leu371).

A number of other drugs that also stabilize microtubules have been isolated from diverse natural sources over the last few years. Most of these bind competitively with Taxol, and several attempts have been made to identify a "common pharmacophore" that would describe the common properties of their interactions with the protein. Although it is clear that most of the drugs do bind in the same region as Taxol, they show a great deal of chemical diversity. Thus it is not too surprising that the first structure of one of these, epothilone-A, bound to tubulin *(24)* suggests more of a "distributed pharmacophore." In Fig. 9, Taxol and epothilone-A are superimposed in their tubulin-bound conformations. This comparison makes it clear that these two compounds present quite different funtionality to the protein. There is an active search for derivatives of these drugs with enhanced clinical and synthetic properties that also provides significant information on their binding modes. Further structural and synthetic chemical work should generate a fuller understanding of the complex landscape of the Taxol-binding site and the range of interactions available to drugs that bind in this region.

The mechanism of microtubule stabilization remains somewhat obscure. As pointed out earlier, Taxol sits adjacent to the M-loop, and one might expect that it serves to stabilize a particular conformation of the loop that strengthens the interprotofilament interaction. Figure 10 shows the location of Taxol with respect to the M-loop and interprotofilament contacts. In α-tubulin, the conformation of the M-loop is stabilized by the long S9-S10 loop *(see* Fig. 1). In the β-monomer a similar stabilizing function may be played by Taxol and Taxol-like compounds *(6,7,23)*. It has also been suggested that

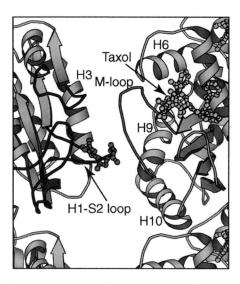

Fig. 10. Detail of the interface between protofilaments, as seen from the inside of the microtubule. Residues of the M-loop on one side make contacts mainly with those in the H1-S2 loop (dark blue) and helix H3 on the other. A cluster of residues on the H1-S2 loop, shown in purple ball-and-stick style, varies among tubulin isotypes, possibly contributing to different interprotofilament bond strength and different dynamics for microtubules formed with different isotypes. To view this figure in color, see the insert and the companion CD-ROM.

some motion of the intermediate and nucleotide-binding domains with respect to each other might account for the curvature of protofilaments in the GDP-bound state, and that Taxol might inhibit such motion *(25)*. The position of the binding site, as well as structural work discussed later that visualizes such a motion, suggest that Taxol may act as a key to block this motion and keep the protofilaments in the straight conformation. The way that epothilone-A binds to tubulin, and presumably other drugs that bind in the same region, support the idea that a combination of these two effects is likely to account for the microtubule stabilization.

7. CONFORMATIONAL CHANGES

Tubulin is known to exist in at least two main conformations: straight when bound to GTP or fixed within the microtubule lattice, and curved when bound to GDP. (*see* ref. (30) for a recent description of a third conformational state.) The latter is most obvious during microtubule depolymerization as protofilaments bend and peel off from microtubule ends. This curved, low energy form of tubulin can also be obtained from GDP-tubulin dimers that, in the presence of stabilizing divalent cations, form intriguing ring structures. The rings correspond closely to protofilaments as if they have been curved radially from the microtubule so that the "bumpy" side of tubulin, which in the microtubule corresponds to the inside surface, faces the outside of the ring *(26)*. In addition, cellular factors and antimitotic agents that act by depolymerizing microtubules can induce the formation of rings. A number of studies have used the atomic structure of tubulin to model the binding and effect of these ligands. In general, tubulin dimers in the rings appear kinked at both the dimer–dimer interface and, to the same or to a lesser extent, at the monomer–monomer interface; with each agent giving rise to particular subtleties in the structures that may reflect their distinctive mechanisms of action.

For example, Sackett and coworkers *(27)* have recently shown that the tubulin-crypto-phycin 1 ring is formed by eight dimers. Averaging of electron microscopy images of frozen-hydrated rings and docking of the crystal structure indicated the presence of 13° intradimer bends and 32° bends between dimers, indicating that the interdimer interface may be significantly more flexible than the intradimer interface.

More precise information on the conformational flexibility of tubulin has come from the first X-ray crystal structures of the protein, in complex with either the stathmin homolog RB3 or RB3 and colchicine *(28,29)*. These results clearly show a rotation of the intermediate domain with respect to the nucleotide-binding domain in both α- and β-monomers. This is the type of motion that had been postulated to account for the cur-vature of protofilaments in the GDP-bound form *(25)*. Recently a similiar, but not iden-tical, conformation has been described for GDP–tubulin without additional ligands (30).

8. CONCLUSION

The structure of tubulin and the model for the microtubule explain a number of the mysteries that have fascinated cell biologists and biophysicists since the discovery of microtubules. For example, the different nucleotide exchangeabilities between α- and β-tubulin and the coupling of polymerization and hydrolysis are a direct consequence of the position of the nucleotides at the intermonomer interfaces. The structure also provides some important clues about the effect of nucleotide hydrolysis on microtubule disassembly and gives a basis for understanding the stabilizing effect of Taxol and related drugs.

REFERENCES

1. Mitchison T, Kirschner M. Dynamic instability of microtubule growth. Nature 1984;312:237–242.
2. Carlier MF, Hill T, Chen YD. Interference of GTP hydrolysis in the mechanism of microtubule assem-bly: an experimental study. Proc Nat Acad Sci USA 1984;81:771–775.
3. Ludueña RF. The multiple forms of tubulin: different gene products and covalent modifications. Int Rev Cyt 1998;178:207–275.
4. Desai A, Mitchison TJ. Microtubule polymerization dynamics. Annu Rev Cell Dev Biol 1997;13:83–117.
5. Jordan A, Hadfield JA, Lawrence NJ, McGown AT. Tubulin as a target for anti-cancer drugs - agents which interact with the mitotic spindle. Med Res Rev 1998;18:259–296.
6. Nogales E, Downing K, Amos L, Löwe J. Tubulin and FtsZ form a distinct family of GTPases. Nature Struct Biol 1998;5:451–458.
7. Nogales E, Wolf SG, Downing KH. Structure of the αβ tubulin dimer by electron crystallography. Nature 1998;391:199–203.
8. Löwe J, Li H, Downing KH, Nogales E. Refined structure of alpha beta-tubulin at 3.5 Å resolution. J Mol Biol 2001;313:1045–1057.
9. Amos LA, Baker TS. The three-dimensional structure of tubulin protofilaments. Nature 1979;279:607–612.
10. Sander C, Schneider R. Database of homology-derived structures and the structural meaning of sequence alignment. Proteins 1991;9:56–68.
11. Löwe J, Amos LA. Crystal structure of the bacterial cell division protein FtsZ. Nature 1998;391:203–206.
12. Jenkins C, Samudrala R, Anderson I, et al. Genes for the cytoskeletal protein tubulin in the bacterial genus Prosthecobacter. Proc Nat Acad Sci USA 2002;99:17,049–17,054.
13. Oliva MA, Cordell SC, Lowe J. Structural insights into Fts Z protofilament formation, Nat Struct Mol Biol. 2004;11:1243–1250.

14. Menendez M, Rivas G, Diaz JF, Andrew JM. Control of the Structural stability of the tubulin dimer by one high affinity bound magnesium ion at nucleotide N-site, J Biol Chem 1998;273:167–176.

15. Dai K, Mukherjee A, Xu Y, Lutkenhaus J. Mutations in FtsZ that confer resistance to SulA affect the interaction of FtsZ with GTP. J Bacteriol 1994;175:130–136.

16. Anders KR, Botstein D. Dominant-lethal alpha-tubulin mutants defective in microtubule depolymerization in yeast. Mol Biol Cell 2001;12:3973–3986.

17. Inclan YF, Nogales E. Structural models for the self-assembly and microtubule interactions of gamma-, delta- and epsilon-tubulin. J Cell Sci 2001;114:413–422.

18. Nogales E, Whittaker M, Milligan RA, Downing KH. High resolution model of the microtubule. Cell 1999;96:79–88.

19. Li H, DeRosier D, Nicholson W, Nogales E, Downing K. Microtubule structure at 8 Å resolution. Structure 2002;10:1317–1328.

20. Kikkawa M, Okada Y, Hirokawa N. 15 Å resolution model of the monomeric kinesin motor, KIF1A. Cell 2000;100:241–252.

21. Han Y, Sablin EP, Nogales E, Fletterick RJ, Downing KH. Visualizing a new binding site of ncd-motor domain on tubulin. J Struct Biol 1999;128:26–33.

22. Kingston DGI. Taxol: the chemistry and structure-activity relationships of a novel anticancer agent. Trends Biotechnol 1994;12:222–227.

23. Snyder JP, Nettles JH, Cornett B, Downing KH, Nogales E. The binding conformation of Taxol in beta-tubulin: A model based on electron crystallographic density. Proc Nat Acad Sci USA 2001;98:5312–5316.

24. Nettles JH, Li HL, Cornett B, Krahn JM, Snyder JP, Downing KH. The binding mode of epothilone-A on αβ-tubulin by electron crystallography. Science 2004;305:866–869.

25. Amos L, Löwe J. How Taxol stabilizes microtubule structure. Chem Biol 1999;6:R65–R69.

26. Nicholson WV, Lee M, Downing KH, Nogales E. Cryoelectron microscopy of GDP-tubulin rings. Cell Biochem Biophys 1999;31:175–185.

27. Watts NR, Cheng NQ, West W, Steven AC, Sackett DL. The cryptophycin-tubulin ring structure indicates two points of curvature in the tubulin dimer. Biochemistry 2002;41:12,662–12,669.

28. Gigant B, Curmi PA, Martin-Barbey C, et al. The 4 angstrom X-ray structure of a tubulin : stathmin-like domain complex. Cell 2000;102:809–816.

29. Ravelli RBG, Gigant B, Curmi PA, et al. Insight into tubulin regulation from a complex with colchicine and a stathmin-like domain. Nature 2004;428:198–202.

30. Wang HW, Nogales E. Nucleotide-dependent bending flexibility of tubulin regulates microtubule assembly. Nature. 2005,435:911–915.

10 Destabilizing Agents

Peptides and Depsipeptides

Lee M. Greenberger and Frank Loganzo

CONTENTS

SUMMARY

Numerous peptides bind to tubulin and inhibit the function of microtubules. The molecular classes discussed in this chapter are dolastatins, cryptophycins, hemiasterlins, phomopsin, ustiloxins, diazonamides, tubulysins, and vitilevuamide. These natural product agents bind to tubulin at a site that is similar but distinct compared with vinblastine and depolymerize microtubules. Many of these agents are potent inhibitors of cell division that have been tested in clinical trials as antitumor agents. These agents are excellent research tools to understand the molecular basis of tubulin function. Further exploration is needed to define their optimal utility in the clinic.

Key Words: Dolastatins; cryptophycins; hemiasterlins; phomopsin; ustiloxins; diazonamides; tubulysins; vitilevuamide; tubulin; vinblastine; drug binding site.

1. INTRODUCTION

Numerous small molecules bind to tubulin, alter the stability of microtubules, and inhibit microtubule-mediated cell division (*1*). These molecules have received tremendous attention as two classes, the vinca alkaloids and particularly the taxanes, have been broadly used to treat cancer (*2*). The clinical data suggest that inhibition of microtubule function by antimitotics has meaningful therapeutic value. However, although effective in some patients, vinca alkaloids and taxanes are rarely used to cure patients, and resistance either at the onset or during the course of therapy occurs with high frequency.

From: *Cancer Drug Discovery and Development: The Role of Microtubules in Cell Biology, Neurobiology, and Oncology* Edited by: Tito Fojo © Humana Press, Totowa, NJ

This has stimulated an intensive search for new antimicrotubule agents that has occurred in the last 20 yr *(1,3)*. Indeed, the structural and functional diversity of new small molecule inhibitors of microtubules is so high that combined with the heterogeneity of cancers, more than a decade has been required to even begin to evaluate the utility of many of these molecules. A most outstanding example is in the area of peptide and depsipeptide inhibitors of microtubules. By definition, these agents contain amide bonds or in the case of depsipeptides, contain an ester linkage with amide bonds. All of these agents were initially isolated as natural products. These agents are the subject of this chapter.

Prerequisite to understanding the mechanism of action of peptide-based inhibitors of microtubules, some basic biochemistry of microtubules must be briefly reviewed (for a detailed excellent review, *see* Jordan *[1]*). Microtubules are highly dynamic polymers whose essential element is the α-/β-tubulin heterodimer. Under certain conditions, the dimer binds in a head to tail fashion to form linear protofilaments. About 13 protofilaments associate in parallel to a cylindrical axis to form the microtubule of 25 nm diameter and up to micrometers in length. The polymer is initially formed from a short microtubule nucleus. Thereafter, the microtubule can elongate slowly or shorten in a rapid fashion; this is done by adding or removing tubulin dimers from the existing polymer, respectively. The transition between these two behaviors is known as dynamic instability. Although both ends of the microtubule can grow or shorten, the changes in length at the "plus end" are more dramatic than those at the "minus end."

The polymerization of tubulin is dependent upon GTP hydrolysis. Both α- and β-tubulin bind one molecule of GTP. The GTP bound to α-tubulin, located in the intradimer region, is nonexchangeable (N-site) and requires harsh conditions (i.e., urea) to extract it. However, the GTP bound to β-tubulin, located in the interdimer region, is exchangeable (E-site). Hydrolysis of GTP is required at the E-site for tubulin polymerization. Upon formation of the protofilament, the E-site becomes N-site and contains GDP. Only the accessible end of the microtubule, the so-called cap, contains GTP. In the event the GTP bound cap is lost, rapid depolymerization of the microtubule occurs.

After chromosome replication during cell division, the sister chromatids become attached to the microtubule and move to the metaphase plate. Subsequently, during anaphase the sister chromatids move to the opposite poles; this is coordinated with widening of the distance between the poles. Both events are mediated by the spindle microtubules, which are highly dynamic compared with microtubules in the cytoplasm present during interphase. All antimicrotubule agents induce a blockage at the G_2/M phase of the cell cycle during the metaphase/anaphase transition by inhibiting the microtubule-dependent movement of sister chromatids to the opposite pole of the cell as well as inhibiting the lengthening of the interpolar distance. At low stoichiometric ratios compared with tubulin, all antimitotics act in a similar manner. They inhibit dynamic instability, causing improper tensioning of the mitotic apparatus and/or insufficient microtubule contraction required to move the chromosomes to the pole. When the mitotic apparatus is upset, this signals a cascade of phosphorylation- and caspase-dependent events leading to apoptosis *(4,5)*.

Despite these similar features, different classes of antimitotic agents can be easily distinguished based on their phenotypic effect on microtubules at high concentrations as well as by their distinct binding sites within tubulin. The first class of agents, the microtubule stabilizing agents, increase the polymer mass when present at equal concentrations compared with tubulin. These agents, which include taxanes, epothilones, discodermalide, eluterobins, and laulimalide have been reviewed elsewhere *(1,6)*. In contrast, the microtubule destabilizing agents decrease the polymer mass when present at equal concentrations compared with tubulin. Three classes of depolymerizing agents have been distinguished from each other based on binding analysis with purified tubulin. The classes are

1. The vinca alkaloids, or agents that competitively inhibit vinca binding to microtubules;
2. The colchicine class, or agents that competitively inhibit colchicine binding to tubulin; and
3. Peptides and depsipeptides.

All agents in the latter class inhibit the binding of vinca alkaloids to tubulin, but do so in a noncompetitive manner and contain a peptide-based core structure. A comprehensive review of peptide-based inhibitors, including discussion of structure activity relationships, was published in 2002 (7). This chapter incorporates new knowledge and focuses on the biological features of peptide-based antimitotic drugs. For detailed analysis on the structure-activity relationships (SARs) of the peptide-based inhibitors, the reader is referred elsewhere (7,8).

1.1. Evaluation of Candidate Antimicrotubule Agents

The activity and mechanism of action of peptide-based molecules can be compared in several ways. From the perspective of activity, the agents are typically evaluated for effects on

1. Purified tubulin and/or microtubules;
2. Cell growth in tissue culture;
3. Human tumor growth in animal models; and
4. Patients with tumors.

The antitumor activity using models that fail to respond to already approved clinical antimitotics are a major determinant for stimulating clinical evaluation of novel antimitotics (3). From a mechanistic perspective, agents are evaluated for their ability to

1. Compete with vinca or colchicine binding to tubulin;
2. Alter tubulin morphology;
3. Alter GTP exchange;
4. Alter dynamic instability or tubulin polymerization; and
5. Enhance colchicine interaction with tubulin.

In a few cases, attempts have been made to define the drug interaction site with tubulin on a molecular level.

1.1.1. Interaction With Tubulin Enriched Preparations

Assays with purified tubulin are useful for determining the type of interaction with the test agent on a macromolecular and molecular level. On a macromolecular level, all peptide-based inhibitors reduce the amount of microtubules. They do so by either

1. Binding and blocking the ability of the dimer to form microtubules;
2. Inducing depolymerization of an existing microtubule; and/or
3. Inducing the formation of aberrant microtubules that cannot polymerize to form microtubules.

The determination of depolymerizing activity of a candidate agent is often assessed by its ability to block tubulin polymerization. As tubulin polymerization is typically monitored by an increase in turbidity of the solution, more care must be taken to insure that any changes in turbidity actually reflect the formation and /or elongation of micro-tubules and not the simple aggregation of tubulin that can be induced by high concentrations of depolymerizing agents. On a molecular level, the nature of the interaction with tubulin can be determined. In this assay, if the test agents competitively inhibit the binding of known agents, the prediction is that both agents bind to the same site within tubulin. In contrast, noncompetitive inhibitors bind to a distinct site that allosterically alters the binding of the known agent. There are two other unique features that distinguish the molecular interaction of most of the peptide-based inhibitors of microtubules. First, like the vinca alkaloids, most peptidyl inhibitors of microtubules prevent GTP exchange. This does not occur with paclitaxel but does with drugs that bind the colchicine site in tubulin *(7)*. Second, they stabilize or enhance colchicine binding to tubulin *(7)*.

1.1.2. Effect in Cellular Assays

Cellular assays have been used to examine the effect of the test agent on microtubules *in situ*. This can be done using fluorescence methods to evaluate the microtubule structure (using antibodies to tubulin) and status of the cell cycle (using staining methods for nuclear material) (i.e., ref. *9*). Alternatively, the ability of candidate agents to partition tubulin into the cytosol vs pellet fraction can be done. In this case, agents that depolymerize microtubules shift tubulin into a soluble form that is enriched in the cytosol, whereas agents that stabilize tubulin cause the opposite effect (i.e., ref. *10*). Beyond this, the ability of a candidate agent to overcome resistance to clinically useful antimitotic agents is typically evaluated. In this case, investigators have used cells that are resistant to paclitaxel, docetaxel, or vinca alkaloids This phenotype occurs in cells derived from patient's tumors or, more frequently, induced by selecting sensitive cells in tissue culture for survival in paclitaxel, vinca alkaloids, or other agents *(11,12)*. It is generally assumed that novel antimitotic agents that overcome resistance to taxanes or vinca alkaloids will be an advantage in the clinic. However, this assumption must be made with caution for several reasons. Typically, cells selected for resistance to paclitaxel in the laboratory overexpress the drug efflux pump P-glycoprotein, have mutations or alteration in the expression of tubulin isoforms *(12)*, or have altered apoptotic mechanisms *(11)*. However, the clinical relevance of each of these mechanisms has been clearly established in only a few tumor types *(13,14)*. Much of the confusion regarding resistance resides with the fact that even low-level resistance to anticancer drugs may be meaningful as these agents are typically given to patients at or near the maximum tolerated dose (MTD). Hence, even a small change in resistance would allow the tumor cells to survive at the MTD. Beyond this, the ability to detect small changes in resistance markers in a heterogeneous tumor population is poor. For example, it has been observed that even relatively small changes in resistance in tissue culture (19-fold resistance to paclitaxel), can be associated with a complete lack of response to taxanes in animals *(15)*.

1.1.3. Effects in Animal Models

In vivo assays with any novel anticancer agent are almost always required before justifying clinical trials in cancer patients. Before the 1980's, the most commonly used animal model involved implanting mouse leukemia cells in the peritoneum of mice and determining if the life-span of these animals was extended when given experimental agents *(16)*. However, as these models used mouse tumors, and drugs were often injected intraperitonially (and therefore, could achieve unusually high concentrations in the region of the tumor), these assays were not optimal for most purposes. This system has been replaced with the human tumor xenograft assay *(17)*, which remains the most common assay used to date. In this case, human tumor cells or tumor fragments are implanted subcutaneously (into the flanks) in an immunocompromised "nude" mouse. The tumor-bearing animal is treated with the test agent given, most commonly, by the intravenous or intraperitoneal route of administration. Those test agents that are tolerated (i.e., do not cause death or maximum weight loss > 20%) and inhibit tumor growth are considered of further interest. Indeed, some tumors are highly responsive to vinca alkaloids or taxanes *(18,19)* and short courses of therapy dramatically inhibit tumor growth. If the same robust response occurred with high frequency in humans, or even if in vivo activity in a particular histological tumor type correlated with activity in the same human cancer histology (which it usually does not *[20]*), there would be less urgent need for new agents. Therefore, new animal models are under development *(21)*. It is important to recognize that tumors implanted subcutaneously are morphologically different than those implanted orthotopically (within the same organ from which the tumor was derived), and the response to candidate antitumor agents in the orthotopic model may more closely mimic the clinical scenario *(22–24)*. Beyond this, drug exposure in mice is typically higher than that achieved in humans. A lower response is observed in animal models when a clinically meaningful drug exposure is achieved *(25)*.

1.1.4. Effect in Clinical Trials

Novel antimicrotubule agents have been evaluated in cancer patients who typically have failed existing therapies. Particularly, in trials conducted beyond 1995, the new antimitotic agents have been tested in patients who had solid tumors that are refractory to taxane-based therapy. With this background information in place, the biological activity of peptide-based antimitotic agents can be compared. A broad comparison is summarized in Table 1. A detailed description of each type of peptide-based drug is described later.

2. DOLASTATIN ANALOGS

2.1. Origin

Dolastatin 10 and 15 were initially isolated from the sea hare *Dolabella auricularia* found in the Indian Ocean *(26,27)* (Fig. 1). Subsequently, other dolastatin analogs were found that interacted with tubulin, including dolastatin H and isodolastatin H, isolated from *D. auricularia* found in the Pacific Ocean *(28)* as well as malevamide D *(29)*, symplostatin 1 *(30)* and 3 *(31)*, isolated from marine cyanobacterium *Sympolca* spp. As symplostatin 1 and 3 as well as dolastatin 10 *(32)* have been isolated from cyanobacteria and because sea hares feed on cyanobacteria, all of these compounds may be metabolites derived from cyanobacteria. The complete synthesis of dolastatin 10 *(33)* and dolastatin 15 *(34)* has been achieved. This has allowed the synthesis of many new dolastatin analogs including the dolastatin 10 analog, TZT-1027 (also known as auristatin PE and soblidotin) *(35–37)* and the dolastatin 15 analogs, LU103793 (cemadotin) *(38)* and ILX651 (tasidotin) *(39,173)*. A review of the SAR of dolastatins can be found elsewhere *(7)*.

2.2. Interaction With Tubulin

Dolastatin 10 is one of the most potent dolastatin analogs based on its ability to interact with tubulin and inhibit cell growth. Because other analogs of dolastatin 10, including

Table 1
Peptide and Depsipeptide Inhibitors of Microtubules: Overview

Class	Effect on purified tubulin IC$_{50}$ (μM)	Effect on cell growth IC$_{50}$ (nM)[a]	Dose needed to inhibit tumor growth in animals (mg/kg)[b]	Highest clinical status (MTD)	References
Dolastatin 10	1–3	0.12 (60)	0.15–0.45	Phase II (~0.4 mg/m^2)	35
Dolastatin 15	23	1.6 (60)	>5.25	–	43
LU103793	7	0.48 (3)	0.5–3	Phase II (2.5 mg/m^2)	38
Symplostatin 1	–	1.4 (4)	–	–	42
Symplostatin 3	–	7.1 (2)	0.5–2	–	31
TZT-1027 (auristatin PE)	1–2	0.49 (60)	–	Phase II (2 mg/m^2)	81
Cryptophycin 1	1–5	0.021 (5)	10–30	–	91,96
Cryptophycin 52	2.75	0.046 (9)	5	Phase II (1.5 mg/m^2)	166
Hemiasterlin	0.98	12.3 (6)	–	–	46
HTI-286	1	2.5 (18)	0.2–1.6	Phase II (1.5 mg/m^2)	15
Phomopsin A	3	7000 (1)	–	–	41
Ustiloxin A	0.7	4300 (8)	–	–	151
Tubulysin D	–	0.04 (5)	–	–	162
Vitilevuamide	2	188 (4)	0.03	–	50
Moroidin and	3	–	–	–	161
Celogentin B	0.8	–	–	–	
Diazonamide A	~1	5 (60)	–	–	49

[a]Value in parentheses indicates number of cell lines evaluated.
[b]All drugs administered intravenously except dolastatin 15. Dose range, where known, are the minimum efficacious and MTDs. Animal models include human tumor (xenograft) data except for vitilevuamide, which was analyzed only in the mouse P388 lymphocytic leukemia.

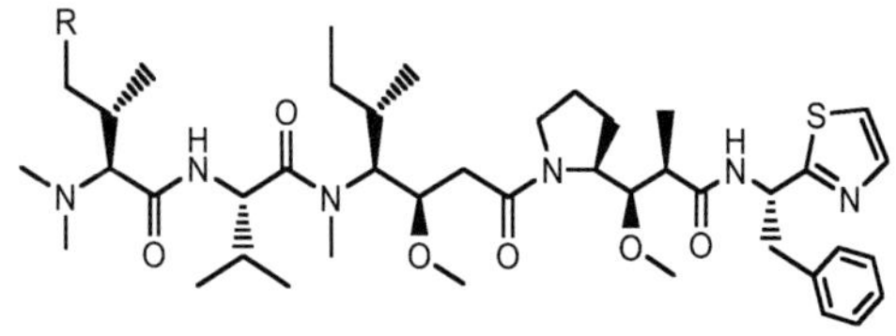

dolastatin 10:   R = H
symplostatin 1: R = CH$_3$

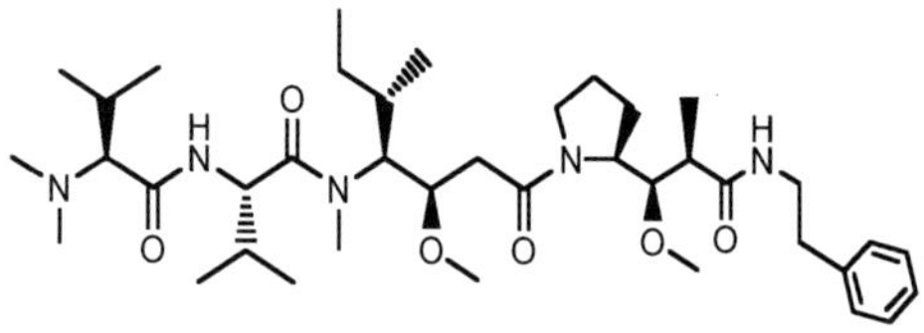

TZT-1027
(auristatin PE)

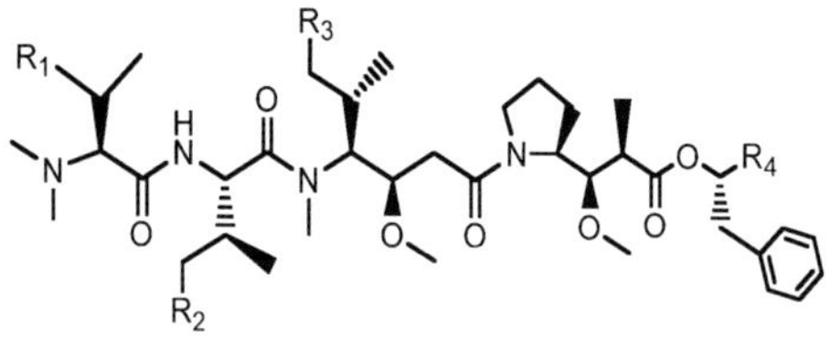

	R$_1$	R$_2$	R$_3$	R$_4$
symplostatin 3:	H	H	H	CO$_2$H
malevamide D	CH$_3$	H	H	CH$_2$OH
isodolastatin H:	H	CH$_3$	CH$_3$	CH$_2$OH

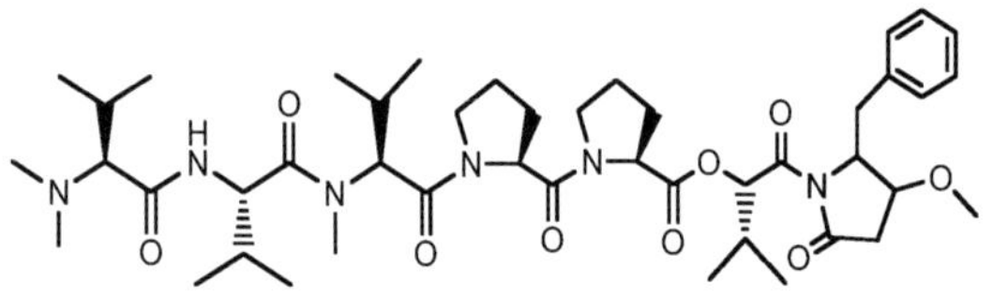

dolastatin 15

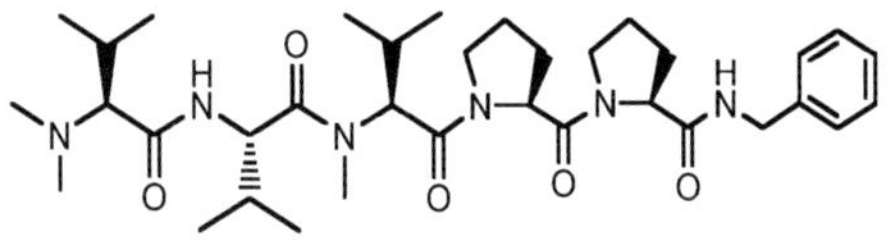

LU103793 (Cemadotin) 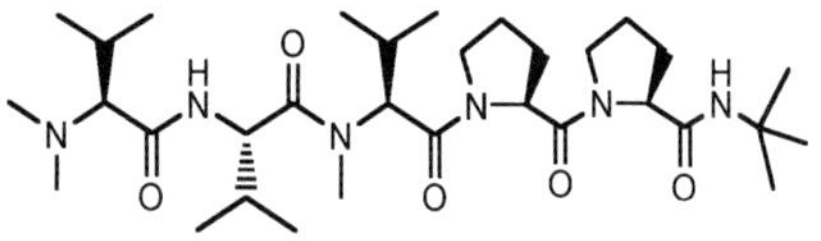   ILX651 (Tasidotin)

Fig. 1. Chemical structure of dolastatin analogs.

dolastatin 15, symplostatins, TZT-1027 and LU103793 are highly related in structure to dolastatin 10 (Fig. 1), it is likely that all of them interact with tubulin in a qualitatively similar manner. Dolastatin 10 *(40,41)*, TZT-1027 *(40)*, and symplostatin 1 *(42)* are potent inhibitors of tubulin polymerization in the presence of microtubule-associated proteins (MAPs) (range 2–3 μ*M*) or absence of MAPs (range 0.6–3 μ*M*) compared with dolastatin 15 (23 μ*M*) or the dolastatin 15 analog, LU 103793 (range = 1–7 μ*M*) *(38,43,44)*. Similar to phomopsin A, dolastatin 10 and TZT-1027 inhibit the binding of vincristine and vinblastine, but not colchicine, to tubulin *(41,45)*. The inhibition of vinblastine binding to tubulin is noncompetitive for dolastatin 10 (K_i = 1.4 μ*M*) *(41)* and TZT-1027 *(40)*, whereas the inhibition of another peptidyl inhibitor of tubulin, hemiasterlin, competitively inhibits dolastatin10 binding to tubulin *(41,46)*. Dolastatin 10 *(47)* and TZT-1027 *(40)* have a high affinity single binding site for tubulin (K_d = 26 n*M* and 200 n*M*). A low affinity site (~10 μ*M*) may also exist for these agents *(40)*, but the analysis is confounded, at least for dolastatin 10, by the aggregation of tubulin that is induced at high concentrations of the drug *(47)*. The affinities of dolastatin 15 ($K_d = 30$ μM) *(48)* and cemadotin ($K_d = 19.4$ μM) for tubulin *(44)* are significantly higher than dolastatin 10 or TZT-1027. Dolastatin10 binding to tubulin is strongly inhibited by some peptide-based inhibitors of tubulin such as hemiasterlin *(46)*, cryptophycin *(46)*, or phomopsin A *(49)*. It is moderately inhibited by the peptide vitilevuamide *(50)* and not inhibited by dolastatin 15 or the peptide diazonamide A *(49)*. Consistent with the inhibition of microtubule assembly, like other peptides, dolastatin 10 and TZT-1027 inhibited tubulin-dependent GTP hydrolysis and nucleotide exchange *(40,45)*. As with phomopsin A, dolastatin 10 enhanced colchicine binding to tubulin *(45)*.

It has been suggested that dolastatin 10 induces an abortive attempt at microtubule formation. In particular, 10 μM or higher dolastatin 10 induces the formation of oligomers of tubulin that in the electron microscope appear as rings, broken rings, and aggregated protein *(29,47)*. Similar effects are observed with LU103793 *(38)*, hemiasterlin *(46)*, and cryptophycin 1 *(51)*. The ring diameter induced by dolastatin 10 is estimated to be approx 45 nM *(52)*.

2.3. Effects on Cells

Dolastatin 10 and TZT-1027 are extremely potent inhibitors of tumor cell growth *(35)*. For example, in the National Cancer Institute (NCI) cell based screen, the average IC$_{50}$ value over 60 cell lines is 120 pM (range 1pM–100 nM) for dolastatin 10 *(7,35)*. TZT-1027 is approximately fourfold less potent using the same assay system *(35)*. Similar IC$_{50}$ values have been obtained for symplostatin 1 *(42)*. Generally, dolastatin 15 *(53)* and its analogs, LU 103973 *(38)* and symplostatin 3 *(31)*, are 10–100-fold less potent than dolastatin 10. The extraordinary potency for dolastatin 10 is related to its ability to be retained in cells. Hamel and colleagues found that cells accumulate the same amount of dolastatin 10 or vinblastine after 24 h, despite the fact that cells are incubated in 20-fold less dolastatin (50 pM) compared with vinblastine (1 nM) *(54)*. Vinblastine reaches peak levels rapidly (20 min) and then decays rapidly ($t_{1/2} = 10$ min), whereas dolastatin 10 reaches peak levels within 6 h and remains high. This coincides with a very high-affinity for tubulin compared with vinblastine, although the relative affinity for efflux pumps (e.g., P-glyco-protein) may also help to explain the relative retention of dolastatin 10. In fact, dolastatin 10 (and symplostatin 1) are substrates for P-glycoprotein based on the crossresistance profile in P-glycoprotein overexpressing cells and verapamil's ability to reverse the phenotype *(42,55)*. However, P-glycoprotein-mediated resistance to dolastatin 10 is not as profound as vinblastine in cells that are moderately or highly resistant to vinblastine *(15,55)*.

2.4. Effect in Tumor Models

The in vivo activity of dolastatin 10 is documented. In human small cell lung cancer (SCLC) or ovarian carcinoma models, 450 µg/kg dolastatin 10 given intravenously as a single dose or two doses 10 d apart markedly inhibits tumor formation *(56,57)*. Dolastatin 15 has weaker activity compared with dolastatin 10 in the ovarian carcinoma model when given at an equitoxic dose *(57)*. Symplostatin 1, given as a 1–3 mg/kg single intravenous dose, is active in murine colon 38 and murine mammary 16C models *(42)*. It is poorly tolerated and mice are slow to recover from toxicity. However, the compound retains activity with better safety when the dose is divided and given multiple times *(42)*.

TZT-1027 is an effective inhibitor of tumor growth in animal models. It is active in the P388 leukemia and other mouse tumor models *(36)*. TZT-1027 appears to have a wider therapeutic window (defined as the difference between minimum effective dose [MED] and MTD) compared with dolastatin 10, as the MED and MTD for TZT-1027 are approx 0. 5 mg/kg and 3 mg/kg, respectively whereas the MED and MTD for dolastatin 10 are 0.125 mg/kg and 0.5 mg/kg, respectively *(36,37)*. In solid tumor models, potent activity is observed with TZT-1027 in the nude or SCID mice models. It is effective in a human pancreatic carcinoma *(58)*, Waldenstrom's macroglobulinemia *(59)*, and a human diffuse large cell lymphoma *(60)* models. When given at the MTDs, TZT-1027 has superior efficacy compared with dolastatin 10 in the Waldenstrom's macroglobulinemia and lymphoma models.

Analogs of TZT-1027, auristatin E and monomethylauristatin E, have also been conjugated to antibodies. This allows the toxin to be selectively targeted to antigens expressed on tumor cells. When monomethyl auristatin E is conjugated to an antibody that recognizes CD 30, which is selectively expressed on lymphomas, it is more efficacious than the antibody given alone *(61)*. Analogs of auristatin E conjugated to cBR96 (specific to Lewis Y on carcinomas) are also highly efficacious *(62)*.

The mechanism of action of TZT-1027 in tumors has been explored. The compound has cytotoxic as well as prominent antivascular effects that are distinct from other antimicrotubule agents studied. In particular, Otani et al. *(63)* found that TZT-1027 causes hemorrhaging of the vascular system within an advanced tumor derived from a murine colon 26 adenocarcinoma, followed by a blockage of tumor blood flow. Effects occur within 1 h after treatment, are mainly in the periphery of the tumor, and are tumor-selective. Consistent with the effects on blood flow, TZT-1027 has a marked effect on cultured human umbilical vein endothelial cells. The effects are more prominent than with vincristine. Natsume and colleagues *(64)* report that TZT-1027 causes prominent destruction of the tumor vasculature in Lewis lung carcinoma cells implanted subcutaneously in mice and that the expression of genes, including those involved with angiogenesis, are altered following treatment with TZT-1027. TZT-1027 also retains good activity in early and late stage tumors derived from a lung cell line that is transfected with a vascular endothelial growth factor, whereas other antimicrotubule agents such as vincristine or paclitaxel, or another antivascular/antimicrotubule agent, combrestatin, are not effective under all conditions. The antivascular effects may help explain why dolastatin 10 is one of the most effective antitumor agents when combined with a *Clostridium novyi*, a bacteria that propagates in anoxic, avascular areas *(65)*. In this case, the bacteria may be more effective because the anoxic field (i.e., the periphery) is increased by dolastatin 10.

2.5 Clinical Trials

The following dolastatin analogs have entered into clinical trials as anticancer agents: dolastatin 10, TZT-1027, LU103793, and ILX651. Two phase I trials with dolastatin 10 were done *(66,67)*. Dolastatin 10 was given intravenously every 21 d as a bolus dose. The MTD was 300–455 µg/m^2; lower tolerated doses were achieved in heavily pretreated patients. In these studies as well as in the phase II studies, dose-limiting toxicity was granulocytopenia. In one study, some patients developed new or increased symptoms of mild peripheral sensory neuropathy *(67)*. Little or no activity was observed with dolastatin 10 when given to patients with a variety of cancers in separate phase II studies (Table 2). Five different dosing schedules were evaluated in phase I trials with LU103793 *(68–71)*. In the

Table 2
Phase II Clinical Trial Results With Peptidyl Antimicrotubule Agents

Tumor type	N^a	Dose (mg/mg^2)	Schedule[b]	CR[c]	PR[d]	SD[e]	Reference
Dolastatin 10							
Ovarian cancer	28	0.4	q3w	0	0	7	167
Colorectal cancer	14	0.3–0.45	q3w	0	0	4	168
NSCLC[f]	19	0.4	q3w	0	0	3	169
Melanoma	12	0.4	q3w	0	0	ND	170
Prostate cancer	16	0.4–0.45	q3w	0	0	3	171
LU103793							
Breast cancer	34	2.5	qd × 5 every 3 wk	0	0	7	72
Melanoma	80	2.5	qd × 5 every 3 wk	1	3	15	73
ILX651							
Melanoma	–	–	qd × 5 every 3 wk	–	–	–	In progress
Cryptophycin 52							
NSCLC	26	1.5	qw × 2 every 3 wk	0	0	10	108
HTI-286							
NSCLC	–	1.5	q3w	–	–	–	In progress

[a]N, number of patients.
[b]q3w, once every 3 wk; qd × 5 every 3 wk = every day for 5 d with 3 wk cycle; qw × 2 every 3 wk = days 1 and 8 every 3 wk.
[c]CR, complete response.
[d]PR, partial response.
[e]SD, stable disease.
[f]NSCLC, nonsmall cell lung cancer.

first studies, where 2.5–10 mg/m^2 LU103793 was given as a 5-min IV infusion repeated every 3 wk or weekly for 4 wk (68), cardiovascular toxicity (mainly severe hypertension) was found. This was predicted by the preclinical safety studies done in rats and dogs. Dose-limiting hypertension also occurred when 10–27.5 mg/m^2 LU103793 was given as a 24 h infusion every 3 wk (69). However, hypertension was substantially reduced if 2.5–17.5 mg/m^2 was given as a 5 d continuous infusion (70), 2.5–10 mg/m^2 was given as a 24 h infusion every week (68) or 0.5–3 mg/m^2 was given as a 5 min infusion daily on days 1–5 every 3 wk (71). In the latter schedules, neutropenia, peripheral edema, and liver function test abnormalities were dose-limiting (68,70,71).

Phase II clinical trials with LU103793 were conducted in advanced breast cancer and melanomas (72,73) where 2.5 mg/m^2 LU 103793 was given over 5 min for 5 consecutive days every 3 wk. Minimal antitumor activity was observed (Table 2).

TZT-1027 has been evaluated in Phase I trials on 3 schedules by a 1 h intravenous administration: days 1 and 8 every 3 wk (74), day 1 every 3 wk (75), and days 1, 8, and 15 every 3 wk (76). Dose limiting toxicity was observed at approx 2–3 mg/m^2 and was associated with neutropenia and other symptoms. A few partial responses and stable diseases have been observed in these trials. The recommended dose for further studies is

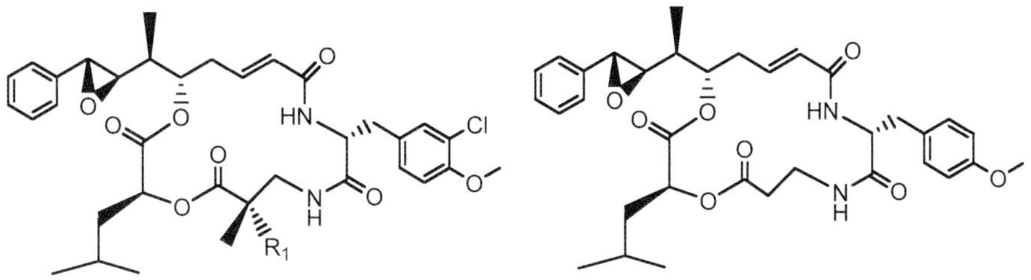

cryptophycin 1: R_1 = H   cryptophycin 24
cryptophycin 52: R_1 = CH_3 (Arenastatin A)

Fig. 2. Chemical structure of cryptophycin analogs.

2.4 mg/m^2 given days 1 and 8 every 3 wk, 2.7 mg/m^2 given once every 3 wk, and 1.8 mg/m^2 in a weekly × 3 regimen.

ILX651 (tasidotin) is a synthetic analog of dolastatin 15 that has broad-spectrum antitumor activity in tumor xenograft models *(39)*. Unlike LU103793, ILX651 does not have cardiotoxicity at myleosuppressive doses in animals. ILX651 is being evaluated in Phase I trials on 3 schedules by IV administration: days 1, 8, and 15 every 4 wk *(39)*, every other day for 3 d every 3 wk *(77)*, and daily administration for 5 d every 3 wk *(78)*. Dose limiting toxicity occurred at 62, 46, and 36 mg/m^2/d in each of these trials, respectively, and was associated with myelosuppression. Unlike LU103793, no major cardiovascular toxicity has been reported in all trials. Approximately, 25% stable disease has been reported for the last 2 trials as well as one partial response in the second trial, and one complete response in the third trial. A phase II trial in melanoma was initiated in October 2003.

3. CRYPTOPHYCINS

3.1. Origin

One of the first cryptophycins (cryptophycin A, also known as cryptophycin 1) was isolated from the terrestrial cyanobacteria *Nostoc* spp. (strain 53789) by investigators at Merck *(79)*. It was originally found to be active against fungi, including strains of *Cryptoccoccus* (Fig. 2). It was subsequently found in another strain of *Nostoc* spp. (strain GSV 224) and the relative and absolute stereochemistry of cryptophycin 1 and six other analogs (cryptophycin B–G) were reported *(80)*. Half-way around the world, Kobayashi and colleagues *(81,82)* reported almost simultaneously the discovery and absolute stereostructure of a potent cryptophycin analog designated arenastatin A, which was isolated from the Okinawan marine sponge *Dysidea arenaria* (Fig. 2). In a short period of time, more than 18 new cryptophycin analogs were reported *(83)* and a numbering system replaced the alphabetic designation of cryptophycin analogs. Subsequently, Lilly has evaluated more than 450 cryptophycin analogs *(84)*. The total synthesis of cryptophycins has been achieved *(85,86)*. The most intensively studied cryptophycins to date are cryptophycin 1, cryptophycin-24 (arenastatin A), and cryptophycin 52 (LY355703). Cryptophycin 52 was designed to reduce the susceptibility of cryptophycin 1 to hydrolysis *(83)*. For further details on analogs of cryptophycins and structure activity relationships, the reader is referred elsewhere *(83,87,88)*.

3.2. Interaction With Tubulin

Cryptophycin 1 *(89–92)* and cryptophycin 52 *(93–95)* interact directly with tubulin. These cryptophycins inhibit tubulin polymerization (cryptophycin-1:$IC_{50} \approx$ 1–5 μM *(89,90,96)*; cryptophycin 52:$IC_{50} \approx$ 1–3 μM. *(94,95)*. They block the binding of vinblastine to tubulin *(89–93)*. In contrast, cryptophycin stabilizes the binding of colchicine to tubulin *(90)*. The interaction of cryptophycin 1 with vinblastine binding is noncompetitive ($K_i = 3.9$ μM), although it is a competitive inhibitor of dolastatin 10 binding ($K_i = 2.1$ μM) to tubulin *(90)*. Cryptophycin 52 also binds to tubulin ($K_d = 3$ μM) *(93,95)* although the type of inhibition of cryptophycin 52 with vinblastine could not be determined *(93)*. Like other peptidyl inhibitors, cryptophycin-1 blocks the hydrolysis of GTP by isolated tubulin *(96)*. However, it did not block the binding of colchicine to tubulin *(96)*.

The interaction of cryptophycin 1 with tubulin was originally suggested to be irreversible as the inhibitory effect of >0.5 μM on tubulin polymerization was sustained after the compound was removed *(96)*. However, as 25 nM cryptophycin 1 inhibits microtubule dynamicity by 50% (without a change in the polymer mass) *(94)*, this substoichiometric amount of cryptophycin compared with tubulin suggests that it binds reversibly and with high affinity to the ends of microtubules ($K_d = 47$ nM) *(94,97)*. Panda reported that cryptophycin 52 binding to microtubules is also poorly reversible, but the binding is not covalent as the complex is denatured with urea or by boiling *(93)*.

Cryptophycin 1 causes aberrant assembly of tubulin that can be detected by gel filtration *(89,90)*, electron microscopy *(89,90,96)*, fluorescence correlation spectroscopy *(52)*, but not by turbidity *(90,96)*. In the electron microscope, cryptophycin 1 induces the formation of ring-like aggregates *(89,90,96)* that are distinct from the spiral-shaped aggregates or twisted ribbons induced by vinblastine *(98)* and colchicine *(99)*, respectively. The ring has a diameter of approx 24 nm and a mass of approx 0.80 Mda, indicating that the ring is made up of eight αβ dimers *(51)*. This is distinct from rings induced by dolastatin 10 or hemiasterlin that are made up of 13–14 heterodimers and are approx 44 nm in diameter *(52)*. The ring is formed by both a bend in the intradimer contact as well as the interdimer contact *(51)*. Consistent with the sustained effects described earlier, the rings are very stable compared with dolastatin 10 (intermediate) and hemiasterlin (low) *(52)*. Cryptophycin 52 also induces the self-association of tubulin in nine-dimer single rings that has been suggested to be formed from peeling off of protofilaments within the microtubule and from soluble tubulin *(95)*.

3.3. Effect on Cells

Cryptophycin 1 and arenastatin A are the most potent peptide-like antimitotic agent inhibitors of tumor cell growth known ($IC_{50} = 0.02$ nM) *(91)*. Cryptophycin 52 also potently inhibits growth in most cell lines ($IC_{50} = 11–37$ nM) *(84,94)*, with the exception of MDA-MB-231 ($IC_{50} = 232$ nM) *(84)*, THP-1 ($IC_{50} = 148$ nM) *(100)* and H-125 cells ($IC_{50} = 2.96$ μM) *(100)*. Treatment of cells with relatively high concentrations of cryptophycin 1 or 52 results in depletion of cellular microtubules *(91,94)*. However, when used at or near the IC_{50} needed to inhibit cell growth, no marked change in the mass of microtubules is found although G_2/M arrest and induction of apoptosis occur *(84,94,101,102)*. Unlike vinblastine, which induces reversible effect on microtubules in cells, the effects of cryptophycins persist for many hours *(91)*. The potency of cryptophycin 1 and 52 are minimally decreased compared with paclitaxel or vinca alkaloids in cells that had high level expression of P-glycoprotein or MRP-1 *(84,91)*.

3.4. Effect in Tumors

Approximately 10–30 mg/kg cryptophycin 1 administered intravenously on every day or every other day for 5–8 cycles inhibit tumor growth 75–100% in a variety of mouse and human tumor models *(80,83,103)*. No significant weight loss is noted. Five mg/kg cryptophycin 52 administered IV on alternate days for 5 or 6 cycles produce tumor growth delays (~5–15 d) in the human MX-1 breast carcinoma, SW-2 SCLC, H82 SCLC, and Calu-6 nonsmall cell lung cancer (NSCLC) xenograft mouse models *(104)*. Cryptophycin 52 (0.3 mg/kg IV given on days 7, 9, and 11) is also efficacious in rats bearing the 13762 mammary carcinoma *(104)*. Additive and supra-additive effects are noted in human tumor xenograft models when cryptophycin 52 is given with a variety of cytotoxic anticancer agents or radiation *(104,105)*. It is obvious that the extraordinary potency of cryptophycin 1 in tissue culture systems does not translate into high potency in vivo. This suggests that the molecule may be unstable, highly bound to plasma protein, or highly metabolized.

3.5. Clinical Trials

Phase I clinical trials with cryptophycin 52 have been reported *(106,107)*. The drug was formulated in a cremophor-containing vehicle (presumably because of limited solubility of the agent) and therefore, patients were treated prophylactically to avoid hypersensitizing reactions (HSRs) that have been observed with other anticancer agents formulated in cremophor (i.e., taxanes *[2]*). The agent was given as a 2 h infusion on either day 1 *(106)*, days 1 and 8 *(107)*, or days 1, 8, and 15 *(106)* on a 3–4 wk cycle. In the single dose schedule, dose limiting toxicity occurred at 1.7 mg/m^2 cryptophycin 52 or higher and was associated with grade 3 peripheral neuropathy and myalgia *(106)*. Grade 3 cardiac arrhythmia, due to bradycardia (14% of all patients) and grade 2 hypertension associated with HSRs, was reported. Dose-limiting neurological toxicity was also observed in patients given a weekly schedule of 1.1–1.84 mg/m^2 cryptophycin 52 *(106,107)*. Some evidence of antitumor activity was noted in these preliminary trials. In one trial, one patient (4%) had a partial response and 5 had stable disease (20%) *(107)*. In the trial run by Sessa, only one objective response was observed and 31% had stable disease *(106)*. One Phase II trial has been reported with cryptophycin 52 given as 1.5 mg/m^2 on days 1 and 8 every 3 wk in patients with refractory advanced NSCLC *(108)* (Table 2). Of the 26 patients, no objective response was seen, although 40% had evidence of disease stabilization. Neurological toxicity in the form of peripheral neuropathy and constipation required that the dose be reduced to 1.125 mg/m^2 in many patients.

4. HEMIASTERLINS

4.1. Origin

Hemiastelin is a tripeptide made up of trimethyltryptophan, tert-leucine, and *N*-methyl homo vinylogous valine. Hemiasterlins and the related milnamides have been isolated from several marine sponges (Fig. 3). Hemiasterlins have been independently isolated from *Hemiasterella minor* found in South Africa *(109)* as well as *Cymbastela* spp. *(110)*, *Auletta* spp. and *Siphonochalina* spp. all collected in Papua New Guinea *(111)*. The original material, hemiasterlin and hemiasterlin A, B, *(110)* and C *(111)* have been identified. Milnamide A and D, isolated from Milne Bay, Papua New Guinea, have been isolated from *Auletta* cf. *constricta* and *Cymbastela* spp. *(112,113)*. Criamides A

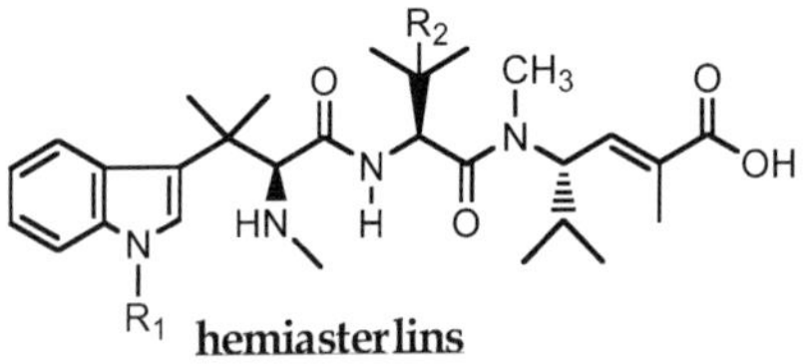

hemiasterlins

hemiasterlin: $R_1 = R_2 = CH_3$

A : R_1=H, R_2=CH_3

B: R_1=R_2=H

C: R_1=CH_3, R_2= H

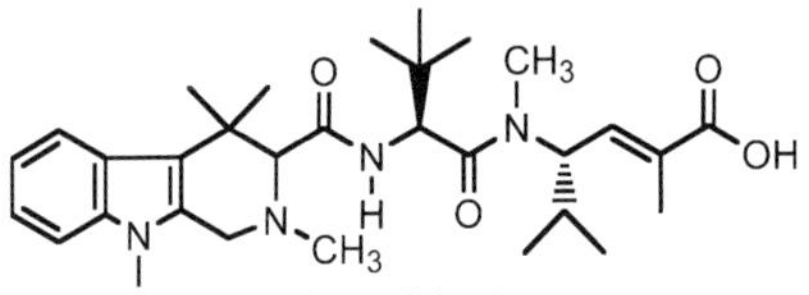

milnamide A

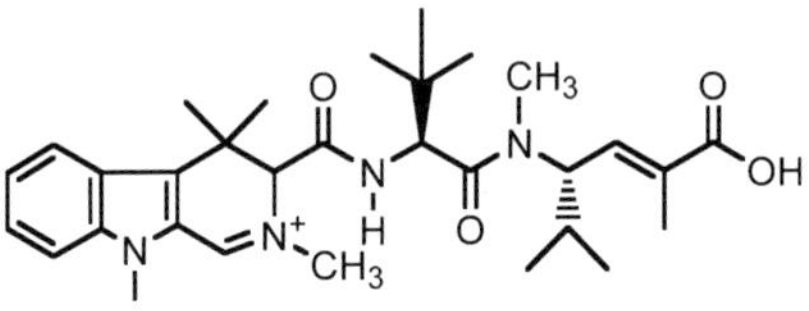

milnamide D

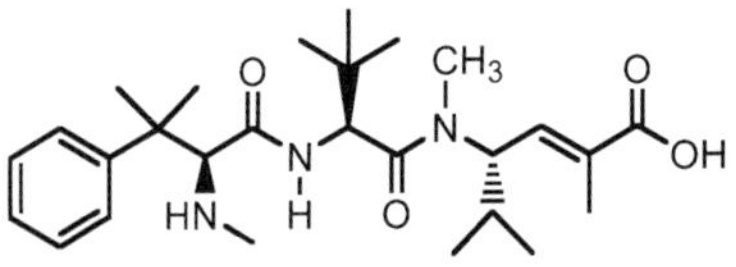

HTI-286

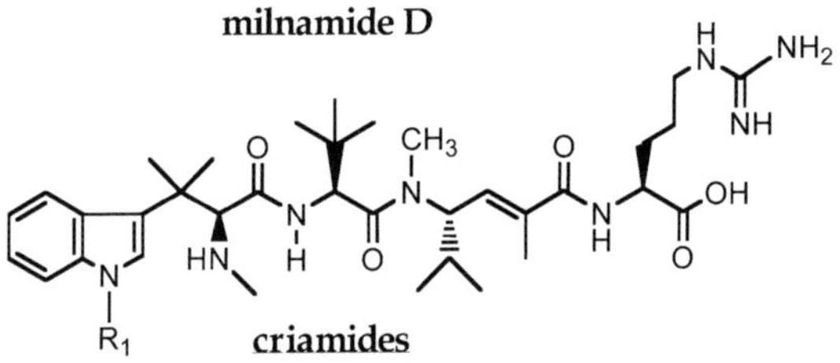

criamides

A : R_1 = H

B: R_1= CH_3

Fig. 3. Chemical structure of hemiasterlin analogs.

and B, which are tetrapeptides closely related to milnamides and hemiasterlins *(110)* as well as the four residue cyclic depsipeptides geodiamolides (*109,110,113–115*) and jaspamide (jasplakinolide) *(113)* have also been isolated from some of the same sponges. Jaspamide (*112,116–118*), geodiamolides A–F and TA *(114)*, criamide B *(110)*, milnamide D *(113)*, and hemiasterlins *(110,111)* are potent inhibitors of tumor cell growth in tissue culture. Jaspamide and the closely related chondramides interact with actin and will not be discussed further *(118,119)*.

Both the hemiasterlins and milnamides inhibit microtubule assembly from purified tubulin, but the hemiasterlins are 1–2-orders of magnitude more potent compared with the milnamides in cellular cytotoxicity assays *(113)*. These facts, combined with the initial in vivo activity of hemiasterlins and criamides *(110)* prompted the need for more substantial amounts of these natural products. Andersen and colleagues *(120)* solved the problem and were the first to develop a total synthesis of hemiasterlin. Other routes for the partial *(121)* or complete *(122)* synthesis of hemiasterlin have also been reported.

Andersen's *(123)* and Zask's laboratories *(124)* have synthesized numerous analogs of hemiasterlin. It has significantly improved the understanding of hemiasterlin SAR and is reviewed elsewhere *(8)*. One of the hemiasterlin analogs synthesized by Andersen was designated SPA-110. This analog was provided to Wyeth for additional testing. It was found to have an excellent profile in both cellular and animal assays *(15)*. This compound is being developed at Wyeth as an anticancer agent and is designated HTI-286 (hemiasterlin tubulin inhibitor also known as taltobulin).

4.2. Interaction With Tubulin

Consistent with the initial report that hemiasterlin depolymerizes microtubules in cells (125), Hamel's laboratory found that hemiasterlin inhibits tubulin assembly (IC_{50} = 0.98 μM), is a noncompetitive and competitive inhibitor of vinblastine and dolastatin 10 binding to tubulin, respectively (K_i = 7 μM and 2 μM), stabilizes colchicine binding to tubulin, and interferes with GTP exchange on tubulin (46). Based on these and other data, hemiasterlin, like other peptidyl antimicrotubule drugs, is suggested to bind at the *Vinca*-peptide domain on β-tubulin (46).

Hemiasterlin, when present at or above stoichiometric concentrations compared with tubulin, induces tubulin oligomers with ring-like structures (46,52). Unlike dolastatin-10, but similar to cryptophicin-1, the formation of the oligomer is not correlated with an increase in turbidity development. Further analysis indicates that the single-walled ring induced by hemiasterlin is similar to that induced by dolastatin 10 (44.6 nm in diameter) and larger than that induced by cryptophycin-1 (24 nm in diameter) (52). Beyond this, the rings induced by hemiasterlin have the least stability compared with those formed with cryptophycin-1 or dolastatin-10.

The hemiasterlin analog, HTI-286 has similar properties compared with hemiasterlin. It is a potent inhibitor of tubulin assembly (IC_{50} ≈ 1 μM) (15) and like hemiasterlin, induces depolymerization of preassembled microtubules (126). Using fluorescence methods, HTI-286 has high affinity to tubulin (K_d ~ 100 nM) that is similar to hemiasterlin (126). HTI-286 induces the aggregation of tubulin corresponding to a ring structure consisting of 13 tubulin dimers (126), which is very close to that observed for dolastatin 10 (52). However, the oligomers induced by HTI-286 are less stable than those induced with hemiasterlin when analyzed by size-exclusion chromatography (126).

Recently, two photoaffinity probe analogs of HTI-286 have been synthesized and their interaction with tubulin has been studied (127,128). Both photoaffinity probes label tubulin. Photolabeling is inhibited by dolastatin-10 and vinblastine but is either unaffected or enhanced by paclitaxel or colchicine. These data are consistent with the idea that the vinblastine /peptide binding site is distinct from the colchicine and taxane binding domains in tubulin. However, unlike the model that places the peptide binding site in β-tubulin and overlaps the vinblastine binding site (45), both photoaffinity probes crosslink predominately, if not exclusively, within α-tubulin. The labeling domain for one probe has been localized to residues 314–339 and corresponds to loop 8/helix 10 of α-tubulin (129). Based on electron micrographic crystal structure of zinc induced tubulin sheets (130), this region is known to have longitudinal interactions with β-tubulin across the interdimer interface as well as lateral interactions with adjacent protofilaments. In particular, helix H10 of α-tubulin interacts across the interdimer interface with H6 and H6-7 of β-tubulin. As helix H6 of β-tubulin contains a vinblastine photoaffinity labeling domain localized to residues 175–213 of β-tubulin (131), a new model of the hemiasterlin and vinblastine binding sites has been proposed (Fig. 4). The close proximity of the vinblastine and hemiasterlin labeling sites would help explain 1) the noncompetitive interaction of these two molecules with tubulin, 2) why vinblastine and peptidyl antimitotic drugs influence GTP exchange (which is in contact with the N-terminus of helix H6 in β-tubulin), and 3) how peptides of similar but distinct structure alter the degree of interdimer bend (51) and thereby influence the size of rings induced by peptides (52). Further identification of the labeling domain within α-tubulin of the second probe should help refine the model. So far, the crosslinking domain has been

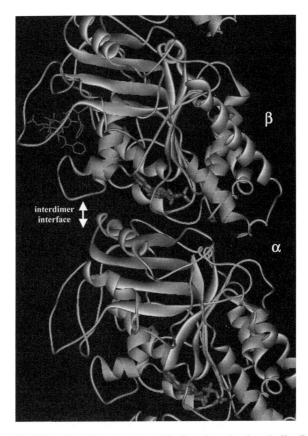

Fig. 4. Model of the HTI-286 and vinblastine photolabeling domains in tubulin. Peptide mapping was used to localize the region in contact with HTI-286 photoprobe 1 (green) and a vinblastine photoaffinity probe (yellow) sites. The binding sites for taxol (located in β-tubulin) exchangeable GDP (in β-tubulin) and nonexchangeable GTP (in α-tubulin) using the ball and stick model are shown. Structural model for tubulin was taken from PDB accession code 1JFF. From ref. *129.* To view this figure in color, see the insert and the companion CD-ROM.

mapped to residues 204–280 of α-tubulin and is therefore distinct from, but possibly close to, the contact site by the first photoaffinity probe *(128).*

Residues 314–339 of α- tubulin are located in almost exactly the same location as residues 321–339 *(132)* or 303–335 *(133),* which crosslink with the N-terminus of stathmin. This is intriguing as stathmin/OP18 *(134)* is a highly conserved 149 amino acid cytosolic phosphoprotein that directly interacts with tubulin, induces depolymerization of microtubules in vitro and in intact cells, and can be crosslinked primarily to α-tubulin *(135).* Although most of the molecule binds along the entire longitudinal length of the 2-β dimer based on cocrystallization experiments *(136),* the N-terminus of the molecule (which was not visualized in the crystal) is believed to bind to the cap located on the plus-end of the microtubule and stimulate microtubule plus-end catastrophe *(137).* These findings raise the possibility that hemiasterlin, and perhaps the peptidyl antimitotic agents in general, bind the stathmin binding site in tubulin and thereby induce microtubule shortening and depolymerization.

4.3. Effect on Cells

Roberge's laboratory was the first to demonstrate that hemiasterlin and hemiasterlin A induce mitotic arrest in cultured cell lines, abnormal spindle formation, and microtubule depolymerization *(125)*. Hemiasterlin causes abnormal spindle formation (at low concentrations) and microtubule depolymerization (at high concentrations) as detected by immunofluoresence microscopy. Consistent with tubulin being the primary target of hemiasterlin, tubulin is the only specifically labeled protein detected when whole cells or a crude cytosolic extract is incubated with an HTI-286 photoaffinity probe *(138)*.

Hemiasterlins potently inhibit the proliferation of cultured tumor cells. In general, the reported IC_{50} of hemiasterlin, hemiasterlin A, B, and C for growth inhibition of various tumor cell lines ranges from approx 0.3–20 nM *(110,111,125)*. The synthetic hemiasterlin analog, HTI-286, inhibits the growth of human tumor cell lines with IC_{50}'s from 0.2 to 7 nM (mean = 2.5 nM ± 2.1 in 18 cell lines) *(15)*. Among natural and synthetic hemiasterlin analogs, there is a direct correlation of cytotoxicity with antimitotic activity *(123)*, supporting the observation that the primary cellular effects of hemiasterlins are due to mitotic arrest. Cell-cycle analyses confirms that HTI-286 also causes mitotic arrest after 24 h exposure to low nanomolar concentrations of drug and causes apoptosis after prolonged exposure *(15)*.

HTI-286 was selected for further development as it was consistently efficacious in cell lines that were resistant to microtubule-stabilizing agents or vinca alkaloids. Resistance to taxanes was associated with overexpression of the drug transporter P-glycoprotein (MDR1/ABCB1) *(15)* or mutations in tubulin (Loganzo, personal observation). HTI-286 also effectively inhibits the growth of cells that overexpress the drug transporters MRP1 (ABCC1) and MXR (ABCG2) *(15)*, which are also implicated in resistance to various chemotherapeutic agents *(14)*.

To better understand the frequency and potential mechanisms of resistance to HTI-286, tumor cell lines have been selected for resistance to HTI-286 by chronic exposure to the drug. KB-3-1 epidermoid carcinoma cells and 1A9 ovarian carcinoma cells develop 4–23-fold resistance to HTI-286 *(9,139)* as well as resistance to most of the antimitotic agents that bind to the vinca or vinca-peptide domain. No significant resistance is observed to colchicine (although the 1A9-selected cells are resistant to podophyllotoxin). Although no resistance to polymerizing agents is observed in the KB-HTI-resistant cells, 1A9-HTI-resistant cells have approx 5–10-fold sensitivity to these agents. The mechanistic basis of resistance in the two cell lines is complex. In a series of independently selected 1A9-selected cells, the microtubules are more stable and associated with point mutations in β-tubulin (S172A) or α-tubulin (I384V or S165P and R221H). In contrast, in the KB-HTI-resistant cells, a point mutation was found in α-tubulin (A12S). The position of the point mutations in tubulin may have a common theme. Serine β172 and alanine α12 contact the E-site and N-site GTP binding site in α- and β-tubulin, respectively. It is speculated that alterations in serine α 165 and arginine α 221 may indirectly influence GTP hydrolysis and interdimer interaction, respectively.

Other phenotypic changes are detected in KB-HTI-resistant cells. Before the selection step where a mutation in tubulin is found, ATP-dependent low cellular accumulation of HTI-286 is noted. This effect, which is lost in the revertant cell line, suggests that an ABC-transporter is overexpressed in these cells *(14)*. However, previous transporters associated with resistance to chemotherapeutic drugs (i.e., MDR1, MRP1, MRP3, or MXR) *(14,140)* have been ruled out.

4.4. Effect in Tumors

Hemiasterlin was originally reported to improve the survival of mice implanted with murine P388 tumor cells *(110)*. Additional animal studies were conducted at Wyeth when synthetic HTI-286 became available. HTI-286, given intravenously, inhibits the growth of subcutaneously implanted tumors derived from human melanoma, colon, breast, and prostate carcinoma (Table 3) *(15)*. The minimally effective dose of HTI-286 needed to inhibit growth of tumors that lack P-glycoprotein expression is approx 0.2–0.5 mg/kg when administered intravenously on days 1, 5, and 9 *(15)*. HTI-286 is also effective when administered on multiple frequent (daily or every other day) or intermittent (weekly, every other week) cycles. Consistent with its ability to kill tumor cells that overexpress P-glycoprotein in tissue culture, HTI-286 also inhibits the growth of tumors in animals that overexpress P-glycoprotein and are insensitive to paclitaxel and vincristine (Table 3) *(15)*. As P-glycoprotein is expressed at high levels at the luminal face of cells that line the gut wall, paclitaxel is not orally bioavailable, and exposure to oral paclitaxel can be enhanced by coadministration with a P-glycoprotein inhibitor or in MDR knockout animals *(140)*. In contrast, it is not surprising that oral administration of HTI-286 to athymic mice inhibits tumor growth *(15)*. As HTI-286 is highly soluble in saline, the drug can be administered to animals and patients without solubilizing agents such as Cremophor or polysorbate-80, which are commonly used for taxanes as well as cryptophycin 52. This appears to be a benefit as no HSRs have been observed in patients given HTI-286.

4.5. Clinical Trials

HTI-286 is the first hemiasterlin analog to be evaluated in humans as an anticancer agent. A phase I trial in patients with advanced malignant solid tumors was initiated in early 2002 to identify the dose-limiting toxicities and maximally tolerated dose of HTI-286 *(141)*. Initially, 35 patients were enrolled and evaluated. The drug was administered intravenously in saline for 30 min once in every 3 wk at doses ranging from 0.06 to 2 mg/m^2. Toxicities were neutropenia, nausea/emesis, alopecia, and pain. Dose-limiting toxicities were limb and chest pain, hypertension, and neutropenia. The dose selected for the phase II efficacy study was 1.5 mg/m^2. Patients are currently being evaluated in a phase II single-agent trial of HTI-286 (Table 2). Additional clinical trials with HTI-286, both as monotherapy as well as in combination with other standard-of-care agents, are in progress or planned.

5. PHOMOPSIN AND USTILOXINS

5.1. Origin

The cyclic hexapeptide, phomopsin A was isolated from the lupin seed or maize infected with the fungus *Phomopsin leptostromiformis* (Fig. 5) *(142)*. In animals it induced a fatal liver disorder *(143)* similar to those animals that ingest the infected herb and developed lupin poisoning (lupinosis). The purified material was then shown to induce mitotic arrest in cells *(144)*. The structure of phomopsin A as well as another

Table 3
Effects of HTI-286, Paclitaxel, and Vincristine on the Growth of Human Tumor Xenografts
in Athymic Mice

Tumor type	N	T/C (%)[a] Day 7	Day 14	Day 21
Tumors with little or no detectable levels of P-glycoprotein expression				
LOX melanoma				
HTI-286	5	17 ± 1	3 ± 1	–
Paclitaxel	2	12 ± 4	3 ± 2	–
Vincristine	4	17 ± 6	6 ± 2	–
KB-3-1 epidermoid cancer				
HTI-286*	2	6 ± 5	3 ± 1	20 ± 3
Paclitaxel	1	12	5	5
Vincristine	1	11	9	9
LOVO colon cancer				
HTI-286	1	88	35	20
Paclitaxel	1	17	6	8
Vincristine	3	34 ± 8	37 ± 15	45 ± 15
MCF-7 breast cancer				
HTI-286	1	15	6	27
Paclitaxel	1	41	3	0
Vincristine	2	33	16	14
Tumors with low to high levels of p-glycoprotein expression				
KB-8-5 epidermoid cancer[b]				
HTI-286	3	21 ± 4	9 ± 4	47[c]
Paclitaxel	2	72 ± 9	77 ± 10	109[c]
Vincristine	2	88 ± 16	84 ± 1	81[c]
SW-620-W colon cancer[b]				
HTI-286	5	19 ± 4	9 ± 4	1[c]
Paclitaxel	3	60 ± 10	73 ± 5	55[c]
Vincristine	3	81 ± 6	63 ± 3	59 ± 3
DLD-1 colon cancer[b]				
HTI-286	4	37 ± 8	37 ± 8	54 ± 12
Paclitaxel	2	71 ± 1.5	73 ± 21	101 ± 4
Vincristine	2	123 ± 10	125 ± 6	153 ± 24
HCT-15 colon cancer[b]				
HTI-286	4	39 ± 12	35 ± 9	38 ± 7
Paclitaxel	3	103 ± 8	93 ± 11	109 ± 7
Vincristine	3	61 ± 26	55 ± 14	59[c]

[a]Tumor cells were implanted into the flanks of nude mice. T/C, percentange of tumor size in treatment group vs control on days 7, 14, and 21 after drug dosing. Values shown are mean ± standard errors (where available). N = number of independent experiments. HTI-286 was given at 1.5 mg/kg IV or 2 mg/kg IV (indicated by *), except LOX given at 1.0 mg/kg. Paclitaxel was given at 60 mg/kg IV. Vincristine was given at 0.8–1 mg/kg IP. All drugs were given on day 1, 5, and 9 to tumors that had an established size of approx 100 mg.

[b]The level of P-glycoprotein expression in SW-620-W (previously reported as MX1W), KB-8-5, DLD-1, and HCT-15 were approximately +, ++, +++, and ++++, respectively; further details can be found in Sampath et al. *(172)*.

[c]Control tumor grew too large at this time-point in one experiment in this group and no standard error could be computed.

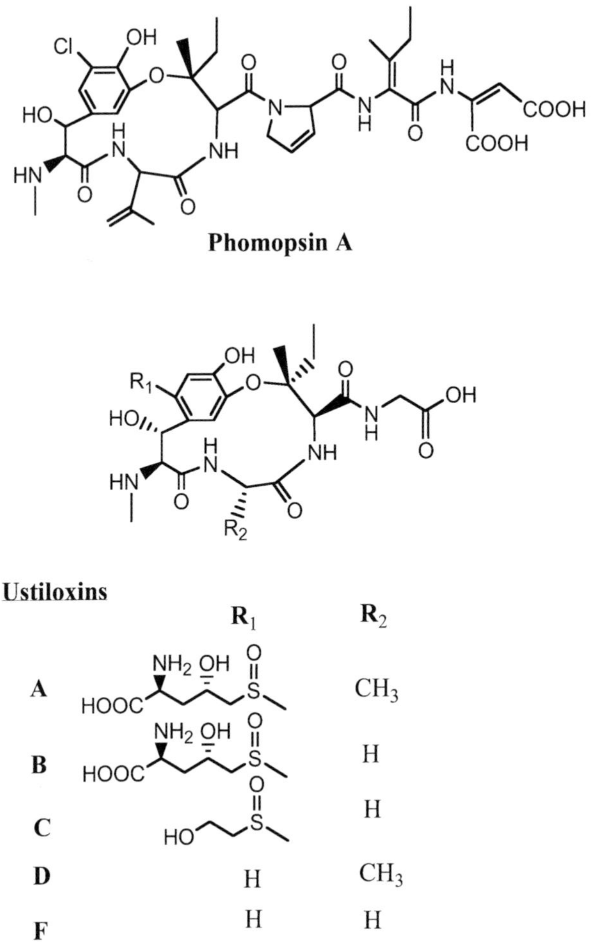

Phomopsin A

Ustiloxins

	R₁	**R₂**
A		CH₃
B		H
C		H
D	H	CH₃
F	H	H

Fig. 5. Chemical structure of phomopsin A and ustiloxin A–D and F.

isolated variant without a halogen at C-14, phomopsin B was solved by Culvenor and colleagues *(145)*. The ustiloxins A–D and F were isolated from rice infected with the fungus *Ustilaginoidea virens (146,147)* (Fig. 5). These agents also cause a disease similar to lupinosis when given to mice *(148)*. Ustiloxins share a highly related core group with phomopsins. Total synthesis of ustiloxin D has been recently achieved *(149,150)*.

5.2. Effect in Biological Assays

Phomopsin A and ustiloxins A and B inhibit polymerization of microtubules and depolymerized existing microtubules derived from purified tubulin *(144,151)*. These particular analogs are potent and effects are achieved at substoichiometric concentrations that are 5–10-fold less than the concentration of tubulin in the reaction. Phomopsin A appears to bind to a distinct site within tubulin that either overlaps with the vinca alkaloids binding site or a distant site that has an allosteric effect on the vinca alkaloid binding site. Consistent with this:

1. Phomopsin A inhibits the binding of vinca alkaloids *(41,152)* in a noncompetitive ($K_i = 2.8$ µ*M*, respectively) *(41)*,

2. Like vinca alkaloids and dolastatin10, phomopsin A stabilizes the ability of tubulin to bind colchicine *(45,152)*, and

3. Like dolastatin 10 (and less so, vinblastine), phomopsin A inhibits tubulin-dependent GTP hydrolysis and nucleotide exchange *(45)*.

Ustiloxins are likely to behave in a similar fashion compared with phomopsin, but have not been well-studied *(151)*. Despite the potency in a cell-free system, the phomopsins and ustiloxins are relatively nonpotent cytotoxic agents (IC_{50}'s typically in the 1–10 μM range) usually requiring 1000-fold higher concentrations (or more) to kill cells compared with other dolastatin, hemiasterlin, paclitaxel, or vinblastine *(41,146,151)*. The basis for the effect is unknown, but may be because of the inability of the drug to penetrate the cell membrane or attributed to metabolism of the agent. Paradoxically, the minimum lethal dose for sheep is extremely low, approx 10 µg/kg, when given subcutaneously, and is inconsistent with the tumor cell data *(143)*.

6. DIAZONAMIDES

6.1. Origin

Diazonamide A and B were originally isolated from the ascidian *Diazona angulata* (initially incorrectly identified as *Diazona chinensis*) collected from the ceilings of small caves located along the northwest coast of Siquijor Island, Philippines *(153)*. The structure of these two agents was initially identified by Lindquist *(153)*, based on chemical modification of the isolated natural product, and subsequently revised *(154,155)* (Fig. 6). Complete chemical synthesis of the diazonamide A has been achieved by two laboratories *(155–157)*, although many laboratories have achieved complete synthesis of the incorrectly identified structure (for review, *see* Ritter and Carreira *[158]*).

6.2. Interaction With Tubulin

Diazonamide A is a potent inhibitor of MAP-induced or glutamate-induced polymerization of tubulin (IC_{50} = 1–2 μM) *(49)*. These values are approx twofold higher than that obtained with dolastatin 10. Diazonamide A also inhibits GTP hydrolysis. However, unlike dolastatin 10, dolastatin 15, hemiasterlins, or cryptophycins, diazonamide A does not inhibit vinblastine or GTP binding to tubulin. It also does inhibit the binding of dolastatin 10 to tubulin or stabilize colchicine binding to tubulin. These data suggest that this agent binds to a unique site in tubulin.

6.3. Effect on Cells

Diazonamide A is a potent inhibitor of cell tumor cell growth in vitro. The mean IC_{50} with 60 cell lines is approx 5 nM *(49)*. A primary target of diazonamide is likely to be the microtubule as it arrests cells in the G_2/M phase of the cell cycle and induces loss of microtubules in cells when given at or above the IC_{50} needed to inhibit cell growth *(49)*. Diazonamide B is less cytotoxic than diazonamide A *(153)*. The activity of diazonamide in tumor models has not been reported, but chemical synthesis insures that supplies of it and analogs will be available to perform such experimentation in the future.

6.4. Moroidin and Celogentins

Moroidin was originally isolated from the leaves and stalks of the bush, *Laportea moroides*, which is found in the rain forests of Eastern Australia (Fig. 7) *(159)*. The unique

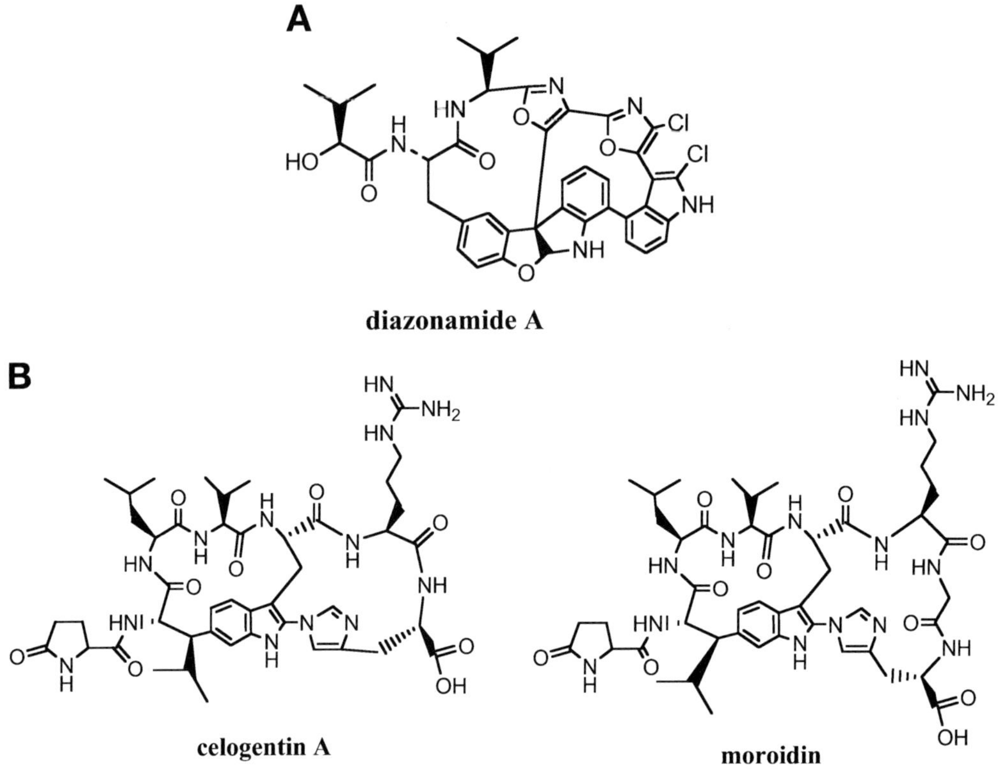

A

diazonamide A

B

celogentin A   **moroidin**

Fig. 6. (**A**) Chemical structure of diazonamide. (**B**) Chemical structure of celogentin A and moroidin.

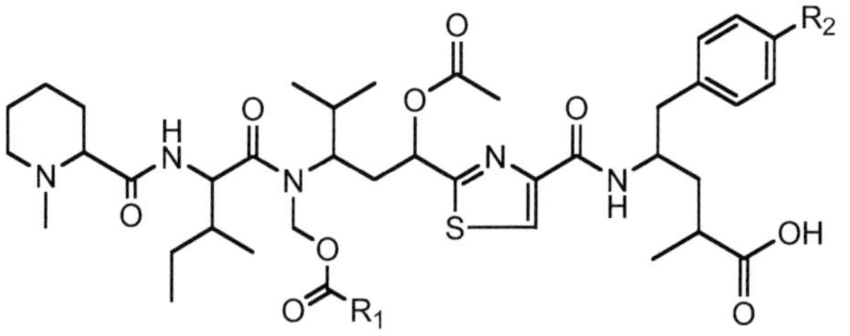

	R₁	R₂
Tubulysin A	$CH_2-CH(CH_3)_2$	**OH**
B	$CH_2-CH_2-CH_3$	**OH**
D	$CH_2-CH(CH_3)_2$	**H**
E	$CH_2-CH_2-CH_3$	**H**

Fig. 7. Chemical structure of tubulysins.

bicyclic peptide made up of eight amino acids was initially suggested to be one of the active ingredients that induced intense pain upon being stung by hairs that grow on the bush. Indeed, intradermal injection into human forearms of moroidin (it can be only imagined who might have been the volunteer) produces pain and redness *(159)*. Moroidin has also been isolated from the seeds of *Celosia argentea* by Morita and coworkers who were the first to show that the compound inhibits tubulin polymerization *(160)*. Subsequently,

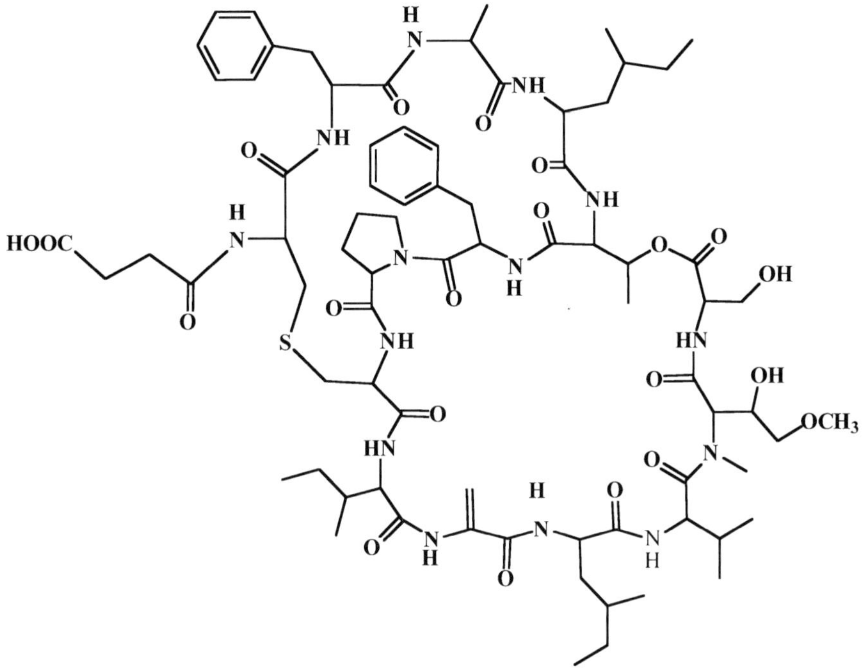

Fig. 8. Chemical structure of vitilevuamide.

three related analogs, designated celogentins A, B, and C were isolated from the seeds of *C. argentea (161)*. Moroidin, celogentins A, B, and C inhibit microtubule assembly (the IC_{50}s are 3, 0.8, 20, 30 μM, respectively).

7. TUBULYSINS

Tubulysins are a family of molecules isolated from the culture broths of a variety of myxobacteria of the genera *Archangium, Angiococcus, Cystobacter,* and *Stigmatella. (162)* (Fig. 8). They share some structural similarity with dolastatins and hemiasterlins. When examined in 5 cell lines, tubulysins A, B, D, and E are potent inhibitors of growth (IC_{50} ~ 0.02–0.4 n*M*). Tubulysin D is the most potent analog (IC_{50} = 0.043 nM across 5 cell lines). When cells are incubated with tubulysin D (at 10-fold the IC_{50} needed to inhibit the growth of cells) multipolar spindle apparatuses are observed. At very high concentrations of tubulysin A (50 ng/mL), microtubule staining becomes diffuse, which is suggestive of a depolymerizing effect of the agent.

8. VITILEVUAMIDE

The bicyclic 13 amino acid peptide, vitilevuamide was isolated from two marine ascidians, *Didemnum cuculiferum* and *Polysyncranton lithostrotum (50)* (Fig. 8). It inhibits the polymerization of purified tubulin (IC_{50} = 2 μM) and at high concentrations, causes aggregation of tubulin. Like other peptidyl agents, vitilevuamide inhibits vinblastine binding (in a noncompetitive manner), has no effect on colchicine binding to tubulin, and prevents the decay of the colchicine–tubulin complex. The agent does inhibit GTP binding to tubulin when tested at concentrations up to 80 μM. This suggests that vitilevuamide binds to the vinca-peptide site within tubulin but has unique properties compared with all petidyl antimicrotubule agents. Vitilevuamide inhibits the growth of a variety of tumor cells lines (IC_{50} range = 6–311 nM) associated with G_2/M cell-cycle arrest. At doses of 12–30 mg/kg given on days 1,5, and 9 by intraperitoneal administration, vitilevuamide significantly increases the life-span of mice implanted with P388 leukemia cells.

9. SUMMARY

A wide variety of potent peptide and depsipeptide natural products that interfere with the function of microtubules have been discovered in the last 20 yr. With the efforts of some very talented chemists, a few of these natural products have been synthesized in quantities sufficient for clinical evaluation and extensive analog work has provided a better understanding of their interaction with tubulin. On a molecular level, the peptidyl agents described here bind to a unique site in tubulin that specifically influences vinblastine binding. However, because most peptidyl agents noncompetitively inhibit vinblastine binding to tubulin, the peptidyl site is either completely separate or overlaps with a part of the vinca site. More work is needed to define the peptide binding site and understand how these agents interact with tubulin on a molecular level. This will lead to a better understanding of how tubulin and microtubules function in general.

On a phenotypic level, many of these agents are potent inhibitors of cell growth that demonstrate a profound ability to inhibit tumor proliferation in animal models. The excellent preclinical profile should translate into clinical utility in cancer patients. This has not happened yet. How can this be? First of all, it must be recognized that the mechanism of action of each peptidyl inhibitor can be subtly distinct (if they indeed only interact with tubulin) and the metabolism of each drug may be very different. Therefore, until each agent is tested in multiple tumor types, schedules, and doses, only then may a clinically useful peptide emerge. In the event that clinical utility is not achieved, it would seem that the balance of efficacy vs toxicity must be tipped unfavorably. As P-glycoprotein is expressed in bone marrow progenitor cells *(163)*, and clinically-tested peptidyl agents have a weaker interaction with P-glycoprotein compared with antimitotic agents in clinical use, peptidyl agents may induce more profound neutropenia than approved antimitotic agents. In any case, there may be ways to overcome this in the future. First, it may be possible to selectively enhance the anticancer effect of these drugs by altering apoptotic machinery (e.g., ref. *5*) or angiogenic systems *(65)* unique to the cancer cell environment. Second, subtoxic doses given metronomically may be more effective *(164)*. Third, efforts must be made to understand why a few patients respond extremely well to these agents. The answer may lie in the tumor. Alternatively, resistance to dose-limiting side effects such as neutropenia or cardiovascular toxicity (observed with many peptidyl agents) could play an important role. Finally, conjugation of these highly toxic molecules to antibodies may selectively target tumor cells *(61,165)*. There is a great deal of work yet to be done with these intriguing molecules. The diversity of these natural products, their unique interaction with tubulin, and their possible utility in treating cancer indicates that continued study of these molecules will be valuable in the future.

ACKNOWLEDGMENTS

The authors thank Dr. Arie Zask for his review of chemical structures and Ms. Randy Simon for her editorial comments. The authors acknowledge all those who were not cited in this review in an effort to condense the work, but have contributed to the field.

REFERENCES

1. Jordan MA. Mechanism of action of antitumor drugs that interact with microtubules and tubulin. Curr Med Chem Anti-Cancer Agents 2002;2:1–17.
2. Rowinsky EK, Tolcher AW. Antimicrotubule agents, In: Devita VT Jr, Hellman S, Rosenberg SA, eds. Cancer Principles and Practice. Lippincott, Philadelphia: Williams and Wilkins; 2001; 431–452.
3. Kavallaris M, Verrills NM, Hill BT. Anticancer therapy with novel tubulin-interacting drugs. Drug Resist Updat 2001;4:392–401.
4. van Loo G, Saelens X, van Gurp M, MacFarlane M, Martin SJ, Vandenabeele P. The role of mitochondrial factors in apoptosis: a Russian roulette with more than one bullet. Cell Death Differ 2002;9:1031–1042.
5. Altieri DC. Survivin, versatile modulation of cell division and apoptosis in cancer. Oncogene 2003;22:8581–8589.
6. He L, Orr GA, Horwitz SB. Novel molecules that interact with microtubules and have functional activity similar to Taxol. Drug Discov Today 2001;6:1153–1164.
7. Hamel E, Covell DG. Antimitotic peptides and depsipeptides. Curr Med Chem Anti-Cancer Agents 2002;2:19–53.
8. Andersen RJ, Roberge M. HTI-286. A synthetic analog of the antimitotic natural product hemiasterlin, In: Cragg GM, Kingston DGI, Newman DJ, eds. Natural Products. Philadelphia: CRC Press; 267–280.
9. Loganzo F, Hari M, Annable T, et al. Cells resistant to HTI-286 do not overexpress P-glycoprotein but have reduced drug accumulation and a point mutation in α-tubulin. Mol Cancer Ther 2004;3:1319–1327.
10. Giannakakou P, Sackett DL, Kang YK, et al. Paclitaxel-resistant human ovarian cancer cells have mutant beta-tubulins that exhibit impaired paclitaxel-driven polymerization. J Biol Chem 1997;272:17,118–17,125.
11. Dumontet C. Mechanisms of action and resistance to tubulin-binding agents. Expert Opin Invest Drugs 2000;9:779–788.
12. Orr GA, Verdier-Pinard P, McDaid H, Horwitz SB. Mechanisms of taxol resistance related to microtubules. Oncogene 2003;22:7280–7295.
13. Bradshaw DM, Arceci RJ. Clinical relevance of transmembrane drug efflux as a mechanism of multidrug resistance. J Clin Oncol 1998;16:3674–3690.
14. Gottesman MM, Fojo T, Bates SE. Multidrug resistance in cancer: role of ATP-dependent transporters. Nat Rev Cancer 2002;2:48–58.
15. Loganzo F, Discafani CM, Annable T, et al. HTI-286, a synthetic analogue of the tripeptide hemiasterlin, is a potent antimicrotubule agent that circumvents P-glycoprotein-mediated resistance in vitro and in vivo. Cancer Res 2003;63:1838–1845.
16. Schabel FM. Animal models as predictive systems in cancer chemotherapy, In: Nineteenth Annual Clinical Conference on Cancer, 1974, M D Anderson Hospital and Tumor Institute, ed. Year Book Medical Publishers: Chicago. 1975;325–355.
17. Ovejera A. The use of human tumor xenografts in large-scale drug screening, In: Kalimann RF, ed. Rodent Tumor Model in Experimental Cancer Therapy. NY:Pergamon Press;1987;218–220.
18. Rose WC. Taxol: a review of its preclinical in vivo antitumor activity. Anticancer Drugs 1992;3: 311–321.
19. Dykes DJ, Bissery MC, Harrison SD Jr, Waud WR. Response of human tumor xenografts in athymic nude mice to docetaxel (RP 56976, Taxotere). Invest N Drugs 1995;13:1–11.
20. Johnson JI, Decker S, Zaharevitz D, et al. Relationships between drug activity in NCI preclinical in vitro and in vivo models and early clinical trials. Br J Cancer 2001;84:1424–1431.
21. Van Dyke T, Jacks T. Cancer modeling in the modern era: progress and challenges. Cell 2002;108:135–144.
22. Hoffman RM. Orthotopic metastatic mouse models for anticancer drug discovery and evaluation: a bridge to the clinic. Invest N Drugs 1999;17:343–359.
23. Killion JJ, Radinsky R, Fidler IJ. Orthotopic models are necessary to predict therapy of transplantable tumors in mice. Cancer Metastasis Rev 1998;17:279–284.
24. Onn A, Isobe T, Itaska S, et al. Development of an orthotopic model to study the biology and therapy of primary human lung cancer in nude mice. Clin Cancer Res 2003;9:5532–5539.
25. Kerbel RS. Human tumor xenografts as predictive preclinical models for anticancer drug activity in humans: better than commonly perceived-but they can be improved. Cancer Biol Ther 2003;2: S134–S139.

26. Pettit GR, Kamano Y, Herald CL, et al. The isolation and structure of a remarkable marine animal antineoplastic constituent: dolastatin 10. J Am Chem Soc 1987;109:6883–6885.
27. Pettit GR, Kamano Y, Dufresne C, Cerny RL, Herald CL, Schmidt JM. et al. Isolation and structure of the cytostatic linear depsipeptide dolastatin 15. J Org Chem 1989;54:6005–6006.
28. Sone H, Shibata T, Fujita T, Ojika M, Yamada K. Dolastatin H and Isodolastatin H, potent cytotoxic peptides from the sea hare *Dolabella auricularia*: isolation, stereostructures, and synthesis. J Am Chem Soc 1996;118:1874–1880.
29. Li Y, Kobayashi H, Hashimoto Y, et al. Interaction of marine toxin dolastatin 10 with porcine brain tubulin: competitive inhibition of rhizoxin and phomopsin A binding. Chem Biol Interact 1994;93:175–183.
30. Harrigan GG, Luesch H, Yoshida WY, et al. Symplostatin 1: A dolastatin 10 analogue from the marine cyanobacterium Symploca hydnoides. J Nat Prod 1998;61:1075–1077.
31. Luesch H, Moore RE, Paul VJ, Mooberry SL, Corbett TH. Symplostatin 3, a new dolastatin 10 analogue from the marine cyanobacterium Symploca sp. VP452. J Nat Prod 2002;65:16–20.
32. Luesch H, Moore RE, Paul VJ, Mooberry SL, Corbett TH. Isolation of dolastatin 10 from the marine cyanobacterium Symploca species VP642 and total stereochemistry and biological evaluation of its analogue symplostatin 1. J Nat Prod 2001;64:907–910.
33. Pettit GR, Singh SB, Hogan F, et al. The absolute configuration and snythesis of natural (-) dolastatin 10. J Am Chem Soc 1990;111:5463–5465.
34. Pettit GR, Herald DL, Singh SB, Thornton TJ, Mullaney JT. Antineoplastic agents. 220. Synthesis of natural (-)-dolastatin 15. J Am Chem Soc 1991;113:6692–6693.
35. Pettit GR, Srirangan JK, Barkoczy J, et al. Antineoplastic agents 337. Synthesis of dolastatin 10 structural modifications. Anti-cancer Drug Des 1995;10:529–544.
36. Miyazaki K, Kobayashi M, Natsume T, et al. Synthesis and antitumor activity of novel dolastatin 10 analogs. Chem Pharm Bull (Tokyo) 1995;43:1706–1718.
37. Kobayashi M, Natsume T, Tamaoki S, et al. Antitumor activity of TZT-1027, a novel dolastatin 10 derivative. Jpn J Cancer Res 1997;88:316–327.
38. de Arruda M, Cocchiaro CA, Nelson CM, et al. LU103793 (NSC D-669356): a synthetic peptide that interacts with microtubules and inhibits mitosis. Cancer Res 1995;55:3085–3092.
39. Mita AC, Hammond LA, Bonate PL, et al. Phase I and pharmacokinetic study of the tasidotin (ILX651), a third-generation dolastatin-15 analogue, administered weekly for 3 weeks every 28 days in patients with advanced solid tumors. Clin Cancer Res 2006;12:5207–5215.
40. Natsume T, Watanabe J, Tamaoki S, Fujio N, Miyaska K, Kobayashi M. Characterization of the interaction of TZT-1027, a potent antitumor agent, with tubulin. Jpn J Cancer Res 2000;91:737–747.
41. Bai R, Pettit GR, Hamel E. Dolastatin 10, a powerful cytostatic peptide derived from a marine animal. Inhibition of tubulin polymerization mediated through the vinca alkaloid binding domain. Biochem Pharmacol 1990;39:1941–1949.
42. Mooberry SL, Leal RM, Tinley TL, Luesch H, Moore RE, Corbett TH. et al. The molecular pharmacology of symplostatin 1: a new antimitotic dolastatin 10 analog. Int J Cancer 2003;104:512–521.
43. Bai R, Friedman SJ, Pettit GR, Hamel E. Dolastatin 15, a potent antimitotic depsipeptide derived from *Dolabella auricularia*. Interaction with tubulin and effects of cellular microtubules. Biochem Pharmacol 1992;43:2637–2645.
44. Jordan MA, Walker D, de Arruda M, Barlozzari T, Panda D. Suppression of microtubule dynamics by binding of cemadotin to tubulin: possible mechanism for its antitumor action. Biochemistry 1998;37:17,571–17,578.
45. Bai RL, Pettit GR, Hamel E. Binding of dolastatin 10 to tubulin at a distinct site for peptide antimitotic agents near the exchangeable nucleotide and vinca alkaloid sites. J Biol Chem 1990;265:17,141–17,149.
46. Bai R, Durso NA, Sackett DL, Hamel E. Interactions of the sponge-derived antimitotic tripeptide hemiasterlin with tubulin: comparison with dolastatin 10 and cryptophycin 1. Biochemistry 1999;38:14,302–14,310.
47. Bai R, Taylor GF, Schmidt JM, et al. Interaction of dolastatin 10 with tubulin: induction of aggregation and binding and dissociation reactions. Mol Pharmacol 1995;47:965–976.
48. Cruz-Monserrate Z, Mullaney JT, Harran PG, Pettit GR, Hamel E. Dolastatin 15 binds in the vinca domain of tubulin as demonstrated by Hummel-Dreyer chromatography. Eur J Biochem 2003;270:3822–3828.

49. Cruz-Monserrate Z, Mullaney JT, Harran PG, Pettit GR, Hamel E. Diazonamide A and a synthetic structural analog: disruptive effects on mitosis and cellular microtubules and analysis of their interactions with tubulin. Mol Pharmacol 2003;63:1273–1280.

50. Edler MC, Fernandez AM, Lassota P, Ireland CM, Barrows LR. Inhibition of tubulin polymerization by vitilevuamide, a bicyclic marine peptide, at a site distinct from colchicine, the vinca alkaloids, and dolastatin 10. Biochem Pharmacol 2002;63:707–715.

51. Watts NR, Cheng N, West W, Steven AC, Sackett DL. The cryptophycin-tubulin ring structure indicates two points of curvature in the tubulin dimer. Biochemistry 2002;41:12,662–12,669.

52. Boukari H, Nossal R, Sackett DL. Stability of drug-induced tubulin rings by fluorescence correlation spectroscopy. Biochemistry 2003;42:1292–1300.

53. Steube KG, Grunicke D, Pietsch T, Gignac SM, Pettit GR, Drexler HG. Dolastatin 10 and dolastatin 15: effects of two natural peptides on growth and differentiation of leukemia cells. Leukemia 1992;6:1048–1053.

54. Verdier-Pinard P, Kepler JA, Pettit GR, Hamel E. Sustained intracellular retention of dolastatin 10 causes its potent antimitotic activity. Mol Pharmacol 2000;57:180–187.

55. Toppmeyer DL, Slapak CA, Croop J, Kufe DW. Role of P-glycoprotein in dolastatin 10 resistance. Biochem Pharmacol 1994;48:609–612.

56. Kalemkerian GP, Ou X, Adil MR, et al. Activity of dolastatin 10 against small-cell lung cancer in vitro and in vivo: induction of apoptosis and bcl-2 modification. Cancer Chemother Pharmacol 1999;43:507–515.

57. Aherne GW, Hardcastle A, Valenti M, et al. Antitumour evaluation of dolastatins 10 and 15 and their measurement in plasma by radioimmunoassay. Cancer Chemother Pharmacol 1996;38:225–232.

58. Mohammad RM. Bryostatin 1 induces differentiation and potentiates the antitumor effect of Auristatin PE in a human pancreatic tumor (PANC-1) xenograft model. Anti-Cancer Drugs 2001;12:735–740.

59. Mohammad RM, Limvarapuss C, Wall NR, et al. A new tubulin polymerization inhibitor, auristatin PE, induces tumor regression in a human Waldenstrom's macroglobulinemia xenograft model. Int J Oncol 1999;15:367–372.

60. Mohammad RM, Pettit GR, Almatchy VP, Wall N, Varterasian M, Al-Kabit A. Synergistic interaction of selected marine animal anticancer drugs against human diffuse large cell lymphoma. Anticancer Drugs 1998;9:149–156.

61. Francisco JA, Cerveny CG, Meyer DL, et al. cAC10-vcMMAE, an anti-CD30-monomethyl auristatin E conjugate with potent and selective antitumor activity. Blood 2003;102:1458–1465.

62. Doronina SO, Toki BE, Torgov MY, et al. Development of potent monoclonal antibody auristatin conjugates for cancer therapy. Nat Biotechnol 2003;21:778–784.

63. Otani M, Natsume T, Watanabe JI, et al. TZT-1027, an antimicrotubule agent, attacks tumor vasculature and induces tumor cell death. Jpn J Cancer Res 2000;91:837–844.

64. Natsume T, Nakamura T, Koh Y, Kobayashi M, Sajio N, Nishio K. Gene expression profiling of exposure to TZT-1027, a novel microtubule-interfering agent, in non-small cell lung cancer PC-14 cells and astrocytes. Invest New Drugs 2001;19:293–302.

65. Dang LH, Bettegowda C, Huso DL, Kinzler KW, Vogelstein B. Combination bacteriolytic therapy for the treatment of experimental tumors. Proc Natl Acad Sci USA 2001;98:15,155–15,160.

66. Madden T, Tran HT, Beck D, et al. Novel marine-derived anticancer agents: a phase I clinical, pharmacological, and pharmacodynamic study of dolastatin 10 (NSC 376128) in patients with advanced solid tumors. Clin Cancer Res 2000;6:1293–1301.

67. Pitot HC, McElroy EA Jr, Reid JM, et al. Phase I trial of dolastatin-10 (NSC 376128) in patients with advanced solid tumors. Clin Cancer Res 1999;5:525–531.

68. Wolff I, Bruntsch U, Cavalli F, de Jonk J, von Broen IM, Sessa C. Phase I clinical and pharmacokinetic study of the dolastatin analogue LU 103793 on a weekly x 4 schedule. Ann Oncol 1996;7(suppl 5):124.

69. Mross K, Berdel WE, Fiebig HH, Velagapudi R, von Broen IM, Unger, C. Clinical and pharmacologic phase I study of Cemadotin-HCl (LU103793), a novel antimitotic peptide, given as 24-hour infusion in patients with advanced cancer. Ann Oncol 1998;9:1323–1330.

70. Supko JG, Lynch TJ, Clark JW, et al. A phase I clinical and pharmacokinetic study of the dolastatin analogue cemadotin administered as a 5-day continuous intravenous infusion. Cancer Chemother Pharmacol 2000;46: 319–328.

71. Villalona-Calero MA, Baker SD, Hammond L, et al. Phase I and pharmacokinetic study of the water-soluble dolastatin 15 analog LU103793 in patients with advanced solid malignancies. J Clin Oncol 1998;16:2770–2779.

72. Kerbrat P, Dieras V, Pavlidis N, Ravaud A, Wanders J, Fumoleau P. Phase II study of LU 103793 (dolastatin analogue) in patients with metastatic breast cancer. Eur J Cancer 2003;39:317–320.

73. Smyth J, Boneterre ME, Schellens J, et al. Activity of the dolastatin analogue, LU103793, in malignant melanoma. Ann Oncol 2001;12:509–511.

74. De Jonge MJ, Madretsma S, Van der Gaast A, et al. TZT-1027, a novel dolastatin 10 derivative: Phase I and pharmacologic study of day 1 and 8 IV administration every 3 weeks in patients (pts) with advanced solid tumors. Proc Am Soc Clin Oncol 2003;22:153.

75. Schoffski P, Thate B, Beutel G, et al. Phase I evaluation of the 3-weekly administration of TZT-1027 in patients with solid tumors. Proc Am Soc Clin Oncol 2003;22:211.

76. Yamamoto N, Andoh Kawahara M, Fukuokam M, Niitani H. Phase I study of TZT-1027, an inhibitor of tubulin polymerization, given weekly × 3 as a 1-hour intravenous infusion in patients (pts) with solid tumors. Proc Am Soc Clin Oncol 2002;21, abstract 420.

77. Eder J, Schwartz RE, Hirsch CF, et al. ILX651, a third generation dolastatin 15: Results of a phase I dose escalating and pharmacokinetic study of ILX-651 administered as a 30 minute IV infusion every other day for 3 days every three weeks. Proc Am Soc Clin Oncol 2003;22:205.

78. Ebbinghaus S, Rubin E, Hersh E, et al. A phase I study of ILX-651 administered intravenously daily for five days every three weeks in patients with advanced solid tumo. Proc Am Soc Clin Oncol 2003;22:129.

79. Schwartz RE, Hirsch CF, Sesin DF, et al. Pharmaceuticals from cultured algae. J Ind Microbiol 1994;1990:5.

80. Trimurtulu G, Ohtani I, Patterson GML, et al. Total structures of cryptophycins, potent antitumor depsipeptides from the blue-green alga *Nostoc* sp. strain GSV 224. J Am Chem Soc 1994;116:4729–4737.

81. Kobayashi H, Aoki S, Ohyabu N, Kurosu M, Wang W, Kitagawa I. Arenastatin A, a potent cytotoxic depsipeptide from the Okinawan marine sponge *Dysidea arenaria*. Tetrahedron Lett 1994;35:7969–7972.

82. Kobayashi H, Kurosu M, Ohyabu N, Wang W, Fujii S, Kitagawa I. The absolute stereostructure of arenastatin A, a potent cytotoxic depsipeptide from the Okinawan marine sponge Dysidea arenaria. Chem Pharm Bull (Tokyo) 1994;10:2196–2198.

83. Golakoti T, Ogino J, Heltzel CE, et al. Structure determination, conformational analysis, chemical stability studies, and antitumor evaluation of the cryptophycins. Isolation of 18 new analogs from Nostoc sp. strain GSV 224. J Am Chem Soc 1995;117:12,030–12,049.

84. Wagner MM, Paul DC, Shih C, Jordan MA, Wilson L, Williams DC. In vitro pharmacology of cryptophycin 52 (LY355703) in human tumor cell lines. Cancer Chemother Pharmacol 1999;43:115–125.

85. Rej R, Nguyen D, Go B, Fortin S, Lavallee JF. Total synthesis of cryptophycins and their 16-(3-phenylacryloyl) derivatives. J Org Chem 1996;61:6289–6295.

86. Salamonczyk GM, Han K, Guo Zw Z, Sih CJ. Total synthesis of cryptophycins via a chemoenzymatic approach. J Org Chem 1996;61:6893–6900.

87. Subbaraju GV, Golakoti T, Patterson GML, Moore RE Three new cryptophycins from Nostoc sp. GSV 224. J Nat Prod 1997;60:302–305.

88. Eggen M, Georg GI. The cryptophycins: their synthesis and anticancer activity. Med Res Rev 2002; 22:85–101.

89. Kerksiek K, Mejillano MR, Schwartz RE, Georg GI, Himes RH. Interaction of cryptophycin 1 with tubulin and microtubules. FEBS Lett 1995;377:59–61.

90. Bai R, Schwartz RE, Kepler JA, Pettit GR, Hamel E. Characterization of the interaction of cryptophycin 1 with tubulin: binding in the Vinca domain, competitive inhibition of dolastatin 10 binding, and an unusual aggregation reaction. Cancer Res 1996;56:4398–4406.

91. Smith CD, Zhang X, Mooberry SL, Patterson GM, Moore RE. Cryptophycin: a new antimicrotubule agent active against drug-resistant cells. Cancer Res 1994;54:3779–3784.

92. Mooberry SL, Taoka CR, Busquets L. Cryptophycin 1 binds to tubulin at a site distinct from the colchicine binding site and at a site that may overlap the vinca binding site. Cancer Lett 1996;107: 53–57.

93. Panda D, Ananthnarayan V, Larson G, Shih C, Jordan MA, Wilson L. Interaction of the antitumor compound cryptophycin 52 with tubulin. Biochemistry 2000;39:14,121–14,127.

94. Panda D, DeLuca K, Williams D, Jordan MA, Wilson L. Antiproliferative mechanism of action of cryptophycin 52: kinetic stabilization of microtubule dynamics by high-affinity binding to microtubule ends. Proc Natl Acad Sci USA 1998;95:9313–9318.

95. Barbier P, Gregoire C, Devred F, Sarrazin M, Peyrot V. In vitro effect of cryptophycin 52 on micro-tubule assembly and tubulin: molecular modeling of the mechanism of action of a new antimitotic drug. Biochemistry 2001;40:13,510–13,519.

96. Smith CD, Zhang X. Mechanism of action cryptophycin. Interaction with the Vinca alkaloid domain of tubulin. J Biol Chem 1996;271:6192–6198.

97. Panda D, Himes RH, Moore RE, Wilson L, Jordan MA. Mechanism of action of the unusually potent microtubule inhibitor cryptophycin 1. Biochemistry 1997;36:12,948–12,953.

98. Himes RH, Kersey RN, Heller-Bettinger I, Sampson FE. Action of the vinca alkaloids vincristine, vinblastine, and desacetyl vinblastine amide on microtubules in vitro. Cancer Res 1976;36: 3798–3802.

99. Andreu JM, Wagenknecht T, Timasheff SN. Polymerization of the tubulin-colchicine complex: rela-tion to microtubule assembly. Biochemistry 1983;22:1556–1566.

100. Chen BD, Nakeff A, Valeriote F. Cellular uptake of a novel cytotoxic agent, cryptophycin 52, by human THP-1 leukemia cells and H-125 lung tumor cells. Int J Cancer 1998;77:869–873.

101. Mooberry SL, Busquets L, Tien G. Induction of apoptosis by cryptophycin 1, a new antimicrotubule agent. Int J Cancer 1997;73:440–448.

102. Drew L, Fine RL, Do TN, Douglas GP, Petrylak DP. The novel antimicrotubule agent cryptophycin 52 (LY355703) induces apoptosis via multiple pathways in human prostate cancer cells. Clin Cancer Res 2002;8:3922–3932.

103. Corbett TH, Valeriote FA, Demchik L, et al. Discovery of cryptophycin-1 and BCN-183577: examples of strategies and problems in the detection of antitumor activity in mice. Invest N Drugs 1997;15:207–218.

104. Menon K, Alvarez E, Forler P, et al. Antitumor activity of cryptophycins: effect of infusion time and combination studies. Cancer Chemother Pharmacol 2000;46:142–149.

105. Teicher BA, Forler P, Menon K, Phares V, Amsrud T, Shih C. Cryptophycin 52 and cryptophycin 55 in sequential and simultaneous combination treatment regimens in human tumor xenografts. In Vivo 2000;14:471–480.

106. Sessa C, Weigang-Korler K, Pagani O, et al. Phase I and pharmacological studies of the cryptophycin analogue LY355703 administered on a single intermittent or weekly schedule. Eur J Cancer 2002;38:2388–2396.

107. Stevenson JP, Sun W, Gallagher M, et al. Phase I trial of the cryptophycin analogue LY355703 admin-istered as an intravenous infusion on a day 1 and 8 schedule every 21 days. Clin Cancer Res 2002;8:2524–2529.

108. Edelman MJ, Gandara DR, Hausner P, et al. Phase 2 study of cryptophycin 52 (LY355703) in patients previously treated with platinum based chemotherapy for advanced non-small cell lung cancer. Lung Cancer 2003;39:197–199.

109. Talpir R, Benayahu Y, Kashman Y, Pannell L, Schleyer M. Hemiasterlin and geodiamolide TA; two new cytotoxic peptides from the marine sponge hemiasterella minor (Kirkpatrick). Tetrahedron Lett 1994;35:4453–4456.

110. Coleman JE, Dilip De Silva E, Kong F, Andersen RJ, Allen TM. Cytotoxic peptides from the marine sponge Cymbastela sp. Tetrahedron 1995;51:10,653–10,662.

111. Gamble WR, Durso NA, Fuller RW, et al. Cytotoxic and tubulin-interactive hemiasterlins from Auletta sp. and Siphonochalina spp. sponges. Bioorg Med Chem 1999;7:1611–1615.

112. Crews P, Farias JJ, Emrich R, Keifer PA. Milnamide A, an unusual cytotoxic tripeptide from the marine sponge Auletta cf. constricta. J Org Chem 1994;59:2932–2934.

113. Chevallier C, Richardson AD, Edler MC, Hamel E, Harper MK, Ireland CM. A new cytotoxic and tubulin-interactive milnamide derivative from a marine sponge Cymbastela sp. Org Lett 2003;5:3737–3739.

114. deSilva ED, Andersen RJ, Allen TM. Geodimolides C to F, new cytotoxic cyclodepsipeptides from the marine sponge Pseudaxinyssa Sp. Tetrahedron Lett 1990;31:489–492.

115. Coleman JE, Van Soest R, Andersen RJ. New geodiamolides from the sponge Cymbastela sp. col-lected in Papua New Guinea. J Nat Prod 1999;62:1137–1141.

116. Zabriskie TM, Klocke JA, Ireland CM, et al. Jaspamide, a modified peptide from a Japis sponge, with insecticidal and antifungal activity. J Am Chem Soc 1986;108:3123–3124.

117. Crews P, Manes LV, Boehler M. Jasplakinolide a cyclodepsipeptide from the marine sponge Jaspis Sp. Tetrahedron Lett 1986;27:2797–2800.

118. Senderowicz AM, Kaur G, Sainz E, et al. Jasplakinolide's inhibition of the growth of prostate carcinoma cells in vitro with disruption of the actin cytoskeleton. J Natl Cancer Inst 1995;87:46–51.

119. Sasse F, Kunze B, Gronewold TM, Reichenbach H. The chondramides: cytostatic agents from myxobacteria acting on the actin cytoskeleton. J Natl Cancer Inst 1998;90:1559–1563.

120. Andersen RJ, Coleman JE, Piers E, Wallace DJ. Total synthesis of (-)-hemiasterlin, a structurally novel tripeptide that exhibits potent cytotoxic activity. Tetrahedron Lettt 1997;38:317–320.

121. Reddy R, Jaquith JB, Neelagiri VR, Saleh-Hanna S, Durst T. Asymmetric synthesis of the highly methylated tryptophan portion of the hemiasterlin tripeptides. Org Lett 2002;4:695–697.

122. Vedejs E, Kongkittingam C. A total synthesis of (-)-hemiasterlin using N-Bts methodology. J Org Chem 2001;66:7355–7364.

123. Nieman J, Coleman J, Wallace D, et al. Synthesis and antimitotic / cytotoxic activity of hemiasterlin analogs. J Nat Prod 2003;66:183–199.

124. Zask A, Birnberg G, Cheung K, et al. Synthesis and biological activity of analogs of the antimicrotubule agent HTI-286. Proc Am Assoc Cancer Res 2002;43:737.

125. Anderson HJ, Coleman JE, Andersen RJ, Roberge M. Cytotoxic peptides hemiasterlin, hemiasterlin A and hemiasterlin B induce mitotic arrest and abnormal spindle formation. Cancer Chemother Pharmacol 1997;39:223–226.

126. Krishnamurthy G, Cheng W, Lo MC, et al. Biophysical characterization of the interactions of HTI-286 with tubulin heterodimer and microtubules. Biochemistry 2003;42:13,484–13,495.

127. Nunes M, Kaplan J, Wooters J, et al. Two photoaffinity analogues of the tripeptide, hemiasterlin, exclusively label alpha-tubulin. Biochemistry 2005;44:6844–6857.

128. Hari M, Nunes M, Zask A, et al. Hemiasterlin analogs exclusively label α-tubulin at the interdimerface and specifically block subtilisin digestion of α-tubulin.Proc Am Assoc Cancer Res 2004;45:359.

129. Lowe J, Li H, Downing KH, Nogales E. Refined structure of alpha beta-tubulin at 3.5 Å resolution. J Mol Biol 2001;313:1045–1057.

130. Nogales E, Whittaker M, Milligan RA, Downing KH. High-resolution model of the microtubule. Cell 1999;96:79–88.

131. Rai SS, Wolff J. Localization of the vinblastine-binding site on beta-tubulin. J Biol Chem 1996;271: 14,707–14,711.

132. Muller DR, Schindler P, Towbin H, et al. Isotope-tagged cross-linking reagents. A new tool in mass spectrometric protein interaction analysis. Anal Chem 2001;73:1927–1934.

133. Wallon G, Rappsilber J, Mann M, Serrano L. Model for stathmin/OP18 binding to tubulin. EMBO J 2000;19:213–222.

134. Cassimeris L. The oncoprotein 18/stathmin family of microtubule destabilizers. Curr Opin Cell Biol 2002;14:18–24.

135. Larsson N, Marklund U, Gradin HM, Brattsand G, Gullberg M. Control of microtubule dynamics by oncoprotein 18: dissection of the regulatory role of multisite phosphorylation during mitosis. Mol Cell Biol 1997;17:5530–5539.

136. Gigant B, Curmi PA, Martin-Barbey C, et al. The 4 Å X-ray structure of a tubulin:stathmin-like domain complex. Cell 2000;102:809–816.

137. Steinmetz MO, Kammerer RA, Jahnke W, Goldie KN, Lustig A, van Oostrum J. Op18/stathmin caps a kinked protofilament-like tubulin tetramer. EMBO J 2000;19:572–580.

138. Nunes M, Kaplan J, Loganzo F, Zask A, Ayral-Kaloustial S, Greenberger LM. Two photoaffinity analogs of HTI-286, a synthetic analog of hemiasterlin, interact with alpha-tubulin. Eur J Cancer 2002;38:S119.

139. Poruchynsky MS, Kim JH, Nogales E, Loganzo F, Greenberger LM. Tumor cells resistant to a microtubule-depolymerizing hemiasterlin analog, HTI-286, have mutations in α- or β-tubulin and increased microtubule stability. Proc Am Assoc Cancer Res 2003;44:2nd ed., 535.

140. Borst P, Elferink RO. Mammalian ABC transporters in health and disease. Annu Rev Biochem 2002;71:537–592.

141. Ratain MJ, Undevia S, Janisch L, et al. Phase I and pharmacological study of HTI-286, a novel antimicrotubule agent: correlation of neutropenia with time above theshold plasma concentration. Proc Am Soc Clin Oncol 2003; abstract 516.

142. Culvenor CC, Beck AB, Clarke M, et al. Isolation of toxic metabolites of *Phomopsis leptostromiformis* which produce lupinosis. Aust J Biol Sci 1977;30:269–278.

143. Jago MV, Peterson JE, Payne AL, Campbell DG. Lupinosis: response of sheep to different doses of phomopsin. Aust J Exp Bio Med Sci 1982;60:239–251.

144. Tonsing EM, Steyn PS, Osborn M, Weber K. Phomopsin A, the causitive agent of lupinosis, interacts with microtubules in vivo and in vitro. Eur J Cell Biol 1984;35:156–164.

145. Mackay MF, Van Donkelaar A, Culvenor CJ. The X-ray structure of phomopsin A, a hexapeptide mycotoxin. J Chem Soc Chem Commun 1986;1219–1221.

146. Koiso Y, Li Y, Iwasaki S, et al. Ustiloxins, antimitotic cyclic peptides from false smut balls on rice panicles caused by *Ustilaginoidea virens*. J Antibiot (Tokyo) 1994;47:765–773.

147. Koiso Y, Morisaki N, Yamashita Y, et al. Isolation and structure of an antimitotic cyclic peptide, ustiloxin F: chemical interrelation with a homologous peptide, ustiloxin B. J Antibiot (Tokyo) 1998;51:418–422.

148. Nakamura K, Izumiyama N, Ohtsubo K, et al. Lupinosis-like lesions in mice caused by ustiloxin, produced by *Ustilaginoieda virens*: a morphological study. Nat Toxins 1994;2:22–28.

149. Tanaka H, Sawayama AM, Wandless TJ. Enantioselective total synthesis of ustiloxin D. J Am Chem Soc 2003;125:6864–6865.

150. Cao B, Park H, Joullie MM. Total synthesis of ustiloxin D. J Am Chem Soc 2002;124:520–521.

151. Li Y, Koiso Y, Kobayashi H, Hashimoto Y, Iwasaki S. Ustiloxins, new antimitotic cyclic peptides: interaction with porcine brain tubulin. Biochem Pharmacol 1995;49:1367–1372.

152. Lacey E, Edgar JA, Culvenor CC. Interaction of phomopsin A and related compounds with purified sheep brain tubulin. Biochem Pharmacol 1987;36:2133–2138.

153. Lindquist N, Fenical W, van Duyne GD, Clardy J. Isolation and structure determination of diazonamides A and B, unusual cytotoxic metabolites from the marine ascidian Diazona chinesis. J Am Chem Soc 1991;113:2303–2304.

154. Li J, Burgett WG, Esser L, Amezcua C, Harran PG. Synthesis of nominal diazoniamides- part 2: on the true structure and origin of natural isolates. Angew Chem Int Ed 2001;40:4765–4770.

155. Li J, Jeong S, Esser L, Harran PG. Total synthesis of nominal diazonamides—part 1: convergent preparation of the structure proposed for (-)—diazonamide A. Agnew Chem Int Ed 2001;40: 4765–4770.

156. Nicolaou KC, Bella M, Chen DY, Huang X, Ling T, Snyder SA. Total synthesis of diazonamide A. Angew Chem Int Ed 2002;41:3495–3499.

157. Nicolaou KC, Bheema Rao P, Hao J, et al. The Second Total Synthesis of Diazonamide A. Angew Chem Int Ed 2003;42: 1753–1758.

158. Ritter T, Carreira EM. The diazonamides: the plot thickens. Angew Chem Int Ed 2002;41:2489–2495.

159. Leung TW, Williams DH, Barna JCJ, Foti S, Oelrichs PB. Structural studies on the peptide morisin from *Laportea moroides*. Tetrahedron 1986;42:3333–3348.

160. Morita H, Shimbo K, Shigemori H, Kobayashi J. Antimitotic activity of moroidin, a bicyclic peptide from the seeds of *Celosia argentea*. Bio Med Chem Lett 2000;10:469–471.

161. Kobayashi J, Suzuki H, Shimbo K, Takeya K, Morita H. Celogentins A-C, new antimitotic bicyclic peptides from the seeds of *Celosia argentea*. J Org Chem 2001;66:6626–6633.

162. Sasse F, Steinmetz H, Heil J, Hofle G, Reichenbach H. Tubulysins, new cytostatic peptides from myxobacteria acting on microtubuli. Production, isolation, physico-chemical and biological properties. J Antibiot (Tokyo) 2000;53:879–885.

163. Chaudhary PM, Roninson IB. Expression and activity of P-glycoprotein, a multidrug efflux pump, in human hematopoietic stem cells. Cell 1991;66:85–94.

164. Kerbel RS, Klement G, Pritchard KI, Kamen B. Continuous low-dose anti-angiogenic/ metronomic chemotherapy: from the research laboratory into the oncology clinic. Ann Oncol 2002;13:12–15.

165. Dubowchik GM, Walker MA. Receptor-mediated and enzyme-dependent targeting of cytotoxic anticancer drugs. Pharmacol Ther 1999;83:67–123.

166. Shih C, Teicher BA. Cryptophycins: a novel class of potent antimitotic antitumor depsipeptides. Curr Pharm Des 2001;7:1259–1276.

167. Hoffman MA, Blessing JA, Lentz SS, Gynecologic Oncology Group S. A phase II trial of dolastatin-10 in recurrent platinum-sensitive ovarian carcinoma: a Gynecologic Oncology Group study. Gynecol Oncol 2003;89:95–98.

168. Saad ED, Kraut EH, Hoff PM, et al. Phase II study of dolastatin-10 as first-line treatment for advanced colorectal cancer. Am J Clin Oncol 2002;25:451–453.

169. Krug LM, Miller VA, Kalemkerian GP, et al. Phase II study of dolastatin-10 in patients with advanced non-small-cell lung cancer. Ann Oncol 2000;11:227–228.

170. Margolin K, Longmate J, Synold TW, et al. Dolastatin-10 in metastatic melanoma: a phase II and pharmokinetic trial of the California Cancer Consortium. Invest N Drugs 2001;19:335–340.

171. Vaishampayan U, Glode M, Du W, et al. Phase II study of dolastatin-10 in patients with hormone-refractory metastatic prostate adenocarcinoma. Clin Cancer Res 2000;6:4205–4208.
172. Sampath D, Discafani CM, Loganzo F, et al. MAC-321, A novel taxane with greater efficacy than paclitaxel and docetaxel in vitro and in vivo. Mol Cancer Ther 2003;2:873–994.
173. Ray A, Okauneva T, Manna T, et al. Mechanism of action of the microtubule-targeted antimitotic depsipeptide tasidotin (formerly ILX651) and is major metabolite tasidotin C-carboxylate. Cancer Res 2007;67:3767–3776.

11

Molecular Features of the Interaction of Colchicine and Related Structures with Tubulin

Susan L. Bane

CONTENTS

SUMMARY

Colchicine is one of the oldest known antimicrotubule drugs. It exerts its biological effects by binding to a single site on the β-subunit of the tubulin heterodimer. The resulting colchicine–tubulin complex substiochiometrically inhibits tubulin assembly and suppresses microtubule dynamics. A large number of molecules with significant structural diversity interact with the colchicine site on tubulin; literally hundreds of potential colchicine site ligands have been synthesized and tested in the hopes of finding a better clinical agent. In spite of the wealth of data, an understanding of the structure–activity relationship for these colchicine site ligands remains elusive. Colchicine site drugs are believed to act as a common mechanism, which has been studied extensively in the case of colchicine but much less studied for other ligands. In this review, the molecular mechanisms by which colchicine and closely related structures interact with tubulin are explored. Thermodynamic, kinetic, and structure–activity analyses as well as more recent structural information about the ligand–receptor complex are discussed.

Key Words: Antimitotic; colchicine; combretastatin; mechanism; microtubules; podophyllotoxin; structure–activity relationships; tubulin.

From: *Cancer Drug Discovery and Development: The Role of Microtubules in Cell Biology, Neurobiology, and Oncology* Edited by: Tito Fojo © Humana Press, Totowa, NJ

1. INTRODUCTION

Drugs that affect microtubule dynamics are among the most widely used agents in cancer chemotherapy *(1–4)*. These substances interact with tubulin, a 100,000 Da heterodimer that is the major protein component of cellular microtubules. Tubulin contains at least three distinct drug-binding sites, known as the colchicine, Taxol, and vinblastine-binding sites. Both the Taxol-binding site and the vinblastine-binding site have been exploited for drugs that are used in clinical oncology *(5–7)*, but there is not yet a colchicine site drug that is in routine use for cancer treatment. The discovery of the impressive antivascular activity of the colchicine site ligand combretastatin A-4 phosphate *(8–10)* sparked renewed interest in searching for potential anticancer drugs that exert their activity by binding to tubulin at this site.

A large number of molecules with significant structural diversity interact with the colchicine site on mammalian tubulin. Figure 1 illustrates some of these structures. Similarities between the natural products colchicine, podophyllotoxin *(11)*, combretastatin A-4 *(8,12)*, phenstatin *(13)*, and steganacin *(11)* are apparent, but common structural themes in other colchicine site ligands such as 2-methoxyestradiol *(14)*, nocodazole *(15)*, curacin A *(16)*, chalcones such as MDL-27048 *(17)*, and additional examples shown in Fig. 1 *(18–21)* are not obvious.

A substance is considered to be a colchicine site ligand if it meets the minimum requirements of inhibiting in vitro microtubule assembly and [^{3}H]colchicine binding to tubulin *(22)*. Competitive inhibition of colchicine binding to tubulin has been demonstrated for some colchicine site ligands *(23–26)*, but in most studies competitive inhibition is assumed but not explicitly demonstrated. It is also implicitly assumed in many studies that inhibition of colchicine binding to tubulin is the result of occupancy of the same binding pocket within the protein. This assumption is not necessarily true. The colchicine C ring analog tropolone methyl ether (TME) (Fig. 2) blocks colchicine binding to tubulin but does not inhibit the binding of podophyllotoxin to tubulin *(27)*, which seems to suggest that podophyllotoxin and colchicine share only part of the binding site.

It is probably more accurate to say that these molecules associate with a colchicine-binding *region* of tubulin rather than the colchicine-binding *site*, although the latter term will be retained for convenience. Tubulin binding by these ligands results in the same final effect, inhibition of tubulin assembly, presumably by the same general mechanism, production of a protein conformation that is ineffective in forming a stable microtubule lattice. The details of the binding interactions between colchicine site ligands and tubulin are unknown and likely to be different for various ligands. It is therefore perhaps fruitless to attempt to define a single pharmacophore for the agents that bind to the colchicine site on tubulin. There are probably multiple pharmacophores. A major challenge for the future is to determine which structural families should be considered together and which should be considered separately when molecular mechanisms are proposed.

The molecular mechanism by which colchicine and closely related structures bind to tubulin and subsequently inhibit tubulin polymerization has been extensively investigated *(28–30)*. Quantitative studies of colchicinoid–tubulin associations have shed some light on the molecular features of the interaction *(31–33)*. Far less is known about the molecular mechanisms for the other classes of colchicine site ligands. This review will focus on colchicine and closely related structural molecules. Several reviews that include greater detail on other colchicine site drugs have been published recently

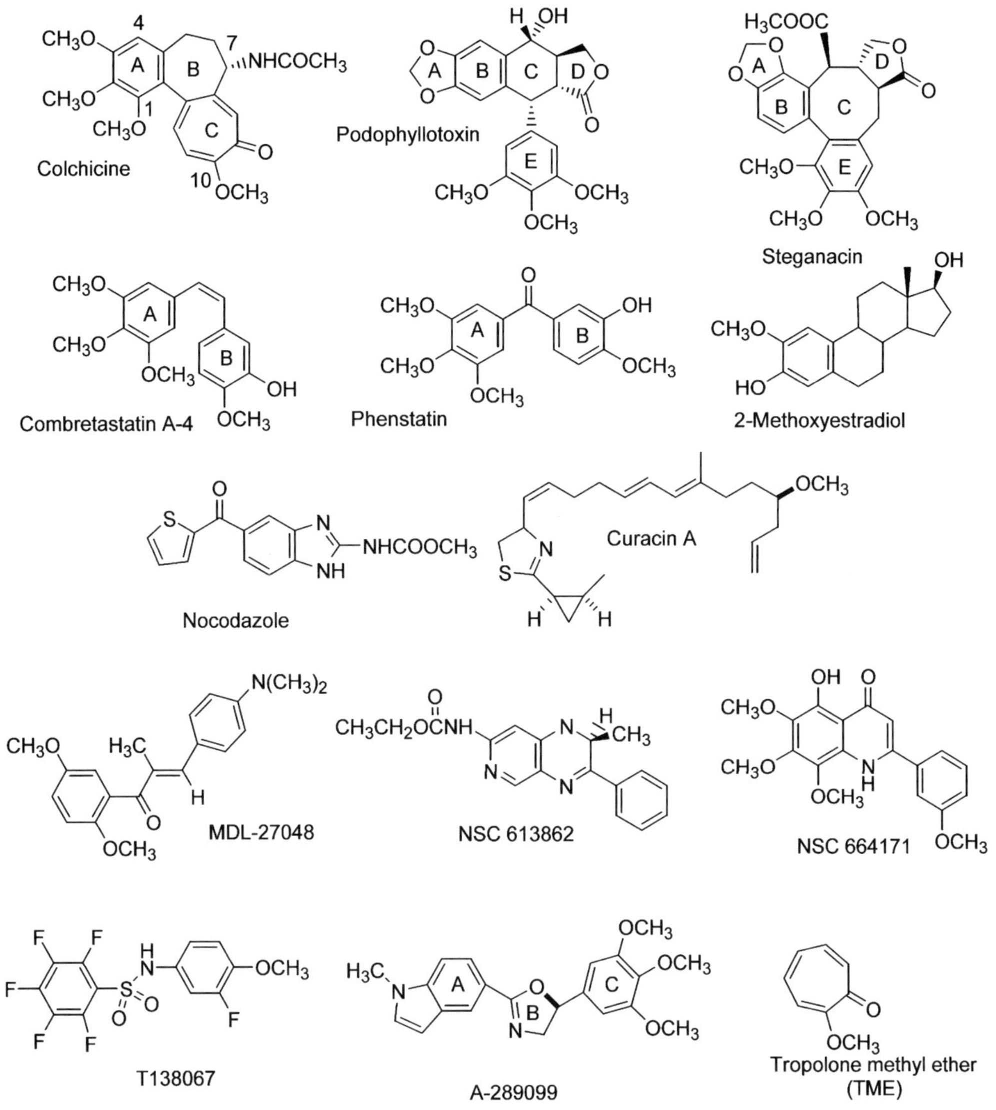

Fig. 1. Structures of colchicine and colchicine site ligands. The standard numbering system and designation of the rings for colchicine is indicated. Standard designation for the rings of some of the other molecules is also shown.

(1,8,10,17,34–39). Many of the important findings about colchicine are likely to be relevant to other colchicine site drugs. However, the colchicine findings also illustrate that similarities in structure do not necessarily imply identical mechanisms.

2. COLCHICINE AND DISEASE

Colchicine itself is the oldest drug in this family of antimicrotubule agents. Although used for more than two millennia for various ailments, its cellular target was discovered just four decades ago. Taylor and coworkers *(40,41)* demonstrated simultaneously that the protein tubulin was both the core protein of the microtubule and the colchicine-binding

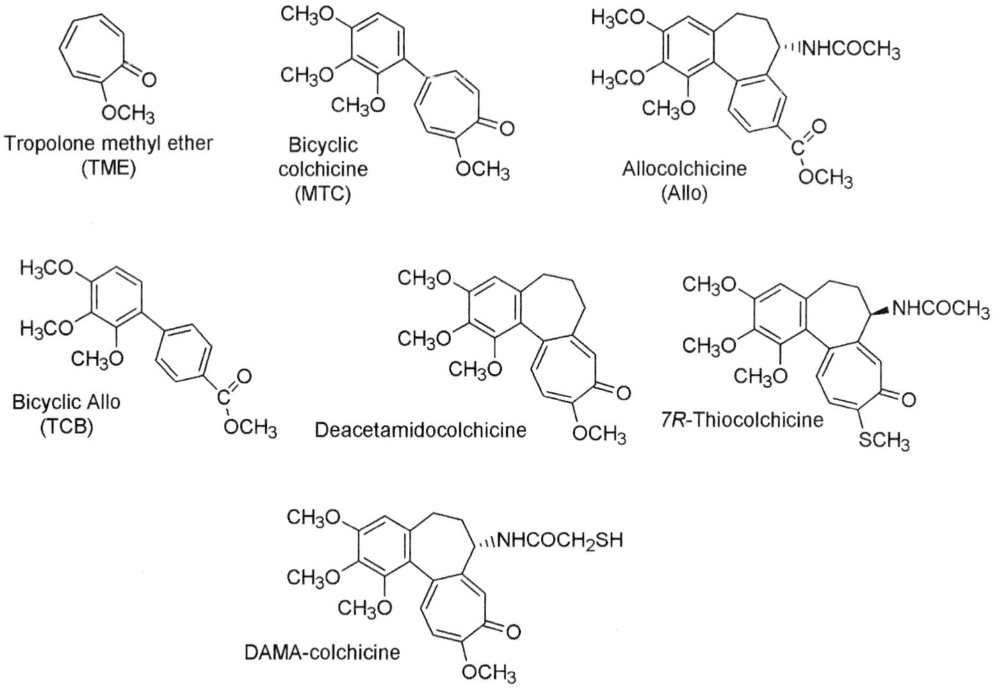

Fig. 2. Structures of some colchicine site ligands.

protein in the cell. The specificity of colchicine for tubulin is remarkable: colchicine's effect on cellular processes appears to be entirely dependent on its interaction with tubulin. This association is also ubiquitous in mammals. Colchicine-binding activity frequently has been used to identify a protein as tubulin, to assess the activity of the protein, and to specifically disrupt microtubule-mediated processes in cells. The ubiquity and specificity of colchicine-binding activity in tubulin from higher organisms has intrigued researchers to search for an endogenous substance that regulates microtubule behavior through the colchicine site. Although a few candidates have been suggested, none has been widely accepted *(22)*.

Colchicine itself is not used in cancer chemotherapy, presumably because of its high toxicity *(3)*. However, colchicine is clinically important in treatment of other diseases. It remains an important drug in the management of acute gout and is often used in many less common disease states, such as familial Mediterranean fever, amyloidosis, and Behcet's disease. Le Hello *(42)* has documented the use of colchicine in more than 50 disorders and diseases, which are chronicled in a thorough review of the pharmacology of colchicine and its use in medicine.

3. COLCHICINE'S EFFECTS ON MICROTUBULE ASSEMBLY AND DYNAMICS

It is well established that the effect of colchicine on the assembly properties of tubulin is owing to the tubulin–colchicine complex (TC) and not to free colchicine *(43)*. Colchicine binding to tubulin induces a conformational change in the protein that alters its association with unliganded tubulin within the microtubule *(44)* in ways that are not well understood. Detailed discussions of the effects of colchicine and other antimicrotubule drugs on microtubule assembly and dynamicity have been published recently *(32,45,46)*, so discussion here will be brief.

The effects of TC on the bulk properties of tubulin in vitro are to substoichiometrically inhibit polymerization of the protein and, to a lesser extent, induce disassembly of the steady state polymer. It is known that TC is incorporated into the end of the microtubule and that the affinity of TC for the microtubule end is about the same as the affinity of guanosine 5′-triphosphate (GTP)-tubulin for the microtubule end *(43,47)*. The difference in the protein conformation between GTP-tubulin and TC affects the longitudinal and lateral protein–protein contacts. As a result, microtubule ends that incorporate TC have a lower affinity for additional tubulin dimers.

The ability of a prospective colchicine site drug to inhibit in vitro microtubule assembly is used as a standard measure of the molecule's potential effect on a cell. It was long thought that inhibition of microtubule assembly by TC was the cause of colchicine's cytotoxicity. However, it is now believed that the effect of TC on in vivo microtubule dynamics is the therapeutically relevant activity of the drug *(3)*. TC potently decreases the rate of growth and shortening of individual microtubules and increases the amount of time the microtubule resides in the attenuation state *(48)*. Precisely how colchicine binding alters the tubulin conformation to produce these effects on tubulin assembly is unclear. Taxol and vinblastine also suppress microtubule dynamics by interacting with the end of a microtubule, yet their tubulin-binding sites and their effects on the bulk properties of microtubules are very different from one another and from colchicine *(32,49,50)*. It therefore seems that more than one type of conformational readjustment in the protein can lead to the same final result.

An additional complication is that different colchicine site ligands can induce different effects on tubulin conformation. For example, there is significant evidence that conformation of tubulin when it is complexed with podophyllotoxin is different from that of tubulin in TC. Colchicine binding to tubulin enhances a weak intrinsic GTPase activity in the protein, but podophyllotoxin binding to tubulin inhibits this GTPase *(51,52)*. Tubulin bound to colchicine can assemble to form sheet-like structures rather than microtubules *(53)*, and the thermodynamic parameters of TC assembly are very similar to that of pure tubulin *(54)*. TC copolymerizes with tubulin, but the tubulin–podophyllotoxin complex does not; in fact, tubulin–podophyllotoxin inhibits both tubulin assembly into microtubules and TC assembly into polymers *(55)*. Podophyllotoxin but not colchicine is able to stabilize ring oligomers of avian erythrocyte tubulin to form crystals suitable for low-resolution electron diffraction *(56)*.

4. THE MECHANISM OF COLCHICINE BINDING TO TUBULIN

The interaction of colchicine with tubulin has several unusual and intriguing features. The association is very slow for a small molecule binding to a protein: the second order rate constant is in the range of $100/M/s$ at $37°C$ *(57)*. The resulting complex is noncovalent but dissociates extremely slowly ($\sim15 \times 10^{-6}/s$) *(58)*. Certain optical properties of the colchicine–tubulin complex are very unusual and have resisted satisfactory explanation. Specifically, the low energy absorption band of colchicine bound to tubulin displays absorption, fluorescence, and circular dichroic spectra that have not been precisely duplicated in the absence of the protein. Many of the earlier investigations into the mechanism of colchicine binding to tubulin were reviewed in the 1990s *(22,28–30,59)*.

4.1. Kinetics

A thorough review of the kinetics of colchicine and analogues binding to tubulin has been published recently *(59)*, so the treatment here will be brief. Colchicine binding to tubulin is a two step process: a rapid equilibrium between the ligand and the protein to form a low affinity complex, which is followed by a slow, unimolecular step to form the stable colchicine/tubulin complex. It is generally accepted that colchicine associates with tubulin through one of its conjugated rings in the first, low affinity complex. The second ring binds in the slow step, which requires a protein conformational change to form the final complex. A substituent on the C-7 carbon (*see* Fig. 1 for numbering system) raises the activation energy of the second step, indicating that there might be a steric interaction between the C-7 substituent and the protein during the unimolecular step. However, the B ring itself does not affect the activation energy of the process *(60)*.

The kinetics of colchicine binding to tubulin can be observed by monitoring the increase in colchicine fluorescence or quenching of tubulin fluorescence as a function of time. Under pseudo-first order conditions, two phases corresponding to two parallel reactions are observed in the kinetic data for colchicine binding to brain tubulin: a fast phase that accounts for most of the amplitude of the fluorescence change, and a slow phase of smaller amplitude. (Each phase of the association consists of the fast and slow steps described earlier.) Ludueña and coworkers have shown that the two kinetic phases observed when colchicine binds to brain tubulin are the result of the isotype composition of the tubulin. The slow phase of the association is owing to the lower rate constant for colchicine binding to $\alpha\beta$-III tubulin than to tubulin containing the other β-isotypes *(61–63)*.

4.2. Equilibrium Studies

The slow binding of colchicine to tubulin to form a nearly irreversible complex makes measuring equilibrium binding parameters difficult. Andreu and coworkers *(58)* used kinetic measurements to calculate a standard free energy of binding change of about –10 kcal/mol at 37°C, which corresponds to an association constant of ~$1 \times 10^7 M$. The enthalpy change for colchicine binding to brain tubulin has been determined from temperature dependence of the kinetic data and from calorimetry *(64)* to be around –5 to –6 kcal/mol. It was noted in the calorimetry studies that the bicyclic derivative of colchicine 2-methoxy-5-(2,3,4-trimethoxyphenyl) -2,46-cycloheptatrien-1-on (MTC) (Fig. 2) binds to tubulin with lower affinity than colchicine but with the same enthalpy change *(64)*. These data led to the conclusion that neither the B ring of colchicine nor its substituents make contact with the protein in the ground state.

There are small but real differences in the colchicine binding constants of purified tubulins with different β-isotypes *(65)*, but the differences are apparently too small to be observed in unfractionated tubulin. Whether the differences in the colchicine-binding properties of tubulin isotypes are relevant in vivo remains an open question *(66)*. It is not clear how the sequence differences in the isotypes affect the colchicine-binding site, as most of the sequence variation between the various isotypes is found in the C-termini of each subunit. Bhattacharyya and coworkers *(67)* have observed that removal of the C-termini from tubulin in unfractionated brain tubulin affects the association of colchicine to tubulin; the affinity of the protein for colchicine is decreased, and the

$$ M \bullet TL \overset{K_i}{\rightleftharpoons} M + TL \overset{K_b}{\rightleftharpoons} M + T + L \overset{K_g}{\rightleftharpoons} M \bullet T $$

liganded tubulin binding to the end of the microtubule	association of ligand with tubulin	unliganded tubulin binding to the end of the microtubule

Fig. 3. Timasheff's mechanism for the equilibria present at steady state. M—microtubule, T—unliganded tubulin, L—ligand, TL—ligand–tubulin complex. K_g—normal growth equilibrium constant, K_b—ligand–tubulin association constant, K_i—microtubule inhibition constant.

drug/tubulin association is no longer pH dependent. These investigators postulate that "tail-body" interactions between the C-terminal amino acids and the rest of the protein may allosterically affect the colchicine-binding site *(65,68)*.

Timasheff, Andreu, and coworkers *(69–72)* have carried out detailed experiments designed to elucidate the molecular interaction between tubulin and colchicine. They measured equilibrium binding constants and inhibition of tubulin assembly for a series of biphenyls and bridged biphenyls. Equilibrium-binding experiments yielded association constants for the ligands with unassembled tubulin (K_b). The microtubule inhibitory potential of these same ligands was assessed by measuring the I_{50} for pure tubulin, which is the total concentration of ligand required to reduce the extent of tubulin self-assembly to 50% of its value in the absence of added inhibitor. As the TC and not free colchicine is responsible for inhibition of microtubule assembly *(43)*, a simple model for the events in the process was constructed to calculate the "microtubule inhibition constant" K_i, which represents the binding constant for the ligand–tubulin complex for the end of the microtubule. The K_i provides a quantitative assessment for the inhibitory potency of a ligand (Fig. 3).

Another parameter was also calculated for these studies. The extent of ligand binding to tubulin at the I_{50} concentration, r, gives the fraction of unliganded tubulin bound to ligand at the I_{50}, which is inversely proportional to K_i. The r-value is used in the subsequent discussion, as it is a more intuitive expression. Table 1 lists the results for a few of the ligands *(71)*. As expected, colchicine binds with high affinity to tubulin (K_b is large) and the colchicine/tubulin complex is highly potent as an inhibitor of tubulin assembly (r is small). However, TME yields some unexpected results. The affinity of TME for tubulin is very low, and the I_{50} value is very high, but the r-value is nearly as small as that found for colchicine. In other words, TME does not bind as well as colchicine to tubulin, but once formed the TME/tubulin complex is nearly as effective as the colchicine–tubulin complex as an inhibitor of tubulin assembly. The bicyclic analog of allocolchicine, 2,3,4-trimethoxy-4'-carbomethoxy 1,1'-biphenigl (TCB) (Fig. 2), has an affinity for tubulin that is intermediate between colchicine and TME, and its I_{50} value is also intermediate between the two ligands. However, its r-value is very large: 40% of the tubulin dimers must be liganded with TCB to inhibit polymerization to the same extent as the other two ligands.

This is a dramatic illustration that tubulin binding and polymerization inhibition are not linked thermodynamically; that is, the efficacy of the drug–tubulin complex is not necessarily a direct function of binding site occupancy. Inhibition of tubulin assembly is a "post-binding phenomenon" *(71)*, dependent on the ability of the particular ligand to induce an "inactive" conformation in tubulin, not on the affinity of the ligand for the binding site.

Table 1
Tubulin Binding and Polymerization Inhibition by Selected Colchicine Site Ligands

Ligand	K_a (× 10^{-6}/M)	I_{50} (μM)	r (%)
Colchicine	16	0.42	1.9
TME[a]	0.00035	310	4.2
TCB[a]	0.1	19	40

[a]See Fig. 2 for structure.

From the series of compounds studied, Timasheff, Andreu, and coworkers formed some general conclusions. They postulated that the interaction between tubulin and one of the oxygens on the C ring is responsible for the extent of the conformational change that leads to inhibition of tubulin assembly. The B ring serves as a scaffold, fixing the substituents on the C ring in a particular locus within the ligand–tubulin complex, which may or may not be optimal for inducing a conformational change to an inhibitory species. The A ring primarily serves as an anchor and affects the conformation of the protein only by how it affects the orientation of the C ring in the binding site (72).

5. STRUCTURE–ACTIVITY RELATIONSHIPS

Literally hundreds of colchicine analogs have been prepared and tested for antimicrotubule activity and/or cytotoxicity. A great deal of the structure–activity relationship literature for colchicinoids was published in the 1980s and 1990s and has been reviewed (22,73–77). This review concentrates on literature published after 1990.

5.1. Three-Dimensional-Quantitative Structure–Activity Relationships

A few attempts to define a pharmacophore for the colchicine site on tubulin have been made through evaluation of quantitative structure–activity relationships (QSAR) in three-dimensions (3D). Two groups have used comparative molecular field analysis (CoMFA) (78). Briefly and greatly simplified, in this method the molecules of interest are aligned such that common structural features thought to be important for the activity are in the same region of space. A correlation between the electrostatic and steric fields of these molecules is sought mathematically, and the results of the calculations are displayed graphically. An attractive feature of this method is that the predictive ability of the model is also evaluated in the analysis. Macdonald and coworkers used CoMFA on a series of ligands, primarily combretastatin derivatives, with inhibition of [^{3}H] colchicine binding to tubulin or inhibition of tubulin assembly as the biological variables. The two biological variables yielded similar CoMFA models (79). Lee and coworkers performed CoMFA on a series of colchicine site ligands with a great deal of structural diversity, including colchicinoids, quinolines, and naphthyridinones, and used inhibition of tubulin assembly as the biological variable (80). The best results were obtained when the alignment required the aR-conformation of the A and C rings (Section 5.2) and for B ring substituent be oriented in the same direction on all the pertinent molecules. Each investigation produced a model with high correlation and good predictive ability, but the CoMFA fields appear quite different in the models from the different groups.

Polanski *(81,82)* reported using self-organized neural networks to develop a 3D-QSAR for colchicinoids and biphenyls, using inhibition of tubulin polymerization as the biological variable. The neural network approach is said to diminish one of the main criticism of the CoMFA method, the "alignment rule," in which the superimposition of molecules for analysis requires the operator to choose the molecular similarities within the set. The predictive ability of the model reported was quite low; however, the range of biological activities for this molecular set was small (less than one order of magnitude). It is possible that this type of model would provide better results with a data set possessing a greater range of biological activities.

Ducki et al. *(83)* used a "5D-QSAR" approach to evaluate a series of combretastatin and chalcone analogs. In this approach, operator bias is reduced by allowing for multiple conformations and orientations of the ligand molecules to be explored, and multiple *quasi*-atomic receptor models that can accommodate ligand-dependent induced fit are considered. Inhibition of tubulin polymerization data obtained in a single lab was used as the biological variable. Three receptor models were developed: one for the chalcone series, one for the combretastatin series, and a third model that combined molecules of the other two sets. The chalcone model had good predicative power, whereas the predictive power of the combretastatin model was moderate. The model generated from both structural classes, however, had lower predictive ability. The authors speculate that multiple ligand classes may not be well accommodated in the 5D-QSAR approach.

5.2. Axial Chirality

The tricyclic colchicinoids can exist in two atropisomeric forms. The relative amount of each atropisomer in solution is affected by the substituent on the C-7 carbon. Deacetamidocolchicine (Fig. 2) exists in solution in a racemic mixture of the two atropisomers *(80,84,85)*. Natural 7S-colchicine, also known as (−)-colchicine, exists in solution exclusively as the aR atropisomer, whereas unnatural 7R-colchicine, also known as (+)-colchicine, exists in solution as the aS atropisomer *(74,86)*. (It should be noted that in papers published before 1999, the axial chirality of natural (−)-colchicine was incorrectly designated aS *[87]*.) The C-7 amide is pseudoequatorial in the observed conformer for each molecule, which is presumably the molecular feature that dictates the axial chirality in the colchicine series (Fig. 4).

Changing the substitution pattern on the A or C rings of 7S-colchicine has not been observed to lead to the appearance of the aS atropisomer *(88–91)*. It is therefore generally assumed that the 7S-colchicinoids and 7R-colchicinoids are exclusively in the aR and aS axial configurations, respectively. When the C ring is aromatic rather than a tropone, however, both atropisomers are frequently observed in solution, and the ratio of atropisomers depends on solvent as well as the nature of the C-7 substituent *(92)*.

The 7R-isomers of colchicinoids are invariably less active than the 7S-isomers *(75)*. 7R-Colchicine is about 1/3 as active as natural 7S-colchicine as an inhibitor of [3H]colchicine binding to tubulin and about 40-fold less cytotoxic *(86)*. The 7R-isomer of thiocolchicine (Fig. 2) is 15-fold less potent in polymerization assays and 29-fold less cytotoxic than the 7S-isomer *(90)*. The activities of a series of 7-*O*-substituted deacetamidothiocolchicines were recently investigated *(91)*. The 7R-isomers were 4- to 12-fold less active than the 7S-isomers, further supporting the idea that the colchicine-binding site is stereoselective *(93)*. It is interesting to note that the axial chirality of the colchicinoid is

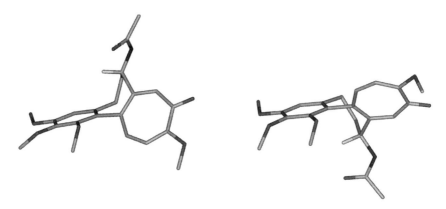

Fig. 4. Structures of (7S,aR)-(−)-colchicine (left) and (7R,aS)-(+)-colchicine (right), illustrating the atropisomers of the colchicine ring system. Note the C-7 amide is in a pseudoequatorial position in each structure. Color key: green—carbon, red—oxygen, blue—nitrogen, white—C-7 hydrogen. The remaining hydrogen atoms are omitted for clarity. To view this figure in color, see the insert and the companion CD-ROM.

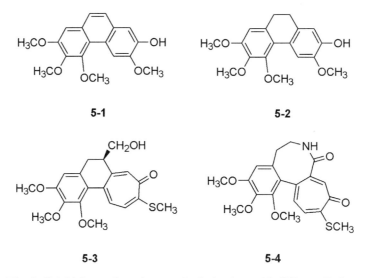

Fig. 5. Colchicine and combretastatin derivatives with different B rings.

much less detrimental to tubulin-binding activity than changing the substitution pattern of the tropone (Section 5.3).

The magnitude as well as the sign of the dihedral angle between the two rings is important in tubulin binding. Phenanthrenes and dihydrophenanthrenes with aromatic ring substitution patterns found in combretastatin A-4 (Fig. 5, **5-1** and **5-2**) are reported to be inactive as inhibitors of assembly, indicating that a planar or nearly planar arrangement of the A and C rings does not result in an active molecule in this series *(94)*. Thiocolchicine analogs possessing a 6-membered B ring have been prepared and tested. The crystal structure of one of these shows a*S* axial chirality and a dihedral angle of about 30° between the two rings (Fig. 5, **5-3**). This molecule binds rapidly to the colchicine site on tubulin *(95)* and inhibits tubulin assembly better than colchicine does *(88)* in spite of having the "wrong" axial chirality in the crystal. It was proposed that the a*R* atropisomer is the active conformation of the molecule. Circular dichroic spectra

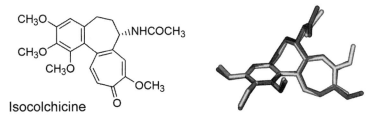

Isocolchicine

Fig. 6. Isocolchicine vs colchicine. The two-dimensional structure of isocolchicine is shown on the left. On the right is an overlay of the crystal structures of colchicine (red) and isocolchicine (blue). Rotations about the single bonds of some substituents were performed to improve the clarity of the illustration. All hydrogens are omitted on both structures. To view this figure in color, see the insert and the companion CD-ROM.

were interpreted to infer that both atropisomers of **5-3** are present in solution and inter-coversion of the atropisomers is rapid *(89)*.

A thiocolchicine analog with an 8-membered B ring was prepared and tested by Berg and coworkers *(96)* (Fig. 5, **5-4**). Two atropisomeric forms of the molecule were formed and were separated chromatographically. The atropisomer that corresponds to the axial chirality of natural a*R*-colchicine is about 10-fold less active than colchicine in inhibiting tubulin polymerization. The dihedral angle between the A and C rings in the crystal structure is 73°, which is about 20° larger than the 53° dihedral angle typically found in colchicinoids or the 49–60° angles found in methylated bridged biphenyls *(97)*. The other atropisomer was reported to be inactive at reasonable concentrations.

The colchicine-binding site on tubulin therefore seems fairly tolerant of variations in the dihedral angle between the A and C rings, provided that the correct axial chirality is maintained. High activity is found in molecules with angle sizes varying from 30° to about 60° *(98,99)*. Molecules with angles larger or smaller than these values are less active.

5.3. C Ring Derivatives

The structure–activity relationships within the C ring derivatives of colchicine and aromatic colchicinoids are not well understood. Isocolchicine is nearly 1000-fold less active than colchicine as an inhibitor of microtubule assembly and as a ligand for the colchicine-binding site, even though the crystal structures of the two molecules are virtually super imposable (Fig. 6) *(100)*. In fact, the C ring isoforms of all colchicinoids that have been tested are so much less active than the corresponding colchicine derivatives *(76)* that their activities are rarely reported anymore. Yet the C-10 position on colchicine can accommodate a wide variety of substituents with little change in the molecule's activity. Thiocolchicines show activity equal to or slightly more than that of the corresponding colchicines *(101)*. Replacing the methoxy group of colchicine with sterically and electronically diverse substituents such as halogens, azide, amines, ethers, and alkyl groups (including *tert*-butyl) usually resulted in only minor change in polymerization and [³H]colchicine-binding inhibition *(102)*.

A further complication is that the structure–activity relationship for the substituted tropone is affected by the nature of the connection between the A and C rings. The bicyclic colchicine derivative MTC binds to tubulin with 10-fold less affinity than colchicine. But if the methoxy group on the tropone ring is replaced by a chlorine atom, the bicyclic molecule becomes essentially inactive as an inhibitor of tubulin assembly,

Table 2
Inhibition of In Vitro Microtubule Assembly by Selected Colchicinoids

(I)   (II)   (III)

I_{50} (μM)

Substituent (R_1)	Colchicine (I, R_2 = NHAc)	MTC (II)	Combretatropone (III)	Deacetamidocolchicine (I, R_2 = H)
OCH$_3$	5.1	6.9	54	4.2
CH$_2$CH$_3$	4	63	160	NT
Cl	4	>500	300	3.7

whereas the colchicine analog becomes more potent *(103,104)*. The substitution is similarly deleterious for the combretatropone series, but does not significantly affect activity in the deacetamidocolchicine series (Table 2) *(105)*.

As only polymerization data were obtained, it cannot be determined whether the differences in activity are owing to differing affinity of the ligands for the protein, differing affects of the ligand on the conformation of the protein, or a combination of the two. However, this example does serve to further illustrate the potential pitfalls in rational drug design for the colchicine-binding site on tubulin.

5.4. A Ring Derivatives

A ring derivatives of colchicine show little structural diversity, primarily owing to the difficulty in preparing such compounds. Most colchicine derivatives are semisynthetic, and it is difficult to severely modify the A ring substitution pattern without destroying the tropone ring *(22)*. More variety in the A ring substitution pattern can be obtained from totally synthetic molecules such as bicyclic colchicinoids *(106)*, combretastatin derivatives *(107,108)*, and biaryls and bridged biphenyls *(38,109)*, but to date most of these derivatives have retained the trimethoxyphenyl ring of colchicine. Removing one of the methoxy groups or replacing a methyl group with another alkyl or an acyl group is normally detrimental to activity in these series *(108,110)*. Recently, more variety of A ring substituents was explored for combretastatin A-4 *(111)*. Replacing all three methoxy groups with methyl groups resulted in a molecule that is slightly more active than combretastatin A-4 as an inhibitor of microtubule assembly (Fig. 7, **7-1**), although it is less cytotoxic than combretastatin A-4 in both cell lines examined. There seems to be some link between the substitution patterns on the aromatic rings in this series. The fluorine-substituted derivative **7-2** also inhibits microtubule polymerization more potently than combretastatin A-4, but a combination of the two modifications (**7-3**) results in decreased activity in both polymerization and cytotoxicity assays.

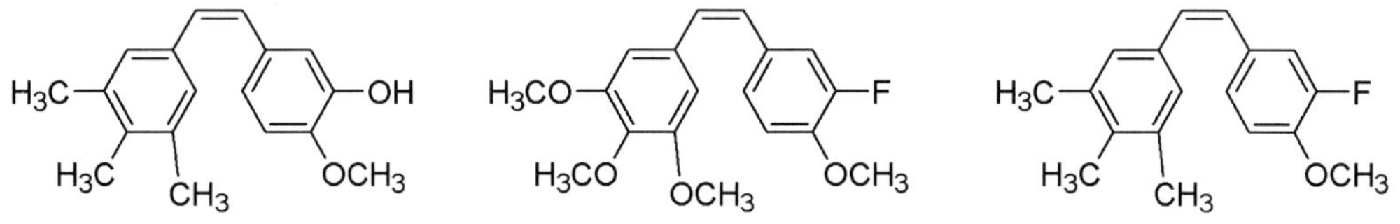

7-1   **7-2**   **7-3**

Fig. 7. A ring derivatives of combretastatin A-4.

6. THE COLCHICINE-BINDING SITE ON TUBULIN

A tremendous boost to understanding the molecular mechanisms of antimictubule agents resulted from the first publication of a 3D structure for tubulin *(112)*. The tubulin structure was obtained by electron crystallography from Zn-sheets of the protein stabilized with Taxol, thus the location of the Taxol-binding site on the protein could be unequivocally identified. The other two drug-binding sites can each be localized to a general region of the 3D-structure of the protein using photoaffinity labeling and other biochemical data *(113,114)*. It seems unlikely that more precise information about the colchicine-binding site will emerge from further study of these sheets, as it has not been possible to create a similar tubulin assemblage with colchicine occupying its binding site.

More recently, a 3.5 Å resolution structure of tubulin–colchicine complexed with the stathmin-like domain (SLD) of RB3 was solved by X-ray diffraction *(115)*. The conformation of tubulin in this structure is very similar to that found in Zn sheets in the terminal regions of the protein, but differs significantly in the "intermediate" domain. Colchicine is found in this domain, in a region at the interface of the α- and β-subunits. Most of the contacts between colchicine and tubulin are in the β-subunit, but the loop T5 from α-tubulin also makes contact with the ligand in this structure. The colchicine-binding site is deep within the protein in this structure; the T7 loop and H8 helix must be displaced from their positions in the Zn-sheet structure to form a binding site for colchicine. The conformational differences in the β-subunits of the two tubulin structures extend from the αβ-interface across the surface of the protein involved in longitudinal interactions between protofilaments to the GTP-binding site at the locus of lateral interactions between dimers in a protofilament.

It is interesting to compare the characteristics of the colchicine-binding site in the tubulin/colchicine: RB3-SLD structure with those predicted from biochemical studies. Photoaffinity labeling performed with [³H]colchicine identified two β-tubulin peptides: β1–41 and β214–241. Figure 8 shows a stereoview of the X-ray structure in which the two peptides are highlighted. The β214-241 peptide forms part of the colchicine-binding site, but the β1–41 peptide does not, although the C terminal amino acids of this peptide are in the vicinity of the colchicine-binding site. Earlier photoaffinity labeling studies used an aromatic azide attached to the C-7 substituent of colchicine. Labels attached to the C-7 substituent of colchicine labeled the α-subunit exclusively when a long spacer was included, whereas both α- and β-tubulin were labeled when no atoms were incorporated between the amine and the aromatic azide *(116,117)*. The crystal structure can accommodate these results: the C-7 substituent is near the exterior of the binding site, oriented toward the α-subunit. Experiments in which sulfhydryl groups

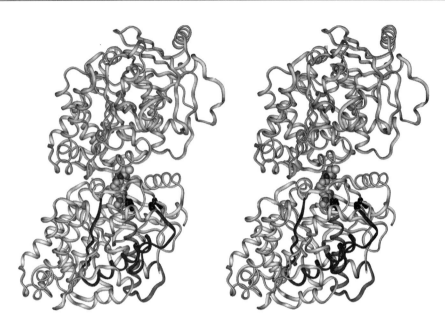

Fig. 8. Stereoview of one of the tubulin-colchicine complexes from the crystal structure of tubulin/ colchicine:stathmin-like domain complex (PDB 1SA0, *[110]*). The backbone of the α-subunit is shown in light blue and the backbone of the β-subunit is colored yellow. DAMA-colchicine (Fig. 2) is colored by atom type (green—carbon, red—oxygen, blue—nitrogen, yellow—sulfur; hydrogen atoms are not shown). Cys-241 (left) and Cys-356 (right) are shown in black. The backbone of the β-subunit peptide 1–41 is shown in dark blue and the backbone of the β-subunit peptide 214–241 is shown in pink. To view this figure in color, see the insert and the companion CD-ROM.

were covalently modified support the location of the trimethoxyphenyl A ring of colchicine very close to Cys-241. Colchicine affinity labels in which the methyl group at C-2 or C-3 is replaced with a chloroacetyl group form covalent bonds with Cys-241 primarily and Cys-356 secondarily *(118)*. The intramolecular crosslink between Cys-241 and Cys-356 is strongly inhibited by colchicine binding to tubulin, but is unaffected by the presence of the C ring analog TME *(119)*.

\Cabral and coworkers *(120)* have created and studied many Chinese hamster ovary cells that are resistant to tubulin-binding drugs. The primary mechanism by which tubulin mutations in these cells produce drug resistance is through alterations of amino acids outside the binding sites, which result in altered microtubule assembly properties but not drug-binding properties. In one notable exception, a mutation in β-tubulin (A254V) caused transfected cells to be hypersensitive to colchicine *(121)*. This activity was attributed to an increase in affinity for colcemid by the mutant tubulin. The mutated residue is located on helix 8, about 5 Å from tubulin-bound colchicine. Although the affected amino acid is not on the interior of the binding site defined by X-ray crystallography, it is reasonable to propose that the mutation allosterically rather than sterically affects the shape of the binding site.

An unexpected discovery in the crystal structure is that the colchicine-binding site is deep in the protein (Fig. 9). This finding is in harmony with the noncovalent but essentially irreversible nature of the colchicine–tubulin complex—clearly a significant conformational change in the protein would be necessary for the complex to dissociate. The origin of colchicine fluorescence in the colchicine–tubulin complex can

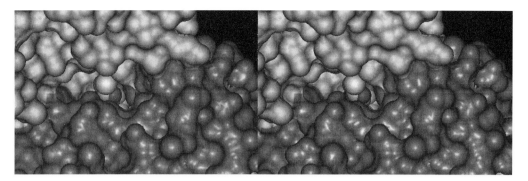

Fig. 9. Stereoview of the colchicine-binding site of tubulin/colchicine:stathmin-like domain complex (PDB 1SA0). The surface of the protein was rendered without hydrogen atoms. The α-subunit is colored blue and the β-subunit is light brown. DAMA-colchicine is colored by atom type. To view this figure in color, see the insert and the companion CD-ROM.

also be rationalized. Intrinsic colchicine fluorescence is greatly enhanced on tubulin binding, possibly by rigidifying the molecule and thereby decreasing the loss of excited state energy by vibrational motion *(122)*. The depth of the colchicine-binding site in this structure might be an indication of a rigid binding site. But how colchicine can access such a deep pocket in the protein? Kinetic data indicate that the initial encounter of the ligand and the protein forms a low-affinity complex and tubulin conformational changes occur in a subsequent unimolecular step. How colchicine binding to a pocket on the exterior of the protein would trigger the requisite conformational adjustment is currently unclear.

The unusual absorption, circular dichroic, and resonance Raman spectra of colchicine in its complex with tubulin are also not apparently explained by the structure. A stacking interaction between the C ring of colchicine and an aromatic amino acid in the binding site has been suggested to explain these data *(22)*, but no such interaction is found in the X-ray structure. Finally, a second structure was solved in which tubulin–podophyllotoxin was complexed with RB3-SLD. Both aromatic rings of podophyllotoxin in the structure were found to overlap with the A and C rings of tubulin-bound colchicine, which does not seem to agree with data from competition experiments. Differences in tubulin conformation that would account for the differences in the activities of the two drugs were not apparent in the two structures.

In spite of the unanswered questions, the X-ray structure of tubulin–colchicine is proving useful in picturing how structurally diverse ligands can bind to the same region of the protein. Nguyen et al. *(123)* used the X-ray structure of the complex to design a pharmacophore model for a diverse set of colchicine site ligands, ranging from colchicine itself to curacin A. Correlations between pharmacophoric elements of the ligands were obtained in 3D by aligning each ligand to colchicine and podophyllotoxin, followed by refinement using the molecular volume and electrostatic surface from the X-ray structure. Pharmacophoric points were derived from recurring tubulin–ligand interactions, and overall pharmacophore for the colchicine-binding site consists of a total of seven points. Each ligand possesses five or six of the elements of the overall pharmacophore. The authors intend to use this pharmacophore model to design ligands with superior tubulin-binding activity.

7. CONCLUSION

Colchicine continues to confound and intrigue investigators even after nearly 40 yr of studying its interactions with tubulin. Structure–activity, mechanistic and the recent protein structural studies have yielded significant inroads into understanding the molecular features of colchicine activity, although some important aspects of the mechanism remain unclear. Perhaps the most fundamental question unanswered by current data is how the colchicine-binding site can accommodate quite disparate ligand structures, yet be exquisitely sensitive to minor variations within a single structural class. The advent of a crystal structure of colchicine–tubulin, albeit complexed with RB3-SLD, should greatly assist in unraveling the current structure–activity questions. It is not unreasonable to hope that we are approaching a time when *de novo* design of a colchicine site drug is achievable.

ACKNOWLEDGMENTS

Dr. Barbara M. Poliks created all of the 3D-illustrations. Thanks to Drs. Jack Correia, Jim Snyder, and Rudy Ravindra for critiquing parts of this work.

REFERENCES

1. Jordan A, Hadfield JA, Lawrence NJ, McGown AT. Tubulin as a target for anticancer drugs: Agents which interact with the mitotic spindle. Med Res Rev 1998;18:259–296.
2. Checchi PM, Nettles JH, Zhou J, Snyder JP, Joshi HC. Microtubule-interacting drugs for cancer treatment. Trends Pharmacol Sci 2003;24:361–365.
3. Jordan MA, Wilson L. Microtubules as a target for anticancer drugs. Nature Rev Cancer 2004;4: 253–265.
4. Giannakakou P, Sackett D, Fojo T. Tubulin/microtubules: Still a promising target for new chemotherapeutic agents. J Natl Cancer Inst 2000;92:182–183.
5. Rowinsky EK, Donehower RC. Drug-Therapy - Paclitaxel (Taxol). N Engl J Med 1995;332:1004–1014.
6. Altaha R, Fojo T, Reed E, Abraham J. Epothilones: A novel class of non-taxane microtubule-stabilizing agents. Curr Pharm Des 2002;8:1707–1712.
7. Kruczynski A, Hill BT. Vinflunine, the latest Vinca alkaloid in clinical development—A review of its preclinical anticancer properties. Crit Rev Oncol Hematol 2001;40:159–173.
8. Cirla A, Mann J. Combretastatins: from natural products to drug discovery. Nat Prod Rep 2003;20: 558–564.
9. Chaplin DJ, Hill SA. The development of combretastatin A4 phosphate as a vascular targeting agent. Int J Rad Oncol Biol Phys 2002;54:1491–1496.
10. Marx MA. Small-molecule, tubulin-binding compounds as vascular targeting agents. Expert Opin Ther Patents 2002;12:769–776.
11. Sackett DL. Podophyllotoxin, steganacin and combretastatin: natural products that bind at the colchicine site of tubulin. Pharmacol Ther 1993;59:163–228.
12. Tozer GM, Kanthou C, Parkins CS, Hill SA. The biology of the combretastatins as tumour vascular targeting agents. Int J Exp Pathol 2002;83:21–38.
13. Pettit GR, Toki B, Herald DL, et al. Antineoplastic agents. 379. Synthesis of phenstatin phosphate. J Med Chem 1998;41:1688–1695.
14. Lakhani NJ, Sarkar MA, Venitz J, Figg WD. 2-methoxyestradiol, a promising anticancer agent. Pharmacotherapy 2003;23:165–172.
15. Xu K, Schwarz PM, Luduena RF. Interaction of nocodazole with tubulin isotypes. Drug Dev Res 2002;55:91–96.
16. Wipf P, Reeves JT, Day BW. Chemistry and biology of curacin A. Curr Pharm Des 2004;10: 1417–1437.
17. Lawrence NJ, McGown AT. The chemistry and biology of antimitotic chalcones and related enone systems. Curr Pharm Des 2005;11:1679–1693.

18. Barbier P, Peyrot V, Sarrazin M, Rener GA, Briand C. Differential effects of ethyl 5-amino-2-methyl-1,2-dihydro-3-phenylpyrido [3,4-b] pyrazin-7-yl carbamate analogs modified at position C-2 on tubulin polymerization, binding, and conformational changes. Biochemistry 1995;34:16,821–16,829.

19. Peyrot V, Leynadier D, Sarrazin M, et al. Mechanism of binding of the new antitumor drug MDL 27048 to the colchicine site on tubulin: equilibrium studies. Biochemistry 1992;31:11,125–11,132.

20. Li Q, Woods KW, Claiborne A, et al. Synthesis and biological evaluation of 2-indolyloxazolines as a new class of tubulin polymerization inhibitors. Discovery of A-289099 as an orally active antitumor agent. Bioorg Med Chem Lett 2002;12:465–469.

21. Tahir SK, Nukkala MA, Mozny NAZ, et al. Biological activity of A-289099: An orally active tubulin-binding indolyloxazoline derivative. Mol Cancer Ther 2:227–233.

22. Hastie SB. Interactions of colchicine with tubulin. Pharmacol Ther 1991;51:377–401.

23. Kelleher JK. Tubulin binding affinities of podophyllotoxin and colchicine analogs. Mol Pharmacol 1977;13:232–241.

24. Bhattacharyya B, Wolff J. Promotion of fluorescence upon binding of colchicine to tubulin. Proc Natl Acad Sci USA 1974;71:2627–2631.

25. Banerjee A, Bhattacharyya B. Colcemid and colchicine binding to tubulin: similarity and dissimilarity. FEBS Lett 1979;99:333–336.

26. Ray K, Bhattacharyya B, Biswas BB. Role of B-ring of colchicine in its binding to tubulin. J Biol Chem 1981;256:6241–6244.

27. Cortese F, Bhattacharyya B, Wolff J. Podophyllotoxin as a probe for the colchicine binding site of tubulin. J. Biol. Chem. 1977;252:1134–1140.

28. Guha S, Bhattacharyya B. The colchicine-tubulin interaction: A review. Curr Sci 1997;73:351–358.

29. Hamel E. Antimitotic natural products and their interactions with tubulin. Med Res Rev 1996;16: 207–231.

30. Hamel E. Interactions of tubulin with small ligands. In: Microtubule Proteins, Avila J, ed. CRC Press, Inc., Boca Raton, FL, 1990; pp 89–191.

31. Correia JJ. Effects of antimitotic agents on tubulin-nucleotide interactions. Pharmacol Ther 1991;52: 127–147.

32. Correia JJ, Lobert S. Physiochemical aspects of tubulin-interacting antimitotic drugs. Curr Pharm Des 7:1213–1228.

33. Timasheff SN, Andreu JM, Na GC. Physical and spectroscopic methods for the evaluation of the interactions of antimitotic agents with tubulin. Pharmacol Ther 1991;52:191–210.

34. Lee KH. Novel antitumor agents from higher plants. Med Res Rev 1999;19:569–596.

35. Canel C, Moraes RM, Dayan FE, Ferreira D. Podophyllotoxin. Phytochemistry 2000;54:115–120.

36. Bacher G, Beckers T, Emig P, Klenner T, Kutscher B, Nickel B. New small-molecule tubulin inhibitors. Pure Appl Chem 2001;73:1459–1464.

37. Hamel E. Interactions of antimitotic peptides and depsipeptides with tubulin. Biopolymers 2002;66: 142–160.

38. Li Q, Sham HL. Discovery and development of antimitotic agents that inhibit tubulin polymerisation for the treatment of cancer. Expert Opin Ther Patents 2002;12:1663–1702.

39. Lee KH. Current developments in the discovery and design of new drug candidates from plant natural product leads. J Nat Prod 2004;67:273–283.

40. Borisy GG, Taylor EW. The mechanism of action of colchicine. Binding of colchicine-^{3}H to cellular protein. J Cell Biol 1967;34:525–533.

41. Weisenberg RC, Borisy GG, Taylor EW. The colchicine-binding protein of mammalian brain and its relation to microtubules. Biochemistry 1968;7:4466–4479.

42. Le Hello C. The pharmacology and therapeutic aspects of colchicine. In: The Alkaloids, Cordell GA, ed. Academic Press, New York, vol 53, 2000; pp 287–352.

43. Skoufias DA, Wilson L. Mechanism of inhibition of microtubule polymerization by colchicine—Inhibitory potencies of unliganded colchicine and tubulin colchicine complexes. Biochemistry 1992; 31:738–746.

44. Sackett DL, Varma JK. Molecular mechanism of colchicine action - Induced local unfolding of β-tubulin. Biochemistry 1993;32:13,560–13,565.

45. Wilson L, Jordan MA. Drug Interactions with Microtubules. In: Microtubules, Hyams JS, Lloyd CW, eds. Wiley-Liss, New York, 1994; pp 59–83.

46. Wilson L, Panda D, Jordan MA. Modulation of microtubule dynamics by drugs: A paradigm for the actions of cellular regulators. Cell Struct Funct 1999;24:329–335.

47. Vandecandelaere A, Martin SR, Engelborghs Y. Response of microtubules to the addition of colchicine and tubulin-colchicine: Evaluation of models for the interaction of drugs with microtubules. Biochem J 1997;323:189–196.

48. Panda D, Daijo JE, Jordan MA, Wilson L. Kinetic stabilization of microtubule dynamics at steady state in vitro by substoichiometric concentrations of tubulin-colchicine complex. Biochemistry 1995;34:9921–9929.

49. Hamel E. Evaluation of antimitotic agents by quantitative comparisons of their effects on the polymerization of purified tubulin. Cell Biochem Biophys 2003;38:1–21.

50. Sackett DL. Vinca site agents induce structural-changes in tubulin different from and antagonistic to changes induced by colchicine site agents. Biochemistry 1995;34:7010–7019.

51. David-Pfeuty T, Simon C, Pantaloni D. Effect of antimitotic drugs on tubulin GTPase activity and self-assembly. J Biol Chem 1979;254:11,696–11,702.

52. Lin CM, Hamel E. Effects of inhibitors of tubulin polymerization on GTP hydrolysis. J Biol Chem 1981;256:9242–9245.

53. Andreu JM, Timasheff SN. Tubulin bound to colchicine forms polymers different from microtubules. Proc Natl Acad Sci USA 1982;79:6753–6756.

54. Andreu JM, Wagenknecht T, Timasheff SN. Polymerization of the tubulin-colchicine complex: relation to microtubule assembly. Biochemistry 1983;22:1556–1566.

55. Saltarelli D, Pantaloni D. Polymerization of the tubulin-colchicine complex and guanosine 5′-triphosphate hydrolysis. Biochemistry 1982;21:2996–3006.

56. Schonbrunn E, Phlippen W, Trinczek B, et al. Crystallization of a macromolecular ring assembly of tubulin liganded with the anti-mitotic drug podophyllotoxin. J Struct Biol 1999;128:211–215.

57. Garland DL. Kinetics and mechanism of colchicine binding to tubulin: evidence for ligand-induced conformational changes. Biochemistry 1978;17:4266–4272.

58. Diaz JF, Andreu JM. Kinetics of dissociation of the tubulin-colchicine complex: complete reaction scheme and comparison to thermodynamic measurements. J Biol Chem 1991;266:2890–2896.

59. Engelborghs Y. General features of the recognition by tubulin of colchicine and related compounds. Eur Biophys J 1998;27:437–445.

60. Pyles EA, Hastie SB. Effect of the B-ring and the C-7 substituent on the kinetics of colchicinoid tubulin associations. Biochemistry 1993;32:2329–2336.

61. Banerjee A, Luduena RF. Distinct colchicine binding-kinetics of bovine brain tubulin lacking the type-III isotype of β-tubulin. J Biol Chem 1991;266:1689–1691.

62. Banerjee A, Luduena RF. Kinetics of colchicine binding to purified β-tubulin isotypes from bovine brain. J Biol Chem 1992;267:13,335–13,339.

63. Banerjee A, Engelborghs Y, D' Hoore A, Fitzgerald TJ. Interactions of a bicyclic analog of colchicine with β-tubulin isoforms αβII, αβIII, αβIV. Eur J Biochem 1997;246:420–424.

64. Menendez M, Laynez J, Medrano FJ, Andreu JM. A thermodynamic study of the interaction of tubulin with colchicine site ligands. J Biol Chem 1989;264:16,367–16,371.

65. Banerjee A, Luduena RF. Kinetics of colchicine binding to purified beta-tubulin isotypes from bovine brain. J Biol Chem 1992;267:13,335–13,339.

66. Luduena RF. Multiple forms of tubulin: Different gene products and covalent modifications. In: International Review of Cytology - a Survey of Cell Biology, vol 178, 1998; pp 207–275.

67. Mukhopadhyay K, Parrack PK, Bhattacharyya B. The carboxy terminus of the α-subunit of tubulin regulates its interaction with colchicine. Biochemistry 1990;29:6845–6850.

68. Pal D, Mahapatra P, Manna T, et al. Conformational properties of alpha-tubulin tail peptide: Implications for tail-body interaction. Biochemistry 2001;40:15,512–15,519.

69. Andreu JM, Gorbinoff MJ, Medrano FJ, Rossi M, Timasheff SN. Mechanism of colchicine binding to tubulin: tolerance of substituents in ring C′ of biphenyl analogues. Biochemistry 1991;30:3777–3786.

70. Medrano FJ, Andreu JM, Gorbunoff MJ, Timasheff SN. Roles of the ring C oxygens in the binding of colchicine to tubulin. Biochemistry 1991;30:3770–3777.

71. Perez-Ramirez B, Gorbunoff MJ, Timasheff SN. Linkages in tubulin-colchicine functions: The role of the ring C (C′) oxygens and ring B in the controls. Biochemistry 1998;37:1646–1661.

72. Andreu JM, Perez-Ramirez B, Gorbunoff MJ, Ayala D, Timasheff SN. Role of the colchicine ring A and its methoxy groups in the binding to tubulin and microtubule inhibition. Biochemistry 1998;37:8356–8368.

73. Shi Q, Chen K, Morris-Natschke SL, Lee KH. Recent progress in the development of tubulin inhibitors as antimitotic antitumor agents. Curr Pharm Des 1998;4:219–248.
74. Brossi A, Yeh HJC, Chrzanowska M, et al. Colchicine and its analogs—recent findings. Med Res Rev 1988;8:77–94.
75. Boye O, Brossi A. Tropolonic *Colchicum* alkaloids and all congeners. In: The Alkaloids, Brossi A, ed. Academic Press, New York, vol 41, 1992; pp 125–176.
76. Capraro HG, Brossi A. Tropolonic *Colchicum* alkaloids. In: The Alkaloids, Brossi A, ed. Academic Press, New York, vol XXIII, 1984; pp 1–70.
77. Lee KH. Anticancer drug design based on plant-derived natural products. J Biomed Sci 1999;6:236–250.
78. Cramer RD, Patterson DE, Bunce JD. Comparative Molecular-Field Analysis (CoMFA) .1. Effect of shape on binding of steroids to carrier proteins. J Am Chem Soc 1988;110:5959–5967.
79. Brown ML, Rieger JM, Macdonald TL. Comparative molecular field analysis of colchicine inhibition and tubulin polymerization for combretastatins binding to the colchicine binding site on β-tubulin. Bioorg Med Chem 2000;8:1433–1441.
80. Zhang SX, Feng J, Kuo SC, et al. Antitumor agents. 199. Three-dimensional quantitative structure–activity relationship study of the colchicine binding site ligands using comparative molecular field analysis. J Med Chem 2000;43:167–176.
81. Polanski J. The non-grid technique for modeling 3D QSAR using self-organizing neural network (SOM) and PLS analysis: Application to steroids and colchicinoids. SAR QSAR Environ Res 2000;11:245–261.
82. Polanski J. Self–organizing neural network for modeling 3D QSAR of colchicinoids. Acta Biochim Pol 2000;47:37–45.
83. Ducki S, Mackenzie G, Lawrence NJ, Snyder JP. Quantitative structure-activity relationship (5D-QSAR) study of combretastatin-like analogues as inhibitors of tubulin assembly. J Med Chem 2005;48:457–465.
84. Berg U, Deinum, J., Lincoln, P., and Kvassman, J. (1991) Stereochemistry of colchicinoids. Enantiomeric stability and binding to tubulin of desacetamidocolchicine and desacetamidoisocolchicine. *Bioorg. Chem.* 19, 53–65.
85. Brossi A, Boye O, Muzaffar A, et al. aS,7S-absolute configuration of natural (−)-colchicine and allo-congeners. FEBS Lett 1990;262:5–7.
86. Roesner M, Capraro HG, Jacobson AE, et al. Biological effects of modified colchicines. Improved preparation of 2-demethylcolchicine, 3-demethylcolchicine, and (+)-colchicine and reassignment of the position of the double bond in dehydro-7-deacetamidocolchicines. J Med Chem 1981;24: 257–261.
87. Berg U, Bladh H. The absolute configuration of colchicine by correct application of the CIP rules. Helv Chim Acta 1999;82:323–325.
88. Sun L, Hamel E, Lin CM, Hastie SB, Pyluck A, Lee KH. Antitumor Agents .141. Synthesis and biological evaluation of novel thiocolchicine analogs - N-acyl-(substituted benzyl)deacetylthiocolchicines, and N-aroyl-(substituted benzyl)deacetylthiocolchicines, and (substituted benzyl) deacetylthiocolchicines as potent cytotoxic and antimitotic compounds. J Med Chem 1993;36: 1474–1479.
89. Sun L, McPhail AT, Hamel E, et al. Antitumor Agents .139. synthesis and biological evaluation of thiocolchicine analogs 5,6-dihydro-6(S)-(acyloxy) and 5,6-dihydro-6(S)- (aroyloxy)methyl -1,2,3-trimethoxy-9-(methylthio)-8H-cyclohepta[a]naphthalen-8-ones as novel cytotoxic and antimitotic agents. J Med Chem 1993;36:544–551.
90. Shi Q, Verdier-Pinard P, Brossi A, Hamel E, Lee KH. Antitumor agents—CLXXV. Anti-tubulin action of (+)-thiocolchicine prepared by partial synthesis. Bioorg Med Chem 1997;5:2277–2282.
91. Shi Q, VerdierPinard P, Brossi A, Hamel E, McPhail AT, Lee KH. Antitumor agents .172. Synthesis and biological evaluation of novel deacetamidothiocolchicin-7-ols and ester analogs as antitubulin agents. J Med Chem 1997;40:961–966.
92. Shi Q, Chen K, Chen X, et al. Antitumor agents. 183. Syntheses, conformational analyses, and anti-tubulin activity of allothiocolchicinoids. J Org Chem 1998;63:4018–4025.
93. Yeh HJC, Chrzanowska M, Brossi A. The importance of the phenyl-tropolone 'aS' configuration in colchicine's binding to tubulin. FEBS Lett 1988;229:82–86.
94. Lin CM. Singh SB, Chu PS, et al. Interactions of tubulin with potent natural and synthetic analogs of the antimitotic agent combretastatin—a structure-activity study. Mol Pharmacol 1988;34:200–208.

95. Lincoln P, Nordh J, Deinum J, Angstrom J, Norden B. Conformation of thiocolchicine and 2 B-ring-modified analogs bound to tubulin studied with optical spectroscopy. Biochemistry 1991;30: 1179–1187.

96. Berg U, Bladh H, Svensson C, Wallin M. The first colchicine analogue with an eight membered B-ring. Structure, optical resolution and inhibition of microtubule assembly. Bioorg Med Chem Lett 1997;7:2771–2776.

97. Edwards DJ, Pritchard RG, Wallace TW. Fine-tuning of biaryl dihedral angles: structural characterization of five homologous three-atom bridged biphenyls by X-ray crystallography. Acta Cryst B 2005;61:335–345.

98. Banwell MG, Peters SC, Greenwood RJ, Mackay MF, Hamel E, Lin CM. Semisyntheses, X-ray crystal-structures and tubulin-binding properties of 7-oxodeacetamidocolchicine and 7-oxodeacetami-doisocolchicine. Aust J Chem 1992;45:1577–1588.

99. Buttner F, Bergemann S, Guenard D, Gust R, Seitz G, Thoret S. Two novel series of allocolchicinoids with modified seven membered B-rings: design, synthesis, inhibition of tubulin assembly and cyto-toxicity. Bioorg Med Chem 2005;13:3497–3511.

100. Hastie SB, Williams RC, Puett D, Macdonald TL. The binding of isocolchicine to tubulin - mechanisms of ligand association with tubulin. J Biol Chem 264:6682–6688.

101. Kerekes P, Sharma PN, Brossi A, Chignell CF, Quinn FR. Synthesis and biological effects of novel thiocolchicines. 3. Evaluation of N-acyldeacetylthiocolchicines, N-(alkoxycarbonyl)deacetylthio-colchicines, and O-ethyldemethylthiocolchicines. New synthesis of thiodemecolcine and antileukemic effects of 2-demethyl- and 3-demethylthiocolchicine. J Med Chem 1985;28:1204–1208.

102. Staretz ME, Hastie SB. Synthesis and tubulin binding of novel C-10 analogs of colchicine. J Med Chem 1993;36:758–764.

103. Andres CJ, Bernardo JE, Yan Q, Hastie SB, Macdonald TL. Combretatropones—Hybrids of combre-tastatin and colchicine—synthesis and biochemical evaluation. Bioorg Med Chem Lett 1993;3: 565–570.

104. Hahn KM, Humphreys WG, Helms AM, Hastie SB, Macdonald TL. Structural requirements for the binding of colchicine analogs to tubulin—the role of the C-10 substituent. Bioorg Med Chem Lett 1991;1:471–476.

105. Janik ME, Bane SL. Synthesis and antimicrotubule activity of combretatropone derivatives. Bioorg Med Chem 2002;10:1895–1903.

106. Deveau AM, Macdonald TL. Practical synthesis of biaryl colchicinoids containing 3′,4′-catechol ether-based A-rings via Suzuki cross-coupling with ligandless palladium in water. Tetrahedron Lett 2004;45:803–807.

107. Cushman M, Nagarathnam D, Gopal D, Chakraborti AK, Lin CM, Hamel E. Synthesis and evaluation of stilbene and dihydrostilbene derivatives as potential anticancer agents that inhibit tubulin poly-merization. J Med Chem 1991;34:2579–2588.

108. Cushman M, Nagarathnam D, Gopal D, He HM, Lin CM, Hamel E. Synthesis and evaluation of analogs of (Z)-1-(4-methoxyphenyl)-2-(3,4,5-trimethoxyphenyl)ethene as potential cytotoxic and antimitotic agents. J Med Chem 1992;35:2293–2306.

109. Banwell MG, Cameron JM, Corbett M, et al. Synthesis and tubulin-binding properties of some AC-ring and ABC-ring analogs of allocolchicine. Aust J Chem 1992;45:1967–1982.

110. Boye O, Brossi A, Yeh HJC, Hamel E, Wegrzynski B, Toome V. Natural-products - antitubulin effect of congeners of N-acetylcolchinyl methyl-ether - synthesis of optically-active 5-acetamido-deaminocolchinyl methyl-ether and of demethoxy analogs of deaminocolchinyl methyl-ether. Can J Chem 1992;70:1237–1249.

111. Gaukroger K, Hadfield JA, Lawrence NJ, Nolan S, McGown AT. Structural requirements for the interaction of combretastatins with tubulin: how important is the trimethoxy unit? Org Biomol Chem 2003;1:3033–3037.

112. Nogales E, Wolf SG, Downing KH. Structure of the αβ tubulin dimer by electron crystallography. Nature 1998;391:199–203.

113. Downing KH. Structural basis for the interaction of tubulin with proteins and drugs that affect micro-tubule dynamics. Annu Rev Cell Dev Biol 2000;16:89–111.

114. Downing KH, Nogales E. New insights into microtubule structure and function from the atomic model of tubulin. Eur Biophys J 1998;27:431–436.

115. Ravelli RBG, Gigant B, Curmi PA, et al. Insight into tubulin regulation from a complex with colchicine and a stathmin-like domain. Nature 2004;428:198–202.

116. Williams RF, Mumford CL, Williams GA, et al. A photoaffinity derivative of colchicine: 6-(4′-azido-2′-nitrophenylamino)hexanoyldeacetylcolchicine. Photolabeling and location of the colchicine-binding site on the α-subunit of tubulin. J Biol Chem 1985;260:13,794–13,802.

117. Floyd LJ, Barnes LD, Williams RF. Photoaffinity labeling of tubulin with (2-nitro-4-azidophenyl) deacetylcolchicine: direct evidence for two colchicine binding sites. Biochemistry 1989;28:8515–8525.

118. Bai RL, Covell DG, Pei XF, et al. Mapping the binding site of colchicinoids on β-tubulin - 2-chloroacetyl-2-demethylthiocolchicine covalently reacts predominantly with cysteine 239 and secondarily with cysteine 354. J Biol Chem 2000;275:40,443–40,452.

119. Luduena RF, Roach MC. Tubulin sulfhydryl-groups as probes and targets for antimitotic and antimicrotubule agents. Pharmacol Ther 1991;49:133–152.

120. Hari M, Wang YQ, Veeraraghavan S, Cabral F. Mutations in α- and β-tubulin that stabilize microtubules and confer resistance to Colcemid and vinblastine. Mol Cancer Ther 2003;2:597–605.

121. Wang YQ, Veeraraghavan S, Cabral F. Intra-allelic suppression of a mutation that stabilizes microtubules and confers resistance to colcemid. Biochemistry 2004;43:8965–8973.

122. Bhattacharyya B, Wolff J. Immobilization-dependent fluorescence of colchicine. J Biol Chem 1984;259:11,836–11,843.

123. Nguyen TL, McGrath C, Hermone AR, et al. A common pharmacophore for a diverse set of colchicine site inhibitors using a structure-based approach. J Med Chem 2005;48:6107–6116.

12 Antimicrotubule Agents That Bind Covalently to Tubulin

Dan L. Sackett

SUMMARY

The binding of most microtubule inhibitors (MTI) is reversible. Most of these inhibit the polymerization of tubulin to microtubules (MT), but an important group promotes polymerization and stabilizes MT. A subset of MTI are not reversible due to covalent adduct formation with tubulin. Some of these agents are quite specific in the amino acid residue targeted for reaction, while others are less specific, though still targeting MT function in cells. Almost all reactive MTI cause loss of MT, but one reactive agent is known that stabilizes MT. Many reactive MTI target cysteines in tubulin, especially in β-tubulin, although MTI are known that target lysine and other residues. Reactive MTI include organic, inorganic, natural, and synthetic compounds, environmental toxicants as well as potential therapeutics.

Key Words: Tubulin; sulfhydryl; cysteine; heavy metals; mercury; cisplatin; lysine.

1. INTRODUCTION

Most microtubule (MT) inhibitors (MTI) bind reversibly either to the tubulin dimer, inhibiting assembly, or to the MT polymer, inhibiting disassembly. However, not all MTI interact by reversible interactions; some mediate covalent modifications of tubulin that result in inhibition of MT function. To date, all of these compounds inhibit assembly of tubulin to MT, but in principle such compounds could also be MT stabilizers. At one level, virtually any molecule that can react with proteins can probably inhibit MT polymerization from purified tubulin. Indeed, just adding sufficient HCl will do the same thing. Much useful information has been obtained about structural requirements for tubulin polymerization by using general reagents that react with specific groups on proteins

From: *Cancer Drug Discovery and Development: The Role of Microtubules in Cell Biology, Neurobiology, and Oncology* Edited by: Tito Fojo © Humana Press, Totowa, NJ

(hence more specific than HCl). These reagents react with any such residue on essentially any protein on which that group is exposed. Some such studies will be mentioned in this review. But the main focus is on molecules that react covalently with tubulin with some specificity, as opposed to other proteins. In addition to reacting with tubulin, molecules considered here inhibit assembly of MT from purified protein as well as in cells. Many of these covalent MTI are of interest precisely because they are active in a cellular—and whole organism—context, and some are in clinical development or current use.

Covalent MTI are a diverse set of molecules, though not as numerous as the reversibly binding MTI. Many to most of these reactive compounds form covalent bonds with the sulfhydryl moiety of cysteine residues. Of these compounds, some show remarkable residue specificity among the many cysteines of tubulin. Others react with multiple cysteines, though perhaps preferentially with particular residues, especially at low concentrations that select for the most reactive cysteines. Most reversible MTI bind to β-tubulin, and this is also true of these covalent MTI. However, it is certainly not true of all of them, at least in reactions with purified tubulin. Many clearly react with both α- and β-tubulin, and at least one is specific to α-tubulin. Some react with other proteins, but their reaction with tubulin appears to drive the biological response (e.g., mitotic arrest).

It is perhaps remarkable that the biological activity, measured as the range of cytotoxic IC_{50} or MT polymerization-inhibitory concentrations with purified tubulin or MT subunits, is not remarkably different for these irreversible inhibitors compared with the more well-known reversible inhibitors. In fact, many of the compounds reviewed here require higher concentrations to achieve their MTI effect. This may be because of two things. First, many reversible binders nonetheless bind with very high-affinity. Colchicine is an example. Once formed, the colchicine–tubulin complex, though noncovalent, is essentially irreversible on a time-scale long in terms of molecular diffusion (i.e., many tens of minutes). Others bind with high intrinsic affinity, whereas remaining easily displaceable such as taxol. Second, many covalent binders are slow to form complexes, possibly reflecting the need for conformational changes in tubulin to expose the reactive residues.

The compounds are grouped by reactive site on tubulin. First are compounds known to react with cysteine. These fall into two categories, based on the current knowledge of their reaction site: those with a specific, known residue of reaction, and those with possibly broader reactivity at unknown cysteine residues, some of which may react with a specific, but not yet known residue. Second are compounds that react at noncysteine residues. The first group here shows high site specificity on tubulin. The final group of compounds react at multiple (and not always known) amino acid residues, yet appear to produce typical MTI effects in cells and in purified systems, suggesting a functional specificity that belies their apparent chemical nonspecificity.

1.1. Terminology/Methodology/Mechanisms

It is probably useful to clarify a number of points before proceeding. These have to do with terminology used to describe tubulin structures and preparations of tubulin and other MT proteins, methodology used to quantitate sulfhydryls, and a discussion of mechanisms of action of anti-MT agents. In terminology, several terms will be used that require some clarification. The first is the term "MT," which is used to mean a microtubule or many microtubules. In biological terms, of course, the singular is almost never relevant: MT act in groups, frequently large arrays, as in the mitotic spindle.

Preparations of purified MT proteins are often referred to as MTP or tubulin, which are not identical. Preparations of MT proteins that consist of the tubulin core with all of the associated proteins (microtubule associated proteins; MAPs) that copolymerize, are termed microtubule protein, or MTP. In its usual preparation, MTP contains tubulin (85–95%) and a collection of MAPs that include structural MAPs that regulate assembly (such as MAP1, MAP2, and tau) as well as ATPase motor proteins (such as kinesin and dynein). It is perhaps worth noting that although "MTP" contains the term "MT," the preparation is usually depolymerized and contains no MT, until it is deliberately induced to assemble. Tubulin purified away from the MAPs is referred to as "purified tubulin" or just "tubulin." Tubulin so purified consists of depolymerized α-β dimers until it is deliberately induced to polymerize. When the term "tubulin" is used it is assumed to be non-polymerized dimer, not monomer or MT, unless specified as polymerized tubulin or MT. If only one monomer is intended, it will be referred to as α- or β-tubulin.

There are multiple genes for mammalian α- and β-tubulin, expressed in a tissue- and developmental stage-specific manner. This isotype diversity is discussed in the chapter by Luduena (Chapter 40). Isotype differences have little relevance to covalent MTI, at least as currently known, except for one. The β-III isotype, expressed in neural tissue, has a serine at position 239 instead of the cysteine that all other β-isotypes contain. As β239cys is one of the most reactive cysteines in tubulin, this isotype difference becomes relevant in discussing the MTI that target cysteines.

Another source of chemical diversity in tubulins is post-translational modification. Both α- and β-tubulin undergo many post-translational modifications (see the chapter by Luduena, Chapter 40). Here, the most significant post-translational modifications are detyrosination of the α-tubulin carboxyl terminus, and acetylation of lysine near the amino terminus. Both of these modifications are associated with stable, low-turnover MT (such as axonal or flagellar MT), as opposed to dynamic MT that lack these modifications such as mitotic MT.

The principal methodology that may require comment is the quantiation of sulfhydryls, most commonly accomplished with dithiobis-(nitrobenzoic acid) (DTNB). A free sulfhydryl or thiol group on a protein (more correctly the thiolate form) can react with the disulfide group in DTNB, forming a protein-disulfide-linked nitrobenzoic acid group and a free nitrobenzoic acid thiolate anion, whose absorbance at 410 nm can be used to quantitate the release of anion, and hence the protein sulfhydryls that reacted with DTNB (1).

The detailed mechanism of action of anti-MT agents is often ambiguous or ambiguously reported. As MT polymerization requires the (usually reversible) assembly of tubulin dimers, anti-MT agents could target the dimer only, preventing assembly, or the polymer only, stabilizing the polymerized form and preventing disassembly. Other possibilities exist: e.g., a compound could bind to the dimer, preventing assembly into MT as well as binding to MT, causing disassembly. Indeed, several compounds are known that show this combination of actions such as vinblastine. When assayed with MTP or purified tubulin, it may be possible to distinguish between

1. Binding to the dimer and thereby preventing assembly and consequently causing loss of MT polymer in a reversible system and
2. Binding to preformed MT and consequently directly inducing depolymerization of MT.

This distinction is often not made in the literature and may be difficult or impossible to infer after the fact.

When MTI are added to cells, the mechanistic details are usually not accessible, in principle, and the experimental observables are either changes in the dynamics of the MT, or much more commonly, a change in the number density or mass of MT (e.g., by immunofluorecent staining of cells), or in the fraction of tubulin in polymer (MT) form, assayed by Western blots of pelletable MT vs supernatant dimeric tubulin. In any case, the observable is usually loss- or gain-of-MT, but the detailed path to that end is not clear. Therefore, a statement that treatment of cells with agent X results in loss of MT does not imply that the agent acts directly on MT polymer as opposed to acting on the dimer, disfavoring polymerization.

2. CYSTEINE REACTIVE REAGENTS

Near neutral pH, sulfhydryl groups are the most reactive residues in proteins, and hence, it is perhaps not surprising that most small antimitotic molecules that react with tubulin do so through reactions at cysteine residues *(2)* This is not limited to tubulin; cysteine-reactive compounds have been developed as candidate clinical agents that target other proteins as well *(3,4)*.

2.1. Tubulin Cysteine Residues

The tubulin heterodimer has 20 cysteine residues and in native tubulin all are reduced and reactive (as with DTNB), although with varying reactivities *(5–7)*. Binding of many reversible ligands to tubulin alters the reaction of some or many sulfhydryl residues, because of steric protection by the ligand, and/or to allosteric changes in the structure of the tubulin dimer *(6–10)*. MT polymerization has long been known to be sensitive to sulfhydryl modification. Oxidation or alkylation of sulfhydryls was shown by many laboratories to prevent polymerization of purified MTP as well as purified tubulin *(11–13)* These early studies demonstrated that modification of as few as two sulfhydryls could prevent MT assembly.

Many cys-reactive drugs show significant selectivity for particular residues on tubulin. This appears to be partially because of details of the structure of the small molecules, reflected in structure-activity studies that demonstrate the significance of alterations to portions of the molecule distinct from the sulfhydryl-reactive moiety. However, it is also because of intrinsic reactivity differences between different cysteine residues in the protein *(5)*. This is reflected in the ability of some "general sulfhydryl-active" compounds to exhibit residue and activity-specific reactivity. Examples are the preference of mono-bromobimane for reactivity with βcys239 *(10)* and the ability of *N*-ethylmaleimide-blocked tubulin to selectively interfere with (–)-end assembly *(14)*. The numerous publications on the selectivity of bifunctional sulfhydryl reagents (summarized in ref. *9*) presumably reflect a combination of site reactivity and especially site separation.

It is important to note that the reactivity of cysteine residues with sulfhydryl-directed reagents is not a function only, or even primarily, of the degree of exposure of a given residue to the solvent. This is well-illustrated by studies of the reactivity of particular cysteine residues of tubulin. Surface exposed cysteines can be made nearly unreactive (for example with iodoacetamide) by neighboring acidic residues (aspartic or glutamic acid), and more buried cysteines can be made quite reactive by the presence of neighboring basic groups such as lysine *(5)*. Cysteine reactivity (for example with DTNB) is often taken in the literature as the same as solvent exposure, and of course a soluble reactant cannot react with a cysteine residue if that residue is not at least transiently exposed to the solvent. However, the *average* position of that cysteine residue may be somewhat or even considerably buried, as revealed by the crystal structure, for example.

2.2. Tubulin βcys239

Most of the cysteine residues in tubulin are buried in the electron crystallographic model *(15)* of the tubulin dimer *(6)*. Indeed, four of the five fastest reacting cysteines in tubulin are buried *(5)*. The issue of solvent exposure is well-illustrated by the case of βcys239. This cysteine residue reacts rapidly with general sulfhydryl-reactive agents such as monobromobimane *(10)*, iodoacetamide, or N-ethylmaleimide *(5)*. It is also the site of reaction of several site-specific MTI, as discussed later. The position of this residue is shown in several views in (Fig. 1). As shown in the upper two panels of Fig. 1, βcys239 is located near the α-β intradimer contact, and is not exposed on the surface of the protein. The buried status of the residue is better illustrated in the space-filling models in the two lower panels. The lower left panel shows the whole dimer with the water-exposed surface shown in blue. The sulfur atom of βcys239 is shown in yellow and is barely visible through a small "window" in the β surface. This is shown in an enlarged view in the lower right panel, slightly turned. The diameter of the opening is about 6.4 Å, shown by the dotted line. Thus, the residue can be considered solvent exposed in that a water molecule with diameter of approx 2.8 Å could contact it. However, the high reactivity of this residue to molecules much larger than water must be because of transient opening of the protein structure (breathing) and/or electrostatic activation *(5,6)*.

2.3. Specific Site Agents

Specific site agents are those demonstrated to react preferentially with one residue/site. A number of agents have been shown to react with considerable, if not exclusive, specificity with a particular cysteine residue. These all apparently react with βcys239. The structures of the various agents are not obviously related, and all interfere with colchicine binding or are interfered with by colchicine binding.

2.3.1. ARYL-PENTAFLUOROSULFONAMIDES

The closely related aryl-pentafluorosulfonamide agents T138067 and T113242 (Fig. 2) react covalently with tubulin, alkylating βcys239. Colchicine, but not vinblastine competes for binding of these agents to purified tubulin. Treatment of cells with T138067 disrupts MT polymerization, leading to G2/M blockade, induction of apoptosis, and an increase in chromosomal ploidy in surviving cells. IC_{50} in cell culture is 10^{-8}–10^{-7} M, and is not substantially affected by the multidrug resistance (MDR) phenotype *(16,17)*. Exposure of MCF7 human breast carcinoma cells to radioactive T138067 followed by sodium dodecyl sulfate (SDS) electrophoresis revealed a single radioactive band, corresponding to tubulin, and shown to be tubulin by immunoprecipitation with anti-β tubulin antibodies. Further studies with β-isotype-specific antibodies revealed binding to βII and βIV, the major β-isotypes in these cells. Some labeling was found of the small amount of βI, but none could be found with βIII *(16)* Thus, in the cell this compound appears to react only with tubulin, and specifically β-tubulin. The reaction with purified tubulin demonstrated that this is largely with βcys239 *(10,16)*. This explains the lack of reaction with βIII, as this isotype has a serine, not a cysteine,

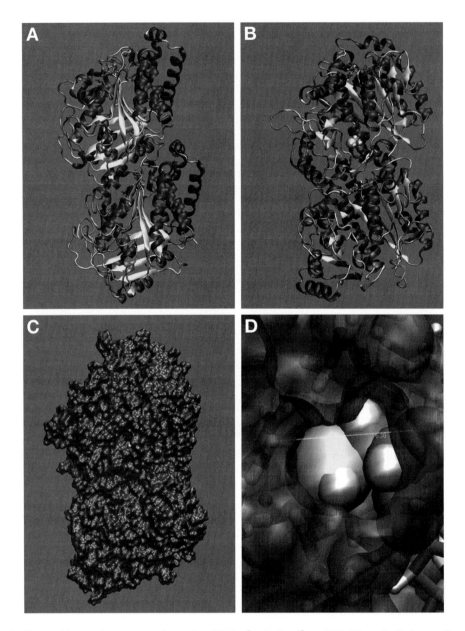

Fig. 1. The position and exposure of cysteine 239 in β-tubulin (βcys239). The tubulin heterodimer is shown with β-tubulin on top and α-tubulin below, based on the electron crystallographic model JFF1 *(15)*. GDP in α and GTP in β are shown in ball-and-stick representation and βcys239 is shown in space-filling format near the α-β intradimer interface. **Panel A** shows the dimer from the "side," with the dimer surface exposed on the microtubule surface to the right, and the surface exposed in the MT lumen to the left. **Panel B** shows the dimer in A rotated 90° to the right. The sulfur in βcys239 is more visible as a yellow sphere centered left-right and near the intradimer interface. **Panel C** shows the same view as B, but in space-filling mode. The outer surface of the dimer is colored blue and represents the surface of closest approach of a water molecule rolling over the dimer. The cysteine βcys239 can be seen in color through a small "window." **Panel D** shows a close in view of the "window." The sulfur of βcys239 can be seen in yellow. The diameter of the opening is approx 6.5 Å, allowing passage of water (2.8 Å diameter), but restrictive to larger molecules such as DTNB. To view this figure in color, see the insert and the companion CD-ROM.

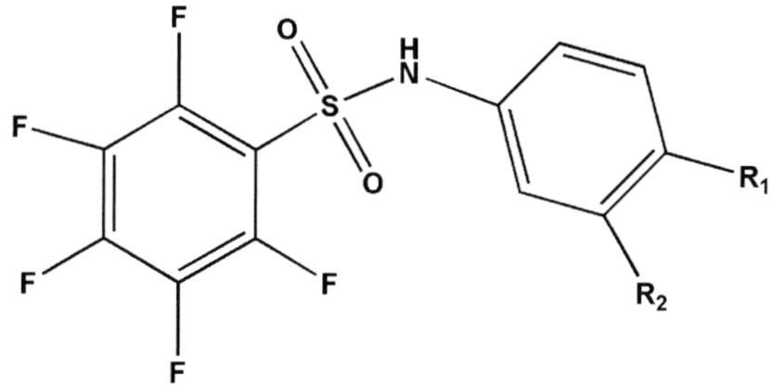

	R₁	R₂
T138067	-O-CH₃	-F
T113242	-N-(CH₃)₂	-H

Fig. 2. The structures of the aryl-pentafluorosulfonamide agents T138067 and T113242.

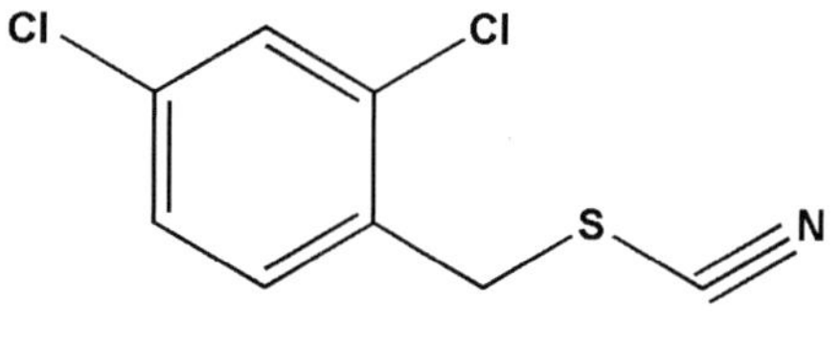

DCBT

Fig. 3. The structure of DCBT, dichlorobenzylthiocyanate.

at this position. The reaction is not totally specific to βcys239, as alklyation of βcys12 is also observed, even under mild reaction conditions *(10)*. Nonetheless, at large excess of drug to tubulin (60-fold excess) and long incubation time (8 h at 37°C), reaction is limited to β, and is largely limited to βcys239 *(16)*. This appears to be the site important in the cellular action of these compounds.

2.3.2. DICHLOROBENZYL THIOCYANATE

Chlorinated derivatives of benzyl thiocyanate have been shown to induce mitotic arrest and disrupt intracellular microtubules *(18)*. This activity was most studied with 2,4-dichlorobenzyl thiocyanate (DCBT) (Fig. 3), and shown to be because of alkylation of tubulin sulfhydryls *(19)*. With equimolar DCBT and tubulin, reaction is mostly with β-tubulin, but a higher drug:tubulin ratio (5:1) resulted in nearly equal reaction with both subunits. Significantly, when cells were exposed to radioactive DCBT and extracts examined by SDS electrophoresis, the most prominent band corresponded to β-tubulin, although multiple other proteins were also labeled, including one corresponding to α-tubulin *(19)*. Subsequently, it was shown that the site of reaction on tubulin is βcys239 at low drug:tubulin ratios (1:5) *(20)*. Tubulin polymerization is inhibited by low ratios of DCBT to tubulin, indicating that βcys239 is essential for MT assembly. Preincubation of tubulin with colchicine prevents reaction with DCBT, but DCBT does not prevent colchicine binding, indicating that βcys239 is not essential for colchicine binding *(20)*.

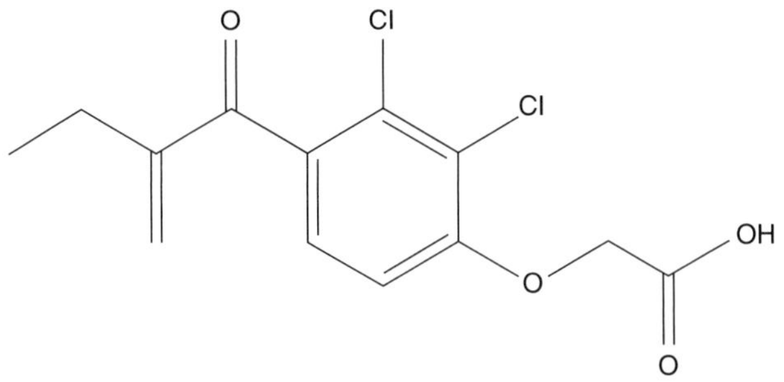

Ethacrynic acid

Fig. 4. The structure of ethacrynic acid.

2.4. Cysteine-Reactive Compounds of Unknown Site Specificity

A number of compounds are known that interfere with microtubule function by reaction with unknown or multiple tubulin sulfhydryls. In fact, virtually any compound that can react with sulfhydryls can inhibit MT polymerization. However, as detailed earlier, most of those compounds do not meet the criteria for inclusion here: specificity to tubulin in reaction and/or biological inpact. Those compounds included here are probably not exhaustive of the category, but were chosen either because of clinical importance as a therapy, or alternatively, clinical importance as a toxicant found in food or the environment. The former category includes organic natural products and analogs, whereas the latter includes both inorganic heavy metal salts and organometalic compounds. Although somewhat arbitrary, the discussion of the latter group will be separated into inorganic and organometalic compounds, even though the active metal may be the same, as with mercury salts and methylmercury. Equally arbitrarily, discussion of unsaturated aldehydes which can react with cysteines will be defered until a later section whose main topic is aliphatic aldehydes, which react with lysines.

2.4.1. ORGANIC COMPOUNDS–ETHACRYNIC ACID

Ethacrynic acid (Fig. 4) is a sulfhydryl reactive diuretic, under investigation for use as an ocular hypotensive agent in the treatment of glaucoma. It is thought that its usefulness in this context may be because of alteration of MT in trabecular meshwork cells, leading to increased aqueous humor outflow (21,22). It has long been recognized as an anti-MT agent, which can inhibit colchicine binding to tubulin and whose inhibitory action can be prevented (or partially reversed) by added sulfhydryl agents such as mercaptoethanol (23). Reaction of the tubulin sulfhydryl is likely with the α, β-unsaturated carbonyl of ethacrynic acid, and reduction of either component of this moiety causes loss of activity (22). Exposure of purifed tubulin or MTP to ethacrynic acid inhibited polymerization in a dose- and preincubation time-dependent manner, accompanied by loss of titrable sulfhydryl residues. Loss of approx two sulfhydryl residues corresponded to 50% inhibition of polymerization (24). Studies with the bifunctional sulfhydryl agent N,N′-ethylenebis(iodoacetamide) indicated that ethacrynic acid does not interact with βcys239 or βcys354, but may interact with cys residues 12, 201, or 211 (25). Studies with tubulin prepolymerized to steady state, demonstrated that ethacrynic

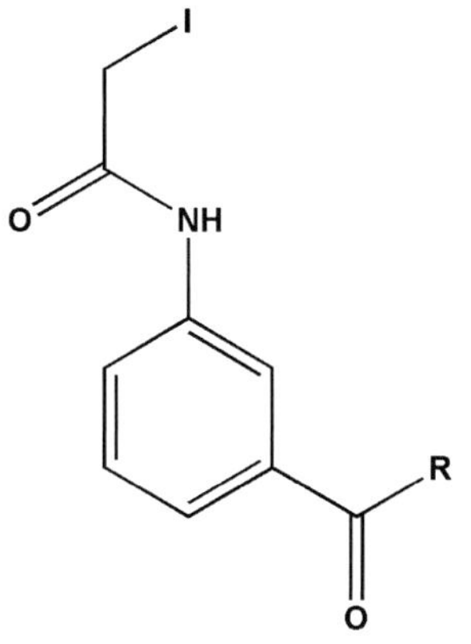

	R
IAABE	-O-C$_2$H$_5$
IAABU	-NH-CO-NH$_2$

Fig. 5. The structure of the haloacetamidobenzoyl ethyl ester IAABE and the haloacetamidobenzoy-lurea IAABU.

acid could induce disruption of MT in a concentration range (0.1–1 mM) similar to that for inhibition of polymerization *(26)*. Recent studies have highlighted derivatives of ethacrynic acid with increased potency in cytoskeletal modulation but lower cytotoxic potential, although studies with MTP or tubulin were not included *(22)*.

2.4.2. HALO-ACETAMIDOBENZOYL—ETHYL ESTERS OR -UREAS

Halogenated derivatives of acetamido benzoyl ethyl ester and acetamido benzoylurea have been reported to be cytotoxic agents acting through inhibition of MT polymerization. 3-haloacetamido benzoylureas exhibited in vitro cytotoxicity with IC$_{50}$ values in the 10–100 nM range, and were not affected by MDR phenotype. MT were lost from treated cells, resulting in mitotic arrest followed by apoptosis. The chloro analog was less active than the bromo analog, and 3-iodoacetamido benzoylurea (3-IAABU) *(see* Fig. 5) was the most active *(27,28)*. Preincubation with compound inhibited assembly of tubulin dimer but did not induce disassembly of preassembled MT, indicating binding to the dimer. 3-IAABU interfered with colchicine binding but not with vinblastine. It did not displace bound GTP/GDP but did inhibit GTPase activity.

The 3-haloacetamido benzoyl ethyl esters showed similar activity with tubulin and 3-iodoacetamido benzoyl ethyl ester (3-IAABE) was the most active *(see* Fig. 5). It inhibited in vitro polymerization of tubulin by reaction with tubulin sulfhydryls. Exposure of cells to radioactive 3-IAABE resulted in labeling of multiple proteins, but the major peak on SDS gels corresponded in molecular weight to tubulin and was bound by an ion exchange resin, as expected for tubulin *(29,30)*. Further incubation resulted in loss of this band, presumably because of cellular turnover of alkylated tubulin.

There is no direct evidence that these compounds react with β-tubulin, or with any particular amino acid residue. Interference with colchicine binding and the reactivity of the nearby βcys239 suggest that this may be the site of reaction, but the example of the arylchloroethylureas (Subheading 3.1.1.), which apparently do not react with this cysteine despite similar indirect evidence, suggest caution in this interpretation.

Interestingly, exposure to 3-IAABE resulted in blockade of G1/S transition as well as mitotic arrest in human leukemia/lymphoma cells as well as early induction of apoptosis. Similar treatment with 3-IAABU showed G2/M arrest with later progression to apoptosis (29,30). The G1/S arrest was also observed in cells, which had taken up the isolated and purified covalent tubulin drug complex, purified following reaction of tubulin with 3-IAABE. These results suggest a role for tubulin/MT in the G1/S transition, as has also been observed in cells treated with taxol (31–33). The reason that a G1/S arrest is observed with 3-IAABE and not 3-IAABU is not clear.

2.4.2.1. Inorganic Compounds. Heavy metal salts have been long known to induce alterations of cellular structure that have been shown to be because of disruption of the cytoskeleton and MT in particular. A wide variety of metals have been shown to perturb MT, although some only at very high concentrations (34). Aluminum has been reported to increase MT polymerization (35), but most other metals both inhibit tubulin polymerization by binding to the dimer and induce depolymerization of preformed MT by binding to MT. Notable in this regard are mercury and arsenic.

Inorganic mercury (Hg^{2+}) has long been appreciated to be a neurotoxicant, and has also been shown to be genotoxic because of disruption of MT function (36,37). Tubulin assembly is inhibited by stoichiometric or slightly substoichiometric Hg^{2+} salts, and is independent of the anion (37–39). Preformed MT are depolymerized by addition of Hg^{2+} at similar concentrations (38), and this also occurs in cultured cells (40). Inhibition of polymerization is prevented by preaddition of sulfhydryl agents (100–500-fold of β-mercaptoethanol or dithiothreitol (DTT) over Hg^{2+}), and is reversible upon such addition after exposure of tubulin to the Hg^{2+} salt (36,38). Interestingly, complexation of the Hg^{2+} with chelators such as ethylenediamine tetra acetic acid (EDTA), ethylene glycol–bis (2–aminoethylether)–N, N, N', N'–tetraacetic acid (EGTA), or nitrilotriacetic acid (NTA) does not prevent the inhibition of MT assembly (40). Chelators may potentiate the action, reducing the concentration required for inhibition (39), although this has not always been observed (37). Hg–EDTA appears to inhibit the interaction of GTP with the E-site of β-tubulin, possibly resulting in the inhibition of polymerization (39).

The action of Hg^{2+} and Hg-chelator complexes on MT is most likely because of reaction with sulfhydryl(s), consistent with the protection by sulfhydryl agents, though the residues involved are not known. From the stoichiometry of inhibition (~1 Hg per dimer), it is likely that the reaction is with one or two cys residues. As the presence of metal chelators of high-affinity (Hg–EDTA has a log_{10} stability constant of approx 21.5 (38) do not prevent reaction with tubulin, the tubulin site must have very high-affinity for Hg^{2+}. A simple way to account for this is to posit that the site of reaction involves two nearby cys residues, which react with one Hg^{2+} ion (Dithiol–Hg complexes have a log_{10} stability constant of approx 40 (39). The resulting high-affinity of the site may explain how Hg^{2+} can target tubulin in a cell filled with other sulfhydryl-containing proteins and peptides. The identity of these vicinal cys residues on tubulin remains undefined.

It may seem that the requirement that the two cysteine residues be close enough to bond simultaneously to Hg would narrow the possibilities considerably, but it is not the case. Both α- and β-tubulin contain sequence-vicinal cysteines. Thus, α-cysteines 315 and 316 or β-cysteine 129 and 131 are possible. However, binding of tubulin to arsenical chromatography media, specific for vicinal dithiol proteins, showed that the tubulin interaction with the column was significantly weakened by denaturation after adsoption onto the column. This suggested that the vicinal cysteines responsible for interaction with the arsenic on the column (likely the same pair as bond with Hg^{2+}) were not close in sequence, but were close because of the native fold of the protein *(41)*. If one considers those cysteines whose sulfurs are <10 Å apart as likely candidates for the reactive cysteine pair, the number of possibilities is surprisingly large: 15 of the 20 cysteines in tubulin are close enough to another cysteine residue in the electron crystallographic model *(15)* that their sulfur atoms are separated by <10 Å. Thus, identification of the reactive cysteines will require direct experimental evidence.

Inorganic arsenic has been used medically for several millennia, and has recently become a focus of interest because of its efficacy in treatment of acute promyelocytic leukemia, despite the lack of a clear mechanism of action *(42–44)*. A direct effect on MT is one possible mechanisms being explored, and indeed, inorganic arsenite (As^{3+}) has been shown to alter MT, both in cultured cells and in purified in vitro systems. Some reports have examined the effect of sodium arsenite ($NaAsO_2$) *(45)*, but most of the studies reviewed here have used arsenic trioxide (As_2O_3), the compound in use clinically.

Treatment of a variety of cell lines (including myeloid leukemia cells as well as lines from nonsmall cell lung cancer, neuroblastoma, ovarian, cervical, and breast cancer cell lines) with As_2O_3 results in accumulation in G2/M, confirmed to be early mitotic arrest by microscopy *(46,47)* and other markers *(47,48)*. Effects on cellular MT were reported in some lines with sodium arsenite *(45)* and As_2O_3 *(47)*. Reported effects on purified tubulin vary. As_2O_3 has been reported to promote polymerization of purified tubulin and to stabilize polymerized tubulin against cold-induced depolymerization *(47)*. On the other hand, other reports indicate inhibition of GTP- and paclitaxel-induced polymerization, although not depolymerization of preformed MT *(46,49)*. One of these papers reports that As_2O_3 reaction results in loss of free sulfhydryls in tubulin *(49)*. The other report suggests that As_2O_3 inhibits polymerization by inhibiting GTP binding to tubulin. This is suggested to be because of crosslinking of two cysteine residues (βcys12 and βcys213) that are known to be near each other when the GTP site is unoccupied *(46)*. There is no direct evidence that these are the residues of reaction, and the data of Carre et al. *(49)* indicate that reaction occurs with more than two cysteine residues (reaction with ~20% of tubulin SH groups, or four cysteines, results in 60–80% reduction in polymerization). The identity of the cysteine residues reacting with As_2O_3 remains to be determined. It will be interesting to see if these residues are the same as those involved in Hg^{2+}-based reaction with tubulin. In addition, the role of MT in the cellular effects of As_2O_3 remains unanswered. This is particularly relevant to understanding the basis of the efficacy of As_2O_3 in acute promyelocytic leukemia. It is possible that a β-isotype reaction bias may be an important factor to investigate in this regard.

2.4.2.2. Organometalic Compounds. Organometal compounds are common naturally occurring environmental toxicants as well as industrial products and environmental contaminants, and have been the cause of significant human poisoning. A number have been shown to cause mitotic arrest in cultured cells and to interfere with MT polymerization. The most studied compound, and the cause of most human poisonings is methylmercury. Other organomercurials have also been considered as well as organocompounds of arsenic, tin, and lead.

Methylmercury (MeHg) is the most common organic mercurial, but ethylmercury and phenylmercury have also been studied in cell culture systems. All have been shown to cause accumulation of cells in mitosis (50) because of disruption of MT (40,51). MeHg is a potent neurotoxin as well as being toxic to dividing cells. It easily crosses the placenta and this leads to embryotoxic and teratogenic effects. MeHg has been the cause of multiple mass human poisonings because of environmental contamination as well as deliberate use of MeHg, e.g., as a seed fungicide in agriculture (52). The sensitivity of MT to low concentrations of MeHg may account for the developmental toxicity seen after these events. The high-affinity of organomercurials for sulfhydryls leads to inhibition of sulfhydryl enzymes, and this sulfhydryl reactivity is clearly the reason for MT sensitivity, although particular sulfhydryl residues on tubulin that mediate this sensitivity have not been identified.

MeHg cytotoxicity in cultured mammalian cells has been demonstrated. The IC_{50} for MeHg is about 1/10 that for Hg^{2+}, and MeHg does not cause membrane damage as much as Hg^{2+}. Rather, exposure to MeHg results in specific loss of MT (40). Concentrations of MeHg that cause loss of most MT do not affect vimentin or actin filaments in PtK-2 cells (53). A cell concentration of MeHg of >0.6 µg/mg protein leads to loss of MT, and essentially all MT are lost at a concentration of approx 1.6 µg Hg/mg of protein (54). This corresponds to an intracellular concentration of MeHg of approx 1 µM (assuming an intracellular protein concentration of 200 mg/mL).

Loss of MT because of exposure to MeHg selectively affects dynamic MT with tyrosinated α-tubulin, whereas the less dynamic MT with acetylated α-tubulin are resistant to the effects of MeHg, as with other MT-destabilizing agents (55). All Mt are not equally sensitive: platelet MT are more sensitive to MeHg than those of lymphocytes (56). This may reflect differential isotype expression, as platelets express an unusual β-isotype (βVI). This is similar to the isotype expressed in avian erythrocytes, the tubulin from which was previoiusly shown to differ in its sensitivity to sulfhydryl reagents compared with brain tubulin (57). In that case, sensitivity to mono- and bifunctional iodoacetamide reaction was reduced compared with brain, but it is possible that the differential sensitivity is reversed for other sulfhydryl agents. In a number of cell types, MeHg exposure leads to mitotic accumulation, and this has been shown to lead to induction of apoptosis in PC12 and HeLa cells (58). The sensitivity of P19 mouse embryonal carcinoma cells to MeHg is altered by induction of neuronal differentiation, possibly because of altered tubulin isotype expression, or by induced expression of particular MAPs (55,59).

Studies with purified MTP or pure tubulin have shown that MeHg inhibits polymerization with an IC_{50} about half the tubulin concentration, or about twice the IC_{50} for Hg^{2+}. Both MeHg and Hg^{2+} cause depolymerization of existing MT, but at higher concentrations than required for inhibition of polymerization (40). In a study comparing polymerization inhibition to cysteine status, it was found that 15, free –SH groups were measurable with DTNB per tubulin dimer and all could react with MeHg in the unpolymerized dimeric protein. Reaction with ≥2 –SH groups per dimer prevented polymerization (60) MeHg can also react with polymerized MT (and induce depolymerization) with a stoichiometry of 1 MeHg per tubulin dimer (60,61).

Organolead compounds are second in toxicity only to organomercurials. The most studied is triethyllead, Et_3Pb, a degradation product of the now discontinued antiknock fuel additive tetraethyllead. Exposure of cultured mammalian cells (Wi38 human fibroblasts and rat kangaroo PtK-1) to micromolar levels of Et_3Pb results in loss of interphase MT as well as MT of the mitotic spindle. Notably, this effect is reversible by washing out the agent. Curiously, little increase in mitotic index was observed, and instead lack of cytokinesis resulted in accumulation of multinucleate cells *(62)*. Different cell types display differential sensitivity to Et_3Pb. Leukemic lymphocytes are reported to be more sensitive to Et_3Pb than normal lymphocytes, possibly because of an increased pool of unpolymerized tubulin in these cells *(63)*.

In studies with purified MTP or tubulin, Et_3Pb was found to prevent polymerization and induce disassembly of MT polymers. Inhibition of polymerization was observed even at ratios as low as 1:30, Et_3Pb to tubulin. This also resulted in partial disassembly of preformed MT. Inhibition is reversed by addition of 100-fold excess of glutathione or other monothiol compounds *(62,64)*. Surprisingly, it was also reported that simply gel filtering tubulin that had been made unpolymerizable by reaction with Et_3Pb sufficed to restore polymerization levels to untreated levels *(64)*, although this may explain the easy reversibility of cell effects by washout. Direct assay of tubulin sulfhydryl content of porcine brain tubulin detected 18 sulfhydryls per tubulin dimer, and revealed that exposure to Et_3Pb reduced this by 2. Even when present in 100-fold molar excess, Et_3Pb caused the same reduction, suggesting a specific interaction with two distinct thiol groups *(62,64)*. Further indication of the specificity of this interaction was provided by a lack of interaction of Et_3Pb with actin, which possesses reactive thiols. No evidence for loss of thiols was found, and no change was observed in actin polymerization *(64)*.

3. NONCYSTEINE REACTIVE COMPOUNDS

Studies with general, amino acid residue-specific- reagents demonstrated some time ago that polymerization of tubulin was sensitive to reaction with other amino acids in addition to cysteine. Tryptophan residues were implicated by the inhibition of polymerization (and colchicine binding) by reaction of at least 4–5 of the 8 trp residues with 2-hydroxy-5-nitrobenzyl bromide *(65)*. Positively charged residues were implicated by several studies. Carbamoylation of lysine residues, in both α and β-tubulin, resulted in inhibition of assembly, indicating that some lysines in one or the other or both subunits were important for assembly *(66)*. The importance of lysine residues is examined later. 2,3-butanedione reacts with arginine residues, and was found to inhibit polymerization (but not colchicine binding) after reaction with as few as 3 arg residues in tubulin *(67)*. Histidine residues can be modified by reaction with diethylpyrocarbonate or by photooxidation with methylene blue, and both treatments were shown to inhibit polymerization *(68)*. Inhibition required modification of not more than 3 his residues, and was partially reversible by hydroxylamine, as expected for his modification. A later study showed that the hydroxylamine-resistant component of diethylpyrocarbonate inhibition was because of reaction with the N-terminal methionine *(69)*. Despite the possibilities for side reactions with unintended residues, these studies clearly showed that amino acids other than cysteine play a critical role in the process of MT polymerization. Since these studies were reported, a number of new compounds have been described, whose reaction with tubulin is often much more specific.

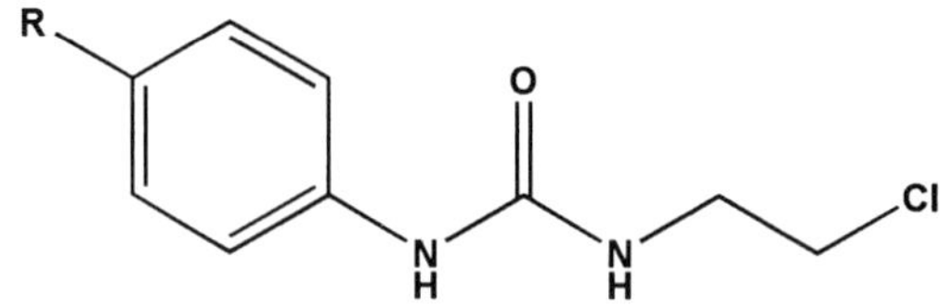

	R
ICEU	-I
tBCEU	-*tert*-butyl

Fig. 6. The structures of arylchloroethylureas ICEU and tBCEU.

3.1. Residue-Specific Compounds

3.1.1. ARYLCHLOROETHYLUREAS

A series of 1-aryl-3-(2-chloroethyl) ureas (CEUs) have been prepared and studied to obtain a combination of features of known alklyating agents: the aromatic moiety of nitrogen mustards such as chlorambucil and the non-nitrosated core of aliphatic nitrosoureas. The resulting compounds such as 4-tert-butyl-[3-(2-chloroethyl)ureido]phenyl (tBCEU) (*see* Fig. 6) showed enhanced cytotoxicity (IC$_{50}$ 1–10 μM for more active agents) compared with the nitrogen mustard or nitrosourea parent compounds, are not mutagenic, and do not show reduced effectiveness in cell lines with acquired resistance to other compounds mediated by several different clinically relevant mechanisms *(70)*. Although derived from DNA alkylating agents, the CEUs are MT disrupting agents. They have lower alkylating activity (measured with nitrobenzylpyridine), do not cause DNA damage, and do not react with the thiol of glutathione or glutathione reductase *(71)*. They do induce G2/M arrest, increase in the mitotic index, and lead to loss of cellular MT. Incubation of cells with radioactive tBCEU, followed by SDS gel electrophoresis and autoradiography, revealed reaction with a small number of bands, the major one corresponding to tubulin. Two dimensional gel separation indicated that the labeled tubulin was β-tubulin. Further evidence of reaction with β-tubulin was revealed by a time-dependent shift of β-tubulin to a band of higher mobility on Western blots following exposure of cells to tBCEU and other active CEUs *(70)*. The shifted band was only found in the soluble tubulin, not MT, fraction from treated cells, and was not found following treatment with inactive analogs lacking the chloro-substituent such as tBEU. Possibly the increased mobility in SDS gels is because of intramolecular crosslinking as has been shown with bifunctional cys-reactive probes *(9)*, but it is not obvious how such a crosslink would arise from this monofunctional probe. Induction of the shifted band was prevented by pretreatment of cells with colchicine but not vinblastine, suggesting alkylation near the colchicine site. Exposure of neuroblastoma cells to tBCEU followed by Western blotting with a βIII-specific antibody revealed no shift in position. As βIII has a serine rather than a cysteine, at position 239, it was proposed that active CEUs act by covalent reaction with βcys239.

Direct examination of the β-tubulin in the shifted band following reaction with either ICEU or tBCEU revealed that none of the cysteines were modified. The alkylated

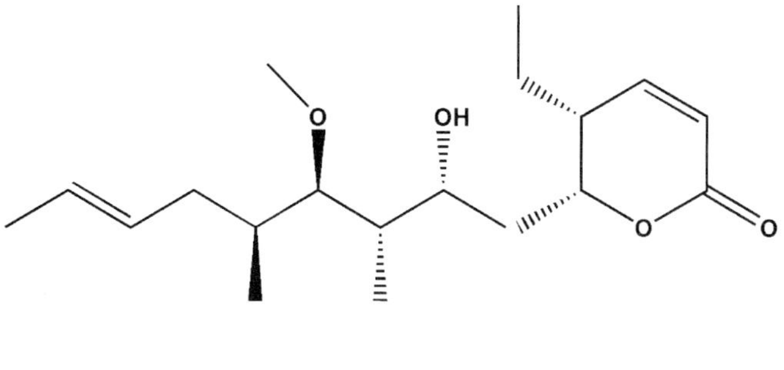

Pironetin

Fig. 7. The structure of pironetin.

β-tubulin had a more basic isoelectric point than the unmodified protein, which also was not consistent with a cysteine reaction. Mass spectrometry of proteolytic fragments of the alkylated protein showed that the adduct was contained in residues 175–213, and fragmentation sequencing demonstrated linkage to glutamic acid 198 through an ester bond *(72)*. This reaction is similar to the established alkylation of an acidic residue through an ester linkage by chlorambucil, a parent drug of the CEUs *(73)*. βglu198 is located near the α-β intradimer interface, near the colchicine site, and may play a role in the conformational changes in the dimer that accompany nucleotide hydrolysis *(15)*. The lack of reaction with tubulin preincubated with colchicine or containing serine rather than cysteine at position 239 (i.e., βIII isotype) is proposed to be because of conformational changes secondary to those changes *(72)*.

A number of analogs contribute to increased understanding of the mechanism of these agents. Substantial modification of the aryl group is possible without major loss of cytotoxicity or tubulin reactivity, whereas extension of the chloroalkyl group causes significant loss of activity *(71)*. A "metabolically stabilized" analog, ICEU (Fig. 6), was produced to circumvent inactivation of tBCEU by hepatic cytochrome P450, and shown to be similar to tBCEU in cytotoxicity and tubulin alkylation *(74)*. CEUs with branched chloroalkyl moieties have been produced by substitution at the carbon β to the chlorine *(75)*. This can lead to enhanced cytotoxicity and more rapid tubulin alkylation, but also reveals the stereoselectivity of the binding site, as the addition introduces an asymmetric center, and one enantiomer showed more activity than the other *(75)*.

3.1.2. LYSINE-REACTIVE AND -SPECIFIC—PIRONETIN

Pironetin is a natural product isolated from *Streptomyces* culture broth *(76)*. It has shown plant growth regulatory activity, immunosuppressive, and more recently, antimitotic and antitumor activity *(77)*. Antimitotic activity is reported to be because of inhibition of MT assembly because of binding to α-tubulin through Michael addition to αlys352. Pironetin is interesting not only because its activity is because of reaction with α-tubulin as opposed to β-tubulin, but also because of its relatively simple structure, consisting of one pyran residue and an alkyl chain (*see* Fig. 7). Despite the relatively simple structure and reaction mode, pironetin shows remarkable specificity for tubulin.

Pironetin is cytotoxic to cultured mammalian cells, exhibiting an IC$_{50}$ of 20–30 nM against a variety of tumor cell lines including HeLa, A2780, HL60, and K-NRK cells

(77,78). Growth inhibition is because of mitotic arrest, also demonstrated in normal rat fibroblast 3Y1 cells. Treatment of these cells results in disruption of MT, but not actin filaments, and is slowly reversible (~12 h) following removal of pironetin; the slow turnover is suggested to be because of turnover of the protein *(77,78)*. Treatment of HL60 cells with pironetin leads to phosphorylation of Bcl2 and induction of apoptosis indicated by DNA laddering *(78)*. Pironetin showed moderate activity against leukemia cell line P388 in mice, accompanied by significant weight loss *(77)*.

Pironetin and a number of analogs have been synthesized *(79,80)*. Structure-activity studies revealed that saturation of the α-β unsaturated lactone, reversal of chirality at the 7-position bearing a hydroxyl, and some alterations of the terminal portion of the alkyl chain resulted in loss of activity. Demethylation caused little loss of activity, and esterification of the 7-hydroxyl resulted in moderate loss of activity *(77,80)*.

Pironetin inhibits polymerization of MTP or pure tubulin and induces disassembly of preformed MT. Preincubation with pironetin inhibits vinblastine binding and enhances colchicine binding. Surface plasmon resonance measurements of tubulin–pironetin binding yielded a K_d of approx 0.3 µ*M*, compared with approx 2 µ*M* for vinblastine *(78,80)*. It is not obvious what this number means in the context of a covalent interaction (i.e., where k_{off} presumably = 0), as pironetin is shown to be.

The pironetin binding site and binding mode have been revealed by use of biotin-pironetin, a biotinylated linker arm esterified to pironetin through the 7-hydroxyl. Treatment of cells with biotin-pironetin resulted in loss of MT and SDS gel blots of treated cells reveal covalent linkage to one band corresponding to tubulin (50 kD). Sequencing of peptides with bound biotin-pironetin gave α-tubulin sequences and implicated residues 270–370 as the site of binding. Alanine scanning of all cysteine and lysine residues in this region showed that only loss of αlys352 prevented binding of biotin-pironetin. This indicates that pironetin binds to α-tubulin through reaction of the ε-amino group of αlys352, probably with the α-β- unsaturated bond in the pyran group of pironetin. αlys352 is located in the interdimer contact interface in a cavity formed by α-helices 8 and 10 and β-sheet8, which faces the GTP binding site on the β-subunit of the adjacent dimer. αlys352 is near the highly conserved αglu254, and it is proposed that pironetin binding to lys352 perturbs glu254 and interferes with its normal role in hydrolysis of GTP on the adjacent β-subunit *(81)*.

3.1.3. LYSINE REACTIVE–NONSPECIFIC

Aldehydes are reactive molecules that can show significant antimitotic activity because of inhibition of MT polymerization *(82)*. In normal biological circumstances, the concentration of reactive aldehydes is low, but may become significant in some contexts. Acetaldehyde is produced by hepatic metabolic oxidation of ethanol and may be involved in loss of MT-dependent hepatic function seen upon prolonged exposure to alcohol *(83)*. In experiments with cultured hepatocytes, significant loss of MT was observed upon 3 h exposure to 5 m*M* ethanol, as acetaldehyde accumulated in the media to 150–300 µ*M*. Inhibiting alcohol dehydrogenase prevented MT loss, showing that this was because of acetaldehyde production and not to alcohol presence *(84)*. In another context, acetaldehyde and acrolein are significant components of the volatile fraction of cigarette smoke. Both compounds caused dose-dependent loss of adhesion and viability, accompanied by disruption of MT and disorganization of intermediate filaments and actin filaments in human gingival fibroblasts *(85)*, and lung fibroblasts *(86)*.

A number of studies have addressed the mechanism of inhibition of MT polymerization by simple aldehydes. Acetaldehyde forms stable adducts to tubulin as well as unstable ones (Schiff bases), which can be stabilized by reduction. After short reaction times, inhibition of polymerization is observed but is reversed by simple gel filtration *(87)*. Longer reaction times lead to formation of stable adducts, and this inhibition is not reversed by gel filtration *(87,88)*. Stable adduct formation occurred with time in the absence of added reductant, and occurred preferentially on α-tubulin (α/β ratio as high as 3 at low acetaldehyde concentrations), if the reaction was with tubulin dimer. Reaction with MT was reduced compared with reaction with dimer, and showed no preferential reaction with α, suggesting the existence of a highly reactive lysine (HRL) exposed on α in the dimer but protected in the polymer. Similar reactions were observed with brain or liver tubulin *(88,89)*.

Studies of the reaction of tubulin and formaldehyde in the presence of the reductant cyanoborohydride demonstrated that tubulin polymerization is prevented (without effect on GTP or colchicine binding) by reaction with a highly reactive lysine (HRL) in α-tubulin *(90,91)*. This residue was shown to be αlys394, found in a highly evolutionarily conserved cluster of basic residues, suggested to be required for assembly. Furthermore, reactivity of this residue is reduced approx 10-fold by protonation of a residue with pK approx 6.3, presumably a histidine, and possibly αhis393, in the same basic cluster *(92)*. These residues are located on the solvent-exposed surface of helix 11 of α-tubulin, on the outer surface of the MT. Although it is not clear exactly how these positive charged residues participate in polymerization, it may be by interaction with the negatively charged C-terminus. It is possible that the acidic groups of the unstructured C-terminus interact with the basic groups of the lysine cluster, reducing charge–charge repulsion between dimers and hence, facilitating polymerization *(92–95)*. Hence, reaction with lysine(s) may reduce this interaction and inhibit polymerization.

It is presumably αlys394 that is the HRL with which acetaldehyde reacts, though this has not been explicitly demonstrated. As expected from the results with formaldehyde and αlys394, acetaldehyde adduction on the HRL of the dimer inhibits polymerization, whereas reaction with polymerized tubulin (in which the HRL is protected) did not inhibit subsequent depolymerization or repolymerization *(96)*. Inhibition is distinctly substoichiometric. Low mole fractions (as low as 5%) of tubulin adducted on this lysine completely prevented polymerization of added, unreacted tubulin. This level of HRL-bound aldehyde can be achieved by reaction of tubulin with concentrations as low as 50 μM acetaldehyde *(97)*.

Of course, acetaldehyde can react with other proteins as well. Among them, actin also forms unstable and stable adducts with acetaldehyde under nonreducing conditions, and actin monomer is more reactive than polymerized actin, similar to the case with tubulin. Interestingly, reaction with unpolymerized actin does not prevent its subsequent polymerization *(98)*.

Unsaturated aldehydes such as 4-hydroxynonenal also inhibit MT polymerization, although probably because of reaction with sulfhydryls not lysines *(99)*. 4-hydroxynonenal, a major product of lipid peroxidation, causes loss of MT in 3T3 fibroblasts at concentrations as low as 10 μM. Polymerization of purified tubulin was inhibited in a dose-dependent manner, accompanied by loss of up to 4 sulfhydryls per dimer and inhibition of colchicine binding. Addition of cysteine to the reaction prevented inhibition of polymerization *(100)*. The related unsaturated aldehyde 2-nonenal showed similar MT inhibition in cells and with purified tubulin. However, the saturated nonanal had much less effect on cells and a reduced activity on tubulin–colchicine binding,

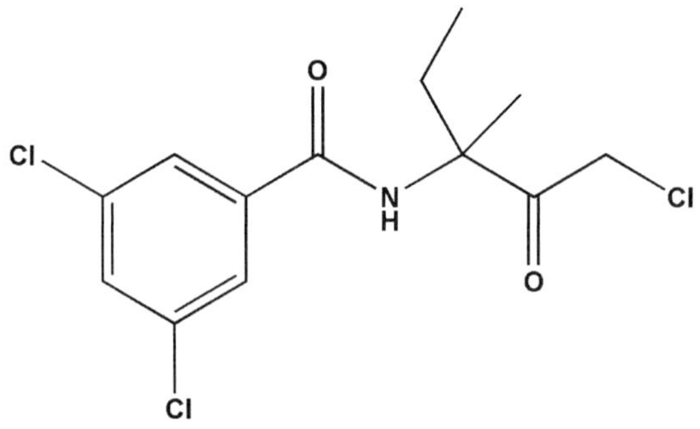

RH-4032

Fig. 8. The structure of the benzamide RH-4032.

while the aromatic benzaldehyde was inactive *(101)*. Nonanal inhibited MTP polymerization significantly more potently than the other aldehydes, however *(99)*, possibly by reacting with the HRL on α-tubulin.

4. UNKNOWN REACTIVITY

4.1. Unknown Reactivity—Specific Target

The benzamide RH-4032 (Fig. 8) is a potent anti-MT agent in plant cells that covalently reacts with both plant and mammalian tubulin *(102)*. It inhibited tobacco root elongation, caused accumulation of metaphase-arrested cells, and induced loss of MT. Incubation with bovine brain tubulin resulted in loss of polymerizability in a dose- and preincubation time-dependent manner. The presence of an α-chloroketone in the structure of RH-4032 suggested the possibility that it might react with tubulin. Incubation of brain tubulin with radiolabeled RH-4032 followed by SDS gel electrophoresis resulted in covalent labeling of the β-tubulin band. A similar incubation with tobacco cells followed by SDS gel electrophoresis resulted in labeling of essentially one band, corresponding to β-tubulin. Binding was saturable, was consistent with one binding site, and was inhibited by the structurally related compounds pronamide and zarilamide, and by *N*-phenylcarbamates related to chlorprophan. Binding was not inhibited by the dinitroaniline herbicide trifluralin, or the phosphoric amide herbicide amiprophos-methyl. The lack of inhibition by these compounds is consistent with the location of their binding site, thought to be on α-tubulin *(103,104)*. RH-4032 is unusual among covalently reacting anti-MT agents in that it is much more potent (100–1000-fold) in biological assays compared with presumably noncovalent, presumably reversible, structurally related agents (pronamide and zarilamide) that presumably bind to the same site *(102)*.

RH4032 contains a chloroketone moiety, known to be reactive with sulfhydryls. This, combined with the specificity of the reaction (β only), suggests that a cysteine residue, possibly βcys239 could be the targeted residue. Further work will be required to establish this.

4.2. Unknown Reactivity—Multiple Sites

4.2.1. PHOTODYNAMIC PROBES

A number of photosensitizers have been described as potential agents in the photodynamic therapy (PDT) of cancer. This treatment is based on the retention of these

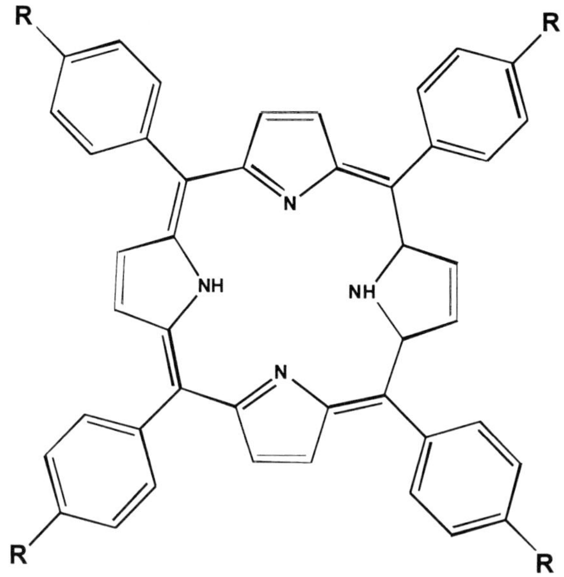

	R =
TPPS$_0$	-H
TPPS$_4$	-SO$_3$H

Fig. 9. The structure of the photodynamic probes TPPS$_0$ and TPPS$_4$.

compounds by tumor cells and on the generation of reactive singlet oxygen and other reactive oxygen species upon exposure of the compound to light. Reactive oxygen species damage occurs to cell membranes, organelles such as the lysosome, and nucleus as well as cytoplasmic proteins (105). Some compounds, including porphyrins, phthalocyanines, and others, cause damage to MT, resulting in mitotic arrest and disruption of the cytoskeleton. This may be because of covalent modification(s) of tubulin, some of which have been identified.

Endothelial cells exposed to the porphrin Photofrin show light-dependent loss of MT that is reversible at sublethal light doses but irreversible at higher doses. No effect was observed on F-actin organization or cellular ATP levels (106). Cellular exposure to tetrasulfonated aluminum phthalocyanine, tetrahydroxy- and monosulfonated m-tetraphenylporphine resulted in light-dependent accumulation in mitosis, mostly metaphase, with abnormal mitotic spindles. Mitotic accumulation was increased more by water-soluble compounds than with lipophilic ones (107). Similar results with tetra(4-sulfonatophenyl)porphine (tetraphenylporphine tetrasulfonate, TPPS$_4$; see Fig. 9) were potentiated by previous treatment of the cells with nocodazole, suggesting that the target of damage is the unpolymerized fraction of tubulin (108,109). Combination of PDT using TPPS$_2$ with previous vincristine or taxol treatment resulted in an enhanced in vivo antitumor effect in a mouse mammary tumor model, whereas no enhancement was observed upon combining vincristine with PDT with Photofrin (110).

Variation of the extent of sulfonation (TPPS$_n$, with $n = 0, 1, 2, 3,$ or 4, see Fig. 9) indicated that the disulphonated compound was most active, and that sulfonation on adjacent phenyl rings (producing an amphipatic compound), rather than opposite phenyl rings (producing a hydrophilic compound) enhanced activity. It was suggested that a spacing of 1.2 nm between oxygen atoms is optimal for photosensitization in these and other compounds and it was proposed that tubulin is the main target molecule *(111)*.

In addition to these light-dependent effects, light-independent inhibition of MT polymerization using purified tubulin has been observed with $TPPS_{2-4}$ as well as with mesotetra (4N–methylpyridyl) porphine (T_4MPyP). The mitotic index also increased in lymphoblasts exposed to $TPPS_4$ in the dark *(112)*. A later study *(113)* found that MT damage mediated by T_4MPyP was proportional to the light dose given. As in many other studies, this study showed that increased mitotic index following sublethal light dose was reversible. Higher light doses (90% killing) resulted in loss of MT, but no increase in mitotic index.

Other agents have also been found to induce light-dependent mitotic arrest. Thiazine dyes such as methylene blue and toluidine blue caused severe depletion of cellular MT upon light exposure *(114)*. The carbocyanine dye 3,3′-dihexylcarbocyanine iodide concentrates in mitochondria and the endoplasmic reticulum, but nonetheless mediates a large loss of cellular MT upon exposure to light. Treatment of purified MTP with dye and light causes loss of polymerization that is not rescued by additional GTP, indicating that the protein is the target, not the nucleotide. Dye with no light had no effect *(115)*.

In a few cases, studies have revealed the nature of the covalent modifications of tubulin. Exposure of rat or human fibroblasts to light following loading with the photosensitizer Hypocrellin A resulted in oxidation of a set of proteins different from those oxidized upon exposure to H_2O_2. Both α- and β-tubulin were prominent targets of PDT-induced oxidation, and this oxidation was on tyrosine residues, as indicated by a tyramine labeling specific to oxidized tyrosine *(116)*. In a study with $TPPS_4$, inhibition of MTP polymerization was proportional to light dose, and amino acid analysis indicated that oxidation of only a few histidine and cysteine residues was responsible for inhibition *(105)*. Purpurin-18-mediated PDT of HL60 cells resulted in oxidation of α-tubulin as well as β-actin and some chaperones. Oxidation was detected by the appearance of carbonyl groups on the proteins, which could be because of oxidation of side chains of lysine, arginine, proline, or threonine residues *(117)*.

4.3. Cisplatin

Cis-dichlorodiammine-platinum (II) (cisplatin, CDDP) is a widely used antitumor agent whose activity is widely accepted because of reaction with DNA. Dose-limiting toxicities include peripheral neuropathy, and this led to the investigation of reaction with tubulin or microtubules as a possible origin of this. *Cis-* or *trans*-dichlorodiammine-platinum(II) were found to inhibit assembly of MTP or purified tubulin, with somewhat higher inhibition with tubulin: 40 minutes contact of 1.2 mg/ml protein with 250 μM CDDP at 27°C resulted in 50% inhibition of MTP polymerization and 80% inhibition of tubulin polymerization *(118)*. Increased inhibition was found with increased contact time or increased temperature (although reaction still occurred at 4°C). DTNB assay yielded 20 sulfhydryls per tubulin dimer; this was reduced by 1–2 sulfhydryls per dimer following incubation with 20–60-fold excess of CDDP for 1 h at 27°C, which resulted in 100% inhibition of polymerization. Direct assay by atomic absorption spectrophotometry following dialysis revealed 1.5–2 platinums bound per dimer. Significantly increasing reaction time resulted in further loss of sulfhydryls. It was concluded that CDDP reacts with tubulin sulfhydryls and that one adduct was sufficient to block polymerization. Only small effects were seen on colchicine or vinblastine binding following CDDP reaction. A possible relation to the neurotoxic effects of CDDP was suggested as the previously published platinum concentration found in nerves of patients with neuropathies was similar to those resulting in approx 50% inhibiton of MT polymerization (118).

A direct alteration of MT assembly was reported in a study of rats given ip CDDP (119). Tubulin, purified from testes of exposed rats showed lower cold-induced depolymerization compared with control rats. Overnight exposure of bovine brain tubulin to CDDP, followed by a cycle of polymerization purification, produced tubulin that also depolymerized more slowly in the cold, and produced shorter MT than control tubulin. This effect was partially reversed by diethyldithiocarbamate, which cleaves protein sulfhydryl–platinum complexes. Carboplatin showed a reduced ability to alter MT assembly and disassembly (119). Another study found CDDP to inhibit in vitro polymerization of tubulin, and to induce formation of stable rings (120). Nuclear magnetic resonance studies reveal a spectral signature indicative of a complex of CDDP with GTP. It is suggested that reaction occurs with N^7 of GTP, as (Met-N^7)-GTP substitution in tubulin before reaction with CDDP fails to produce the nuclear magnetic resonance signature. It is not clear why the proposed CDDP–GTP complex cannot exchange with free GTP, but a structure is proposed for the intermediate, and this intermediate is proposed to account for the increased GTP hydrolysis observed (120). These three studies all report anti-MT activity of CDDP, though by rather different mechanisms, and with differing functional consequences. It is not clear how to rationalize these divergent results. Future work, informed by these studies, may reconcile the differences and define the anti-MT action of CDDP and its role in cytotoxicity and clinical application of this drug.

5. MICROTUBULE STABILIZERS

All of the previous compounds act as MT destabilizers, and induce loss of MT in treated cells. An important exception to this is cyclostreptin, which irreversibly stabilizes cellular MT and induces cell cycle arrest due to covalent reaction with β-tubulin (121). Cyclostreptin reacts with MT in a 1:1 stoichiometry to tubulin, with reaction distributed between two nearby sites—one on the outside and one on the inside of the MT. The outside site is Thr220, located in the outside part of a pore through the MT wall, while the inside site, Asn228, is on the lumenal of the pore at the paclitaxel binding site. Reaction with unpolymerized dimer only labels Thr220 (121). It will be interesting to see if cyclostreptin is unique or if other agents are found that are also covalent MT stabilizers.

ACKNOWLEDGMENTS

The author wishes to thank Dr. David Sept, Washington University, for the figures included in Fig. 1. This work was supported in part by the Intramural program of the National Institute of Child Health and Human Development, NIH, Bethesda, MD.

REFERENCES

1. Lundblad RL. Techniques in Protein Modification. 169–173. Boca Raton, CRC Press, 2000.
2. Han KK, Delacourte A, Hemon B, Chemical modification of thiol group(s) in protein: application to the study of anti-microtubular drugs binding. Comp Biochem Physiol B 1987;88:1057–1065.

 3. Izbicka E, Tolcher AW. Development of novel alkylating drugs as anticancer agents. Curr Opin Invest Drugs 2004;5:587–591.
 4. Casini A, Scozzafava A, Supuran CT. Cysteine-modifying agents: a possible approach for effective anticancer and antiviral drugs. Environ Health Perspect 2002;110(suppl 5):801–806.
 5. Britto PJ, Knipling L, Wolff J. The local electrostatic environment determines cysteine reactivity of tubulin. J Biol Chem 2002;277:29,018–29,027.
 6. Roychowdhury M, Sarkar N, Manna T, et al. Sulfhydryls of tubulin. A probe to detect conformational changes of tubulin. Eur J Biochem 2000;267:3469–3476.
 7. Sackett, unpublished results.
 8. Chaudhuri AR, Seetharamalu P, Schwarz PM, Hausheer FH, Luduena RF. The interaction of the B-ring of colchicine with α-tubulin: a novel footprinting approach. J Mol Biol 2000;303:679–692.
 9. Luduena RF, Roach MC. Tubulin sulfhydryl groups as probes and targets for antimitotic and antimicrotubule agents. Pharmacol Ther 1991;49:133–152.
10. Kim YJ, Pannell LK, Sackett DL. Mass spectrometric measurement of differential reactivity of cysteine to localize protein-ligand binding sites. Application to tubulin-binding drugs. Anal Biochem 2004;332:376–383.
11. Mellon MG, Rebhun LI. Sulfhydryls and the in vitro polymerization of tubulin. J Cell Biol 1976;70:226–238.
12. Kuriyama R, Sakai H. Role of tubulin-SH groups in polymerization to microtubules. Functional-SH groups in tubulin for polymerization. J Biochem (Tokyo) 1974;76:651–654.
13. Lee YC, Yaple RA, Baldridge R, Kirsch M, Himes RH. Inhibition of tubulin self-assembly in vitro by fluorodinitrobenzene. Biochim Biophys Acta 1981;671:1–77.
14. Phelps KK, Walker RA. NEM tubulin inhibits microtubule minus end assembly by a reversible capping mechanism. Biochemistry 2000;39:3877–3885.
15. Lowe J, Li H, Downing KH, Nogales E. Refined structure of αβ-tubulin at 3.5 A resolution. J Mol Biol 2001;313:1045–1057.
16. Shan B, Medina JC, Santha E, et al. Selective, covalent modification of β-tubulin residue Cys-239 by T138067, an antitumor agent with in vivo efficacy against multidrug-resistant tumors. Proc Natl Acad Sci USA 1999;96:5686–5691.
17. Ziegelbauer J, Shan B, Yager D, Larabell C, Hoffmann B, Tjian R. Transcription factor MIZ-1 is regulated via microtubule association. Mol Cell 2001;8:339–349.
18. Abraham I, Dion RL, Duanmu C, Gottesman MM, Hamel E. 2,4-dichlorobenzyl thiocyanate, an antimitotic agent that alters microtubule morphology. Proc Natl Acad Sci USA 1986;83:6839–6843.
19. Bai R, Duanmu C, Hamel E. Mechanism of action of the antimitotic drug 2,4-dichlorobenzyl thiocyanate: alkylation of sulfhydryl group(s) of β-tubulin. Biochim Biophys Acta 1989a;994:12–20.
20. Bai RL, Lin CM, Nguyen NY, Liu TY, Hamel E. Identification of the cysteine residue of β-tubulin alkylated by the antimitotic agent 2,4-dichlorobenzyl thiocyanate, facilitated by separation of the protein subunits of tubulin by hydrophobic column chromatography. Biochemistry 1989b;28:5606–5612.
21. Gills JP, Roberts BC, Epstein DL. Microtubule disruption leads to cellular contraction in human trabecular meshwork cells. Invest Ophthalmol Vis Sci 1998;39:653–658.
22. Shimazaki A, Suhara H, Ichikawa M, et al. New ethacrynic acid derivatives as potent cytoskeletal modulators in trabecular meshwork cells. Biol Pharm Bull 2004;27:846–850.
23. Mallevais ML, Delacourte A, Luyckx M, Cazin M, Brunet C, Lesieur D. Antimicrotubular effects of ethacrynic acid. Methods Find Exp Clin Pharmacol 1984;6:675–677.
24. Xu S, Roychowdhury S, Gaskin F, Epstein DL. Ethacrynic acid inhibition of microtubule assembly in vitro. Arch Biochem Biophys 1992;296:462–467.
25. Luduena RF, Roach MC, Epstein DL. Interaction of ethacrynic acid with bovine brain tubulin. Biochem Pharmacol 1994;47:1677–1681.
26. O'Brien ET, Lee RE 3rd, Epstein DL. Ethacrynic acid disrupts steady state microtubules in vitro. Curr Eye Res 1996;15:985–990.
27. Jiang JD, Wang Y, Roboz J, Strauchen J, Holland JF, Bekesi JG. Inhibition of microtubule assembly in tumor cells by 3-bromoacetylamino benzoylurea, a new cancericidal compound. Cancer Res 1998a;58:2126–2133.
28. Jiang JD, Davis AS, Middleton K, et al. 3-(Iodoacetamido)-benzoylurea: a novel cancericidal tubulin ligand that inhibits microtubule polymerization, phosphorylates bcl-2, and induces apoptosis in tumor cells. Cancer Res 1998b;58:5389–5395.
29. Davis A, Jiang JD, Middleton KM, et al. Novel suicide ligands of tubulin arrest cancer cells in S-phase. Neoplasia 1999;1:498–507.

30. Jiang JD, Denner L, Ling YH, et al. Double blockade of cell cycle at G1–S transition and M phase by 3- Iodoacetamido benzoyl ethyl ester, a new type of tubulin ligand. Cancer Res 2002;62: 6080–6088.

31. Donaldson KL, Goolsby GL, Wahl AF. Cytotoxicity of the anticancer agents cisplatin and taxol during cell proliferation and the cell cycle. Int J Cancer 1994;57:847–855.

32. Trielli MO, Andreassen PR, Lacroix FB, Margolis RL. Differential Taxol-dependent arrest of transformed and nontransformed cells in the G1 phase of the cell cycle, and specific-related mortality of transformed cells. J Cell Biol 1996;135:689–700.

33. Blagosklonny MV, Darzynkiewicz Z, Halicka HD, et al. Paclitaxel induces primary and postmitotic G1 arrest in human arterial smooth muscle cells. Cell Cycle 2004;3:1050–1056.

34. Graff RD, Reuhl KR. Cytoskeletal Toxicity of Heavy Metals in Toxicology of Metals (Ed.: Liu Chang), pp. 639–658 Boca Raton: Lewis Pub.1996.

35. Macdonald TL, Humphreys WG, Martin RB. Promotion of tubulin assembly by aluminum ion in vitro. Science 1987;236:183–186.

36. Stoiber T, Bonacker D, Bohm KJ, et al. Disturbed microtubule function and induction of micronuclei by chelate complexes of mercury(II). Mutat Res 2004;563:97–106.

37. Bonacker D, Stoiber T, Wang M, et al. Genotoxicity of inorganic mercury salts based on disturbed microtubule function. Arch Toxicol 2004;78:575–583.

38. Keates RA, Yott B. Inhibition of microtubule polymerization by micromolar concentrations of mercury (II). Can J Biochem Cell Biol 1984;62:814–818.

39. Duhr EF, Pendergrass JC, Slevin JT, Haley BE. HgEDTA complex inhibits GTP interactions with the E-site of brain β-tubulin. Toxicol Appl Pharmacol 1993;122:273–280.

40. Miura K, Inokawa M, Imura N. Effects of methylmercury and some metal ions on microtubule networks in mouse glioma cells and in vitro tubulin polymerization. Toxicol Appl Pharmacol 1984;73:218–231.

41. Hoffman RD, Lane MD. Iodophenylarsine oxide and arsenical affinity chromatography: new probes for dithiol proteins. Application to tubulins and to components of the insulin receptor-glucose transporter signal transduction pathway J Biol Chem 1992;267:14,005–14,011.

42. Larochette N, Decaudin D, Jacotot E, et al. Arsenite induces apoptosis via a direct effect on the mitochondrial permeability transition pore. Exp Cell Res 1999;249:413–421.

43. Fojo T, Bates S. Arsenic trioxide (As(2)O(3)): still a mystery. Cell Cycle 2002;1:183–186.

44. Karlsson J, ORa I, Porn-Ares I, Pahlman S. Arsenic trioxide-induced death of neuroblastoma cells involves activation of Bax and does not require p53. Clin Cancer Res 2004;10:3179–3188.

45. Li W, Chou IN. Effects of sodium arsenite on the cytoskeleton and cellular glutathione levels in cultured cells. Toxicol Appl Pharmacol 1992;114:132–139.

46. Li YM, Broome JD. Arsenic targets tubulins to induce apoptosis in myeloid leukemia cells. Cancer Res 1999;59:776–780.

47. Ling YH, Jiang JD, Holland JF, Perez-Soler R. Arsenic trioxide produces polymerization of microtubules and mitotic arrest before apoptosis in human tumor cell lines. Mol Pharmacol 2002;62: 529–538.

48. Halicka HD, Smolewski P, Darzynkiewicz Z, Dai W, Traganos F. Arsenic trioxide arrests cells early in mitosis leading to apoptosis. Cell Cycle 2002;1:201–209.

49. Carre M, Carles G, Andre N, et al. Involvement of microtubules and mitochondria in the antagonism of arsenic trioxide on paclitaxel-induced apoptosis. Biochem Pharmacol 2002;63:1831–1842.

50. Umeda M, Saito K, Hirose K, Saito M. Cytotoxic effect of inorganic, phenyl, and alkyl mercuric compounds on HeLa cells. Jpn J Exp Med 1969;39:47–58.

51. Miura K, Suzuki K, Imura N. Effects of methylmercury on mitotic mouse glioma cells. Environ Res 1978;17:453–471.

52. Elhassain SB. The many faces of methylmercury poisoning. J Toxicol Clin Toxicol 1983;19:875–906.

53. Sager PR. Selectivity of methyl mercury effects on cytoskeleton and mitotic progression in cultured cells. Toxicol Appl Pharmacol 1988;94:473–486.

54. Sager PR, Doherty RA, Olmsted JB. Interaction of methylmercury with microtubules in cultured cells and in vitro. Exp Cell Res 1983;146:127–137.

55. Graff RD, Falconer MM, Brown DL, Reuhl KR. Altered sensitivity of posttranslationally modified microtubules to methylmercury in differentiating embryonal carcinoma-derived neurons. Toxicol Appl Pharmacol 1997;144:215–224.

56. Durham HD, Minotti S, Caporicci E, Chakrabarti S, Panisset JC. Sensitivity of platelet microtubules to disassembly by methylmercury. J Toxicol Environ Health 1996;48:57–69.

57. Luduena RF, Roach MC, Jordan MA, Murphy DB. Different reactivities of brain and erythrocyte tubulins toward a sulfhydryl group-directed reagent that inhibits microtubule assembly. J Biol Chem 1985;260:1257–1264.

58. Miura K, Koide N, Himeno S, Nakagawa I, Imura N. The involvement of microtubular disruption in methylmercury-induced apoptosis in neuronal and nonneuronal cell lines. Toxicol Appl Pharmacol 1999;160:279–288.

59. Hunter AM, Brown DL. Effects of microtubule-associated protein (MAP) expression on methylmercury-induced microtubule disassembly. Toxicol Appl Pharmacol 2000;166:203–213.

60. Vogel DG, Margolis RL, Mottet NK. Analysis of methyl mercury binding sites on tubulin subunits and microtubules. Pharmacol Toxicol 1989;64:196–201.

61. Vogel DG, Margolis RL, Mottet NK. The effects of methyl mercury binding to microtubules. Toxicol Appl Pharmacol 1985;80:473–486.

62. Zimmermann HP, Faulstich H, Hansch GM, Doenges KH, Stournaras C. The interaction of triethyl lead with tubulin and microtubules. Mutat Res 1988;201:293–302.

63. Stiakaki E, Stournaras C, Dimitriou H, Kalmanti M. High sensitivity of leukemic peripheral blood lymphocytes to triethyllead action. Biochem Pharmacol 1997;54:1371–1376.

64. Faulstich H, Stournaras C, Doenges KH, Zimmermann HP. The molecular mechanism of interaction of Et3Pb+ with tubulin. FEBS Lett 1984;174:128–131.

65. Maccioni RB, Seeds NW. Involvement of tryptophan residues in colchicine binding and the assembly of tubulin. Biochem Biophys Res Commun 1982;108:896–903.

66. Mellado W, Slebe JC, Maccioni RB. Tubulin carbamoylation. Functional amino groups in microtubule assembly. Biochem J 1982;203:675–681.

67. Maccioni RB, Vera JC, Slebe JC. Arginyl residues involvement in the microtubule assembly. Arch Biochem Biophys 1981;207:248–255.

68. Lee YC, Houston LL, Himes RH. Inhibition of the self-assembly of tubulin by diethylpyrocarbonate and photooxidation. Biochem Biophys Res Commun 1976;70:50–57.

69. Levison BS, Wiemels J, Szasz J, Sternlicht H. Ethoxyformylation of tubulin with [3H]diethyl pyrocarbonate: a reexamination of the mechanism of assembly inhibition. Biochemistry 1989;28: 8877–8884.

70. Legault J, Gaulin JF, Mounetou E, et al. Microtubule disruption induced in vivo by alkylation of β-tubulin by 1-aryl-3-(2-chloroethyl)ureas, a novel class of soft alkylating agents. Cancer Res 2000; 60:985–992.

71. Mounetou E, Legault J, Lacroix J, C-Gaudreault R. Antimitotic antitumor agents: synthesis, structure-activity relationships, and biological characterization of N-aryl-N′-(2-chloroethyl)ureas as new selective alkylating agents. J Med Chem 2001;44:694–702.

72. Bouchon B, Chambon C, Mounetou E, et al. Alkylation of β-tubulin on Glu 198 by a microtubule disrupter. Mol Pharmacol 2005;68:1415–1422.

73. Harris RB, Wilson IB. Irreversible inhibition of bovine lung angiotensin I-converting enzyme with p-[N,N-bis(chloroethyl)amino]phenylbutyric acid (chlorambucil) and chlorambucyl L-proline and with evidence that an active site carboxyl group is labeled J Biol Chem 1982;257:811–815.

74. Petitclerc E, Deschesnes RG, Cote MF, et al. Antiangiogenic and antitumoral activity of phenyl-3-(2-chloroethyl) ureas: a class of soft alkylating agents disrupting microtubules that are unaffected by cell adhesion-mediated drug resistance. Cancer Res 2004;64:4654–4663.

75. Mounetou E, Legault J, Lacroix J, C–Gaudreault R. A new generation of N-aryl-N′-(1-alkyl-2-chloroethyl)ureas as microtubule disrupters: synthesis, antiproliferative activity, and β-tubulin alkylation kinetics. J Med Chem 2003;46:5055–5063.

76. Kobayashi S, Tsuchiya K, Harada T, et al. Pironetin, a novel plant growth regulator produced by Streptomyces spp. NK10958. I. Taxonomy, production, isolation and preliminary characterization. J Antibiot (Tokyo) 1994;47:697–702.

77. Kondoh M, Usui T, Kobayashi S, et al. Cell cycle arrest and antitumor activity of pironetin and its derivatives. Cancer Lett 1998;126:29–32.

78. Kondoh M, Usui T, Nishikiori T, Mayumi T, Osada H. Apoptosis induction via microtubule disassembly by an antitumour compound, pironetin Biochem J 1999;340:411–416.

79. Keck GE, Knutson CE, Wiles SA. Total synthesis of the immunosupressant (–)-pironetin (PA48153C). Org Lett 2001;3:707–710.

80. Watanabe H, Watanabe H, Usui T, Kondoh M, Osada H, Kitahara T. Synthesis of pironetin and related analogs: studies on structure-activityrelationships as tubulin assembly inhibitors. J Antibiot (Tokyo) 2000;53:540–545.

81. Usui T, Watanabe H, Nakayama H, et al. The anticancer natural product pironetin selectively targets Lys352 of α-tubulin. Chem Biol 2004;11:799–806.

82. Sentein P. Action of glutaraldehyde and formaldehyde on segmentation mitoses. Inhibition of spindle and astral fibres, centrospheres blocked. Exp Cell Res 1975;95:233–246.

83. Israel Y. Covalent binding of acetaldehyde to liver tubulin: a step in the right direction. Hepatology 1989;9:161–162.

84. Kawahara H, Matsuda Y, Takada A. Effects of ethanol on the microtubules of cultured rat hepatocytes. Alcohol Alcohol 1987;1(suppl):307–311.

85. Poggi P, Rota MT, Boratto R. The volatile fraction of cigarette smoke induces alterations in the human gingival fibroblast cytoskeleton. J Periodontal Res 2002;37:230–235.

86. Nakamura Y, Romberger DJ, Tatel, et al. Cigarette smoke inhibits fibroblast proliferation and chemotaxis. Am J Respir Crit Care Med 1995;151:1497–1503.

87. McKinnon G, Davidson M, De Jersey J, Shanley B, Ward L. Effects of acetaldehyde on polymerization of microtubule proteins. Brain Res 1987;416:90–99.

88. Jennett RB, Sorrell MF, Johnson EL, Tuma DJ. Covalent binding of acetaldehyde to tubulin: evidence for preferential binding to the α-chain. Arch Biochem Biophys 1987;256:10–18.

89. Jennett RB, Sorrell MF, Saffari-Fard A, Ockner JL, Tuma DJ. Preferential covalent binding of acetaldehyde to the α-chain of purified rat liver tubulin. Hepatology 1989;9:57–62.

90. Szasz J, Burns R, Sternlicht H. Effects of reductive methylation on microtubule assembly. Evidence for an essential amino group in the α-chain. J Biol Chem 1982;257:3697–3704.

91. Sherman G, Rosenberry TL, Sternlicht H. Identification of lysine residues essential for microtubule assembly. Demonstration of enhanced reactivity during reductive methylation J Biol Chem 1983;258:2148–2156.

92. Szasz J, Yaffe MB, Elzinga M, Blank GS, Sternlicht H. Microtubule assembly is dependent on a cluster of basic residues in α-tubulin. Biochemistry 1986;25:4572–4582.

93. Pal D, Mahapatra P, Manna T, et al. Conformational properties of α-tubulin tail peptide: implications for tail-body interaction. Biochemistry 2001;40:15,512–15,519.

94. Sackett DL, Bhattacharyya B, Wolff J. Tubulin subunit carboxyl termini determine polymerization efficiency. J Biol Chem 1985;260:43–45.

95. Sackett DL. Structure and function in the tubulin dimer and the role of the acidic carboxyl terminus. Subcell Biochem 1995;24:255–302.

96. Smith SL, Jennett RB, Sorrell MF, Tuma DJ. Acetaldehyde substoichiometrically inhibits bovine neurotubulin polymerization. J Clin Invest 1989;84:337–341.

97. Smith SL, Jennett RB, Sorrell MF, Tuma DJ. Substoichiometric inhibition of microtubule formation by acetaldehyde-tubulin adducts. Biochem Pharmacol 1992;44:65–72.

98. Xu DS, Jennett RB, Smith SL, Sorrell MF, Tuma DJ. Covalent interactions of acetaldehyde with the actin/microfilament system. Alcohol Alcohol 1989;24:281–289.

99. Miglietta A, Olivero A, Gadoni E, Gabriel L. Effects of some aldehydes on brain microtubular protein. Chem Biol Interact 1991;78:183–191.

100. Olivero A, Miglietta A, Gadoni E, Gabriel L. 4-Hydroxynonenal interacts with tubulin by reacting with its functional –SH groups. Cell Biochem Funct 1990;8:99–105.

101. Olivero A, Miglietta A, Gadoni E, Gabriel L. Aldehyde-induced modifications of the microtubular system in 3T3 fibroblasts. Cell Biochem Funct 1992;10:19–26.

102. Young DH, Lewandowski VT. Covalent binding of the benzamide RH-4032 to tubulin in suspension-cultured tobacco cells and its application in a cell-based competitive-binding assay. Plant Physiol 2000;124:115–124.

103. Anthony RG, Hussey PJ. Double mutation in eleusine indica α-tubulin increases the resistance of transgenic maize calli to dinitroaniline and phosphorothioamidate herbicides. Plant J 1999;18:669–674.

104. Morrissette NS, Mitra A, Sept D, Sibley LD. Dini troanilines bind alpha–tubulin to disrupt microtubules. Mol Biol Cell 2004;15:1960–1968.

105. Berg K, Moan J. Lysosomes and microtubules as targets for photochemotherapy of cancer. Photochem Photobiol 1997;65:403–409.

106. Sporn LA, Foster TH. Photofrin and light induces microtubule depolymerization in cultured human endothelial cells. Cancer Res 1992;52:3443–3448.

107. Berg K, Moan J. Mitotic inhibition by phenylporphines and tetrasulfonated aluminium phthalocyanine in combination with light. Photochem Photobiol 1992;56:333–339.

108. Berg K. The unpolymerized form of tubulin is the target for microtubule inhibition by photoactivated tetra(4-sulfonatophenyl)porphine. Biochim Biophys Acta 1992;1135:147–153.

109. Berg K, Steen HB, Winkelman JW, Moan J. Synergistic effects of photoactivated tetra(4-sulfonatophenyl)porphine and nocodazole on microtubule assembly, accumulation of cells in mitosis and cell survival. J Photochem Photobiol B 1992;13:59–70.
110. Ma LW, Berg K, Danielsen HE, Kaalhus O, Iani V, Moan J. Enhanced antitumour effect of photodynamic therapy by microtubule inhibitors. Cancer Lett 1996;109:129–139.
111. Winkelman JW, Arad D, Kimel S. Stereochemical factors in the transport and binding of photosensitizers in biological systems and in photodynamic therapy. J Photochem Photobiol B 1993;18:181–189.
112. Boekelheide K, Eveleth J, Tatum AH, Winkelman JW. Microtubule assembly inhibition by porphyrins and related compounds. Photochem Photobiol 1987;46:657–661.
113. Juarranz A, Villanueva A, Diaz V, Canete M. Photodynamic effects of the cationic porphyrin, mesotetra(4N-methylpyridyl)porphine, on microtubules of HeLa cells. J Photochem Photobiol B 1995;27:47–53.
114. Stockert JC, Juarranz A, Villanueva A, Canete M. Photodynamic damage to HeLa cell microtubules induced by thiazine dyes. Cancer Chemother Pharmacol 1996;39:167–169.
115. Lee C, Wu SS, Chen LB. Photosensitization by 3,3′-dihexyloxacarbocyanine iodide: specific disruption of microtubules and inactivation of organelle motility. Cancer Res 1995;55:2063–2069.
116. Sakharov DV, Bunschoten A, van Weelden H, Wirtz KW. Photodynamic treatment and H2O2-induced oxidative stress result in different patterns of cellular protein oxidation. Eur J Biochem 2003;270:4859–4865.
117. Magi B, Ettorre A, Liberatori S, et al. Selectivity of protein carbonylation in the apoptotic response to oxidative stress associated with photodynamic therapy: a cell biochemical and proteomic investigation. Cell Death Differ 2004;11:842–852.
118. Peyrot V, Briand C, Momburg R, Sari JC. In vitro mechanism study of microtubule assembly inhibition by cis-dichlorodiammine-platinum(II). Biochem Pharmacol 1986;35:371–375.
119. Boekelheide K, Arcila ME, Eveleth J. cis-diamminedichloroplatinum (II) (cisplatin) alters microtubule assembly dynamics. Toxicol Appl Pharmacol 1992;116:146–151.
120. Tulub AA, Stefanov VE. Cisplatin stops tubulin assembly into microtubules. A new insight into the mechanism of antitumor activity of platinum complexes. Int J Biol Macromol 2001;28:191–198.
121. Buey RM, Calvo E, Barasoain I, et al. Cyclostreptin binds covalently to microtubule pores and lumenal taxoid binding sites. Nature Chem Biol 2007;3:117–125.

13 Microtubule Stabilizing Agents

Susan Band Horwitz and Tito Fojo

SUMMARY

The microtubule cytoskeleton continues to be both an effective and a validated target in the therapy of cancer. Beginning with the vinca alkaloids almost 50 years ago and encouraged by the broad activity of first taxol® (paclitaxel) and then taxotere® (docetaxel) numerous investigators have identified structurally diverse compounds that interact with the tubulin/microtubule system displaying antimitotic and anticancer properties. Paclitaxel was the first of a class of agents with a binding site on the microtubule polymer that act principally by stabilizing microtubules. Other members of this class of compounds referred to as the microtubule-stabilizing agents include the epothilones, eleutherobin and discodermolide, natural products that are mechanistically similar but structurally unrelated to paclitaxel. With an emphasis on paclitaxel, the original and best-characterized member of the microtubule-stabilizing class, this chapter will review the compounds that comprise this diverse and interesting class of anti-cancer agents.

Key Words: Microtubule-stabilizing agents; paclitaxel (Taxol®); taxotere (Docetaxel®); epothilones; discodermolide; microtubules; tubulin.

From: *Cancer Drug Discovery and Development: The Role of Microtubules in Cell Biology, Neurobiology, and Oncology* Edited by: Tito Fojo © Humana Press, Totowa, NJ

1. INTRODUCTION

The microtubule cytoskeleton is both an effective and a validated target in the therapy of cancer. In numerous studies, structurally diverse compounds have been shown to interact with the tubulin/microtubule system displaying antimitotic and anticancer properties. These tubulin/microtubule interacting compounds can be divided into two major classes: The first class includes those agents that bind preferentially to α/β-tubulin heterodimers and thus inhibit microtubule polymerization. The second class consists of those agents with a binding site on the microtubule polymer that act principally by stabilizing microtubules. The first class exemplified by the vinca alkaloids, are reviewed elsewhere in this text (*see* Chapter 10). Paclitaxel is the prototype for the second class that is inclusive of the microtubule-stabilizing agents (Fig. 1). Other members of the microtubule-stabilizing class of compounds include mechanistically similar but structurally unrelated natural products, such as the epothilones, eleutherobin, and discodermolide. This chapter will review the compounds that include the microtubule-stabilizing class of agents. Emphasis will be placed on paclitaxel, the original microtubule-stabilizing agent of which the authors have the greatest understanding.

2. THE STRUCTURE AND FUNCTION OF THE MICROTUBULE CYTOSKELETON

2.1. Microtubule Dynamics and Function

In eukaryotes, microtubules are essential to a diverse range of cellular functions including mitosis, meiosis, motility, maintenance of cell shape, and intracellular trafficking of macromolecules and organelles *(1–3)*. Microtubules are hollow cylindrical tubes formed by the self-association of α/β-tubulin heterodimers into polymers (*see* Chapter 3). The tubulin heterodimers associate in a head-to-tail fashion to form protofilaments, which in turn associate in a lateral manner to form hollow microtubules. The arrangement of protofilaments in a parallel array imparts polarity to the structure. The β-subunits of the tubulin dimer are exposed at the "plus end" of the polymer, and the α-subunits at the "minus end." In cells, the minus end of a microtubule is usually associated with the microtubule-organizing center near the nucleus. From this anchored position, the microtubules radiate outward so that the plus ends are near the periphery of the cell. γ-Tubulin, a protein highly homologous to the α/β-tubulins, is localized at the microtubule-organizing center, and plays an important role in microtubule nucleation by interacting with α-tubulin *(2)*.

Microtubules are highly dynamic structures characterized by a nonequilibrium behavior termed dynamic instability (*see* Chapter 3). In this process, microtubules undergo rapid stochastic transitions between growth and shrinkage, as a result of the association and dissociation, respectively, of tubulin dimers from the microtubule ends. The transition from growing to shrinking is referred to as a catastrophe, whereas the transition from shrinkage to growth is called a rescue. GTP binding and hydrolysis at the exchangeable or E-site of β-tubulin appears to be crucial for this dynamic instability (GTP also binds to α-tubulin, but at the nonexchangeable or N-site). Microtubule assembly requires that GTP be bound to β-tubulin, so that it can be hydrolyzed on addition of the tubulin dimer to the elongating microtubule. After hydrolysis, the guanine nucleotide becomes nonexchangeable, and so microtubules are

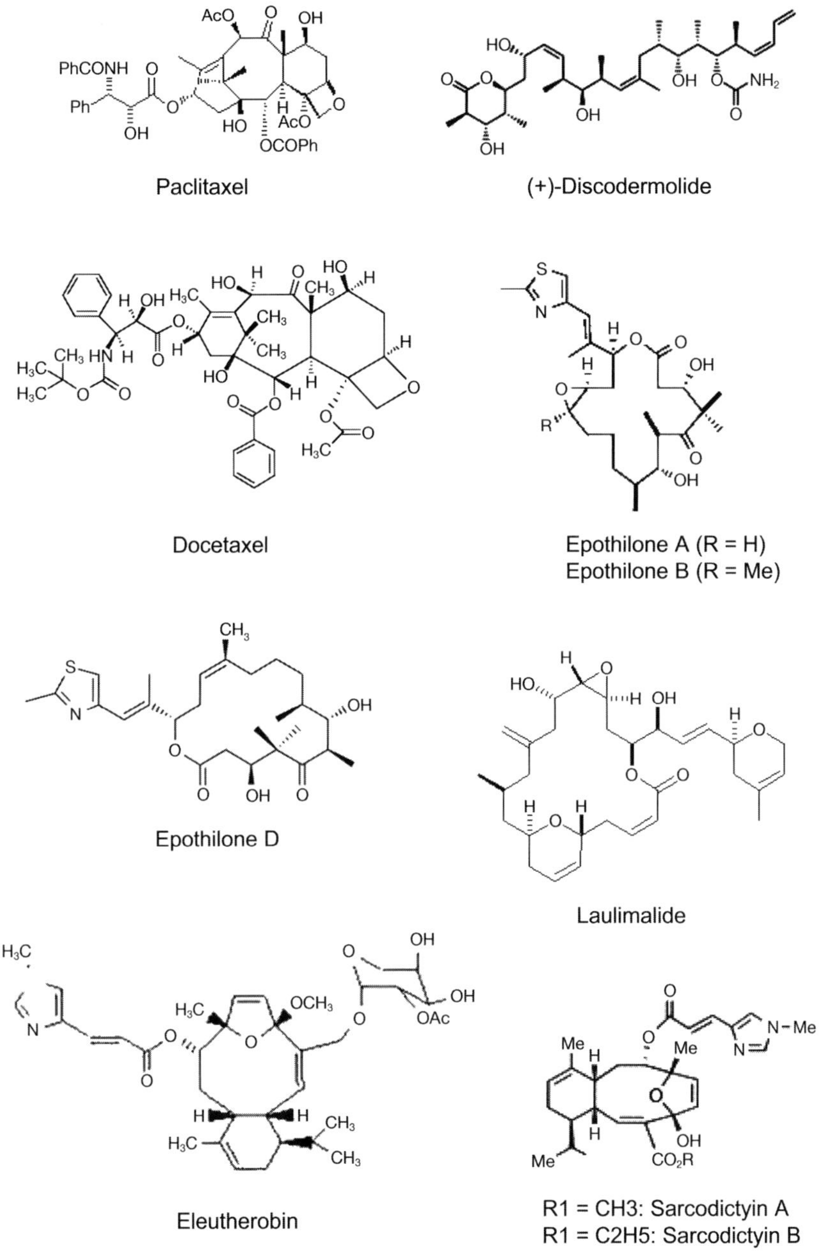

Fig. 1. Structure of microtubule stabilizing agents.

mostly made up of (GTP: α-tubulin/GDP: β-tubulin), with the growing end capped with GTP (or GDP·Pi): β-tubulin. In the GTP-cap model, microtubules, which are inherently unstable, are stabilized by GTP (or GDP·Pi)-tubulin at the growing ends. When the GTP cap is lost, the microtubules rapidly depolymerize, with the protofilaments peeling outward.

There are numerous proteins that can interact with microtubules and/or free tubulin dimers and participate in the regulation of microtubule dynamics (*see* Chapter 4). The best characterized of these regulatory proteins are the microtubule-associated proteins (MAPs), that stabilize microtubules by decreasing the frequency and duration of catastrophes and/or increasing the frequency and duration of rescues. In addition to these proteins that associate with and stabilize microtubules, other proteins, such as stathmin, may regulate microtubule dynamics. Unlike MAPs, stathmin binds exclusively to tubulin dimers and not to microtubules, thereby increasing the catastrophe rate and promoting depolymerization. In addition, the activities of many of these microtubule-stabilizing and microtubule-destabilizing proteins can be regulated by cyclic phosphorylation and dephosphorylation, often in a cell cycle-dependent manner.

3. THE MECHANISM OF ACTION OF PACLITAXEL AND OTHER MICROTUBULE-STABILIZING COMPOUNDS

In 1971 Monroe Wall, Mansukh Wani, and their colleagues *(4)* published a landmark study describing the isolation of a compound from the bark of the yew tree (*Taxus brevifolia*). They determined its structure and demonstrated its cytotoxic activity against KB (HeLa) cells in culture. The publication came nearly a decade after the first samples of the yew tree were collected as part of a joint project of the Department of Agriculture and the National Cancer Institute to search for new plant products with antitumor activity. In 1967, Dr. Wall gave the compound the name taxol based on its source and because it was an alcohol (the drug is now known by the generic name paclitaxel and the trade name Taxol®[Bristol Myers Squibb]). Paclitaxel was first recognized as a microtubule-targeting agent in 1979 when studies showed it was able to increase the rate and extent of microtubule assembly in vitro and to stabilize microtubules in vitro and in cells *(5,6)*.

The first few experiments demonstrated that nanomolar concentrations of paclitaxel inhibited the replication of HeLa cells by blocking cells in metaphase (Fig. 2). After 18 h in 250 nM paclitaxel, essentially all cells had replicated their DNA, and had a 4N DNA content, however they were blocked in metaphase. Although previously studied drugs such as colchicine and the Vinca alkaloids had been shown to block cells in mitosis, paclitaxel-treated cells reorganized their microtubules so that distinct microtubule bundles could be seen (Fig. 3). Paclitaxel was also shown in mouse fibroblasts to both block cell replication by arresting cells in the G2/M phase of the cell cycle, and to stabilize cytoplasmic microtubules. The latter were visualized by transmission electron microscopy and indirect immunofluorescence microscopy as bundles that radiated from a common site (or sites) after cells were treated with 10 µM taxol for 22 h at 37°C (Fig. 4). In the experiments demonstrating stability against cold-induced depolymerization, untreated cells that were kept at 4°C for 16 h lost their microtubules, whereas cells that were pretreated with paclitaxel continued to display their microtubules and bundles of microtubules in the cold.

The formation of *stable microtubule bundles*, now recognized as diagnostic of a microtubule-stabilizing compound and *a hallmark of drug binding to microtubules*, suggested paclitaxel was able to enhance microtubule assembly and stabilize existing microtubules. These hypotheses were confirmed by several experiments including early studies that showed that paclitaxel was able to stabilize microtubules against cold-induced depolymerization; as well as studies demonstrating that paclitaxel was able to augment microtubule assembly when added to microtubules at apparent steady state by promoting both the

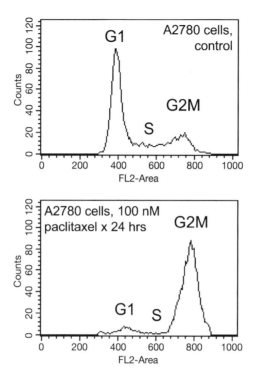

Fig. 2. Paclitaxel-induced G2M arrest. Propidium iodide staining, FACS (fluorescence activated cell sorting) analysis.

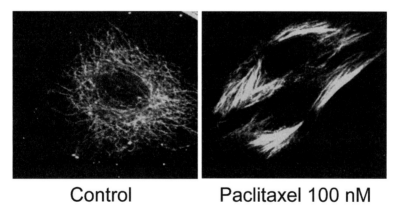

Fig. 3. Paclitaxel-induced tubulin polymerization *in vitro*. To view this figure in color, see the insert and the companion CD-ROM.

elongation of existing microtubules and spontaneous nucleation of new microtubules (Fig. 5) *(7)*. Using purified calf brain tubulin, microtubule polymerization was assessed by monitoring the increase in absorption at 350 nm. In the absence of paclitaxel, poly-merization was seen to occur after a 3–4 min lag period. However, when the experiment was repeated in the presence of paclitaxel, the lag period observed in the absence of drug was eliminated, indicating paclitaxel enhanced the initiation phase of microtubule

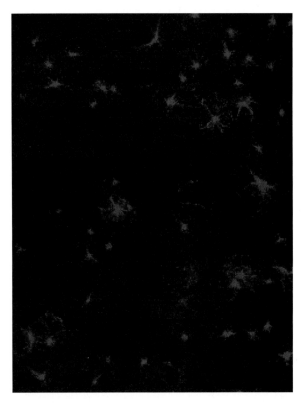

Fig. 4. Paclitaxel-induced mitotic arrest with aster formation *in vivo*. To view this figure in color, see the insert and the companion CD-ROM.

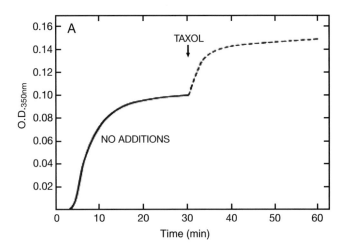

Fig. 5. Additional microtubule assembly occurs when paclitaxel is added at a final concentration of 5 μM after apparent steady state has been reached at 30 minutes. [Schiff and Horwitz, 1981].

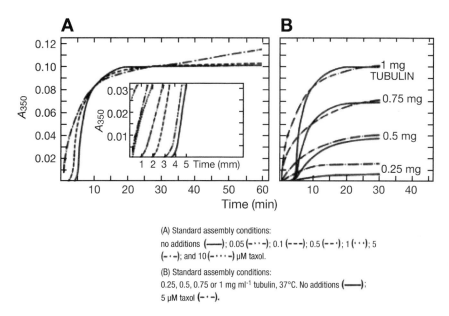

Fig. 6. (A) Paclitaxel causes a dose-dependent decrease in the lag time for assembly. (B) Paclitaxel decreased the lag time independently of tubulin concentration. [Schiff et al, 1979]

polymerization *(5,8)* (Fig. 6). Furthermore, paclitaxel was able to bring about tubulin polymerization even at cold temperatures and in the absence of MAPs and GTP that are required for polymerization in the absence of paclitaxel. The most striking observation was that the microtubules formed in the presence of paclitaxel were resistant to cold (4°C) and Ca^{2+} induced depolymerization, conditions previously shown to depolymerize microtubules. By enhancing polymerization, paclitaxel alters the in vitro kinetics such that a paclitaxel concentration of 5 μM reduces the critical concentration of microtubule protein necessary for microtubule assembly by a factor of 20 from 0.2 to less than 0.01 mg/mL. Paclitaxel also affects the structure of the microtubule polymer by reducing the number of protofilaments from an average of 13 under normal circumstances to an average of 12 *(9)*.

Whereas these data strongly suggested that paclitaxel could interact with microtubules, convincing evidence as well as accurate stoichiometry was provided by sedimentation assays assessing the binding of [³H]paclitaxel to microtubule protein *(10)*. Under assembly conditions both podophyllotoxin and vinblastine were able to inhibit the binding of [³H]paclitaxel to microtubule protein. However, unlike unlabeled paclitaxel, which competitively displaced [³H]paclitaxel from microtubules, podophyllotoxin, vinblastine, and $CaCl_2$ did not. Podophyllotoxin and vinblastine were able to reduce the mass of sedimented paclitaxel-stabilized microtubules, but the specific activity of bound [³H]paclitaxel in the pellet remained constant. Collectively this data was interpreted as a competition between paclitaxel on the one hand and either podophyllotoxin or vinblastine on the other, although not for a single binding site, but rather for different forms of tubulin: the dimeric or soluble form and the polymeric or microtubule form. Finally it was noted that steady-state microtubules assembled with GTP or a GTP analog (GPCPP), bound [³H]paclitaxel with approximately the same stoichiometry as microtubules assembled in the presence of [³H]paclitaxel. These observations led to the conclusion that paclitaxel bound specifically and reversibly to the polymerized form of tubulin with a stoichiometry approaching unity and that the paclitaxel-binding site was present in intact microtubules in the β-tubulin subunit *(11)*. This conclusion has received support from numerous subsequent experiments and today it is known that *paclitaxel binds at a specific site on β-tubulin in the microtubule* as confirmed by electron crystallography *(12)* (*see* Chapter 9). There is no evidence that paclitaxel binds to soluble tubulin dimer.

In vivo, high concentrations of paclitaxel increase the mass of microtubule polymer, and also induce microtubule bundle formation in interphase cells. This observation has become a hallmark of paclitaxel binding and is now recognized as a characteristic of microtubule stabilization *(6)*. However, the formation of microtubule bundles occurs only above a threshold concentration of drug. At paclitaxel concentrations less than 10 nM, when only a fraction of the paclitaxel-binding sites on the microtubule are occupied, there is no obvious effect on polymer mass; instead the principal drug effect is suppression of microtubule dynamics *(13,14)*. Because low concentrations of vinblastine, a microtubule-destabilizing drug, have a similar effect on microtubule dynamics as low-paclitaxel concentrations, it has been suggested that both the microtubule stabilizing and destabilizing drug classes block mitosis by decreasing the dynamics of spindle microtubule. However, the two drug classes exhibit different mitotic effects at low concentrations *(15)*. Microtubule-stabilizing drugs, including paclitaxel, the epothilones, and discodermolide, produce aneuploid populations of cells in the absence of a sustained mitotic block *(15,16)*. In contrast, destabilizing drugs such as colchicine, vinblastine, and nocodazole, do not lead to aneuploidy at these low concentrations. Aneuploidy appears to occur as a result of cells exiting from an aberrant mitosis, as the stabilizing drugs induce multipolar spindles. These studies imply that paclitaxel, and most likely other microtubule stabilizing drugs exert their effects by alternate mechanisms, depending on the drug concentration *(16)*.

4. THE INTERACTION OF PACLITAXEL AND MICROTUBULES

Lacking a high-resolution structure of tubulin, early studies used photoaffinity labeling to understand the nature of the interaction between paclitaxel and its target *(11,17–21)*. Initial studies used [³H]-paclitaxel as a direct photoaffinity-label and demonstrated that paclitaxel binds specifically to the β-subunit of tubulin *(11)*. However, because of low photoincorporation, a detailed analysis of the paclitaxel-binding site was not possible. A more detailed definition of the contact sites between paclitaxel and β-tubulin became possible when a series of paclitaxel analogs bearing photoreactive groups at defined positions around the taxane nucleus became available. Initially, analogs with photoreactive groups at the C-2′, C-3′, or C-7 positions were found to be specifically incorporated into β-tubulin following photoactivation. Subsequent studies used chemical and enzymatic digestion to isolate photolabeled fragments that were then subjected to N-terminal amino acid sequencing, making it possible to identify residues in close proximity to the paclitaxel-binding site. Studies using a paclitaxel analog where the photoreactive arylazide group was incorporated into the C-13 side-chain of paclitaxel ([³H]3′-[p-azidobenzamido]paclitaxel), led to the isolation of a photolabeled peptide containing amino-acid residues 1–31 of β-tubulin *(17)*. Similarly, a peptide containing amino acid residues 217–233 of β-tubulin was identified as the region interacting with the 2-benzoyl group by using an analog with the photoreactive group attached to the B ring of the paclitaxel nucleus ([³H]2-[m-azidobenzoyl]Taxol) *(18)*. Finally, specific photocrosslinking to Arg282 was observed when a photoreactive benzophenone (BzDC) substituent was attached to the C-7 hydroxyl group of the C ring *(20)*. The usefulness of this approach became apparent after the structure of tubulin was determined by electron crystallography. Excellent agreement was seen between the binding site predicted using the various photoaffinity analogs and that determined by electron crystallography. (*See* Chapter 9.)

5. NONTAXANE MICROTUBULE-STABILIZING DRUGS

The success of paclitaxel and then docetaxel (Taxotere® [Sanofi Aventis]) in the treatment of patients with a wide range of malignancies prompted an extensive search for additional natural products that could stabilize microtubules. Using a variety of different screens, several natural products with unique structures unrelated to that of the taxanes, were identified and reported to stabilize microtubules (Fig. 1) *(22–24)*. The first and most extensively characterized, were the epothilones A and B, founding members of a class of compounds isolated from the fermentation broth of a soil *Myxobacterium*, *Sorangium cellulosum* strain 90. Epothilones A and B are naturally produced by *S. cellulosum* by the action of a hybrid nonribosomal peptide synthetase 1/polyketide synthase together with an epoxidase.

Inactivation or deletion of the epoxidase gene precludes the synthesis of epothilones A and B and results in the production of epothilones C and D as the end products of the nonribosomal peptide synthetase/polyketide synthase. These novel polyketide natural products were found to polymerize tubulin, form microtubule bundles, and arrest cells in mitosis *(22)*. Compared with paclitaxel, both epothilones were reported to be more potent in promoting microtubule assembly in vitro, with epothilone B more potent than epothilone A, as well as epothilone D. Whereas isolation of the epothilones hinted at the existence of additional agents with microtubule stabilizing properties, the isolation of discodermolide affirmed their widespread occurrence in nature. Discodermolide is a C24:4, trihydroxylated, octamethyl, carbamate-bearing fatty acid lactone originally isolated from a Caribbean sponge, *Discodermia dissoluta*, that like the epothilones was reported to induce microtubule assembly in vitro more rapidly than paclitaxel as well as cause microtubule bundling and mitotic arrest *(23,25,26)*.

Similarly, first eleutherobin, a diterpene glycoside isolated from an *Eleutherobia* species of soft, red colored coral found near Western Australian, and then the closely related sarcodictyins were shown to stabilize microtubules with potency similar to that of paclitaxel *(27,28)*. Finally the laulimalides, complex macrolide compounds isolated from the marine sponge *Cacospongia mycofijiensis* were also reported to possess microtubule-stabilizing activity *(29)*. Undoubtedly others have yet to be discovered and it is likely that their characterization will reveal an increasingly complex mechanism of action. Already it is known that whereas the epothilones, discodermolide, and eleutherobin are all competitive inhibitors of the binding of [3H]-paclitaxel to microtubules, laulimalide is not *(26,30)*. This has been interpreted as evidence that the former drugs interact at the same or an overlapping binding domain as paclitaxel on β-tubulin, with laulimalide binding at a different site on the microtubules. Based on a computational search, the binding site for laulimalide has been proposed to be in α-tubulin *(31)*. Additional evidence cited in support of this includes the observation that unlike epothilone A, laulimalide was able to enhance microtubule assembly synergistically with paclitaxel, as would be predicted if the two drugs bound at different sites on the polymer *(32)*. Although most would agree that this evidence supports the concept that laulimalide binds at a different site on the microtubules, it must be noted that the combination of paclitaxel and discodermolide also has been shown to exhibit a synergistic cytotoxic interaction in human carcinoma cell lines *(33)*. In addition, synergism between paclitaxel and discodermolide has been demonstrated in an ovarian cancer model in nude mice *(34)*. As discodermolide is a competitive inhibitor of the binding of [^{3}H]-paclitaxel to microtubules, synergism in cytotoxic activity does not require distinct binding sites *(33)*.

In addition to its ability to stabilize microtubules, discodermolide may have a second activity that is responsible for its synergy with paclitaxel. There is evidence that discodermolide induces a senescence phenotype at its IC$_{50}$ concentration, which paclitaxel does not do at the same concentration *(35)*. As regards the concept of similar or overlapping binding sites, it should be noted that emerging electron crystallographic evidence might challenge the extent to which the description of the binding sites as overlapping is stretched. As has been shown by electron crystallography, compared with paclitaxel, the epothilones exploit the tubulin-binding pocket in a unique and independent manner *(36)* *(see* Chapter 9). Indeed, it could be argued that one should expect differences in the way that these diverse compounds interact with microtubules.

Although all of these agents share the ability to stabilize mammalian microtubules, this might be too crude or imprecise an assay. From other lines of investigation it is known that differences can be found when microtubules from other sources are used in stabilization assays. For example, unlike paclitaxel, the epothilones promote the in vitro assembly of yeast tubulin *(37)*. This discrepancy might be explained by sequence variation in three β-tubulin residues, K19A, V23T, and D26G, which make contact with the 3′-benzamidophenyl group of paclitaxel, but not with the epothilones.

Identifying/characterizing the binding site on the microtubule is not only of academic interest, but will be of value in the future design of analogs or even synthetic compounds targeting microtubules. Structural information obtained from experimental approaches such as electron crystallography will be invaluable. In addition, for a macromolecule as large and complex as the microtubule, additional information including structure activity correlations and cross-resistance data are also deemed helpful. However, findings from diverse experimental approaches may eventually have to be reconciled. For example, like paclitaxel, 2-*m*-azido baccatin III, a paclitaxel analog that lacks the C-13 side-chain but has a *m*-azido benzoyl group at the C-2 position, promotes microtubule assembly in the absence of GTP, stabilizes microtubules, and competitively inhibits the binding of [3H]-Taxol to the microtubule protein *(38)*. Although 2-*m*-azido baccatin III is certainly not as potent as paclitaxel, the fact that it retained all of the basic properties of paclitaxel was taken as evidence that the C-13 side-chain is not required for the biological activity of a taxane. Based on this, a common pharmacophore model for paclitaxel and epothilone was proposed with the macrolide ring system of the epothilones overlapping the taxane ring system and the thiazole side-chain of the epothilones corresponding to the C-2 side-chain of 2-*m*-azido baccatin III, binding in the pocket formed by His227 and Asp224 *(38)*. This model of the epothilone:tubulin interaction was similar to one of two potential models based on β-tubulin mutations identified in epothilone-resistant cells *(39,40)*. However, neither of these agrees fully with a model based on electron crystallographic data, underscoring the need to reconcile evidence obtained from different experimental sources *(36)* *(see* Chapter 9).

6. HOW ALTERATIONS IN MICROTUBULE STABILITY/DYNAMICS CAN AFFECT DRUG SENSITIVITY

The selection of drug-resistant cell lines usually has, as a principal goal, understanding mechanisms of drug tolerance. However, such selections can also help discern the factors involved in intrinsic drug sensitivity as well as provide insight into how drugs work. In the case of the microtubule stabilizing agents, selection of drug-resistant cell lines have delivered information on all these fronts. As investigators have sought to understand how cells develop resistance, their understanding of normal microtubule function, stability, and dynamics has grown.

For drugs such as the microtubule-stabilizing agents that likely have a defined "binding site" on the microtubule, one could envisage mutations at these loci interfering with drug binding and conferring resistance. Such mutations have been discussed previously in this chapter and will be referred to below *(39–42)*. But remarkably, in a majority of the resistant cells lines, mutations that map to putative drug-binding sites and interfere with drug binding have not been identified (*see* Chapter 14). Instead the mutations or adaptations that have been identified affect intrinsic properties of the microtubule—stability/dynamics—and in turn drug sensitivity. Two models have been proposed to help us understand the relationship between resistance to microtubule-targeting agents and cellular microtubule stability/dynamics. Although it can be argued that absolute proof is lacking for either of these models, data accumulated since they were first proposed attest to their utility. At a minimum they provide a very rational framework for thinking about resistance to microtubule-targeting agents, and also for considering how microtubules function normally.

The first hypothesis advanced by Cabral and coworkers *(43–45)* in the mid-to-late 1980s and discussed in detail elsewhere in this book (*see* Chapter 14), envisioned that under normal circumstances microtubule stability or polymerization was maintained within a limited range. According to this hypothesis, intrinsic sensitivity to a microtubule-stabilizing agent depends on the basal level of microtubule stability/polymerization. A cell with a higher percent of polymerized tubulin is more sensitive to a microtubule-stabilizing agent. One can envision that in this case the equilibrium is already shifted to a more stable state, making it easier for the stabilizing agent to achieve its goal or that as the paclitaxel-binding site is present only on microtubules (polymerized tubulin), and not on tubulin dimers, a cell with a more stable polymer presents more sites for drug binding and would be more sensitive to paclitaxel. Conversely, cells with less stable microtubules should be resistant to polymer-binding drugs like paclitaxel, but demonstrate increased sensitivity to agents such as vinblastine and colchicine that target tubulin dimers. Thus, according to this model, cells resistant to a microtubule-stabilizing agent contain "hypostable" microtubules in which the equilibrium between the dimer and polymer is shifted toward the former. The model also offers a potential explanation for the observation that some paclitaxel-resistant cell lines require low-paclitaxel concentrations for normal growth. In these drug-dependent cells, polymer stability is perturbed to such an extent—actually the polymer is so hypostable—that normal cell function is compromised, and the cells require low concentrations of paclitaxel to stabilize their microtubules in order to survive.

A second model by Wilson and Jordan evolved from the observation that low concentrations of both microtubule-stabilizing and -destabilizing drugs inhibit microtubule *dynamics* without affecting the microtubule polymer mass *(13,46,47)* (*see* Chapter 3).

According to this view, microtubule dynamics, like microtubule stabilization affect drug sensitivity: increased dynamics confer resistance to microtubule stabilizing agents, whereas reduced dynamics bestow tolerance to destabilizing compounds. Therefore, in paclitaxel-resistant cell lines, the equilibrium between hypodynamic and highly dynamic microtubules is shifted toward the latter, antagonizing the effects of paclitaxel *(14,46–48)*. This increase in microtubule dynamics provides a survival advantage to a cell challenged with a microtubule-stabilizing drug such as paclitaxel. Evidence for the latter has come from several sources including studies quantifying the dynamics of individual rhodamine-labeled microtubules by digital time-lapse microscopy *(48)*. Parental A549 human lung carcinoma cells, and paclitaxel-resistant A549-T12 and -T24 cell lines, selected by continuous exposure of A549 cells to increasing concentrations of paclitaxel were examined. The A549-T12 and A549-T24 cell lines were found to be 9- and 17-fold resistant, respectively, to paclitaxel. Both resistant cell lines exhibited *increased dynamic instability* compared with the parental A549 cell line. As noted earlier, increased microtubule dynamics would be predicted to provide a survival advantage to a cell challenged with paclitaxel. Furthermore, both resistant cell lines were also found to be dependent on low concentrations of paclitaxel for growth. In the absence of 2 n*M* paclitaxel to stabilize the hyperdynamic microtubules, the resistant cells were blocked in the G2/M phase of the cell cycle.

7. RESISTANCE TO MICROTUBULE-STABILIZING AGENTS

Several potential mechanisms can be envisaged by which a cell may develop resistance to microtubule-stabilizing agents. Included among these would be changes that alter microtubule stability/dynamics rendering a cell resistant to a microtubule-targeting agent, mutations that alter drug binding or changes in signaling pathways that modulate the cellular response to drug exposure. Although the evidence accumulated to date has been garnered principally with paclitaxel, and to a lesser extent with the epothilones, it is likely that these changes/mechanisms apply broadly to varying extents to all microtubule-stabilizing agents. For the purpose of this chapter the authors propose to subdivide these mechanisms of drug resistance as follows:

1. Alterations in the absolute or relative expression levels of tubulin isotypes (*see* Chapters 5–7).
2. Tubulin mutations that affect either binding of regulatory proteins or GTP or that impact longitudinal/lateral interactions and can alter microtubule dynamics.
3. Post-translational modifications that modify regulatory protein binding.
4. Altered expression and post-translational modifications of proteins that regulate the dynamics/stability of the microtubules by interacting with soluble tubulin or microtubules.
5. Altered drug binding to the microtubule.
6. Alterations in signaling pathways.

7.1. Alterations in the Absolute or Relative Expression Levels of Tubulin Isotypes

In vitro analysis of β-tubulin isotypes purified by immunoaffinity from bovine brain tubulin (β-tubulin composition: 3% class I, 58% class II, 25% class III, and 13% class IV) has identified *inherent* differences in the assembly properties, microtubule dynamics, and drug interactions of the various β-tubulin isotypes *(49–54)* (*see* Chapter 6). Initial studies examined the differences in assembly and microtubule dynamics. Because differences were observed using purified tubulin devoid of MAPs, it suggested that the differences were *inherent* to the various tubulin isotypes *(53)*. Subsequent in vitro studies examining the effect of paclitaxel on microtubule dynamics demonstrated that it could be modulated by the β-tubulin isotype composition. These studies eventually formed the basis for the notion that altered cellular expression of β-tubulin isotypes could be an important determinant of cellular sensitivity to paclitaxel. The initial studies demonstrated differences between βIII-tubulin compared with βII-, βIV-, or unfractionated tubulin when microtubule assembly was examined *(49,51,52)*. Compared with other isotypes, βIII-tubulin required the highest critical concentration of tubulin for assembly, exhibited a distinct delay in nucleation and proceeded at a slower rate. In addition, the dynamicity of microtubules containing only βIII-tubulin was more than double that of microtubules formed from βII- and βIV-tubulin, so that microtubules made up exclusively of βIII-tubulin were less stable *(53)*. Consistent with this, the dynamics of microtubules made up of βIII-tubulin were considerably less sensitive to the suppressive effects of paclitaxel than the dynamics of microtubules assembled from βII or unfractionated tubulin; although not significantly different from microtubules assembled from βIV-tubulin *(54)*. Finally, using Chinese hamster ovary (CHO) cells overexpressing either βI- or βIII-tubulin, microtubule dynamic instability during interphase was analyzed by microinjection of rhodamine-labeled tubulin and time-lapse fluorescence microscopy. Somewhat unexpectedly, when dynamic instability was assessed in the absence of paclitaxel, no differences were found between the two β-tubulin-overexpressing cell types. However, when the same analysis was conducted in 150 nm paclitaxel, dynamic instability was suppressed only 12% in cells overexpressing βIII-tubulin compared with a 47% suppression in cells with similar levels of βI-tubulin. These results suggest that by reducing the ability of paclitaxel to suppress microtubule dynamics, overexpression of β III-tubulin induces paclitaxel resistance *(55)*.

Although collectively these in vitro studies suggest that microtubule dynamics and the effects of paclitaxel on this process could be affected by the β-tubulin isotype composition, care must be exercised in reaching this conclusion. Two reasons for this restrain are, first, that the β-tubulin isotype composition of brain tubulin is unlike that of most cancer cells, and second, uncertainty regarding the α-tubulin isotype composition. Regarding the first concern, whereas bovine brain is widely used as a source of mammalian tubulin for in vitro studies, its tubulin composition is probably not representative of most cancer cells. Specifically, unlike brain tubulin that has 58% βII-tubulin, βI is the major β-tubulin isotype in most cancer cells. Regarding the second uncertainty, bovine brain tubulin has three major α-tubulin isotypes, α1, α2, and α4 all of which undergo extensive post-translational modifications. Because the α-tubulin isotype composition of the immunoaffinity-purified β-tubulin isotypes is not known, it is not possible to predict how this might have impacted the published data. For example, the effect of a variable such as the preferential association that could occur between specific α- and β-isotypes cannot be ascertained.

Despite the concerns regarding the use of purified bovine brain tubulin, it appears that at least some of the observations might be relevant and correct. The accumulated data imply that altered expression of some β-tubulin isotypes, especially class III (and possibly IVa), may be correlated with paclitaxel sensitivity *(56)*. This data has been gathered from (1) drug-resistant cell lines, (2) drug naive cells that have not undergone drug selection before, and (3) from transfection and antisense experiments.

Examining cells selected for resistance to antimitotic drugs, numerous investigators have reported altered expression of β-tubulin isotypes (Table 1).

In drug-selected cell lines the evidence includes:

1. About two- to threefold increase in class III and IVa-isotypes in paclitaxel-resistant A549-T12 cells and about fourfold increase of the same isotypes in the A549-T24 cell lines *(57)*.
2. A twofold increase in class IVa β-tubulin messenger RNA (mRNA) and protein level in a K562 erythroleukemia cell line that is ninefold resistant to paclitaxel *(56)*.
3. About threefold increase in total α- and β-tubulin, and a fourfold increase in class III protein with a ninefold increase in RNA expression in paclitaxel-resistant DU-145 human prostate carcinoma cells *(58)*.
4. A correlation between paclitaxel but not vinblastine, vincristine, and rhizoxin sensitivity and the absolute levels of mRNA expression for βIII-tubulin in 17 cancer cell lines from the National Cancer Institute-Anticancer Drug Screen *(59)*.

As expected in these cancer cell lines, βI was the major tubulin isotype accounting for 85–99% of all the β-tubulin mRNA, but after βI, βIII mRNA was the next most predominant mRNA expressed, with levels ranging from 0.5 to 14%.

Additional data in drug naive cells includes:

1. A correlation among brain cell lines between intrinsic levels of class III β-tubulin and paclitaxel sensitivity. Specifically, the two cell lines with elevated class III protein were approx 5.5-fold less sensitive to paclitaxel, than the cell line with no detectable class III β-tubulin *(60)*.
2. The paclitaxel sensitivity of HT29-D4 human colon adenocarcinoma cells depends on their differentiation status *(61)*. Whereas in undifferentiated cells microtubule bundling occurs in the presence of paclitaxel, on differentiation there is a selective increase in class III β-tubulin mRNA and protein, and the microtubules fail to bundle.
3. An apparent correlation between epidermal growth factor receptor (EGFR)vIII and HER2 expression, increases in class IVa β-tubulin and decreases paclitaxel sensitivity; with suppressed paclitaxel-induced polymerization in cells expressing EGFRvIII and HER2 compared with cells expressing wild-type EGFR. *(62,63)*.

These latter studies suggest that by modulating β-tubulin isotype levels, some oncogenes may affect drug sensitivity.

Although together the studies described earlier support the thesis that isotype composition might be important in drug sensitivity, they can be criticized for providing principally correlative rather than direct evidence. To counter this criticism, several investigators have sought to modulate specific isotype levels in cells by either using antisense oligonucleotides or overexpressing protein. In the paclitaxel-resistant A549-T24 cells described earlier, the antisense strategy was modestly successful in bringing about a 40–50% decrease in both class III mRNA and protein levels, and a 39% increase in sensitivity to paclitaxel *(64)*. However, in several drug-naive cell lines, transfection experiments were unsuccessful *(65–67)*. Although this lack of success could indicate that isotype composition is not important in drug sensitivity, one could argue that given the obstacles that had to be surmounted, a negative result is not surprising, and indeed may have been anticipated. For example, the transfection strategy sought to increase the levels of specific isotypes but could not avoid compensatory changes in the expression levels of other β-tubulin

Table 1
Nomenclature, Tissue Distribution, C-Terminus Sequence and Post-Translational Modifications
of Human Tubulin Isotypes

Isotype	Human gene	Tissue expression[a]	C-terminus sequence[b]	Post-translational modifications[c]
α				
Kα1	TUBA1/k-α1	Widely expressed	MAALEKDYEEVGVDS VEGEGEEEGEEY	glutamylation (E445)
bα1	TUBA3/b-α1	Mainly in brain	MAALEKDYEEVGVHSVE GEGEEEGEEY	gylcylation (E445)
3	TUBA2	Testis specific	LAALEKDYEEVGVDS VEAEAE-EGEEY	detyrosination/ tyrosination
4	TUBA4	Brain muscle	MAALEKDYEEVGIDS YEDEDE—GEE-	removal of penultimate
6	TUBA6	Widely expressed	MAALEKDYEEVGADS ADGEDE—GEEY	glutamate acetylation
8	TUBA8	Heart, muscle testis	LAALEKDYEEVGTD SFEEENE—GEEF	(K40)
β				
I	HM40/ TUBB	Constitutive	YQDATAEEEEDFGEEA EEEA	glutamylation (E445)
II	Hβ9/TUBB2	Major neuronal lung	YQDATADEQGEFEE EEGEDEA	gylcylation (E445,E447, E448)
III	Hβ4/TUBB4	Minor neuronal testis	YQDATAEEEGEMYEDD EEESEAQGPK	phosphorylation of βIII isotype (S444)
IVa	Hβ5/TUBB5	Brain specific	YQDATAEQGEFEE EAEEEVA[d] YQDATAEEGEFEEE AEEEVA	
IVb	Hβ2	Major testis	YQDATAEEEGEFEEE AEEEVA	
V	5β/βV	Uterus adeno- carcinoma	YQDATANDGEEAFED EEEEIDG	
VI	Hβ1/TUBB1	Blood	FQDAKAVLEEDEEVTE EAEMEPEDKGH	

[a]Tubulin isotype distribution in human tissues has been performed mainly by RT-PCR, Western blotting, and immunohistochemistry.

[b]Amino acids differing from isotype 1 (k-α1) for α-tubulins or from isotype I (HM40/TUBB) for β-tubulins are highlighted in black.

[c]Position and amino acid modified by posttranslational modification are indicated in brackets and refer to studies performed on mammalian tubulin.

[d]Two βIVa-tubulin sequences with distinct C-termini were found in the NCBI protein database. The upper C-terminus sequence was found in human brain and lower sequence was found in a human oligodendroglioma and in mouse brain.

isotypes that might occur. Furthermore, the major isotype in the transfected cells, βI-tubulin, is the least studied of the isotypes and whether βIII- or βIV-isotypes can alter the dynamics of microtubules made up predominantly of βI-tubulin has not been established. In this regard, the levels of overexpression achieved following transfection may not have been sufficient to alter microtubule dynamics and affect drug sensitivity. Finally, the evidence in both the drug resistant and drug-naive cell lines notwithstanding, it is possible that the isotype composition is not *solely* responsible, but instead requires additional isotype-specific proteins for the full effect. One possibility would be MAPs that might be coordinately regulated. MAPs are known to bind to the highly divergent C-terminal regions of tubulin, and although not yet proven, MAPs that are isotype-specific or that have a preference for one isotype over another may exist. These would be coordinately expressed both in drug naive and in drug-resistant cell lines and could be downregulated using an antisense strategy explaining the observed experimental results. This hypothesis has yet to be proven and one must remember that the differences in assembly and microtubule dynamics observed using purified brain tubulin were seen using tubulin preparations devoid of MAPs. The lack of MAPs in these experiments has been used as evidence in support of the notion that the differences are *inherent* to the various tubulin isotypes.

Although most of the emphasis has been placed on the β-tubulin isotypes, drug sensitivity could also be potentially affected by *alterations in the α-tubulin isotype composition*. Although additional information will need to be gathered to draw firm conclusions, preliminary data suggest that at least in vitro, tubulin enriched by immunoaffinity purification in the tyrosinated α1, α2-isotypes assembles faster than tubulin containing the nontyrosinated forms *(68)*. In separate experiments, NCI-H460/T800, a paclitaxel resistant lung carcinoma cell line, was reported to express high levels of multidrug resistance 1 gene (MDR1) as well as the k-α1-tubulin *(69)*. That the overexpression of k-α1-tubulin may have a role in drug tolerance was supported by a 2.5-fold increase in paclitaxel resistance when k-α1-tubulin was overexpressed in parental H460 cells, and conversely an increase in paclitaxel sensitivity when the levels of k-α1-tubulin were downregulated 45–51% using antisense RNA.

7.2. Tubulin Mutations That Affect Either Binding of Regulatory Proteins or GTP or That Impact Longitudinal/Lateral Interactions and Can Alter Microtubule Dynamics

As discussed elsewhere in greater detail, studies dating back to the 1980s had shown altered migration of α- and β-tubulin by two-dimensional gel electrophoresis in CHO cell lines isolated by single-step selection, for resistance to paclitaxel (*see* Chapter 14). Additional characterization led to the conclusion that many of the paclitaxel-resistant cell lines contained a less stable microtubule polymer, and that as discussed earlier, this less-stable polymer with its reduced microtubule assembly was responsible for the resistant phenotype. Furthermore, *some* of the resistant cell lines were shown to be paclitaxel-dependent. Compared with other resistant cell lines that were not paclitaxel-dependent and could grow in drug-free medium, the paclitaxel-dependent cell lines exhibited even *lower* levels of microtubule assembly. Consistent with the notion that the resistance was mediated at least in part by the existence of a less stable microtubule polymer, many of the paclitaxel-resistant cell lines were shown to be hypersensitive to microtubule-destabilizing drugs that bind to free tubulin dimers, such as vinblastine or colchicine. Based on these observations, the authors predicted that mutations would be identified in these cell lines, a forecast that was subsequently confirmed when sequence analysis of class I β-tubulin revealed a cluster of mutations at leucines 215, 217, and 228 *(70)*. The authors concluded that these mutations conferred resistance by altering microtubule dynamics/stability. When subsequently the electron crystallographic model of paclitaxel-stabilized microtubules became available, it was concluded that the effect of the mutations was mediated by affecting the lateral/longitudinal interactions important for microtubule assembly. By destabilizing the microtubules, the mutations counteract the stabilizing effects of paclitaxel, and as discussed earlier, reduce paclitaxel affinity by presenting less binding sites. Importantly, using a tetracycline-regulated expression system, the authors subsequently showed in CHO cells that low-level expression of a β-tubulin containing any one of these mutations could confer paclitaxel resistance *(70)*.

Subsequent studies with paclitaxel as well as other microtubule-stabilizing agents have extended these observations. For example, three epothilone-resistant A549 and HeLa cell lines cross-resistant to the taxanes were found by sequence analysis of class I β-tubulin, the predominant β-tubulin isotype, to harbor single point mutations at amino acids 173 (Pro to Ala), 292 (Gln to Glu), and 422 (Tyr to Tyr/Cys) *(24)*. These mutations are found near the M-loop (a region involved in stabilizing the lateral contacts between adjacent protofilaments), at the nucleotide-binding site (important for the hydrolysis of GTP), and in the C-terminus (a region important for the binding of MAPs). It is now believed that by decreasing the endogenous stability of the microtubule, these mutations counteract the activities of microtubule-stabilizing drugs and hence confer resistance. Consistent with this hypothesis, the cell lines have been shown to be more sensitive to microtubule-destabilizing drugs such as vinblastine and colchicines *(24)*. Furthermore, studies in other models have established that these mutations are not confined to β-tubulin. For example, in the paclitaxel-selected A549-T12 cells described earlier, sequencing of class I β-tubulin, again the predominant β-tubulin isotype, did not uncover any mutations. However, when the predominant α-tubulin isotype, K-α1, was sequenced a heterozygous point mutation was found at residue 379 (Ser to Ser/Arg) *(71)*. Expression of both the wild-type and mutated α-tubulins in this heterozygous cell line was subsequently confirmed by mass spectrometry *(72)*. Exactly how this mutation impacts drug resistance remains to be elucidated, but its location near the C-terminus of α-tubulin is close to the proposed sites of interaction for both MAP4 and stathmin, providing possible explanations. This latter suggestion might also apply to an A549 cell line selected to high levels of resistance with a Val195Phe mutation in α-tubulin *(73)*. Finally, we would also note here that the development of drug-resistance by acquiring mutations that affect polymer stability does not appear to be confined to the microtubule stabilizing agents, as similar observations have been seen with the destabilizing hemiasterlins described elsewhere in this book *(42,74)* (*see* Chapter 10). In these drug-selected cell lines, the mutations instead of destabilizing the polymer, lead to enhanced stabilization and in turn drug resistance. Thus, a similar mechanism is used by cells to withstand both stabilizing and destabilizing agents.

7.3. Post-Translational Modifications to Tubulin

The structural diversity of the tubulin protein family is further increased by extensive post-translational modifications (*see* Chapter 5). With the exception of α-tubulin acetylation that occurs on Lys40, all post-translational modifications occur in the C-termini of both α- and β-tubulin. Because several MAPs interact with the C-terminal region of tubulin, it has been suggested that the interaction of MAPs and in turn microtubule stability/dynamics might be regulated by post-translational modifications. In vitro evidence in support of this concept includes (1) Data showing that the level of polyglutamylation of α- and β-tubulins can affect the interaction of both structural (tau, MAP-2) and motor MAPs (kinesin) with microtubules *(75–77)* and (2) Studies demonstrating that MAP-2-stimulated microtubule assembly could be inhibited by the removal of phosphate from βIII-tubulin using protein phosphatase 2A *(78)*. However, despite this in vitro data, there is still very little evidence that altered post-translational modifications can affect the sensitivity of cells toward any tubulin-directed antimitotic agent.

7.4. Altered Expression and Post-Translational Modifications of Proteins That Regulate the Dynamics/Stability of the Microtubules by Interacting With Soluble Tubulin or Microtubules

Previous sections have alluded to the potential role that proteins, such as MAPS that regulate microtubule stability/dynamics, might have on modulating drug sensitivity. In discussing variables other than isotype composition that might affect drug sensitivity, the possibility that MAPs might be coordinately regulated was raised. It has been suggested, although not yet proven that MAPs that are isotope-specific or that have a preference for one isotype over another may exist and might be coordinately expressed both in drug naive and in drug-resistant cell lines. Similarly, in the previous section, the possibility that the interaction of MAPs and in turn microtubule stability/dynamics might be regulated by post-translational modifications of the C-terminal region of tubulin was raised.

The discussion has thus far focused principally on proteins that associate with and stabilize polymerized microtubules. However, proteins also exist that can interact with tubulin dimers and could modulate cellular sensitivity. Examples of proteins that regulate the stability/dynamics of cellular microtubules in opposite ways include MAP4, a protein that can stabilize microtubules and stathmin, a cytoplasmic protein that destabilizes microtubules. These are discussed later.

Turning first to the family of microtubule stabilizing proteins, it is now known that it includes a large and diverse group of proteins *(see* Chapter 4). For example, because neurons maintain a very stable microtubule network, they possess a repertoire of microtubule-stabilizing proteins, the most abundant of which is MAP2 *(79)*. In cells other than neurons, MAP4 is the predominant human non-neuronal MAP, and its function as a microtubule-stabilizing protein is regulated by phosphorylation *(80,81)*. MAP4 alters microtubule dynamics by *associating with microtubules* and increasing the rescue frequency (the transition from shrinkage to growth), as microtubules undergo stochastic transitions between growth and shrinkage. Phosphorylation of MAP4 results in a loss of this microtubule-stabilizing activity. By comparison to the family of microtubule stabilizing proteins, the family of microtubule destabilizing proteins is less expansive. One member is stathmin, a soluble cytoplasmic protein that *binds tubulin dimers* and facilitates microtubule catastrophes (the transition from growing to shrinking) *(82–84) (see* Chapter 4). As with MAP4, stathmin's function is regulated by phosphorylation, so that its destabilizing activity is abrogated when stathmin is fully phosphorylated *(85,86)*. MAP4 has no effect on the catastrophe frequency; and similarly, stathmin does not influence the rescue frequency. However, given their effect on the stochastic transitions between growth and shrinkage, the levels or activity of these proteins would be expected to modulate drug sensitivity. Down regulation or inactivation of MAP4 and/or overexpression or activation of stathmin should increase the dynamicity of microtubules and decrease their stability. If a cancer cell acquired such changes they could provide a mechanism of resistance to microtubule-stabilizing drugs such as paclitaxel, whereas inversely enhancing the potency of microtubule-depolymerizing drugs like the vinca alkaloids. Indeed, the intrinsic levels of proteins such as MAP4 and stathmin could modulate the inherent cellular sensitivity to microtubule-stabilizing and microtubule-destabilizing drugs.

Evidence that MAP4 expression might modulate the cellular sensitivity to microtubule interacting drugs such as paclitaxel, was originally provided by studies examining the paclitaxel sensitivity of cells with wild-type and mutant p53. The observation that unlike most chemotherapy agents, paclitaxel retained activity in cells harboring a mutant p53 was the catalyst for these experiments (87–89). Evidence from the NCI anticancer drug screen confirmed this, establishing that unlike the majority of chemotherapeutic agents, paclitaxel activity was indifferent to the p53 status of cells (90). One possible explanation for this emerged from studies in *murine cells*. Whereas MAP4 was actively transcribed in cells harboring a mutant p53, MAP4 was transcriptionally repressed in the presence of wild-type p53 (87–89). Additional studies in *murine fibroblasts* found that inactivation of p53 increased paclitaxel sensitivity and decreased vinblastine sensitivity (87); whereas induction of p53 by DNA-damaging agents resulted in decreased paclitaxel sensitivity and increased vinblastine sensitivity (88). These observations have been explained by a modulation of MAP4 levels. In untreated cells or following a DNA-damaging agent, wild-type p53 represses MAP4 transcription leading to less stable microtubules and reduced paclitaxel sensitivity. The transcriptional repression of MAP4 is released in cells lacking a wild-type p53 (or those with a mutant p53) resulting in higher levels of MAP4. More stable microtubules in these cells lacking a wild-type p53 function result in enhanced paclitaxel sensitivity as more sites are available for drug binding and greater stabilization is achieved starting from a more stable microtubule. Although this straightforward thesis may prove correct, the data in murine cells have been considered preliminary and in need of supporting data. Additional evidence in *human cells* supporting this hypothesis has been gathered and includes:

1. Observations in paclitaxel-resistant human ovarian carcinoma cells where paclitaxel sensitivity was shown to correlate with MAP4 phosphorylation and its dissociation from microtubules (91),
2. Findings in vinblastine resistant CCRF-CEM cells where the expression of nonphosphorylated forms of MAP4 was shown to be increased (92), and
3. Data in C127 breast cancer cells and patient samples demonstrating activation of p53 and repression of MAP4 in normal and malignant tissues in patients treated with a DNA-damaging agent (93).

Similarly, there is evidence that stathmin levels might modulate the cellular sensitivity to microtubule-interacting drugs. As indicated previously, the expression level of stathmin should affect the dynamicity/stability of microtubules and influence drug sensitivity. Evidence supporting this has been presented in K562 erythroleukemia stably transfected with a stathmin-antisense construct. Consistent with the increased stability of microtubules that would be expected in a cell with low-stathmin levels, the authors reported enhanced inhibition of growth and clonogenicity when the antisense containing constructs were treated with low concentrations of paclitaxel; and increased vinblastine resistance compared with control mock-transfected cells *(83,94)*. Similarly, overexpression of stathmin increased the vindesine and vincristine sensitivity of SBC-3 human lung carcinoma cells; although it did not significantly decrease paclitaxel or docetaxel sensitivity, a result that is not consistent with the hypothesis advocated in this section *(95)*. Finally as would be expected, stathmin inhibited in vitro paclitaxel-induced microtubule polymerization *(96)*. In their entirety, these data indicate that in a cancer cell, overexpression of stathmin could reduce drug sensitivity by opposing the microtubule-stabilizing effect of microtubule-polymerizing agents such as paclitaxel, whereas increasing sensitivity to microtubule-destabilizing agents.

That both stathmin and MAP-4 might interact to affect drug sensitivity has been investigated in the parental and paclitaxel-resistant A549 cells discussed previously. Compared with the parental cells, stathmin protein levels in the A549-T12 and -T24 resistant cell lines were increased approximately twofold, a change that would be predicted to antagonize paclitaxel action. Furthermore, when the phosphorylation status of stathmin was monitored in parental and drug-resistant cells, the latter were seen to have acquired additional adaptations. In the drug-naive parental cells the addition of increasing concentrations of paclitaxel-shifted stathmin from the active, nonphosphorylated form to the inactive, fully phosphorylated protein; a change that would enhance paclitaxel sensitivity by allowing for unopposed microtubule stabilization. In contrast, in the paclitaxel-resistant cell lines the increase in phosphorylation did not occur. Coordinate changes in MAP4 were also reported, with only the active, nonphosphorylated form found in the drug-naive parental cell line, whereas in the resistant A549-T24 cells the inactive, phosphorylated protein was predominantly expressed. A partial adaptation was noted in the less-resistant A549-T12 cells with expression of both the phosphorylated and the nonphosphorylated forms of MAP4. These changes in stathmin and MAP4, two tubulin-/microtubule-regulatory proteins acting in concert would be expected to increase the dynamicity/instability of the microtubules in the paclitaxel-resistant A549 cells and contribute to the resistant phenotype.

Given the nearly two-dozen proteins that have been shown to stabilize and destabilize microtubules *(see* Chapter 4), one can envision how complex a formula might be needed to predict the outcome of their interplay on the stability of the microtubule. Although only a subset of these proteins will be expressed in a given cell at any given time, the regulation of microtubule dynamics is likely to involve proteins other than stathmin and MAP4. The majority of these other proteins have not been explored. For example, E-MAP-115 is a MAP expressed in cells of epithelial origin, whose expression appears to be linked to the degree of differentiation. In HT29-D4 colon adenocarcinoma cells only low levels of E-MAP-115 can be detected in undifferentiated cells, and these levels increase during differentiation; the MAP4 levels do not change during differentiation *(61)*. Furthermore, consistent with its function as a microtubule-stabilizing MAP, overexpression of E-MAP-115 in MCF-7 and HeLa cells has been shown to increase their sensitivity to paclitaxel *(97)*.

7.5. Altered Drug Binding to the Microtubule

Surprisingly, although acquired mutations affecting drug binding to its primary target have been reported for several chemotherapeutic agents, a small number of drug selected cell lines exist that have acquired such mutations following selections with microtubule interacting agents. The reasons for this are beyond the scope of this chapter, but likely include:

1. The imperfect fit of drugs that have evolved in nature to target other tubulins and have been adapted as cancer chemotherapeutics to target human tubulins;
2. The adaptability of the drug interacting sites;
3. The relative intolerance of tubulin for structural changes;
4. The availability of numerous other mechanisms of resistance as discussed in this chapter that might arise more readily.

Nevertheless, several examples of altered drug binding have been reported in drug-resistant cell lines selected with microtubule-stabilizing agents. Examples of this mechanism of drug tolerance were reported following the isolation of paclitaxel-resistant human ovarian carcinoma cell lines from an A2780 subclone designated A2780 (1A9). Two independent sublines, 1A9PTX10 and 1A9PTX22, were isolated and shown to be 20- to 30-fold resistant to paclitaxel *(41)*. Interestingly, these paclitaxel-resistant cells retained their sensitivity to epothilone B and to 2-*m*-azido-benzoyl-Taxol, two microtubule-stabilizing drugs that are considerably more potent than paclitaxel, and in the case of the epothilones likely have a (somewhat) separate binding site within the drug-binding pocket. Unlike the cell lines described previously that harbor tubulin mutations that alter the dynamicity/stability of the microtubules, these cell lines were not found to be paclitaxel-dependent. Consistent with a stable, inheritable change, such as an acquired mutation, the resistant phenotype was stable even after the cells had been passaged in drug-free medium for up to 3 yr. The total tubulin content of the resistant cells was similar to that of the drug-naive parental cells. Furthermore, unlike the cell lines with the tubulin mutations that altered the dynamicity/stability of the microtubules, the fraction of tubulin that was polymerized was similar in the parental and resistant cells, suggesting that microtubule dynamics/stability was not altered. However, tubulin purified from the resistant cells polymerized poorly or not at all in the presence of paclitaxel when assayed in vitro, suggesting that the acquired changes (mutations) abrogated paclitaxel-binding. When the sequence of βI, the major β-tubulin isotype, was determined, 1A9PTX10 cells were found to harbor a Phe270Val substitution, whereas 1A9PTX22 cells had acquired an Ala364Thr. Molecular modeling studies demonstrated that Phe270 is close to the region of tubulin in contact with the baccatin ring system of paclitaxel *(98,99)* (*see* Chapter 9). It is possible that substituting the phenyl ring with the less bulky side-chain of valine could disrupt paclitaxel binding in the mutant tubulin. In addition, epothilone A and B were used in similar drug selections again starting with the A2780 (1A9) subclone *(39,40)*. The resultant epothilone-resistant cell lines were shown to have impaired epothilone- and paclitaxel-induced tubulin polymerization and to also harbor acquired mutations. In the epothilone A selected subline, a Thr274Ile substitution was found, whereas in the epothilone B isolate an Arg282Glu change was detected. Molecular modeling suggested that the Thr274Ile substitution could disrupt a hydrogen bond between the threonine hydroxyl and the C7-hydroxyl of the epothilones that might stabilize the epothilone–tubulin interaction *(39,40)*. Arginine 282 had been previously shown to be the site of photoincorporation of 7-BzDC paclitaxel *(20)*.

7.6. Alterations in Signaling Pathways

The number of crucial cellular proteins that bind to, interact with or are transported by the microtubules continues to increase *(100–102)*. Among these are numerous key proteins that mediate signaling pathways, and existing data has shown that microtubule-interacting agents can interfere with these interactions *(39,40,100,103,104)*. The best documented example of a signaling pathway intertwined with microtubules is the mitogen-activated protein kinase (MAPK) pathway whose components include the extra-cellular signal-regulated kinases (ERK)1 and ERK2. Several investigators have documented activation of ERK-signaling following microtubule disruption, a response that might be seen as adaptive *(105–107)*. That the activation of the MAPK pathway might be beneficial, is supported by the additive toxicity seen when paclitaxel was combined with the MEK inhibitor, U0126 *(107)*. This study documented this interaction, but more importantly highlighted that the degree of activation of the MAPK pathway determines whether the interaction between paclitaxel and a MEK inhibitor is additive/synergistic, or antagonistic. The enhanced paclitaxel cytotoxicity when a MEK inhibitor is coadministered may involve more than one mechanism and this must be further clarified. In addition to the often-stated "inhibition of the survival-signaling function of the MAPK pathway," synergism may also occur owing to enhanced microtubule polymerization, as it has been proposed that microtubule stabilization is *inhibited* by MAPK activation *(105)*. Although this latter explanation may provide a microtubule specific link, it is not surprising that pathways implicated in other settings as promoting cellular survival have also been said to modulate sensitivity to microtubule interacting agents. For example, overexpression of a catalytically active subunit of PI-3 kinase (PI3k) in ovarian cancer cells confers paclitaxel resistance, and this tolerance can be reversed with a selective PI3k inhibitor *(108)*. And similarly, overexpression of EGFRvIII, a receptor variant of the EGFR gene with a deletion encompassing exons 2-7 has been reported to increase paclitaxel tolerance. The EGFRvIII variant, the most common alteration of the EGFR gene, is associated with constitutive activation of the pERK (MAP kinase) *(62)* and the phosphatidylinositol 3-kinase pathways (PI3k/AKT) *(109)*, consistent with increased cellular survival.

Other signaling pathways that may interact directly or indirectly with microtubule function include the Erb pathways. For example, in breast cancer cells ErbB2 overexpression inhibits Cdc2 activation and paclitaxel-induced cell death, by deregulating the G2/M cell-cycle checkpoint *(110)*. A possible explanation for this observation has been provided by the demonstration of elevated levels of inhibitory phosphorylation of Cdc2 on tyrosine 15 in ErbB2-overexpressing breast cancer cells and primary tumors *(111)*.

Finally as discussed previously, oncogenic growth factor signaling may represent a novel mechanism for the modulation of tubulin isotypes *(63)*. Although intellectually attractive, this hypothesis will require validation in cell models and in human tumors that express oncogenic forms of receptor tyrosine kinases.

8. PACLITAXEL RESISTANCE IN PATIENTS

Understanding the mechanisms of resistance to paclitaxel in the clinic is best considered a work in progress. Both intrinsic and acquired resistance occurs in the clinic, and to what extent the mechanisms responsible for these presentations overlap is uncertain. In the attempt to understand clinical drug resistance, the field is still in the information-gathering phase, and the data gathered to date is preliminary at best. Given these caveats potential mechanisms that have been investigated can be summarized as follows:

1. Alterations in tubulin isotype composition;
2. Alterations in tubulin/microtubule regulatory proteins;
3. P-glycoprotein mediated resistance;
4. Tubulin mutations.

8.1. Alterations in Tubulin Isotype Composition in Tumors

Given the data discussed previously as to the potential role that differential isotype expression might have in paclitaxel sensitivity, a limited number of studies have examined the isotype expression profile in clinical samples. The first and most convincing of these examined the β-tubulin isotype expression profile using reverse transcription-polymerase chain reaction and were undertaken in paclitaxel-sensitive and -resistant human ovarian carcinomas *(57)*. As conventionally defined clinically, a tumor was deemed to be resistant to paclitaxel if disease progression had occurred during treatment, or there had been a relapse within 6 mo of successful treatment. Compared with primary untreated ovarian tumors the paclitaxel-resistant tumors displayed significant increases in class I (3.6-fold), class III (4.4-fold), and class IVa (7.6-fold) β-tubulin. These findings were confirmed and extended in 41 patients with a diagnosis of advanced ovarian cancer. The β-tubulin isotype expression was evaluated by semiquantitative and real-time polymerase chain reaction and by immunohistochemistry. A statistically significant upregulation of class III β-tubulin was detectable at both the mRNA and protein level in the resistant subset *(112)*.

In similar studies, the prognostic value of the isotype expression pattern in tumors was examined in 19 patients with nonsmall-cell lung cancer receiving taxane-based regimens. Patient samples were stained with antibodies directed against total β-tubulin, as well as classes I, II, and III β-tubulin isotypes. As expected, all tumors stained with the antibody against the class I-tubulin isotype. Furthermore, a majority of the tumor samples expressed class II and class III, although the percent of positive cells varied significantly between tumors. The authors noted that the progression-free survival of patients whose tumors expressed high levels of the class III-tubulin isotype was shorter (41 d) than that of patients whose tumors had low levels (288 d, $p = 0.02$). Furthermore, the tumor in two of the nine patients (22%) whose tumors had high levels of class III-tubulin responded to chemotherapy compared with 6 among the 10 patients (60%) whose tumors had low levels of expression (Fisher exact test: $p = 0.11$). Although not conclusive, these data suggest that high expression of class III-tubulin by tumor cells is associated with poor prognosis in patients with NSCLC receiving a taxane-based regimen *(113)*. Although these studies examined clinical samples, the authors would note that in mouse xenografts established from 12 human ovarian carcinoma specimens obtained before or after the paclitaxel therapy, no correlation was observed between β-tubulin mRNA expression and paclitaxel sensitivity *(59)*.

8.2. Alterations in Tubulin/Microtubule Regulatory Proteins in Human Cancers

Based on the in vitro data implicating proteins that stabilize and destabilize microtubules with drug resistance, the levels of such proteins have been examined in clinical samples. In a recent phase 1 clinical study using a sequential doxorubicin/vinorelbine regimen, a partial correlation was observed with induction of p53 and decreased MAP4 expression both in peripheral blood mononuclear cells and in tumors. This observation is consistent with the in vitro studies demonstrating transcriptional regulation of MAP4 by wild-type p53 *(93)*. A larger number of studies have examined stathmin mRNA levels in a wide range of malignancies. These have documented elevated stathmin mRNA levels in breast carcinoma cells from patients with more aggressive disease and in acute leukemias, lymphomas, as well as a diverse group of carcinomas *(114–117)*.

8.3. P-Glycoprotein Mediated Resistance

Although paclitaxel has been shown to be effective in a broad range of cancers including ovarian, breast, and lung cancer, a subset of solid tumors have proven refractory to paclitaxel therapy. Solid tumors that are intrinsically resistant to paclitaxel and that express P-glycoprotein such as colon cancer and renal cell carcinoma may be resistant by virtue of the expression of this transporter. Although the fact that clinical responses to epothilone B, which is not a substrate for P-glycoprotein, have been reported in these patients is consistent with this thesis, but clear evidence for this is lacking *(118,119)*. Indeed in women with advanced ovarian cancer no statistically significant changes of MDR-1 expression were noticed between those with chemotherapy sensitive tumors and those whose tumors were clinically resistant to chemotherapy, either at the mRNA or protein level *(112)*.

8.4. Tubulin Mutations in Human Tumors

To date seven β-tubulin pseudogenes have been reported *(120–122)*. Their existence makes it difficult to use genomic DNA as the starting point for analyzing the precise nucleotide sequence of β-tubulin *(123,124)*. With these constraints, an analysis of 62 human breast cancer tumors did not result in any β-tubulin mutations *(125)*. Similarly, when 30 paclitaxel-resistant specimens (nine ovarian cancers, nine ovarian cancer cell lines, and 12 ovarian cancer xenografts in nude mice) were analyzed, a very high degree of sequence conservation in class I β-tubulin was noted, without detectable mutations in class I β-tubulin *(126)*. No mutations were detected when the genomic sequence of Class I -tubulin from a series of 29 patients with resected lung tumors (15 male, 14 female, median age 67 yr) was examined *(127)*. Although further studies may be forthcoming the data to date suggest that mutations will not likely be found in clinical samples *(128–131)*.

ACKNOWLEDGMENTS

We thank our many wonderful colleagues, who have worked with us over the years, for their interest and contributions to studies on microtubule targeting agents and drug resistance. This work was supported in part by USPHS Grants CA 39821 (SBH), CA 77263 (SBH), and the National Foundation for Cancer Research (SBH). As a member of the intramural program of the NCI all support for TF has come from the NCI.

REFERENCES

1. Desai A, Mitchison TJ. Microtubule polymerization dynamics. Annu Rev Cell Dev Biol 1997;13: 83–117.
2. Oakley BR. An abundance of tubulins. Trends Cell Biol. 2000;10:537–542.
3. Sharp DJ, Rogers GC, Scholey JM. Microtubule motors in mitosis. Nature 2000;407:41–47.
4. Wani MC, Taylor HL, Wall ME, Coggon P, McPhail AT. Plant antitumor agents. VI. The isolation and structure of taxol, a novel antileukemic and antitumor agent from Taxus brevifolia. J Am Chem Soc 1971;93:2325–2327.
5. Schiff PB, Fant J, Horwitz SB. Promotion of microtubule assembly in vitro by taxol. Nature 1979;277:665–667.
6. Schiff PB, Horwitz SB. Taxol stabilizes microtubules in mouse fibroblast cells. Proc Natl Acad Sci USA, 1980;77:1561–1565.
7. Schiff PB, Horwitz SB. Taxol assembles tubulin in the absence of exogenous guanosine 5'–triphosphate or microtubule-associated proteins. Biochemistry 1981;20:3247–3252.

8. Horwitz SB, Parness J, Schiff PB, Manfredi JJ. Taxol: a new probe for studying the structure and function of microtubules. Cold Spring Harb Symp Quant Biol 1982;46 Pt 1:219–226.

9. Diaz JF, Valpuesta JM, Chacon P, Diakun G, Andreu JM. Changes in microtubule protofilament number induced by Taxol binding to an easily accessible site. Internal microtubule dynamics. J Biol Chem 1998;273:33803–33810.

10. Parness J, Horwitz SB. Taxol binds to polymerized tubulin in vitro. J Cell Biol 1981; 91:479–487.

11. Rao S, Horwitz SB, Ringel l. Direct photoaffinity labeling of tubulin with taxol. J Natl Cancer Inst 1992;84:785–788.

12. Lowe J, Li H, Downing KH, Nogales E. Refined structure of alpha beta-tubulin at 3.5 A resolution. J Mol Biol 2001;313:1045–1057.

13. Jordan MA, Toso RJ, Thrower D, Wilson L. Mechanism of mitotic block and inhibition of cell proliferation by taxol at low concentrations. Proc Natl Acad Sci USA 1993;90:9552–9556.

14. Derry WB, Wilson L, Jordan MA. Substoichiometric binding of taxol suppresses microtubule dynamics. Biochemistry 1995;34:2203–2211.

15. Chen JG, Horwitz SB. Differential mitotic responses to microtubule-stabilizing and destabilizing drugs. Cancer Res 2002;62:1935–1938.

16. Torres K, Horwitz SB. Mechanisms of Taxol-induced cell death are concentration dependent. Cancer Res 1998;58:3620–3626.

17. Rao S, Krauss NE, Heerding JM, et al. 3'-(p-azidobenzamido)taxol photolabels the N–terminal 31 amino acids of beta-tubulin. J Biol Chem 1994;269:3132–3134.

18. Rao S, Orr GA, Chaudhary AG, Kingston DG, Horwitz SB. Characterization of the taxol binding site on the microtubule. 2–(m-Azidobenzoyl)taxol photolabels a peptide (amino acids 217–231) of beta-tubulin. J Biol Chem 1995;270:20235–20238.

19. Orr GA, Rao S, Swindell CS, Kingston DG, Horwitz SB. Photoaffinity labeling approach to map the Taxol-binding site on the microtubule. Methods Enzymol 1998;298:238–252.

20. Rao S, He L, Chakravarty S, Ojima I, Orr GA, Horwitz SB. Characterization of the Taxol binding site on the microtubule. Identification of Arg(282) in beta-tubulin as the site of photoincorporation of a 7-benzophenone analogue of Taxol. J Biol Chem 1999;274:37990–37994.

21. Rao S, Aberg F, Nieves E, Horwitz SB, Orr GA. Identification by mass spectrometry of a new alpha-tubulin isotype expressed in human breast and lung carcinoma cell lines. Biochemistry 2001; 40:2096–2103.

22. Bollag DM, McQueney PA, Zhu J, et al. Epothilones, a new class of microtubule-stabilizing agents with a taxol-like mechanism of action. Cancer Res 1995;55:2325–2333.

23. ter Haar E, Kowalski RJ, Hamel E, et al. Discodermolide, a cytotoxic marine agent that stabilizes microtubules more potently than taxol. Biochemistry 1996;35:243–250.

24. He L, Orr GA, Horwitz SB. Novel molecules that Interact with microtubules and have functional activity similar to Taxol. Drug Discov Today 2001;6:1153–1164.

25. Hung DT, Chen J, Schreiber SL. (+)-Discodermolide binds to microtubules in stoichiometric ratio to tubulin dimers, blocks taxol binding and results in mitotic arrest. Chem Biol 1996;3:287–293.

26. Kowalskr RJ, GrannakaKou P, Gunasekera SP, Longley RE, Day BW, Hamel E. The microtubule-stabilizing agent discodermolide competitively inhibits the binding of paclitaxel (Taxol) to tubulin polymers, enhances tubulin nucleation reactions more potently than paclitaxel, and inhibits the growth of paclitaxel-resistant cells. Mol Pharmacol 1997;52:613–622.

27. Long BH, Carboni JM, Wasserman AJ, et al. Eleutherobin, a novel cytotoxic agent that induces tubulin polymerization, is similar to packlitaxel (Taxol). Cancer Res., 1998;58:1111–1115.

28. Hamel E, Sackett DL, Vourloumis D, Nicolaou KC. The coral-derived natural products eleutherobin and sarcodictyins A and B: effects on the assembly of purified tubulin with and without microtubule-associated proteins and binding at the polymer taxoid site. Biochemistry, 1999;38:5490–5498.

29. Mooberry SL, Tien G, Hernandez AH, Plubrukarn A, Davidson BS. Laulimalide and isolaulimalide, new paclitaxel-like microtubulestabilizing agents. Cancer Res 1999;59:653–660.

30. Pryor DE, O'Brate A, Bilcer G, et al. The microtubule stabilizing agent laulimalide does not bind in the taxoid site, kills cells resistant to paclitaxel and epothilones, and may not require its epoxide moiety for activity. Biochemistry 2002;41:9109–9115.

31. Pineda O, Farras J, Maccari L, Manetti F, Botta M, Vilarrasa J. Computational comparison of microtubule-stabilising agents laulimalide and peloruside with taxol and colchicine. Bioorg Med Chem Lett 2004;14:4825-4829.

32. Gapud EJ, Bai R, Ghosh AK, Hamel E. Laulimalide and paclitaxel: a-comparison of their effects on tubulin assembly and their synergistic action when present simultaneously. Mol Pharmacol 2004; 66:113–121.

33. Martello LA, McDaid HM, Regl DL, et al. Taxol and discodermolide represent a synergistic drug combination in human carcinoma cell lines. Clin Cancer Res 2000;6:1978–1987.

34. Huang GS, Lopez-Barcons L, Freeze BS, et al. Potentiation of taxol efficacy and by discodermolide in ovarian carcinoma xenograft-bearing mice. Clin Cancer Res 2006;12:298–304.

35. Klein LE, Freeze BS, Smith AB, Horwitz SB. The microtubule stabilizing agent discodermolide is a potent inducer of accelerated cell senescence. Cell Cycle 2005;4:501–507.

36. Nettles JH, Li H, Cornett B, Krahn JM, Snyder JP, Downing KH. The binding mode of epothilone A on alpha,beta-tubulin by electron crystallography. Science 2004;305:866–869.

37. Bode CJ, Gupta ML, Reiff EA, Suprenant KA, Georg GI, Himes RH. Epothilone and paclitaxel: unexpected differences in promoting the assembly and stabilization of yeast microtubules. Biochemistry 2002;41:3870–3874.

38. He L, Jagtap PG, Kingston DG, Shen HJ, Orr GA, Horwitz SB. A common pharmacophore for Taxol and the epothilones based on the biological activity of a taxane molecule lacking a C-13 side chain. Biochemistry 2000;39:3972–3978.

39. Giannakakou P, Gussie R, Nogales E, et al. A common pharmacophore for epothilone and taxanes: molecular basis for drug resistance conferred by tubulinmutations in human cancer cells. Proc Natl Acad Sci USA 2000;97:2904–2909.

40. Giannakakou P, Sackett DL, Ward Y, Webster KR Blagosklonny MV, Fojo T. p53 is associated with cellular microtubules and is transported to the nucleus by dynein. Nat Cell Biol 2000;2:709–717.

41. Giannakakou P, Sackett DL, Kang YK, Zhan Z, Buters JT, Fojo T, Poruchynsky MS. Paclitaxel-resistant human ovarian cancer cells have mutant betatubulins that exhibit impaired paclitaxel-driven polymerization. J Biol Chem 1997;272:17118–17125.

42. Hari M, Loganzo F, Annable T, et al. Paclitaxel-resistant cells have a mutation in the paclitaxel-binding region of beta-tubulin (Asp26Glu) and less stable microtubules. Mol Cancer Ther 2006;5:270–278.

43. Cabral FR, Brady RC, Schibler MJ. A mechanism of cellular resistance to drugs that interfere with microtubule assembly. Ann NY Acad Sci 1986;466:745–756.

44. Cabral F, Barlow SB. Mechanisms by which mammalian cells acquire resistance to drugs that affect microtubule assembly. FASEB J 1989;3:1593–1599.

45. Minotti AM, Barlow SB, Cabral F. Resistance to antimitotic drugs in Chinese hamster ovary cells correlates with changes in the level of polymerized tubulin. J Biol Chem 1991;266:3987–3994.

46. Wilson L, Jordan MA. Microtubule dynamics: taking aim at a moving target. Chem Biol 1995;2:569–573.

47. Jordan MA, Wilson L. Microtubules and actin filaments: dynamic targets for cancer chemotherapy. Curr Opin Cell Biol 1998;10:123–130.

48. Goncalves A, Braguer D, Kamath K, et al. Resistance to Taxol in lung cancer cells associated with increased microtubule dynamics. Proc Natl Acad Sci USA 2001;98:11737–11742.

49. Banerjee A, Roach MC, Trcka P, Luduena RF. Increased microtubule assembly in bovine brain tubulin lacking the type III isotype of beta-tubulin. J Biol Chem 1990;265:1794–1799.

50. Banerjee A, Roach MC, Trcka P, Luduena RF. Preparation of a monoclonal antibody specific for the class IV isotype of beta-tubulin. Purification and assembly of alpha beta II, alpha beta III, and alpha beta IV tubulin dimers from bovine brain. J Biol Chem 1992;267:5625–5630.

51. Lu Q, Luduena RF. Removal of beta III isotype enhances taxol induced microtubule assembly. Cell Struct. Funct 1993;18:173–182.

52. Lu Q, Luduena RF. In vitro analysis of microtubule assembly of isotypically pure tubulin dimers. Intrinsic differences in the assembly properties of alpha beta II, alpha beta III, and alpha beta IV tubulin dimers in the absence of microtubule-associated proteins. J Biol Chem 1994;269:2041–2047.

53. Panda D, Miller HP, Banerjee A, Luduena RF, Wilson L. Microtubule dynamics in vitro are regulated by the tubulin isotype composition. Proc Natl Acad Sci USA 1994;91:11358–11362.

54. Derry WB, Wilson L, Khan lA, Luduena RF, Jordan MA. Taxol differentially modulates the dynamics of microtubules assembled from unfractionated and purified beta-tubulin isotypes. Biochemistry 1997;36:3554–3562.

55. Kamath K, Wilson L, Cabral F, Jordan MA. BetaIII-tubulin induces paclitaxel resistance in association with reduced effects on microtubule dynamic instability. J Bioi Chem 2005;280:12902–12907.

56. Jaffrezou JP, Dumontet C, Derry WB, et al. Novel mechanism of resistance to paclltaxel (Taxol) in human K562 leukemia cells by combined selection with PSC 833. Oncol Res 1995;7:517–527

57. Kavallaris M, Kuo DY, Burkhart CA, et al. Taxol-resistant epithelial ovarian tumors are associated with altered expression of specific beta-tubulin isotypes. J Clin Invest 1997;100:1282–1293.

58. Ranganathan S, Benetatos CA, Colarusso PJ, Dexter DW, Hudes GR. Altered beta-tubulin isotype expression in paclitaxel-resistant human prostate carcinoma cells. Br J Cancer 1998a;77:562–566.

59. Nicoletti Ml, Valoti G, Giannakakou P, et al. Expression of beta-tubulin isotypes in human ovarian carcinoma xenografts and in a sub-panel of human cancer cell lines from the NCI-Anticancer Drug Screen: correlation with sensitivity to microtubule active agents. Clin Cancer Res 2001;7:2912–2922.

60. Ranganathan S, Dexter DW, Benetatos CA, Hudes GR. Cloning and sequencing of human betaIII-tubulin cDNA: induction of betaIII isotype in human prostate carcinoma cells by acute exposure to antimicrotubule agents. Biochim Biophys Acta 1998b;1395:237–245.

61. Carles G, Braguer D, Dumontet C, et al. Differentiation of human colon cancer cells changes the expression of beta-tubulin isotypes and MAPs. Br J Cancer 1999;80:1162–1168.

62. Montgomery RB, Moscatello DK, Wong AJ, Cooper JA, Stahl WL. Differential modulation of mitogen-activated protein (MAP) kinase/extracellular signal-related kinase kinase and MAP kinase activities by a mutant epidermal growth factor receptor. J Biol Chem 1995;270:30562–30566.

63. Montgomery RB, Guzman J. O'Rourke DM, Stahl WL. Expression of oncogenic epidermal growth factor receptor family kinases induces paclitaxel resistance and alters beta-tubulin isotype expression. J Biol Chem 2000;275:17358–17363.

64. Kavallaris M, Burkhart CA, Horwitz SB. Antisense oligonucleotides to class III beta-tubulin sensitize drug-resistant cells to Taxol. Br J Cancer 1999;80:1020–1025.

65. Burkhart CA, Kavallaris M, Band Horwitz S. The role of beta-tubulin isotypes in resistance to antimitotic drugs. Biochim. Biophys Acta 2001;1471:01–09.

66. Blade K, Menick DR, Cabral F. Overexpression of class I, II or IVb beta-tubulin isotypes in CHO cells is insufficient to confer resistance to paclitaxel. J Cell Sci 112 1999;Part 13:2213–2221.

67. Ranganathan S, McCauley RA, Dexter DW, Hudes GR. Modulation of endogenous beta-tubulin isotype expression as a result of human beta(lll) cDNA transfection into prostate carcinoma cells. Br J Cancer 2001;85:735–740.

68. Banerjee A, Kasmala LT. Differential assembly kinetics of alpha-tubulin isoforms in the presence of paclitaxel. Biochem. Biophys Res Commun 1998;245:349–351.

69. Han EK, Gehrke L, Tahir SK, et al. Modulation of drug resistance by alpha-tubulin in paclitaxelresistant human lung cancer cell lines. Eur J Cancer 2000;36:1565–1571.

70. Gonzalez-Garay ML, Chang L, Blade K, Menick DR, Cabral F. A beta-tubulin leucine cluster involved in microtubule assembly and paclitaxel resistance. J Biol Chem 1999;274:23875–23882.

71. Martello LA, Verdier-Pinard P, Shen H-J, He L, Torres K, Orr GA, Horwitz SB. Elevated levels of microtubule destabilizing factors in a Taxolresistant/dependent A549 cell line with an alpha-tubulin mutation. Cancer Res 2003;63:1207–1213.

72. Verdier-Pinard P, Wang F, Martello LA, Orr GA, Horwitz SB. Analysis of tubulin isotypes and mutations from taxol-resistant cells by combined isoelectrofocusing and mass spectrometry. Biochemistry 2003;42:5349–5357.

73. Yang CP, Verdier-Pinard P, Wang F, et al. A highly epothilone B-resistant A549 cell line with mutations in tubulin that confer drug dependence. Mol Cancer Ther 2005;6:987–995.

74. Poruchynsky MS, Kim JH, Nogales E, et al. Tumor cells resistant to a microtubule-depolymerizing hemiasterlin analogue, HTI-286, have mutations in alpha- or beta-tubulin and increased microtubule stability. Biochemistry 2004;43:13944–13954.

75. Boucher D, Larcher JC, Gros F, Denoulet P. Polyglutamylation of tubulin as a progressive regulator of in vitro interactions between the microtubule-associated protein Tau and tubulin. Biochemistry 1994;33:12471–12477.

76. Larcher JC, Boucher D, Lazereg S, Gros F, Denoulet P. Interaction of kinesin motor domains with alpha- and beta-tubulin subunits at a tau-independent binding site. Regulation by polyglutamylation. J Biol Chem 1996;271:22117–22124.

77. Bonnet C, Boucher D, Lazereg S, et al. Differential binding regulation of microtubule-associated proteins MAP1A, MAP1B, and MAP2 by tubulin polyglutamylation. J Biol Chem 2001;276:12839–12848.

78. Khan IA, Luduena RF. Phosphorylation of beta III-tubulin. Biochemistry 1996;35:3704–3711.

79. Shafit-Zagardo B, Kalcheva N. Making sense of the multiple MAP-2 transcripts and their role in the neuron. Mol Neurobiol 1998;16:149–162.

80. Chapin SJ, Lue CM, Yu MT, Bulinski JC. Differential expression of alternatively spliced forms of MAP4: a repertoire of structurally different microtubule-binding domains. Biochemistry 1995;34:2289–2301.

81. Chang W, Gruber D, Chari S, et al. Phosphorylation of MAP4 affects microtubule properties and cell cycle progression. J Cell Sci 2001;114:2879–2887.

82. Belmont LD, Mitchison TJ. Identification of a protein that interacts with tubulin dimers and increases the catastrophe rate of microtubules. Cell 1996;84:623–631.

83. Iancu C, Mistry SJ, Arkin S, Atweh GF. Effects of stathmin inhibition on the mitotic spindle. Cancer Res 2000;60:3537–3541.

84. Cassimeris L. The oncoprotein 18/stathmin family of microtubule destabilizers. Curr Opin Cell Biol 2002;14:18–24.

85. Marklund U, Larsson N, Gradin HM, Brattsand G, Gullberg M. Oncoprotein 18 is a phosphorylation-responsive regulator of microtubule dynamics. EMBO J 1996;15:5290–5298.

86. Horwitz SB, Shen HJ, He L, et al. The microtubule-destabilizing activity of metablastin (p19) is controlled by phosphorylation. J Biol Chem 1997;272:8129–8132.

87. Zhang CC, Yang JM, White E, Murphy M, Levine AJ, Hait WN. The role of MAP4 expression in the sensitivity to paclitaxel and resistance to vinca alkaloids in p53 mutant cells. Oncogene 1998;16:1617–1624

88. Zhang CC, Yang JM, Bash-Babula J. et al. DNA damage increases sensitivity to vinca alkaloids and decreases sensitivity to taxanes through p53-dependent repression of microtubule-associated protein 4. Cancer Res 1999;59:3663–3670.

89. Murphy M, Hinman A, Levine AJ. Wild-type p53 negatively regulates the expression of a microtubule-associated protein. Genes Dev 1996;10:2971–2980.

90. Weinstein JN, Myers TG, O'Connor PM, et al. An information-intensive approach to the molecular pharmacology of cancer. Science 1997;298:343–349.

91. Poruchynsky MS, Giannakakou P, Ward Y, et al. Accompanying protein alterations in malignant cells with a microtubule-polymerizing drug-resistance phenotype and a primary resistance mechanism. Biochem. Pharmacol 2001;62:1469–1480.

92. Kavallaris M, Tait AS, Walsh BJ, et al. Multiple microtubule alterations are associated with Vinca alkaloid resistance in human leukemia cells. Cancer Res 2001;61:5803–5809.

93. Bash-Babula J, Toppmever D, Labassi M, et al. A Phase I/pilot study of sequential doxorubicin/vinorelbine: effects on p53 and microtubule-associated protein 4. Clin Cancer Res 2002;8:1057–1064.

94. Iancu C, Mistry SJ, Arkin S, Wallenstein S, Atweh GF. Effects of stathmin inhibition on the mitotic spindle. J Cell Sci 2001;114:909–916.

95. Nishio K, Nakamura T, Koh Y, Kanzawa F, Tamura T, Saijo N. Oncoprotein 18 overexpression increases the sensitivity to vindesine in the human lung carcinoma cells. Cancer 2001;91:1494–1499

96. Larsson N, Segerman B, Gradin HM, Wandzioch E, Cassimeris L, Gullberg M. Mutations of oncoprotein 18/stathmin identify tubulin-directed regulatory activities distinct from tubulin association. Mol Cell Biol 1999;19:2242–2250.

97. Gruber D, Faire K, Bulinski JC. Abundant expression of the microtubule-associated protein, ensconsin (E-MAP-115), alters the cellular response to Taxol. Cell Motil. Cytoskeleton 2001;49:115–129.

98. Downing KH, Nogales E. New insights into microtubule structure and function from the atomic model of tubulin. Eur Biophys J 1998;27:431–436.

99. Downing KH. Structural basis for the interaction of tubulin with proteins and drugs that affect microtubule dynamics. Annu Rev Cell Dev Biol 2000;16:89–111.

100. Gundersen GG, Cook TA. Microtubules and signal transduction. Curro Opin Cell Biol 1999;11:81–94.

101. Hirokawa N, Takemura R. Kinesin superfamily proteins and their various functions and dynamics. Exp Cell Res 2004 301:50–59.

102. Palmer KJ, Watson P, Stephens DJ. The role of microtubules in transport between the endoplasmic reticulum and Golgi apparatus in mammalian cells. Biochem Soc Symp 2005;72:1–13.
103. Hollenbeck P. Cytoskeleton: Microtubules get the signal Curr. Biol 2001;11:R820–R823.
104. Cardone L, de Cristofaro T, Affaitati A, et al. A-kinase anchor protein 84/121 are targeted to mitochondria and mitotic spindles by overlapping amino-terminal motifs J Mol Biol 2002;320:663–675.
105. Shinohara-Gotoh Y, Nishida E, Hoshi M, Sakai H. Activation of microtubule-associated protein kinase by microtubule disruption in quiescent rat 3Y1 cells. Exp Cell Res 1991;193:161–166.
106. Schid-Alliana A, Menou L, Manie S, et al. Microtubule integrity regulates src-like and extracellular signal-regulated kinase activities in human pro-monocytic cells. Importance for interleukin-1 production. J Biol Chem 1998;273:3394–3400.
107. McDaid HM, Horwitz SB. Selective potentiation of paclitaxel (taxol)-induced cell death by mitogen-activated protein kinase kinase inhibition in human cancer cell lines. Mol Pharmacol 2001;60:290–301.
108. Hu L, Hofmann J, Lu Y, Mills GB, Jaffe RB. Inhibition of phosphatidylinositol 3'-kinase increases efficacy of paclitaxel in in vitro and in vivo ovarian cancer models. Cancer Res 2002;62:1087–1092.
109. Moscatello DK, Holgado-Madruga M, Ernlet DR, Montgomery RB, Wong AJ. Constitutive activation of phosphatidylinositol 3-kinase by a naturally occurring, mutant epidermal growth factor receptor. J Biol Chem 1998;273:200–206.
110. Yu D, Jing T, Liu B, et al. Overexpression of ErbB2 blocks Taxol-induced apoptosis by upregulation of p21Cip1, which inhibits p34Cdc2 kinase Mol Cell 1998;2:581–591.
111. Tan M, Jing T, Lan KH, Neal C, et al. Phosphorylation on tyrosine-15 of p34(Cdc2) by ErbB2 inhibits p34(Cdc2) activation and is involved in resistance to taxol-induced apoptosis. Mol Cell 2002;9:993–1004.
112. Mozzetti S, Ferlini C, Concolino P, et al. Class III beta-tubulin overexpression is a prominent mechanism of paclitaxel resistance in ovarian cancer patients. Clin Cancer Res 2005;11:298–305.
113. Dumontet C, Isaac S, Souquet PJ, et al. Expression of class III beta tubulin in non-small cell lung cancer is correlated with resistance to taxane chemotherapy. Bull Cancer 2005;2:E25–30.
114. Hanash SM, Strahler JR, Kuick R, Chu EH, Nichols D. Identification of a polypeptide associated with the malignant phenotype in acute leukemia. J Biol Chem 1988;263:12813–12815.
115. Nylander K, Marklund U, Brattsand G, Gullberg M, Roos G. Immunohistochemical detection of oncoprotein 18 (Op18) in malignant lymphomas. Histochem J 1995;27:155–160.
116. Bieche I, Lachkar S, Becette V, et al. Overexpression of the stathmin gene in a subset of human breast cancer. Br J Cancer 1998;78:701–709.
117. Curmi PA, Nogues C, Lachkar S, et al. Overexpression of stathmin in breast carcinomas points out to highly proliferative tumours. Br J Cancer 2000; 82:142–150.
118. Calvert PM, O'Neill V, Twelves C, et al. A Phase I Clinical and Pharmacokinetic Study of EP0906 (Epothilone B), Given Every Three Weeks, in Patients with Advanced Solid Tumors. Proc Am Soc Clin Oncol 20 2001 (Abstract 429).
119. Zhuang SH, Menefee M, Kotz H, et al. A phase II clinical trial of BMS-247550 (ixabepilone), a microtubulestabilizing agent in renal cell cancer. (2004) Vol 22, No 14S (July 15 Supplement) (Abstract 4550).
120. Wilde CD, Crowther CE, Cowan NJ. Diverse mechanisms in the generation of human beta-tubulin pseudogenes. Science 1982;217:549.
121. Wilde CD, Crowther CE, Cripe TP, Gwo-Shu Lee M, Cowan NJ. Evidence that a human beta-tubulin pseudogene is derived from its corresponding mRNA. Nature 1982;297:83–84.
122. Lee MG, Lewis SA, Wilde CD, Cowan NJ. Evolutionary history of a multigene family: an expressed human beta-tubulin gene and three processed pseudogenes. Cell 1983;33:477–487.
123. Kelley MJ, Li S, Harpole DH. Genetic analysis of the beta-tubulin gene, TUBB, in non-small-cell lung cancer. J Natl Cancer Inst 2001;93:1886–1888.
124. Tsurutari J, Komiya T, Uejima H, et al. Mutational analysis of the beta-tubulin gene in lung cancer. Lung Cancer 2002;35:11–16.
125. Hasegawa S, Miyoshi Y, Egawa C, et al. Mutational analysis of the class I beta-tubulin gene in human breast cancer. Int J Cancer 2002;101:46–51.
126. Sale S, Sung R, Shen P, et al. Conservation of the class I beta-tubulin gene in human populations and lack of mutatlons in lung cancers and paclitaxel-resistant ovarian cancers. Mol Cancer Ther 2002;1:215–225.

127. Kohonen-Corish MR, Qin H, Daniel JJ, et al. Lack of beta-tubulin gene mutations in early stage lung cancer. Int J Cancer 2002;101:398–399.

128. Diaz JF, Andreu JM. Assembly of purified GDP-tubulin into microtubules induced by taxol and taxotere: reversibility, ligand stoichiometry, competition. Biochemistry 1993;32:2747–2755.

129. Horwitz SB. Mechanism of action of taxol. Trends Pharmacol Sci 1992;13:134–136.

130. Horwitz SB, Lothstein L, Manfredi JJ, et al. Taxol: mechanisms of action and resistance. Ann NY Acad Sci 1986;466:733–744.

131. Manfredi JJ, Parness J, Horwitz SB. Taxol binds to cellular microtubules. J Cell Biol 1982;94:688–696.

14 Mechanisms of Resistance to Drugs That Interfere with Microtubule Assembly

Fernando Cabral

Contents

Summary

Patient relapse during or following chemotherapy is a complex problem that potentially involves suboptimal drug dosing, changes in pharmacokinetics, sequestration of cancer cells, and genetic changes in the tumor cells themselves. This review focuses on possible mechanisms of drug resistance caused by mutations in cancer cells, and critically discusses evidence from cell culture models in support of each of these mechanisms.

Key Words: Drug resistance; paclitaxel; colcemid; vinblastine; tubulin; mutations; isotypes.

1. INTRODUCTION

Antimitotic drugs are important agents for treating a number of medical conditions including gout, Chediak-Higashi Syndrome, Familial Mediterranean Fever, cutaneous fungal infections, helminthiasis, and others (1). Their best known use, however, is in the treatment of cancer. The *vinca* alkaloids, vinblastine, and vincristine, are first line drugs for acute lymphocytic leukemia, Hodgkin's lymphoma, testicular carcinoma, Ewing's sarcoma, hepatoblastoma, pheochromocytoma, and Wilm's tumor. More recently, taxanes including paclitaxel and docetaxel have become front line drugs for treating breast, ovarian, head and neck, and lung carcinomas (2). The importance of these drugs in cancer therapy derives in part from the fact that they inhibit the microtubule cytoskeleton, a unique target for cancer drugs. Because they do not target DNA synthesis, structure,

From: *Cancer Drug Discovery and Development: The Role of Microtubules in Cell Biology, Neurobiology, and Oncology* Edited by: Tito Fojo © Humana Press, Totowa, NJ

or replication like most other cancer drugs, antimitotic drugs are effective choices for use in combination chemotherapy. As a result, much current effort is being devoted to find new drugs that target microtubules, and to modify existing drugs to enhance their therapeutic index.

Despite the effectiveness of these agents, their ability to cure cancer is limited by the problem of drug resistance. Many patients who initially respond to chemotherapy eventually relapse and their tumors become refractory to further treatment with the drugs that previously worked so well. Patients may relapse for a variety of reasons. Temporary forms of resistance may include the migration of cancer cells into areas such as the central nervous system that are not well-penetrated by the drugs. In such cases, the patient can be treated with radiation or intrathecal drug administration to reach these "hidden" cells. Other temporary forms of resistance might include changes in the way the drug is metabolized, excreted, or delivered to the tumor cells. Again, however, it may be possible to overcome this resistance by altering the manner by which the drug is administered or by switching to an analog of the drug with different pharmacological properties. A more permanent form of resistance occurs when there are genetic changes in the tumor cells themselves that exclude the drug from entry or interfere with the action of the drug. It is this last form of resistance that will be covered by this review. Since much of the work in this area has been covered in previous reviews *(3–6)*, the focus here will be to provide a framework for understanding why different results have been reported by various laboratories, and to summarize some of the most notable contributions that have appeared after the earlier reviews were written.

2. SELECTIONS FOR DRUG RESISTANCE

The central premise for studying drug resistance is that understanding the underlying mechanisms should make it possible to prevent resistance or to circumvent it when it occurs. But how does one uncover these mechanisms? Studies involving patients are too problematic: the patient population is too heterogeneous, the drugs may be given suboptimally and/or in combination with a variety of other drugs, the tumor cells are often difficult to obtain and are contaminated with other nontumor cells from the patient, and pretreatment controls are not always available. Animals represent a more genetically defined population and are more experimentally tractable, but still suffer from a number of drawbacks that include expense, a mixture of temporary and permanent forms of resistance, and difficulties in growing and analyzing tumor cells ex vivo. Most research laboratories have therefore turned to cultured cell lines to carry out their studies. This approach, however, can still present problems. Many mammalian cell lines suffer from unstable genomes, slow growth, and an inability to grow from a single cell. To avoid these limitations, we have chosen to carry out studies using Chinese hamster ovary (CHO) cells. These cells have the advantage of a relatively fast doubling time (12 h), a stable genome that ensures that any mutants isolated will not be rapidly lost, and a very high cloning efficiency that gives the ability to start selections with a homogeneous cell population. Moreover, many laboratories have selected mutant cell lines from CHO cells, making this the most genetically well-characterized mammalian cell line available *(7)*. Although not of human origin, the nearly identical amino acid sequences of tubulin between the two species suggests that similar mutations and mechanisms of resistance will be found in CHO cells.

Once an appropriate cell line is chosen, it is necessary to formulate a protocol to isolate drug resistant mutants. Ideally, the cells should be recloned to be sure one starts with a genetically homogeneous population, and the sensitivity of the cells to a given drug should be established by measuring the cloning efficiency of the cells in various concentrations of the drug. Because a typical spontaneous mutation frequency for drug resistance is on the order of 10^{-6} to 10^{-5} (8,9), the drug concentration used for mutant selection should, in most cases, be able to reduce the cloning efficiency to those levels. The selection is then quite simply to plate a sufficient number of cells in a lethal concentration of the drug, wait for resistant colonies to appear, isolate each colony individually, grow the cells, and analyze the phenotypes. This kind of single step procedure gives only low resistance (two- to fivefold), but also provides the maximum likelihood that single gene mutations will be responsible for the drug resistance, and that biochemical and phenotypic differences between the mutant and parental cells will be linked to the mutation.

For many (but not all) human cancer cell lines, this kind of single step procedure is unworkable because of their inability to grow from single cells. In these cases, multistep selection schemes are often used in which cells are treated with a toxic concentration of the drug, survivors are regrown, and the cells are then treated with a higher concentration of the same drug. This procedure is often repeated through many cycles yielding cell populations that can be 100 or even 1000-fold resistant. The final cell population is usually analyzed directly.

Each of these procedures has advantages and limitations. Because mutations occur spontaneously during the growth of cells (10), the short, single step procedure has the advantage that a single mutation is likely to be responsible for the drug resistance phenotype. On the other hand, the low level of resistance makes it more difficult to clearly measure differences in drug sensitivity. The multistep procedure gives more robust resistance that is easily measured, but suffers from two major limitations. Because the selections may start with a heterogeneous cell population, and because the cells are grown over a period of months during the selection, the final cell population is likely to have numerous genetic and phenotypic differences compared with the starting cell population. Such complex phenotypes make it very difficult to sort out, which changes are directly responsible for drug resistance, which may be secondary changes related to drug resistance, and which are coincidental and have no relationship to drug resistance.

A second disadvantage of multistep procedures is that they are inherently biased; i.e., not all mechanisms are capable of producing high resistance. Thus, cell lines that are several hundredfold resistant to an antimitotic drug are unlikely to have changes in tubulin as their primary mechanism of resistance. Tubulin is a highly conserved, essential protein that does not tolerate major changes in structure. As a result, it is only able to acquire subtle mutations that increase resistance to drugs a few fold. To acquire high resistance, other changes are needed, the most common of which is P-glycoprotein mediated multidrug resistance (MDR). Because it is a nonessential (at least in cell culture) membrane protein, P-glycoprotein can be mutated and/or amplified to give very high levels of resistance in mutant cells. Single step mutants may also acquire resistance by MDR or tubulin based mechanisms, but they are not biased toward MDR as are the multistep procedures and seldom produce cell lines in which multiple mechanisms of resistance coexist.

Regardless of the selection method used, it should be recognized that finding a mutation or biochemical change in a drug resistant cell line does not prove that the alteration is responsible for resistance. To establish proof, it is necessary to demonstrate further changes in the same gene upon reversion of the resistance, or show that wild-type cells transfected with the mutant, but not the wild-type, gene become drug resistant. Our laboratory has used both approaches to demonstrate that mutations in tubulin genes are sufficient to confer resistance in wild-type cells *(11–15)*.

3. MECHANISMS OF DRUG ACTION

Microtubules are essential structures in all eukaryotic cells. They exist as cytoplasmic filaments that serve as highways for the transport of vesicles into and out of the cell, they are needed to maintain the structure and organization of the endoplasmic reticulum and the Golgi apparatus, and they are the major structural component of the axonemes responsible for ciliary and flagellar motility. Their most essential function, however, is to segregate sister chromatids before cell division. Thus, drugs that interfere with microtubule assembly are frequently called spindle poisons or antimitotic drugs because cells treated with these agents form dysfunctional spindles and become blocked in mitosis. Most known antimitotic drugs act by binding to αβ-tubulin heterodimers, the building blocks for microtubule assembly, but they bind to multiple sites and act by differing mechanisms. Drugs like colchicine, vinblastine, and vincristine inhibit microtubule formation and bind to one of two sites: the colchicine site located on β-tubulin near the intradimer interface *(16)*, or the *vinca* site located near the interdimer interface *(17–19)*. More recently, a third class of drugs that includes paclitaxel, docetaxel, and epothilones has been described that promotes microtubule assembly and binds to β-tubulin at the taxane-binding site *(17)*.

The mechanisms by which these drugs act to inhibit microtubule function are discussed in another chapter in this volume. Functionally, the drugs can be divided into two groups: those that inhibit the formation of microtubules, and those that promote microtubule assembly and stabilize the filaments to disassembly. Microtubules are known to be very dynamic structures that frequently transit from states of growth to rapid disassembly *(20)*. Microtubule inhibitory drugs like colchicine, vinblastine, and vincristine bind to αβ-tubulin heterodimers and poison the growth of microtubules *(21)*; whereas microtubule stabilizing drugs like paclitaxel and epothilones bind to polymerized tubulin and inhibit microtubule disassembly *(22)*. The fact that drugs that inhibit or promote microtubule assembly are equally toxic argues that this dynamic behavior is crucial to the function of the microtubule cytoskeleton *(23)*.

4. MECHANISMS OF RESISTANCE

Once a drug resistant cell line has been isolated, the hard work begins: namely, finding the genetic mutation or biochemical change that is causing resistance. To facilitate the search, it is useful to consider some theoretical mechanisms by which cells can acquire resistance to antimitotic drugs. Figure 1 shows a simplified scheme depicting the plasma membrane and the intracellular microtubule machinery. When an antimitotic drug (D) is added to the culture medium (or serum in the case of an in vivo tumor), it is able to diffuse through the lipid bilayer and enter the cell because most of these drugs are quite hydrophobic. Once in the cell, the drug can either be pumped back out by P-glycoprotein mediated MDR (site 1) or bind to its intracellular target, tubu-

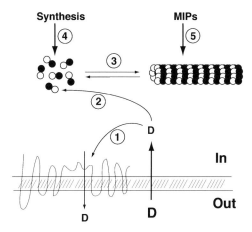

Fig. 1. Hypothetical mechanisms of resistance to antimitotic drugs. D, drug; MIPs, microtubule interacting proteins. Numbers refer to various mechanisms that are described in the text.

lin (site 2). The binding of the drug to tubulin can then influence the distribution of the protein between the soluble and polymer pools (site 3). If the intracellular drug concentration is sufficiently high, the steady state assembly of tubulin may be disrupted to such an extent that mitosis is blocked and the cells are killed. The concentration of drug needed to produce these effects could potentially be influenced by the composition and/or amount of tubulin present in the cell (site 4), or by the presence or activity of microtubule interacting proteins (MIPs) (site 5). In the following sections, the involvement of each of these sites of action in mammalian cell drug resistance will be discussed in more detail.

4.1. P-Glycoprotein Mediated MDR (Site 1)

The phenomenon of MDR has been studied for more than 30 yr. The most common form of MDR is mediated by a low specificity membrane pump, called P-glycoprotein that spans the membrane 12 times and expels a large repertoire of hydrophobic, weakly cationic compounds from the cell (24). Given that many of the most powerful drugs for treating cancer are substrates for P-glycoprotein, MDR is widely believed to be a major cause for the failure of chemotherapy. A number of recent clinical observations, however, have begun to question the prevalence of this mechanism. For example, it has been recognized that cancer patients resistant to Adriamycin are still sensitive to paclitaxel therapy even though both drugs are substrates for P-glycoprotein (25). Moreover, patients resistant to paclitaxel are frequently sensitive to the structural analog docetaxel, making MDR an improbable cause of the resistance (26).

Discrepancies between the expected prevalence of MDR based on in vitro studies, and its actual prevalence found by clinical experience are likely to be due in part to the multistep selections used to obtain drug resistant cells in culture, a procedure that is biased toward the MDR phenotype as described in Subheading 2 of this review. The prevalence of MDR in single step selections, on the other hand, appears to depend on the drug being used. Early selections for paclitaxel resistance yielded cell lines with changes in tubulin rather than MDR; and in one large study in CHO cells less than 10% of resistant cell lines had the MDR phenotype as judged by cross resistance to unrelated drugs (27). Although a subsequent study in human cells reported a higher incidence of MDR (40%), it still appeared less frequently than other mechanisms *(28)*. In the case of drugs such as vinblastine and colchicine that inhibit microtubule assembly, on the other hand, MDR is the predominant mechanism of resistance even in single step selections where it is found in approx 80% of mutant cell lines *(29)*. Why is MDR so much more prevalent in selections using these latter drugs? Although it is possible that these drugs have higher affinity for P-glycoprotein, a simpler argument is that tubulin mutations that cause paclitaxel resistance are much more common than mutations that cause resistance to drugs that inhibit microtubule assembly. A rationale for this explanation will be discussed later.

4.2. Altered Drug Binding (Site 2)

Even if a drug is able to accumulate within a cell, it must still be able to bind to tubulin in order to affect microtubule assembly and produce cytotoxicity. Thus, it stands to reason that mutations in tubulin that reduce drug-binding affinity should produce drug resistance. In fact, some of the earliest mutants resistant to an antimitotic drug, benomyl, were isolated in *Aspergillus nidulans* and were found to have defects in drug binding *(30)*. Altered drug binding was also reported in CHO cells selected for resistance to colchicine, but subsequent sequencing of tubulin from the mutant cells failed to reveal any mutations *(31,32)*. In contrast, other CHO mutants resistant to colchicine, colcemid, and griseofulvin were shown to have altered tubulin yet exhibited no evidence for altered drug binding *(33,34)*.

In order to make sense of these conflicting results, it is useful to remember that reduced drug binding should produce a recessive phenotype. Thus, in *A. nidulans* which expresses a single β-tubulin gene during hyphal growth and can propagate as a haploid, mutations in tubulin that reduce drug binding will produce cells that are drug resistant. In mammalian cells, however, which are diploid and express multiple tubulin genes *(35)*, a single gene mutation that reduces drug binding will generally not produce resistance because other wild-type tubulins are still present to bind the drug with normal affinity and inhibit microtubule assembly. Indeed, as will be discussed under Subheading 4.3, most mammalian cell lines with alterations in tubulin have properties that are inconsistent with altered drug binding as a mechanism of resistance.

Nevertheless, several recent publications have appeared reporting mutations in tubulin that affect drug binding in mammalian cells. In one report, β-tubulin mutations were found in cells selected in multiple steps to 24-fold resistance to paclitaxel *(36)*. The proximity of the mutations to the drug-binding site and the observation that the cells were much less resistant (1.4- to 3-fold) to epothilone B, another drug that stabilizes microtubules and binds to the same site, prompted the authors to conclude that the mutations acted by inhibiting the binding of paclitaxel to tubulin. It is noteworthy, however, that when the authors looked at expression levels of the mutant allele, they found almost no coexpression of the wild-type allele, even though it was clearly present at the DNA level.

This observation suggests a plausible scenario for the generation of these cells. Early steps in the selection may have allowed the survival of a population of cells with tubulin mutations that confer low resistance to paclitaxel and epothilone B, but increased sensitivity to vinblastine, a phenotype that is characteristic of changes in microtubule assembly (*see* Subheading 4.3).

Some of these mutations may have affected paclitaxel binding in addition to altering microtubule assembly, but altered binding could not contribute to the phenotype because of the coexpression of the wild-type allele. On further selection in paclitaxel, mutations that confer resistance solely by altering microtubule assembly would be lost because they are incapable of conferring a high level of resistance. Cells with mutations that additionally affect drug binding, however, could survive at the higher drug concentrations by now allowing the drug binding phenotype to contribute to the resistance, provided the wild-type allele was lost. Thus, in order for the drug-binding phenotype to be seen, the cells had to become functionally haploid. The same authors reported similar results in cells selected for resistance to epothilones A and B but did not indicate whether these cells were also functionally haploid *(37)*. A second laboratory has described the isolation of human lymphoblastoid cells 115-fold resistant to indanocine, a drug that binds to the colchicine site. These cells were found to also exhibit 40-fold cross-resistance to vinblastine, a drug that binds to a different site *(38)*. Despite this drug cross-resistance pattern, the authors favored a mechanism based on altered drug binding based, in part, on the proximity of the mutation to the colchicine-binding site. Although the authors did not measure mutant expression, sequencing indicated the absence of heterozygosity at the mutant locus, indicating that these cells were also functionally haploid for the major β-tubulin gene.

Based on these limited studies, it appears that drug-binding mutations represent a major mechanism of resistance in haploid organisms, but are not common in mammalian cells when single step selections are used. In multistep selections, on the other hand, selective pressure for resistance to relatively high drug concentrations results in mutant cells with multiple genetic changes; i.e., a tubulin mutation that affects microtubule assembly and drug binding coupled with loss of a wild-type tubulin gene and, in some cases, increased activity of P-glycoprotein or other mechanisms that limit intracellular accumulation of the drug.

Parenthetically, it should also be pointed out that it is possible to isolate mutations that increase drug binding. Unlike mutations that decrease drug binding, these mutations should behave in a dominant fashion and produce increased drug sensitivity even when wild-type subunits are present. Direct selections for these mutants, however, are problematic because they are the cells that are dying in drug resistance selections. Recently, a revertant of a colcemid resistant cell line was described with a mutation (A254V) in helix 8 of β-tubulin, close to the colchicine binding site. Unlike most mutations that have been analyzed, transfection of β-tubulin containing A254V into wild-type CHO cells, increased colcemid sensitivity four- to fivefold but did not change sensitivity to paclitaxel or vinblastine *(15)*. Moreover, it did not alter the assembly of the microtubule cytoskeleton, consistent with a mechanism based on changes in drug binding. Similarly, an L215I β-tubulin mutation created by site-directed mutagenesis and transfected into CHO cells increased the sensitivity of the cells to paclitaxel, but not to epothilone A or colcemid. Expression of this mutant β-tubulin also did not perturb microtubule assembly and thus appeared to be acting by enhancing the binding of paclitaxel *(39)*.

4.3. Altered Microtubule Assembly (Site 3)

Some of the earliest mammalian cell mutants resistant to antimitotic drugs were shown to have alterations in tubulin by two-dimensional gel electrophoresis *(27,33,40–42)*. These mutants were selected in a single step from a cloned CHO cell line and displayed a consistent phenotype that suggested a resistance mechanism based on changes in microtubule assembly. For example, it was commonly seen that cells

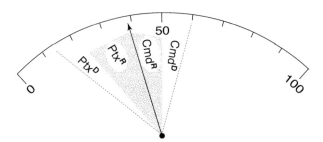

Fig. 2. The extent of microtubule assembly in wild-type and mutant CHO cell lines. Solid arrow, the level of assembly in wild-type cells; shaded area, levels of assembly associated with drug resistance and normal growth; clear areas inside the dotted lines, levels of assembly associated with drug-dependent growth; areas outside the dotted lines, levels of assembly associated with lethality.

selected for resistance to a drug that inhibits microtubule assembly were cross resistant to other drugs that inhibit assembly, but were more sensitive to paclitaxel, a drug that promotes assembly. Conversely, cells resistant to paclitaxel were more sensitive to drugs that inhibit microtubule assembly. Resistance of the cells in these single step selections was low (two- to fivefold) and the resistance was dominant in somatic cell hybrids. Also, alterations in both α- and β-tubulin were found among cells selected for resistance to any given drug. Together, these properties argued against a mechanism of resistance involving changes in drug binding.

Even more telling, however, was the isolation of CHO mutants that were not only resistant to paclitaxel, but also dependent on the drug for proliferation *(27,43)*. Clearly, reduced drug binding could not explain this phenotype. Instead, a model was proposed in which the mutations affect microtubule assembly in a manner that opposes the action of the drug *(34,44)*. The model is based on the premise that tubulin in mammalian cells is not fully polymerized at steady state and that the extent of polymerization provides a read-out for the stability of the microtubules. Cells proliferate normally so long as microtubules remain dynamic and microtubule stability stays within acceptable limits (shaded area, Fig. 2). Changes in stability will manifest as changes in the distribution of tubulin between the soluble (heterodimer) and polymer (microtubule) pools. Drugs such as colcemid that destabilize microtubule assembly will cause treated cells to make less microtubule polymer (a lower percentage polymerized tubulin [%P]), whereas drugs like paclitaxel that stabilize microtubules will lead to a higher %P. The model predicts that mutations in tubulin that confer resistance to microtubule destabilizing drugs make microtubules more stable, thereby increasing %P to the upper limits of the shaded region in Fig. 2. This increased stability explains why colcemid resistant cells are frequently more sensitive than wild-type cells to the stabilizing effects of paclitaxel.

Conversely, mutations that confer resistance to paclitaxel make microtubules less stable and decrease %P to the lower limits of the shaded region. Cells with these destabilized microtubules will therefore tolerate more paclitaxel but will show increased sensitivity to colcemid. The model also predicts the existence of drug-dependent mutants *(27,43)*. Paclitaxel-dependent cells have tubulin mutations that decrease %P lower than the shaded region; i.e., the microtubules are too unstable to function normally. Addition of any agent that stabilizes microtubules would then increase %P back into the shaded area, allowing the cells to carry out their microtubule mediated functions. Colcemid-dependent mutants should also exist, but have not been selected directly. However, a colcemid-dependent phenotype can be created by transfection and overexpression of a mutant tubulin that confers colcemid resistance *(45)*.

To evaluate whether drug resistant cells conform to this model, an assay was developed that allows an accurate and precise measure of the extent of tubulin assembly in wild-type and mutant cell lines *(46)*. With this assay, it was found that wild-type CHO cells have about 38% of their cellular tubulin in the polymer fraction. All paclitaxel resistant CHO cells that have been examined to date have values for %P that are lower than 38%; all CHO cells resistant to drugs that inhibit microtubule assembly have %P values greater than 38%. Also consistent with the model, paclitaxel-dependent cell lines have values for %P as low as 12–15% *(47)*. The transition of resistance into dependence (*see* Fig. 2) was established by examining cell lines that appeared to be only partially dependent on the presence of a selecting drug for growth *(27)*. For paclitaxel, this transition occurs at approx 22%P *(47)*. In the case of colcemid, a transition point of approx 55%P was estimated by noting that colcemid resistant cells seldom have values more than 50%P *(14)* and that a transfected cell line overexpressing tubulin with a colcemid resistance alteration had a value of 57%P and showed signs of partial colcemid dependence *(47)*. Thus, examination of a variety of drug resistant mutant cell lines has given a good quantitative picture of the limits on microtubule stability that can be tolerated by a mammalian cell line before functionality of the microtubules is lost.

Although the actual numbers may vary from cell line to cell line, it is anticipated that most cells can tolerate a sizable increase or decrease in microtubule assembly before toxicity is encountered. By isolating revertants of a colcemid resistant mutant with a microtubule stabilizing D45Y mutation, it has also been possible to identify tubulin mutations such as βQ292H that decrease microtubule stability even more radically *(15)*. Wild-type cells transfected with a β-tubulin complementary DNA (cDNA) containing the Q292H mutation under the control of an inducible promoter are unable to proliferate even in the presence of the microtubule stabilizing drug, paclitaxel. These cells have values for %P that are below the lower dotted line in Fig. 2. Thus, drug dependence is only seen within a limited range of microtubule disruption, and some mutations cannot be corrected by adding the appropriate drug to counteract their effects on microtubule assembly.

Although the studies just described were carried out in CHO cells, it is likely that similar mechanisms of resistance occur in human cells. An early study in a human small cell lung cancer cell line identified a paclitaxel resistant mutant with partial drug dependence, an increased sensitivity to vinca alkaloids, and an altered α-tubulin in isoelectric focusing gels *(48)*. More recent publications using single step and multistep selections to relatively low levels of resistance have reported α- and β-tubulin mutations that act by altering microtubule assembly in a variety of human and rodent cell lines (summarized in Table 1).

It might be anticipated that mutations affecting microtubule assembly should affect multiple regions of the tubulin molecule. This expectation is supported by early observations that alterations in both α- and β-tubulin were associated with resistance to any given antimitotic drug *(27,29,33,40–42,49)*. The earliest sequenced mutations found in paclitaxel and epothilone resistant mammalian cells were located near the paclitaxel binding site but were reported to affect drug binding *(36,37)*. It was surprising, therefore, that the first 12 sequences from paclitaxel resistant CHO mutants with clear alterations in microtubule assembly, indicated various amino acid substitutions at L215, L217, and L228 of β-tubulin *(13)*, a small region close to the paclitaxel binding site (Fig. 3).

Table 1
Drug Resistance Mutations in Mammalian Cells

Mutation	Phenotype	Mechanism	Location	References
αA12S	HTI-286[R]	↑MT stability	H1	95
αE55K	Cmd[R]	↑MT stability	H1/S2 loop	14
αS165P	HTI-286[R]	↑MT stability	H4/S5 loop	96
αR221H	HTI-286[R]	↑MT stability	H6/H7 loop	96
αH283Y	Cmd[R]	↑MT stability	M loop	14
αS379R	Ptx[R]	↓MT stability	S10	79
αA383V	Cmd[R]	↑MT stability	H11	14
αI384V	HTI-286[R]	↑MT stability	H11	96
αR390C	Cmd[R]	↑MT stability	H11	14
βD45Y	Cmd[R]	↑MT stability	H1/S2 loop	14
βV60A	Ptx[R]	↓MT stability	H1/S2 loop	15
βS172A	HTI-286[R]	↑MT stability	T5 loop	96
βP173A	Epo[R]	↓MT stability	T5 loop	97
βC2911F	Cmd[R]	↑MT stability	H6	14
βL215F	Ptx[R]	↓MT stability	H6/H7 loop	13
βL215H	Ptx[R]	↓MT stability	H6/H7 loop	13
βL215R	Ptx[R]	↓MT stability	H6/H7 loop	13
βL215I	Ptx[SS]	↑drug binding	H6/H7 loop	39
βL217R	Ptx[R]	↓MT stability	H6/H7 loop	13
βD224N	Cmd[R]	↑MT stability	H7	14
βL228F	Ptx[R]	↓MT stability	H7	13
βL228H	Ptx[R]	↓MT stability	H7	13
βA231T	Epo[R]	↓MT stability	H7	98
βS234N	Cmd[R]	↑MT stability	H7	14
βL240I	Vcr[R]	↑MT stability	H7	99
βA254V	Cmd[SS]	↑drug binding	H8	15
βF270V	Ptx[R]	↓drug binding	M loop	36
βT274I	Epo[R]	↓drug binding	M loop	37
βR282Q	Epo[R]	↓drug binding	M loop	37
βQ292H	Lethal	↓↓MT stability	H9	15
βQ292E	Epo[R]	↓MT stability	H9	97,98
βK350N	Cmd[R], Ind[R]	↑MT stability, ↓drug binding	S9	14,38
βA364T	Ptx[R]	↓drug binding	S9/S10 loop	36
βY422C	Epo[R]	↓MT stability	H12	97

Cmd, colcemid; Epo, epothilone; Ptx, paclitaxel; Ind, indanocine; Vcr, vincristine; HTI, hemiasterlin; R, resistance; SS, supersensitivity.

Data from several laboratories including the author's own, however, subsequently identified residues in both α- and β-tubulin outside of this area that are associated with paclitaxel resistance and with resistance to other antimitotic drugs (*see* Table 1). Thus, the reason for the initial cluster of mutations at L215, L217, and L228 remains unexplained. One attractive hypothesis is that this region plays an especially crucial role in maintaining tubulin in an assembly competent conformation, and so mutations or paclitaxel binding in this area can more easily influence microtubule assembly. In support of this notion, a recent publication comparing the structures of soluble (assembly incompetent or "curved" conformation) tubulin and filamentous (assembly competent or "straight" conformation)

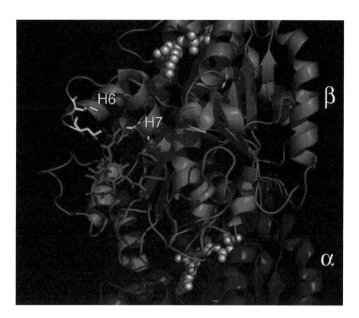

Fig. 3. Structure of an αβ-tubulin heterodimer. The model was drawn with MacPymol (pymol.source-forge.net) based on published atomic coordinates 1JFF *(94)*. Note that only a portion of the α-subunit (blue) is shown. The β-subunit is labeled to show helix 6 (H6) and helix 7 (H7) as well as residues L215 (green), L217 (cyan), and L228 (yellow). GTP and GDP are shown as pink spheres; docetaxel (an analog of paclitaxel) is shown in red. To view this figure in color, see the insert and the companion CD-ROM.

tubulin noted that the greatest changes occurred in the positions of helix 6, helix 7, and the H6-H7 loop, the exact regions where the cluster of mutations was found *(16)*.

It should be noted that a growing list of mutations are reported to be associated with drug resistance in various cell lines, but finding a mutation in a resistant cell line does not prove that the mutation is responsible for resistance. To accomplish this, loss of the mutation in a drug sensitive revertant or demonstration of drug resistance in wild-type cells transfected with the mutant gene is necessary. This, however, has only been done for a few of the reported mutations. For example, it was shown that the altered β-tubulin in a colcemid-resistant cell line was lost or further altered at high frequency in revertants of colcemid-resistant and paclitaxel-resistant CHO cell lines *(11,15,27,50)*. More recently, a tetracycline regulated expression system has been used to demonstrate that β-tubulin mutations identified in drug-resistant CHO cells are able to confer resistance in a tetracycline regulated manner when transfected into wild-type CHO cells *(12–15,51)*.

4.4. Altered Tubulin Synthesis (Site 4)

Mammalian cells express multiple tubulin genes whose protein products are highly homologous but have unique sequences that differ most notably in the last 15 amino acids *(35)*. Interestingly, these C-terminal sequences are well-conserved across verte-brate species and has led to the classification of 7 distinct β-tubulin isotypes *(52)*. Moreover, many of these distinct gene products or isotypes are expressed in a tissue-specific manner, fueling the hypothesis that different isotypes carry out different func-tions *(53)*. Transfection experiments, however, have indicated that all isotypes coassemble into all microtubules *(54)*. On the other hand, some in vitro studies have uncovered

small, but potentially significant differences in drug binding or assembly among purified, or partially purified, tubulin isotypes (55). Thus, it is reasonable to conclude that
cells could potentially alter the synthesis of specific β-tubulin isotypes as a way to gain
resistance to antimitotic drugs.

Evidence for such a mechanism came first from a series of multistep mutants selected
for moderate resistance to paclitaxel (reviewed in ref. 56). These studies reported
increased expression of βII-tubulin in mouse macrophage cells (57), βI–βII,βIII, and
βIVa in human lung cancer cells and ovarian tumors (58), and βIVa in human leukemia
cells (59). Although these studies were consistent with the idea that altered synthesis
of specific tubulin isotypes might constitute a mechanism of drug resistance, they only
provided a correlation, not a proof. Moreover, it seemed unlikely that so many different
isotypes could be contributing to drug resistance. Some experimental evidence also
argued against such a mechanism. One study used single step selections to obtain paclitaxel resistant human sarcoma cells (28). Resistant clones had reduced rather than elevated expression of specific tubulin isotypes, and other drug-sensitive clones also
exhibited reduced expression of these isotypes making it unlikely that altered expression was tied to the drug-resistance phenotype. A second laboratory reported increased
expression of βIII and βIVa-tubulin in human prostate carcinoma cells selected for
resistance to estramustine (60), a drug that inhibits microtubule assembly and binds to
a site that is distinct from other antimitotic drugs (61). It seemed unlikely that increased
expression of the same isotypes could confer resistance to two drugs with opposing
mechanisms of action and separate binding sites.

In an effort to clear up the conflicting data and provide direct evidence for a role of
tubulin expression in drug resistance, CHO cells were transfected with cloned cDNAs
for βI-, βII-, and βIVb-tubulin under the control of a tetracycline-regulated promoter
(12). The results showed that overexpression of these isotypes was insufficient to alter
sensitivity to paclitaxel. Transfection of a cDNA containing a mutation in the βI-tubulin
gene, on the other hand, produced clear resistance to the drug. A later publication
reported the effects of βIII-tubulin overexpression (62). This isotype is normally found
in appreciable amounts only in neuronal and Sertoli cells (55) but is also expressed in
many tumor cells for reasons that are as yet unclear (63). When expressed in CHO cells,
it was found to efficiently assemble into microtubules and produce no obvious phenotype at low levels of expression. When expressed at 50% of total tubulin, however, it
caused a small reduction in polymerized tubulin and conferred weak (1.5–2X) resistance to paclitaxel. At still higher levels of expression (80% of total tubulin), it became
toxic and interfered with mitotic progression. This toxicity was not reversed by
paclitaxel and so resistance in those cells could not be measured. It thus appears that
βIII-tubulin is able to confer very weak resistance to paclitaxel but does not appear to
be capable of conferring the high levels of resistance reported in the studies described
earlier in this section.

More recently, a mouse βV-tubulin cDNA was transfected into CHO cells with quite
interesting results (51). Unlike earlier transfections, even low expression of βV-tubulin
reduced polymerized tubulin and conferred resistance to paclitaxel. Higher levels of
expression (50% of total tubulin) produced a dramatic reduction in cellular microtubules that could be reversed by paclitaxel. These cells were approx twofold resistant
to the drug but also required the drug for growth. In the absence of paclitaxel, the cells
exhibited mitotic defects similar to the paclitaxel-dependent cells described earlier.

As pointed out in an earlier review *(35)*, β-tubulin isotypes can be divided into two groups: outside of their C-termini, βI, βII, and βIV have very similar amino acid sequences and are widely distributed among vertebrate tissues. A second group consisting of βIII, βV, and βVI is much more heterogeneous in sequence and its members exhibit tissue-specific expression (neuronal tissue for βIII; hematopoietic cells for βVI), or wide distribution but low abundance in the case of βV *(55)*. It is interesting that overexpression of members of the first group have little or no effect on microtubule assembly or drug resistance whereas overexpression of two members of the second group have small to dramatic effects on the same properties. One possible explanation for these differences is that members of the first group are used to build the basic microtubule structure whereas members of the second group are used in various tissues to modulate the properties of the microtubule cytoskeleton.

Altered synthesis of some β-tubulin isotypes thus appears to be a viable mechanism for changing the assembly properties of microtubules and conferring weak resistance to drugs such as paclitaxel. Mechanistically, these isotypes appear to act in manner similar to mutant forms of βI-tubulin that confer paclitaxel resistance; i.e., they produce a dominant phenotype, reduce microtubule assembly, and confer only low resistance to the drug. It will be interesting to determine which amino acid differences in βIII and βV relative to βI are responsible for these effects.

4.5. Alterations in MIPs (Site 5)

In previous sections it has been seen that changes in microtubule assembly frequently accompany the drug resistance phenotype and thus, it makes sense to suggest that tumor cells might use a variety of mechanisms to modulate microtubule assembly and thereby gain resistance to chemotherapy. It has long been recognized from in vitro studies that cells produce a variety of proteins that cofractionate with polymerized tubulin and that promote the assembly of purified tubulin. The earliest such proteins, called microtubule-associated proteins (MAPs) are large fibrous polypeptides that form a stable association with intact microtubules *(64)*. In the intervening years, many additional proteins have been discovered that bind stably or transiently to microtubules or to soluble tubulin heterodimers *(65,66)*. Collectively, all of these proteins (including the MAPs) are referred to as MIPs.

The role of MIPs in drug resistance is still uncertain. Although many of these proteins have been shown to affect microtubule assembly in vitro, it should be recognized that they may have very different roles in living cells. For example, structural MAPs have been repeatedly shown to promote microtubule assembly in vitro and to alter microtubule dynamics, yet in vivo studies have yielded conflicting results. In an early microinjection study, tau, a brain-specific MAP was reported to cause microtubule bundles and to make the microtubules more resistant to nocodazole-induced disassembly based on a fluorescence assay in which the time required for microtubule disassembly in high concentrations of the drug was measured *(67)*. The time required for microtubule dissolution, however, may not be a valid measure of cellular resistance to a drug. What is needed is a measurement of cloning efficiency of the cells in various concentrations of the drug. Using this as an assay, a later publication reported that overexpression of neither MAP 4, a ubiquitous high molecular weight MAP nor tau had any effect on level of polymerized tubulin or drug resistance in CHO cells even though the tau rearranged microtubules into bundles that depolymerized slowly *(68)*. The absence of any effect with overexpression of MAP 4 was consistent with lack of any phenotype in *Drosophila* with a genetic knockout of a MAP similar to MAP 4 *(69)*, and in CHO cells microinjected with an antibody that prevented MAP 4 from binding to microtubules *(70)*. The lack of effects of these manipulations on microtubule assembly and drug resistance could potentially be explained by assuming that the function of these MAPs is redundant with other MAPs in the cell. Even if true, however, it would not change the fact that mutations or changes in expression of these MAPs are unlikely to affect drug sensitivity. In contrast to these studies, others have reported changes in drug sensitivity in response to overexpression of MAP 4 *(71,72)*. The reasons for these discrepancies are uncertain but could include differences in cell lines or differences in the assays used to measure sensitivity to antimitotic drugs.

In addition to the structural MAPs, several other tubulin interacting proteins have been shown to affect microtubule assembly in vitro. These include stathmin, a protein that binds two tubulin heterodimers and acts by sequestering tubulin and/or promoting microtubule catastrophe *(73)*, mitotic centromere associated kinesin (MCAK), a kinesin-like protein that binds to microtubule ends to promote disassembly *(74)*, and katanin, a protein that severs intact microtubules *(75)*. Each of these proteins can desta- bilize microtubules and perhaps thereby confer resistance to paclitaxel and/or enhanced sensitivity to drugs like colcemid, but there is currently little evidence that they are actu- ally involved in mechanisms of resistance. The use of antisense RNA to reduce stath- min in erythroleukemia cells has been linked to increased microtubule assembly, resistance to vinblastine, and increased sensitivity to paclitaxel, a phenotype consistent with increased stabilization of microtubules *(76)*.

Conversely, the use of transfection to increase stathmin levels has been reported to decrease microtubule polymer and confer resistance to both paclitaxel and vinblastine *(77)*. The unexpected resistance to vinblastine was attributed to a decrease in the ability of the cells to enter mitosis when stathmin was overexpressed; but it should be noted that another laboratory reported increased sensitivity to vindesine in stathmin overex- pressing cells, an outcome that is more consistent with the observed decrease in micro- tubule polymer *(78)*. In addition to these transfection studies, a direct selection for paclitaxel resistance has yielded cells with increased active stathmin *(79)*; but the cells also have a point mutation in α-tubulin making it unclear whether it is the change in stathmin that is responsible for drug resistance. Thus, there are tantalizing indications that MIPs may be able to confer resistance to antimitotic drugs. Inconsistencies in the reported data, however, and the large number of tubulin interacting proteins that have yet to be examined, suggest there is still much work to be done in this area.

5. MECHANISMS THAT OCCUR MOST FREQUENTLY

Although a variety of mechanisms are theoretically capable of producing resistance to antimitotic drugs, the mechanism seen most commonly depends upon the experimen- tal conditions used to select mutants. Thus, decreased drug binding is a major mecha- nism of resistance in lower eukaryotes, but is not commonly seen in mammalian cells unless accompanied by additional changes that lead to mutant tubulin being almost the only tubulin produced. It has also been seen that multistep selections are frequently biased for the survival of mutants with MDR, whereas the predominant mechanism of resistance in single step selections depends on which drug is used. Analysis of hundreds of single step mutants in CHO cells has demonstrated that about 80% of mutants selected for resistance to drugs that destabilize microtubules are multidrug-resistant, whereas >90% of mutants selected for resistance to drugs that stabilize microtubules have tubulin alterations. The reason for this large difference in which mechanism predominates is not clear, but one simple explanation is that mutations that destabilize microtubules (resistance to stabilizing drugs) are much more frequent than mutations that stabilize microtubules (resistance to destabilizing drugs). Of course, these numbers will vary considerably based on the actual drug chosen for mutant selection. For example, many newer drugs under development are able to circumvent the MDR pathway and in selections using those drugs, it is expected that tubulin mutations will be the most common.

Of the mutants that are not multidrug resistant, our experience indicates that virtually all have mutations in either α- or β-tubulin that affect microtubule assembly. To date, changes have not been found in drug binding, tubulin synthesis, or MIPs in any cell line selected for resistance to an antimitotic agent. Moreover, transfection of mutant tubulin cDNA has invariably been able to confer resistance in wild-type cells indicating that the mutant tubulin alone is sufficient to confer resistance. Thus, the fact that changes in tubulin isotype expression or stathmin can produce drug resistance under a defined set of experimental conditions does not necessarily mean that those mechanisms will be found at an appreciable frequency under selective conditions for mutant isolation.

6. CLINICAL AND BIOLOGICAL SIGNIFICANCE

Understanding how cells acquire resistance to a drug used in chemotherapy is clearly of importance to learn how to prevent patient relapse, circumvent relapses when they occur, and design diagnostic tests for the early detection of drug resistant tumor cells. The question that has not yet been adequately answered is what mechanism is most commonly associated with clinical resistance to chemotherapy. For a large number of tumors of epithelial origin, inherent resistance to a wide range of antitumor antibiotics is seen because they derive from tissues that are naturally high in P-glycoprotein. For tumors that are initially sensitive to drugs but acquire resistance during or after therapy, however, the cause of resistance is much less clear. Because patients are usually treated with a combination of drugs having diverse mechanisms of action, it would make sense that MDR, which can confer simultaneous resistance to many drugs, should be very common; but, as mentioned earlier, many patients who fail combination chemotherapy remain sensitive to drugs such as paclitaxel even though accumulation of these drugs in the cell should be limited by P-glycoprotein. This suggests that mutations in tubulin might represent a major mechanism of resistance to antimitotic drugs in patients, just as it does in cell culture. Although tubulin mutations frequently give only low resistance (e.g., two- to fivefold), this should be sufficient to protect a tumor cell from chemotherapy because the drugs are usually given to patients at concentrations close to the maximum tolerated dose.

A number of recent studies have begun to directly address whether tubulin mutations are commonly found in tumors from patients who have relapsed on paclitaxel therapy, but to date no mutation that alters the coding sequence of the major βI-tubulin has been found. An initial study examining the sequence of β-tubulin in nonsmall-cell lung tumors found numerous changes that appeared to correlate with patient response to paclitaxel (80). The results were troubling, however, because they suggested a high incidence of tubulin mutations in tumors that had not received treatment with any antimitotic drug. In fact, a later study failed to replicate the findings and showed that the authors of the original work had been sequencing pseudogenes (81). Not surprisingly, numerous other studies have reported the absence of tubulin mutations in tumors from nontreated patients (82–87). Two studies, however, examined tumors from patients that failed paclitaxel therapy but again found no tubulin mutations (88,89). Although there are a number of possible technical and experimental problems that could explain these negative results, a likely conclusion is that tubulin mutations do not account for patient relapse under current conditions of therapy.

The paradox thus exists that the resistance mechanisms that appear to be most common in tissue culture are not commonly found in an in vivo situation. How does one explain this result? It is likely that the lack of tubulin alterations in vivo simply means that other mechanisms of relapse are more common. Although these other mechanisms could also be cellular, the fact that they are not commonly seen in tissue culture, makes this unlikely. Rather, the author proposes that patient relapse occurs most frequently because of noncellular, pharmacological changes. For example, a patient could relapse because of increased metabolism or excretion of one or more of the drugs used in combination chemotherapy. This would explain why patients who fail paclitaxel chemotherapy are often responsive to docetaxel even though the two drugs have the same mechanism of action and are both extruded from the cell by P-glycoprotein (26). Once the relevant pharmacological factors are identified, it should be possible to alter the manner in which the drug is administered, or switch to a related drug with different pharmacokinetic properties, to reduce the number of relapses that are seen. Even with optimized drug administration, however, relapses are likely to still occur because of the cellular mechanisms described in this review.

Beyond the obvious importance of understanding mechanisms of resistance for clinical reasons, the isolation of cell lines with mutant tubulin has provided unique insights into microtubule assembly and structure. One of the most exciting discoveries in microtubule biology in recent years was the elucidation of the crystal structure for tubulin (90). Although this has given an unprecedented glimpse into microtubule structure (91), the structural model is in need of biological and genetic confirmation. Moreover, the dynamic aspects of microtubule assembly need to be integrated with structural information in order to get a true understanding of microtubule regulation. The existence of a library of mutant tubulins will be crucial for understanding how structure supports microtubule function. Already, a number of mutations in α- and β-tubulin have been shown to localize to regions that were predicted by the structural model to form lateral and longitudinal contact sites (see Table 1). These mutations have frequently been found to affect microtubule polymer levels, and in one case, has also been shown to affect microtubule dynamics (92).

Further analysis of how specific mutations affect the assembly and dynamic behavior of microtubules should begin to give a detailed molecular view of how tubulin function is affected by its assembly. For example, a recent study has used mutants to probe how much stability or instability is consistent with essential microtubule function in living cells and to further elucidate how tubulin expression is affected by microtubule assembly (47). Other mutations have been shown to affect residues in β-tubulin that are near drug binding sites and that may decrease affinity for the drug (36,37) or in some cases increase the affinity (15,39). This information will be useful in designing new drugs with increased efficacy. Additional phenotypes have been produced by looking for revertants of drug resistant mutants. Such revertants may have second mutations in the same gene that counteract the effects of the first mutation (15), have second mutations that increase drug binding affinity

(15), or have mutations that produce unstable tubulin *(93,100)*. These latter mutations should be helpful to further dissect the folding pathway for tubulin and to reveal the signals that target assembly defective forms of tubulin for degradation.

ACKNOWLEDGMENTS

The author wishes to thank the US Public Health Service for its generous support of his research and his students and fellows who have stimulated much of the thinking and carried out most of the experiments that led to the conclusions summarized in this review.

REFERENCES

1. Hardman JG, Limbird LE, Molinoff PB, Ruddon RW, Gilman AG. The Pharmacological Basis of Therapeutics. 9th ed. McGraw-Hill, New York, 1996.
2. Mekhail TM, Markman M. Paclitaxel in cancer therapy. Expert Opin Pharmacother 2002;3:755–766.
3. Cabral F. Factors determining cellular mechanisms of resistance to antimitotic drugs. Drug Resistance Updates 2000;3:1–6.
4. Casazza AM, Fairchild CR. Paclitaxel (Taxol): mechanisms of resistance. Cancer Treatment Res 1996;87:149–171.
5. Dumontet C, Sikic BI. Mechanisms of action of and resistance to antitubulin agents: microtubule dynamics, drug transport, and cell death. J Clin Oncol 1999;17:1061–1070.
6. Orr GA, Verdier-Pinard P, McDavid H, Horwitz SB. Mechanisms of Taxol resistance related to microtubules. Oncogene 2003;22:7280–7295.
7. Gottesman MM. Molecular Cell Genetics, John Wiley, NY, 1985.
8. Siminovitch L. On the nature of hereditable variation in cultured somatic cells. Cell 1976;7:1–11.
9. Goldie JH, Coldman AJ. A mathematical model for relating the drug sensitivity of tumors to their spontaneous mutation rate. Cancer Treatment Rep 1979;63:1727–1733.
10. Luria SE, Delbruck M. Mutations of bacteria from virus sensitivity to virus resistance. Genetics 1943;28:491–511.
11. Cabral F, Abraham I, Gottesman MM. Revertants of a Chinese hamster ovary cell mutant with an altered β-tubulin: evidence that the altered tubulin confers both colcemid resistance and temperature sensitivity on the cell. Mol Cell Biol 1982;2:720–729.
12. Blade K, Menick DR, Cabral F. Overexpression of class I, II, or IVb β-tubulin isotypes in CHO cells is insufficient to confer resistance to paclitaxel. J Cell Sci 1999;112:2213–2221.
13. Gonzalez-Garay ML, Chang L, Blade K, Menick DR, Cabral F. A β-tubulin leucine cluster involved in microtubule assembly and paclitaxel resistance. J Biol Chem 1999;274:23,875–23,882.
14. Hari M, Wang Y, Veeraraghavan S, Cabral F. Mutations in α- and β-tubulin that stabilize microtubules and confer resistance to colcemid and vinblastine. Mol Cancer Ther 2003;2:597–605.
15. Wang Y, Veeraraghavan S, Cabral F. Intra-allelic suppression of a mutation that stabilizes microtubules and confers resistance to colcemid. Biochemistry 2004;43:8965–8973.
16. Ravelli RBG, Gigant B, Curmi PA, et al. Insight into tubulin regulation from a complex with colchicine and a stathmin-like domain. Nature 2004;428:198–202.
17. Downing KH. Structural basis for the interaction of tubulin with proteins and drugs that affect microtubule dynamics. Ann Rev Cell Dev Biol 2000;16:89–111.
18. Rai SS, Wolff J. Localization of the vinblastine-binding site on β-tubulin. J Biol Chem 1996;271:14,707–14,711.
19. Gigant B, Wang C, Ravelli RBG, et al. Structural basis for the regulation of tubulin by vinblastine. Nature 2005;435:519–522.
20. Mitchison TJ, Kirschner MW. Some thoughts on the partitioning of tubulin between monomer and polymer under conditions of dynamic instability. Cell Biophys 1987;11:35–55.
21. Wilson L. Action of drugs on microtubules. Life Sci 1975;17:303–310.
22. Manfredi JJ, Parness J, Horwitz SB. Taxol binds to cellular microtubules. J Cell Biol 1981;94:688–696.
23. Jordan MA, Wilson L. Microtubules and actin filaments: dynamic targets for cancer chemotherapy. Curr Opin Cell Biol 1998;10:123–130.

24. Ambudkar SV, Dey S, Hrycyna CA, Ramachandra M, Pastan I, Gottesman MM. Biochemical, cellular, and pharmacological aspects of the multidrug transporter. Annu Rev Pharmacol Toxicol 1999;39: 361–398.

25. Rowinsky EK, Donehower RC. Paclitaxel (Taxol). N Engl J Med 1995;332:1004–1014.

26. Verschraegen CF, Sittisomwong T, Kudelka AP, et al. Docetaxel for patients with paclitaxel-resistant Mullerian carcinoma. J Clin Oncol 2000;18:2733–2739.

27. Schibler M, Cabral F. Taxol-dependent mutants of Chinese hamster ovary cells with alterations in α- and β-tubulin. J Cell Biol 1986;102:1522–1531.

28. Dumontet C, Duran GE, Steger KA, Beketic-Oreskovic L, Sikic BI. Resistance mechanisms in human sarcoma mutants derived by single-step exposure to paclitaxel (Taxol). Cancer Res 1996;56: 1091–1097.

29. Schibler MJ, Barlow SB, Cabral F. Elimination of permeability mutants from selections for drug resistance in mammalian cells. FASEB J 1989;3:163–168.

30. Davidse LC, Flach W. Differential binding of methyl benzimidazole-2-yl carbamate to fungal tubulin as a mechanism of resistance to this antimitotic agent in mutant strains of Aspergillus nidulans. J Cell Biol 1977;72:174–193.

31. Ling V, Aubin JE, Chase A, Sarangi F. Mutants of Chinese hamster ovary (CHO) cells with altered colcemid-binding efficiency. Cell 1979;18:423–430.

32. Keates RAB, Sarangi F, Ling V. Structural and functional alterations in microtubule protein from Chinese hamster ovary mutants. Proc Natl Acad Sci USA 1981;78:5638–5642.

33. Cabral F, Sobel ME, Gottesman MM. CHO mutants resistant to colchicine, colcemid or griseofulvin have an altered β-tubulin. Cell 1980;20:29–36.

34. Cabral F, Barlow SB. Mechanisms by which mammalian cells acquire resistance to drugs that affect microtubule assembly. FASEB J 1989;3:1593–1599.

35. Sullivan KF. Structure and utilization of tubulin isotypes. Ann Rev Cell Biol 1988;4:687–716.

36. Giannakakou P, Sackett DL, Kang Y-K, et al. Paclitaxel-resistant human ovarian cancer cells have mutant β-tubulins that exhibit impaired paclitaxel-driven polymerization. J Biol Chem 1997;272: 17,118–17,125.

37. Giannakakou P, Gussio R, Nogales E, et al. A common pharmacophore for epothilone and taxanes: molecular basis for drug resistance conferred by tubulin mutations in human cancer cells. Proc Natl Acad Sci USA 2000;97:2904–2909.

38. Hua XH, Genini D, Gussio R, et al. Biochemical genetic analysis of indanocine resistance in human leukemia. Cancer Res 2001;61:7248–7254.

39. Wang Y, Yin S, Blade K, Cooper G, Menick DR, Cabral F. Mutations at Leucine 215 of β-tubulin affect paclitaxel sensitivity by two distinct mechanisms. Biochemistry 2006;45:185–194.

40. Barlow SB, Cabral F. Alterations in microtubule assembly caused by the microtubule-active drug LY195448. Cell Motil Cytoskeleton 1991;19:9–17.

41. Cabral F, Abraham I, Gottesman MM. Isolation of a taxol-resistant Chinese hamster ovary cell mutant that has an alteration in α-tubulin. Proc Natl Acad Sci USA 1981;78:4388–4391.

42. Schibler MJ, Cabral F. Maytansine-resistant mutants of Chinese hamster ovary cells with an alteration in α-tubulin. Cancer J Biochem Cell Biol 1985;63:503–510.

43. Cabral F. Isolation of Chinese hamster ovary cell mutants requiring the continuous presence of taxol for cell division. J Cell Biol 1983;97:22–29.

44. Cabral F, Brady RC, Schibler MJ. A mechanism of cellular resistance to drugs that interfere with microtubule assembly. Ann NY Acad Sci 1986;466:745–756.

45. Whitfield C, Abraham I, Ascherman D, Gottesman MM. Transfer and amplification of a mutant β-tubulin gene results in colcemid dependance: use of the transformant to demonstrate regulation of β-tubulin subunit levels by protein degradation. Mol Cell Biol 1986;6:1422–1429.

46. Minotti AM, Barlow SB, Cabral F. Resistance to antimitotic drugs in Chinese hamster ovary cells correlates with changes in the level of polymerized tubulin. J Biol Chem 1991;266:3987–3994.

47. Barlow SB, Gonzalez-Garay ML, Cabral F. Paclitaxel-dependent mutants have severely reduced microtubule assembly and reduced tubulin synthesis. J Cell Sci 2002;115:3469–3478.

48. Ohta S, Nishio K, Kubota N, et al. Characterization of a taxol-resistant human small-cell lung cancer cell line. Jpn J Cancer Res 1994;85:290–297.

49. Cabral F, Wible L, Brenner S, Brinkley BR. Taxol-requiring mutant of Chinese hamster ovary cells with impaired mitotic spindle assembly. J Cell Biol 1983;97:30–39.

50. Wang Y, Cabral F. Paclitaxel resistance in cells with reduced β-tubulin. Biochim Biophys Acta 2005;1744:245–255.
51. Bhattacharya R, Cabral F. A ubiquitous β-tubulin disrupts microtubule assembly and inhibits cell proliferation. Mol Biol Cell 2004;15:3123–3131.
52. Lopata MA, Cleveland DW. In vivo microtubules are copolymers of available β-tubulin isotypes: localization of each of six vertebrate β-tubulin isotypes using polyclonal antibodies elicited by synthetic peptide antigens. J Cell Biol 1987;105:1707–1720.
53. Fulton C, Simpson PA. In: Cell Motility Goldman R, Pollard T, Rosenbaum J, eds. NY: Cold Spring Harbor Press; 1976:987–1006.
54. Joshi H C, Cleveland DW. Diversity among tubulin subunits: toward what functional end. Cell Motil Cytoskel 1990;16:159–163.
55. Luduena RF. Multiple forms of tubulin: different gene products and covalent modifications. Int Rev Cytol 1998;178:207–275.
56. Burkhart CA, Kavallaris M, Horwitz SB. The role of β-tubulin isotypes in resistance to antimitotic drugs. Biochim Biophys Acta 2001;1471:O1–O9.
57. Haber M, Burkhart CA, Regl DL, Madafiglio J, Norris MD, Horwitz SB. Altered expression of M beta 2, the class II beta-tubulin isotype, in a murine J774.2 cell line with a high level of taxol resistance. J Biol Chem 1995;270:31,269–31,275.
58. Kavallaris M, Kuo DYS, Burkhart CA, et al. Taxol-resistant epithelial ovarian tumors are associated with altered expression of specific beta-tubulin isotypes. J Clin Invest 1997;100:1282–1293.
59. Jaffrezou JP, Dumontet C, Derry WB, et al. Novel mechanism of resistance to paclitaxel (Taxol) in human K562 leukemia cells by combined selection with PSC 833. Oncol Res 1995;7:517–527.
60. Ranganathan S, Dexter DW, Benetatos CA, Chapman AE, Tew KD, Hudes GR. Increase of β_{III}- and β_{IVa}-tubulin in human prostate carcinoma cells as a result of estramustine resistance. Cancer Res 1996;56:2584–2589.
61. Laing N, Dahllof B, Hartley-Asp B, Ranganathan S, Tew KD. Interaction of estramustine with tubulin isotypes. Biochemistry 1997;36:871–878.
62. Hari M, Yang H, Zeng C, Canizales M, Cabral F. Expression of class III β-tubulin reduces microtubule assembly and confers resistance to paclitaxel. Cell Motil Cytoskeleton 2003;56:45–56.
63. Katsetos CD, Herman MM, Mork SJ. Class III beta-tubulin in human development and cancer. Cell Motil Cytoskeleton 2003;55:77–96.
64. Olmsted JB. Non-motor microtubule-associated proteins. Curr Opin Cell Biol 1991;3:52–58.
65. Cassimeris L, Spittle C. Regulation of microtubule-associated proteins. Int Rev Cytol 2001;210:163–226.
66. Gundersen GG, Cook TA. Microtubules and signal transduction. Curr Opin Cell Biol 1999;11:81–94.
67. Drubin DG, Kirschner MW. Tau protein function in living cells. J Cell Biol 1986;103:2739–2746.
68. Barlow SB, Gonzalez-Garay ML, West RR, Olmsted JB, Cabral F. Stable expression of heterologous microtubule associated proteins in Chinese hamster ovary cells: evidence for differing roles of MAPs in microtubule organization. J Cell Biol 1994;126:1017–1029.
69. Pereira A, Doshen J, Tanaka E, Goldstein LS. Genetic analysis of a Drosophila microtubule-associated protein. J Cell Biol 1992;116:377–383.
70. Wang XM, Peloquin JG, Zhai Y, Bulinski JC, Borisy GG. Removal of MAP4 from microtubules in vivo produces no discernible phenotype at the cellular level. J Cell Biol 1996;132:349–358.
71. Nguyen H-L, Charl S, Gruber D, Lue C-M, Chapin SJ, Bulinski JC. Overexpression of full- or partial-length MAP4 stabilizes microtubules and alters cell growth. J Cell Sci 1997;110:281–294.
72. Zhang CC, Yang J-M, Bash-Babula J, et al. DNA damage increases sensitivity to vinca alkaloids and decreases sensitivity to taxanes through p53-dependent repression of microtubule-associated protein 4. Cancer Res 1999;59:3663–3670.
73. Cassimeris L. The oncoprotein 18/stathmin family of microtubule destabilizers. Curr Opin Cell Biol 2002;14:18–24.
74. Walczak CE. Microtubule dynamics and tubulin interacting proteins. Curr Opin Cell Biol 2000;12:52–56.
75. Quarmby L. Cellular Samurai: katanin and the severing of microtubules. J Cell Sci 2000;113:2821–2827.
76. Iancu C, Mistry SJ, Arkin S, Wallenstein S, Atweh GF. Effects of stathmin inhibition on the mitotic spindle. J Cell Sci 2001;114:909–916.
77. Alli E, Bash-Babula J, Yang J-M, Hait WN. Effect of stathmin on the sensitivity to antimicrotubule drugs in human breast cancer. Cancer Res 2002;62:6864–6869.

78. Nishio K, Nakamura T, Koh Y, Kanzawa F, Tamura T, Saijo N. Oncoprotein 18 overexpression increases the sensitivity to vindesine in human lung carcinoma cells. Cancer 2001;91:1494–1499.

79. Martello LA, Verdier-Pinard P, Shen H-J, et al. Elevated levels of microtubule destabilizing factors in a taxol-resistant/dependent A549 cell line with an α-tubulin mutation. Cancer Res 2003;63: 1207–1213.

80. Monzo M, Rosell R, Felip E, et al. Paclitaxel resistance in non-small-cell lung cancer associated sith beta-tubulin gene mutations. J Clin Oncol 1999;17:1786–1793.

81. Kelley MJ, Li S, Harpole DH. Genetic analysis of the β-tubulin gene, TUBB, in non-small-cell lung cancer. J Nat Cancer Inst 2001;93:1886–1888.

82. Achiwa H, Sato S, Shimizu S, et al. Analysis of beta-tubulin gene alteration in human lung cancer cell lines. Cancer Lett 2003;201:211–216.

83. Maeno K, Ito K, Hama Y, et al. Mutation of the class I beta-tubulin gene does not predict response to paclitaxel for breast cancer. Cancer Lett 2003;198:89–97.

84. Urano N, Fujiwara I, Hasegawa S, et al. Absence of beta-tubulin gene mutation in gastric carcinoma. Gastric Cancer 2003;6:108–112.

85. de Castro J, Belda-Iniesta C, Cejas P, et al. New insights in beta-tubulin sequence analysis in non-small cell lung cancer. Lung Cancer 2003;41:41–48.

86. Hasegawa S, Miyoshi Y, Egawa C, et al. Mutational analysis of the Class I beta-tubulin gene in human breast cancer. Int J Cancer 2002;101:46–51.

87. Tsurutani J, Komiya T, Uejima H, et al. Mutational analysis of the beta-tubulin gene in lung cancer. Lung Cancer 2002;35:11–16.

88. Lamendola DE, Duan Z, Penson RT, Oliva E, Seiden MV. Beta tubulin mutations are rare in human ovarian carcinoma. Anticancer Res 2003;23:681–686.

89. Sale S, Sung R, Shen P, et al. Conservation of the class I beta-tubulin gene in human populations and lack of mutations in lung cancers and paclitaxel-resistant ovarian cancers. Mol Cancer Ther 2002;1:215–225.

90. Nogales E, Wolf SG, Downing KH. Structure of the αβ tubulin dimer by electron crystallography. Nature 1998;391:199–203.

91. Nogales E, Whittaker M, Milligan RA, Downing KH. High-resolution model of the microtubule. Cell 1999;96:79–88.

92. Goncalves A, Braguer D, Kamath K, et al. Resistance to Taxol in lung cancer cells associated with increased microtubule dynamics. Proc Natl Acad Sci USA 2001;98:11,737–11,742.

93. Boggs B, Cabral F. Mutations affecting assembly and stability of tubulin: evidence for a non-essential β-tubulin in CHO cells. Mol Cell Biol 1987;7:2700–2707.

94. Lowe J, Li H, Downing KH, Nogales E. Refined structure of αβ-tubulin at 3.5 A resolution. J Mol Biol 2001;313:1045–1057.

95. Loganzo F, Hari M, Annable T, et al. Cells resistant to HTI-286 do not overexpress P-glycoprotein but have reduced drug accumulation and a point mutation in alpha-tubulin. Mol Cancer Ther 2004;3:1319–1327.

96. Poruchynsky MS, Kim JH, Nogales E, et al. Tumor cells resistant to a microtubule-depolymerizing hemiasterlin analogue, HTI-286, have mutations in alpha- or beta-tubulin and increased microtubule stability. Biochemistry 2004;43:13,944–13,954.

97. He L, Yang CH, Horwitz SB. Mutations in β-tubulin map to domains involved in regulation of micro-tubule stability in epothilone-resistant cell lines. Mol Cancer Ther 2001;1:3–10.

98. Verrills NM, Flemming CL, Liu M, et al. Microtubule alterations and mutations induced by desoxye-pothilone B: implications for drug-target interactions. Chem Biol 2003;10:597–607.

99. Kavallaris M, Tait AS, Walsh BJ, et al. Multiple microtubule alterations are associated with Vinca alkaloid resistance in human leukemia cells. Cancer Res 2001;61:5803–5809.

100. Wang Y, Tian G, Cowan NJ, Cabral F. Mutations affecting β-tubulin folding and degradation. J Biol Chem 2006;281:13,628–13,635.

15 Resistance to Microtubule-Targeting Drugs

Paraskevi Giannakakou and James P. Snyder

SUMMARY

As essential components of cell shape, signaling, movement and division, microtubules (MTs) are crucial to normal cellular functions and survival. Beginning with vincristine in the 1950's numerous agents targeting the microtubules have been identified and developed as anti-cancer agents. Their use in the clinic has at times led to complete regression of tumors, but unfortunately in too many cases, regressions have not occurred or have been followed by the recurrence of tumor that is then usually resistant to the agent that appeared initially to be effective. These clinical observations have driven a body of work that has sought to understand why some cells are resistant to begin with while others become resistant after drug treatment. Numerous approaches have been used to identify the molecular changes responsible for this differential sensitivity, and once identified these have been used to understand at a structural level how resistance is conferred. These studies have helped to elucidate mechanisms of drug resistance and also enhance our understanding of how microtubule-targeting drugs interact with tubulin and the microtubules.

Key Words: Microtubules; drug-resistance; taxanes; vinca alkaloids; colchicine; tubulin mutations.

From: *Cancer Drug Discovery and Development: The Role of Microtubules in Cell Biology, Neurobiology, and Oncology* Edited by: Tito Fojo © Humana Press, Totowa, NJ

1. INTRODUCTION

Microtubules (MTs) are major dynamic structural components in cells that are essential for the development and maintenance of cell shape, cell signaling, movement, and division. They are cytoskeletal polymers built by self-association of αβ-tubulin dimers that exist in a state of dynamic equilibrium between the polymer and its soluble dimer form. This constant shortening and elongation is necessary for MTs to function within a cell.

Drugs that bind to either tubulin or MTs form one of the most effective classes of anticancer agents. This class of drugs is large, and continues to expand. The majority of these compounds are natural products, but they have remarkably diverse chemical structures. Tubulin-targeting agents are divided into two main groups according to their effects on microtubule polymer mass. The first group is made up of microtubule-destabilizing drugs that bind preferentially to tubulin dimers and inhibit tubulin assembly. These include the *Vinca* alkaloids, cryptophycins, and colchicine. The second group is made up of microtubule-stabilizing drugs that bind preferentially to the microtubule polymer, enhance tubulin polymerization and include the taxanes (such as Taxol™ and Taxotere), epothilones, eleutherobins, laulimalide, and discodermolide. Although all these drugs bind to distinct sites on the microtubule or the tubulin dimer, they all affect microtubule dynamics, block mitosis at the metaphase/anaphase transition, and consequently induce cell death (for review, *see* ref. *1*). As cancer cells are more dynamic and divide at much higher rates than normal cells, the drugs are most toxic to cancer cells.

Currently, there are five known drug-binding sites on tubulin: four that are well-characterized and a fifth that is not yet thoroughly characterized. They have names assigned depending on which drug was originally found to bind the site. The **taxane**-binding site on β-tubulin (Fig. 1) is shared by drugs that stabilize MTs and bind preferentially at the microtubule polymer *(2–6)*. The prototype of this class of drugs is Taxol™ *(7)*, but newer members include the epothilones *(8)*, discodermolide *(9)*, eleutherobin, and the sarcodictyins *(10)*.

All of the other well-characterized binding sites, with the exception of the fifth binding site, are shared by drugs that bind preferentially to unpolymerized tubulin, inhibiting tubulin assembly. These destabilizing agents either form covalent crosslinks to tubulin cysteine residues, like Cys-β239, such as the small molecules 2,4-dichlorobenzyl thiocyanate *(11)* and T138067 *(12)*; bind tubulin at the **colchicine** site (Fig. 1) such as the combretastatins *(13)*, curacins *(14)*, 2-methoxyestradiol (2ME2) *(15)*, and the podophylotoxins *(16)*; bind tubulin at the *Vinca* domain (Fig. 1), such as maytansin and rhizoxin *(17,18)*; or locate in α-tubulin as do the hemiasterlins *(19)*, which also bind at the *Vinca* domain *(20)*, and perhaps the cryptophycins (for review, *see* ref. *21*). In the case of drugs that bind at the *Vinca* alkaloid or the colchicine sites, it should also be noted that although they show a preference for unpolymerized tubulin, they can also bind MTs to a lesser extent *(1)*.

The fifth binding site, on tubulin, has been recently identified as the location where the microtubule-stabilizing drug laulimalide binds *(22)*. This is the first report of the existence of a second drug-binding site other than the taxane site on the microtubule polymer that promotes microtubule stability. Recently, a microtubule-stabilizing natural product derived from a New Zealand marine sponge, peluroside, was also found to compete with laulimalide for this site *(23)*. Future studies are warranted to identify the exact location of this novel binding site.

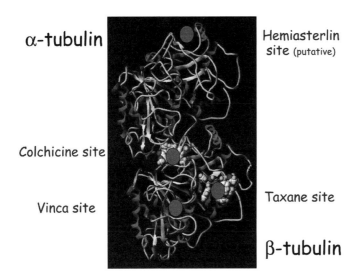

Fig. 1. An overview of some of the ligand-binding sites on the αβ-tubulin dimer. Both microtubule stabilizing (taxane-binding site) and destabilizing (*Vinca*-, colchicine-binding sites), ligands are represented. Hemiasterlin has been shown to bind at the *Vinca*-domain as well as in α-tubulin. Red circles represent approximate locations rather than site identity. To view this figure in color, see the insert and the companion CD-ROM.

2. CLINICAL APPLICATIONS OF MICROTUBULE-TARGETING DRUGS

The use in medicine of natural products containing microtubule-targeting drugs dates back to antiquity. In more recent years, the microtubule-destabilizing compounds have been widely used in the treatment of human disease, even before the protein tubulin was discovered as their target. Today, microtubule-destabilizing drugs are being used as anti-fungal and antihelminthic agents *(24)*, whereas colchicine is also being used for treatment of both familial Mediterranean fever *(25)* and gout *(26)*. However, the main use of microtubule-targeting drugs is in clinical oncology.

The *Vinca* alkaloids vinblastine and vincristine, the first antitubulin agents released for clinical trials in 1961, were approved by 1966 for the treatment of child-hood acute leukemia. However, the mechanism of action of these *Vinca* alkaloids was not elucidated until later. In fact, in a case report published in 1968, where intrathecal administration of high doses of vincristine caused the death of childhood leukemia patient, postmortem staining of the spinal cord revealed the presence of argentophilic strands and rhombohedral crystals *(27)*. At the time, the report authors suggested that the mechanism by which vincristine produced these neuronal changes might lie in a decrease in DNA and RNA synthesis. Since then, the understanding of the mechanism of action of these drugs has deepened significantly, and it is now known that these histological findings were likely owing to the formation of paracrystaline tubulin aggregates, which are often observed when tubulin is exposed to high concentrations of *Vinca* alkaloids. Although the *Vinca* alkaloids had been used successfully for the treatment of both hematological malignancies and solid tumors, interest in the development of new agents that target MTs declined until the introduction of Taxol™ into clinical oncology.

Arguably the most effective anticancer agent introduced since cisplatin, Taxol's remarkable activity in a broad range of malignancies reignited interest in tubulin and MTs as chemotherapeutic targets. Taxol™, the prototype of the microtubule-stabilizing drugs, was originally isolated in 1971 from the bark of the Pacific yew tree, *Taxus brevifolia (28)*, becoming the first microtubule-stabilizing natural product described. Taxol's unique mechanism of action, first reported in 1979 *(7)*, generated intense interest in its clinical development. In the ensuing years, Taxol™ was moved into the clinic, where its oncological response rates have been continually improved. So far, the taxanes Taxol™ and Taxotere have been approved by the Food and Drug Administration (FDA) for the treatment of ovarian, breast, prostate, head and neck, and nonsmall-cell lung cancer (NSCLC). The authors expect that these clinical indications will continue to expand *(29,30)*.

3. DRUG RESISTANCE

The clinical success of Taxol™, together with the fact that its antitumor spectrum appears to be the broadest of any class of anticancer agents, make MTs arguably the single best anticancer drug target identified to date. However, the emergence of drug resistant tumor cells has limited Taxol's ability to cure disease. Anticancer drug resistance is best defined as a state of insensitivity or decreased sensitivity of cancer cells to drugs that would ordinarily cause cell death. This resistance can be either intrinsic or acquired. Intrinsic resistance is defined as a state of insensitivity to initial therapy in response to a drug or combination of drugs. On the other hand, acquired drug resistance is defined as a state whereby a population of cancer cells that were initially sensitive to a drug undergoes a change toward insensitivity. Acquired drug resistance is the most common reason for the failure of drug treatment in cancer patients with initially sensitive tumors, and as such, is presently responsible for the majority of deaths from cancer. The gravity of the drug resistance problem in clinical oncology has led to intense scientific efforts to understand the molecular mechanisms that lead to drug resistance, as well as to identify the ways to overcome it.

In the case of Taxol™, several mechanisms of resistance have been described. With the exception of P-glycoprotein (Pgp)-mediated multidrug resistance (MDR) *(31,32)* resulting in decreased intracellular drug accumulation, all other mechanisms appear to involve alterations in tubulin. Such alterations in Taxol™-resistant cancer cells include: (1) differential expression levels of β-tubulin isotypes *(33–35)*, (2) increase in microtubule dynamics *(36)*, decreased ability of Taxol™ to suppress microtubule dynamics in cells overexpressing βIII-tubulin *(37)*; and most importantly, (3) the presence of tubulin mutations that impair the drug's ability to interact with tubulin *(38–41)*.

Research on tubulin mutations is not new. The tubulin protein, first isolated in 1968 as "the colchicine-binding protein" *(42)*, has since been the subject of constant discoveries not only regarding its wide array of cellular functions, but also the variety of effects that different microtubule-targeting drugs have on it. The original discoveries of relationships between tubulin mutations and drug resistance actually began in the late 1970s. In fact, the first study reporting that tubulin mutations can confer resistance to a microtubule-targeting drug, benomyl, found this resistance to be because of a decrease in binding affinity in cells from the fungal strain *Aspergillus nidulans (43)*. Following this initial discovery, two later studies described a reduction in the binding affinity of radiolabeled colcemid to tubulin *(44)*, as well as an altered electrophoretic mobility of β-tubulin obtained from Chinese hamster ovary cells resistant to either colcemid or colchicine, respectively *(45)*. However, no specific tubulin mutations were identified in these drug resistant cells.

In the ensuing years, discoveries of drug resistance to antimitotic agents owing to alterations or mutations in tubulin continued to amass in studies involving lower organisms such as the yeast *Schizosaccharomyces pombe (46)* and the myxoamoebae *Physarum* regarding resistance to benzimidazol *(47)*, as well as in organisms resistant to benomyl such as *Neurospora crassa (48)* and *Saccharomyces cerevisiae (49)*. However, it was not until 1997 when acquired tubulin mutations conferring Taxol™ resistance in human cancer cells were described for the first time *(38)*.

Since then, several other groups have described acquired tubulin mutations in human cancer cell lines, which confer resistance to either microtubule-stabilizing or microtubule-destabilizing drugs, making tubulin mutations one of the most frequently encountered mechanisms of antitubulin-drug resistance. In the end, what this means is that no matter where an organism is on the phylogenetic tree, it may be capable of acquiring mutations in its tubulin genes as a mechanism that allows these organisms to survive in otherwise toxic circumstances, such as in the presence of the powerful tubulin-targeting drugs. Thus, the acquisition of mutations that alter cellular MTs' response to drug binding appears to be a simple but universal mechanism for an organism's survival in the presence of these drugs.

In this chapter, the focus will be on the discoveries involving drug resistance through acquired tubulin mutations in human cancer cells, as well as the role of these mutations in the cell's sensitivity to other microtubule-targeting drugs belonging to both similar and distinct drug-binding classes. The authors believe that this mechanism of drug resistance occurs very frequently and almost invariably when intracellular drug accumulation is not impaired, as is evidenced by the numerous ongoing reports of acquired tubulin mutations in response to *in vitro* selection with different microtubule-targeting drugs. In this chapter the attempt will be to thoroughly describe all separate tubulin mutations that can confer drug resistance, as identified to date in human cancer cell lines, according to each drug's binding site on tubulin.

4. TAXANE-BINDING SITE TUBULIN MUTATIONS

This section will cover tubulin mutations identified in human cancer cell lines selected with either Taxol™ or the epothilones.

4.1. Mutations in Response to Selection With Taxol™

The taxanes, **Taxol™** and its semisynthetic analog **Taxotere** (Fig. 2), were among the most important new additions to the cancer chemotherapeutic arsenal after 1990. The taxanes bind preferentially and with high affinity to the β-subunit of the tubulin dimer, along the entire length of the microtubule. Electron crystallographic studies on Taxol™ complexed with tubulin have allowed the precise identification of the amino acids, which include the taxane-binding pocket on β-tubulin *(6)* and revealed the location of the site at the inside lumen of the microtubule *(50)*. At high drug concentrations there is one molecule of Taxol™ bound on every molecule of tubulin dimer in a microtubule, which results in increased tubulin polymerization both in vitro and in cells at approximately a 1:1 stoichiometry. However, even at lower concentrations of Taxol™

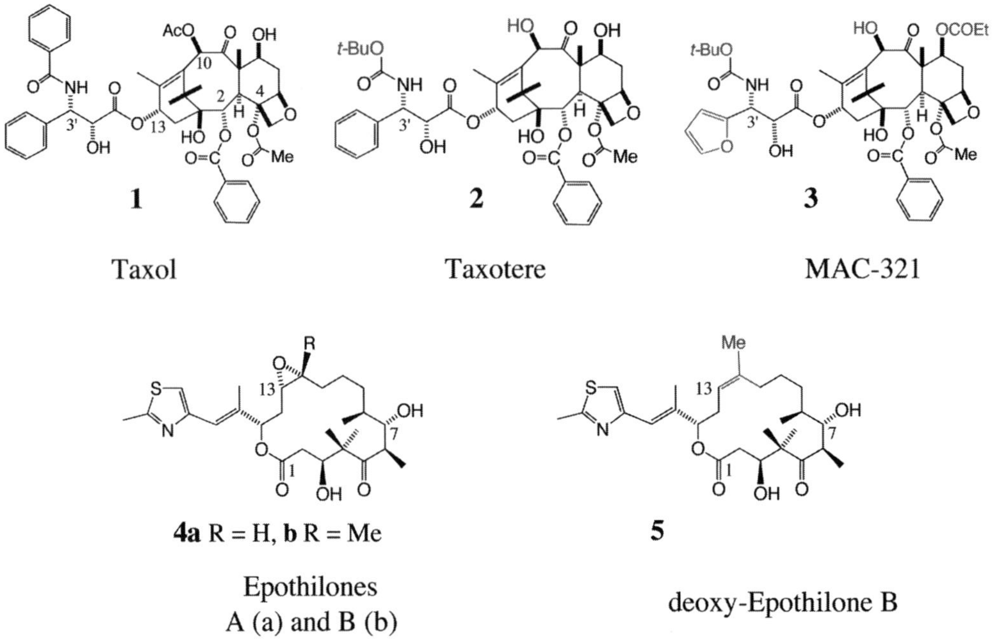

Fig. 2. Structures of taxane-binding site microtubule stabilizing drugs. To view this figure in color, see the insert and the companion CD-ROM.

where one molecule of Taxol™ is bound to every few hundred tubulin molecules along the microtubule, microtubule function can still be affected by a reduction in microtubule dynamics, which ultimately leads to microtubule stabilization, mitotic arrest, and cell death (for review, *see* ref. *51*).

Research on the mechanisms involved in taxane resistance has revealed that sequence alterations of single amino acids at the taxane-binding site can have a significant impact on the drug's ability to bind tubulin. As mentioned earlier, this work has provided the first identification of acquired β-tubulin mutations in Taxol™-resistant human cancer cell lines *(38)*. Two distinct β-tubulin mutations in two human ovarian cancer clones were identified, namely clones 1A9/PTX10 and 1A9/PTX22, selected independently with Taxol™ from the parental 1A9 human ovarian cancer cells. Sequence analysis of the predominantly expressed β-tubulin isotype (class I/gene HM40) revealed that both mutations were single nucleotide substitutions at residues **βPhe270Val** in 1A9/PTX10 cells, and **βAla364Thr** in 1A9/PTX22 cells *(38)* (*see* Figure 3 and Table 1). As a result, impaired Taxol™-induced tubulin polymerization was observed in the two clones, which were found to exhibit a 30-fold resistance to Taxol™. To exclude other potential cellular alterations that could contribute to the observed Taxol™ resistance the mutant tubulins from each clone were purified and in vitro tubulin polymerization experiments with Taxol™ were performed. The purified tubulins from clones 1A9/PTX10 and 1A9/PTX22 failed to polymerize in vitro in the presence of Taxol™, in contrast to the 1A9 wild-type (wt) parental cell tubulin, which was readily polymerized by Taxol™ under the same conditions. These results provided direct evidence that these specific acquired mutations were indeed responsible for the Taxol™-resistance, owing to a compromised drug-binding site on tubulin.

However, these two β-tubulin mutant clones exhibited very low cross-resistance to epothilone B (EpoB) (about two- to threefold), even though epothilone competes with Taxol™ for the same binding site on β-tubulin. Consistent with this finding, EpoB was still able to induce polymerization of these two isolated mutant tubulins in vitro. Thus, it appears that although the taxane-binding site is the same for the two drugs, the specific amino acids necessary for binding of each drug to this pocket are distinct as determined by a comparison of epothilone A (EpoA) and Taxol™ bound to β-tubulin *(52)* (Fig. 11).

Subsequent reports have described other acquired β-tubulin mutations in human breast cancer MDA-MB-231 cells *(41)* and human epidermoid cancer KB-31 cells *(40)*, following Taxol™ selections (Table 1). In the first case, a mutation was described in the class I β-tubulin gene at residue **βGlu198Gly** located near the α/β interphase *(41)*. These β-tubulin mutant MDA-MB-231 breast cancer cells displayed 17-fold resistance to Taxol™, some cross-resistance to Taxotere and the epothilones and did not express Pgp. Although the authors of this report did not isolate the mutant tubulin from their resistant cells to perform in vitro binding assays, they did, however, introduce exogenous wt class I β-tubulin into these resistant cells, thereby demonstrating partial reversal of the resistant phenotype, suggesting that this mutant residue was indeed responsible for the taxane-resistant phenotype seen in these cells. In a second study, Hari et al. *(40)* reported the identification of a novel point mutation **βAsp26Glu** in class I β-tubulin. This mutation is at the N-terminus of β-tubulin, which forms part of the taxane-binding pocket *(6)*, and confers an 18-fold resistance to Taxol™ with minimal cross-resistance to EpoB and the Taxol™ analog milataxel (MAC-321), yet 10-fold cross-resistance to Taxotere. Interestingly, these β-tubulin mutant KB3-1 cells are also drug-dependent, as they require the presence of low concentrations of Taxol™ in the tissue culture medium for optimal growth. This finding would be consistent with the presence of less stable MTs as has been reported in other studies of drug-resistant cell lines displaying a combination of Taxol™-resistant and Taxol™-dependent phenotypes *(36,53)*.

Along the same lines, two related studies have described the characteristics of yet another Taxol™ selected human cancer cell line, the NSCLC A549 cells *(36,53)*. In these studies two clones derived from the A549 parental cells were isolated and characterized, namely A549-T12 and A549-T24. The A549-T12 cells were found to be nine-fold resistant to Taxol™ and did not express P-gp, whereas the A549-T24 cells were 17-fold resistant to Taxol™ and expressed low levels of P-gp. Sequence analysis revealed no alterations in the class I β-tubulin sequence; however, an α-tubulin mutation **αSer379Arg** was detected in the kα1-isotype of α-tubulin *(53)* (Table 1). Interestingly, the ability of Taxol™ to induce tubulin polymerization in these cells has not been impaired, suggesting that the presence of this mutation in α-tubulin is not likely to affect the binding of Taxol™ to its target site. The A549-Taxol™ resistant cells are also drug-dependent and were found to have increased MTs dynamics *(36)* as well as alterations in the microtubule-associated proteins (MAPs) MAP4 and stathmin. As a result, it has been postulated that α-tubulin, and hence the identified **αSer379Arg** mutation, may play a role in the binding of these regulatory proteins to MTs, thus ultimately regulating microtubule dynamics. However, there is no evidence that the presence of this mutation plays a role in the observed Taxane-resistance, as no in vitro data using this particular purified mutant have been gathered. Similarly, the mutation has not been reintroduced into cells harboring wt tubulin to observe what effects it alone

Table 1
α- and β-Tubulin Mutations in Drug-Resistant Human Cell Lines

Line and selecting agent	Tubulin mutation	Fold resistance to selecting agent[a]	Fold cross-resistance
Taxol[b]			
1A9PTX10 (ovarian carcinoma)	βPhe270Val	24 (38)	Docetaxel 4.2 (98) MAC321 3.4 (98) EpoA 8.9 (98) EpoB 2.8 (38) TacA 2.3 (99) TacE 4.8 (99) Vinblastine 0.5 (38) Laulimalide 1.5 (22) Noscapine 1 (100)
1A9PTX22 (ovarian carcinoma)	βAla364Thr	24 (38)	Docetaxel 4.4 (98) MAC321 5 (98) EpoA 2.3 (98) EpoB 1.4 (38) TacA 2.1 (99) TacE 12 (99) Vinblastine 0.4 (38) Laulimalide 1.6 (22) Noscapine 1 (100)
MDA.MB231/K20T (breast cancer)	βGlu198Gly	19 (41)	Vinblastine 1.0 (41)
KB 3-1/KB-15-PTX/099 (epidermoid carcinoma)	βAsp26Glu	18 (40)	Taxotere 10 (40)
A549-T24 (NSCLC)	αSer379 Ser/Arg	17 (53)	Vinblastine 1.4 (53) Colchicine 1.3 (53)
Epothilones[b]			
1A9/A8 (ovarian carcinoma)	βThr274Ile	Epo A 40 (57) 55 (22)	Taxol™ 10 (57) 4.1 (98) 8.0 (99) 7.6 (22) Docetaxel 7 (57) 1.2 (98) MAC321 1.6 (1) EpoB 25 (57) 38 (22) TacA 2.9 TacE 4.2 Laulimalide 2.4 (3)
1A9/B10 (ovarian carcinoma)	βArg282Gln	Epo B 24 (57)	Taxol™ 6.5 (57) 9.4 (22) Docetaxel 5 (6) EpoA 57 (57) 74 (22) Laulimalide 3.8

(Continued)

Table 1 *(Continued)*

Line and selecting agent	Tubulin mutation	Fold resistance to selecting agent[a]	Fold cross-resistance
A549.epoB40 (NSCLC)	βGln292Glu	EpoB 95 *(59)*	Taxol™ 22 *(59)* 21 *(98)* Taxotere 13 *(9)* 17 *(1)* MAC321 10.7 *(1)* EpoA 72 Vinblastine 0.5 Colchicine 0.6
KB-C5/0 (epidermoid carcinoma)	βThr274IPro	Epo A 45 *(58)*	Epo B 8 Taxol™ 98 Discodermolide 1.3 Vinblastine 0.7 Demecolcine 0.7 Doxorubicine 1.6 5-FU 1.2 Paraplatine 1.2 *(58)*
KB-D4/40 (epidermoid carcinoma)	βThr274IPro	Epo A 71 *(58)*	Epo B 9 Taxol™ 200 Discodermolide 0.5 Vinblastine 0.2 Demecolcine 0.5 Doxorubicine 1.2 5-FU 0.9 Paraplatine 3.8 *(58)*
dEpoB30 (CCRF-CEM leukemia cells)	βAla231Ala/Thr	dEpoB 21 *(60)*	Taxol™ 16 EpoB No Test Vinblastine 0.28 Colchicine 0.78
dEpoB60 (CCRF-CEM leukemia cells)	βAla231Ala/Thr	dEpoB 60 *(60)*	Taxol™ 26 EpoB No Test Vinblastine 0.58 Colchicine 0.88
dEpoB140 (CCRF-CEM leukemia cells)	βAla231Ala/Thr	dEpoB 173 *(60)*	Taxol™ 7 EpoB 12 Vinblastine 0.79 Colchicine 0.88
dEpoB300 (CCRF-CEM leukemia cells) Indanocine[b]	βAla231Ala/Thr βGln292Gln/Glu	dEpoB 307 *(60)*	Taxol™ 467 EpoB 77 Vinblastine 0.25 Colchicine 0.85
Leukemia CEM-178	βLys350Asn	Indanocine 115 *(69)*	Taxol™ 1 Vinblastine 40 Colchicine 31 Fludarabine 2.3 Doxorubicin 1.9 Cytochalasin B 0.6 *(69)*

(Continued)

Table 1 *(Continued)*

Line and selecting agent	Tubulin mutation	Fold resistance to selecting agent[a]	Fold cross-resistance
Vincristine[b]			
CEM/VCR-R (CCRF-CEM leukemia cells)	βLeu240Leu/Ile	Vincristine 22,600 *(76)*	
Hemiasterlin[b]			
1A9/HTIβS172A (ovarian carcinoma)	βSer172Ser/Ala	HTI-286 52 *(77)*	Hemiasterlin 9 Dolastatin-10…10 Cryptophysin 4 Vincristine 20 Vinorelbine 78 Taxol™ 0.12 Epo A 0.13 Doxorubicin 0.6 *(77)*
1A9/HTIαI384V (ovarian carcinoma)	αIle384Ile/Val	HTI-286 65 *(77)*	Hemiasterlin 7 Dolastatin-10…7 Cryptophysin 4 Vincristine 27 Vinorelbine 122 Taxol™ 0.15 Epo A 0.13 Doxorubicin 0.17 *(77)*
1A9/HTIαS165P; R221H (ovarian carcinoma)	αSer165Pro αArg221His	HTI-286 89 *(77)*	Hemiasterlin 6 Dolastatin-10…10 Cryptophysin 7 Vincristine 13 Vinorelbine 186 Taxol™ 0.32 Epo A 0.24 Doxorubicin 0.74 *(77)*

[a]Fold resistance is calculated as the ration of the drug's IC_{50} ($IC50$ is the drug concentration required to inhibit the cell growth of 50% of cells) against the resistant cell lines over the drug's IC_{50} against the respective parental cell line.

[b]Selecting agent.

would have on Taxol™ resistance. Additional studies probing the exact function of this mutation are required before conclusions can be made on the exact nature of the Taxol™-resistant phenotype obtained in these cells.

5. STRUCTURAL ANALYSIS OF MUTATIONS IN RESPONSE TO SELECTION WITH TAXOL™

To increase the understanding of the role of the reported tubulin gene mutations in resistance to Taxol™ and other microtubule-targeting drugs that bind at or near the tax-ane site, the authors have performed detailed structural analyses using the protein-ligand complexes of Taxol™ (PTX) *(54,55)* as well as EpoA *(52)* from the computationally refined electron crystallographic structures. In this section, the authors will describe the

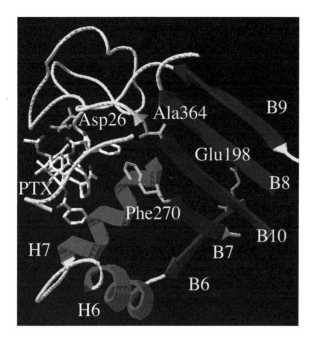

Fig. 3. The Asp26, Phe270, Ala364 (green), and Glu198 (gold) residues are mutated following drug selection with Taxol™. The first three surround the ligand Taxol™ at short distances, whereas Glu198 is 17–18 Å distant from the taxane-binding center. Taxol™ is overlayed by a portion of the M-loop (yellow), whereas helix H-7 is at the back of the hydrophobic-binding pocket. To view this figure in color, see the insert and the companion CD-ROM.

results obtained for Taxol™ and the other microtubule-stabilizing drugs that bind to the taxane site. The structure of each of these drugs is given in Fig. 2.

Four mutations in β-tubulin, exhibiting a range of 18- to 55-fold resistance, arose in response to selection with Taxol™ (PTX or paclitaxel). Three of these were found to be clustered in the taxane-binding site, in direct contact with bound drug (Fig. 3). The first two mutations (**βPhe270Val** and **βAla364Thr**) were observed in 1A9 ovarian cancer cells *(38)* and are closely associated with Taxol™ in the microtubule protein. Phe270 is in van der Waals contact with Taxol's™ C-3′ phenyl group, whereas Ala364 resides in a five-residue hydrophobic cluster at the bottom of the tubulin-binding pocket housing the ligand. In Fig. 4, the side-chains immediately associated with Taxol™ (Phe270, Pro272, Pro358, and Leu 361) are depicted in gold, whereas Ala364 is shown in magenta. The mutation of β364 from a nonpolar alanine to a polar threonine can be expected to cause reorganization of this cluster, with consequences for binding affinity with the ligand.

The **βAsp26Glu** change in KB-3-1 epidermal cells observed in response to Taxol™-selection *(40)* is similarly accompanied by direct contact between protein and Taxol™. In particular the CH_2 of the Asp side chain is at the van der Waals boundary with respect to two CH centers of the phenyl ring of Taxol™'s C-3′ benzamido group. The same methylene abuts one methyl group of the *t*-butyl group of taxotere (Fig. 2) docked in the same site (Fig. 5). The steric resistance between drugs and protein side chain permits a hydrogen bond between the Asp26 carboxylate and taxotere's NH, but only a longer range electrostatic interaction for Taxol™ *(40)*. Replacement of Asp26 with Glu causes 18-fold resistance to Taxol™, but also results in a decrease in microtubule stability,

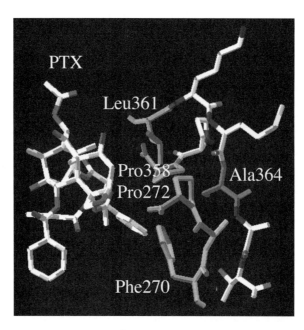

Fig. 4. The Ala364 cluster representing the βAla364Thr mutation seen in the ovarian carcinoma 1A9 cells. Although Ala364 (magenta) is 6–7 Å distant from the bound Taxol™ ligand (at left, CPK colors), it resides in the bottom of a local pocket consisting of four hydrophobic residues (gold) that form part of the hydrophobic basin in which the entire Taxol™ molecule resides. To view this figure in color, see the insert and the companion CD-ROM.

likely responsible for the drug-dependent nature of these cells. Importantly, this mutation creates only three- to fivefold resistance to taxotere, as well as to a furan-containing analog, MAC-321 (Fig. 2). One of the outcomes of extending the Asp chain by an extra methylene unit (CH_2) in Glu is to bring the negatively charged carboxylate functionality (i.e., CO_2^-) in closer contact with the NHCO centers of the ligands. Taxotere and MAC-321 persist in a productive hydrogen bond with the lengthened Glu side-chain, but severe steric interactions of the same Glu conformation with Taxol™ appear to force this drug up and out of the binding pocket (yellow, Fig. 5). As a result, it has been proposed that in the case of Taxol™ an alternative Glu side-chain conformation is adopted; one that does not contribute to ligand binding. The relative resistance to the drug is thereby rationalized *(40)*.

Although the **βGlu198Gly** mutation observed by Wiesen and colleagues *(41)* in MDA-MB 231/K20T cells is 17–18 Å distant from the bound ligand, it causes 19-fold resistance to Taxol™. In this case, interactions within the protein-binding site are complex. Glu198 resides on the β-sheet strand B4, making hydrogen bonds with residues on B5 and the short helix H8. The cluster of nonbonded contacts resides on one side of the longer helix H7, whereas Taxol™ is found on the opposite side (Glu198 green; Fig. 6). Examination of contact residues between the β-strands and helix H7 reveals no obvious electrostatic or hydrogen bonding network between them. Thus, although mutation of Glu198 to Gly can be expected to destabilize the structure around this center, the mode of transmission of this effect across H7 to influence Taxol™ binding is not obvious. A more detailed examination of the dynamics of secondary structure in

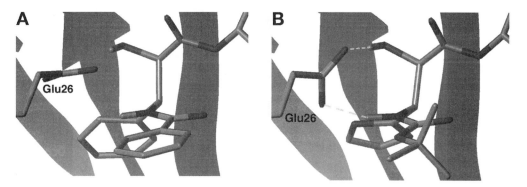

Fig. 5. Binding models of Taxol™ and taxotere in the taxane-binding pocket of the βAsp26Glu mutant seen in the epidermoid carcinoma KB-31 cells. **(A)** The partial view illustrates the CH_2 of Asp in van der Waals contact with the phenyl ring of Taxol™ (magenta) and the *t*-Bu group of taxotere (teal). The CO_2 of Asp makes a hydrogen bond to the NH of taxotere (and MAC-321; not shown) but only electrostatic interaction with Taxol™; **(B)** Mutation to Glu results in the persistence of the H-bond to taxotere (not shown), but severe steric contact with Taxol™ (original placement, magenta) causing the drug to be pushed back and out of the binding site (yellow). To view this figure in color, see the insert and the companion CD-ROM.

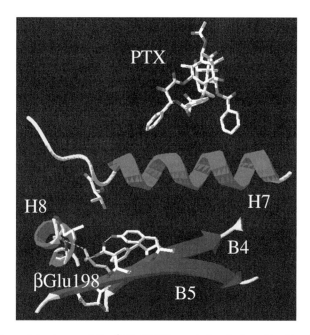

Fig. 6. Representation of the locus of the βGlu198Gly mutation induced by Taxol™-selection in the MDA-MB-231 breast cancer cells. Glu198 (green) on strand B4 makes hydrogen bonds (dashed yellow) with residues on helix H8 and strand B5. However, the glutamate residue is separated from PTX (at top) by helix H7. To view this figure in color, see the insert and the companion CD-ROM.

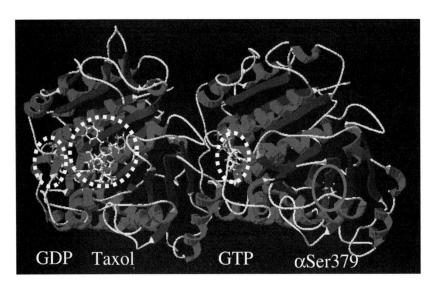

Fig. 7. The αβ-tubulin dimer illustrating the spatial relationships of GDP, GTP, Taxol™ (dotted circles), and αSer379 (magenta solid circle) in the context of the Taxol™-induced αSer379Arg mutation in the A549 lung cancer cells. αSer379 and Taxol™ reside approx 50 Å apart. To view this figure in color, see the insert and the companion CD-ROM.

this region of the protein would be required to pinpoint the exact cause of the resistance produced by this mutation.

An interesting **αSer379Arg** mutation was reported to appear in the α-tubulin of A549-T12 lung cells on exposure to Taxol™, yet no mutation was found in β-tubulin *(53)*. These mutant cells are dependent on Taxol™ for growth and the corresponding MTs display increased dynamicity when compared with the parental cells. Figure 7 illustrates that this substitution appears in strand B10, far from either the intra- or interdimer interface of the tubulin subunits, as well as from the GDP and GTP centers. However, the altered amino acid is near a site of interaction with both a MAP4 and stathmin. Martello and coworkers *(53)* proposed that the increased microtubule instability is related to the **αSer379** mutation, and that the cell dependence on Taxol™ could be the result of compensatory stability provided by the drug. An alternative explanation is that Taxol™ engages in low affinity binding in α-tubulin near Ser379. This possibility is suggested by a labeling study using the ureido taxane derivative [^{3}H]-TaxAPU that photolabels residues 281–304 on helix H7 and the loop between B7 and H7 in the α-subunit of the dimer. As shown by the spatial representation in Fig. 8, the **αSer379** residue (in orange) resides on strand B10, separated from the labeled peptide by strand B7. The shortest separations between α379 and the residues forming the loop are 6.5–8 Å. However, were Taxol™ to bind in this region, it could readily bridge the labeled and mutated centers. Future studies are warranted to resolve this question.

5.1. Mutations in Response to Selection With Epothilones

Epothilones A and B are soil bacteria-derived microtubule-targeting drugs that were first identified in 1995 as part of a screening program for compounds with microtubule stabilizing activity *(8)*. Although epothilones are structurally distinct from Taxol™, they compete

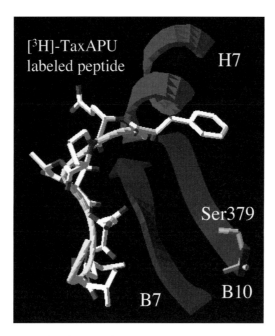

Fig. 8. The αSer379Arg mutation is located on strand B10 in α-tubulin. The Ser379 residue (orange) is located near the loop (yellow) linking B7 and H7. A peptide spanning the loop and H7 (α-281–304) is labeled by a radioactive ureido taxane derivative ([^{3}H]-TaxAPU) that also labels partial peptide sequences in β-tubulin (β-217–229). To view this figure in color, see the insert and the companion CD-ROM.

for the same binding site and exert similar microtubule-stabilizing activity. However, the epothilones maintain activity against Pgp-expressing multidrug-resistant cells as well as Taxol™-resistant β-tubulin mutant cells *(56)*. These characteristics, along with the epothilones' increased water solubility and ease of production in large quantities through bacterial fermentation, led to epothilones being moved quickly into the clinic. Currently, several independent clinical trials are being carried out with EpoB and its analogs in parallel.

Investigations on potential mechanisms of epothilone resistance began soon after the discovery of epothilones. To date, the only solid mechanism of epothilone resistance that has been described occurs through acquisition of β-tubulin mutations. The authors have isolated two epothilone-resistant human ovarian cancer cell lines, namely 1A9/A8 and 1A9/B10, selected with EpoA and B, respectively *(57)*. These epothilone-resistant sublines exhibit impaired epothilone-and Taxol™-driven tubulin polymerization, caused by individual acquired β-tubulin mutations in each clone: **βThr274Ile** in 1A9/A8 cells and **βArg282Gln** in 1A9/B10 cells *(57)*. Interestingly, both these mutations are located near the Taxol™-binding site in the atomic model of αβ-tubulin, explaining why these cells are also cross-resistant to Taxol™ (7- to 10-fold) (Table 1). By using molecular modeling guided by the mutation data and the activity-profile of several MT-stabilizing drugs against these cell lines with mutant tubulins, the authors were able to propose a common pharmacophore that is shared by both the epothilones and the taxanes; and in turn model epothilone binding onto the atomic model of tubulin *(57)*. However, recent electron crystallographic studies of EpoA complexed with tubulin *(52)* suggest that although the two compounds occupy common space, they do not

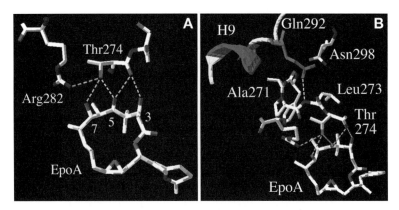

Fig. 9. Environment around the C-3 to C-7 oxygen substituents of EpoA bound to β-tubulin. **(A)** Residues βThr274 and βArg282 provide a hydrogen bond network (dashed yellow) that anchors this sector of the ligand to the protein. Thr274Ile and Arg282Gln tubulin mutations found in the 1A9 human ovarian cancer cells selected with EpoA and EpoB, respectively, disrupt the network. **(B)** The amide terminus of βGln292 (magenta) resides in a polar pocket and H-bonds to the backbone NH of Leu273 that is linked to the crucial Thr274 anchor. The βGln292Glu mutation found in the A549 lung cancer cells, is thought to perturb this anchoring. To view this figure in color, see the insert and the companion CD-ROM.

present a common tubulin pharmacophore (Fig. 9). The following section discusses this point in more detail in the context of cross-resistance.

Interestingly, an alteration in the same residue, **βThr274Pro,** has been identified in another cell line, the EpoA-selected human epidermoid carcinoma KB-31 cells *(58)*. In this case, the amino acid substitution was from Thr to Pro, instead of Thr to Ile as in 1A9 cells. The **βThr274Pro** mutation conferred 45-fold resistance to EpoA, eightfold resistance to EpoB and significant cross-resistance to Taxol™ (96-fold) (Table 1). In addition, this mutation was associated with drug dependence for optimal growth. The fact that two different human cancer cell lines, selected by two different research groups, acquire mutations at the same residue in response to epothilone selection suggests that this specific residue is very important for epothilone binding to tubulin and that it may prove to be a "hot spot" for acquired tubulin mutations following epothilone treatment.

Shortly after, two additional studies reported the presence of acquired β-tubulin mutations in the NSCLC A549 and HeLa cells *(59)*, as well as in the human leukemia CCRF-CEM cells *(60)* (Table 1). In the first study, a **βGln292Glu** mutation has been identified in the A549 cells conferring 70- to 95-fold resistance to the epothilones and significant cross-resistance to Taxol™ (22-fold) and Taxotere (13-fold). Although two mutations have been identified in epothilone-selected HeLa cells, at **βPro173Ala** and **βTyr422Tyr/Cys,** they did not seem to confer significant resistance to the epothilones (1.5- to 2.5-fold) or the taxanes (three- to sixfold), suggesting that these particular residues are not critical for drug binding *(59)*. In the second study, the CCRF-CEM human leukemia cells were selected with the potent EpoB analog, desoxyepothilone B (dEpoB) *(60)*. After selecting with increasing concentrations of dEpoB at 30, 60, 140, and 300 n*M*, four distinct clones were isolated: dEpoB30, dEpoB60, dEpoB140, and dEpoB300, respectively. The four clones had acquired the same β-tubulin mutation at **βAla231Ala/Thr,** whereas the dEpoB140 and dEpoB300 clones had acquired an additional mutation at **βGln292Gln/Glu** *(60)*. The presence of the **βAla231Ala/Thr** mutation was associated with 21- and 60-fold resistance to dEpoB for clones dEpoB30 and dEpoB60 respectively, and cross-resistance to Taxol™ (16- and 26-fold, respectively). The presence of the additional mutation at **βGln292Gln/Glu** in clones dEpoB140 and 300, was associated with higher degrees of resistance to the selecting agent dEpoB. All clones exhibited impaired drug-induced tubulin polymerization, suggesting that these mutations do play a role in drug binding to tubulin.

6. STRUCTURAL ANALYSIS OF MUTATIONS IN RESPONSE TO SELECTION WITH EPOTHILONES

In 1A9/A8 ovarian cancer cells selected with EpoA, a **βThr274Ile** mutation occurred, thereby conferring 40- to 55-fold resistance to this antimitotic agent. EpoB exhibited corresponding cross-resistance of 25- to 38-fold. In a previous analysis of the basis for cross-reactivity and resistance to microtubule-stabilizing drugs, a β-tubulin-binding model was created based on a single Taxol™-Epo pharmacophore. In this model, it was postulated that the C-7 OH of the epothilones forms a hydrogen bond in the vicinity of Thr274 *(57)*. The conversion of this threonine to isoleucine was proposed to disrupt the H-bond, thus causing drug resistance through decreased binding affinity. A recent electron crystallography model of the binding of EpoA to β-tubulin confirms the premise in this initial model, but indicates that the situation is somewhat more complex *(52)*. Figure 9 illustrates that Thr274 and Arg282 of tubulin are jointly engaged in a network of hydrogen bonds with the oxygens at C-3, C-5, and C-7 on the EpoA ligand. The mutation replacing Thr with Ile on residue **β274** not only eliminates the βThr274-mediated tubulin–Epo interaction, but obviates the β274Thr–β282Arg interaction, whereas impacting on the β282Arg-mediated tubulin–Epo contacts as well. The close association of Thr274 and Arg282 implies mutuality. That is, mutation of Arg282 would be predicted to disengage Thr274 in a reciprocal manner. This viewpoint is fully realized in the observation that in EpoB-selected 1A9/B10 ovarian cancer cells, a mutation of **βArg282Gln** confers 24- to 53-fold resistance to that ligand, and 57- to 74-fold cross-resistance to EpoA. This amino acid change not only shrinks the side chain by two heavy atoms (i.e., by $-CH_2-CH_2-$), eliminating a direct hydrogen bond, but also removes the positive charge, and damps any longer-range electrostatic stabilization as well.

Apart from the Arg282Gln mutation mentioned earlier, the one other significant acquired EpoB alteration was observed in A549 lung cell tubulin at **βGln292Glu**. As illustrated by Fig. 9, Gln292 (magenta) resides on helix H9 at a distance of 7–9 Å from the nearest bound epothilone atom. However, the terminal amide of this Gln residue hydrogen bonds to the Leu273, which is adjacent to Thr274, the amino acid residue in direct contact with the ligand. Substitution of a negatively charged Glu for the Gln can be expected to strengthen the first H-bond to Leu273; however, there are most likely other consequences as well. The terminus of side chain 292 resides in a highly polar pocket, which includes carbonyl groups from Phe270, Ala271, and Asn298. The electrostatic repulsion occasioned by the negative carboxylate can be expected to reorganize this pocket and translate its effects outward to Thr274. A reasonable interpretation of the observed 95-fold EpoB resistance is that the network of hydrogen bonds associating with C-3 to C-7 of the epothilone (Fig. 10) is perturbed sufficiently, so that ligand binding is considerably weakened. Theoretically, this mutation should also influence

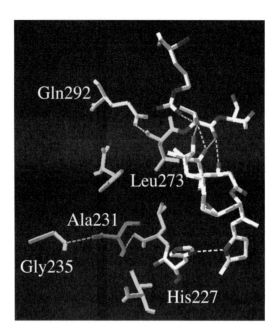

Fig. 10. At lower left, Ala231 (magenta) surrounded by a hydrophobic pocket (green) is subjected to the Ala231Thr mutation, which presumably perturbs the His227 interaction with dEpoB (represented by EpoA at their binding pocket). Above, the second dEpoB-induced mutations in βGln292 (yellow) that it is H-bonded to Leu273 (green) (*see* Fig. 9). This βGln292Glu mutation may potentially influence the binding of the drug to Thr274 in the binding pocket. To view this figure in color, see the insert and the companion CD-ROM.

Taxol™ binding, as this ligand similarly interacts with Thr274. Indeed, it was found that the EpoB mutant's Taxol™ cross-resistance is measured at 20- to 22-fold (Table 1).

An important dEpoB-induced mutation that is cross-resistant to EpoB has been reported in B140 leukemia cells when exposed to 30–140 n*M* concentrations of the drug: **βAla231Thr** *(60)*. This alanine is not in direct contact with the ligand, but lies on helix H7, deep in the β-tubulin binding pocket at a distance of 7–8 Å from the ligand and surrounded largely by hydrophobic residues (i.e., Val23, Leu228, Gly235, Phe270, and Leu273). Given that helix H7 is regarded as being central to the conformation of tubulin *(61)*, and that the Ala231Thr mutant retains the ability to bind Taxol™, Verrills and coworkers strongly favored decreased microtubule stability over reduced drug binding as the cause of this selective resistance. However, H7 is similarly one of the structural elements that forms one wall of the hydrophobic taxane pocket, as illustrated by the subset of residues from His227 to Gly235 that bracket Ala231 (magenta), as shown in Fig. 10. A change in this residue may perturb the normal interaction of His227 with dEpoB, as seen in this spatial model of drug-pocket interactions and the mutated residue. Not displayed are a number of water molecules that most certainly reside in the space between the bottom of the ligand and the basin of the binding site.

As can be seen in the model, one edge of the Ala231 cluster is populated by His227, which provides an anchor to the epothilone thiazole side chain *(52)*. The replacement of alanine with threonine is envisioned to result in two alterations of the environment around the ligand. First, addition of a polar OH group to the CH_3 of Ala to give the CH_2OH in Ser can be expected to perturb both the small pool of water between the ligand and the tubulin protein, thus influencing the shape of the hydrophobic cluster. Second, the somewhat extended serine side chain is in a position to compete with the epothilone thiazole moiety for hydrogen bonding to His227. Both these actions can be seen as deleterious for dEpoB binding and therefore responsible for the considerable degree of the resistance observed.

Higher concentrations of dEpoB (300 n*M*) cause the appearance of yet another, second **βGln292Glu** mutation, which is identical to that observed in A549 lung cells. Highlighted in yellow in Fig. 10, the Gln292 location, hydrogen bonded to Leu273, is readily perceived as a partner of the Ala231 to Ser change in its ability to further disrupt EpoD binding. Complementing this Epo resistance (EpoB 77-fold; dEpoB 307-fold) is an even greater impact on Taxol™ (467-fold) of these combined mutations. This result was interpreted by Verrills to be a mutational effect on drug binding, rather than on microtubule stability, and also appeared to imply that the two drugs do not bind in an identical manner *(60)*. The recently disclosed EpoA-tubulin structure confirms the latter insight *(52)*, but dispels the notion of a common PTX-Epo pharmacophore (*see* Subheading 6.1). Qualitatively, the potent double mutant combination is precisely the type of structural response one might imagine a cell would marshal against high doses of drug.

6.1. Structural Analysis of Cross-Resistance Between Taxanes and Epothilones

The separate evaluations of taxanes and epothilones, sets the stage for examining cross-resistance between the two drug classes. A key biostructural concept is that although the two drugs occupy common receptor volume, they do not engage in a common β-tubulin pharmacophore. Figure 11 illustrates the point by simultaneously displaying aspects of the overlapped binding sites for the Taxol™–tubulin *(54,55)* and EpoA–tubulin *(52)* models. A preliminary examination of the topologies of the ligand structures 1 and 4 (Fig. 2) puts this in perspective. Taxol™ (1) is made up of a rigid four-ring baccatin core supporting three critical flexible side chains at C-2, C-4, and C-13. Two of these chains terminate in phenyl rings. EpoA (4a), on the other hand is a structural composite of a flexible 16-membered lactone ring and a single side chain terminating in a methylated thiazole ring.

A number of attempts have been made to understand how structures of such diversity might occupy the taxane-binding site to produce a common biological response; namely efficient stabilization of MTs. All of these efforts have adopted the essence of the definition of a pharmacophore proposed by Marshall: the "3D arrangement of functional groups essential for recognition and activation (of a receptor)" *(62)*. Thus, four separate "common pharmacophore" models separately superimposed the Epo thiazole ring on the C-3′, *(57,63)*, C-3′ benzamido *(64)*, and C-2 benzoyl *(57,65)* phenyl rings of Taxol™. However, the same studies differed completely on the relative positioning of other functional groups. They either simply overlapped the entire taxane baccatin core and the Epo macrocycle *(65)* or sought specific functional group superpositions. For example, Giannakakou, Gussio, and colleagues *(57)* proposed two distinct models, each of which brought a minimum of five different functionalities into spatial match.

In spite of these efforts to explain drug potency and mutation-related resistance profiles, the electron crystallographic structures of Taxol™ and EpoA strongly suggest that a common pharmacophore is behind neither the similar activities of the drugs nor the mutation-based resistance. Cross-resistance is best understood in this context. Figure 3, for example, shows that the thiazole ring of EpoA is not coincident with any

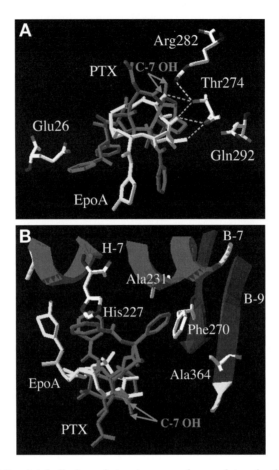

Fig. 11. Taxol™ and EpoA tubulin-bound structures overlap as determined by the corresponding electron crystallographic structures. Only their respective C-7 OH groups occupy a common location. **(A)** Residues Thr274, Arg282, and Gln292 mutate in response to epothilones and are cross resistant to Taxol™. Mutation of Glu26 against Taxol™ shows minimal cross-resistance to EpoB; **(B)** Taxol™ induced mutations at Phe270 and Ala364 experience marginal cross-resistance against epothilones, whereas the Ala231 mutation against dEpoB shows cross-resistance to Taxol™. To view this figure in color, see the insert and the companion CD-ROM.

of the rings in Taxol™. Furthermore, a careful examination of common structural features that promote the binding of the ligands to β-tublin makes it clear that the single substituent moiety common to 1 and 4 is the C-7 OH group *(52)*. This same functional group in both ligands interacts with Thr274 and is the basis for the characterization of the residue as a "hot spot." Figures 9 and 11 also provide an explanation for the significant cross-resistance between Taxol™ and the epothilones against the **Arg282Gln** and **Gln292Glu** mutations, given that these share a network of noncovalent interactions with Thr274. The same picture similarly rationalizes the lack of cross-resistance for the epothilones with respect to the **Asp26Glu** mutation. Whereas the C-3′ benzamido ring of Taxol™ is in direct contact with the Asp26, the closest atoms of the thiazole ring and the epoxide center of EpoA are 5–8 Å away from the side chain and, most likely, are shielded from it by water molecules.

Three mutations are found in the hydrophobic basin that makes up the taxane-binding pocket. They are in intimate contact with Taxol™, and all confer 15- to 50-fold resistance to the drug. Two of them, **Phe270Val** and **Ala364Thr** (Fig. 4) show very little cross-resistance to EpoB (two- to threefold). Figure 11B illustrates that these residues reside on B-7 and B-9, respectively, well away from the epothilone ligand (5–8 Å) shown shifted to the left in the diagram. The **Ala231Thr** exchange, causing a 16-fold response against Taxol™, is similarly well separated from EpoA (7–9 Å) (Fig. 11), but it nonetheless elicits a 21-fold resistance against dEpoB. The latter can be understood from the discussion previously, which proposes that mutant Thr231's residence on H-7 (Fig. 10) can influence the orientation of His227 on the same helix and, thereby, the His227-thiazole interaction. Finally, it was speculated earlier that mutations at Pro173 and Thr422 are unlikely to be involved in ligand binding, as they result in marginal resistance of 1.5- to 6-fold. In fact, both residues are found between 18–30 Å distant from the closest atoms of both 1 and 4 in the electron crystallographic models.

6.2. The Role of Tubulin Isotypes and Loss of Tubulin Heterozygosity in the Development of Microtubule-Targeting Drug-Induced Resistance

To date, there are seven β-tubulin isotypes described in mammalian cells (66). Although the exact role of each of these isotypes has yet to be defined, it seems that all of them are incorporated into the microtubule polymer and contribute to the overall cellular microtubule function. Interestingly, however, all tubulin mutations identified so far occur at the major β-tubulin isotype (class βI/gene HM40) (38), the expression of which accounts for 80–95% of total tubulin mRNA in a subset of cancer cell lines from the NCI60 human cancer cell collection (67). From a mechanistic standpoint, one wonders why the other β-tubulin isotypes in their wt sequence do not "take over" expression-wise, in order for the cell to escape the toxic effects of microtubule-targeting drugs that use class I β-tubulin as a target isotype. The authors' experiences, together with recurring data in the literature, suggest authors that the cell's first response is to adapt to the otherwise lethal effects of the drug by acquiring tubulin mutations in sites that are important for drug–tubulin interactions. To the best of the authors' knowledge, these mutations do not impair any other major function of cellular tubulin besides its drug response, and exert only a subtle effect on microtubule stability.

This cellular behavior of acquiring mutations *vs* substituting isotypes suggests that the class I β-tubulin isotype is not functionally redundant and its role in the cell simply cannot be replaced by the other isotypes. The latter constitutes a compelling hypothesis for why this is the single isotype identified in which sequence alterations occur to block the action microtubule-targeting drugs.

For the most part, the tubulin mutations described in this chapter appear to be homozygous (i.e., only mutant tubulin residues expressed). However, even in the case of heterozygous mutations (i.e., both wt and mutant tubulin residues expressed), when only one allele is mutated and the other remains wt, significant levels of drug resistance (10-fold or higher) are still obtained, and impaired drug-induced tubulin polymerization is observed consistent with compromised drug binding. This result is understandable if one realizes that in cells one molecule of drug bound to every few hundred tubulin dimers in the microtubule is sufficient to affect tubulin function (51). Consequently, even if only every other tubulin molecule to which the drug binds is mutated, this is apparently sufficient to confer significant levels of resistance to those drugs.

Although the selection process instigates mutations that confer resistance, continued selection of these clones in the presence of higher drug concentrations causes the level of resistance to increase, as has been reported in several cases. To investigate the mechanism by which this increased resistance occurs, the authors studied the temporal variations of wt- and mutant tubulin genes in the Taxol™- and epothilone-resistant 1A9 clones *(38,58)*. The authors observed that the early-step isolates, which were 10- to 15-fold drug resistant were heterozygous for the main tubulin isotype (HM40), whereas the late-step isolates, which were 30- to 50-fold resistant did not express the wt tubulin nor retain the wt sequence at the genomic level *(68)*. This result suggested that either a second identical mutation in the wt β-tubulin gene had occurred in the late-step isolates or that this wt was altogether lost. Detailed investigation of the resistance mechanism using single nucleotide polymorphism and fluorescence *in situ* hybridization analyses, revealed that loss of heterozygosity (LOH) for the wt β-tubulin allele had occurred in the late-step isolates so that only the remaining mutant allele is present at the genomic level leading to expression of the mutant β-tubulin only (Fig. 12) *(68)*. This result confirms that although acquisition of a tubulin mutation may be the first response of a cell to drug selection, increased selection pressure makes the cell further adapt by loss of the remaining wt allele, which leads to increased levels of drug resistance (Fig. 13). This is the first report of tubulin LOH conferring drug-resistance and such a mechanism is consistent with the genomic instability of cancer cells together with the cell's ability to adapt to external stress factors, such as prolonged exposure to toxic chemotherapeutic drugs. Despite the fact that there are other tubulin isotypes present in cells, the main isotype HM40 is the focus of cellular remodeling through mutation. Future studies are warranted to determine whether tubulin LOH in conjunction with tubulin mutations is a more universal mechanism of resistance to microtubule-targeting drugs.

6.3. Colchicine-Binding Site Tubulin Mutations and Structural Analysis

Antimitotic drugs that inhibit the binding of colchicine to tubulin appear to bind at a common site called the colchicine site. These agents are for the most part relatively simple structurally (Fig. 14), especially when they are compared with those binding at the taxane (Fig. 2) or the *vinca*-binding sites (Fig. 15). However, they are still structurally diverse. Although the prototype of this class is colchicine, it includes podophylotoxin, 2ME2, and indanocine, among others. Colchicine-binding site antitubulin drugs inhibit microtubule function by causing microtubule depolymerization. Although colchicine is used for the treatment of gout disease and Mediterranean fever *(25,26)*, neither colchicine nor colchicine-binding site drugs are currently FDA approved for the treatment of cancer. Interestingly, however, a few colchicine-site drugs like the combretastatins and 2ME2 are currently undergoing clinical trials as inhibitors of tumor angiogenesis.

The first report of acquired tubulin mutations in human cancer cells, in response to selection with colchicine-binding site drugs, was demonstrated with indanocine as the selecting agent *(69)*. Indanocine is a colchicine-site binding agent, as it is known to be competitive with colchicine when bound to β-tubulin *(70)*. In this study, human T-lymphoblatoid CEM cells selected with indanocine exhibited a drug-resistant phenotype owing to an acquired β-tubulin mutation (*HM40* gene) in residue 350 leading to a Lys to Asn amino acid substitution. The homozygous **βLys350Asn** mutation conferred

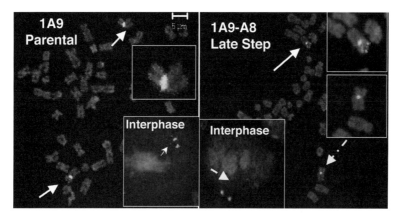

Fig. 12. Fluorescence *in situ* hybridization analysis of 1A9 cells Depicts tubulin LOH in the drug-resistant 1A9-A8 cells. Metaphase spreads from both cell lines were hybridized with the bacterial artificial chromosome (BAC) clone RP11-506k6 (orange) containing the β-tubulin gene *HM40*, and a centromeric probe for chromosome 6 (green), followed by counterstaining with nucleic acid stain Sytox Blue (blue). The parental 1A9 cells display two copies of chromosome 6 (white arrows) as evidenced by the chromosome 6 centromeric probe staining. Both 6 chromosomes displayed staining for the BAC clone indicating two copies of the β-tubulin gene *HM40*. In the late-step 1A9-A8 cells however, only one chromosome 6 stained for the BAC clone, although both copies of the chromosome 6 were present (green staining). The white arrows indicate chromosome 6, as evidenced by the green centromeric staining, and the BAC hybridization at the tips of the chromosome. In 1A9-A8 the yellow dashed arrow depicts the chromosome 6 that has lost the chromosomal region of 6p25. Insets display either a magnification of chromosome 6 in metaphase, or interphase. Scale bar = 5 µm. (Figure displayed with permission from ref. *68*.) To view this figure in color, see the insert and the companion CD-ROM.

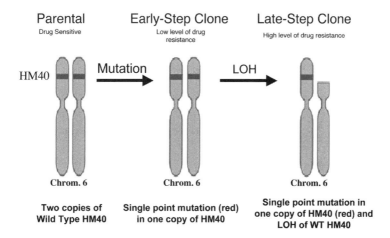

Fig. 13. Schematic representation of Evolution of Microtubule-Targeting Drug-Induced Resistance. (Figure displayed with permission from ref. *68*.) To view this figure in color, see the insert and the companion CD-ROM.

6

Indanocine

7

Colchicine

8

Podophyllotoxin

9

2-methoxy-estradiol

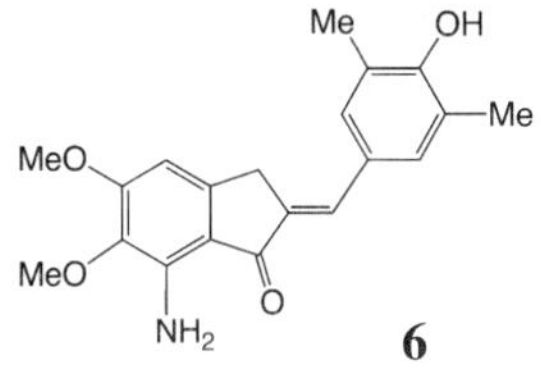

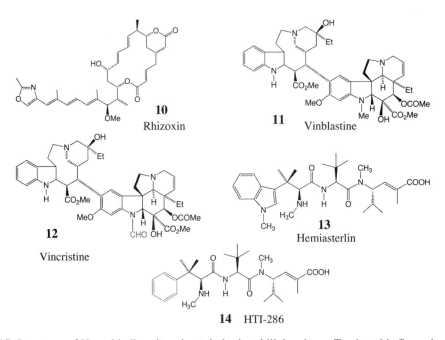

Fig. 14. Structures of colchicine-binding site microtubule destabilizing drugs.

10

Rhizoxin

11 Vinblastine

12

Vincristine

13

Hemiasterlin

14 HTI-286

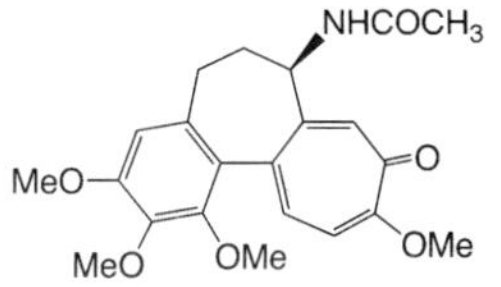

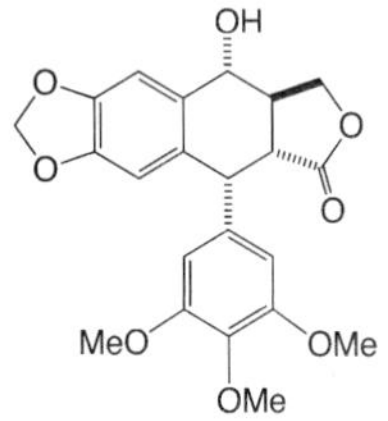

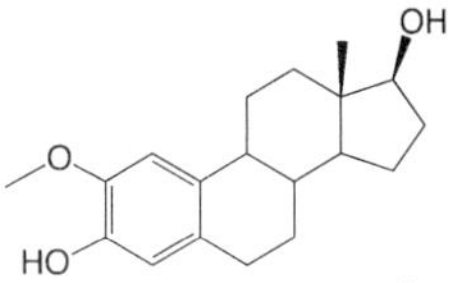

Fig. 15. Structures of *Vinca*-binding site microtubule-destabilizing drugs. To view this figure in color, see the insert and the companion CD-ROM.

115-fold resistance to indanocine and 30- to 40-fold cross-resistance to colchicine and vinblastine, respectively, whereas no cross-resistance was obtained with Taxol™ *(69)*. In addition comparison of the parental wt with the β-tubulin mutant resistant clone revealed impaired indanocine-induced tubulin depolymerization in the latter, although there were no differences in the cold-induced tubulin depolymerization between the two

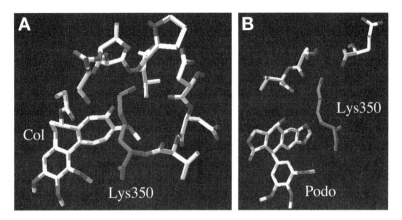

Fig. 16. The colchicine and podophyllotoxin binding sites as determined by X-ray crystallography. The βLys350Asn mutation is depicted by Lys350 (purple) in direct contact with the ligands. (**A**) The hydrophobic –(CH$_2$)$_4$– moiety of Lys350 is presented to the colchicine (a) and podophyllotoxin (b) ligands, whereas the terminal –NH$_3^+$ cation resides in a polar pocket. To view this figure in color, see the insert and the companion CD-ROM.

cell lines. These results strongly suggest compromised drug binding to tubulin. A qualitative understanding of the effect of this amino acid substitution is obtained by examination of the recently divulged tubulin–stathmin complexes separately incorporating colchicine and another colchicine-site drug, podophyllotoxin, near the α,β-tubulin dimer interface *(71)*. The colchicine-binding site is depicted in Fig. 16A, showing that Lys350 (magenta) is in van der Waals contact with colchicine. Surprisingly, the cationic NH$_3^+$ terminus of this amino acid is not directed at the C=O center of the tropolone ring of the colchicine ligand (Ring C), though it is within range and could make a productive hydrogen bond. The model of the protein complex shows this cationic head to be surrounded by a cluster of carbonyl groups from the protein instead (Fig. 16A). Thus, Lys350 interacts with colchicine (7) hydrophobically by means of the CH$_2$CH$_2$CH$_2$CH$_2$ component of its side chain. The corresponding podophyllotoxin (8)-lysine interaction in the corresponding complex (Fig. 16B) shows identical behavior. Given the structural similarity of 6, 7, and 8, it would seem safe to project that indanocine (6) will most likely interact with Lys350 in a similar fashion. If so, mutation to the shorter Asp residue will eliminate the charge-dipole interactions in the polar clusters shown at the top of Fig. 16A,B, whereas presenting a polar rather than nonpolar chain of atoms to the bound ligands. The combination of events would seem sufficient to rationalize resistance to both indanocine and colchicine as well as vinblastine (*see* Section 7). It should be noted that based on the unrefined electron crystallographic structure of the tubulin dimer *(6)*, Hua and colleagues *(69)* have suggested that Lys350 and Lys352 may sequester a Mg^{2+} cation and assist the neutralization of nonexhangeable GTP.

6.4. Vinca-*Binding Site Tubulin Mutations*

Antitubulin compounds that inhibit the binding of radiolabeled vinblastine and vincristine (Fig. 15) to tubulin share a common binding site described as the "*Vinca* domain." The prototypes of this class are the *Vinca* alkaloids, vinblastine, and vincristine, which were isolated more than 40 yr ago from the leaves of the periwinkle *Catharanthus roseus*. These compounds were initially studied because of their hypoglycemic activities, but were discovered to have antileukemic effects and cause bone marrow suppression *(72,73)*. Since then they have been widely used clinically for the treatment of leukemias, lymphomas, and some solid malignancies. The clinical success of these two natural products (vinblastine and vincristine) together with the elucidation of their mechanism of action on cellular MTs, have facilitated the development of several semisynthetic derivatives notably vindesine, vinorelbine, and vinflunine, which are now used in the clinic for the treatment of cancer *(74)*. Several other naturally occurring microtubule-interfering compounds have been identified that bind β-tubulin at the *Vinca*-domain including the marine-sponge derived halichondrins, spongistatin, dolastatins (isolated from the sea hare *Dolabella auricularia*), and cryptophycins (isolated from the blue-green algae *Nostoc sp.*) (for review, *see* ref. *75*). These agents depolymerize MTs and are currently at various stages of clinical development for the treatment of cancer.

Acquired tubulin mutations in human cancer cells for the "*Vinca*-domain" have been reported in response to selections with vincristine *(76)* and the hemiasterlin (HTI) analog HTI-286 *(77,78)*. In the first study, the human leukemia cell lines CCRF-CEM has been selected with vincristine and exhibited very high levels (22,600-fold) of drug-resistance to the selecting agent. This cell line overexpressed Pgp, which accounted for a major part of the resistant phenotype; however, the Pgp modulator verapamil was able to restore only partial sensitivity to vincristine. These data led to the hypothesis that other cellular alterations might be present contributing to the overall resistant phenotype. Sequencing of β-tubulin (HM40) revealed the presence of a heterozygous β-tubulin (HM40) mutation at residue **βLeu240Ile** *(76)*. In addition, multiple other microtubule-related and other cytoskeletal alterations have been identified in this cell line *(76,79)*. These observations preclude the ability to draw a firm conclusion on the importance of this mutation, especially as no direct studies assessing the role of this point mutation in resistance to *Vinca* alkaloids have been performed.

In the second study, 1A9 human ovarian carcinoma cells were selected with a synthetic analog of hemiasterlin, HTI-286. Several HTI-resistant 1A9 clones were isolated exhibiting a 57- to 89-fold resistance to HTI, significant cross-resistance (3- to 186-fold) to "*Vinca*-domain" binding drugs, minimal two- to fourfold cross-resistance to colchicine-site drugs and collateral sensitivity to taxane-site microtubule stabilizing drugs *(77)*. Sequencing of both α- (Kα1 isotype) and β-tubulin (HM40 isotype) in these clones revealed distinct acquired tubulin mutations in both subunits as follows: several HTI-resistant cell clones harbored only one heterozygous mutation at **βSer172Ala,** whereas others harbored either one heterozygous α-tubulin mutation at residue **αIle384Val** or two homozygous tubulin mutations at residues **αSer165Pro** and **αArg221His**. In all cases the presence of these mutations has been associated with increased microtubule stability. This finding together with the location of these mutations and the cross-resistance profile of the HTI-resistant cells led to the formation of a model whereby the drug resistance phenotype is attributed to the increased microtubule stability rather than to the reduced binding affinity of HTI to tubulin, although the latter has not been tested.

7. STRUCTURAL ANALYSIS OF MUTATIONS IN RESPONSE TO SELECTION WITH *VINCA*-BINDING SITE DRUGS

Many studies have attempted to identify the tubulin-binding site of the *Vinca* alkaloids vinblastine and vincristine. Figure 1 is very conservative in depicting it in a small region of β-tubulin near the interface between αβ-tubulin dimers. Current evidence

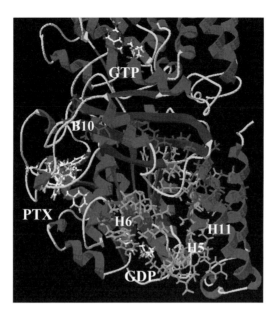

Fig. 17. Photoaffinity labeled peptides and mutations around the *Vinca* site. β-Tubulin (lower half of protein) displaying peptides labeled by rhizoxin (magenta) and vinblastine (cyan). Taxol™ (PTX-left), GTP (in α-tubulin), and GDP (in β-tubulin) are displayed in CPK colors. To view this figure in color, see the insert and the companion CD-ROM.

suggests the binding locus for these compounds is either much more expansive or not well defined. In 1993, an attempt to locate the binding region took advantage of the observation that rhizoxin is a potent inhibitor of binding of vinblastine to tubulin. An azidodansyl derivative of the rhizoxin was used as a photoaffinity label, in order to identify interaction with residues β363–379 *(80)*. In Figs. 17 and 18 these residues are magenta colored, and the peptide defines a belt across the upper part of the β-tubulin subunit. Shortly thereafter, a photoactive and fluorescently labeled analog of vinblastine was used to identify a second peptide sequence from β175–213 *(81)*. Shown in Figs. 17 and 18 in cyan, the depiction gives the misleading impression that there is overlap between the two peptides identified by the photoaffinity probes. However, the rhizoxin-determined magenta peptide sequence is confined largely to B10 and H11, whereas the vinblastine-determined cyan peptide sequence runs from the loop leading to H5 through B6 and H6. The expanse covered by these peptides is so broad that it is not possible to define a local site that might span some of the secondary structure between them. However, by combining photoaffinity and mutation data, a more specific binding locus can be conceived (*see* below).

In addition to the binding site information obtained through photolabeling, two acquired mutations have been reported. One study of rhizoxin resistance discovered a mutant gene corresponding to **βAsn100Ile** in *Aspergillus nidulans (82,83)*. The asparagine is not a part of the vincristine-related cyan peptide, but falls 4–6 Å behind it in the view displayed. The most prominent feature in the vicinity of Asn100 is Tyr406, the aromatic ring, which is directly over the side chain's amide group, as shown in Fig. 19A. Exchange of Asn for the similarly sized Ile should cause no steric conflict in the pocket, but the change of polarity can be expected to have an influence, if rhizoxin is nearby.

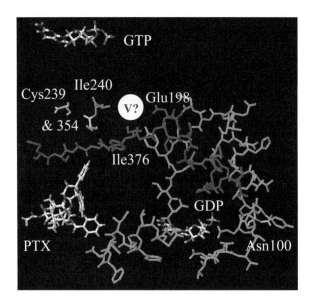

Fig. 18. Closeup of the photoaffinity labeled peptides and mutations around the *Vinca* site. The labeled peptides in β-tubulin colored as in Fig. 17 are complemented with mutations Asn100Ile (orange), Ile240 (green), residues Cys239 and Cys354 (yellow), and labeled peptide residues Glu198 and Ile376 (red). The circled V represents a possible *Vinca*-binding site. To view this figure in color, see the insert and the companion CD-ROM.

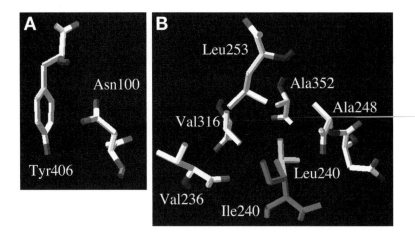

Fig. 19. The local environments of the βAsn100Ile and βLeu240Ile mutations. **(A)** The Asn100 amide group lies directly over the face of Tyr406's aromatic ring. **(B)** The terminal methyl groups of Leu240 (green) are located in a hydrophobic pocket. Mutation to Ile (magenta) appears to pull the residue somewhat out of the pocket. To view this figure in color, see the insert and the companion CD-ROM.

More recently, CEM VCR-R human leukemia cells were cultured in the presence of vincristine, after which extraordinary resistance (22,600-fold) developed *(84)*. Subsequently, a **βLeu240Ile** mutation has been associated with the latter (Leu green, Fig. 19B) *(76)*. Like the Asn100 mutant, this residue is not included in either peptide derived by photolabeling. Not surprisingly, Leu240 is centered in a hydrophobic pocket including residues Val236, Ala248, Leu253, Val316, and Ala352 (Fig. 19B). Although mutation to Ile (magenta) is conservative, the change in branching of the side chain pulls four carbons of the unit somewhat out of the pocket. Not only can this cause rearrangement of the hydrophobic cluster, but the steric effects near the pocket's base are predicted to interfere with backbone atoms, as well. Along similar lines, Kavallaris and coworkers argued that the Leu to Ile switch occasions short contacts between Ileu240 and Ala248 on the loop linking helices H7 and H8 in the protein. As these interactions operate near the intradimer surface of the tubulin αβ-heterodimer, it was speculated that coupled displacements of the helices might affect the longitudinal interactions between the subunits. Thus, in turn, microtubule stability could be affected.

In combination with the peptide labeling data, the βLeu240Ile mutant information is potentially diagnostic of a more specific locale for *Vinca* alkaloid binding. In this region of the protein, two sectors of the labeled peptides are within 4–6 Å. One of these is indicated in Fig. 18 by the two residues Glu198 and Ile376 colored red, belonging to each of the peptide sequences (in magenta and cyan), respectively. These two amino acids are within 6–8 Å of mutant Leu240Ile (green). Two additional pieces of information are exhibited in Fig. 18. First, the amino acids Cys239 and Cys354 (yellow) are intimately associated with the mutated center. These residues have been crosslinked to suitable colchicine derivatives *(85)*. Second, the latter is in the same region of the protein where Taxol™ binds. These spatial connections may reflect the observations that vinblastine seems to protect the colchicine site from decay, even though vinblastine and colchicine do not affect each other's binding to tubulin *(86)*. This collection of evidence motivates us to speculate that the *Vinca* alkaloids may bind near the white circle with the "**V?**" seen in Fig. 18. This view would be consistent with the observation that vinblastine exhibits 40-fold cross-resistance against the **βLys350Asn** mutation induced by indanocine that was previously described *(69)*. It is worth noting that despite the existing data and although all of the earlier proposals are reasonable, the precise role played by vincristine and vinblastine in both the induction of mutations and the resulting resistance will remain unclear until more detailed information on the precise binding center of the *Vinca* drugs is obtained.

8. HEMIASTERLIN: RESISTANCE FROM α-TUBULIN

Hemiasterlin (Fig. 15) is a member of a small family of cytotoxic tri- and tetrapeptides isolated from marine sponges. The compound is a more potent in vitro cytotoxin and antimitotic agent than either Taxol™ or vincristine *(87)*. A synthetic analog, HTI-286 (Fig. 15), overcomes resistance mediated by drug efflux pumps including Pgp (MDR1), multidrug-resistance-associated protein 1 (MRP1), and mitoxantrone-resistance gene (MXR) *(88,89)*.

Attempts to locate the hemiasterlin-binding site have utilized two-labeled benzophenone photoaffinity probes of HTI-286, both of which exhibit similar tubulin affinity (apparent KD = 0.2–1.1 μM), potent inhibition of cell growth (IC$_{50}$, 1–22 nM), and exclusive labeling of α-tubulin *(19,90,91)*. Probe 1 labels α-tubulin both in cytosolic extracts of whole cells and in the pure form. Deconvolution of the latter identifies residues α314–338 as the target, a region that corresponds to strand B8 and helix H10 (Fig. 20, cyan). The α-tubulin peptide tagged by probe 2 in bovine brain and HeLa cell

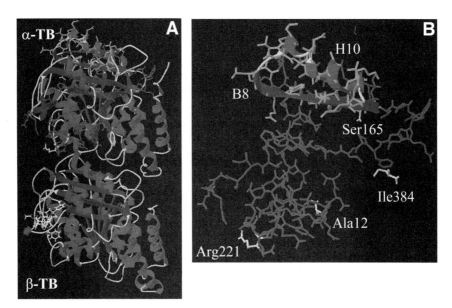

Fig. 20. Photoaffinity labeled peptides and mutations around the α-tubulin hemiasterlin-binding site. **(A)** Benzophenone photoaffinity probes of HTI-286 (**13**) label peptides that cover the α-subunit cap; probe 1 (314–388, cyan), probe 2 (204–280 magenta). **(B)** Four mutations induced by HTI-286 are pictured in yellow. Arg221 falls within the magenta labeled peptide sequence. Ala12, Ser165, and Ile384 are independent of the peptides. To view this figure in color, see the insert and the companion CD-ROM.

tubulin preparations (residues α204–280) is considerably more expansive, but complementary to probe 1 (Fig. 20, magenta). Together the labeled peptides cap α-tubulin at its northern end. Ligand binding in this sector provides opportunities to alter wt tubulin behavior associated with longitudinal interactions with β-tubulin across the interdimer interface, lateral interactions with adjacent protofilaments, and the ability to cross-link to stathmin, a protein that induces depolymerization of MTs. These phenomena are, in turn, intimately affiliated with the extent of microtubule stabilization in the cell.

Further biological evidence complementary to the photoprobe studies, is provided by the identification of tubulin mutants from cells raised in the presence of HTI-286; namely the 1A9-HTI[R] cells *(77)*. Derived from the 1A9 human ovarian carcinoma cells, this cell line is 57- to 66-fold resistant to **HTI-286** and exhibits a **βSer172Ala** mutation in the HM40 β-tubulin isotype. Two other 1A9-HTI[R] cells lines with no mutations in the β-subunit display **αSer165Pro/Arg221His** and **αIle384Val** replacements, respectively, in the Kα-1 α-tubulin isotype *(77)*. The presence of these α-tubulin mutations is associated with 65- to 89-fold drug resistance. In a separate study using the human epidermoid carcinoma KB3-1 cells, a KB3-1-HTI resistant clone revealed a fourth α-tubulin mutation: **Ala12Ser** *(78)*. However, this mutation was only partially responsible for the resistant phenotype as these cells also displayed significantly impaired intracellular drug accumulation, indicating overexpression of a drug-efflux pump conferring drug resistance. Figures 20–22 display the corresponding amino acids in different settings. Residue αArg221 falls within the magenta peptide (α204–280) identified by photolabeling; however, this residue, similar to residues Ile384 and Ala12, appears to be at a distance from the putative hemiasterlin-binding site. It might appear from Fig. 20B that the latter

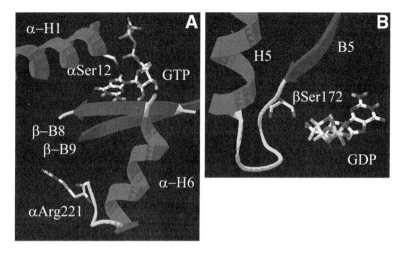

Fig. 21. Local environments of the αAla12Ser, βSer172Ala, and βArg221His mutations. **(A)** The αSer12 mutation on helix αH1 appears to be within hydrogen bonding distance of GTP. The latter, in turn, is in direct contact with α-H6 and the connected loop on which resides Arg221. **(B)** Ser172 on the loop connecting β-H5 and β-B5 is within a few Å of GDP and the channel that separates the α- and β-tubulin subunits. To view this figure in color, see the insert and the companion CD-ROM.

amino acid resides near the magenta peptide, however, closest contacts are 5–7 Å. As depicted by Figs. 20 and 21A, mutant Ser12 (Ala12 is the wt shown in the Fig. 19) is not only near the tubulin dimer interface, but is also in direct contact with the nonexchangable N-site GTP *(55)*. In particular, the serine OH is in a position to make a hydrogen bond with either the oxygen of the 5′-α-phosphate, the oxygen of the sugar or the nitrogen of the guanine base. Furthermore, although Arg221 is distant in space from GTP (14–15 Å), it is linked through helix H6 and the distal loop on which it resides. It seems unlikely that mutations at positions α12 and α221 can confer resistance by ablating the binding of hemiasterlin analogs, but one possibility for mutual interaction between them is shown in Fig. 21A. The αSer12/GTP couple can potentially perturb the orientation of the helix H6, which may in turn alter the conformation of the loop that harbors Arg221. The plausibility of this long range interaction is strengthened by the observation that the NH_2 of GTP's guanine base is within hydrogen bonding distance of the side chain of Asn204 at the end of H6.

With respect to the **βSer172Ala** mutation, *(77)* Fig. 21B illustrates that the residue is located on the loop between β-strand B5 and helix H5. Strikingly, like the αSer12 mutation, it is situated within 5 Å of the nucleotide, in this case exchangeable GTP at the interdimer interface. If this residue is one of the factors that indeed causes resistance to HTI-286, mutation most probably transmits conformational information from the β-subunit to the α-subunit to prevent the microtubule destabilizing action of the hemiasterlin drugs. The speculation carries the caveat that any accompanying resistance operates without interfering with GTP/GDP dynamics.

Finally, the **αSer165Pro** change has been suggested to operate by a train of events similar to that depicted for mutation centers αSer12 and αArg221, this time across the tubulin dimer interface *(77)*. Figure 22 intimates that αSer165 can engage in nonbonded contact with Gln256 on helix α-H8, a key part of the longitudinal interface between

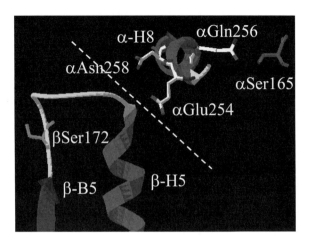

Fig. 22. A potential interaction train for αSer165Pro and βSer172Ala across the αβ-tubulin dimer interface (dotted line); αSer165 (magenta), βSer172 (green). αGln256 is a residue required for hydrolysis of GTP in β-tubulin on polymerization. To view this figure in color, see the insert and the companion CD-ROM.

dimers. A half turn away, αAsn258 interacts with helix β-H5 and the β-B5 connected loop on which βSer172 resides. Furthermore, α-H8 includes Glu254, a residue required for hydrolysis of GTP in β-tubulin on polymerization (92). Conceptually, then, two additional centers susceptible to mutation in response to HTI-286 may well engage in mutual long-range interaction in a sector crucial for regulation of the dimer–microtubule equilibrium.

An insightful and provocative analysis of the role of mutations distant from the tubulin binding site of an effective drug has been presented by Poruchynsky and colleagues (77). It centers on the principle that mutations to a cytotoxic drug must still allow tubulin self-assembly and assembly regulation to ensure cell survival. Two cases are cited. First, resistance to ligands that bind in sites not directly involved in dimer aggregation can be mediated by mutations within the same sites without detrimental influence on the tubulin–microtuble equilibrium. The taxane cleft mutations against Taxol™ analogs and epothilones (Fig. 2) discussed earlier are pertinent examples. Second, antimitotic drugs that operate from a site of tight sequence conservation require an alternative mutation strategy to achieve resistance. Such amino acid interchanges are predicted to be localized distant from the binding site and to counter drug action by a mechanism other than drug-binding inhibition. Thus, effective mutations are regarded to be those that influence microtubule dynamics opposite to the drug's effect; stabilizing mutations for depolymerization and destabilizing mutations for microtubule stabilizers. Poruchynsky and colleagues argue that HTI-286 induced mutations of the first type should act to increase the stability of MTs, confer general resistance to depolymerizing agents, and increase sensitivity to stabilizing ligands. Indeed, as described earlier, the tubulin mutations in 1A9-HTI[R] resistant cells are accompanied by increased microtubule stability. They similarly display cross-resistance to *Vinca* alkaloids (10- to 186-fold), dolostatin-10 (7- to 10-fold), cryptophycin-1 (four- to sevenfold), colchicines, and podophyllotoxin (two- to fourfold), but none to Taxol™, Taxotere (2), epothilones (4), and other microtubule stabilizing agents. By contrast, these HTI-resistant cells were 2- to 14-fold more sensitive to all microtubule-stabilizing drugs tested. The resistant cells were equally sensitive to adriamycin (one- to fivefold), a drug that targets DNA. In sum, these cytotoxicity profiles suggest that at least three of the mutations identified in the α- and β-subunits of the tubulin dimer (βSer172Ala, αIle384Val, and αSer165Pro) alter the stability of MTs rather than interfering with HTI-286 binding. The principle is intriguing and may apply similarly to the colchicine, *Vinca,* and laulimalide *(22)* sites as well as to other proteins. However, as mentioned earlier, additional studies using isolated tubulins harboring the aforementioned mutations are required to elucidate the role of these mutations in drug-binding *vs* microtubule stability.

9. CONCLUSION

Clinical drug resistance is a major barrier to overcome before chemotherapy can become curative for most cancer patients. In the majority of cases, drug resistance eventually develops and is universally fatal. If this could be prevented or at least overcome, the impact for cancer therapy would be substantial. Rational attempts to tackle clinical drug resistance need to be based on the understanding of the molecular mechanisms involved. These mechanisms are likely to be complex and multifactorial, given the high level of genomic instability and mutations seen in cancer cells, as well as the fact that a combination of chemotherapy agents, each one having a different cellular target, are generally administered clinically. All of these factors together with the high degree of tumor heterogeneity seen in patients allow the cancer cell many escape routes to survival.

Molecular insights into mechanisms of drug resistance are often achieved through studies of experimental models of induced drug resistance. In the case of taxanes and other microtubule-targeting agents these models primarily include cancer cell lines with induced-resistance phenotypes following drug selection. Although, the role of acquired tubulin mutations in clinical drug resistance is not yet clear, acquisition of tubulin mutations in cultured cells is one of the most frequently occurring mechanisms of resistance to microtubule-targeting agents. Molecular and structural studies of these acquired tubulin mutations have significantly deepened the understanding of the basic biology of tubulin and MTs and have revealed important information on drug–tubulin interactions. This knowledge has provided attractive alternative treatments to overcome resistance by using structurally related compounds whose activity is not affected by the presence of specific tubulin mutations.

As microtubule-targeting drugs will continue to play a major role in anticancer therapy, and resistance to these drugs will continue to be an important clinical issue until minimized, it is hoped that the knowledge gained from these preclinical studies will lead to more effective cancer therapies. Room for optimism is found in current treatments of AIDS, another disease plagued by drug resistance owing to mutations in two important clinical targets; namely reverse transcriptase and HIV protease. Drug cocktails targeting different viral proteins appear often to be effective at causing significant and long-term reduction in viral load *(93)*. A similar strategy has been less successful in the case of bacterial infection. Within this regime, mutations of a variety of bacterial proteins are sufficiently rapid that monodrug therapies are consistently playing catch-up *(94,95)*. In some instances, control of clinical infection would appear to be threatened by the potential development of resistance to last-resort vancomycin, a powerful antibiotic without deep backup *(96,97)*. Thus, it is not yet clear that, even if a detailed understanding of mutation in tumor-causing systems is gained, general and widely applicable remedies for the range of diseases that fall under the cancer umbrella can be devised. However, without deeper insights into the operation of mutant-driven resistance, the opportunities for rational intervention are considerably diminished whatever the long-term prospects may be.

ACKNOWLEDGMENT

We thank Dr. Jim Nettles (Emory University) for providing Fig. 6 and Ms Kathleen Kite-Powell for text editing.

REFERENCES

1. Jordan MA. Mechanism of Action of Anittumor Drugs that Interact with Microtubules and Tubulin. Curr Med Chem 2002;2:1–17.
2. Rao S, Horwitz SB, Ringel I. Direct photoaffinity labeling of tubulin with taxol™. J Natl Cancer Inst 1992;84:785–788.
3. Rao S, Krauss NE, Heerding JM, et al. 3′-(p-azidobenzamido)taxol™ photolabels the N-terminal 31 amino acids of beta-tubulin. J Biol Chem 1994;269:3132–3134.
4. Rao S, Orr GA, Chaudhary AG, Kingston DG, Horwitz SB. Characterization of the taxol™ binding site on the microtubule. 2-(m- Azidobenzoyl)taxol™ photolabels a peptide (amino acids 217-231) of beta- tubulin. J Biol Chem 1995;270:20,235–20,238.
5. Rao S, He L, Chakravarty S, Ojima I, Orr GA, Horwitz SB. Characterization of the Taxol™ binding site on the microtubule. Identification of Arg(282) in beta-tubulin as the site of photoincorporation of a 7-benzophenone analogue of Taxol™. J Biol Chem 1999;274:37,990–37,994.
6. Nogales E, Wolf SG, Downing KH. Structure of the alpha beta tubulin dimer by electron crystallography. Nature 1998;391:199–203.
7. Schiff PB, Fant J, Horwitz SB. Promotion of microtubule assembly in vitro by taxol™. Nature 1979; 277:665–667.
8. Bollag DM, McQueney PA, Zhu J, et al. Epothilones, a new class of microtubule-stabilizing agents with a taxol™- like mechanism of action. Cancer Res 1995;55:2325–2333.
9. ter Haar E, Kowalski RJ, Hamel E, et al. Discodermolide, a cytotoxic marine agent that stabilizes microtubules more potently than taxol™. Biochemistry 1996;35:243–250.
10. Long BH, Carboni JM, Wasserman AJ, et al. Eleutherobin, a novel cytotoxic agent that induces tubulin polymerization, is similar to paclitaxel (Taxol™). Cancer Res 1998;58:1111–1115.
11. Bai RL, Lin CM, Nguyen NY, Liu TY, Hamel E. Identification of the cysteine residue of beta-tubulin alkylated by the antimitotic agent 2,4-dichlorobenzyl thiocyanate, facilitated by separation of the protein subunits of tubulin by hydrophobic column chromatography. Biochemistry 1989;28:5606–5612.
12. Shan B, Medina JC, Santha E, et al. Selective, covalent modification of beta-tubulin residue Cys-239 by T138067, an antitumor agent with in vivo efficacy against multidrug-resistant tumors. Proc Natl Acad Sci USA 1999;96:5686–5691.
13. Pettit GR, Singh SB, Schmidt JM, Niven ML, Hamel E, Lin CM. Isolation, structure, synthesis, and antimitotic properties of combretastatins B-3 and B-4 from Combretum caffrum. J Nat Prod 1988;51:517–527.
14. Blokhin AV, Yoo HD, Geralds RS, Nagle DG, Gerwick WH, Hamel E. Characterization of the interaction of the marine cyanobacterial natural product curacin A with the colchicine site of tubulin and initial structure-activity studies with analogues. Mol Pharmacol 1995;48:523–531.
15. D'Amato RJ, Lin CM, Flynn E, Folkman J, Hamel E. 2-Methoxyestradiol, an endogenous mammalian metabolite, inhibits tubulin polymerization by interacting at the colchicine site. Proc Natl Acad Sci USA 1994;91:3964–3968.
16. Wilson L. Properties of colchicine binding protein from chick embryo brain. Interactions with vinca alkaloids and podophyllotoxin. Biochemistry 1970;9:4999–5007.
17. Lin CM, Hamel E, Wolpert-DeFilippes MK. Binding of maytansine to tubulin: competition with other mitotic inhibitors. Res Commun Chem Pathol Pharmacol 1981;31:443–451.

18. Bai RL, Pettit GR, Hamel E. Binding of dolastatin 10 to tubulin at a distinct site for peptide antimitotic agents near the exchangeable nucleotide and vinca alkaloid sites. J Biol Chem 1990;265: 17,141–17,149.

19. Nunes M, Kaplan J, Loganzo F, Zask A, Ayral-Kaloustian S, Greenberger LM. Two photoaffinity analogs of HTI-286, a synthetic analog of hemisasterlin, interact with alpha-tubulin. Eur J Cancer 2002;38:S119.

20. Bai R, Durso NA, Sackett DL, Hamel E. Interactions of the sponge-derived antimitotic tripeptide hemiasterlin with tubulin: comparison with dolastatin 10 and cryptophycin 1. Biochemistry 1999;38:14,302–14,310.

21. Hamel E, Covell DG. Antimitotic peptides and depsipeptides. Curr Med Chem Anti-Canc Agents 2002;2:19–53.

22. Pryor DE, O'Brate A, Bilcer G, et al. The microtubule stabilizing agent laulimalide does not bind in the taxoid site, kills cells resistant to paclitaxel and epothilones, and may not require its epoxide moiety for activity. Biochemistry 2002;41:9109–9115.

23. Pineda O, Farras J, Maccari L, Manetti F, Botta M, Vilarrasa J. Computational comparison of microtubule-stabilising agents laulimalide and peloruside with taxol™ and colchicine. Bioorg Med Chem Lett 2004;14:4825–4829.

24. McKellar QA, Scott EW. The benzimidazole anthelmintic agents—a review. J Vet Pharmacol Ther 1990;13:223–247.

25. Ben-Chetrit E, Levy M. Colchicine: 1998 update. Semin Arthritis Rheum 1998;28:48–59.

26. Borstad GC, Bryant LR, Abel MP, Scroggie DA, Harris MD, Alloway JA. Colchicine for prophylaxis of acute flares when initiating allopurinol for chronic gouty arthritis. J Rheumatol 2004;31:2429–2432.

27. Schochet SS Jr, Lampert PW, Earle KM. Neuronal changes induced by intrathecal vincristine sulfate. J Neuropathol Exp Neurol 1968;27:645–658.

28. Wani MC, Taylor HL, Wall ME, Coggon P, McPhail AT. Plant antitumor agents. VI. The isolation and structure of taxol™, a novel antileukemic and antitumor agent from Taxus brevifolia. J Am Chem Soc 1971;93:2325–2327.

29. Rowinsky EK. The development and clinical utility of the taxane class of antimicrotubule chemotherapy agents. Annu Rev Med 1997;48:353–374.

30. Rowinsky EK. On pushing the outer edge of the outer edge of paclitaxel's dosing envelope. Clin Cancer Res 1999;5:481–486.

31. Ling V. Charles F. Kettering Prize. P-glycoprotein and resistance to anticancer drugs. Cancer 1992;69:2603–2609.

32. Horwitz SB, Cohen D, Rao S, Ringel I, Shen HJ, Yang CP. Taxol™: mechanisms of action and resistance. J Natl Cancer Inst Monogr 1993;55–61.

33. Jaffrezou JP, Dumontet C, Derry WB, et al. Novel mechanism of resistance to paclitaxel (Taxol™) in human K562 leukemia cells by combined selection with PSC 833. Oncol Res 1995;7:517–527.

34. Kavallaris M, Kuo DY, Burkhart CA, et al. Taxol™-resistant epithelial ovarian tumors are associated with altered expression of specific beta-tubulin isotypes. J Clin Invest 1997;100:1282–1293.

35. Haber M, Burkhart CA, Regl DL, Madafiglio J, Norris MD, Horwitz SB. Altered expression of M beta 2, the class II beta-tubulin isotype, in a murine J774.2 cell line with a high level of taxol™ resistance. J Biol Chem 1995;270:31,269–31,275.

36. Goncalves A, Braguer D, Kamath K, et al. Resistance to Taxol™ in lung cancer cells associated with increased microtubule dynamics. Proc Natl Acad Sci USA 2001;98:11,737–11,742.

37. Kamath K, Wilson L, Cabral F, Jordan MA. beta III-tubulin induces paclitaxel resistance in association with reduced effects on microtubule dynamic instability. J Biol Chem 2005;280: 12,902–12,907.

38. Giannakakou P, Sackett DL, Kang YK, et al. Paclitaxel-resistant human ovarian cancer cells have mutant beta-tubulins that exhibit impaired paclitaxel-driven polymerization. J Biol Chem 1997;272: 17,118–17,125.

39. Gonzalez-Garay ML, Chang L, Blade K, Menick DR, Cabral F. A beta-tubulin leucine cluster involved in microtubule assembly and paclitaxel resistance. J Biol Chem 1999;274:23,875–23,882.

40. Hari M, Loganzo F, Annable T, et al. Paclitaxel resistant cells have a mutation in the paclitaxel binding region of beta-tubulin (Asp26Glu) and less stable microtubules. Mol Cancer Ther 2006;5:2, 270–278.

41. Wiesen KM, Xia S, Horwitz SB. Exogenous class I beta-tubulin sensitizes Taxol™-resistant MDA cells. Proc Amer Assoc Cancer Res 2002;43:788.

42. Weisenberg RC, Borisy GG, Taylor EW. The colchicine-binding protein of mammalian brain and its relation to microtubules. Biochemistry 1968;7:4466–4479.

43. Davidse LC, Flach W. Differential binding of methyl benzimidazol-2-yl carbamate to fungal tubulin as a mechanism of resistance to this antimitotic agent in mutant strains of Aspergillus nidulans. J Cell Biol 1977;72:174–193.

44. Ling V, Aubin JE, Chase A, Sarangi F. Mutants of Chinese hamster ovary (CHO) cells with altered colcemid-binding affinity. Cell 1979;18:423–430.

45. Cabral F, Sobel ME, Gottesman MM. CHO mutants resistant to colchicine, colcemid or griseofulvin have an altered beta-tubulin. Cell 1980;20:29–36.

46. Yamamoto M. Genetic analysis of resistant mutants to antimitotic benzimidazole compounds in Schizosaccharomyces pombe. Mol Gen Genet 1980;180:231–234.

47. Burland TG, Schedl T, Gull K, Dove WF. Genetic analysis of resistance to benzimidazoles in Physarum: differential expression of beta-tubulin genes. Genetics 1984;108:123–141.

48. Orbach MJ, Porro EB, Yanofsky C. Cloning and characterization of the gene for beta-tubulin from a benomyl-resistant mutant of Neurospora crassa and its use as a dominant selectable marker. Mol Cell Biol 1986;6:2452–2461.

49. Thomas JH, Neff NF, Botstein D. Isolation and characterization of mutations in the beta-tubulin gene of Saccharomyces cerevisiae. Genetics 1985;111:715–734.

50. Nogales E, Whittaker M, Milligan RA, Downing KH. High-resolution model of the microtubule. Cell 1999;96:79–88.

51. Jordan MA, Wilson L. Microtubules as a target for anticancer drugs. Nat Rev Cancer 2004;4:253–265.

52. Nettles JH, Li H, Cornett B, Krahn JM, Snyder JP, Downing KH. The binding mode of epothilone A on alpha,beta-tubulin by electron crystallography. Science 2004;305:866–869.

53. Martello LA, Verdier-Pinard P, Shen HJ, et al. Elevated levels of microtubule destabilizing factors in a Taxol™-resistant/dependent A549 cell line with an alpha-tubulin mutation. Cancer Res 2003;63:1207–1213.

54. Snyder JP, Nettles JH, Cornett B, Downing KH, Nogales E. The binding conformation of Taxol™ in beta-tubulin: a model based on electron crystallographic density. Proc Natl Acad Sci USA 2001;98: 5312–5316.

55. Lowe J, Li H, Downing KH, Nogales E. Refined structure of alpha beta-tubulin at 3.5 A resolution. J Mol Biol 2001;313:1045–1057.

56. Kowalski RJ, Giannakakou P, Hamel E. Activities of the microtubule-stabilizing agents epothilones A and B with purified tubulin and in cells resistant to paclitaxel (Taxol™(R)). J Biol Chem 1997;272:2534–2541.

57. Giannakakou P, Gussio R, Nogales E, et al. A common pharmacophore for epothilone and taxanes: molecular basis for drug resistance conferred by tubulin mutations in human cancer cells. Proc Natl Acad Sci USA 2000;97:2904–2909.

58. Mehdi N, Lartigot M, Loretan J, Altman KH, Fabbro D, Wartmann M. Asingle point mutation in Human b-tubulin isoform M40 is associated with Epothilone-A resistance in two KB-31 Epidermoid carcinoma cell Lines. Proc Am Assoc Cancer Res 2001;42:920.

59. He L, Yang CP, Horwitz SB. Mutations in beta-tubulin map to domains involved in regulation of microtubule stability in epothilone-resistant cell lines. Mol Cancer Ther 2001;1:3–10.

60. Verrills NM, Flemming CL, Liu M, et al. Microtubule alterations and mutations induced by desoxyepothilone B: implications for drug-target interactions. Chem Biol 2003;10:597–607.

61. Amos LA, Lowe J. How Taxol™ stabilises microtubule structure. Chem Biol 1999;6:R65–R69.

62. Marshall G, Barry C, Bosshard h, Dammkoehler R. The conformation Parameter in drug Design: The Active Analog Approach. In: Computer-Assisted Drug Design, Olson EC, Christoffersen RE, eds. Washington, Amer Chem Soc, 1979; pp 205–226.

63. Wang M, Xia X, Kim Y, et al. A Unified and Quantitative Receptor Model for the Microtubule Binding of Paclitaxel and Epothilone. Org Lett 1999;1:43–46.

64. Ojima I, Chakravarty S, Inoue T, et al. A common pharmacophore for cytotoxic natural products that stabilize microtubules. Proc Natl Acad Sci USA 1999;96:4256–4261.

65. He L, Jagtap PG, Kingston DG, Shen HJ, Orr GA, Horwitz SB. A common pharmacophore for Taxol™ and the epothilones based on the biological activity of a taxane molecule lacking a C-13 side chain. Biochemistry 2000;39:3972–3978.

66. Luduena RF. Multiple forms of tubulin: different gene products and covalent modifications. Int Rev Cytol 1998;178:207–275.

67. Nicoletti MI, Valoti G, Giannakakou P, et al. Expression of beta-tubulin isotypes in human ovarian carcinoma xenografts and in a sub-panel of human cancer cell lines from the NCI- Anticancer Drug Screen: correlation with sensitivity to microtubule active agents. Clin Cancer Res 2001;7:2912–2922.

68. Wang Y, O'Brate A, Zhou W, Giannakakou P. Resistance to Microtubule-Stabilizing Drugs Involves two Genetic Events: Mutation in ··-tubulin in one allele Followed by Loss of the second allele. Cell Cycle 2005;4:1847–53.

69. Hua XH, Genini D, Gussio R, et al. Biochemical genetic analysis of indanocine resistance in human leukemia. Cancer Res 2001;61:7248–7254.

70. Leoni LM, Hamel E, Genini D, et al. Indanocine, a microtubule-binding indanone and a selective inducer of apoptosis in multidrug-resistant cancer cells. J Natl Cancer Inst 2000;92:217–224.

71. Ravelli RB, Gigant B, Curmi PA, et al. Insight into tubulin regulation from a complex with colchicine and a stathmin-like domain. Nature 2004;428:198–202.

72. Johnson IS, Wright HF, Svoboda GH, Vlantis J. Antitumor principles derived from Vinca rosea Linn. I. Vincaleukoblastine and leurosine. Cancer Res 1960;20:1016–1022.

73. Cutts JH, Beer CT, Noble RL. Biological properties of Vincaleukoblastine, an alkaloid in Vinca rosea Linn, with reference to its antitumor action. Cancer Res 1960;20:1023–1031.

74. Duflos A, Kruczynski A, Barret JM. Novel aspects of natural and modified vinca alkaloids. Curr Med Chem Anti-Cancer Agents 2002;2:55–70.

75. Zhou J, Giannakakou P. Targeting microtubules for cancer chemotherapy. Curr Med Chem Anti-Cancer Agents 2005;5:65–71.

76. Kavallaris M, Tait AS, Walsh BJ, et al. Multiple microtubule alterations are associated with Vinca alkaloid resistance in human leukemia cells. Cancer Res 2001;61:5803–5809.

77. Poruchynsky MS, Kim JH, Nogales E, et al. Tumor cells resistant to a microtubule-depolymerizing hemiasterlin analogue, HTI-286, have mutations in alpha- or beta-tubulin and increased microtubule stability. Biochemistry 2004;43:13,944–13,954.

78. Loganzo F, Hari M, Annable T, et al. Cells resistant to HTI-286 do not overexpress P-glycoprotein but have reduced drug accumulation and a point mutation in alpha-tubulin. Mol Cancer Ther 2004;3:1319–1327.

79. Verrills NM, Walsh BJ, Cobon GS, Hains PG, Kavallaris M. Proteome analysis of vinca alkaloid response and resistance in acute lymphoblastic leukemia reveals novel cytoskeletal alterations. J Biol Chem 2003;278:45,082–45,093.

80. Sawada T, Kobayashi H, Hashimoto Y, Iwasaki S. Identification of the fragment photoaffinity-labeled with azidodansyl-rhizoxin as Met-363-Lys-379 on beta-tubulin. Biochem Pharmacol 1993;45:1387–1394.

81. Rai SS, Wolff J. Localization of the vinblastine-binding site on beta-tubulin. J Biol Chem 1996;271:14,707–14,711.

82. Takahashi M, Kobayashi H, Iwasaki S. Rhizoxin resistant mutants with an altered beta-tubulin gene in Aspergillus nidulans. Mol Gen Genet 1989;220:53–59.

83. Takahashi M, Matsumoto S, Iwasaki S, Yahara I. Molecular basis for determining the sensitivity of eucaryotes to the antimitotic drug rhizoxin. Mol Gen Genet 1990;222:169–175.

84. Haber M, Norris MD, Kavallaris M, et al. Atypical multidrug resistance in a therapy-induced drug-resistant human leukemia cell line (LALW-2): resistance to Vinca alkaloids independent of P-glycoprotein. Cancer Res 1989;49:5281–5287.

85. Bai R, Covell DG, Pei XF, et al. Mapping the binding site of colchicinoids on beta-tubulin. 2-Chloroacetyl-2-demethylthiocolchicine covalently reacts predominantly with cysteine 239 and secondarily with cysteine 354. J Biol Chem 2000;275:40,443–40,452.

86. Gupta S, Bhattacharyya B. Antimicrotubular drugs binding to vinca domain of tubulin. Mol Cell Biochem 2003;253:41–47.

87. Anderson HJ, Coleman JE, Andersen RJ, Roberge M. Cytotoxic peptides hemiasterlin, hemiasterlin A and hemiasterlin B induce mitotic arrest and abnormal spindle formation. Cancer Chemother Pharmacol 1997;39:223–226.

88. Nieman JA, Coleman JE, Wallace DJ, et al. Synthesis and antimitotic/cytotoxic activity of hemiasterlin analogues. J Nat Prod 2003;66:183–199.

89. Loganzo F, Discafani CM, Annable T, et al. HTI-286, a synthetic analogue of the tripeptide hemiasterlin, is a potent antimicrotubule agent that circumvents P-glycoprotein-mediated resistance in vitro and in vivo. Cancer Res 2003;63:1838–1845.

90. Nunes M, Kaplan J, Wooters J, et al. A photoaffinity analog of the tripeptide hemiasterlin labels alpha-tubulin within residues. Proc Am Assoc Cancer Res 2004;45:314–338.

91. Hari M, Nunes M, Zask A, et al. Hemiasterlin analogs exclusively label -tubulin at the interdimer region and specifically block subtilisin digestion of -tubulin. Proc Am Assoc Cancer Res 2004;45.

92. Nogales E. Structural insights into microtubule function. Annu Rev Biochem 2000;69:277–302.

93. Mayhew CN, Sumpter R, Inayat M, et al. Combination of inhibitors of lymphocyte activation (hydroxyurea, trimidox, and didox) and reverse transcriptase (didanosine) suppresses development of murine retrovirus-induced lymphoproliferative disease. Antiviral Res 2005;65:13–22.

94. Kollef MH. Gram-negative bacterial resistance: evolving patterns and treatment paradigms. Clin Infect Dis 2005;40(suppl 2):S85–S88.

95. Waterer GW. Monotherapy versus combination antimicrobial therapy for pneumococcal pneumonia. Curr Opin Infect Dis 2005;18:157–163.

96. DiazGranados CA, Jernigan JA. Impact of vancomycin resistance on mortality among patients with neutropenia and enterococcal bloodstream infection. J Infect Dis 2005;191:588–595.

97. Palazzo IC, Araujo ML, Darini AL. First report of vancomycin-resistant staphylococci isolated from healthy carriers in Brazil. J Clin Microbiol 2005;43:179–185.

98. Sampath D, Discafani CM, Loganzo F, et al. MAC-321, a novel taxane with greater efficacy than paclitaxel and docetaxel in vitro and in vivo. Mol Cancer Ther 2003;2:873–884.

99. Tinley TL, Randall-Hlubek DA, Leal RM, et al. Taccalonolides E and A: Plant-derived steroids with microtubule-stabilizing activity. Cancer Res 2003;63:3211–3220.

100. Zhou J, Gupta K, Yao J, et al. Paclitaxel-resistant human ovarian cancer cells undergo c-Jun NH2-terminal kinase-mediated apoptosis in response to noscapine. J Biol Chem 2002;277:39,777–39,785.

16 Microtubule Stabilizing Agents in Clinical Oncology

The Taxanes

Chris H. Takimoto and Muralidhar Beeram

CONTENTS

INTRODUCTION
PACLITAXEL
DOCETAXEL
CONCLUSIONS
REFERENCES

SUMMARY

The taxanes are a versatile and important class of drugs that target microtubules. Paclitaxel and docetaxel are some of the most widely used cancer chemotherapeutic agents in clinical oncology. In this chapter, we discuss the most common dose and treatment schedules, clinical efficacy, and the toxicity profiles of these agents in detail.

Key Words: Taxanes; paclitaxel; docetaxel.

1. INTRODUCTION

The taxanes represent one of the most important and clinically useful classes of anti-cancer agents in all of clinical oncology. Initial studies of the taxanes began in 1963 when natural product screening programs demonstrated that extracts of the bark of the Pacific yew tree (*Taxus brevifolia*) had antitumor activity *(1,2)*. Several years later, Monroe Wall and colleagues *(3)* identified paclitaxel as the active component in these crude extracts. Early studies of the taxanes were limited by the difficulties in preparing sufficient quantities of paclitaxel for pharmacological testing and by its poor aqueous solubility. However, further interest in this novel agent was stimulated by pioneering studies by Horowitz and colleagues that characterized paclitaxel's novel mechanism of action as a microtubule stabilizer *(4,5)*. Subsequent studies of related compounds led to the synthesis of docetaxel, a semisynthetic derivative of 10-deacetylbaccatin III, which is an inactive but much more easily isolated taxane precursor found in yew tree needles. The yew tree needles represent a renewable resource that is much more readily available than tree bark *(6)*.

From: *Cancer Drug Discovery and Development: The Role of Microtubules in Cell Biology, Neurobiology, and Oncology* Edited by: Tito Fojo © Humana Press, Totowa, NJ

Paclitaxel and docetaxel are both complex ester derivatives that share a common tax-ane 15-ring structure bound to an oxetan ring. Structurally, paclitaxel and docetaxel dif-fer in the nature of the substitutions on the C13 ester side chain and in the alkyl groups bound to the C10 carbon. These structural differences render docetaxel slightly more water soluble than paclitaxel; however, both taxanes share a common binding site on the β-tubulin subunit that localizes to the inner surface of the microtubule lumen. These binding regions are completely distinct from those used by colchicine and the vinca alkaloids. Taxane binding results in the stabilization of polymerized microtubules by promoting the nucleation and elongation phases of microtubule assembly; thereby reducing the tubulin protein concentration required for microtubule formation *(1,2)*. Docetaxel has about a twofold higher binding affinity for β-tubulin and may have slightly higher in vitro potency. This taxane-induced stabilization alters microtubule dynamics, ultimately interfering with mitotic spindle formation. This disruption can induce p53 expression and inhibit cyclin-dependent kinases resulting in cell cycle arrest in mitosis, identified on fluorescent activated cell sorter analysis as arrest in the G_2M phase of the cell cycle (G_2M arrest). Subsequently, inactivation of the antiapoptotic $Bclx_L$/Bcl-2 proteins occurs and apoptosis ensues. The similarity of action of paclitaxel and docetaxel raises the question of whether complete clinical cross resistance exists between the taxanes. The clear clinical activity of one taxane in cancer patients truly refractory to another has not been unequivocally demonstrated. A detailed discussion of the taxane's mechanism of action is provided in Chapter 13.

Taxane resistance has been associated with enhanced drug efflux from cells mediated by the multidrug resistance gene, *MDR1* that encodes for the membrane-associated drug efflux pump, P-glycoprotein *(7)*. Other less well-characterized transporters may also be important in mediating clinical taxane resistance. In cell lines, mutations in the α- and β-tubulin subunits have been characterized that promote taxane resistance; however, the importance of these mutations in the clinical setting has not been determined *(8,9)*. Finally, the increased expression of antiapoptotic proteins such as Bcl-2 may also con-fer a relative degree of resistance to the taxanes.

2. PACLITAXEL

2.1. Doses/Schedules

Paclitaxel has been administered as an intravenous infusion in schedules ranging in duration from 1-h to as long as 96-h. In early studies of 6 h infusion schedules, the phar-macokinetics of the drug in plasma was characterized by a biphasic process with dose proportional kinetics *(1,2,10,11)*. However, detailed pharmacokinetics analyses of shorter 3 h paclitaxel infusions demonstrate nonlinear pharmacokinetics with saturable drug distribution and elimination. These shorter infusions are associated with higher plasma concentrations of the Cremophore™ vehicle, which may contribute in part to paclitaxel's nonlinear kinetics *(12)*.

Paclitaxel is the most commonly administered at doses ranging from 175 to 200 mg/m^2 infused during 3 or 24 h every 3 wk. Prophylactic administration of glucocorticoids and H1 and H2 antagonists to prevent hypersensitivity reactions is considered routine when administering paclitaxel chemotherapy. Because longer infusion schedules have not demonstrated clear superiority during shorter infusion durations, most oncologists admin-ister 175 mg/m^2 of paclitaxel as a 3-h infusion. In addition, weekly drug administration schedules have been explored with some promising early results. Weekly paclitaxel infusions at doses ranging from 80 to 100 mg/m^2 induce less bone marrow suppression with no apparent loss in antitumor efficacy compared with conventional schedules *(13–17)*. Disadvantages of weekly administration schedules include lack of extensive data in randomized phase III studies, and the requirement for weekly treatment center visits. Higher doses of paclitaxel, often administered with growth factor support, have been referred to as "dose-intense" paclitaxel regimens and this strategy has been extensively explored in clinical trials. In advanced solid tumors, high doses of paclitaxel of up to 250 mg/m^2 given every 3 wk may increase response rates in diseases such as advanced ovarian, breast, nonsmall cell lung (NSCL), and head and neck cancers; but, overall survival (OS) is not improved *(18)*. However, these dose intense paclitaxel regimens are associated with increased drug-related toxicities, and in the final analysis, there are very little compelling data to support the aggressive use of these toxic, dose-intense regimens. In contrast, emerging data on the use of more frequent administration of paclitaxel at standard doses given every 2 wk instead of every 3 wk are quite promising *(19,20)*. These so called "dose-dense" treatment strategies may offer improved efficacy in the adjuvant treatment of breast cancer without substantially increasing drug-related toxicities.

2.2. Toxicities

The major dose limiting toxicity of paclitaxel is myelosuppression, principally neutropenia, with thrombocytopenia and anemia occurring less commonly. When administered on an every 3 wk schedule bloodcounts typically nadir by days 8–10, with recovery occurring by day 21. The degree of previous bone marrow suppressive therapy is predictive of the severity of neutropenia, with heavily pretreated patients showing a substantial increase in risk of more profound neutropenia. In pharmacodynamic studies, the degree of neutropenia was found to correlate with the time that the concentrations of paclitaxel in plasma are higher than threshold plasma concentrations, typically 0.05–0.1 µM *(10,21,22)*. Paclitaxel induced neutropenia is not cumulative, and in most cases the duration of severe neutropenia is brief, lasting less than 7 d even in patients with extensive earlier chemotherapy.

Paclitaxel therapy is associated with a classical peripheral neuropathy that is frequently cumulative in nature. The clinical presentation is one of loss of proprioception combined with diminished soft touch sensation in a stocking and glove distribution, with attenuation of deep tendon reflexes *(23,24)*. The mechanism of paclitaxel induced neuropathy has been attributed to axonal degradation associated with demyelination. Symptoms are commonly dose related with severe peripheral neuropathy being infrequent at doses less than 200 mg/m^2 on an every 3 wk schedule, or at doses less than 100 mg/m^2 on a weekly administration schedule. Higher doses above 250 mg/m^2 or more may be associated with acute presentations of neuropathy beginning as early as 1–3 d after treatment. However, the most common clinical scenario is the slow, cumulative onset of neuropathy occurring after multiple rounds of chemotherapy at doses ranging from 135 to 250 mg/m^2 on an every 3 wk schedule. Risk factors for severe neuropathy include earlier exposure to other neurotoxic drugs (e.g., cisplatin), or the presence of chronic neuropathy because of other conditions including diabetes, neurological disorders, or ethanol abuse.

In early clinical trials of paclitaxel, the incidence of severe hypersensitivity reactions was as high as 30% *(25,26)*. However, the institution of routine prophylaxis with antihistamines, glucocorticoids, and H$_2$-blocking agents has reduced this incidence to 1%

and 2.1%, for the 3 h and 24 h infusions, respectively *(27)*. Acute paclitaxel hypersensitivity reactions are manifested as shortness of breath, airway hyper-reactivity, bradycardia, low blood pressure, body pains, and an urticarial skin rash, frequently beginning within minutes of the first or second drug infusion. Prompt termination of the infusion followed by standard supportive care for acute hypersensitivity reactions is usually sufficient to resolve the episode. Reinitiation of paclitaxel infusions has been successful in some cases following high dose glucocorticoid administration. Minor acute reactions including skin rash and flushing responses may be seen more commonly, but in such cases continued treatment is not usually associated with any severe sequelae. These paclitaxel-associated allergic reactions have been attributed in large part to the Cremophore vehicle used to administer the drug.

Paclitaxel is a mild vesicant, and localized necrosis at sites of extravasation has been reported. One clinical report recommends close observation and the administration of cold packs instead of heat as potential treatment for paclitaxel extravasation events *(28)*. Another relatively infrequent dermatological side effect is a localized skin reaction at the site of previous radiation therapy; the so-called radiation "recall" reaction *(29,30)*.

2.3. Paclitaxel: Clinical Utility

2.3.1. OVARIAN CANCER

Paclitaxel has clinically significant activity in the treatment of a variety of solid tumors, including breast and lung cancers *(1,2)*; however, the earliest clinical studies of paclitaxel were in ovarian cancer. Paclitaxel was initially approved for the treatment of patients with advanced ovarian cancer refractory to first-line platinum-containing chemotherapy. Later studies demonstrated that paclitaxel in combination with either cisplatin or carboplatin was a useful adjuvant therapy for earlier stage ovarian cancer patients. Currently, the treatment of choice for newly diagnosed patients with advanced ovarian cancer is surgical debulking followed by six courses of carboplatin and paclitaxel.

The role of paclitaxel in combination with a platinum agent in the treatment of ovarian cancer was established in several studies. In Gynecological Oncology Group study 111, 386 patients with advanced ovarian cancer and large volume disease were randomized to treatment with 6 courses of cisplatin at 75 mg/m^2 in combination with either cyclophosphamide 750 mg/m^2 or paclitaxel 135 mg/m^2 infused during 24 h *(31)*. All patients had either stage IV or suboptimally debulked stage III disease. The cisplatin-paclitaxel combination was associated with a higher incidence of febrile neutropenia, peripheral neuropathy, myalgias, hypersensitivity, adverse cardiac events, and alopecia; however, the efficacy of the taxane combination arm was also superior. After a median follow-up of 37 mo, the paclitaxel treatment arm showed a significant improvement in objective response rate (73% vs 60%, $p = 0.01$) and in median survival (38 vs 24 mo, $p < 0.001$). Overall, cisplatin-paclitaxel was associated with a 33% more reduction in mortality compared with the cisplatin–cyclophosphamide containing arm.

These data were confirmed by a European–Canadian trial (OV10) comparing the combination of cyclophosphamide 750 mg/m^2 and cisplatin 75 mg/m^2 to paclitaxel 175 mg/m^2 during 3 h and cisplatin 75 mg/m^2 in patients with stage IIB to Stage IV ovarian cancer *(32)*. Once again the paclitaxel containing arm demonstrated a superior response rate (59% vs 45%) and better OS (35.6 vs 25.8 mo) after a median follow-up period of 38.5 mo. These two studies firmly established the role of paclitaxel in combination with a platinum agent as the standard of care for newly diagnosed patients with advanced ovarian cancer. At present, it is still unclear if one platinum derivative is superior than another when given in combination with paclitaxel. However, because cisplatin-paclitaxel regimens may generate more severe toxicities compared with carboplatin combinations *(33,34)*, most oncologists now favor carboplatin dosed to an area-under-the-curve (AUC) of 6–7 in combination with paclitaxel at 175 mg/m^2 during 3 h as the treatment of choice for patients with ovarian cancer.

A Scottish study examined the utility of using docetaxel instead of paclitaxel in combination with carboplatin in advanced ovarian cancer *(35)*. Although the paclitaxel arm generated slightly more neurotoxicity, it was also significantly better tolerated in terms of myelosuppressive and nonhematological toxicities when compared with docetaxel. Thus, paclitaxel remains the most common taxane used in this setting I.

An as yet unanswered question is what is the optimal duration of paclitaxel therapy in the treatment of advanced ovarian cancer? A single randomized Phase III study examined the utility of administering a prolonged maintenance course of paclitaxel chemotherapy to advanced ovarian cancer patients experiencing a complete clinical response during front-line therapy with paclitaxel and carboplatin *(37)*. Completely responding patients were randomized to receive an additional 3 or 12 mo of paclitaxel at 175 mg/m^2 during 3 h. A marked improvement in progression-free survival favoring the prolonged 12 mo maintenance treatment schedule led to early termination of this study. The median progression-free survival was 21 mo vs 28 mo ($p = 0.0023$) for the 3 and 12 mo groups, respectively; however, OS was equivalent at the time of study closure. Toxicities were enhanced by the prolonged drug administration; but overall, these were judged to be clinically manageable. Studies to confirm show the benefit of maintenance paclitaxel therapy in this setting have been designed.

In early stage disease, ovarian cancer patients undergoing complete surgical debulking with no other risk factors have a recurrence risk of less than 10%. Such patients need no additional therapy other than observation. However, patients with any high risk features have a risk of recurrence of about 40%. High risk factors are defined as any of the following: histological grade 3 (poorly differentiated) cancers, tumor on the surface of or extrinsic to the ovary, positive peritoneal cytology, or extra ovarian stage II disease. In early stage high-risk ovarian cancer patients, a European study randomized 925 women to treatment with a platinum containing adjuvant chemotherapy regimen or to observation alone. At 4 yr of follow-up, the recurrence free 5-yr survival was 76% for the treatment arm compared with 65% for the controls, corresponding to a relative risk of 0.64 ($p = 0.001$). Overall 5 yr survival was 82% for the treatment arm and 74% for the control patients (relative risk of 0.67, $p = 0.008$) *(32)*. Based upon these finding, most physicians now favor adjuvant therapy for any early stage ovarian cancer patients with high risk factors.

A Gynecological Oncology Group study (Protocol 157) randomized 457 women with completely resected early stage high-risk ovarian cancer to treatment with three or six cycles of postoperative paclitaxel at 175 mg/m^2 during 3 h in combination with carboplatin dosed at an AUC of 7.5 every 3 wk. No significant differences were seen for the two treatment durations *(38)*. Currently, most patients with completely resected early stage ovarian cancer and any high-risk factors are treated with a minimum of three cycles of adjuvant paclitaxel and carboplatin chemotherapy *(37,38)*.

An important issue in the optimal use of paclitaxel is whether increasing the dose intensity by escalating the taxanes to maximal dose levels with growth factor support provides an overall improvement in therapeutic index. This approach was examined in a randomized study by Omura et al *(39)* comparing dose-intense single agent paclitaxel to standard dose regimens in women with platinum refractory ovarian cancer. Women were randomized to receive 24 h infusions of 175 mg/m^2 or 250 mg/m^2 infusions of paclitaxel, with growth factor support given at the higher dose level. Primary end points were response rate, progression-free survival, and OS. Response rates were significantly better for the higher dose vs the lower (36% vs 27%, respectively; $p = 0.027$), but progression-free survival (5.5 mo vs 4.8 mo, respectively) and median survival (13.1 mo vs 12.3 mo, respectively) were not significantly improved. Furthermore, myelosuppression, neuropathy, and myalgias were substantially more severe on the high dose arm. This study is in complete agreement with many earlier trials of paclitaxel dose-intense regimens conducted in breast, lung, and head and neck cancer patients that consistently demonstrate higher response rates without a survival benefit *(18)*. The lack of improved survival and the increase in drug-related serious toxicities argue strongly against the wide-spread implementation of dose-intense paclitaxel strategies in the palliative treatment of solid tumors.

2.3.2. ADVANCED BREAST CANCER

After its initial success in ovarian cancer, paclitaxel was also subsequently approved for use in advanced breast cancer patients in the second-line setting *(40)*. Currently, the taxanes have an established role in the management of patients with advanced anthracycline-resistant metastatic breast cancer. Paclitaxel has also been approved for the adjuvant treatment of patients with early stage breast cancer with lymph node involvement when administered after surgery in combination with anthracyclines. This indication was based upon randomized studies showing an improved progression-free survival, although a clear benefit in OS was not definitively demonstrated *(41)*. More recently, clinical trials of adjuvant chemotherapy for breast cancer patients have examined more novel taxane-containing regimens by exploring dose-dense and/or dose-intense treatment strategies.

Paclitaxel has been extensively studied as first-line therapy for the treatment of women with metastatic breast cancer. In a randomized study in 209 breast cancer patients previously untreated for metastatic disease, 200 mg/m^2 of paclitaxel infused during 3 h was compared with a nonanthracycline containing regimen consisting of cyclophosphamide, methotrexate, 5-fluorouracil, and prednisone *(42)*. Overall, a nonsignificant trend favoring paclitaxel was seen in the objective response rate (35% vs 29%, $p = 0.37$), and in the median time to tumor progression (6.4 vs 5.3, $p = 0.25$). Crossover between the treatment arms was not permitted and median survival was marginally better for the paclitaxel arm (17.3 vs 13.9 mo, $p = 0.068$). Myelosuppressive toxicities were less common on the paclitaxel arm; however, overall quality of life was similar.

A second study randomized 331 patients with advanced metastatic breast cancer to either 175 mg/m^2 of paclitaxel during 3 h vs single-agent doxorubicin 75 mg/m^2 *(43)*. The primary end points favored the doxorubicin arm over the paclitaxel treatment, with response rates of 41% and 25%, respectively ($p = 0.003$), and progression-free survivals of 7.5 vs 3.9 mo, respectively ($p < 0.001$). Median survival was not significantly different for doxorubicin compared with paclitaxel (18.3 vs 15.6 mo, $p = 0.38$); however, crossover between the two arms was allowed. Doxorubicin was associated with higher bone marrow suppression and more cardiac side effects, whereas myalgias and peripheral neuropathy were more common with paclitaxel. Thus, paclitaxel offered no advantage over single agent doxorubicin as first-line therapy for metastatic breast cancer.

Finally, a large intergroup three arm trial compared single agent doxorubicin vs single agent paclitaxel at 175 mg/m^2 during 3 h vs the combination of both paclitaxel and doxorubicin in 739 patients with advanced untreated metastatic breast cancer *(44)*. Interestingly, paclitaxel response rates (34%) were equivalent to doxorubicin alone (34%, $p = 0.77$), but were inferior to the combination arm (47%, $p = 0.006$). No difference in OS was seen when paclitaxel (22.5 mo) was compared with either doxorubicin (19.1 mo, $p = 0.60$) or the combination (22.4 mo, $p = 0.49$). Not surprisingly, myelosuppressive toxicities were the most pronounced on the combination arm, but no increase in cardiotoxicity was observed when paclitaxel was combined with doxorubicin. Together, these two first-line therapy studies in advanced breast cancer patients suggest that single-agent paclitaxel may have some benefit over nonanthracycline-based chemotherapy, but it offers no advantage over anthracycline-containing regimens.

In combination chemotherapy studies, paclitaxel plus doxorubicin was compared with 5-fluorouracil, doxorubicin, and cyclophosphamide (FAC) in 267 advanced, chemotherapy-naïve breast cancer patients *(45)*. The time to tumor progression (8.3 vs 5.2 mo, $p = 0.034$), response rate (68 vs 55%, $p = 0.032$), and OS (23.3 vs 18.3 mo, $p = 0.013$) all favored the doxorubicin and paclitaxel combination. Tolerability of both arms was similar. A second trail conducted by the European Organization for Research and Treatment of Cancer (EORTC) compared doxorubicin and paclitaxel with doxorubicin and cyclophosphamide as initial treatment for advanced breast cancer in 275 patients *(46)*. In contrast to the first study, the progression-free survival, response rate, and median survival were equivalent for the two arms. The incidence of neutropenic fever (32% vs 9%, $p < 0.001$) was higher for the paclitaxel arm, although other toxicities were comparable. Several similar studies have also examined whether paclitaxel plus the anthracycline epirubicin has any advantage over a nontaxane combination such as epirubicin and cyclophosphamide in women with advanced breast cancer. However, epirubicin–paclitaxel had no clear superiority over any of the nontaxane combination treatment regimens used in these studies *(46,47)*.

For women with HER2/neu expressing tumors, the addition of trastuzumab to single agent paclitaxel provides clear clinical benefit. This was demonstrated in a large study of women with advanced breast cancer treated with chemotherapy with and without the Her2/neu (ErbB2) targeting antibody, trastuzumab *(48)*. In this study design, most women were randomized to treatment with doxorubicin ± trastuzumab; however, if they had previously received any anthracycline treatments, they were randomized to single-agent paclitaxel ± trastuzumab. Overall, 92 patients were treated with paclitaxel and trastuzumab, and 96 received paclitaxel alone. For the paclitaxel-treated subgroup, the median time to tumor progression (6.9 vs 3, $p < 0.001$) and the overall objective response rate (38% vs 16%, $p < 0.001$) favored the paclitaxel + trastuzumab combination. The combination trended toward an improved median survival (22.1 vs 18.4 mo, $p = 0.17$), but it was also associated with an increased risk of cardiotoxicity. However, in the context of the larger study, this subgroup of paclitaxel–trastuzumab treated patients had less overall toxicity than the much larger number of women treated with doxorubicin and trastuzumab.

Unfortunately, there is a paucity of published randomized studies that carefully examine the activity of second-line paclitaxel chemotherapy in patients with anthracycline resistant breast tumors *(47,49)*. One phase III study presented only in preliminary form suggested that second-line paclitaxel was less effective than second-line treatment with cisplatin and etoposide *(49)*. In women previously treated with an anthracycline, paclitaxel had an inferior response rate and time to tumor progression than etoposide-cisplatin. This is in contrast to the numerous phase III studies of docetaxel in the second-line setting showing an advantage over nontaxane containing combinations regimens. (*see* docetaxel discussion later). Thus, docetaxel instead of paclitaxel is typically the favored taxane for the initial treatment of advanced breast cancer patients who are clearly anthracycline refractory.

To summarize, as first-line therapy for advanced metastatic breast cancer patients, single agent paclitaxel is active, but offers no benefit over doxorubicin-based chemotherapy. In combination with an anthracycline, initial therapy with paclitaxel may offer some benefit over nontaxane anthracycline-based regimens, but the total data are not consistent on this point. Clearly, clinical evidence does not support the replacement of doxorubicin with paclitaxel in frontline regimens, but the use of sequential single agent therapy and the use of paclitaxel–anthracycline combinations need further study. One caution regarding the expanded use of paclitaxel combination with doxorubicin is the pharmacokinetic interaction that leads to higher exposures to doxorubicin and its metabolites *(50)*. This may cause more exposures to doxorubicin and enhance response rates but may also increase long-term cardiotoxicity. Current studies are aimed at minimizing this interaction by using sequential instead of concurrent administration of these two agents. Finally, in Her2/neu over expressing patients, paclitaxel and trastuzumab may offer a benefit over paclitaxel alone.

The optimal taxane to use in the treatment of metastatic breast cancer patients is made more complicated by the growing body of data supporting the use of docetaxel in this setting. No pharmacokinetic interaction has been reported between docetaxel and doxorubicin. At present, the use of single agent docetaxel or combination regimens containing docetaxel have largely supplanted paclitaxel as the treatment of choice for anthracycline resistant breast cancer patients. However, direct head to head comparisons between docetaxel and paclitaxel are lacking. In a preliminary but as yet not fully published report on 449 women with anthracycline-refractory metastatic breast cancer, docetaxel had superior time to tumor progression and OS compared with paclitaxel *(46,47)*. Interestingly, some clinical studies suggest a lack of cross-resistance between paclitaxel and docetaxel *(51,52)*, and in clinical practice, the use of both taxanes at some point during the treatment of women with advanced breast cancer is relatively common.

2.3.3. ADJUVANT BREAST CANCER THERAPY

In adjuvant breast cancer trials, paclitaxel is almost always used in sequence with anthracyclines in order to circumvent potential increases in cardiac toxicity induced by the concurrent administration of paclitaxel and doxorubicin. Two conceptually similar trials have been conducted: the Cancer and Leukemia Group B (CALGB) 9344 study and the National Surgical Adjuvant Breast and Bowel Project (NSABP) B-28 study. Both trials compared 4 cycles of standard doxorubicin and cyclophosphamide (AC) chemotherapy with the same AC regimen followed by an additional 4 courses of single-agent paclitaxel given in 3-h.

In the CALGB 9344 study, 3121 women with positive lymph nodes and early stage breast cancer were randomized to four cycle of AC alone, or the same regimen followed by four additional cycles of 175 mg/m^2 of paclitaxel *(41)*. The taxane-containing arm showed a clear benefit in improved overall and disease-free survival (DFS). In the NSABP (B-28) trial, 3060 women with node positive breast cancer were randomized to adjuvant postoperative treatment with 4 cycles of AC or the same treated followed by four additional cycles of paclitaxel at 225 mg/m^2 *(49)*. Once again, the addition of paclitaxel to the treatment regimen provided a significant benefit in DFS (76% vs 72%, $p = 0.008$); but in contrast to the CALGB study, no improvement in OS was demonstrated.

Norton and colleagues *(53)* have extended the concept of dose intensity to explore the more refined strategy of dose-density, which they define as standard doses of taxane chemotherapy administered on a more frequent schedule. In early stage breast cancer patients, the CALGB 9741 study compared a dose-dense combination of doxorubicin, cyclophosphamide followed by paclitaxel administered every 2 wk with the same regimen given on a standard every 3 wk schedule *(20)*. This study also compared sequential doxorubicin, paclitaxel, and cyclophosphamide administered separately as consecutive single agents with the more common coadministration of doxorubicin and cyclophosphamide together followed by additional course of single-agent paclitaxel. Interestingly, the DFS and OS were improved on the dose-dense arms with a 26% and 31% decrease in the relative risk in these two key end points, respectively. Surprisingly, the incidence of serious neutropenia and myelosuppression was less in the dose-dense treatment group, probably because of the mandatory use of prophylactic growth factors. No difference was observed in these end points when the sequential consecutive schedules were compared with the more standard coadministration schedules. Based upon these data, administration of every 2 wk courses of paclitaxel-based chemotherapy is a reasonable option to consider in the adjuvant treatment of breast cancer patients *(54)*.

In the neoadjuvant setting, paclitaxel has been less extensively studied than docetaxel. In a randomized trial, Buzdar and colleagues *(55)* compared preoperative paclitaxel at 250 mg/m^2 with doxorubicin, 5-fluorouracil, and cyclophosphamide in 174 women with early stage breast cancer. The response rates in both arms were comparable for paclitaxel vs the combination (80 vs 79%, respectively) although the pathological complete response rate was lower in the paclitaxel alone arm (8% vs 16%). Longer term follow-up of these patients beyond 2-yr l is necessary. Several other trials of neoadjuvant paclitaxel in combination with anthracyclines are in progress *(56)*.

2.3.4. LUNG CANCER

Paclitaxel is approved in combination with cisplatin chemotherapy for use in the initial treatment of advanced NSCL cancer patients *(57)*. Based upon impressive single agent activity and potential synergistic interactions with platinum derivatives, taxanes in combination with a platinum agent are a common treatment choice for first-line therapy of patients with this disease *(58)*. Currently, a variety of two-drug combination regimens are employed in advanced NSCL cancer, and a worldwide consensus on a specific optimal combination does not exist. In North America, paclitaxel and carboplatin is the most common chemotherapy doublet used in the treatment of patients with NSCL cancer.

The clearest demonstration of the impressive utility of paclitaxel in NSCL cancer was the randomized study by Ranson et al. *(59)* where 157 chemotherapy naïve patients with Stage IIIB or IV disease were randomized to paclitaxel 200 mg/m^2 in 3 h or to the best supportive care. Median survival was significantly improved by chemotherapy (6.8 vs 4.8 mo, $p = 0.0037$), and functional and quality of life assessments also favored the chemotherapy arm. Subsequent studies in advanced NSCL cancer patients have combined paclitaxel with a variety of agents including cisplatin and carboplatin, with generally comparable results in terms of antitumor efficacy *(58)*.

Two of the larger studies that illustrate the current state of knowledge about the use of taxane-platinum combination chemotherapy are the SWOG 9509 and the Eastern Cooperative Oncology Group (ECOG) 1594 studies. In the large SWOG 9509 study, 202 advanced NSCL cancer patients were randomized to treatment with paclitaxel 225 mg/m^2 and carboplatin dosed to an AUC of 6 every 2 wk or to another doublet chemotherapy regimen consisting of vinorelbine and cisplatin *(61)*. When the paclitaxel– carboplatin treatment was compared with the vinorelbine–cisplatin arm, objective response rates (25% vs 28%, respectively) and median survival (8.6 vs 8.1 mo, respectively) were comparable. However, myelosuppressive toxicities, nausea and vomiting, and dose reductions were more common on the nontaxane containing treatment arm. Assessment of quality of life was similar for both therapies. Overall the paclitaxel and carboplatin doublet was favored because of improved convenience and tolerability. In the ECOG 1594 trial, four distinct combination chemotherapy doublets were directly compared: paclitaxel and cisplatin, paclitaxel and carboplatin, gemcitabine and cisplatin, and docetaxel and cisplatin *(62)*. Again no difference in efficacy end points was seen and overall toxicity was generally comparable. OS was 7.8 vs 8.1 vs 8.1 vs 7.4 for the paclitaxel/cisplatin, paclitaxel/carboplatin, gemcitabine/cisplatin, and docetaxel/cisplatin arms, respectively. Response rates were 21% vs 17% vs 22% vs 16%, respectively. Toxicity was judged to be similar on all arms, although severe grade 3 and 4 toxicities were the lowest in the carboplatin and paclitaxel arm *(57)*.

Finally, a large Italian study compared the vinorelbine and cisplatin regimen with paclitaxel and carboplatin and with gemcitabine and cisplatin in a three arm randomized study in advanced NSCL cancer patients. When the vinorelbine and cisplatin, paclitaxel and carboplatin, and gemcitabine and cisplatin arms were compared, no significant differences were seen in response rate (30% vs 32% vs 30%, respectively) or in median survival (9.5 vs 9.9 vs 9.8 mo, respectively) *(63)*. Toxicity was worse on vinorelbine and cisplatin arm, but was comparable for the other two chemotherapy doublets. The paclitaxel–carboplatin therapy was associated with more alopecia that the other two arms, but it was otherwise well tolerated. In summary, these trials together do not strongly favor any one chemotherapy doublet over another for the treatment of newly diagnosed patients with advanced NCSL cancer.

The issue of dose intensity in NSCL cancer was addressed in a Greek trial that randomized 198 advanced NSCL patients to either carboplatin and paclitaxel at 175 mg/m^2 or to the same regimen with a higher dose of paclitaxel of 225 mg/m^2 *(64)*. There was a trend favoring the more dose-intense paclitaxel regimen in terms of response rate (25.6% vs 31.8%) and the median time to tumor progression was significantly improved (4.3 vs 6.4 mo, $p = 0.044$). However, median survival was not significantly better (9.5 vs 11.4 mo, $p = 0.16$). As expected, higher neurotoxicity and myelosuppression was seen on the high dose arm, but the regimens were otherwise well tolerated. Again, there seems to be little benefit to using the more dose-intense regimen on a routine basis.

Finally, considerable clinical research is ongoing on the use of paclitaxel-based chemotherapy in treating NSCL cancer patients in other settings. These include the use of adjuvant chemotherapy after surgical resection of early stage disease, or the concurrent administration of paclitaxel chemotherapy in combination with radiation therapy for locally advanced disease *(57)*.

2.3.5. OTHER TUMORS

Paclitaxel has also been approved for use by the US Food and Drug Administration (US FDA) for the second-line treatment of Kaposi's sarcoma in HIV positive patients *(65,66)*. Paclitaxel also has activity in a variety of additional solid tumor types, including head and neck cancer *(67)*, small cell lung cancer *(68)*, gastric *(69)*, esophageal *(69)*, prostate *(70)*, unknown primary tumors *(71)*, nonovarian gynecological tumors *(72)*, bladder cancers *(73)*, and lymphomas *(74)*. However, its optimal clinical role in the pharmacological management of these diseases is less well defined than it is in breast, lung and ovarian cancers.

Other promising early strategies for improving paclitaxel-based chemotherapy are the development of novel formulations such as nanoparticle paclitaxel. Weekly nanoparticle albumin paclitaxel (ABI-007) has shown preliminary promising activity in a randomized study in advanced breast cancer *(75)*. ABI-007 is designed to improve intracellular tumor drug delivery by exploiting an albumin-specific transport mechanisms in endothelial cells. Advantages of this approach include ease of formulation and the ability to administer the agent without the need for the usual premedication for hypersensitivity reactions. Other modifications in drug delivery systems include paclitaxel poliglumex (Xyotax™, CT-2103), a polyglutamated paclitaxel derivative, which is in phase III testing in lung and ovarian cancer *(76)*.

3. DOCETAXEL

Docetaxel is structurally related to paclitaxel, but differs in the substitutions located at the C10 position on the taxane ring and in the ester side chain present at the C13 position *(6)*. Isolation of the inactive taxane compound, 10-deacetylbaccatin III, from the needles of the yew tree provides a renewable source for the synthetic precursor of docetaxel. The structural differences compared with paclitaxel may explain why docetaxel has a 1.9-fold higher affinity for binding to β-tubulin *(77)*. As a tubulin stabilizer, stoichiometric docetaxel binding blocks the disassembly of tubulin polymers and disrupts mitotic spindle function, ultimately leading to apoptosis. These processes occur at clinically relevant nanomolar concentrations of docetaxel.

Docetaxel is administered intravenously, and its plasma concentrations profiles are characterized by triphasic elimination kinetics with an elimination half-life ranging from 11–19 h. Docetaxel has a relatively large volume of distribution, consistent with its binding to a wide range of tissues. However, it is a substrate for the *MDR1*-mediated P-glycoprotein drug efflux pump, which explains its poor oral bioavailability and its limited penetration into the central nervous system *(78)*. Covariate factors that substantially influence docetaxel clearance include abnormalities in liver function, plasma α-1-acid glycoprotein concentrations, liver dysfunction, and to a lesser extent, age and serum albumin levels *(79)*. Elimination occurs predominantly through the biliary tract with about 70–80% of the administered agent eliminated in the feces. In contrast, only about 10% of the total administered dose of docetaxel is excreted in the urine. Major metabolites of docetaxel have not been well-characterized.

3.1. Doses/Schedules

Docetaxel is commonly administered every 3 wk at doses ranging from 60 to 100 mg/m^2 infused in 1-h, with 75 mg/m^2 being a frequent dose when docetaxel is given as a single agent or in combination with other chemotherapeutic agents *(6,80)*. At higher doses, docetaxel may be less well-tolerated, especially by patients who have received extensive previous bone marrow suppressive therapies. One-hour weekly administration schedules of docetaxel have also been explored, and may induce less myelosuppression compared with conventional schedules.

3.2. Toxicities

Docetaxel's major dose-limiting toxicity is myelosuppression *(6,80)*. Typically, blood counts will nadir on about day 8 of treatment, with full recovery by days 15–21. The extent of previous myelosuppressive therapy is an important determinant for the risk of severe docetaxel-induced myelosuppression. Alopecia is also extremely common and dermatological toxicities can occur in up to 50–75% of patients. Docetaxel skin manifestations can include a pruritic maculopapular rash appearing on the forearms, hands, or feet, and less commonly, desquamation of the palms or the soles of the feet (palmar-plantar erythrodysesthesia) can occur. Unlike paclitaxel, docetaxel therapy can cause edema and fluid accumulation including pleural effusions and ascites. The mechanism of this adverse reaction is not precisely defined, but may relate to increased capillary permeability. Fluid accumulation is often cumulative and resolves slowly after docetaxel discontinuation. Complete resolution may take several months *(81)*. Glucocorticoid premedication may ameliorate this fluid retention. Acute hypersensitivity reactions may arise and have been attributed to either Tween™ 80, the solvent for intravenous docetaxel administration, or to the drug itself. The incidence of allergic reactions is about 31% and can also be substantially controlled by glucocorticoid medication. The clinical manifestations of these allergic reactions often have their onset during the initial course or two of docetaxel therapy and are characterized by dyspnea, bronchospasm, and hypotension that begin within minutes after starting the infusion. Immediate termination of the infusion and antihistamine administration will generally allow the symptoms to resolve within 15 min and in the absence of more severe reactions, treatment often can be reinitiated without serious sequelae. Similar to paclitaxel, docetaxel therapy can cause mild, self-limited myalgias, and arthralgias; however, peripheral neuropathies, diarrhea, and mucositis are much less frequent than with paclitaxel therapy. Elevated liver function tests are uncommon.

3.3. Docetaxel: Clinical Utility

3.3.1. BREAST CANCER

Over the past decade, docetaxel has been extensively studied in women with breast cancer. The activity of single agent docetaxel was initially well documented in the treatment of women with advanced metastatic breast cancer patients refractory to anthracycline-containing therapy. Recent studies have demonstrated considerable clinical utility when docetaxel is combined with anthracyclines for the treatment of patients with earlier stages of this disease.

3.3.1.1. Advanced Metastatic Breast Cancer. Metastatic breast cancer has a very poor long-term prognosis with a median survival after the first documentation of metastatic disease of approx 2 yr. Anthracyclines are now an established component for the initial treatment of early stage and advanced breast cancer patients. As a consequence, clinical trials examining the utility of docetaxel in advanced breast cancer have been conducted within two distinct populations:

1. Patients with no- or minimal-earlier anthracycline exposure, or
2. Patients whose disease has recurred despite previous anthracycline therapy and whose tumors are anthracycline refractory or resistant *(47)*.

In both of these settings, docetaxel has been studied as a single agent and in combination with other therapies. In the majority of these studies, docetaxel appears to improve overall response rates and times-to-disease progression.

3.3.1.2. First-Line Therapy in Anthracycline Naïve Advanced Breast Cancer. Single agent docetaxel at 100 mg/m^2 was compared with doxorubicin monotherapy at 75 mg/m^2 in 326 patients with anthracycline-naïve advanced breast cancer. Both drugs were administered every 3 wk for up to a maximum of seven courses *(82)*. Progressive disease on a previous alkylating agent-containing regimen was required for study enrollment. Docetaxel resulted in a significantly higher response rates than doxorubicin (48% vs 33%, $p = 0.008$), but the median time to tumor progression (6.1 vs 4.9 mo, $p = 0.45$) and median OS (15 vs 14 mo, $p = 0.39$) were not significantly improved. Patients with a poor prognosis with resistant disease (defined as relapsing within 12 mo of adjuvant therapy), liver metastases, or other visceral disease all showed a benefit from docetaxel therapy. Toxic drug-related deaths, mostly because of cardiotoxicity, were more frequent with anthracycline therapy (3%) than with docetaxel (1.2%). Grade 4 neutropenia was similar in both groups; however, doxorubicin caused more grades 3 and 4 nonhematological toxicities. No differences in the quality of life scales were observed. Thus, in contrast to paclitaxel, single-agent docetaxel may offer an advantage over anthracycline single agent therapy in the initial treatment of advanced breast cancer patients.

The combination of docetaxel and an anthracycline was compared with a nontaxane-containing regimen in four randomized studies of first-line therapy in women with metastatic breast cancer. In all these studies, the docetaxel and anthracycline combination produced superior response rates. Three of the studies showed a benefit for the docetaxel regimens in terms of time to tumor progression and a benefit in OS was seen in two of the four studies *(49)*. The *TAX306* study compared the combination of docetaxel 75 mg/m^2 plus doxorubicin at 50 mg/m^2 (AT) with a nontaxane regimen consisting of doxorubicin at 60 mg/m^2 plus cyclophosphamide at 600 mg/m^2 (AC) *(83)*. Treatments were administered every 3-wk for up to 8 courses to 429 anthracycline-naïve metastatic breast cancer patients. Forty-two percent of patients had received previous adjuvant chemotherapy. Overall, AT therapy produced superior response rates (60 % vs 47%, $p = 0.0012$) and median time to tumor progression (8.7 vs 7.4 mo, $p = 0.015$) compared with the AC regimen. Median OS was not statistically significant (22.5 vs 21.5 mo). Both regimens were judged to be tolerable, but the AT arm was associated with slightly more neutropenia (82% vs 69%), febrile neutropenia (6% vs 2%), diarrhea, and asthenia. No differences in cardiotoxicity were reported.

The second randomized phase III study compared the triple combination of docetaxel, doxorubicin, and cyclophosphamide (TAC) with FAC as first-line chemotherapy in 484 patients with advanced breast cancer *(84)*. Thirty nine percent of patients had received previous chemotherapy including 11% with previous history of anthracycline-based chemotherapy, but in all cases the cumulative dose was less than 240 mg/m^2. Significantly better response rates were seen on the TAC arm compared with FAC therapy (55% vs 44%, respectively; $p = 0.023$). No differences were observed in the median time to tumor progression (31 vs 29 wk, respectively; $p = 0.51$) or in the overall median survival (21 vs 22 mo, respectively; $p = 0.93$). Grade 3 and 4 hematological toxicities including febrile neutropenia (29% vs 5%) and nonhematological toxicities were more frequent on the TAC arm.

A third randomized study compared docetaxel in combination with the anthracycline, epirubicin, (ET) with the nontaxane containing three drug combination of 5-fluorouracil, epirubicin, and cyclophosphamide chemotherapy (85). Both regimens were administered every 3-wk to 132 anthracycline-naive advanced breast cancer patients. The ET arm demonstrated a superior response rate (63% vs 34%) and median time to tumor progression (7.8 vs 5.9 mo). The fourth and final study compared the doublet of AT with FAC in 216 metastatic breast cancer patients (86). This study demonstrated a significantly superior response rate (64 % vs 41%, $p = 0.002$), median time to progression (8.1 vs 6.6 mo, $p = 0.002$), and median OS (22.6 vs 16.1 mo, $p = 0.02$) for the AT arm. Taken together, these series of studies suggest that docetaxel and anthracycline combinations may offer modest advantages in efficacy when compared with nontaxane containing regimens in treating newly diagnosed patients with advanced breast cancer.

3.3.1.3. Anthracycline-Resistant Advanced Breast Cancer. Single-agent docetaxel was compared with two separate doublet combination chemotherapy regimens in several randomized studies examining second-line therapy of patients with anthracycline-refractory advanced breast cancer. In the first study of 392 women with advanced breast cancer, all patients had previously had progression of their disease after previous anthracycline-based therapy had been administered for metastatic disease (50%), for adjuvant therapy (21%), or for both (29%) (83). Compared with mitomycin C and vinblastine, single-agent docetaxel generated superior response rates (30% vs 12%, $p < 0.0001$), median time to tumor progression (4.4 vs 2.6 mo, $p = 0.001$) and OS (11.4 vs 8.7 mo, $p = 0.0097$). Febrile neutropenia and the incidence of serious grade 3 or 4 infections were significantly more common for the docetaxel-treated patients, and nonhematological toxicities such as weakness, stomatitis, diarrhea, skin rash, nail changes, and neuropathy were also more frequent. A second study compared single agent docetaxel with methotrexate and 5-fluorouracil in 283 breast cancer patients, all of whom had received previous anthracycline-based chemotherapy (87). In this Scandinavian study, docetaxel again demonstrated superior response rates (42% vs 21%, $p < 0.001$) and median time to tumor progression (6.3 vs 3 mo, $p < 0.001$). Both treatment arms were generally well tolerated with infrequent grade 3 or 4 toxicities in both arms. No differences in quality of life assessments were observed in either of these two trials.

Finally, a third randomized phase III study compared single-agent docetaxel with continuous infusion 5-fluorouracil and vinorelbine in 176 women with anthracycline-refractory breast cancer (85). Objective response rates were equivalent for docetaxel vs 5-fluorouracil and vinorelbine (43% vs 38.9%, respectively; $p = 0.69$), and the overall efficacy of the two regimens were felt to be comparable. However, less febrile neutropenia and treatment related deaths were observed on the docetaxel arm.

These studies in total strongly support the use of docetaxel in the treatment of women with advanced breast cancer. However, an as yet unanswered question is what is the optimal taxane for use in this disease? Studies directly comparing the two taxanes head to head are lacking; however, a growing body of clinical evidence suggests that single-agent docetaxel may offer a slight advantage over paclitaxel as initial therapy for patients with advanced disease (49). As previously mentioned, very preliminary evidence suggests a modest time to tumor progression and OS advantage may exist for docetaxel at 100 mg/M^2 compared with paclitaxel at 175 mg/M^2 administered to women previously treated with anthracyclines on an every 3-wk schedule (88). However, the overall toxicity of these two taxane regimens was also different, with the docetaxel arm experiencing substantially more grade 3 or 4 neutropenia, asthenia, infection, edema, and stomatitis.

Weekly regimens of docetaxel are gaining popularity and have been explored in a variety of studies of women with advanced breast cancer. Because of the substantial toxicities observed with every 3 wk schedules of docetaxel, especially at higher doses, the weekly regimens may be an attractive alternative, especially for more elderly and frail patients. A review of the various weekly docetaxel regimes has been published by Wildiers et al. (89). When administered weekly, docetaxel doses range from 35 to 50 mg/m^2/wk given for 3 of 4 wk or for 6 of 8 wk. Although qualitatively toxicities are unchanged, the severity of observed toxicities, especially myelosuppression, may be modestly reduced with no apparent loss of antitumor effect.

To date, only one phase III trial has compared single-agent docetaxel to a capcitabine-docetaxel combination (90). In this study, 511 patients with anthracycline-resistant metastatic breast cancer were randomized to either capecitabine and docetaxel or to docetaxel alone. All measured efficacy parameters strongly favored the combination including time to tumor progression (6.1 vs 4.2 mo, $p = 0.0001$), median survival (14.5 vs 11.5 mo, $p = 0.0126$), and response rate (42% vs 30%, $p = 0.006$). Overall treatment related toxicities were comparable on both arms with slightly more grade 3 adverse events seen with the combination, whereas the docetaxel alone had more grade 4 events. Dose reductions were required in 65% of patients on the combination, compared with only 36% of patients receiving docetaxel alone.

Another taxane combination in clinical testing is docetaxel plus gemcitabine in advanced breast cancer. Phase II studies of docetaxel administered at 35 mg/m^2/wk for 3 out of 4 wk or as 75–100 mg/m^2 every 4 wk along with gemcitabine administered weekly have yielded response rates of 36–79% in the second-line setting with 5 of the 6 studies to date showing response rates higher than 50%. These individual phase II studies have been reviewed in detail (91,92). A phase III study comparing the docetaxel–gemcitabine doublet with docetaxel–capecitabine is presently ongoing.

3.3.1.4. Adjuvant Therapy of Early Stage Breast Cancer. The documented single-agent activity of taxanes in patients with advanced breast cancer has stimulated the evaluation of these agents in the adjuvant setting. The strategies employed in the clinical testing of these two taxanes have differed. Paclitaxel has been most frequently studied in sequential combination regimens such as doxorubicin and cyclophosphamide followed by paclitaxel (AC followed by paclitaxel) or doxorubicin followed by paclitaxel followed by cyclophosphamide (A→P→C). In contrast, docetaxel has been studied both in sequential (AC followed by docetaxel) and concurrently administered regimens (doxorubicin/ docetaxel and docetaxel/doxorubicin/ cyclophosphamide). As such, the adjuvant clinical trials of docetaxel can be classified into two groups. The first group of trials compared taxane and anthracycline containing regimen with nontaxane containing regimens given in combination or in sequence. These trials addressed the question of whether taxanes are able to change the natural history of early stage breast cancer. The second generation trials included taxanes in all the arms in order to address whether sequential or simultaneous administration may offer more clinical benefit.

The *Breast Cancer International Research Group Trial 001*, an international phase III study in 1491 patients with node-positive breast cancer, compared the combination of TAC with FAC on a once every 3-wk schedule *(93)*. The interim analysis at 33 mo indicates that TAC is significantly superior to FAC resulting in an 8% absolute improvement in the primary end point of DFS (82% vs 74%, $p = 0.0011$) for a relative risk of 0.68 ($p = 0.001$). Furthermore, in a prospectively defined subgroup analysis, women with 1–3 positive nodes had an even lower relative risk of 0.50 ($p = 0.0002$). However, patients with more than 4 positive nodes showed no significant difference in DFS (relative risk of 0.086; $p = 0.33$) or in OS.

Many additional adjuvant trials of docetaxel-based therapy in women with surgically resectable breast cancer are ongoing or have just recently been completed with efficacy data pending. These include the ECOG trial comparing four courses each of AC and AT (doxorubicin/docetaxel) in patients with high-risk node negative breast cancer and in those with 1–3 positive nodes. The study closed in 2001 after accrual of 3200 patients and interim analyses are expected. The *Breast Intergroup Trial (TAX 315)* is comparing docetaxel given in sequence or in combination with doxorubicin in more than 2700 node positive breast cancer patients. Another North American Intergroup Trial (E1199) is addressing the use of taxanes in sequence after 4 courses of AC. The design allows for comparison between paclitaxel and docetaxel in 3-weekly and weekly schedules. Finally, the NSABP B-30 study is a three-arm study randomizing patients to AC for four cycles followed by either four cycles of docetaxel (Arm 1) or four cycles of doxorubicin plus docetaxel (Arm 2) or to docetaxel, doxorubicin and cyclophosphamide (Arm 3). Results from all of these trials are pending.

3.3.1.5. Neoadjuvant Therapy of Early Stage Breast Cancer. Primary surgical treatment of locally advanced breast cancer is associated with a high probability of local recurrence and poor OS. Therefore, alternative therapeutic strategies are essential for patients with locally advanced breast cancer (stages IIB, IIIA, and IIIB), and one such approach is the use of neoadjuvant chemotherapy administered before surgical intervention. In phase II studies, objective clinical responses to neoadjuvant chemotherapy are as high as 90% although most studies report a lesser range of 70–75% *(94)*. However, less than one-half of these are clinical complete responses with pathological complete responses being even rarer, typically less than 20% *(95)*. Multiple clinical trials, with OS and DFS as the primary end points, have evaluated breast cancer patients treated with various combinations of chemotherapy given in the neoadjuvant and adjuvant setting. Emerging data from most of these trials indicate no difference in survival end points when neoadjuvant chemotherapy is compared with standard postoperative adjuvant treatment *(96,97)*.

The largest and most important trial addressing this issue is the NSABP B-18 study *(96)*. A total of 1523 patients with stage IIB and IIIA disease were randomized to receive combination AC in either the adjuvant or neoadjuvant setting. The difference between the two arms in terms of the DFS and OS was not statistically significant. An important secondary observation was patients who achieved a pathological complete response (pCR) with neoadjuvant treatment experienced longer DFS compared with those who did not achieve pCR at the time of surgery.

One important strategy to enhance the pCR rate in subsequent clinical trials has been to add taxanes, particularly docetaxel, to the neoadjuvant treatment of patients with locally advanced breast cancer. Nonrandomized phase II studies of neoadjuvant docetaxel-containing chemotherapy reported highly favorable response rates of 88–100% and clinical CR rate of 33%. As a consequence, several important randomized trials are in progress to evaluate the role of docetaxel in combination with anthracycline-based chemotherapy in patients with operable locally advanced breast cancer.

3.3.1.6. National Surgical Adjuvant Breast and Bowel Project B-27. The NSABP B-27 trial was designed to determine whether the addition of docetaxel to AC would improve DFS and OS *(98)*. Additionally, this trial also assessed whether the addition of preoperative docetaxel resulted in improved clinical and pathological response rates. Overall, 2411 patients were enrolled and preoperative AC followed by docetaxel was found to increase both the clinical and pathological response rates compared with preoperative AC alone. The clinical response rate increased from 85% to 91%, with the complete response rate improving from 40% to 65%. More importantly, the pathological response rate nearly doubled from 13.5% to 25.6% in patients with ER-positive and ER-negative tumors. The median tumor size in this study was 4.5 cm in comparison with 2.2 cm in the NSABP B-18. As B-27 involved eight cycles of therapy, there may have been a selection bias to enroll higher-risk patients with larger tumors. Surprisingly, the pathological response rate (~13%) for AC was the same in both trials, indicating that tumor response to neoadjuvant chemotherapy is independent of size. The dfs and OS analyses are awaited as follow-up is ongoing.

The NSABP B-27 data are corroborated by the results of a neoadjuvant trial of 913 operable breast cancer patients conducted by the German Adjuvant Breast Cancer Group (GEPARDUO) *(99)*. This study compared four cycles of concomitant dose-dense AT (doxorubicin and docetaxel) with a sequentially administered regimen of doxorubicin and cyclophosphamide for four cycles followed by four cycles of docetaxel (AC-T). The sequential AC-T regimen was associated with a superior pCR rate (22.4% vs 11.5%) and objective response rate as well as a higher rate of breast-conserving surgery.

In a Scottish (Aberdeen) trial, 162 patients stage IIIA and IIIB breast cancer were treated with four courses of neoadjuvant cyclophosphamide, vincristine, doxorubicin, and prednisone (CVAP) *(100)*. Patients whose tumors responded to therapy were then randomized to receive 4 additional courses of CVAP using an increased dose of prednisone, or 4 courses of docetaxel at 100 mg/m^2. Patients whose tumors did not respond to therapy were not randomized and were treated only with 4 courses of docetaxel. Upon completion of all chemotherapy, responses were again assessed before surgical intervention. In the patients whose tumors responded to CVAP, docetaxel resulted in higher objective clinical response rates compared with CVAP (85% vs 64%, respectively) with a similar increase in pathological response rates also observed (71% vs 57%, respectively). Furthermore, significantly higher rates of breast conservation were achieved on the docetaxel arm (67% vs 48%, respectively). Docetaxel also offered an apparent survival advantage at 5-yr of follow-up, even though this was not the primary end point of this relatively small study.

In contrast to the previous study, the interim results of another neoadjuvant phase III trial, the Anglo-Celtic Cooperative Oncology Group Trial, showed no benefit to using a concomitantly administered docetaxel plus doxorubicin regimen instead of AC in the neoadjuvant setting *(101)*. A total of 363 women with locally advanced breast cancer were randomized in this study. At the time of analysis, there was a trend toward a higher overall response rates (88% vs 78%) for AT and AC, respectively, ($p = 0.11$). However, there were no differences in rates of pCR (8% vs 12 %, $p = 0.26$), relapse-free survival, or OS. Follow-up continues, although these early results suggest that scheduling may be important in affecting clinical outcomes *(49)*.

Preclinical evidence of supra-additive interactions between paclitaxel and anti-HER-2 antibodies *(102)*, formed the basis for a pivotal trial comparing the combination of trastuzumab and paclitaxel with paclitaxel alone in patients with HER2-overexpressing breast cancers *(48)*. Impressively, the clinical data mirrored the laboratory findings with the demonstration of improved OS, progression-free-survival, and response rate by the combination arm. This landmark trial led to the evaluation of other taxane-based combinations such as docetaxel plus platinum derivatives, and trastuzumab in patients with HER2-positive breast cancer. A second, two-arm, phase III study compared the combination of paclitaxel, carboplatin, and trastuzumab (TPC) with paclitaxel and trastuzumab alone in patients with HER2-positive advanced breast cancer *(103)*. The TPC arm produced a superior response rate (52% vs 36%, $p = 0.04$) and time to tumor progression (1.2 mo vs 6.9 mo, $p = 0.007$), relative to paclitaxel/trastuzumab (TP). These promising results have led to the development and testing of several other docetaxel, trastuzumab regimens including combinations with cisplatin *(49)*.

3.3.2. Lung Cancer

Two randomized studies conclusively demonstrated the utility of single agent docetaxel in previously treated patients with NSCL cancer. In the TAX 317 trial, previously treated lung cancer patients were randomized to either docetaxel 75 mg/m^2 every 21 d or the best supportive care *(104)*. Initially, patients were treated with 100 mg/m^2 of docetaxel, but in this previously treated population, the incidence of severe febrile neutropenia was unacceptable and taxane doses had to be decreased. Nonetheless, median survival was significantly improved by chemotherapy from 4.6–7.0 mo ($p = 0.010$). Quality of life parameters were also superior and at the lower dose of 75 mg/m^2 of docetaxel, and overall toxicities were felt to be tolerable.

A three-arm randomized study treated patients with docetaxel at 100 mg/m^2 (Arm 1), docetaxel 75 mg/m^2 (Arm 2), or either vinorelbine 30 mg/m^2/weekly or ifosfamide 2 g/m^2 × 3 d every 3 wk (both grouped as Arm 3). All patients had previously received at least one line of chemotherapy for advanced NSCL cancer *(105)*, and some patients had been treated with multiple previous treatment regimens. Overall, the 11% response rates generated by the 100 mg/m^2 of docetaxel in Arm 1 was superior to the 6% and 1% response rates seen in Arms 2 and 3, respectively. Median survival was not different for any treatment, but 1 yr survival was significantly better with values of 32%, 32%, and 10% for the 100 mg/m^2 docetaxel, 75 mg/m^2 docetaxel, and the vinorelbine/ ifosfamide arms, respectively. In general, the lower dose Arm 2 of 75 mg/m^2 of docetaxel was favored slightly by the investigators because of its better tolerability than the higher docetaxel dose level. A very interesting subgroup analysis examined the outcome of docetaxel-treated patients who had received previous paclitaxel chemotherapy. Overall, 31% of the patients on the high-dose docetaxel arm and 42% of the patients in the low-dose docetaxel regimen had a history of previous paclitaxel therapy; however, response rates, median survivals, and 1 yr survivals were unaffected by previous paclitaxel therapy.

Thus, docetaxel is an active regimen even in patients whose tumors were previously exposed to paclitaxel chemotherapy. This strongly supports docetaxel's use in the second line setting in NSCL cancer patients even after treatment with front-line paclitaxel chemotherapy.

In untreated patients with advanced lung cancer, docetaxel has clear single-agent activity, similar to paclitaxel. In a randomized phase II trial, 200 patients with advanced NSCL cancer were treated with either docetaxel 100 mg/m^2 every 3 wk or with the best supportive care *(106)*. One-year survival favored docetaxel chemotherapy (25% vs 16%) and at two yr the difference was 12% vs 0%. Despite more chemotherapy related side effects, the quality life assessment significantly favored the docetaxel arm.

Most initial treatment regimens for NSCL cancer patients now favor combination chemotherapy and several studies have examined docetaxel paired with other agents as initial front line treatment. Fossella and colleagues *(107)* randomized 1218 patients with untreated advanced NSCL cancer to three different therapies in the TAX 326 trial: docetaxel 75 mg/m^2 plus cisplatin 75 mg/m^2 every 21 d; docetaxel 75 mg/m^2 plus carboplatin at an AUC of 6 also every 21 d; or a nontaxane doublet consisting of vinorelbine 25 mg/m^2 plus cisplatin 100 mg/m^2 every 28 d. Overall response rates were significantly better for the docetaxel–cisplatin treatments compared with the nontaxane vinorelbine–cisplatin treatment arm (31.6% vs 24.5%, respectively, $p = 0.029$) and median survival was also significantly improved (11.3 vs 10.1 mo, respectively, $p = 0.044$). In this study, the two docetaxel arms were not directly compared; however, the docetaxel–carboplatin arm was not statistically better than the inferior vinorelbine plus cisplatin therapy in terms of response rates and median survival. Overall toxicities were not significantly different and formal quality of life assessments favored the docetaxel treatments. Thus, clear advantages were observed for the docetaxel combination over a nontaxane combination in this study.

As previously mentioned in the discussion on paclitaxel in lung cancer, the ECOG 1594 trial compared four chemotherapy doublets in advanced NSCL cancer patients (paclitaxel/cisplatin, paclitaxel/carboplatin, gemcitabine/cisplatin, and docetaxel/cisplatin), and found no differences in any efficacy end point. Thus, several large randomized studies in untreated advanced NSCL cancer patients support the use of docetaxel plus cisplatin chemotherapy. Currently, in the United States, regulatory approval has been granted to docetaxel and cisplatin as initial front-line therapy for patients with advanced NSCL cancer. However, clear evidence suggesting an advantage of this regimen over paclitaxel plus carboplatin in this setting is lacking. In the North America, most medical oncologists favor use of a paclitaxel and carboplatin as initial therapy for advanced NSCL cancer, with docetaxel reserved for second-line therapy. In the United States, docetaxel also has regulatory approval as single-agent chemotherapy for use in the second-line setting after previous platinum-based chemotherapy.

3.3.3. Prostate Cancer

Clinical progress in developing effective chemotherapeutic regimens for the treatment of advanced hormone refractory prostate cancer has been frustratingly slow *(108)*. After many years of negative studies, the demonstration of a modest clinical benefit for mitoxantrone and glucocorticoids led the US FDA to approve this regimen for the treatment of patients with hormone refractory prostate cancer. Early promising results from nonrandomized clinical trials of docetaxel in this disease led the to conduct of several large randomized phase III trials testing the utility of docetaxel-based chemotherapy in advanced prostate cancer *(109)*. The Southwest Oncology Group (SWOG) trial 9916 randomized 770 patients with advanced hormone refractory prostate cancer to receive docetaxel at 60 mg/m2 and estramustine 280 mg/d for 5 d or mitoxantrone 12 mg/m^2 and prednisone every 21 d *(110)*. OS was the primary end point and in 338 patients treated with the taxane combination, median survival was 17.5 mo in contrast to the 15.6 mo seen in the 336 patients treated with mitoxantrone and steroids ($p = 0.02$). Time to tumor progression was also better for the docetaxel combination (6 mo vs 3 mo, $p < 0.01$). However, neutropenia, nausea and vomiting, and cardiovascular events such as stroke, myocardial infarction, or thrombosis were significantly more common for the docetaxel–estramustine combination.

A second study (TAX 327) compared three different regimens in hormone refractory patients and used survival as the major end point *(111)*. All three treatment arms included 5 mg of prednisone given twice daily; however, the chemotherapy component consisted of docetaxel 75 mg/m^2 every 21 d; weekly docetaxel at 30 mg/m^2/wk given for 4 out of every 6 wk; or mitoxantrone 12 mg/m^2 repeated every 21 d. Consistent with the earlier SWOG study, a significant survival advantage was seen when docetaxel every 3 wk was compared with the mitoxantrone regimen (18.9 vs 16.5 mo, $p = 0.009$). However, the weekly docetaxel arm's median survival of 17.4 mo was not significantly better than the control mitoxantrone regimen ($p = 0.36$). Serious neutropenia, alopecia, diarrhea, neuropathy, and peripheral edema were significantly more common with every 3 wk docetaxel; however, decrements in left ventricular cardiac function were more common with mitoxantrone therapy. Nonetheless, despite the modest increase in toxicities, the overall palliation of cancer related pain and quality of life assessments were significantly better for both docetaxel regimens compared with the mitoxantrone therapy.

Both of these studies suggest that docetaxel has clinically meaningful activity in the treatment of patients with hormone refractory prostate cancer. The different docetaxel arms used in these studies (docetaxel/estramustine and docetaxel/prednisone) have not been directly compared; however, given the across study comparable survival end points, most oncologists are favoring the more tolerable docetaxel and prednisone regimen, thereby avoiding the toxicities of estramustine. The US FDA has approved docetaxel in combination with prednisone for the treatment of patients with advanced prostate cancer *(108)*. Ongoing studies in this disease are now exploring other docetaxel combination therapies in an attempt to improve on this active regimen *(109)*.

3.3.4. OTHERS TUMORS

Similar to paclitaxel, docetaxel also has a broad spectrum of antitumor activity. Although the clinical benefit of docetaxel chemotherapy in the following diseases has not been fully characterized, docetaxel-induced objective responses have been reported in tumors such as head and neck cancer *(112)*, gastric *(113,114)*, esophageal *(114)*, non-Hodgkin's lymphoma *(115)*, ovarian *(114)*, and bladder cancers *(116)*.

4. CONCLUSIONS

The clinical development of the taxanes at the end of the 20th century represents a major advance in the clinical treatment of solid tumors. Their broad range of activity and predictable toxicity profile has led to their widespread use in the palliative treatment of a wide range of tumors, both as single agents and in combination chemotherapy regimens. Further refinements in their use will involve formulation advances such as nanoparticle technology, and improvements in drug delivery systems such as the development of high molecular weight conjugates. Further, taxanes derivatives with unique properties such as of the ability to avoid P-glycoprotein-mediated multidrug resistance and high-oral bioavailability are already in clinical development. Thus, in the near future the utility of this class of agents is likely to continue to expand and benefit a wider range of patients with malignant neoplasms.

REFERENCES

1. Rowinsky EK, Cazenave LA, Donehower RC. Taxol: a novel investigational antimicrotubule agent. J Natl Cancer Inst 1990;82:1247–1259.
2. Rowinsky EK, Donehower RC. Paclitaxel (taxol). N Engl J Med 1995;332:1004–1014.
3. Wall ME, Wani MC. Camptothecin and taxol: discovery to clinic—thirteenth Bruce F. Cain Memorial Award Lecture. Cancer Res 1995;55:753–760.
4. Manfredi JJ, Fant J, Horwitz SB. Taxol induces the formation of unusual arrays of cellular microtubules in colchicine-pretreated J774.2 cells. Eur J Cell Biol 1986;42:126–134.
5. Schiff PB, Fant J, Horwitz SB. Promotion of microtubule assembly in vitro by taxol. Nature 1979;277:665–667.
6. Cortes JE, Pazdur R. Docetaxel. J Clin Oncol 1995;13:2643–2655.
7. Horwitz SB, Cohen D, Rao S, Ringel I, Shen HJ, Yang CP. Taxol: mechanisms of action and resistance. J Natl Cancer Inst Monogr 1993;55–61.
8. Lamendola DE, Duan Z, Penson RT, Oliva E, Seiden MV. β tubulin mutations are rare in human ovarian carcinoma. Anticancer Res 2003;23:681–686.
9. Urano N, Fujiwara Y, Hasegawa S, et al. Absence of β-tubulin gene mutation in gastric carcinoma. Gastric Cancer 2003;6:108–112.
10. Huizing MT, Keung AC, Rosing H, et al. Pharmacokinetics of paclitaxel and metabolites in a randomized comparative study in platinum-pretreated ovarian cancer patients. J Clin Oncol 1993;11:2127–2135.
11. Huizing MT, Vermorken JB, Rosing H, et al. Pharmacokinetics of paclitaxel and three major metabolites in patients with advanced breast carcinoma refractory to anthracycline therapy treated with a 3-hour paclitaxel infusion: a European Cancer Centre (ECC) trial. Ann Oncol 1995;6:699–704.
12. Sparreboom A, van Zuylen L, Brouwer E, et al. Cremophor EL-mediated alteration of paclitaxel distribution in human blood: clinical pharmacokinetic implications. Cancer Res 1999;59:1454–1457.
13. Akerley W. Recent developments in weekly paclitaxel therapy in lung cancer. Curr Oncol Rep 2001;3:165–169.
14. Marchetti P, Urien S, Cappellini GA, Ronzino G, Ficorella C. Weekly administration of paclitaxel: theoretical and clinical basis. Crit Rev Oncol Hematol 2002;44(suppl):S3–S13.
15. Markman M. Weekly paclitaxel in the management of ovarian cancer. Semin Oncol 2000;27:37–40.
16. Seidman AD, Hudis CA, Albanel J, et al. Dose-dense therapy with weekly 1-hour paclitaxel infusions in the treatment of metastatic breast cancer. J Clin Oncol 1998;16:3353–3361.
17. Alberola V, Cortesi E, Juan O. Weekly paclitaxel in the treatment of metastatic and/or recurrent non-small cell lung cancer. Crit Rev Oncol Hematol 2002;44(suppl):S31–S41.
18. Takimoto CH, Rowinsky EK. Dose-intense paclitaxel: deja vu all over again? J Clin Oncol 2003;21:2810–2814.
19. Citron ML. Dose density in adjuvant chemotherapy for breast cancer. Cancer Invest 2004;22:555–568.
20. Citron ML, Berry DA, Cirrincione C, et al. Randomized trial of dose-dense versus conventionally scheduled and sequential versus concurrent combination chemotherapy as postoperative adjuvant treatment of node-positive primary breast cancer: first report of Intergroup Trial C9741/Cancer and Leukemia Group B Trial 9741. J Clin Oncol 2003;21:1431–1439.
21. Gianni L, Kearns CM, Giani A, et al. Nonlinear pharmacokinetics and metabolism of paclitaxel and its pharmacokinetic/pharmacodynamic relationships in humans. J Clin Oncol 1995;13:180–190.
22. Ohtsu T, Sasaki Y, Tamura T, et al. Clinical pharmacokinetics and pharmacodynamics of paclitaxel: a 3-hour infusion versus a 24-hour infusion. Clin Cancer Res 1995;1:599–606.

23. Chaudhry V, Rowinsky EK, Sartorius SE, Donehower RC, Cornblath DR. Peripheral neuropathy from taxol and cisplatin combination chemotherapy: clinical and electrophysiological studies. Ann Neurol 1994;35:304–311.

24. Rowinsky EK, Chaudhry V, Cornblath DR, Donehower RC. Neurotoxicity of Taxol. J Natl Cancer Inst Monogr 1993;15:107–115.

25. Weiss RB, Donehower RC, Wiernik PH, et al. Hypersensitivity reactions from taxol. J Clin Oncol 1990;8:1263–1268.

26. Grem JL, Tutsch KD, Simon KJ, et al. Phase I study of taxol administered as a short i.v. infusion daily for 5 days. Cancer Treat Rep 1987;71:1179–1184.

27. Eisenhauer EA, ten Bokkel Huinink WW, Swenerton KD, et al. European-Canadian randomized trial of paclitaxel in relapsed ovarian cancer: high-dose versus low-dose and long versus short infusion. J Clin Oncol 1994;12:2654–2666.

28. Stanford BL, Hardwicke F. A review of clinical experience with paclitaxel extravasations. Support Care Cancer 2003;11:270–277.

29. Raghavan VT, Bloomer WD, Merkel DE. Taxol and radiation recall dermatitis. Lancet 1993; 341:1354.

30. Shenkier T, Gelmon K. Paclitaxel and radiation-recall dermatitis. J Clin Oncol 1994;12:439.

31. McGuire WP, Hoskins WJ, Brady MF, et al. Cyclophosphamide and cisplatin compared with paclitaxel and cisplatin in patients with stage III and stage IV ovarian cancer. N Engl J Med 1996; 334:1–6.

32. Piccart MJ, Bertelsen K, James K, et al. Randomized intergroup trial of cisplatin-paclitaxel versus cisplatin-cyclophosphamide in women with advanced epithelial ovarian cancer: three-year results. J Natl Cancer Inst 2000;92:699–708.

33. du Bois A, Luck HJ, Meier W, et al. A randomized clinical trial of cisplatin/paclitaxel versus carboplatin/paclitaxel as first-line treatment of ovarian cancer. J Natl Cancer Inst 2003;95:1320–1329.

34. Ozols RF, Bundy BN, Greer BE, et al. Phase III trial of carboplatin and paclitaxel compared with cisplatin and paclitaxel in patients with optimally resected stage III ovarian cancer: a Gynecologic Oncology Group study. J Clin Oncol 2003;21:3194–3200.

35. Vasey PA, Jayson GC, Gordon A, et al. Phase III randomized trial of docetaxel-carboplatin versus paclitaxel-carboplatin as first-line chemotherapy for ovarian carcinoma. J Natl Cancer Inst 2004;96: 1682–1691.

36. Katsumata N. Docetaxel: an alternative taxane in ovarian cancer. Br J Cancer 2003;89(suppl 3): S9–S15.

37. McGuire WP 3rd, Markman M. Primary ovarian cancer chemotherapy: current standards of care. Br J Cancer 2003;89(suppl 3):S3–S8.

38. Thigpen T. First-line therapy for ovarian carcinoma: what's next? Cancer Invest 2004;22(suppl 2): 21–28.

39. Omura GA, Brady MF, Look KY, et al. Phase III trial of paclitaxel at two dose levels, the higher dose accompanied by filgrastim at two dose levels in platinum-pretreated epithelial ovarian cancer: an intergroup study. J Clin Oncol 2003;21:2843–2848.

40. Sparano JA. Taxanes for breast cancer: an evidence-based review of randomized phase II and phase III trials. Clin Breast Cancer 2000;1:32–40; discussion 41–32.

41. Henderson IC, Berry DA, Demetri GD, et al. Improved outcomes from adding sequential Paclitaxel but not from escalating Doxorubicin dose in an adjuvant chemotherapy regimen for patients with node-positive primary breast cancer. J Clin Oncol 2003;21:976–983.

42. Bishop JF, Dewar J, Toner GC, et al. Initial paclitaxel improves outcome compared with CMFP combination chemotherapy as front-line therapy in untreated metastatic breast cancer. J Clin Oncol 1999;17:2355–2364.

43. Paridaens R, Biganzoli L, Bruning P, et al. Paclitaxel versus doxorubicin as first-line single-agent chemotherapy for metastatic breast cancer: a European Organization for Research and Treatment of Cancer Randomized Study with cross-over. J Clin Oncol 2000;18:724–733.

44. Sledge GW, Neuberg D, Bernardo P, et al. Phase III trial of doxorubicin, paclitaxel, and the combination of doxorubicin and paclitaxel as front-line chemotherapy for metastatic breast cancer: an intergroup trial (E1193). J Clin Oncol 2003;21:588–592.

45. Jassem J, Pienkowski T, Pluzanska A, et al. Doxorubicin and paclitaxel versus fluorouracil, doxorubicin, and cyclophosphamide as first-line therapy for women with metastatic breast cancer: final results of a randomized phase III multicenter trial. J Clin Oncol 2001;19:1707–1715.

46. Biganzoli L, Cufer T, Bruning P, et al. Doxorubicin and paclitaxel versus doxorubicin and cyclophos-phamide as first-line chemotherapy in metastatic breast cancer: The European Organization for Research and Treatment of Cancer 10961 Multicenter Phase III Trial. J Clin Oncol 2002;20: 3114–3121.

47. Bernard-Marty C, Cardoso F, Piccart MJ. Use and abuse of taxanes in the management of metastatic breast cancer. Eur J Cancer 2003;39:1978–1989.

48. Slamon DJ, Leyland-Jones B, Shak S, et al. Use of chemotherapy plus a monoclonal antibody against HER2 for metastatic breast cancer that overexpresses HER2. N Engl J Med 2001;344:783–792.

49. Crown J, O'Leary M, Ooi WS. Docetaxel and paclitaxel in the treatment of breast cancer: a review of clinical experience. Oncologist 2004;9(suppl 2):24–32.

50. Gianni L, Vigano L, Locatelli A, et al. Human pharmacokinetic characterization and in vitro study of the interaction between doxorubicin and paclitaxel in patients with breast cancer. J Clin Oncol 1997;15:1906–1915.

51. Ishitobi M, Shin E, Kikkawa N. Metastatic breast cancer with resistance to both anthracycline and docetaxel successfully treated with weekly paclitaxel. Int J Clin Oncol 2001;6:55–58.

52. Valero V, Jones SE, Von Hoff DD, et al. A phase II study of docetaxel in patients with paclitaxel-resistant metastatic breast cancer. J Clin Oncol 1998;16:3362–3368.

53. Norton L. Theoretical concepts and the emerging role of taxanes in adjuvant therapy. Oncologist 2001;6(suppl 3):30–35.

54. Miles D, von Minckwitz G, Seidman AD. Combination versus sequential single-agent therapy in metastatic breast cancer. Oncologist 2002;7(suppl 6):13–19.

55. Buzdar AU, Singletary SE, Theriault RL, et al. Prospective evaluation of paclitaxel versus combina-tion chemotherapy with fluorouracil, doxorubicin, and cyclophosphamide as neoadjuvant therapy in patients with operable breast cancer. J Clin Oncol 1999;17:3412–3417.

56. Estevez LG, Gradishar WJ. Evidence-based use of neoadjuvant taxane in operable and inoperable breast cancer. Clin Cancer Res 2004;10:3249–3261.

57. Simon GR, Bunn PA Jr. Taxanes in the treatment of advanced (stage III and IV) non-small cell lung cancer (NSCLC): recent developments. Cancer Invest 2003;21:87–104.

58. Rigas JR. Taxane-platinum combinations in advanced non-small cell lung cancer: a review. Oncologist 2004;9(suppl 2):16–23.

59. Ranson M, Davidson N, Nicolson M, et al. Randomized trial of paclitaxel plus supportive care ver-sus supportive care for patients with advanced non-small-cell lung cancer. J Natl Cancer Inst 2000;92:1074–1080.

60. Socinski MA, Rosenman JG, Schell MJ, et al. Induction carboplatin/paclitaxel followed by concur-rent carboplatin/paclitaxel and dose-escalating conformal thoracic radiation therapy in unresectable stage IIIA/B nonsmall cell lung carcinoma: a modified Phase I trial. Cancer 2000;89:534–542.

61. Kelly K, Crowley J, Bunn PA Jr, et al. Randomized phase III trial of paclitaxel plus carboplatin ver-sus vinorelbine plus cisplatin in the treatment of patients with advanced non-small-cell lung cancer: a Southwest Oncology Group trial. J Clin Oncol 2001;19:3210–3218.

62. Schiller JH, Harrington D, Belani CP, et al. Comparison of four chemotherapy regimens for advanced non-small-cell lung cancer. N Engl J Med 2002;346:92–98.

63. Scagliotti GV, De Marinis F, Rinaldi M, et al. Phase III randomized trial comparing three platinum-based doublets in advanced non-small-cell lung cancer. J Clin Oncol 2002;20:4285–4291.

64. Kosmidis P, Mylonakis N, Skarlos D, et al. Paclitaxel (175 mg/m^2) plus carboplatin (6 AUC) versus pacli-taxel (225 mg/m^2) plus carboplatin (6 AUC) in advanced non-small-cell lung cancer (NSCLC): a multi-center randomized trial. Hellenic Cooperative Oncology Group (HeCOG). Ann Oncol 2000;11:799–805.

65. Welles L, Saville MW, Lietzau J, et al. Phase II trial with dose titration of paclitaxel for the therapy of human immunodeficiency virus-associated Kaposi's sarcoma. J Clin Oncol 1998;16:1112–1121.

66. Gill PS, Tulpule A, Espina BM, et al. Paclitaxel is safe and effective in the treatment of advanced AIDS-related Kaposi's sarcoma. J Clin Oncol 1999;17:1876–1883.

67. Schrijvers D, Vermorken JB. Role of taxoids in head and neck cancer. Oncologist 2000;5:199–208.

68. Hainsworth JD, Burris HA 3rd, Greco FA. Paclitaxel-based three-drug combinations for the treatment of small cell lung cancer: a review of the Sarah Cannon Cancer Center experience. Semin Oncol 2001;28:43–47.

69. Van Cutsem E. The treatment of advanced gastric cancer: new findings on the activity of the taxanes. Oncologist 2004;9(suppl 2):9–15.

70. Olson KB, Pienta KJ. Recent advances in chemotherapy for advanced prostate cancer. Curr Urol Rep 2000;1:48–56.

71. Greco FA, Gray J, Burris HA 3rd, Erland JB, Morrissey LH, Hainsworth JD. Taxane-based chemotherapy for patients with carcinoma of unknown primary site. Cancer J 2001;7:203–212.
72. Markman M, Fowler J. Activity of weekly paclitaxel in patients with advanced endometrial cancer previously treated with both a platinum agent and paclitaxel. Gynecol Oncol 2004;92:180–182.
73. Hussain SA, James ND. The systemic treatment of advanced and metastatic bladder cancer. Lancet Oncol 2003;4:489–497.
74. Younes A. Paclitaxel-based treatment of lymphoma. Semin Oncol 1999;26:123–128.
75. Ibrahim NK, Desai N, Legha S, et al. Phase I and pharmacokinetic study of ABI-007, a Cremophor-free, protein-stabilized, nanoparticle formulation of paclitaxel. Clin Cancer Res 2002;8:1038–1044.
76. Langer CJ. CT-2103: emerging utility and therapy for solid tumours. Expert Opin Invest Drugs 2004;13:1501–1508.
77. Diaz JF, Andreu JM. Assembly of purified GDP-tubulin into microtubules induced by taxol and tax-otere: reversibility, ligand stoichiometry, and competition. Biochemistry 1993;32:2747–2755.
78. Rowinsky EK, Tolcher AW. Antimicrotuble agents. In: DeVita VT, Hellman S, Rosenberg SA. eds. Cancer: Principles & Practice of Oncology. 7th edition. Philadelphia, Lippincott, Williams & Wilkins, 2005:390–416.
79. Clarke SJ, Rivory LP. Clinical pharmacokinetics of docetaxel. Clin Pharmacokinet 1999;36:99–114.
80. Eisenhauer EA, Vermorken JB. The taxoids. Comparative clinical pharmacology and therapeutic potential. Drugs 1998;55:5–30.
81. Semb KA, Aamdal S, Oian P. Capillary protein leak syndrome appears to explain fluid retention in cancer patients who receive docetaxel treatment. J Clin Oncol 1998;16:3426–3432.
82. Chan S, Friedrichs K, Noel D, et al. Prospective randomized trial of docetaxel versus doxorubicin in patients with metastatic breast cancer. J Clin Oncol 1999;17:2341–2354.
83. Nabholtz J, Falkson G, Campos D, et al. A phase III trial comparing doxorubicin (A) and docetaxel (T) (AT) to doxorubicin and cyclophosphamide (AC) as first-line chemotherapy for MBC. Proc Am Soc Clin Oncol 1999;18:127A.
84. Mackey JR, Paterson A, Dirix LY, et al. Final results of the phase III randomized trial comparing doc-etaxel (T), doxorubicin (A) and cyclophosphamide (C) to FAC as first line chemotherapy (CT) for patients (pts) with metastatic breast cancer (MBC). Proc Am Soc Clin Oncol 2002;21:35A.
85. Bonneterre J, Roche H, Monnier A, et al. Docetaxel vs 5-fluorouracil plus vinorelbine in metastatic breast cancer after anthracycline therapy failure. Br J Cancer 2002;87:1210–1215.
86. Bontenbal M, Braun JJ, Creemers GJ, et al. Phase III study comarpring AT (adriamycin, docetaxel) to FAC (fluorouracil, adriamycin, cyclophosphamide) as first-line chemotherapy (CT) in patients with metastatic breast cancer (MBC). Eur J Cancer 2003;1:5201–5202.
87. Sjostrom J, Blomqvist C, Mouridsen H, et al. Docetaxel compared with sequential methotrexate and 5-flu-orouracil in patients with advanced breast cancer after anthracycline failure: a randomised phase III study with crossover on progression by the Scandinavian Breast Group. Eur J Cancer 1999;35: 1194–1201.
88. Jones SE, Erban J, Overmoyer B, et al. Randomized Phase III study of docetaxel compared with paclitaxel in metastatic breast cancer. J Clin Oncol 2005;23:5542–5551.
89. Wildiers H, Paridaens R. Taxanes in elderly breast cancer patients. Cancer Treat Rev 2004;30: 333–342.
90. O'Shaughnessy J, Miles D, Vukelja S, et al. Superior survival with capecitabine plus docetaxel combination therapy in anthracycline-pretreated patients with advanced breast cancer: phase III trial results. J Clin Oncol 2002;20:2812–2823.
91. Seidman AD. Gemcitabine and docetaxel in metastatic breast cancer. Oncology (Huntingt) 2004;18:13–16.
92. Mavroudis D, Malamos N, Polyzos A, et al. Front-line chemotherapy with docetaxel and gemcitabine administered every two weeks in patients with metastatic breast cancer: a multicenter phase II study. Oncology 2004;67:250–256.
93. Nabholtz J-M, Pienkowski T, Mackey J, et al. Phase III trial comparing TAC (docetaxel, doxorubicin, cyclophosphamide) with FAC (5-fluorouracil, doxorubicin, cyclophosphamide) in the adjuvant treatment of node positive breast cancer (BC) patients: interim analysis of the BCIRG 001 study. Proc Am Soc Clin Oncol 2002;21:36A.
94. Smith IE, Lipton L. Preoperative/neoadjuvant medical therapy for early breast cancer. Lancet Oncol 2001;2:561–570.
95. Heys SD, Eremin JM, Sarkar TK, Hutcheon AW, Ah-See A, Eremin O. Role of multimodality therapy in the management of locally advanced carcinoma of the breast. J Am Coll Surg 1994;179:493–504.

96. Wolmark N, Wang J, Mamounas E, Bryant J, Fisher B. Preoperative chemotherapy in patients with operable breast cancer: nine-year results from National Surgical Adjuvant Breast and Bowel Project B-18. J Natl Cancer Inst Monogr 2001;30:96–102.

97. Scholl SM, Fourquet A, Asselain B, et al. Neoadjuvant versus adjuvant chemotherapy in pre-menopausal patients with tumours considered too large for breast conserving surgery: preliminary results of a randomised trial: S6. Eur J Cancer 1994;30A:645–652.

98. NSABP. The effect of primary tumor response of adding sequential taxotere to adriamcyin and cyclophosphamide: preliminary results from NSABP protocol B-27. Breast Cancer Res Treat 2001;69:210.

99. von Minckwitz G, Raab G, Schuette M, et al. Dose-dense versus sequential adriamcyin / docetaxel combination as preoperative chemotherapy (pCHT) in operable breast cancer (T2-3, N0-2,M0) - primary endpoint analysis of the GEPARDUO-Study. Proc Am Soc Clin Oncol 2002;21:43A.

100. Smith IC, Heys SD, Hutcheon AW, et al. Neoadjuvant chemotherapy in breast cancer: significantly enhanced response with docetaxel. J Clin Oncol 2002;20:1456–1466.

101. Evans TRJ, Gould A, Foster E, Crown JP, Leonard RCF, Mansi JL. Phase III randomised trial of adri-amycin (A) and docetaxel (D) versus A and cyclophosphamide (C) as primary medical therapy (PMT) in women with breast cancer: an ACCOG study. Proc Am Soc Clin Oncol 2002;21:35A.

102. Pegram M, Hsu S, Lewis G, et al. Inhibitory effects of combinations of HER-2/neu antibody and chemotherapeutic agents used for treatment of human breast cancers. Oncogene 1999;18:2241–2251.

103. Robert N, Leyland-Jones B, Asmar L, et al. Phase III comparative study of trastuzumab and pacli-taxel with and without carboplatin in patients with HER-2/neu positive advanced breast cancer. Breast Cancer Res Treat 2002;76(suppl 1):S37.

104. Shepherd FA, Dancey J, Ramlau R, et al. Prospective randomized trial of docetaxel versus best sup-portive care in patients with non-small-cell lung cancer previously treated with platinum-based chemotherapy. J Clin Oncol 2000;18:2095–2103.

105. Fossella FV, DeVore R, Kerr RN, et al. Randomized phase III trial of docetaxel versus vinorelbine or ifosfamide in patients with advanced non-small-cell lung cancer previously treated with platinum-containing chemotherapy regimens. The TAX 320 Non-Small Cell Lung Cancer Study Group. J Clin Oncol 2000;18:2354–2362.

106. Roszkowski K, Pluzanska A, Krzakowski M, et al. A multicenter, randomized, phase III study of doc-etaxel plus best supportive care versus best supportive care in chemotherapy-naive patients with metasta-tic or non-resectable localized non-small cell lung cancer (NSCLC). Lung Cancer 2000;27:145–157.

107. Fossella F, Pereira JR, von Pawel J, et al. Randomized, multinational, phase III study of docetaxel plus platinum combinations versus vinorelbine plus cisplatin for advanced non-small-cell lung can-cer: the TAX 326 study group. J Clin Oncol 2003;21:3016–3024.

108. Gulley J, Dahut WL. Chemotherapy for prostate cancer: finally an advance Am J Ther 2004;11:288–294.

109. Khan MA, Carducci MA, Partin AW. The evolving role of docetaxel in the management of androgen independent prostate cancer. J Urol 2003;170:1709–1716.

110. Petrylak DP, Tangen CM, Hussain MH, et al. Docetaxel and estramustine compared with mitox-antrone and prednisone for advanced refractory prostate cancer. N Engl J Med 2004;351:1513–1520.

111. Tannock IF, de Wit R, Berry WR, et al. Docetaxel plus prednisone or mitoxantrone plus prednisone for advanced prostate cancer. N Engl J Med 2004;351:1502–1512.

112. Posner MR, Lefebvre JL. Docetaxel induction therapy in locally advanced squamous cell carcinoma of the head and neck. Br J Cancer 2003;88:11–17.

113. Haller DG, Misset JL. Docetaxel in advanced gastric cancer. Anticancer Drugs 2002;13:451–460.

114. Ajani JA. Docetaxel for gastric and esophageal carcinomas. Oncology (Huntingt) 2002;16:89–96.

115. Budman DR, Petroni GR, Johnson JL, et al. Phase II trial of docetaxel in non-Hodgkin's lymphomas: a study of the Cancer and Leukemia Group B. J Clin Oncol 1997;15:3275–3279.

116. Hong WK. The current status of docetaxel in solid tumors. An M. D. Anderson Cancer Center Review. Oncology (Huntingt) 2002;6:9–15.

17 Investigational Anticancer Agents Targeting the Microtubule

Lyudmila A. Vereshchagina, Orit Scharf, and A. Dimitrios Colevas

CONTENTS

SUMMARY

This chapter summarizes the preclinical and clinical development to date of investigational anticancer agents whose mechanism of action is thought to be via direct interaction with tubulin or microtubules. All of the compounds discussed are agents discovered or derived from screening natural materials for anti-cancer activity. The underlying theme for pursuit of these agents is that tubulin is a validated anticancer target and that novel interactions between new chemical entities and tubulin may overcome resistance, increase therapeutic index, or alter the spectrum of clinical utility against different cancer types.

Key Words: Epothilone; dolostatin; colchicine; investigational agents.

1. INTRODUCTION

Agents targeting the microtubule have been available in the clinic for more than four decades. Vinca alkaloids, originating from the periwinkle plant, were widely tested in the clinic during the 1960's, and a decade later were incorporated into a number of combination chemotherapy regimens that are still in use today. The four major vinca alkaloids approved for use in various parts of the world are vincristine, vinblastine, vinorelbine, and vindesine. Owing to the unusually diverse spectrum of antitumor effects seen with this class, the vinca alkaloids remain in use as first- and second-line agents against a variety of malignancies. The nonclinical development of the taxanes

From: *Cancer Drug Discovery and Development: The Role of Microtubules in Cell Biology, Neurobiology, and Oncology* Edited by: Tito Fojo © Humana Press, Totowa, NJ

began in the 1960s as well. Because of difficulties in the isolation and purification of paclitaxel from the bark of the Pacific yew tree, it was not until the early 1990s that paclitaxel was widely studied in the clinic. Docetaxel, a semisynthetic taxane, became available for clinical use soon thereafter. Like the vinca alkaloids, taxanes have a diverse spectrum of activity, although the uses of these two classes in the clinic are divergent. Whereas both the vinca alkaloids and taxanes are natural products that bind tubulin, their molecular mechanisms and clinical spectra of efficacy differ in many ways. The tubulin binding sites of each are distinct; vinca alkaloids inhibit microtubule heterodimer polymerization, whereas taxanes stabilize the polymer. However, at low concentrations, both classes of agents can decrease the turnover rate, or "treadmilling," of the tubulin heterodimers through microtubules.

There is little additional knowledge about the association between the site of tubulin binding, effect on microtubule dynamics and polymerization, and the ultimate clinically relevant anticancer effects of these agents. Is the relationship between altered microtubules and induction of apoptosis the most relevant? Or is suppression of antiapoptotic molecules paramount? How does modulation of microtubule dynamics of tumor-associated endothelial cells contribute to these agents' anticancer activity? Why are these agents cancer specific? Whereas there are data that buttress one hypothesis or another, the link between the primary site of action at the microtubule and ultimate anticancer activity remains poorly understood. By extension, the link between the type of effect on the microtubule and spectrum of anticancer activity of a particular compound is perplexing. Therefore, drug discovery and development of agents directed against the microtubule remains empiric. It is no accident that the compounds discussed in this chapter are either natural products or their analogs. Investigators continue to rely on the creativity of evolution, which has provided a diversity of compounds that can screen for activity against this target. Pending a better comprehension of the process resulting in relatively selective anticancer (versus normal tissue) toxicity of these agents, and further understanding of the mechanism enabling each of these agents to have a unique clinically useful spectrum of activity, the investigational antimicrotubule agents in clinical trials are likely to share only two features. They will act on the microtubule, and they will be discovered by drug screening of molecular libraries mostly populated by the largess of nature.

Many of the agents discussed in this chapter exist naturally in some of the most isolated parts of the planet, in minute quantities, and in a highly impure form. A recurrent theme in drug development is the process of screening, isolation, purification, and synthesis of a lead drug candidate, which subsequently is subject to preclinical efficacy and toxicology evaluations. Despite extensive preclinical research, antimicrotubule agents continue to enter the clinic lacking adequate information concerning the relative likelihood of efficacy in a particular cancer type, which would allow clinical development targeted toward a particular tumor type, or with a specific molecularly defined phenotype. Therefore, phase one evaluation is usually a generic study of the human pharmacology of these agents, whereas phase two testing is necessarily broad, across many different schedules and tumor types. The extent of clinical testing may be primarily a result of business decisions concerning the identification of a potential registration niche. Without scientific rationale for pursuit of particular tumor types, there is usually no justification for restricted clinical evaluation.

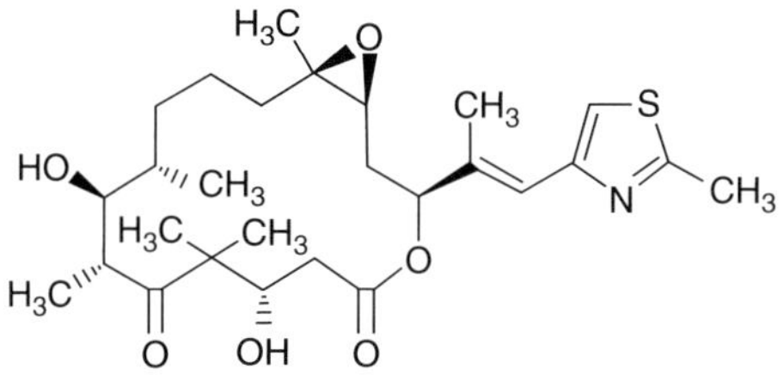

(2) (Novartis Scripps)
Epo B. EPO 906

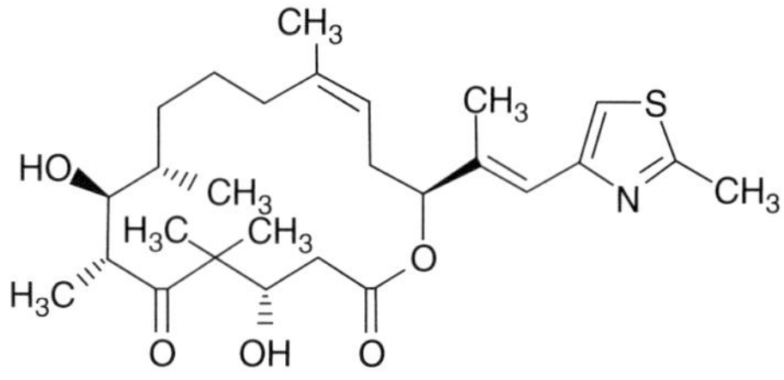

(4) (Kosan Biosciences/MSK)
Epo D, dEpoB, KOS 862

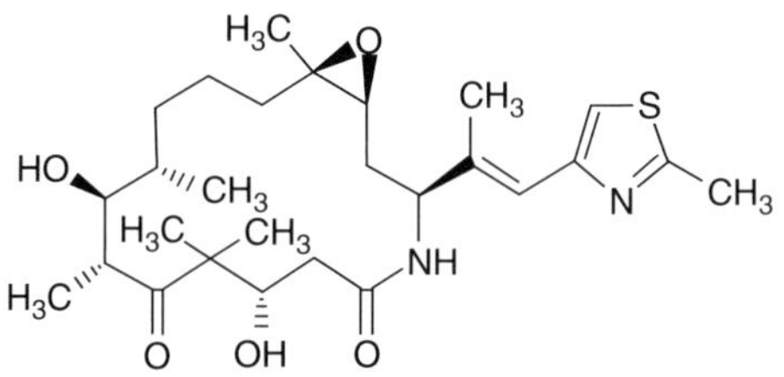

(29) (Bristol-Myers Squibb/GBF)
BMS-247550

Fig. 1. Epothilones currently undergoing clinical trials *(1)*.

2. EPOTHILONES

The epothilones, a relatively new class of nontaxane tubulin polymerization agents, were obtained from the fermentation of the cellulose-degrading myxobacteria *Sorangium cellulosum*. German investigators reported the isolation of epothilones with antifungal and cytotoxic activity from the culture broth of *S. cellulosum* and elucidated the structure of epothilones A and B (Epo B, Fig. 1), which differ only by the presence or absence of a methyl group at the trisubstituted epoxide moiety *(2,3)*. Epothilones C, D (Epo D, 12,13-desoxyepothilone B, Fig. 1), E, and F were originally obtained as minor components in the fermentation process. As a result of ongoing efforts by numerous academic and industrial research groups, the parent compounds and their biologically active analogs were prepared by one or a combination of the following methods: fermentation, biosynthesis, biotransformation, total synthesis, semisynthesis, and combinatorial synthesis *(1)*. The key difference between BMS-247550 and epothilone B is the replacement of the macrolide ring oxygen atom with a nitrogen atom to give the corresponding macrolactam.

Epothilones have a mechanism of action analogous to taxanes, i.e., the stabilization of microtubules through polymerization of tubulin, resulting in mitotic arrest at the G_2/M transition in the cell cycle *(4,5)*. Epothilones compete for the same tubulin-binding site as paclitaxel, but demonstrate significant antitumor activity in taxane-resistant models. To date, epothilones B (EPO 906) and D (KOS 862) and two epothilone B analogs, BMS-247550 and BMS-310705, have been evaluated in human cancer clinical trials. The preliminary safety and efficacy data from some of these trials are described later.

2.1. EPO 906 (Epothilone B, Patupilone)

EPO 906 (Fig. 1) is being clinically developed by Novartis. Two phase 1 trials have evaluated EPO 906 administered as an intravenous (IV) infusion on a weekly (6) or every-21-d (q3w) schedule (7) in patients with advanced solid tumors. On the q3w schedule, the first six dose levels of EPO 906 were administered as a 30-min infusion and subsequent dose levels as a 5–10-min infusion. The starting dose of EPO 906 is 0.3 mg/m^2 in both trials. Diarrhea was dose limiting on the weekly schedule (6 wk on/3 wk off), experienced by 3/5 patients treated at 3.6 mg/m^2 and 3/9 patients at 3 mg/m^2. With a further dose reduction to 2.5 mg/m^2, 1/14 patients experienced dose-limiting toxicity (DLT), and this dose was chosen as the maximum-tolerated dose (MTD). As the severity of the diarrhea often peaked on the fourth wk of treatment, the schedule was changed to 3 wk on/1 wk off with a starting dose of 2.5 mg/m^2, which was the MTD for this schedule as well. Other adverse events (AEs) on the weekly schedule included nausea (44%), fatigue (31%), and vomiting (31%). Hemtological AEs were minimal (grade 3/4 anemia, thrombocytopenia, neutropenia, and leukopenia in 8, 1, 1, and 1% of patients, respectively). Diarrhea was dose limiting at 8 mg/m^2 on the q3w schedule. Other AEs on this schedule included grade 3 fatigue ($n = 4$) and nausea/vomiting ($n = 2$), and grade 2 peripheral neuropathy ($n = 3$).

Antitumor activity of EPO 906 was seen in both trials. Three partial responses (PRs; breast, endometrial, ovarian) and 31 stable disease (9/17 ovarian patients, median duration 16.3 wk) were reported in 60 patients on the weekly schedule. On the q3w schedule, 1 PR (unknown primary), 11 stable diseases and four significant responses (though not reaching criteria for a PR; one breast, three colorectal) have been reported in 36 evaluative patients. Another phase 1 study is evaluating EPO 906 in patients with advanced colon cancer, using three different schedules: q3w as an IV bolus on day 1; continuous 24-h IV infusion (CIV) on day 1; and 16-h CIV on days 1–5 (8). The starting dose is 6.5 mg/m^2. Of 24 enrolled patients, 13, 8, and 3 patients recieved EPO 906 by these three schedules, respectively. The most common AEs were diarrhea (42%), asthenia (17%), and abdominal pain (13%). No dose-limiting diarrhea was reported on the day 1 IV bolus schedule, whereas 2/3 patients on the days 1–5 16-h CIV experienced grade 3 dose-limiting diarrhea. Stable disease was observed in 12 patients mainly on the day 1 bolus IV and the 24-h CIV schedules, including 2 patients with a minor response. These data suggest that the bolus dosing q3w is tolerated better by these patients.

Preliminary pharmacokinetic studies of EPO 906 performed in these trials demonstrate multiphasic clearance with long terminal half-life ($t_{1/2}$) of approx 3.5 d. On the weekly schedule, urinary excretion of EPO 906 was negligible. On the q3w schedule, elimination of EPO 906 was mainly nonrenal, total body clearance was approx 200 mL/min, and volume of distribution at steady state (Vd$_{ss}$) was approx 1000 L.

Two phase 1/2 studies are evaluating EPO 906 in ovarian and nonsmall cell lung cancer (NSCLC). In the ovarian study patients recieve EPO 906 IV once q3w starting at 6.5 mg/m^2, and the MTD has not yet been reached at 10.5 mg/m^2. Of 19 patients evaluated for response, one complete response (CR), one PR, and seven stable diseases were observed. In the NSCLC study, EPO 906 is administered by the same dose and schedule as the ovarian study, and the MTD has not yet been reached at 11.5 mg/m^2. DLTs were observed at 7.5 mg/m^2 (one patient with grade 3 asthenia), and one patient each experienced grade 3 diarrhea at 8 and 8.5 mg/m^2. Only nine patients had grade 3 AEs of any kind. Four PRs were observed in 39 patients.

These results prompted additional trials of EPO 906. All trials are evaluating EPO 906 administered as a 5-min IV infusion on a weekly ×3 q4w or q3w schedules. On the weekly ×3 q4w schedule, EPO 906 has been administered at 2.5 mg/m^2 to 53 patients with advanced epithelial renal cell carcinoma who have had a nephrectomy *(9)*. Grade 4 septic shock was reported in one patient. Grade 3 nonhematological AEs included diarrhea, asthenia, and anemia. In addition, the investigators reported grade 3 neuropathy in 2/15 patients who completed more than six cycles of EPO 906. Three PRs (median duration 4 mo), and 23 stable diseases after four cycles of therapy (durations from 8 to 48 wk) have been reported in 52 evaluative patients. Two other phase 2 trials have evaluated 2.5 mg/m^2 EPO 906 on the weekly ×3 q4w schedule or 6 mg/m^2 EPO 906 on the q3w schedule in patients with advanced colorectal cancer *(10)*. No differences in toxicity between the two regimens were reported in 96 patients evaluative in both trials. Grade 3/4 diarrhea was reported in 29% of patients. Other grade 3 AEs included dehydration (8%), nausea (5%), vomiting (5%), and one case of thrombocytopenia. One PR and six stable diseases were reported in 47 evaluative patients on the weekly schedule. Three PRs and one stable disease were reported in 44 evaluative patients on the q3w schedule.

Two combination phase 1B trials have evaluated EPO 906 in combination with carboplatin in patients with advanced malignancies *(11)*. In one study, EPO 906 was administered to 26 patients as a 5-min IV infusion at doses of 0.5–2.5 mg/m^2 on the weekly ×3 q4w schedule, immediately followed by carboplatin administered as a 30-min IV infusion, AUC = 2. DLTs included diarrhea ($n = 2$) and paresthesias ($n = 1$) in 12 patients treated at the 2.5 mg/m^2 dose level (defined as the MTD). Grade 4 AEs included one case each of eye pain, hyponatremia, respiratory failure, and increased lipase. Grade 3 AEs included neutropenia in one patient at each dose level, suggesting that it may be unrelated to the drug, and hypersensitivity to carboplatin in one patient with a previous reaction to carboplatin. One minimal response in a patient with ovarian cancer and eight stable diseases were reported in 13 evaluative heavily pretreated patients. In another study, EPO 906 was administered as a 5-min IV infusion at doses of 3.6–6 mg/m^2 on the q3w schedule, immediately followed by carboplatin administered as a 60-min IV infusion AUC = 5 or 6 to 37 patients (70% of whom had platinum-sensitive relapsed ovarian cancer). The MTD of EPO 906 in combination with carboplatin AUC = 6 was 4.8 mg/m^2. The most common AEs included diarrhea (84%), fatigue (68%), nausea (60%), and vomiting (43%). Eight patients had CR or PR and three had stable disease in 21 evaluative patients.

2.2. BMS-247550 (Epothilone B Aanalog, Ixabepilone)

BMS-247550 (Fig. 1) is being clinically developed by Bristol-Myers Squibb in collaboration with the National Cancer Institute. Phase 1 clinical trials are evaluating schedules similar to those used with EPO 906. As BMS-247550 is formulated with Cremophor®EL, hypersensitivity reactions have occurred with IV administration of this drug on various schedules *(12,13)*. Therefore, all patients are given 50 mg diphenhydramine (or its equivalent) IV and 50 mg ranitidine (or its equivalent) IV 30 min before BMS-247550 infusion.

Several phase 1 single-agent trials have evaluated BMS-247550 doses ranging from 7.4 to 65 mg/m^2 administered as a 1-h infusion on the q3w schedule. The MTD of BMS-247550 on this schedule is 40–50 mg/m^2 *(13–17)*. Neutropenia and neuropathy were dose limiting in all trials. Other reported grade 3/4 AEs included fatigue, emesis, arthralgia, and myalgia. Antitumor activity seen among 22 patients treated at 50 mg/m^2 and more includes one CR (ovarian post paclitaxel) and three PRs (NSCLC post-taxotere, melanoma in two patients) *(13)*. On another trial, 3/15 patients had minor responses (colon, lung, and melanoma) *(17)*. In addition, responses in taxane-refractory breast cancer (three PRs and one minor response), endometrial cancer (one PR and one minor response), ovarian cancer (one minor response), and melanoma (one minor response) have been reported *(15)*.

Two phase 1 single-agent trials have evaluated BMS-247550 doses ranging from 1 to 30 mg/m^2 administered once weekly in patients with advanced cancer *(18,19)*. The MTD of BMS-247550 administered as a 30-min infusion was 25 mg/m^2 *(20)*. Fatigue was dose limiting in both trials, whereas neutropenia (>5 d) was dose limiting in heavily pretreated patients in one study only *(18,19)*. Other grade 3 AEs included arthralgia and myalgia. As cumulative sensory neuropathy was also observed in both trials, treatment schedules were changed from continuous weekly infusions to a 1-h infusion given weekly ×3 q4w. Promising antitumor activity was seen in both trials. One study reported three PRs (ovarian, colorectal, head, and neck) at 25 mg/m^2 and >50% decrease in tumor markers (three patients with breast and one patient with ovarian cancer) at 20 mg/m^2 *(18)*. Another study reported a minor response in a heavily pretreated patient with melanoma, and tumor marker decrease in three taxane-refractory ovarian cancer patients and one prostate cancer patient *(19)*. BMS-247550 was shown to be orally bioavailable (F ~53%) *(12)*.

Several phase 1 single-agent trials evaluated BMS-247550 doses ranging from 1.5 to 10 mg/m^2 administered as a 1-h infusion daily ×5 on the q3w schedule. The MTD and recommended phase 2 dose (RP2D) of BMS-247550 without granulocyte colony stimulating factor (GCSF) support is 6 mg/m^2/d *(21–23)*. Neutropenia was dose limiting at 8 mg/m^2/d without GCSF. Other grade 3 AEs included fatigue, anemia, abdominal cramping, arthralgias, neuropathic pain, stomatitis, and anorexia. PRs were reported in five patients (two with taxane-pretreated breast cancer, two with taxane-pretreated cervical cancer, and one with taxane-naïve basal cell carcinoma). An additional 3/14 patients with paclitaxel-pretreated ovarian cancer had more than 50% decrease in CA125. The mean terminal $t_{1/2}$ was 16.8 ± 6 h, the Vd_{ss} was 798 ± 375 L, and the clearance was 712 ± 247 mL/min *(21)* on this schedule. This study was amended to establish the RP2D for a daily ×3 q3w schedule with a starting dose of 10 mg/m^2 *(24,25)*. As 3/6 patients treated at this dose level developed grade 4 neutropenia, the dose was reduced to 8 mg/m^2/d, and was escalated to 10 mg/m^2/d in the second cycle for 10/15 patients. Nine of these 10 patients tolerated this dose and experienced no DLTs. Hematological grade 3/4 AEs included neutropenia and thrombocytopenia, and nonhematological grade 3/4 AEs included fatigue, vomiting, nausea, mucositis, hyponatremia, ileus, emesis, and anorexia. There were no objective responses on this schedule, although three patients previously treated with taxol experienced stable disease (renal cell carcinoma, 28 mo; ovarian carcinoma, 14 mo; primary peritoneal mesothelioma, 7 mo). The RP2D of BMS-247550 was established as 8 mg/m^2/d ×3 q3w with escalation to 10 mg/m^2/d on subsequent cycles if tolerated. In a pediatric study, the MTD was exceeded at a dose of 10 mg/m^2/d administered to children with refractory solid tumors, with DLTs of neutropenia and fatigue *(26)*. The 8 mg/m^2/d dose level was expanded, and data are pending. Five patients on this study had stable disease for a median of five cycles. The PK parameters in this pediatric study were similar to adult values.

Preliminary results are available for phase 1 and 1/2 trials evaluating BMS-247550 in combination with carboplatin, estramustine phosphate (EMP), or gemcitabine. In a phase 1 combination trial, 30 patients with advanced solid malignancies were treated with 30 mg/m^2 and 40 mg/m^2 BMS-247550 (1-h infusion) and carboplatin (AUC = 5 and 6; 30-min infusion starting 30 min after completion of the BMS-247550 infusion) administered IV q3w *(27)*. DLTs included febrile neutropenia and thrombocytopenia in one patient, and neutropenia and diarrhea complicated by a cardiac event in another patient at 40 mg/m^2 BMS-247550 and carboplatin AUC = 6. In addition, one case of dose-limiting motor neuropathy was observed at 30 mg/m^2 BMS-247550 and carboplatin AUC = 6. The investigators attributed the unexpected myelotoxicity to an approx 20% increase in carboplatin AUC. Major nonhematological AEs were myalgia 4 d post-treatment and cumulative peripheral sensory neuropathy in patients receiving. three cycles. Three PRs in patients with breast, neuroendocrine, and unknown primary adenocarcinomas have been reported. In a phase 1/2 combination trial, chemotherapy-naïve patients with metastatic prostate cancer were treated with 35–40 mg/m^2 BMS-247550 administered IV q3w and 280 mg estramustine (EMP) given orally 3 times a day for 5 d q3w *(28,29)*. In the phase 1 part, 3/6 patients experienced grade 3/4 neutropenia, and one patient had grade 3 nausea at 40 mg/m^2. Eleven of twelve evaluative patients achieved a more than 50% post-therapy decline in prostate specific antigen (PSA). In addition, seven patients had CRs and three had PRs. In the phase 2 part, patients were randomized to arm A (35 mg/m^2 BMS-247550, 3-h IV on day 1, q3w) or arm B (35 mg/m^2 BMS-247550, 3-h IV on day 2, q3w + 280 mg oral EMP given 3 times a day on days 1–5 + 2 mg coumadin daily) *(28)*. Preliminary data from this study indicate that in arm A (BMS-247550 alone), 18/32 patients had a more than 50% decline in PSA and 6/26 had PRs, whereas in arm B (BMS-247550 + EMP) 22/32 patients had a more than 50% decline in PSA and 8/18 had PRs. Grade 3/4 AEs in both arms included mainly febrile neutropenia, neuropathy, and thrombosis. In a phase 1 study of BMS-247550 and gemcitabine in patients with solid tumors BMS-247550 was administered over 3 h on day 8 with gemcitabine infused over 90 min on days 1 and 8 q3w *(30)*. The MTD was determined at 20 mg/m^2 BMS-247550 and 900 mg/m^2 gemcitabine. No objective responses were observed. However, two patients with ovarian cancer had a more than 50% decrease in their CA125.

Antitumor activity seen on phase 1 trials of BMS-247550 prompted initiation of phase 2 trials using 50 mg/m^2 BMS-247550 administered as a 1-h IV infusion on the q3w schedule. Two phase 2 studies enrolled patients with taxane-refractory or taxane-naïve metastatic breast cancer *(31)*. Grade 3 AEs in seven evaluative taxane-refractory patients included fatigue, sensory neuropathy, proctitis, stomatitis/pharyngitis, thrombocytopenia, and neutropenia. Grade 3 AEs in 19 evaluative taxane-naïve patients included fatigue, myalgia/arthralgia, sensory neuropathy/neuropathic pain, dyspnea, diarrhea, amenorrhea, reversible myocardial ischemia, and febrile neutropenia. Seven out of sixteen patients experienced grade 4 neutropenia. Ten PRs and eight stable diseases were reported in 19 evaluative patients with taxane-naïve breast cancer, and two PRs and three stable diseases were reported in seven evaluative patients with taxane-refractory breast cancer. In order to reduce cumulative neurotoxicity and improve the therapeutic index, both studies were amended to a regimen of 40 mg/m^2 over 3 h q3w. Sixty-five patients with metastatic breast cancer previously treated with an anthracyclin were enrolled on this amended schedule *(32)*. Grade 3/4 neutropenia occurred in 24% of 38 evaluative patients. Grade 3/4 nonhematological AEs occurred in 44% of patients and included myalgia (11%), arthralgia (5%), and fatigue (5%). Grade 3 neuropathy occurred in 16% of patients compared with 37% of patients on the 50 mg/m^2 over-1-h schedule. Twenty-seven PRs have been reported in 61 evaluative patients *(32)*. An additional 49 patients with taxane-resistant metastatic breast cancer, who were also previously treated with an anthracyclin, were enrolled on another study with 40 mg/m^2 BMS-247550 administered over 3 h q3w *(33)*. Grade 3/4 neutropenia and thrombocytopenia occurred in 57% and 8%, respectively, of 49 patients. Grade 3/4 nonhematological AEs other than fatigue and neuropathy were minimal. Grade 3 sensory neuropathy occurred in 12% of patients compared with 25% of patients on the 50 mg/m^2 over-1-h schedule. Six PRs have been reported in 49 evaluative patients. A phase 2 study in patients with gynecological and breast cancer used a BMS-247550 dose of 40 mg/m^2 as a 1-h IV q3w *(34)*. Two PRs were observed in 17 taxane-pretreated gynecological cancer patients (1/14 ovarian, 1/3 endometrial primaries), and 4 PRs were observed in 13 taxane-pretreated breast cancer patients. Neutropenia and neuropathy were the most frequent AEs. Another study in breast cancer patients administered BMS-247550 at 6 mg/m^2 daily ×5 q3w *(35)*. Grade 3/4 AEs included neutropenia (32%), febrile neutropenia (10%), thrombocytopenia (5%), fatigue (15%), diarrhea (12%), nausea/vomiting (5%), constipation (2%), myalgia/arthralgia (2%), and sensory neuropathy (2%; all grades 40%). Of 30 evaluative patients previously treated with taxane, six PRs were observed, and 4 PRs were observed in eight evaluative taxane-naïve patients.

Patients with metastatic renal cell carcinoma received BMS-247550 at 6 mg/m^2 daily ×5 q3w *(36)*. Four PRs were confirmed in 30 patients, and an additional five patients have had minor or mixed reponses. Several post-translational modifications on α-tubulin, such as removal of the C-terminal tyrosine and exposure of glutamic acid as the new C-terminal residue, have been shown to correlate with the extent of microtubule stability.

A phase 2 study evaluated BMS-247550 in patients with NSCLC previously treated with one platinum-based regimen for recurrent or metastatic disease *(37)*. The most common grade 3 nonhematological AEs in 25 evaluative patients included fatigue (20%), and sensory neuropathy/neuropathic pain (20%). Grade 3/4 hematological AEs included febrile neutropenia (10%), neutropenia (50%), and thrombocytopenia (9%). Four PRs were reported in 22 evaluative patients. In order to reduce cumulative neurotoxicity and improve the therapeutic index, the study was amended to a randomized comparison of 40 mg/m^2 over 3 h q3w, vs 6 mg/m^2 administered daily 5 times over 1 h q3w. Eighteen patients were treated at 40 mg/m^2 BMS-247550 administered as a 3-h IV infusion, but owing to the frequency of mucositis and neutropenia, 78 subsequent patients were treated at 32 mg/m^2 *(38)*. At this dose level, grade 3/4 AEs included neutropenia in 20/76 patients, febrile neutropenia in 7/76 patients, sensory neuropathy in 3/76 patients, and grade 3 fatigue in 4/76 patients. In addition, two cases each of grade 3 arthralgia and thrombocytopenia, and one case each of grade 3 myalgia and vomiting were reported. One CR and nine PRs in 76 patients were observed. Median time to progression (TTP) was 1.5 mo. Seventy-four patients were treated at 6 mg/m^2 BMS-247550 administered daily ×5. Grade 3/4 AEs included neutropenia in 10/69 patients and febrile neutropenia in 3/69 patients, and grade 3 AEs included fatigue in 6/69 patients, sensory neuropathy in 4/69 patients, and vomiting in 3/69 patients. In addition, one case each of grade 3 myalgia and thrombocytopenia were reported. One CR and six

PRs in 69 patients were reported. Median time to progression (TTP) was 2 mo. In 21 patients with metastatic gastric cancer previously treated with one taxane-based chemotherapy regimen, grade 3/4 AEs included fatigue (43%), anorexia (29%), nausea/vomiting (24%), sensory neuropathy/neuropathic pain (14%), myalgia/arthralgia (10%), abdominal pain/cramping (10%), diarrhea (5%), febrile neutropenia (5%), and neutropenia (40%) *(39)*. Two PRs and nine stable diseases were reported in 20 evaluative patients.

Thirty-one patients with metastatic and unresectable soft tissue sarcomas were treated with BMS-247550 at 50 mg/m^2 q3w. Twenty-five patients (80%) had grade 3 treatment-related AEs *(40)*. Of these, 19 patients had grade 3 nonhematological AEs. Grade 3/4 neutropenia and leukopenia occurred in 74% patients. One grade 5 AE (aspiration pneumonia resulting in death), which was assessed as not related to BMS-247550, was reported. Twenty-seven percent of patients went off study owing to toxicity (mainly neurotoxicity). Two PRs (one confirmed) and 18 stable diseases were reported in 31 patients. Time to treatment failure was short, with a median of 57 d.

Two patients with advanced colorectal cancer refractory to irinotecan/5-fluorouracil/ leucovorin were initially administered 50 mg/m^2 BMS-247550 as a 1-h IV infusion on the q3w schedule *(41)*. Significant peripheral neuropathy in other studies led to a dose reduction to 40 mg/m^2 BMS-247550 administered as a 3-h IV infusion on the q3w schedule in 23 additional patients. Grade 3/4 AEs included neutropenia (48%), leukopenia (36%), peripheral neuropathy (20%), neutropenic fever (4%), and hypersensitivity reaction (8%). No objective responses were reported in 23 evaluative patients. Median TTP was 11 wk, and median survival was 38 wk.

In patients with hepatobiliary cancer (unresectable or advanced hepatocellular carcinoma [HCC], gallbladder carcinoma, and cholangiocarcinoma), BMS-247550 was administered at 40 mg/m^2 as a 3-h IV infusion on the q3w schedule *(42)*. Grade 3/4 AEs included neutropenia (53%), anemia, febrile neutropenia, fatigue, arthralgia, transaminase elevation, dyspnea, stomatitis, and hypersensitivity (6% each). An unexpected patient death 48 h after the first BMS-247550 administration was assessed as possibly drug-related. Two PRs in patients with HCC and metastatic gallbladder carcinoma lasting for 4 and 4.7 mo, respectively, were reported in 12 evaluative patients. Median survival was 5.7 mo, and progression-free survival (PFS) was 4.3 mo. A study in patients with advanced pancreas cancer using the same schedule reported four confirmed and nine unconfirmed PRs in 54 patients *(43)*. The most common grade 3/4 AEs were neutropenia, leukopenia, nausea, vomiting, neuropathy, and fatigue. Patients with stage IV malignant melanoma were treated with BMS-247550 at 20 mg/m^2 by 1-h infusion weekly ×3 q4w *(44)*. No responses were seen in either untreated or previously treated patients. Median TTP was 8 wk.

When viewed in aggregate, the experience in phases 1 and 2 trials with BMS-247550 to date suggests that reversible bone marrow suppression and cumulative neurotoxicity are the primary DLTs for this agent. Preliminary data suggest that the incidence and severity of neurotoxicity is acceptable at a dose and schedule of 40 mg/m^2 dose as a 3-h infusion q3w. Although neuropathy may be lower on the daily 5 times schedule, definitive statements concerning relative neuropathy rates must await additional data. The bone marrow suppression seen with BMS-247550 is generally within the acceptable range and not considerably different than seen with taxanes.

2.3. KOS-862 (Epothilone D)

KOS-862 (Fig. 1) is being clinically developed by Kosan Biosciences in Hayward, California. A phase 1 trial is evaluating KOS-862 administered as an IV infusion (150 mL/h) q3w *(45)*. Five patients (two colon, one each testicular, HCC, and prostate cancers) have been treated with 9 and 18 mg/m^2 KOS-862 with no DLTs; dose escalation continues. Reported grade 1/2 AEs include emesis and anemia. Preliminary pharmacokinetic studies of KOS-862 performed on this trial demonstrate that the mean ± SD value for AUC at 9 mg/m^2 was 1332 ± 839 h×ng/mL, elimination $t_{1/2}$ was 5–10 h, clearance was 8.4 ± 4.0 L/h/m^2, and Vd$_{ss}$ was 78 ± 31 L/m^2. Another phase 1 trial is evaluating two schedules of KOS-862 administration: a single dose q3w or as doses over 3 consecutive days q3w *(46)*. On the q3w schedule, KOS-862 was administered at 9–185 mg/m^2 to 38 patients with advanced solid malignancies. A dose of 185 mg/m^2 produced DLTs, including grade 3 ataxia with visual disturbances, grade 2 ataxia with visual disturbances/blurry vision, and grade 2 peripheral neuropathy and atypical chest pain. Other grade 3 AEs include cognitive abnormalities, nausea/vomiting/diarrhea, and anemia at the 120 mg/m^2 dose level, cognitive abnormalities, fatigue, nausea/vomiting/diarrhea, and anemia at the 150 mg/m^2 dose level, and grade 3 motor neuropathy at the 185 mg/m^2 dose level. Investigators proposed 120 mg/m^2 as the RP2D on this schedule. On the daily ×3 q3w schedule, doses of 20, 40, and 50 mg/m^2/d KOS-862 were administered to 14 patients with advanced solid malignancies. At the 50 mg/m^2/d dose level, one of two patients had dose-limiting grade 3 ataxia with visual hallucinations and grade 2 neuropathy. At the 40 mg/m^2/d dose level, grade 3 AEs included cognitive abnormalities, motor neuropathy and anemia. No objective responses were reported. Decreases in tumor markers were observed in patients with testicular, ovarian, pancreatic, and breast cancer. No difference in pharmacokinetics, including systemic exposure, was observed between the two schedules. Other phase 1 trials of KOS-862 with alternate schedules are ongoing.

Two phase 1 studies are studying combinations of KOS-862 with gemcitabine or carboplatin *(47,48)*. KOS-862 (50, 60, and 75 mg/m^2; 90-min IV infusion) and gemcitabine (750 mg/m^2; 30-min IV infusion) were administered weekly ×3 q4w to patients with solid tumors. At a KOS-862 and gemcitabine dose of 75 mg/m^2 and 750 mg/m^2, respectively, 2/4 patients (both cholangiocarcinoma) experienced DLT of severe dehydration, diarrhea, and neutropenia. One of these patients had twice the expected maximum plasma concentration (C_{max}) for KOS-862. An intermediate KOS-862 dose level of 60 mg/m^2 produced no DLT in six patients; however, the day-15 dose could not be delivered in 2/6 patients because of thrombocytopenia. The schedule was amended to a q3w cycle (2 wk on, 1 wk off). One PR (unknown primary cancer) and two stable diseases more than 3 mo were observed. KOS-862 was used in combination with carboplatin (KOS-862 at 50 and 75 mg/m^2, 90-min IV on days 1 and 8; carboplatin AUC = 5 on day 1). A DLT of 4-wk delayed recovery of neutrophils was observed in the first patient, and this dose level was expanded to six patients with no other DLTs. At the second dose level, one patient experienced grade 3 neurosensory toxicity. One ovarian cancer patient had a CR, and one HCC patient had stable disease for 21 wk and a 41% drop in α-fetoprotein (AFP).

NSCLC and breast cancer phase 2 trials are underway *(49,50)*. Both trials used a dose and schedule of 100 mg/m^2 90-min IV weekly ×3 q4w. Two out of 10 evaluative breast cancer patients achieved PRs (including one patient with hepatic-metastases who had

Table 1
Clinical Trials of Epothilones

Drug name	Tumor type/schedule	Phase	Activity	Major adverse events (CTC grade 3/4)
EPO 906	Solid (6)/IV wkly	1	3/60 PR (breast, endometrial, ovarian) 25/60 SD	DLT = diarrhea 91 evaluative pts: diarrhea 19%; nausea 9%; vomiting 4%; fatigue 2%; abdominal pain 2% ptt 2%
	Solid (7)/30-min or 5–10-min IV q3w	1	1/36 PR (unknown primary), 4/36 MR (1 breast, 3 colorectal), 11/36 SD	DLT = diarrhea 42 evaluative pts: fatigue 10%; nausea/vomiting 5%
	Colon (8)/IV bolus×1 q3w; 24-h CIV×1; 16-h CIV×5	1	2/24 MR 12/24 SD	DLT = diarrhea in 2/3 pts on the 16-h CIV×5
	Ovarian (51)/10–20-min IV×1 q3w	1/2	1/19 CR 1/19 PR 7/19 SD	DLT = fatigue in 2/9 pts 29 evaluative pt: diarrhea 14%; fatigue 14%; vomiting 7%
	NSCLC (52)/10–20-min IV×1 q3w	1/2	4 PR 9 SD	DLT = asthenia in 1/6; diarrhea in 2/12 pts. 35 evaluative pts: diarrhea 14%
	Renal (9)/5-min IV wkly ×3 q4wks	2	3/52 PR, 23/52 SD	53 evaluative pts: neuropathy 13%; diarrhea 8%; asthenia and anemia 4%; sepsis 2%
	CRC (10)/5-min IV wkly ×3 q4wks 5-min IV q3wks	2	1/47 PR, 6/47 SD 3/44 PR, 1/44 SD	96 evaluative pts: diarrhea 26%; dehydration 8%; nausea 5%; vomiting 5%; thrombocytopenia 1%
EPO 906 + carboplatin	Solid (53)/5-min IV wkly ×3 q4wks + 30-min IV carboplatin	1B	1/28 MR (ovarian), 8/28 SD	DLT = diarrhea in 2/12; paresthesias in 1/12 pts 12 evaluative pts:eye pain 8%; increased lipase 8%
	(11)5-min IV q3wks + 60-min IV carboplatin		8/21 CR+PR 3/21 SD	DLT = 1/6 pts each: elevated liver enzymes; nausea/vomiting; fatigue; nausea; abdominal pain 37 evaluative pts: peripheral neuropathy 3%
BMS-247550	Solid (13)/1-h IV q3w	1	1/22 CR (ovarian), 3/22 PR (NSCLC, 2 melanoma), 11/22 SD	DLT = grade 4 neutropenia; grade 3 neuropathy in 5/6 pts; 3 evaluative pts: arthralgia and myalgia 66%
	Solid (17)/1-h IV q3w	1	3/15 MR (colon, lung, melanoma), 3/15 SD (colon, renal, melanoma)	DLT = grade 4 neutropenia in 4 pts; grade 3/4 emesis and fatigue

(Continued)

Table 1 *(Continued)*

Drug name	Tumor type/schedule	Phase	Activity	Major adverse events (CTC grade 3/4)
	Solid (15)/1-h IV q3w	1	4/40 PR (3 breast, endometrial) 4/40 MR (ovarian, endometrial, breast, melanoma)	DLT = neutropenia in 5/14 pts (16); 49 evaluative pts: neutropenia 30%; fatigue 25%; peripheral neuropathy 10%
	Solid (18)/30-min IV wkly Changed to 1-h IV wkly ×3 q4wks	1	3/36 PR (ovarian, colorectal, head and neck) >50% decrease in tumor markers in 4/36 pts (3 breast, 1 ovarian)	DLT = grade 3 fatigue in 3/4 pts 12 evaluative pts: fatigue 25%; myalgia/arthralgia 17%; sensory neuropathy 8% (Similar to AEs seen with 30-min IV)
	Solid (19)/1-h IV wkly	1	1/16 MR (melanoma) 4/36 decreases in tumor markers (1 prostate, 3 ovarian)	DLT = grade 4 neutropenia in 2/16 pts; grade 3 fatigue in 1/16 pts
	Solid (22)/1-h IV daily ×5 q3w	1	5/27 objective responses (breast, cervical, basal cell), 3/14 >50% decrease in CA125 (ovarian)	DLT = grade 4 neutropenia in 3/3 pts grade 3 fatigue, stomatitis, anorexia
	Solid (25)/1-h IV daily ×3 q3w	1	3/24 SD (ovarian, mesothelioma, renal)	DLT = grade 4 neutropenia in 5/11 pts
BMS-247550 + carboplatin	Solid (27)/1-h IV 3-wkly	1/2	3/30 PR (breast, neuroendocrine, unknown), 16/30 SD	DLT = grade 4 febrile neutropenia, grade 3 thrombocytopenia in 1/2 pts; grade 4 neutropenia, grade 3 diarrhea in 1/2 pts; motor neuropathy in 1/30 pts grade 3/4 ALT 30%; grade 3 arthralgia in 3%; neuropathy in 27% (6/8 pts sensory, 2/8 pts motor)
BMS-247550 + EMP	Prostate (29)/3-h IV on d2 q3w + EMP	1/2 phase 1	1/7 CR 3/7 PR 1/7 SD 11/12 >50% PSA decrease	Gr 3/4 neutropenia in 3/6 pts; grade 3 nausea in 1/6 pts 13 evaluative pts: thrombocytopenia 31%; sensory neuropathy 15%; nausea 15%; thrombosis 15%; rash 8%
	Prostate (28) Arm A: 3-h IV on d1 q3w Arm B: 3-h IV on d2 q3w + EMP	phase 2	6/26 PR >56% pts/arm ≥50% PSA decrease 8./18 PR 69% pts/arm ≥50% PSA decrease	47 evaluative pts: febrile neutropenia 4%; neuropathy 6% 44 evaluative pts: febrile neutropenia 4%; neuropathy 4%; thrombosis 5%

(Continued)

Table 1 *(Continued)*

Drug name	Tumor type/schedule	Phase	Activity	Major adverse events (CTC grade 3/4)
BMS-247550 + gemcitabine	Solid (30)/3-h IV on d8 q3w + gemcitabine	1	5/12 SD (1 leiomyosarcoma, 2 ovarian, 2 uterine)	DLT = neutropenia
BMS-247550	Breast (31)/1-h IV q3w	2	Taxane-refractory 2/7 PR 3/7 SD	Taxane-refractory: 7 evaluative pts: neutropenia 29%; fatigue 14%; sensory neuropathy 14%; proctitis 14%; stomatitis/pharyngitis 14%; thrombocytopenia 14%
			Taxane-naïve 10/19 PR 8/19 SD	Taxane-naïve: 19 evaluative pts: neutropenia 44%; myalgia/arthralgia 26%; sensory neuropathy/neuropathic pain 21%; fatigue 16%; dyspnea 16%; diarrhea 5%; amenorrhea 5%; reversible myocardial ischemia 5%; febrile neutropenia 5%
	Breast (32)/3-h IV q3w	2	27/61 PR (24 confirmed), 21/61 SD	38 evaluative pts: neutropenia 24%; neuropathy 16%; myalgia 11%; arthralgia 5%; fatigue 5%
	Breast (33)/3-h IV q3w	2	6/49 PR 19/49 SD	49 evaluative pts: neutropenia 57%; fatigue 29%; sensory neuropathy 12%; myalgia 8%; thrombocytopenia 8%; febrile neutropenia 6%; nausea 6%; vomiting 6%; constipation 4%; diarrhea 4%
	NSCLC (38)/3-h IV q3w	2	1/76 CR 9/76 PR 27/76 SD	76 evaluative pts: neutropenia 26%; febrile neutropenia 10%; fatigue 5%; sensory neuropathy 4%; arthralgia 3%; thrombocytopenia 3%; myalgia 1%; vomiting 1%
	1-h IV d1-5 q3w		1/69 CR 6/69 PR 21/69 SD	69 evaluative pts: neutropenia 14%; fatigue 9%; neuropathy 6%; febrile neutropenia 4%; vomiting 4%; myalgia 1%; thrombocytopenia 1%
	Breast (35)/IV d1-5 q3w	2	Previous taxane: 6/30 PR 12/30 SD Taxane naïve: 4/8 PR 4/8 SD	42 evaluative pts: Neutropenia 32%; fatigue 15%; diarrhea 12%; febrile neutropenia 10%; thrombocytopenia 5%; nausea/vomiting 5%; constipation 2%; myalgia/arthralgia 2%; sensory neuropathy 2%

(Continued)

Table 1 *(Continued)*

Drug name	Tumor type/schedule	Phase	Activity	Major adverse events (CTC grade 3/4)
	Gynecological (34)/1-h IV q3w	2	6/30 PR (1 ovarian, 1 endometrial, 4 breast)	Neutropenia in 19/42 pts
	Gastric (39)/1-h IV q3w	2	2/20 PR 9/20 SD	21 evaluative pts: fatigue 43%; anorexia 29%; nausea/vomiting 24%; sensory neuropathy/neuropathic pain 14%; myalgia/arthralgia 10%; abdominal pain 10%; diarrhea 5%; febrile neutropenia 5%; neutropenia 40%
	Soft tissue sarcoma (40)/1-h IV q3w	2	2/31 PR (one confirmed), 18/31 SD	31 evaluative pts: neutropenia 48%; leukopenia 36%; sensory neuropathy 19%; myalgia 13%; pain 13%; dyspnea 10%; fatigue 10%; constipation 6%; ileus 6%; hypoxia 6%; rash 6%; nausea 6%; vomiting 6%; arthralgia 6%; pericardial effusion 3%; urinary retention 3%; pleural effusion 3%; stomatitis 3%; syncope 3%; motor neuropathy 3%; anorexia 3%; hypertension 3%
	Colorectal (41)/1-h IV q3w	2	13/23 SD	25 evaluative pts: neutropenia 48%; leukopenia 36%; peripheral neuropathy 20%; hypersensitivity reaction 8%; neutropenic fever 4%;
	Hepatobiliary (42)/3-h IV q3w	2	2/12 PR 6/12 SD	15 evaluativee pts: neutropenia 53%; anemia 6%; febrile neutropenia 6%; fatigue 6%; arthralgia 6%; transaminase elevation 6%; dyspnea 6%; stomatitis 6%; hypersensitivity to Cremophor L 6%
	Pancreas (43)/3-h IV q3w	2	9/54 PR (4 confirmed)	Neutropenia, leukopenia, nausea, vomiting, neuropathy, fatigue
KOS-862	Solid (45)/1-h IV q3w	1	0/5 objective responses	No DLT
	Solid (46)/q3wks	1	7/52 SD	DLT = grade 2/3 ataxia w/visual disturbances in 2/3 pts, atypical chest pain in 1/3 pts

(Continued)

Table 1 *(Continued)*

Drug name	Tumor type/schedule	Phase	Activity	Major adverse events (CTC grade 3/4)
	daily ×3 q3wks			20 evaluative pts: cognitive abnormalities 25%; nausea/vomiting/diarrhea 10%; anemia 15%; fatigue 10% DLT = grade 3 ataxia w/visual hallucinations, grade 2 neuropathy in 1/2 pts 9 evaluative pts: cognitive abnormalities 11%; anemia 11%
KOS-862 + carboplatin	Solid (48)/90-min IV d1 and 8 + carboplain	1	1/12 CR (ovarian) 41% drop in AFP in 1 HCC pt.	DLT = 4-wk delayed recovery of neutrophils in 1 pt.

>50% tumor reduction). Grade 3 AEs include peripheral neuropathy ($n = 3$) and ataxia ($n = 1$). Several patients with neurotoxicity have been successfully retreated with a dose reduction to 75 mg/m^2. In the NSCLC study, 9% of doses required dose reduction, and 6% required dose delay secondary to toxicity. One PR has been reported in the first 35 patients (Table 1).

3. DOLASTATINS

The dolastatin peptides (Fig. 2) were originally isolated from a small sea mollusk *Dolabella auricularia*. Two years later an absolute configuration of the novel pentapeptide dolastatin 10 was determined, enabling the synthesis of a natural compound *(57)*. Dolastatin 10 is a peptide made up of four amino acid residues (dolavaline, valine, dolaisoleuine, dolaproine) linked to a complex primary amine (dolaphenine). Dolastatins possess antimitotic properties and are cytotoxic in a number of cell lines at subnanomolar concentrations *(58)*. Studies of dolastatin 10 isomers and synthetic precursors have shown that the dolaisoleuine amino acid residue is the most critical for inhibition of tubulin polymerization and that modification of the dolaproine or dolavaline moieties decreases its cytotoxicity but not its ability to inhibit tubulin polymerization *(59,60)*. A structurally modified dolastatin 10 analog, auristatin PE (TZT-1027) (Fig. 2), lacks the terminal dolaphenine, which is replaced with a phenethylamine *(61)*. Cemadotin (LU103793) (Fig. 2) is a synthetic water-soluble analog of dolastatin 15, another antiproliferative compound isolated from *D. auricularia*, with a benzylamide moiety as the C-terminal subunit *(62)*. Another synthetic analog of dolastatin 15, ILX651 (LU223631) (Fig. 2), was designed to improve metabolic stability and enhance oral bioavailability of the drug *(63)*. Unlike cemadotin, ILX651 was administered IV or orally without cardiovascular toxicity at myelosuppressive doses in preclinical models.

Dolastatin 10 has been shown to act through inhibition of microtubule assembly and polymerization of tubulin, and thereby causes cells to accumulate in a state of metaphase arrest *(58)*. Dolastatin 10 noncompetitively inhibits vinblastine binding to tubulin, inhibits nucleotide exchange at the exchangeable GTP-binding site of tubulin, stabilizes

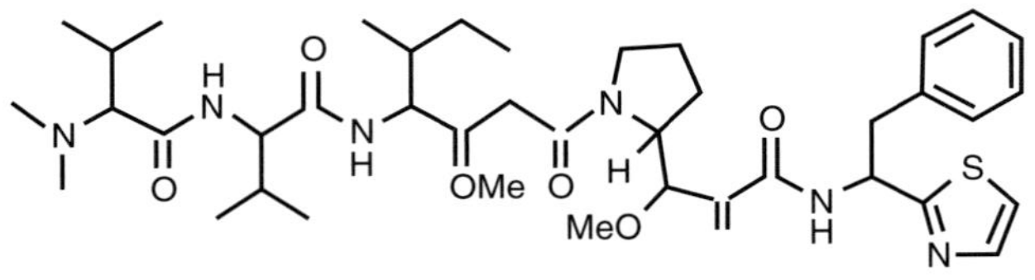

Cemadotin (LU103793) (55)

ILX651 (LU223631)

Dolastatin 10 (54)

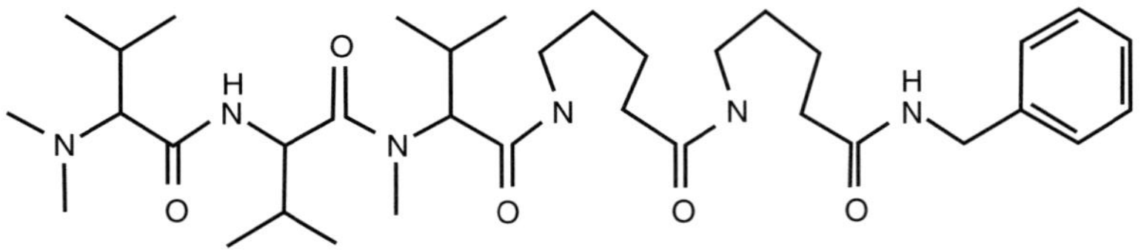

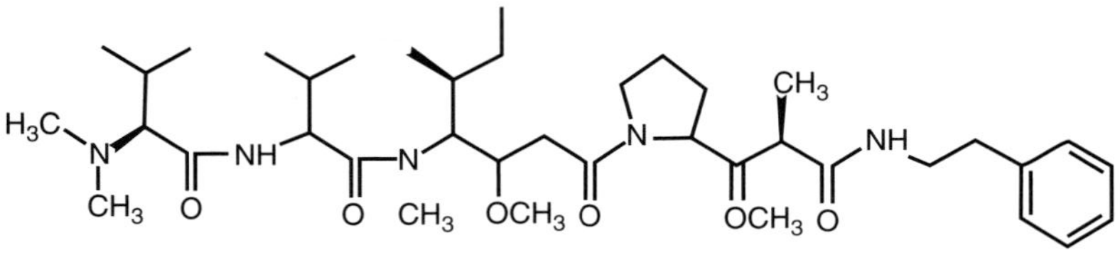

Auristatin PE (TZT 1027) (56)

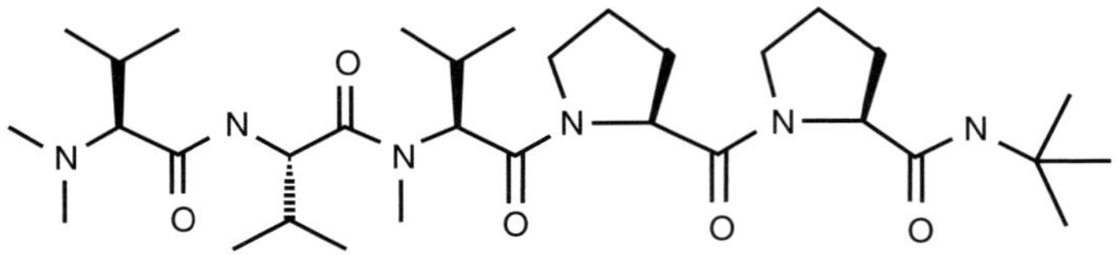

Fig. 2. Dolastatins currently undergoing clinical trials.

436

colchicine's binding, and causes the formation of extensive structured aggregates of tubulin when present in stoichiometric amounts relative to the protein. Originally thought to bind at the Vinca alkaloid site, dolastatin 10 has been shown to competitively inhibit binding of rhizoxin and phomopsin A to tubulin, indicating that dolastatin 10 binds to the rhizoxin/maytansine region on tubulin *(64)*. Dolastatin 10 also induces apoptosis, possibly by phosphorylating and thereby inactivating the bcl-2 protein directly, or as a consequence of microtubule damage *(65,66)*. Studies of auristatin PE have indicated that it inhibits tubulin polymerization similarly to dolastatin 10, but possesses superior antitumor activity *(56,67)*. Similar to dolastatins 10 and 15, cemadotin is highly cytotoxic and inhibits microtubule polymerization *(62)*. Although cellular effects caused by cemadotin are similar to those of vinblastine, cemadotin does not inhibit vinblastine binding to unpolymerized tubulin in vitro. De Arruda and colleagues have also determined that cemadotin does not bind to the colchicine site in tubulin. Subsequent studies have indicated that cemadotin binds to tubulin in two different classes of binding sites: A high-affinity class of sites with a dissociation constant of 19.4 μM, and a low-affinity class of sites with a dissociation constant of 136 μM *(55)*. Some investigators speculated that the low-affinity site is produced by aggregation of tubulin dimers into oligomers in the presence of cemadotin, and suggested that cemadotin suppresses spindle microtubule dynamics through a distinct molecular mechanism by binding at a novel site in tubulin. To date, dolastatin 10, its derivative auristatin PE (TZT-1027), and the two dolastatin 15 analogs ILX651 (LU223631) and cemadotin (LU103793), have been evaluated in cancer clinical trials. The preliminary safety and efficacy data from these trials are described below.

3.1. Dolastatin 10

Dolastatin 10 (Fig. 2) was clinically developed by the National Cancer Institute. Two phase 1 trials have evaluated dolastatin 10 administered as an IV bolus infusion on a q3w in patients with advanced solid tumors *(68,69)*. In one study, 30 patients received dolastatin as bolus doses ranging from 65–455 $\mu g/m^2$ *(69)*. Dose-limiting neutropenia was observed at 455 $\mu g/m^2$ for minimally pretreated patients and at 325 $\mu g/m^2$ for heavily pretreated patients. Nonhematological AEs were generally mild (grade ≤ 1), except for one heavily pretreated patient who experienced grade 3 diarrhea at 325 $\mu g/m^2$, and one patient who experienced grade 2 nausea. Nine of 25 evaluative patients developed new or increased symptoms of peripheral sensory neuropathy (grade 1) at 6 wk. The MTD and the RP2D of dolastatin 10 were 400 $\mu g/m^2$ for minimally pretreated patients and 325 $\mu g/m^2$ for heavily pretreated patients. In another study of 22 patients dolastatin doses ranged from 65 to 300 $\mu g/m^2$ *(68)*. The MTD was reached at the 300 $\mu g/m^2$ level, at which 33% of patients experienced dose-limiting neutropenia. No objective responses were observed on this study. Preliminary pharmacokinetic studies of dolastatin 10 performed in these trials demonstrated a rapid drug distribution with a prolonged plasma elimination phase *(69)*, and mean β and $\gamma\, t_{1/2}$ values of 0.99 and 18.9 h, respectively *(68)*.

In the phase 2 setting, dolastatin 10 has been evaluated in patients with prostate, colorectal, ovarian, NSCLC, renal cell carcinoma, melanoma, breast, and pancreaticobiliary cancers. All phase 2 trials used the same regimen as phase 1 studies, i.e., an IV bolus infusion on the q3w schedule. In 15 patients with hormone-refractory metastatic prostate adenocarcinoma, dolastatin 10 was administered at doses of 400–450 $\mu g/m^2$ *(70)*. Grade 3/4 AEs included neutropenia (53%), neuropathy, anemia, psychiatric AE, liver, and hyperglycemia (one patient each). Three patients had stable disease. In a phase 2 study of dolastatin 10 as first-line treatment for advanced colorectal cancer, 14 patients received doses ranging from 300–450 µg/m^2 *(71)*. Grade 3/4 AEs included neutropenia in 9/42 treatment courses in six patients, anemia in five courses in one patient, creatinine elevation during course eight in one patient, and alkaline phosphatase elevation during one course in one patient. No tumor responses were seen. Thirty patients with recurrent platinum-sensitive ovarian carcinoma were administered 400 µg/m^2 dolastatin 10 *(72)*. Neutropenia was the most common AE with seven cases of grade 4 and three cases of grade 3. Two cases of grade 4 gastrointestinal AEs were reported including one patient with nausea/vomiting and constipation, which resolved with antiemetics and laxatives, and one patient with anorexia. One patient had a pleurodesis for a malignant effusion 8 d after receiving the first dose of dolastatin 10 and died suddenly 4 d later. Autopsy determined that pulmonary embolism was the cause of death, and there were widespread metastases. There were no objective responses, and this study was therefore closed after the first stage of accrual. In 30 patients with advanced renal cell carcinoma, dolastatin 10 was administered at 400 µg/m^2 *(73)*. The most common grade 3/4 AEs included neutropenia, anemia, dyspnea, pleural effusion, fatigue, and constipation. Neurological AEs were mild and did not appear to be cumulative. Three PRs lasting 6, 10.3, and 16.5+ mo have been reported. In patients with stage IIIB or stage IV NSCLC who had not received chemotherapy previously, dolastatin 10 was administered at 400 µg/m^2 *(74)*. Two of 19 evaluative patients developed grade 4 neutropenia, and one patient was hospitalized with fever; no further myelosuppression was noted after a 25% dose reduction. No objective responses were reported. Twelve patients with advanced melanoma who had no previous chemotherapy were administered 400 µg/m^2 dolastatin 10 *(75)*. The only grade 3 AE was neutropenia uncomplicated by infection in four patients. No objective responses were reported. The total systemic clearance and Vd$_{ss}$ were 2.61 ± 1.9 L/h/m^2 and 28.4 ± 13 L/m^2, respectively. In patients with advanced breast cancer, dolastatin 10 was administered at 400 µg/m^2 *(76)*. The most common hematological AE was neutropenia, with grade 3/4 experienced by 62% of patients. Few patients experienced severe nonhematological AEs. One of 21 patients achieved a PR for a duration of 113 d, and four patients maintained stable disease for a median of 87 d. Median survival was 6.3 mo (95%CI: 5.6–23.2 mo), and median TTP was 1.5 mo (95% CI: 1.4–1.9 mo). The same dose and schedule were used in patients with advanced pancreaticobiliary cancer *(77)*. Fifty-nine percent of patients experienced grade 3/4 neutropenia, and 19% developed neutropenic fever. Nonhematological AEs were mostly limited to grade 1/2, with 18% neuropathy (grade 2). There were no objective responses in this study, but 7/27 patients had stable disease (two hepatomas, two cholangiocarcinomas, and one pancreatic carcinoma). In summary, no significant clinical activity of dolastatin 10 as a single-agent was seen in phase 2 trials published to date.

3.2. Auristatin PE (TZT-1027, Dolastatin 10 Analog)

Auristatin PE (TZT-1027) (Fig. 2) is being clinically developed by Daiichi Pharmaceutical Co. LTD, Tokyo, Japan. Auristatin PE (Fig. 2) has been clinically tested in phase 1 trials in Japan, Germany, the Netherlands, and Hungary. In a Japanese study, auristatin PE was administered as a 1-h IV infusion on the weekly ×3 schedule in 40 patients with solid tumors using doses of 0.3–2.1 mg/m^2 *(78)*. Two of four patients treated at the 2.1 mg/m^2 dose level experienced DLTs, including leukopenia, neutropenia, and constipation. Toxicity was acceptable at the 1.8 mg/m² level without DLTs. One PR lasting 183 d has been reported in a patient with a thymoma. The investigators recommended 1.8 mg/m² auristatin PE on the weekly ×3 schedule for further studies.

In a German study, 1.35–3 mg/m² auristatin PE was administered as a 1-h IV infusion on the q3w schedule to 21 patients with solid tumors *(79)*. Three patients experienced DLTs at the 3 mg/m² dose level, including neutropenia, fatigue, and neurotoxicity, whereas other common grade 3/4 AEs included neutropenia, leukopenia, anorexia, alopecia, nausea, fatigue, and constipation. No objective responses were observed. Preliminary pharmacokinetic studies indicate a terminal $t_{1/2}$ of approx 7 h. The investigators recommended 2.7 mg/m² (also the MTD) auristatin PE on the q3w schedule for further clinical trials.

In a study conducted at the Netherlands, 1.35–2.7 mg/m² auristatin PE was administered as a 1-h IV on days 1 and 8 of a q3w schedule *(80)*. DLTs included neutropenia experienced by two patients at 2.7 mg/m², and pain in the infusion arm at 2.4 mg/m² in one patient. The most common nonhematological AEs were fatigue, nausea, vomiting, and diarrhea, mostly grade 1/2. One PR was observed in a heavily pretreated patient with metastatic liposarcoma, lasting more than 54 wk. In addition, eight patients had SD for a median duration of 13 wk. In a Hungarian study, 31 patients with advanced, therapy-resistant NSCLC were treated with escalating doses of auristatin PE of 0.5–3.2 mg/m² q3w *(81)*. The MTD has not yet been reported.

In a phase 1 combination study, auristatin PE was administered with carboplatin to patients with solid tumors *(82)*. Auristatin PE (starting dose 0.4 mg/m²) was administered as a 1-h IV on days 1 and 8 q3w, and carboplatin (AUC = 4 or 5) was administered on day 1 as a 30-min IV before auristatin PE, q3w. Neutropenia, fatigue, and diarrhea were the DLTs in this study (2, 1, and 1 patients, respectively). The most frequent AEs were neutropenia (50%), fatigue (36%), nausea and vomiting (38%), anemia (21%), and peripheral neuropathy (12%). One PR (pancreatic adenocarcinoma) and 7 stable disease were observed in 12 evaluative patients.

3.3. Cemadotin (LU103793, Dolastatin 15 Analog)

Cemadotin (LU103793) (Fig. 2) is being clinically developed by Abbott GmbH and Co. KG (formerly Knoll GmbH/BASF). The early phase 1 trials of cemadotin evaluated 5-min or 24-h IV infusions on various schedules. On the weekly ×4 q5w schedule, cemadotin was administered as a 5-min IV infusion *(83)*. Dose-limiting hypertension and cardiac ischemia were observed at 5 mg/m² cemadotin possibly because of high peak plasma levels. Refractory hypertension prompted the investigators to prolong the infusion time to 24-h based on preclinical testing. Twenty patients with advanced solid tumors were administered 5–12.5 mg/m² cemadotin. In addition to grade 1/2 hypertension already present at 5 mg/m², cardiovascular AEs consisted of grade 1/2 bradycardia and grade 1 angina at 10 and 12.5 mg/m². The MTD for heavily pretreated patients was reached at 10 mg/m² with neutropenia as the DLT. The MTD has not been published for minimally pretreated patients. Other common AEs included liver toxicity, fatigue, phlebitis, fever, nausea/vomiting, and tumor pain. No responses have been observed. Another study administered a 24-h IV infusion of cemadotin at doses of 10–27.5 mg/m²/d to patients with advanced solid tumors. Dose-limiting reversible hypertension at dose levels 20, 25, and 27.5 mg/m² was associated with signs of cardiac ischemia *(84)*. Other grade 3/4 AEs included neutropenia, leukopenia, asthenia, tumor pain, and transient liver enzyme elevation. The best responses were minor tumor regressions in a patient with unknown primary carcinoma and in another patient with liver metastases from a colon cancer. The RP2D was 15 mg/m^2.

Subsequent phase 1 studies examined the influence of the treatment schedule and duration of infusion on cardiovascular toxicity. In one study, doses of 0.5–3 mg/m^2/d cemadotin were administered as a 5-min infusion daily ×5 q3w to 26 patients *(85,86)*. Dose-limiting neutropenia, peripheral edema, and liver function abnormalities were seen at both 2.5 and 3 mg/m^2/d. The RP2D with this schedule was 2.5 mg/m^2/d. Pharmacokinetic parameters in this study were independent of dose and similar on days 1 and 5. Vd_{ss} was 7.6 ± 2 L/m^2, clearance was 0.49 ± 0.18 L/h/m^2, and elimination $t_{1/2}$ was 12.3 ± 3.8 h. In another study, 20 heavily pretreated patients received cemadotin at doses from 0.5–3.5 mg/m^2/d administered as a 5-day continuous IV infusion *(87)*. Reversible dose-related neutropenia was the principal DLT, and 2.5 mg/m^2/d was established as the MTD. No objective responses were observed. Blood drug levels decayed monoexponentially following the end of the infusion, with a mean $t_{1/2}$ of 13.2 ± 4.3 h (n = 14) in all patients. Mean values (n = 14) of the total blood clearance and apparent Vd_{ss} were 0.52 ± 0.09 L/h/m^2 and 9.9 ± 3.3 L/m^2, respectively. There were no cardiovascular AEs observed with the 5-min or 24-h cemadotin infusion in these trials, and the investigators concluded that cardiovascular toxicity appears to be associated with the magnitude of the peak blood levels of cemadotin or its metabolites, whereas myelotoxicity is related to the duration of time that blood levels exceed a threshold concentration.

Two phase 2 trials evaluated 2.5 mg/m^2/d cemadotin administered as a 5-min IV infusion daily ×5 q3w. Thirty-four patients with metastatic breast cancer previously treated with two lines of chemotherapy for advanced disease were enrolled in one study at 18 institutions in Europe *(88)*. Grade 3/4 neutropenia was observed in 18 patients. Other hematological AEs included single cases of anemia and thrombocytopenia. Drug-related nonhematological AEs included grade 3 asthenia, stomatitis, myalgia, and grade 4 serum bilirubin. The main AE was hypertension, occurring in 7/34 patients (one grade 3). No objective responses were reported in 23 evaluative patients. In another study of 80 chemotherapy-naïve melanoma patients, grade 3/4 neutropenia was observed in 19% of patients. Other significant AEs included grade 3 anemia, and a single case of grade 4 thrombocytopenia *(89)*. Nonhematological AEs included asthenia and elevated liver enzymes. One patient each experienced grade 4 cardiac dysrhythmia and grade 3 hypertension. One CR (skin) and three PRs (skin, nodes and liver sites) with a 6 mo median duration (range 3–9.1 mo) were reported in 69 evaluative patients. A study of 17 NSCLC patients reported no objective responses *(90)*. Thirty percent of the patients experienced grade 4 neutropenia.

3.4. ILX651 (LU223631, Dolastatin 15 Analog, Tasidotin)

ILX651 (LU223631) (Fig. 2) is being clinically developed by Genzyme Corp (formerly ILEX™ Oncology Inc). The early phase 1 trials of ILX651 (Fig. 2) evaluated daily ×5 q3w, every-other-day ×3 q3w, and weekly ×3 q4w schedules in patients with advanced solid tumors. On the daily ×5 q3w schedule, ILX651 was administered as a 30-min IV infusion at doses from 2.3–36.3 mg/m^2/d to 36 patients with advanced refractory solid tumors *(91)*. Dose-limiting grade 4 neutropenia was observed in 4/14 patients treated at the 27.3 mg/m^2/d dose level and in 2/2 patients at the 36.3 mg/m^2/d dose level. Other grade 3 AEs included ileus, fever, and AST/ALT elevations. Hypertension was observed in two patients and consisted of transient grade 1 diastolic hypertension. The MTD on this schedule is 27.3 mg/m²/d. One CR in a melanoma patient treated at 15.4 mg/m²/d was reported out of 34 evaluative patients. Two other patients with metastatic melanoma experienced mixed responses of more than 50% reduction of cutaneous nodules. In addition, nine patients had stable disease. Pharmacokinetic studies of ILX651 have shown that the clearance decreased with increasing dose, indicating nonlinear pharmacokinetics.

On the every-other-day ×3 q3w schedule, ILX651 was administered as a 30-min IV infusion at doses from 3.9–45.7 mg/m²/d to 30 patients with advanced refractory solid tumors with no history of significant cardiac disease *(92)*. Dose-limiting transient neutropenia was seen at the 45.7 mg/m²/d dose level in 3/5 patients. The MTD was reached at the 34.4 mg/m²/d level on this schedule. Ten of 32 patients experienced stable disease, with four lasting more than 4 courses (breast, undifferentiated small cell carcinoma, ovarian, and NSCLC for 5, 6, 8, and 14 courses, respectively). Preliminary pharmacokinetic studies of ILX651 in 31 patients indicate that the drug decays from plasma in a biphasic fashion with an effective $t_{1/2}$ ranging from 21 to 55 min. The Vd_{ss} was independent of dose with a mean of 13.2 L/m², area under the concentration-time curve (AUC) was not dose proportional (range 165–2,534 h × ng/mL), and neither was the systemic clearance, which decreased as C_{max} increased, and ranged from 8.6 to 43.2 L/h/m².

On the weekly ×3 q4w schedule, ILX651 was administered as an IV infusion at doses of 7.8–62.2 mg/m²/d to 30 patients with advanced solid malignancies *(93)*. Patients were stratified by previous myelotoxic therapy into minimally and heavily pretreatd cohorts. Dose limiting neutropenia was seen in 2/3 minimally pretreated patients at 62.2 mg/m²/d and in 3 and 2 minimally and heavily pretreated patients, respectively, at the 54.5 mg/m²/d. No objective responses were reported. A 47% decrease was seen in a taxane- and platinum-refractory NSCLC patient lasting 5 mo, and stable disease more than 4 mo was seen in two patients (HCC and renal cancer).

Preliminary results are available from a phase 2 study in melanoma patients, administering ILX651 at 28 mg/m² as a 30-min IV infusion daily ×5 q3w *(94,95)*. Grade 3/4 nondose limiting predicted (day 15) neutropenia and self-limited neutropenia were seen in 2 and 3 patients, respectively. Other AEs were grade 1/2. On dose escalation to 34 mg/m², 3/10 patients experienced dose-limiting neutropenia. Overall, grade 4 neutropenia was seen in 38% of patients. One CR and four PRs (1 unconfirmed) have been observed. At a dose of 28 mg/m², preliminary median PFS was 39 d, and median overall survival was 271 d (Table 2).

4. COLCHICINE-LIKE COMPOUNDS

The classic antimitotic drug colchicine is known to bind tubulin, thereby preventing proper assembly of microtubules in the mitotic spindle and disrupting cell division *(100)*. The binding of colchicine induces partial unfolding of the carboxyl end of an amphipathic helix of β-tubulin, preventing contacts necessary for microtubule assembly. Colchicine, a highly soluble alkaloid, was first isolated from the meadow saffron *Colchicum autumnale* and used as a treatment for gout *(54)*. However, the toxicity profile of colchicine prevented its use in other therapies and prompted a search for simplified, less toxic analogs acting at doses well below their MTDs.

Table 2
Clinical Activity of Single-Agent Dolastatins in Phase 2 Trials

Drug name	Tumor type/schedule	Phase	Activity	Major adverse events (CTC grade 3/4)
Dolastatin 10	Solid (67)/IV bolus q3w	1	0/30 objective responses	DLT = neutropenia 30 evaluative pts: neutropenia 27%; leukopenia 13%; anemia 3%; diarrhea 3%
	Solid (68)/IV bolus q3w	1	0/18 objective responses	DLT = grade 4 neutropenia in 3/9 pts 19 evaluative pts: fatigue 11%
	Prostate (70)/IV bolus q3w	2	3/15 SD	15 evaluative pts: neutropenia 53%; neuropathy 7%
	Colorectal (71)/IV bolus q3w	2	4/13 SD	14 evaluative pts: neutropenia 43%; anemia 7%; creatinine ↑ 7%; alkaline phosphatase ↑ 7%
	Ovarian (72)/IV bolus q3w	2	7/28 SD	30 evaluative pts: neutropenia 33%; nausea/vomiting and constipation 3%; anorexia 3%. 1 death of pulmonary embolism
	RCC (73)/IV bolus q3w	2	3/30 PR, 3/30 SD	30 evaluative pts: neutropenia 47%; leukopenia 17%; anemia 10%; fatigue 7%; constipation 7%; dyspnea 6%; pleural effusion 6%
	NSCLC (74)/IV bolus q3w	2	3/19 SD	19 evaluative pts: neutropenia 11%
	Melanoma (75)/IV bolus q3w	2	0/12 objective responses	12 evaluative pts: neutropenia 33%
	Breast (76)/IV bolus q3w	2	1/21 PR 1/21 SD	21 evaluative pts: neutropenia 62%
	Pancreaticobiliary (77)/IV bolus q3w	2	5/27 SD	28 evaluative pts: neutropenia 59%; fatigue 11%
Auristatin PE (TZT 1027)	Solid (78)/1-h IV wkly ×3	1	1/40 PR	DLT = grade 4 leukopenia, neutropenia and constipation in 1/4 pts, grade 4 neutropenia and grade 3 constipation in 1/4 pts
	Solid (79)/1-h IV q3wks	1	7/21 SD	DLT = grade 4 neutropenia in 3/6 pts 21 evaluative patients: fatigue 38%; insomnia 5%; neurotoxicity 14%

(Continued)

Table 2 *(Continued)*

Drug name	Tumor type/schedule	Phase	Activity	Major adverse events (CTC grade 3/4)
	Solid (80)/1-h IV d1 and 8 q3w	1	1/17 PR (liposarcoma) 8/17 SD	DLT = grade 4 neutropenia in 2/4 pts, pain in infusion arm in 1/7 pts.
	NSCLC (81)/IV q3w	1	0/29 objective responses	29 evaluative pts: neutropenia 3%
Auristatin PE + carboplatin	Solid (82)/1-h IV d1 and 8 q3w + carboplatin	1	1/12 PR (pancreas) 7/12 SD	DLT = neutropenia 2/9, fatigue, diarrhea in 1/9 pts each 13 evaluative pts: neutropenia 36%; nausea/vomiting 2%; anemia 5%
Cemadotin (LU103793)	Solid (83)/5-min IV wkly ×4 q5wks	1	–	DLT = hypertension and cardiac ischemia
	Changed to 24-h IV wkly ×4 q5wks		0/20 objective - responses	DLT = grade 4 neutropenia in 2/4 pts; grade 3 nausea, vomiting and tumor pain
	Solid (84)/24-h IV q3w	1	2/30 MR	DLT = grade 3 hypertension in 10/23 pts 30 evaluative pts: neutropenia 68%; leukopenia 32%; anemia 18%; thrombocytopenia 4%; hypertension 33%; bilirubin 33%; AST/ALT 27%; tumor pain 23%; LDH 23%; asthenia 10%; constipation 10%
	Solid (86)/5-min IV daily × 5 q3w	1	0/26 objective responses	DLT = neutropenia, peripheral edema, and liver function test abnormalities in 6/18 pts
	Solid (87)/5-d CIV q3w	1	0/20 objective responses	DLT = grade 4 neutropenia in 5/20 pts 20 evaluative pts: leukopenia 20%; bilirubin ↑ 30%; AST/ALT 15%; nausea 10%
	Breast (88)/5-min IV daily ×5 q3w	2	7/23 SD	34 evaluativee pts: neutropenia 53%; asthenia 9%; bilirubin ↑ 6%; anemia 3%; thrombocytopenia 3%; stomatitis 3%; myalgia 3%; hypertension 3%
	Melanoma (89)/5-min IV daily × 5 q3w	2	1/69 CR, 3/69 PR, 15/69 SD	Gr 3/4 neutropenia (3%/16%) 80 evaluative pts: leukopenia 17%; anemia 5%; cardiac dysrhythmia 1%; hypertension 1%

(Continued)

Table 2 *(Continued)*

Drug name	Tumor type/schedule	Phase	Activity	Major adverse events (CTC grade 3/4)
ILX651 (LU223631)	Solid (91)/30-min IV daily × 5 q3w	1	1/34 CR (melanoma), 2/34 PR 9/34 SD	DLT = grade 4 neutropenia in 6/16 pts; 36 evaluative pts: AST/ALT 8%; ileus 3%; pyrexia 3%
	Solid (92)/q2d × 3 q3w	1	10/32 SD	DLT = grade 4 neutropenia in 5/16 pts 32 evaluative pts: lymphopenia 31%
	Solid (93)/wkly ×3 q4wks	1	2/30 SD	DLT = neutropenia in 8/17 pts

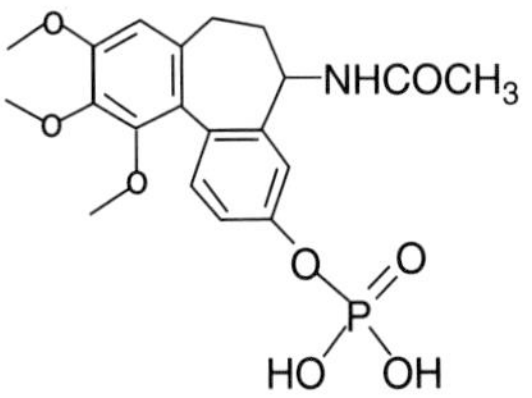

ZD6126 (ANG453) (96)

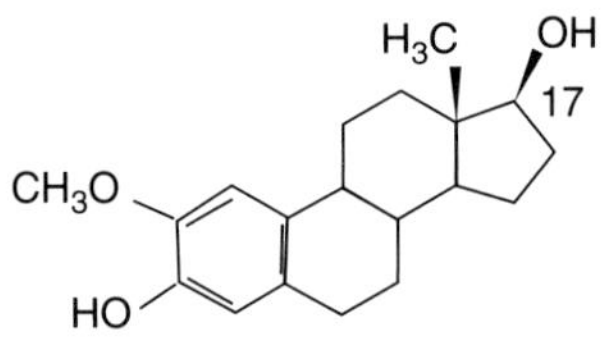

2-methoxyestradiol (Panzem™) (97)

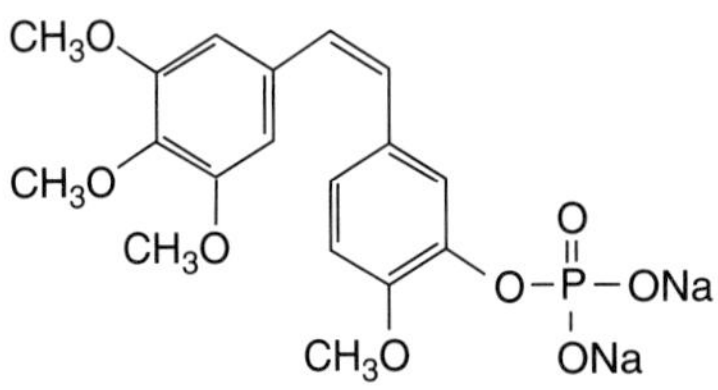

Combretastatin A-4 phosphate (CA-4-P)

(99)

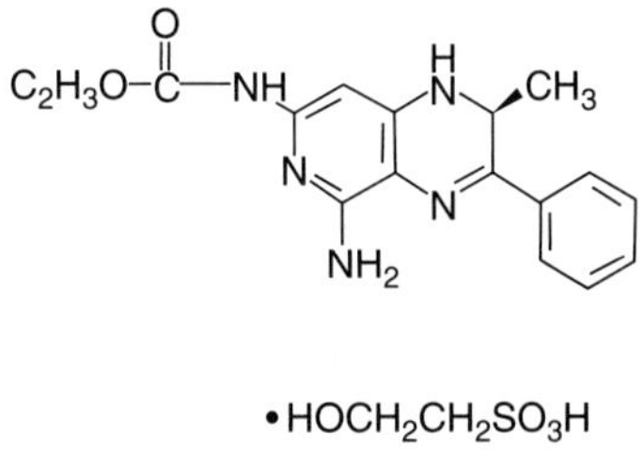

Mivobulin isethionate (CI-980) (98)

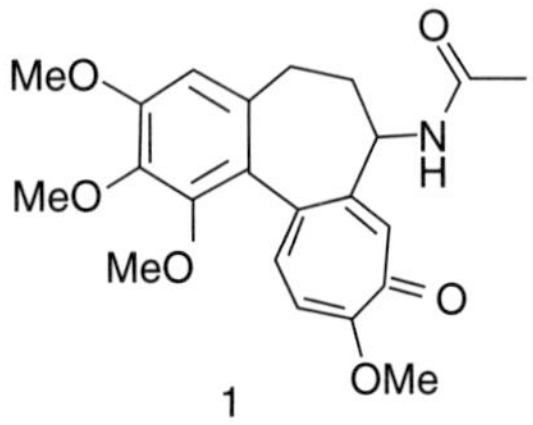

Colchicine (54)

Fig. 3. Colchicine-like compounds currently undergoing clinical trials.

The colchicine analog ZD6126 (ANG453) (Fig. 3) is a synthetic water-soluble phosphate prodrug that is converted in vivo into N-acetylcolchinol (ZD6126 phenol), which binds to the colchicine site on tubulin and causes disruption of microtubules *(101,102)*. In animal models, ZD6126 acts by causing rapid cell shape changes in proliferating immature endothelial cells, leading to exposure of the basal lamina after cell retraction and subsequent loss of endothelial cells *(103)*. This in turn leads to thrombosis and vessel occlusion, resulting in extensive tumor necrosis. The in vivo effects of ZD6126 are highly tumor-selective and rapidly reversible on drug removal.

Another compound that is structurally related to the colchicine antimitotic compound, combretastatin A-4, was originally isolated from the African willow tree *Combretum caffrum (104)*. The combretastatins contain two phenyl rings, which are tilted at 50–60° to each other and linked by a two-carbon bridge, with several methoxy substitutions on the ring system *(99)*. Combretastatin A-4 was shown to interact with tubulin at or near the colchicine-binding site and inhibit tubulin polymerization. However, the clinical development of combretastatin A-4 was hindered by its limited water solubility. Combretastatin A-4 phosphate (CA-4-P) (Fig. 3), which was developed to improve the water solubility, is a prodrug that rapidly dephosphorylates in vivo to the active compound combretastatin A-4 *(105)*. In animal models CA-4-P causes rapid and selective disruption of the tumor vasculature through induction of apoptosis of proliferating endothelial cells, followed by hemorrhagic necrosis of a tumor *(106)*. However, a recent study has indicated that apoptosis is not the major cause of CA-4-P-induced endothelial cell death; instead suggesting that CA-4-P inhibits angiogenesis most likely by signals inhibiting endothelial cell migration *(107)*.

Another antimitotic compound that binds to the colchicine site of tubulin, 2-methoxyestradiol (2-ME, Panzem™) (Fig. 3), is the naturally occurring metabolite of estradiol, which is excreted in the urine *(97)*. 2-ME has been shown to competitively bind to the colchicine site of tubulin, resulting in inhibition of tubulin polymerization or formation of a polymer with altered stability properties, depending on the reaction conditions *(108)*. Several mechanisms of cell growth inhibition by 2-ME have been demonstrated *(97)*. Inhibition of polymerization of tubulin results in mitotic arrest in the G_2/M phase of the cell cycle. Furthermore, 2-ME inhibits angiogenesis *(109)*.

Mivobulin isethionate (CI-980) (Fig. 3) is a synthetic water-soluble antimitotic compound that competitively binds tubulin at the colchicine-binding site and inhibits tubulin polymerization, blocking the formation of the mitotic spindle *(110)*. Cancer cells exposed to mivobulin isethionate accumulate in the M phase of the cell cycle and subsequently die. Preclinical studies have demonstrated that mivobulin isethionate is able to cross the blood–brain barrier *(111)*. Importantly, mivobulin isethionate demonstrated significant antitumor activity in a broad spectrum of murine and human tumor models that were cross resistant to vincristine, cisplatin, vinblastine, navelbine, and doxorubicin and in tumor cell lines exhibiting multidrug resistance owing to P-glycoprotein overexpression *(112)*. In animal studies, activity of mivobulin isethionate was largely independent of the route of drug administration but favored a prolonged treatment schedule.

To date, the colchicine analog ZD6126 (ANG453), combretastatin A-4 phosphate (CA-4-P), 2-ME, and mivobulin isethionate (CI-980) have been evaluated in human cancer clinical trials. The preliminary safety and efficacy data from these trials are described below.

4.1. ZD6126 (ANG453)

ZD6126 (ANG453) (Fig. 3) is being clinically developed by AstraZeneca. Two phase 1 trials are evaluating ZD6126 administered as a 10-min IV infusion on a weekly ×4 (113) or q3w schedule (114) in patients with solid tumors refractory to conventional therapy. On the weekly ×4 schedule, ZD6126 was administered at 5 mg/m^2 (7 patients) and 7 mg/m^2 (five patients) (113). Two cases of hypokalemia were the only grade 3 AEs reported. The MTD has not been reported.

On the q3w schedule, ZD6126 was administered at dose levels of 5–112 mg/m^2 to 31 patients who had received chemotherapy previously (114,115). One patient had asymptomatic reversible grade 2 ischemic changes in ECG and grade 3 elevation in serum troponin I levels with subsequent demonstration of coronary artery disease. Abdominal pain and gastrointestinal symptoms were dose limiting. One patient had a minor response lasting 19 cycles. Dynamic contrast-enhanced magnetic resonance imaging (DCE-MRI) studies have documented reductions in tumor blood flow after ZD6126 administration (115,116). Preliminary pharmacokinetic studies of ZD6126 performed in these trials demonstrate that the drug is rapidly hydrolyzed to ZD6126 phenol, which has an elimination $t_{1/2}$ of 2–3 h.

4.2. Combretastatin A4 Phosphate (CA-4-P)

Combretastatin A-4 phosphate (CA-4-P) (Fig. 3) is being clinically developed by OXiGENE, Inc., Massachusetts, MA. The phase 1 trials of CA-4-P evaluated 10-min or 60-min IV infusions on various schedules. On the q3w schedule, CA-4-P was administered as a 10-min IV infusion at doses of 18, 36, 60, and 90 mg/m^2 and as a 60-min IV infusion at a dose of 60 mg/m^2 to 25 patients with advanced solid tumors (117). DLTs at 90 mg/m^2 included shortness of breath and cardiac ischemia. DLTs at the 10-min infusion of 60 mg/m^2 included pulmonary toxicity, tumor pain, and myocardial ischemia. Tumor pain associated with the infusion of CA-4-P occurred in 10% of cycles at the 60 mg/m^2 dose level on both the 10- and 60-min schedules and completely resolved 24 h after drug administration. The investigators noted the absence of traditional cytotoxic side effects such as myelosuppression, stomatitis, and alopecia. In more than 107 courses of therapy, only one case each of grade 3 anemia, neutropenia, and thrombocytopenia was reported. The investigators concluded that dosages ≤60 mg/m^2 as a 10-min IV infusion define the upper boundary of the MTD. One CR in a patient with refractory metastatic anaplastic thyroid carcinoma treated with a 10-min infusion at 60 mg/m^2 was reported in 25 patients. DCE-MRI indicated that 6/7 patients treated at 60 mg/m^2 had a significant decline in gradient peak tumor blood flow ($p < 0.03$). Pharmacokinetic studies indicate that CA-4-P is rapidly dephosphorylated to combretastatin A4, with a short plasma $t_{1/2}$ of approx 30 min. The mean terminal $t_{1/2}$ was 0.47 h, the mean Vd$_{ss}$ was 7.5 L, and the mean systemic clearance was 24.6 L/h and varied over an eightfold range with a coefficient of variation of 53%.

In a British study, CA-4-P was administered as a 10-min IV infusion on the weekly ×3 q4w schedule to 34 patients at doses of 5–114 mg/m^2 (118). DLTs were reversible ataxia and motor neuropathy at 114 mg/m^2 and vasovagal syncope at 88 mg/m^2. Other drug-related grade 3/4 AEs included pain, fatigue, visual disturbance, hypotension, dyspnea, and fatal ischemia in a patient with previously irradiated bowel. No objective responses were observed, and the RP2D was 68 mg/m^2. Pharmacokinetic studies indicated that CA-4-P was rapidly converted to the active combretastatin A4, which was further metabolized to the CA4 glucuronide, which was the dominant species in the plasma 15 min after the end of infusion.

On the daily ×5 q3w schedule, 6–75 mg/m^2 CA-4-P was administered as a 10-min IV infusion to 37 patients with refractory solid tumors *(119)*. Dose limiting dyspnea occurred in two patients at 75 mg/m^2, as well as DLTs of infustion-associated pain at sites of unknown tumor (2/6 patients). The 75 mg/m^2 dose level was found to be the MTD, and the 52 mg/m^2 was further characterized as the RP2D. A patient with multiple pulmonary metastases had a PR and received eight cycles before progression. In addition, 14 patients maintained stable disease lasting 3–29 cycles. The investigators reported tumor blood flow reduction following CA-4-P therapy in 8/10 patients who were followed by MRI, and three patients demonstrated marked loss of vascularity.

In a phase 1 combination study CA-4-P was administered in combination with carboplatin (carboplatin 30-min IV and CA-4-P 10-min IV 1 h later) *(120)*. At the first dose level of 27 mg/m^2 CA-4-P and carboplatin AUC = 5, 1/6 patients experienced grade 3 dose-limiting neutropenia, and at 36 mg/m^2 CA-4-P and carboplatin AUC = 5, 2/6 patients experienced dose limiting grade 4 thrombocytopenia. As a result of these DLTs, dose excalation was halted and subsequent patients were enrolled at an amended dose level of 36 mg/m^2 CA-4-P and carboplatin AUC = 4. No objective responses were observed in 16 patients, and six patients showed stable disease.

A phase 1b combination trial of CA-4-P in combination with taxanes has been performed (carboplatin or paclitaxel) *(121)*. Twenty-one patients were administered CA-4-P at doses of 27–45 mg/m^2 as a 10-min IV, and paclitaxel (135/175 mg/m^2) or carboplatin (AUC = 4 or 5) were administered 18–22 h later, on a q3w schedule. Grade 3/4 AEs included tumor pain ($n = 2$), increase in liver enzymes ($n = 2$), and one case each of allergy to carboplatin, sensory neuropathy and alopecia related to paclitaxel, lymphopenia, transient dysphasia, and alkaline phosphatase elevation. PRs were seen in 6/9 patients with ovarian cancer and a patient with rapidly progressive renal papillary carcinoma.

4.3. 2-Methoxyestradiol (Panzem)

2-ME (Panzem®) (Fig. 3) is being clinically developed by EntreMed, Inc., Rockville, Maryland. In a phase 1 trial, 2-ME was administered orally at doses of 200–1000 mg/d once daily or at doses of 200–800 mg twice daily for 28 d followed by a 14-d observation period to 31 patients with refractory metastatic breast cancer *(122)*. There were no grade 4 AEs, and the MTD was not reached. No changes in estrogen, luteinizing hormone, or follicle stimulating hormone have been observed. Pharmacokinetic studies indicated that peak serum levels of 2-ME were reached 2–4 h after administration; terminal $t_{1/2}$ was approx 10 h. No drug accumulation was observed with daily administration but was observed throughout the first cycle with every-12-h dosing. Significant interpatient variability and extensive metabolism to 2-methoxyestrone, with 2-methoxyestrone levels significantly higher than those of 2-ME, were noted on both dosing schedules.

In another phase 1 trial, 2-ME was administered orally twice daily at doses 800–6000 mg/d to 20 patients with heavily pretreated solid tumors *(123)*. The drug was well tolerated, and the MTD was not reached. One patient had grade 4 angioedema 42 d into 2-ME treatment at the 3200 mg/d dose level. One patient with clear cell carcinoma of the ovary experienced a PR, with a 67% reduction in the size of pelvic lymph nodes and

78% reduction in CA125 concentrations at the 3200 mg/d dose level. In addition, one patient with prostate cancer, who was treated at the 1600 mg/d dose level, had stable disease and remained on treatment for 350 d.

In a combination phase 1 trial, 2-ME was administered orally at doses of 200–1000 mg/d once daily for 28 d followed by a 13-d observation period in cycle one and continuously thereafter in combination with 35 mg/m² docetaxel administered weekly ×4 q6w to 15 patients with metastatic breast cancer (124). There were no grade 4 AEs, and the MTD was not reached. Grade 3 AEs included fatigue, diarrhea, hand-foot syndrome, and transaminase elevations. The overall response rate was 20% including one CR. The median time to treatment failure was 167 d. As in the single-agent study, no changes in estrogen, luteinizing hormone, or follicle stimulating hormone were observed. Pharmacokinetic studies indicated that 2-ME did not alter docetaxel clearance or dose-normalized AUC, and concurrent docetaxel did not alter 2-ME peak and trough levels. Similarly to the single-agent study, significant interpatient variability in 2-ME clearance and extensive metabolism to 2-methoxyestrone, with 2-methoxyestrone levels significantly higher than those of 2-ME, were reported. No drug accumulation was observed throughout the first cycle in this study. In a randomized, double-blind phase 2 trial, 2-ME was administered orally at doses of 400 or 1200 mg/d once daily to 33 patients with hormone-refractory prostate cancer (125). Grade 2/3 liver function abnormalities, which normalized rapidly after drug discontinuation, were observed in 3/33 patients. A total of eight patients had a decline in PSA of >20% (21–44%). Pharmacokinetic studies reported mean C_{max} values of 2.2 and 5.5 ng/mL on days 1 and 28, respectively, for the 400 mg/d dose, and 2.6 and 9.6 ng/mL on days 1 and 28, respectively, for the 1200 mg/d dose. A new formulation or 2-ME (Panzem NCD) has recently been developed to increase in vivo absorption (126). EntreMed, Inc. is conducting a phase 1b study in advanced cancers, and a phase 2 study in patients with recurring glioblastoma multiforme (127) with this new formulation.

4.4. Mivobulin Isethionate (CI-980)

Mivobulin isethionate (CI-980) (Fig. 3) is being clinically developed by Pfizer, Inc. (formerly Warner-Lambert Co.). Based on the significant antitumor activity mivobulin isethionate showed in preclinical studies, its ability to cross the blood–brain barrier, and independence of the route of administration, mivobulin isethionate was tested in a variety of phase 1 and 2 studies (98,128–142). Phase 2 studies evaluated Mivobulin isethionate in patients with prostate, colorectal, renal cell carcinoma, ovarian, small cell and NSCLC, melanoma, and soft-tissue sarcomas. Activity in these trials was minimal, with one PR lasting 11 wk reported in a patient with colon carcinoma and liver metastases (phase 1) and one PR lasting 27 mo reported in a patient with ovarian cancer (phase 2) (Table 3).

5. SULFONAMIDES

Novel antitumor sulfonamides have been discovered by evaluating sulfonamide-focused libraries and array-based transcriptional profiling (146,147). Previously, sulfonamides have been widely used as antibiotics, insulin-releasing hypoglycemic agents, carbonic anhydrase-inhibitory diuretics, high-ceiling diuretics, and antihypersensitive drugs (143). All sulfonamide derivatives have a common chemical motif of aromatic/heterocyclic sulfonamide (148).

Table 3
Clinical Activity of Single-Agent Colchicine-Like Compounds in Phase 2 Trials

Drug name	Tumor type/schedule	Phase	Activity	Major adverse events (CTC grade 3/4)
ZD6126	Solid (113)/10-min IV wkly ×3	1	–	12 evaluative pts: hypokalemia 17%
	Solid (114)/10-min IV q3w	1	3/31 SD, 1/31 minor response	DLT = abdominal pain and gastrointestinal symptoms. troponin 1 ↑ in 1 pt
CA-4-P	Solid (117)/10 or 60 min IV q3w	1	1/25 CR 3/25 SD	DLT = grade 4 cardiac ischemia, grade 3 shortness of breath, grade 3 pulmonary toxicity, grade 2 tumor pain. 25 evaluative pts: tumor pain 8%; pulmonary 8%; nausea 4%; cardiac/ischemia 4%
	Solid (118)/10 min IV wkly ×3	1	0/34 objective responses	DLT = grade 3/4 vasovagal episode in 1/7 pts, motor neuropathy, ataxia each in 1/6 pts 34 evaluative pts: lymphopenia 3%; vasovagal episode 3%; hypotension 3%; bowel ischemia 3%; fatigue 3%; ataxia 3%; motor neuropathy 3%; vision 3%; abdominal pain or cramping 3%; tumor pain 3%; dyspnea 3%
	Solid (119)/10-min IV daily ×5 q3w	1	1/37 PR 14/37 SD	DLT = tumor pain in 2/6 pts; dyspnea in 2/6 pts
CA-4-P + carboplatin or paclitaxel	Solid (121)/10-min IV q3w + carboplatin or paclitaxel	1b	6/9 PR	21 evaluative pts: tumor pain 10%; GGT↑ 10%; lymphopenia 5%; dysphasia 5%; alk phos 5%
2-Methoxye-stradiol (Panzem)	Breast (122)/once or twice daily q4wks	1	17/22 SD	–
	Solid (123)/twice daily	1	1/20 PR 1/20 SD	20 evaluative pts: angioedema 5%
	Prostate (125)/ once daily	2	–	33 evaluative pts: grade 2/3 liver function abnormalities 9%
Mivobulin isethionate (CI-980)	Solid (98)/24-h IV q21d	1	1/25 PR	DLT = acute reversible neurotoxicity 25 evaluative pts: neutropenia 20%; unconsciousness 12%; nausea 8%; dyspepsia 8%; dizziness 4%; agitation 4%; tremor 4%

(Continued)

Table 3 *(Continued)*

Drug name	Tumor type/schedule	Phase	Activity	Major adverse events (CTC grade 3/4)
	Solid (137,138)/ 72-h IV q21d	1	0/16 objective responses	DLT = neutropenia
	Solid (128)/72-h IV q21d	1 (ped)	1/33 MR 5/33 SD	DLT = myelosuppression, grade 3 reversible cortical toxicity, neurological AEs. 33 evaluative pts: vomiting 3%; myalgia 3%
	Glioma (132)/24-h or 72-h IV	1/2	4/24/SD	DLT = grade 3/4 granulocytopenia
	Glioma (133)/72-h IV q21d	1/2	0/24 objective responses	–
	Colorectal (136)/72-h IV q21d	2	0/14 objective responses	DLT = grade 4 neutropenia in 8/14 pts 14 evaluative pts: granulocytopenia 86%; fatigue 14%; myalgia 7%
	Renal cell (129)/72-h IV q21d	2	0/12 objective responses	12 evaluative pts: neutropeina 33%; fatigue 8%; peripheral neuropathy 8%
	Melanoma (139)/72-h IV q21d	2	2/12 SD	12 evaluative pts: neutropenia 50%; leukopenia 42%
	Prostate (139)/72-h IV q21d	2	4/11 SD	11 evaluative pts: leukopenia 55%; neutropenia 55%; anemia 27%; neutropenic fever; thrombocytopenia; pain; dysrhythmia; neuropathy; thrombosis; infection; hearing loss; hypotension; hypertension; anorexia; nausea; incontinence; alk phos (8% each)
	Melanoma (142)/72-h IV q21d	2	7/24 SD	24 evaluative pts: leukopenia/granulocytopenia 42%; nausea/vomiting 8%; fatigue 8%; anemia 4%; phlebitis/thrombosis 4%; thrombocytopenia 4%
	Ovarian (131)/72-h IV q21d	2	1/16 PR 4/16 SD	16 evaluative pts: granulocytopenia 31%; memory loss 25%; neurosensory 25%; hypoxemia 6%; dyspnea 6%
	Small cell lung (141)/72-h IV q21d	2	0/12 objective responses	12 evaluative pts: 75% neutropenia; 58% anemia
	NSCLC (140)/72-h IV q21d	2	6/14 SD	14 evaluative pts: neutropenia 36%
	Soft tissue (135)/72-h IV q21d	2	6/18 SD	18 evaluative pts: neutropenia 50%; thrombophlebitis 6%

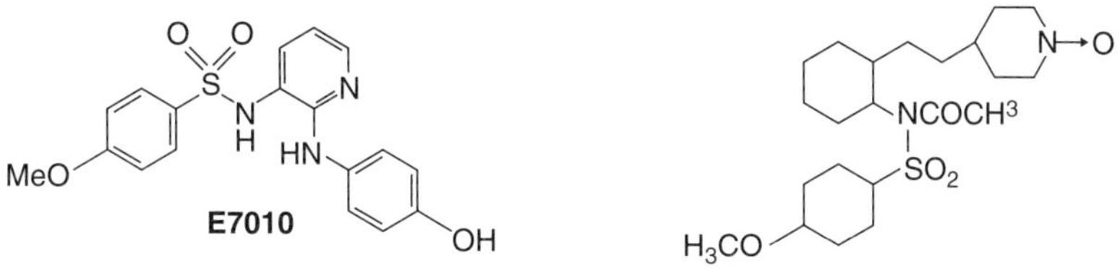

E7010

ABT-751 (E7010) (143)   HMN-214 (144, 145)

Fig. 4. Sulfonamides currently undergoing clinical trials.

5.1. ABT-751 (E7010)

The first orally active antimitotic sulfonamide ABT-751 (E7010) (Fig. 4) was discovered at the Tsukuba Research Laboratories of Eisai Co. LTD, Tokyo, Japan *(147)*. It has a broader antitumor spectrum than vincristine and has different properties from those of vincristine and paclitaxel *(149)*. ABT-751 (E7010) inhibits tubulin polymerization by reversible binding to the colchicine site of β-tubulin, resulting in cell cycle arrest in the M phase. The in vivo activity of ABT-751 (E7010) is similar to that of irinotecan and superior to that of 5-fluorouracil *(150)*.

In the United states, ABT-751 (E7010) (Fig. 4) is being clinically developed by Abbott Laboratories, IL. In Japan, E7010 was clinically developed by Eisai Co. Ltd., Tokyo, Japan. A phase 1 study in Japan explored both single dose and 5-d repeated-dosing (starting dose of 30 mg/m^2/d ×5) of the drug *(151)*. In the single-dose part, E7010 was administered orally to 16 patients with solid tumors at doses ranging from 80–480 mg/m^2. Eleven of 16 patients had lung cancer, and all had had previous chemotherapy. Dose-limiting peripheral neuropathy occurred at 480 mg/m^2. Hematological AEs were mild. Gastrointestinal AEs were also mild, with the exception of one case each of grade 3 nausea/vomiting, anorexia, and epigastric discomfort at 480 mg/m^2, which improved within 3 d of onset. After the safety of E7010 had been confirmed, the drug was administered orally to 41 patients with solid tumors at doses ranging from 30 to 240 mg/m^2 daily ×5. Seventeen of 35 evaluative patients had lung cancer, and 10/35 evaluative patients had gynecological cancer. Thirty-two of 35 evaluative patients had had previous chemotherapy. Both peripheral neuropathy and ileus were dose limiting. Hematological AEs were mild and not dose dependent. Reversible gastrointestinal AEs included nausea/vomiting, anorexia, and epigastric discomfort. The RP2Ds were 320 mg/m^2 for single-dose and 200 mg/m^2/d for a daily ×5 dose. Antitumor activity was limited to one PR in a patient with uterine sarcoma spinal metastases, and mild decrements in CEA and in squamous cell carcinoma antigen in patients with gastric and uterine cancers, respectively. Plasma E7010 levels rapidly increased following administration, with terminal $t_{1/2}$ of 4.4–16.6 h. Total drug recovery in urine 72 h after administration was 77.8 ± 11.4%. No accumulation of the drug was seen on the 5-d schedule.

Another phase 1 study in patients with refractory hematological malignancies administered E7010 orally daily ×7 q3w (100–150 mg/m^2) or daily ×21 q4w (75–200 mg/m^2) *(152)*. One patient experienced a DLT of small bowel obstruction/ileus at the 200 mg/m^2 dose on the q4w schedule. The only grade 3/4 AEs were one episode each of nausea/vomiting and increased alkaline phosphatase/bilirubin. The RP2D was 175 mg/m^2 daily ×21 q4w. One CR was observed in 32 evaluative patients, and four patients had hematological improvement. The same schedules were used in a phase 1 pediatric study in patients with solid tumors *(153)*. DLTs included grade 2 motor neuropathy in 1/6 patients and grade 3 constipation in 1/6 patients. A patient with neuroblastoma receiving 100 mg/m^2 ×7 q3w had stable disease after >24 cycles, and another neuroblatoma patient receiving 165 mg/m^2 ×7 q3w had complete resolution of tumor in all previous disease sites after >14 cycles. A phase 1 pediatric study in patients with neuroblastoma, using doses of 100–250 mg/m^2 on the q3w schedule and 75–165 mg/m^2 on the q4w schedule, defined the MTD as 200 mg/m^2/d (adult equivalent of 375 mg) on the daily ×7 q3w schedule *(154)*. DLTs on the q3w schedule included grade 2 motor neuropathy, grade 3 constipation (one patient each), and asymptomatic decreased left ventricular shortening fraction, neuropathic pain, fatigue, and neutropenia. On the q4w schedule, DLTs included grade 4 neuropathy, vomiting, and dehydration; and grade 3 fatigue, pain, and transaminase elevation.

Two phase 1 trials have evaluated E7010 administered orally on a daily or twice daily ×21 q4w or daily ×7 q3w schedules *(155)*. The starting dose of 25 mg/d or 25 mg/twice-daily was escalated to 250 mg/d or 100 mg/twice-daily. DLTs included hyponatremia and fatigue at 100 mg/twice-daily, fatigue and anorexia at 200 mg/d, and paralytic ileus at the 250 mg/d dose level. Other grade 3/4 AEs included hyponatremia, constipation, abdominal pain, and increased protrombin time in a patient with anorexia on warfarin. A colon cancer patient treated with 50 mg/d experienced a PR lasting 10 cycles. E7010 was rapidly absorbed with a mean time to C_{max} of 1.5 h and a $t_{1/2}$ of approx 6 h.

E7010 was administered orally daily ×7 q3w at 200–300 mg/d to 15 patients, and at doses of 125 and 150 mg/twice-daily to 22 patients *(156, 157)*. DLTs included fatigue, ileus, constipation, and abdominal pain experienced by 2/9 patients at 150 mg/twice-daily, as well as abdominal pain and constipation at the 300 mg/d dose level. The MTD for the daily regimen was 250 mg/d. E7010 was rapidly absorbed with a mean time to C_{max} of approx 1.6 h and a $t_{1/2}$ of approx 4.7 h. Clearance of E7010 was through inactive sulfate and glucuronide metabolites. Four patients had stable disease, including one patient with recurrent anaplastic Astrocytoma.

Several phase 2 studies are evaluating E7010 in breast, colorectal, renal cell cancer, and NSCLC *(158–161)*. All phase 2 studies utilize an oral dose of 200 mg E7010 daily ×21 q4w. Grade 3/4 AEs experienced in these studies include asthenia, constipation, and ileus (breast cancer); asthenia, anorexia, neuropathy, and constipation (colorectal carcinoma); asthenia and ileus (renal cell carcinoma); asthenia, constipation, neuropathy, and paresthesia (NSCLC). Activity in the phase 2 studies thus far has been modest, with one PR and 17 stable disease (less than seven cycles) in renal cell carcinoma; one PR, one minor response, and 14 SD (more than six cycles) in NSCLC; one stable disease (8 cycles) in colorectal carcinoma; and two stable disease (seven and eight cycles) in breast cancer.

5.2. HMN-214

HMN-214 (Fig. 4) is a stilbene derivative with oral bioavailability and prodrug of HMN-176 that inhibits prometaphase spindle formation, resulting in mitotic arrest at the G$_2$/M transition in the cell-cycle and caspase-dependent DNA fragmentation *(162)*. HMN-176 has been shown to interfere with localization of polo-like and cyclin-dependent kinases, in association with profound apoptosis. HMN-176 has demonstrated antitumor activity in a broad spectrum of human xenografts and has shown potent cytotoxic activity against various drug-resistant human tumor cell lines.

HMN-214 is being clinically developed by Nippon Shinyaku Co. Ltd., Japan. Two phase 1 trials are evaluating HMN-214 administered orally on a daily ×5 q4w or a daily

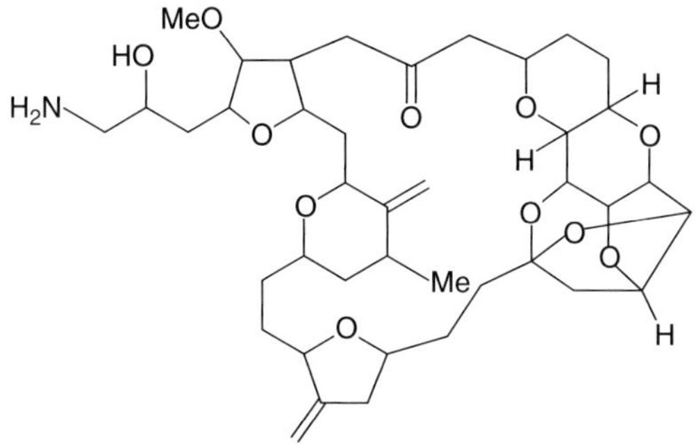

Fig. 5. Halichondrin B analog (E7389, ER-086526, B1939) *(164).*

×21 q4w schedule in patients with advanced solid tumors. Thirty-eight patients stratified according to previous chemotherapy received HMN-214 daily ×5 q4w at doses of 6–48 mg/m^2 *(144)*. In both patient populations, grade 4 neutropenia with fever and prolonged (>5 d) neutropenia were dose limiting. Additional AEs included autonomic neuropathy, myalgia, and electrolyte disturbances. There was one minor response in a patient receiving 24 mg/m^2. Pharmacokinetics studies demonstrate a dose proportional increase in C_{max} and AUC values. The RP2D for lightly pretreated patients was 24 mg/m^2, and accrual continues at 18 mg/m^2 for heavily pretreated patients.

HMN-214 was administered orally daily ×21 q4w at doses of 3–9.9 mg/m^2/d to 33 patients with advanced solid tumors *(163)*. At 9.9 mg/m^2/d, one patient developed grade 3 hyperglycemia and myalgias, and another patient developed grade 3 bone pain. Patients were stratified into lightly pretreated and heavily pretreated cohorts, and treated at a de-escalated dose of 8 mg/m^2/d with no DLTs. Preliminary pharmacokinetic studies of HMN-214 performed on day 1 indicated that the active compound HMN-176 reached C_{max} 2.2 to 6.7 h after oral dosing, and $t_{1/2}$ varied from 11.8–15.8 h. Antitumor activity was seen in a colon cancer patient with a transient decrease (40%) in CEA, and in a breast cancer patient with stable disease for 6 mo (both at the 9.9 mg/m^2/d dose level).

6. MISCELLANEOUS

6.1. Halichondrin B Analog (E7389, ER-086526, and B1939)

The rare marine sponge natural product halichondrin B is an antimitotic agent that exhibits potent anticancer effects in both in vitro and in vivo models of cancer *(164,165)*. It was first isolated from *Halichondria okadai* and later from the unrelated sponges *Axinella carteri* and *Phankella carteri (166,167)*. Limited availability of the natural product prompted efforts to synthesize halichondrin B. The existence of a synthetic route for halichondrin B *(168)* and knowledge that its anticancer activity resides in the macrocyclic lactone C1-C38 moiety *(167)* permitted development of structurally simplified synthetic analogs, such as E7389 (ER-086526, B1939), which retain activity of halichondrin B. E7389 encompasses the biologically active portion of halichondrin B and exhibits similar or identical anticancer properties in preclinical models (Fig. 5).

The mechanism of action of halichondrin B, a noncompetitive inhibitor of vinblastine-tubulin binding, is distinct from all other known classes of tubulin-based agents. Studies with E7389 indicate that its mechanism is similar or identical to that of halichondrin B *(164,165)*. E7389 is a tubulin-binding agent that inhibits microtubule polymerization. Preclinical data show that sub- to low-nanomolar levels of E7389 inhibit cancer cell proliferation through induction of cell-cycle block at G_2/M, disruption of mitotic spindles, and initiation of apoptosis. Human tumor xenograft studies in mice (ovary, breast, colon, and melanoma) demonstrate tumor regressions, remissions, and increased lifespans at dose levels below the MTD *(164)*. Moreover, E7389 showed a wide therapeutic window relative to other cytotoxic anticancer agents. For instance, in the human MDA-MB-435 breast cancer xenograft model, more than 95% tumor suppression occurred over the fourfold dosing range of 0.25–1 mg/kg, with no evidence of toxicity based on body weight losses or decreased water consumption. In contrast, the therapeutic window for paclitaxel in this model is 1.7-fold.

E7389 is being clinically developed by Eisai Research Institute, Andover, Massachusetts, in collaboration with the National Cancer Institute. The first clinical trial in humans is evaluating E7389, administered as a bolus IV infusion over 1–2 min on the weekly ×3 q4w schedule in patients with advanced solid tumors. Patients have been treated at doses up to 2 mg/m^2/wk *(169)*. DLTs consisted of one grade 3 febrile neutropenia and one grade 4 neutropenia. The MTD was 1.4 mg/m^2/ wk. Two PRs (NSCLC and bladder) and three minor responses (NSCLC, breast, and thyroid) were observed in 38 evaluative patients. In addition, 12 patients had stable disease lasting a median of 4 mo (range 2–14). Preliminary pharmacokinetic studies indicated that at the MTD, plasma concentrations of E7389 remained more than those required for cytotoxicity in vitro for more than 1 wk. The data demonstrated a tri-phasic elimination and a prolonged terminal $t_{1/2}$ of 36–48 h.

A second study is using a once q3w schedule, with doses ranging from 0.25 to 4 mg/m^2 *(170)*. All three patients at the 4 mg/m^2 dose level and 2/3 at the 2.8 mg/m^2 dose level developed febrile neutropenia, contributing to DLT at these dose levels, and enrollment continues at the 2 mg/m^2 dose level. No objective responses were observed. One patient with uterine sarcoma previously treated with gemcitabine, paclitaxel, and carboplatin had a 20% reduction in pelvic adenopathy, and 4 patients had stable disease for more than three cycles.

A phase 2 study in patients with refractory breast cancer administered E7389 at 1.4 mg/m^2 weekly ×3 q4w *(171)*. Because of neutropenia on day 15 resulting in a delay in dosing for a considerable number of patients, the schedule was changed to 1.4 mg/m^2 weekly ×2 q3w. Seventy-one and 33 patients were treated on the q4w and q3w schedules, respectively. Twelve PRs (10 confirmed) were observed in 65 evaluative patients on the q4w schedule, and 21 patients had stable disease. Overall survival ranged from 1+ to 253+ days, and median PFS was 7 wk. Forty-six percent of the patients experienced grade 3/4 neutropenia, and 4% experienced grade 3 febrile neutropenia.

6.2. Hemiasterlin Analog HTI-286

HTI-286 is a synthetic analog of the naturally occurring tripeptide hemiasterlin, which was originally derived from marine sponges *Cymbastela* spp, *Hemiasterella minor*, *Siphonochalina* spp, and *Auletta* spp. *(172)*. Hemiasterlins have been shown to noncompetitively inhibit the binding of vinblastine to tubulin, depolymerize microtubules, and arrest cells in the G_2/M phase of the cell cycle *(173)*. In addition, hemiasterlins competitively inhibit the binding of dolastatin 10 to tubulin, suggesting that these two compounds bind to the Vinca-peptide-binding site of tubulin *(174)*. Similar

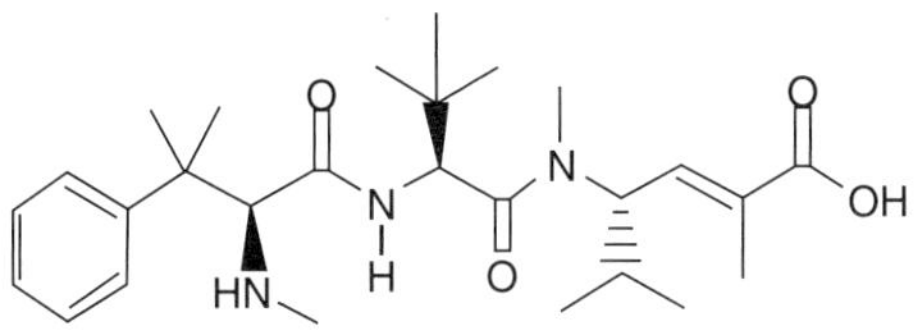

Fig. 6. Hemiasterlin analog HTI-286 *(172)*.

to hemiasterlins, HTI-286 is a potent inhibitor of cell growth that inhibits polymerization of purified tubulin, depolymerizes microtubules, inhibits microtubule assembly, and induces mitotic arrest and apoptosis *(172)* (Fig. 6). In contrast to paclitaxel and vincristine, HTI-286 is a poor substrate for the multidrug resistance protein P-glycoprotein and retains activity against in vitro and in vivo tumor models resistant to several chemotherapeutic agents, including taxanes and vinca alkaloids. In addition, resistance to HTI-286 was not detected in cells overexpressing the drug transporters MRP1 or MXR. Interestingly, mutations in α-tubulin have been found in cells selected for resistance to HTI-286, suggesting HTI-286 resistance may be mediated by mutation in α-tubulin and/or mechanisms not involving MDR1 or MXR *(175,176)*. The activity of HTI-286 against human tumors in xenograft models was independent of the route of administration (*i.e.* IV infusion in saline or oral administration).

HTI-286 is being clinically developed by Wyeth, Cambridge, Massachusetts. A phase 1 trial is evaluating HTI-286 administered as a 30-min IV infusion q3w in patients with metastatic or advanced solid tumors *(177)*. Thirty-five patients were administered doses ranging from 0.06–2 mg/m^2. DLTs included bilateral arm and hand pain at the 1.5 mg/m^2 dose level, and neutropenia, chest pain, and hypertension at the 2 mg/m^2 dose level. Neutropenia also occurred in one patient each at the 0.48 mg/m^2 and 2 mg/m^2 dose levels. The preliminary pharmacokinetic data were characterized by considerable interindividual variability in clearance (15.7 ± 9.3 L/h), Vd$_{ss}$ (177 ± 68 L), and $t_{1/2}$ (9.2 ± 3.8 h), which were independent of dose level and body surface area. The investigators concluded that the best predictors of neutropenia are time above a threshold HTI-286 serum concentration of 0.5 ng/mL and terminal elimination rate constant λ normalized by a dose level. The suggested dose for further testing is 1.5 mg/m^2.

6.3. Cryptophycin 52

The cryptophycins are potent antitumor and antifungal depsipeptides originally isolated from the cyanobacteria (blue-green algae) *Nostoc* species *(179)*. All cyclic cryptophycins consist of a δ-hydroxy acid unit, an α-amino acid unit, a β-amino acid unit, and an α-hydroxy acid unit, connected together in a cyclic sequence *(180)*. In order to improve in vivo hydrolytic stability, the analog cryptophycin 52 (LY355703) was synthesized (Fig. 7). Cryptophycin 52 was chosen from more than 450 synthetic analogs to enter the clinic because of a unique balance of its efficacy and stability. At high concentrations (≥10 times IC$_{50}$), cryptophycin 52 depolymerizes spindle microtubules and blocks cell proliferation at mitosis *(181)*. However, low concentrations of cryptophycin 52 inhibit mitosis owing to suppression of microtubule dynamics, without significantly reducing microtubule polymer mass or mean microtubule length. Five to six molecules of cryptophycin 52 bound per microtubule molecule at its ends are sufficient to decrease

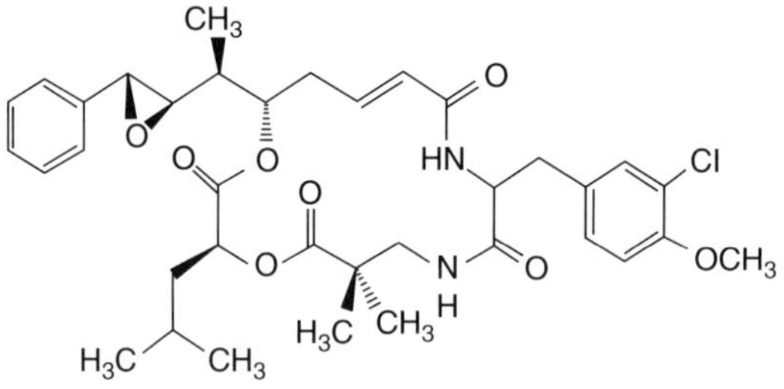

Fig. 7. Cryptophycin 52 (LY355703) *(178)*.

dynamicity by 50%. Cryptophycin 52 noncovalently binds to β-tubulin at a single high-affinity site, inducing a conformational change in tubulin that may be important in the ability of the drug to increase stability at microtubule ends *(182,183)*. Vinblastine inhibits binding of cryptophycin 52 to tubulin, whereas cryptophycin 52 does not inhibit colchicine binding to tubulin, suggesting that cryptophycin 52 interacts with the vinca-binding domain of tubulin. Cryptophycin 52 has demonstrated potent antiproliferative and cytotoxic activity with IC_{50} values for antiproliferative activity in the low picomolar range in both solid and hematological tumor cell lines *(184)*. Cryptophycin 52 is more potent than paclitaxel, vinblastine, and vincristine both in vitro and in human tumor xenograft models, including multidrug-resistant cancer cells overexpressing P-glycoprotein and/or multidrug-resistance protein transport factors.

Cryptophycin 52 is being clinically developed by Eli Lilly and Co., Indiana. Because cryptophycin 52 is formulated with Cremophor EL, hypersensitivity reactions have occurred with IV administration of this drug on various schedules *(178,185)*. Therefore, all patients are given steroid-based prophylaxis *(178,185)* 30–60 min before crypto-phycin 52 infusion. Two phase 1 trials have evaluated cryptophycin 52 administered as a 2-h IV infusion on various schedules in patients with solid tumors. In one study, 25 patients with solid tumors received cryptophycin 52 doses ranging from 0.1–2.22 mg/m² on days 1 and 8 q3w *(178)*. A single case of dose-limiting myalgia at 0.2 mg/m² prompted expansion of the 0.68 mg/m² dose level and further dose escalation up to 2.2 mg/m². At the 2.2 mg/m² dose level, dose-limiting motor neuropathy occurred in 1/1 patient and resulted in exploration of an intermediate dose of 1.84 mg/m². Dose-limiting constipation/ileus occurred in 2/3 patients at that dose level, with severe myalgia in one patient. In an attempt to improve dose intensity and avoid dose-limiting neurotoxicity, the schedule was changed to twice weekly dosing on days 1, 4, 8, and 11 q3w. On this schedule, a starting dose of 0.75 mg/m² was administered to three patients, and 1 mg/m² dose was administered to eight patients. At the 1 mg/m² dose level, dose-limiting constipation/ileus occurred in 2/8 patients despite prophylaxis with an aggressive laxative regimen. Doses of more than 0.75 mg/m² on this schedule were not tolerated as a result of constipation/nausea, suggesting that cryptophycin 52 toxicity is not schedule-dependent and is related to cumulative dose. One PR in a patient with NSCLC treated at 0.1 mg/m² on day 1 and day 8 q3w was reported. Pharmacokinetic studies indicated that cryptophycin 52 was eliminated rapidly with a short terminal $t_{1/2}$ that ranged from 0.8–3.9 h.

In another phase 1 trial, 35 patients with solid tumors were administered crypto-phycin 52 at doses ranging from 0.1 to 1.92 mg/m² on day 1 of a 21-d cycle *(185)*.

Bradycardia was dose limiting in 1/3 patients at 0.68 mg/m^2. A patient experienced hypertension associated with a dose-limiting hypersensitivity reaction, and two additional patients had hypersensitivity reactions at cycle 2, prompting the introduction of prophylactic premedication in all subsequent patients. The dose was escalated to 1.48 mg/m^2, at which dose 1/8 patients experienced dose-limiting cardiac dysrhythmia. At 1.92 mg/m^2, 1/5 patients experienced dose-limiting sensory neuropathy, and 1/5 patients experienced dose-limiting myalgia. Both patients treated at the subsequent dose level of 1.7 mg/m^2 experienced reversible dose-limiting myalgia, prompting expansion of accrual at the 1.48 mg/m^2 dose level, which was recommended for phase 2 studies. The MTD was identified in the dose range of 1.7–1.92 mg/m^2. Owing to the acute dose-related toxicity reported on this regimen, the dose and schedule were changed to 1 mg/m^2 administered weekly ×3 q4w. On this schedule, 1/8 treated patients experienced dose-limiting hypertension. In addition, myalgia occurred in 1/8 patients. Cumulative long-lasting neuroconstipation and neurosensory toxicity precluded the completion cycle in 9 out of 15 cycles. The MTD was not reached, and the clinical development of the weekly regimen was discontinued because a dose below 1 mg/m^2 would provide a dose intensity lower than the RP2D on the q3w schedule. No objective responses were observed in either study.

One of two patients treated at 1.7 mg/m^2 on the q3w schedule experienced neuro-motor and neurosensory AEs with a vulvar vestibulitis syndrome, which disappeared completely after cryptophycin 52 was discontinued (186). This was the first report of vulvar vestibulitis syndrome possibly induced by chemotherapy. The investigators concluded that the neurotoxicity of cryptophycin 52 and the systemic peripheral neuropathy after administration suggest neurological pathogenesis of vulvar vestibulitis and should be considered whenever potential neurotoxic therapy is administered and vulvar pain occurs.

Two phase 2 trials evaluated cryptophycin 52 in patients with stage IIIB or IV NSCLC. In a German study, cryptophycin 52 was administered as first-line therapy at a dose of 1.5 mg/m^2 on days 1 and 8 q3w (187). One patient was hospitalized as a result of grade 4 diarrhea and died from severe lactate acidosis possibly related to cryptophycin 52. Another patient died from a myocardial infarction. No hematological AEs were observed on this trial. Two patients had acute grade 2 hypersensitivity reactions despite prophylaxis. No objective tumor responses were reported, and the trial was suspended. In a United States study, cryptophycin 52 was administered to 26 patients who had received 1–2 chemotherapy regimens previously, including one containing a platinum agent (188). Twenty-five patients were evaluative for AEs and response. The first 12 patients received the drug at a dose of 1.5 mg/m^2 on days 1 and 8 q3w, and eight of those patients experienced grade 3 or greater AEs. Subsequently, the dose was reduced by 20%, and 13 additional patients were treated at the 1.125 mg/m^2 dose level. Four of these 13 patients experienced grade 3 or greater AEs. Grade 4 AEs included one case each of constipation, which appeared to be neurogenic, and dehydration. Grade 3 hematological AEs included decreased hemoglobin and platelets levels, whereas grade 3 nonhematological AEs included pain, neuropathy (including paresthesias and peripheral neuropathy), asthenia/malaise, and nausea/vomiting. No objective responses were observed in this trial. Ten of 25 evaluative patients had stable disease, and median survival was 4.1 mo. The investigators concluded that cryptophycin 52 has only limited activity in NSCLC and suggested that neurotoxicity associated with the drug is related to its peak dose, because toxicity was substantially reduced with a relatively modest dose reduction.

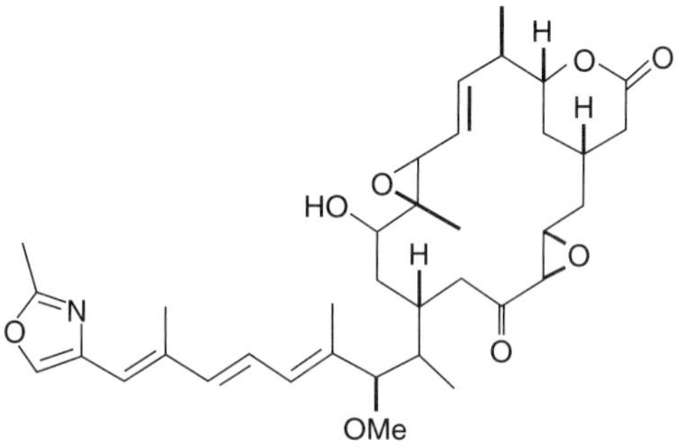

Fig. 8. Rhizoxin *(54).*

A phase 2 study in patients with platinum-resistant ovarian cancer administered cryptophycin 52 at 1.5 mg/m^2 on days 1 and 8 q3w to 25 patients *(189)*. PRs were seen in 3/24 evaluative patients, and 7 patients had stable disease lasting more than 4 mo. Grade 3/4 AEs included anemia, thrombocytopenia, increased creatinine, and hyperbilirubinemia. Median TTP was 5.1 mo.

6.4. Rhizoxin

Rhizoxin is a 16-membered macrolide antitumor and antifungal agent originally isolated from the fungus causing rice seedling blight, *Rhizopus chinensis (190)* (Fig. 8). It has been shown to act through inhibition of microtubule assembly and polymerization of tubulin, which causes cell-cycle arrest in the G$_2$/M phase *(191)*. Rhizoxin interacts with one binding site per molecule of β-tubulin *(192,193)*. It bound to the maytansine binding site on tubulin, separate from the binding site of vinblastine, and demonstrated potent cytotoxic activity in vitro over an exceptionally broad range of concentrations from 10^{-4}–10^{-13} *M (194)*. Rhizoxin demonstrated in vivo activity in human tumor xenograft models including those resistant to vinca alkaloid agents. In animal studies, the antitumor activity of rhizoxin was schedule-dependent as indicated by improved activity after prolonged or repeated drug administration. In addition, rhizoxin has been shown to cause dose-dependent inhibition of angiogenesis and suppress neovascularization *(195)*, possibly through combined inhibition of some functions of endothelial cells responsible for induction of in vivo angiogenesis *(196)*.

Rhizoxin was clinically developed by Fujisawa in collaboration with the National Cancer Institute in the United States. In early clinical studies in the United Kingdom, rhizoxin was administered as a 5-min IV bolus infusion on a q3w schedule to 24 patients with refractory solid tumors *(197,198)*. DLTs at 2.6 mg/m^2 included reversible leukopenia, mucositis, and diarrhea. The RP2D was 2 mg/m^2 on this schedule. In preliminary pharmacokinetic studies, rhizoxin was not detectable in plasma at doses of 0.8 and 1.6 mg/m^2 and was not measurable 10 min after injection at 2 mg/m^2, indicating a very short elimination $t_{1/2}$. Based on preclinical studies that indicated improved antitumor activity of rhizoxin with prolonged administration schedules, and in order to overcome short elimination $t_{1/2}$, subsequent phase 1 studies in the United States evaluated CIV infusions of rhizoxin. In 48 patients with solid tumors, rhizoxin was administered at the starting dose of 1 mg/m^2 as a 3- to 72-h IV infusion *(199)*. Neutropenia was the main DLT. The MTD was 0.8 mg/m^2 administered over 72 h. No objective responses were observed in this study. Based on these data, the investigators developed a mathematical model explaining the schedule-dependent interpatient pharmacodynamic variability in the nadir neutrophil count. In another study, rhizoxin was administered at doses ranging from 0.2 mg/m^2 as a 12-h IV infusion to 2.4 mg/m^2 as a 72-h IV infusion q3w to 19 previously treated patients with solid malignancies *(200)*. At the 2.4 mg/m^2/72 h dose level, a single patient developed dose-limiting neutropenia with fever, mucositis, and diarrhea, and subsequently died. At the 1.2 mg/m^2/72 h dose level, only 1/7 patients experienced dose-limiting neutropenia and fever. A dose of 1.2 mg/m^2 administered over 72 h was determined to be the MTD and RP2D. No objective responses were observed in this study.

The phase 2 trials of rhizoxin were performed by the European Organization for Research and Treatment of Cancer (EORTC) using a 5-min IV bolus infusion on a q3w schedule in patients with melanoma, head and neck, breast, and NSCLC *(201)*. No objective responses were observed in the melanoma and breast studies.

In 32 patients with recurrent and/or metastatic squamous cell head and neck cancer, rhizoxin was administered at the starting dose of 2 mg/m^2 *(202)*. Because of severe tumor pain following rhizoxin administration, the dose was lowered to 1.5 mg/m^2, and the side effect was no longer observed. In total, 9/32 patients treated experienced pain at the tumor site. Hematological AEs included neutropenia and thrombocytopenia, and nonhematological AEs included stomatitis. Two PRs (a large ulcerated cervical recurrence of a tongue carcinoma and a large squamous cell carcinoma of the tongue) lasting 7.5 and 3.5 mo were observed in 25 evaluative patients.

In 31 chemotherapy-naïve patients with advanced NSCLC, rhizoxin was administered at a dose of 2 mg/m^2 *(203)*. Nine of 29 eligible patients were treated surgically, and 3/29 patients received radiotherapy. Drug-related hematological AEs included grade 3/4 leukopenia in 9/118 cycles (8% of cycles) and neutropenia in 18/118 cycles (15% of cycles). In addition, grade 3 anemia was recorded in 1/118 course (<1%). Drug-related nonhematological AEs included grade 3 stomatitis and asthenia/malaise/fatigue in two courses. Four PRs and 13 stable diseases were observed in 27 evaluative patients. The median duration of response was 7 mo (range 6–10.7 mo), and median survival from the start of rhizoxin treatment was 6 mo (range 2–14.7 mo). Rhizoxin was rapidly eliminated from plasma with a median systemic clearance of 8.4 L/min/m^2 and an elimination $t_{1/2}$ of 10.4 min *(204)*. Rhizoxin AUC was higher in patients with a PR or stable disease than in those with progressive disease (median 314 vs. 222 min×ng/mL, $p = 0.03$).

In British phase 2 trials, rhizoxin at a dose of 2 mg/m^2 administered as a 5-min IV bolus infusion on a q3w schedule was evaluated in patients with advanced ovarian, renal, and colorectal cancer *(205)*. In the 17 eligible patients with progressive epithelial ovarian cancer, one patient who had received platinum-based chemotherapy previously had a PR lasting 3 mo in a pelvic mass and para-aortic lymphadenopathy after two cycles of treatment. There were no responses in patients with advanced colorectal cancer or advanced renal cancer.

6.5. T138067 and T900607

T138067 is a novel synthetic antimitotic agent that has been shown to covalently bind to cysteine-239 on β_1, β_2, and β_4 isotypes of tubulin, thereby disrupting microtubule polymerization. Although the apparent association rate constant for T138067 binding to

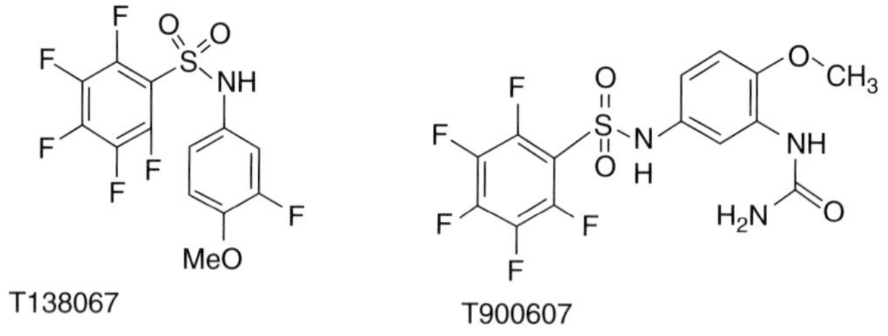

Fig. 9. T138067 *(206)* and T900607 *(207)*.

tubulin is approx 10 times more than that of colchicine, T138067 is more effective in preventing microtubule formation, with an IC_{50} value of approx 2 µM. Shan and colleagues *(208)* suggested that T138067-modified tubulin recruits unmodified heterodimeric tubulin into large, amorphous aggregates, and thus quickly depletes the pool of tubulin available for microtubule formation. Inhibition of tubulin polymerization by T138067 leads to mitotic arrest at the G_2/M transition in the cell cycle followed by apoptosis. T138067 and its analogs, such as T900607, are effective against various tumor cells in culture and mice xenograft models, including those that express the multidrug-resistant phenotype (Fig. 9). T138067 was considerably more toxic than vinblastine, paclitaxel, doxorubicin, and actinomycin D toward multidrug resistant cell lines. In a human T cell leukemia xenograft model, T138067 showed the same degree of efficacy against the drug-sensitive parental cells and vinblastine/paclitaxel-resistant subline. In contrast, paclitaxel and vinblastine showed approx 50% reduced efficacy against multidrug-resistant tumors. The covalent interaction of T138067 with β-tubulin may explain its evasion of cellular multidrug-resistance mechanisms. In vivo and in vitro, T138067 is metabolized through a glutathione S-transferase-mediated metabolic pathway *(209)*. T138067 has the ability to penetrate the blood–brain barrier, which could be partially owing to its lipophilicity. Less lipophilic analogs of T138067 that do not cross the blood–brain barrier are thought to possess fewer potential problems with central neurotoxicity and an increased therapeutic window relative to other members of this class of compounds *(206)*. T900607 is a structural analog of T138067 that differs from T138067 in that it does not cross the blood–brain barrier, has a different tissue distribution profile, and is more water-soluble. Like T138067, T900607 prevents tubulin polymerization by covalently binding to β-tubulin, inhibits the growth of a number of tumor cell lines and tumor growth in human xenograft models, and its activity is not affected by the multidrug-resistant phenotype of the cells *(210–212)*. Additive effects of tumor growth inhibition have been reported with the administration of T900607 and gemcitabine to athymic mice bearing MX-1 mammary tumor xenografts *(211)*.

   T138067 and T900607 are being clinically developed by Tularik Inc., California. Three phase 1 trials have evaluated T138067 administered as a 3-h IV infusion on various schedules in patients with solid tumors. On an q3w schedule, T138067 was administered at dose levels of 11–585 mg/m^2 to 28 patients *(213)*. Grade 4 neurotoxicity (hearing loss, peripheral neuropathy, acute ataxia, dysphoria, lethargy, and tremulousness) was observed in one patient at the 585 mg/m^2 dose level. The authors noted that his exposure was twice the level predicted from linear pharmacokinetics, probably because of reduced hepatic blood flow secondary to reduced cardiac output. No other grade 3/4

AEs were observed, and there was no evidence of cumulative effects. The RP2D was 440 mg/m^2 on the q3w schedule. Pharmacokinetic parameters from 24 patients dosed up to 440 mg/m^2 showed mean clearance of 1.5 ± 0.44 L/h/kg, mean Vd$_{ss}$ of 0.6 ± 0.26 L/kg, and apparent elimination $t_{1/2}$ of 0.5 ± 0.3 h. On a weekly schedule, T138067 was administered at dose levels of 110–440 mg/m^2 to 29 patients *(214, 215)*. In patients with malignancies other than HCC, DLTs included neuropathy and neutropenia at the 440 mg/m^2 dose level, whereas in HCC patients DLTs included thrombocytopenia at the 330 mg/m^2 dose level (grade 1) and neutropenia at the 300 mg/m^2 dose level (grade 3). One PR lasting 15 mo was observed in a patient with HCC at the 110 mg/m^2 dose level. The RP2D of T138067 on the weekly schedule was 330 mg/m^2 for non-HCC patients and 220–300 mg/m^2 for patients with HCC. Pharmacokinetic studies indicate that total body clearance of 51.1 ± 21.7 L/h/m^2 approximates hepatic blood flow, Vd$_{ss}$ of 17 ± 10.6 L/m^2 is slightly smaller than total body water, and the apparent elimination $t_{1/2}$ is short (0.5 ∀ 0.2 h). On a daily ×5 q3w schedule, T138067 was administered at dose levels of 44–250 mg/m^2 to 20 patients *(216)*. DLTs included neutropenia and reversible encephalopathy/hearing. No objective responses and one stable disease were observed in 15 evaluative patients. The MTD and RP2D of T138067 administered daily ×5 q3w was 175 mg/m^2. The pharmacokinetics of T138067 was linear with dose-proportional increases in C_{max} and AUC, and did not change over the 5-d treatment period.

Phase 2 trials of T138067 evaluated the weekly schedule in patients with HCC, NSCLC, and malignant glioma. In 53 patients with unresectable HCC, T138067 was administered as a 3-h IV infusion at a dose of 165 mg/m^2 *(217)*. Thirty-three patients received T138067 as first-line therapy, and 20 patients as second-line therapy. The most common drug-related AEs were fatigue, nausea, vomiting, and diarrhea. No grade 4 AEs were observed. In 21 evaluative first-line patients, two PRs and nine stable diseases were reported. Three of 20 first-line patients had more than 50% reduction in AFP, and one of these three patients had a PR. Of 11 second-line patients, one patient had a PR. The authors concluded that T138067 is well tolerated and active against first- and second-line HCC. In 20 patients with locally advanced or metastatic NSCLC, T138067 was administered at a dose of 330 mg/m^2 *(218)*. All patients had previously received taxane therapy, including Taxol®/carboplatin (six patients), other Taxol® combinations (three patients), and taxotere/irinotecan (one patient). The most common drug-related AEs included neutropenia, nausea, vomiting, and vein irritation. No grade 4 AEs related to T138067 were observed, and no objective responses were observed in 14 evaluative patients. The same dose and schedule were administered to 18 glioma patients (16 patients with glioblastoma multiforme and two patients with anaplastic Astrocytoma) *(219)*. No objective responses were observed. Three patients had stable disease with a median duration of 2.6 mo. AEs were generally mild, with no grade 4 AEs reported and only four patients reporting grade 3 AEs (fatigue, diarrhea, and neurotoxicies consisting of confusion and expressive dysphasia).

A phase 2/3 randomized trial compared overall survival with T138067 vs doxorubicin, in patients with unresectable, chemotherapy-naïve HCC *(220)*. This study was closed early after treating 339 patients (169 on the T138067 arm and 170 patients on the doxorubicin arm) owing to lack of survival benefit in patients treated with T138067.

Three phase 1 trials are evaluating T900607 in patients with refractory solid tumors. In one study, T900607 was administered at doses of 7.5–80 mg/m^2 as a 30- or 60-min

IV infusion on a q3w schedule in 25 patients *(207,221)*. Infusion time and volume were increased from 30 min and 100–250 mL to 60 min and 500 mL, respectively, owing to incidence of vein pain on infusion with T900607 in the first 18 patients. Grade 3 drug-related AEs were anemia, fatigue, troponin elevations, and nausea. The most common drug-related AEs were fatigue, nausea, and diarrhea. One patient had a transient asymptomatic troponin elevation, and one patient had an asymptomatic grade 1 decrease in cardiac ejection fraction and troponin elevation after five cycles of 60 mg/m^2; the ejection fraction recovered to baseline post-treatment. One confirmed PR (95% reduction in tumor and 99% reduction in AFP) lasting for 8+ mo was reported in a patient with HCC treated at the 60 mg/m^2 dose level. The MTD has not yet been reported. T900607 pharmacokinetics was linear over the dose range studied, with dose-proportional increases in C_{max} and AUC. In another study, T900607 was administered at doses of 20–130 mg/m^2 on a weekly schedule to 24 patients *(222)*. Two patients treated at the 130 mg/m^2 dose level experienced DLTs, including abdominal pain and thrombocytopenia. The most common drug-related AEs were nausea, injection site pain/irritation, vomiting, fatigue, and chills. No grade 4 AEs related to T900607 were observed. One patient developed an asymptomatic decline in ejection fraction after six cycles of 60 mg/m^2, and one patient experienced an inferior myocardial infarction after five cycles of 100 mg/m^2. The MTD of T900607 was 100 mg/m^2 on the weekly schedule. The pharmacokinetics of T900607 was linear with proportional increases in C_{max} and AUC. A third study administered T900607 at doses of 15–270 mg/m^2 by 30-min IV infusions q3w *(223)*. At the 270 mg/m^2 dose level, all three patients had grade 3 thrombocytopenia (one of which qualified as a DLT), and the dose level was expanded to two additional patients, one of whom had a fatal DLT (grade 4 troponin elevation, grade 3 leukopenia, and grade 5 myocardial infarction). Patients subsequently enrolled at the 180 mg/m^2 dose level were carefully monitored for cardiac toxicity. One patient had grade 2 atrial fibrillation and grade 3 congestive heart failure; one patient had grade 1 bradycardia; and one patient had palpitations, atrial fibrillation, and died suddenly with a presumed arrhythmia. The fourth patient at this dose level had a 15% decrease in LVEF. The RP2D was subsequently determined at 130 mg/m^2 with only one DLT related to drug (grade 2 anorexia). No objective responses were observed in 20 evaluative patients. Seven patients had stable disease (median duration of 3.9 mo).

In a phase 2 study in patients with chemotherapy-naïve unresectable HCC, T900607 was administered as a weekly 1-h infusion at a dose of 100 mg/m^2 *(224)*. One in 23 evaluative patients had a PR and 13 patients had stable disease. A more than 50% reduction in AFP was observed in 2/15 patients (one of them with PR). No grade 4 AEs were observed, and the most frequent AEs were rigors (38%), pyrexia (32%), nausea (32%), vomiting (21%), anemia (21%), fatigue (15%), and anorexia (12%).

6.6. Cantuzumab Mertansine (SB-408075, huC242-DM1) and BB-10901 (huN901-DM1)

Cantuzumab mertansine (SB-408075, huC242-DM1) (Fig. 10) is a tumor-activated prodrug created by conjugation of approx four molecules of the potent maytansinoid antimitotic agent DM1 to the humanized monoclonal antibody huC242. huC242 specifically binds to the extracellular domain of a tumor-associated carbohydrate epitope of CanAg (a novel glycoform of MUC1) expressed in most colorectal, pancreatic, biliary, and a large proportion of NSCLC (40%), gastric (55%), uterine (45%),

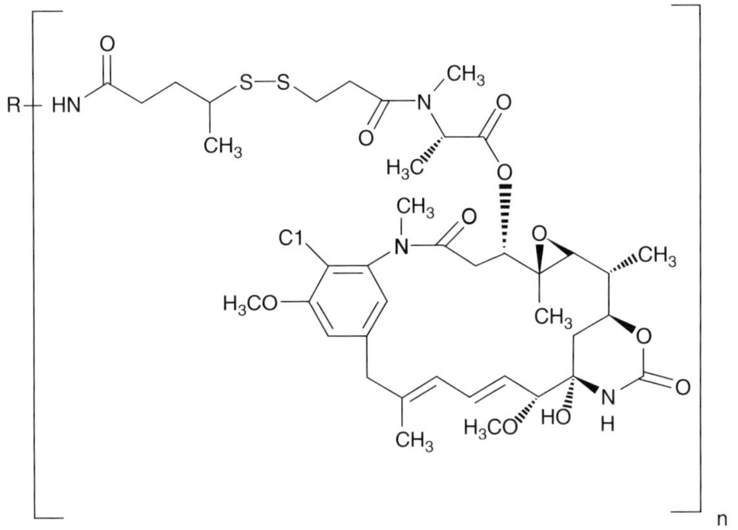

Fig. 10. Schematic of the immunoconjugate structure of cantuzumab mertansine (SB-408075, huC242-DM1) *(225)* and BB-10901 (huN901-DM1).

and bladder (40%) cancers *(225)*. In contrast, CanAg is only minimally expressed on normal tissues. Following CanAg binding, the huC242-DM1 complex is internalized, and the DM1 molecules are released intracellularly by cleavage of the DM1-huC242 disulfide linkage. DM1 is a synthetic derivative of the natural microbial fermentation product and potent but toxic antimicrotubule agent ansamitocin P-3 (maytansinoid). Maytansine was evaluated in phase 1 and 2 clinical trials by the National Cancer Institute in the 1970s *(226–233)*. The development of maytansine was stopped because of severe AEs, including nausea, vomiting, diarrhea, liver function abnormalities, weakness, and lethargy observed in these trials. The conjugation of DM1 to tumor-specific monoclonal antibodies permits the targeted delivery of drug molecules to specific antigen-bearing cells, resulting in enhancement of antitumor activity and reduction of normal tissue toxicity. Cantuzumab mertansine has demonstrated potent cytotoxic activity against colon cancer cell lines in an antigen-specific manner and antitumor efficacy in human colon, pancreatic, and lung tumor xenografts at nontoxic doses *(234,235)*. Maytansinoids, such as maytansine and ansamitocin P-3 derivatives, have been shown to bind to β-tubulin in a reversible manner, competitively inhibit the binding of vinblastine and vincristine, but only marginally affect colchicine binding to tubulin, suggesting interaction with the vinblastine binding site of tubulin *(54,236)*. On binding to tubulin, maytansine causes disassembly of the microtubule and prevents tubulin spiralization. In addition, ansamitocin P-3 competitively inhibits rhizoxin binding to tubulin *(192,193)*, suggesting that rhizoxin and ansamitocin P-3 share the same binding site (the rhizoxin/maytansine binding site), which is distinct from the vinblastine-binding site of tubulin.≠

Cantuzumab mertansine is being clinically developed by ImmunoGen Inc., Massachusetts, MA *(237)*. A phase 1 trial evaluated cantuzumab mertansine administered at doses ranging from 22 to 295 mg/m² as a 30-min IV infusion on a q3w schedule in 37 patients with CanAg-expressing solid malignancies *(225)*. At the 235 mg/m² dose level, 2/16 patients experienced dose-limiting elevations in hepatic transaminases. At the 295 mg/m^2 dose level, 2/3 patients experienced DLTs including grade 3 fatigue and AST elevation in course 1. Other nonhematological grade 3 AEs included asymptomatic elevation of serum lipase in 1/5 patients at 176 mg/m^2, vomiting in 1/16 patients, and peripheral neuropathy in 2/16 patients at 235 mg/m^2 after one and four courses, respectively. Hematological AEs included thrombocytopenia and uncomplicated brief neutropenia. Three of 37 patients experienced hypersensitivity reactions. The symptoms resolved with temporary interruption of cantuzumab mertansine infusion and IV dexamethasone and antihistamine administration. No objective reponses were reported. Seven patients experienced more than 30% decrease in carcinoembryonic antigen levels. The RP2D is 235 mg/m^2 as an IV infusion on the q3w schedule. The results of pharmacokinetic studies indicate that dose, C_{max} and AUC correlated with the severity of transaminase elevation. The mean clearance and terminal elimination $t_{1/2}$ values for cantuzumab mertansine averaged 39.5 ± 13.1 mL/h/m^2 and 41.1 ± 16.1 h, respectively. The mean Vd$_{ss}$ was 1497 ± 584 mL/m^2. In pharmacodynamic studies, strong expression of CanAg was documented in 68% of patients. Another phase 1 study administered cantuzumab mertansine at weekly doses ranging from 40 to 138 mg/m^2 without interruption (238). Thirty-nine patients were treated, and no grade 4 AEs were observed. At the 138 mg/m^2 dose level, 3/6 patients experienced DLTs in the form of grade 3 transaminase elevations and one grade 2 serum bilirubin elevation. Consequently, 115 mg/m^2 was determined as the MTD, and only 2/23 patients treated at this dose level experienced grade 3 AEs (one each elevated lipase and alkaline phosphatase, the latter considered a DLT). One chemotherapy-naïve patient at the 96 mg/m^2 dose level experienced a marked improvement as seen by disappearance of ascites. This patient remained on study for a total of 62 weekly doses. Another patient treated at this dose had a marked decline in CA19-9 associated with stable disease for 5 mo. No objective responses were observed.

DM1 has also been evaluated as an immunoconjugate with humanized murine monoclonal antibody huN901 (239). The huN901 component of a tumor-activated prodrug BB-10901 (huN901-DM1) (Fig. 10), which includes huN901 linked to approx four molecules of DM1, binds to the CD56 antigen expressed on human small-cell-lung cancers (SCLC), neuroblastomas, neuroendocrine tumors, and multiple myelomas. Similar to cantuzumab mertansine, BB-10901 is converted to the active drug inside the target cell, following specific binding to the surface of the tumor cell and internalization. BB-10901 has demonstrated activity in human SCLC xenograft models resulting in cures of mice, in contrast to the clinically used SCLC drugs cisplatin, VP-16, and topotecan, which had only modest effects resulting in tumor growth delays.

BB-10901 (huN901-DM1) is being clinically developed by ImmunoGen Inc., Massachusetts (240). A phase 1 trial is evaluating BB-10901 administered as an IV infusion on a weekly ×4 q6w schedule in patients with CD56-expressing tumors (241). Twenty-four patients, including 15 patients with SCLC and 9 patients with neuroendocrine tumors, have been treated at doses ranging from 5 to 75 mg/m^2. Dose-limiting pancreatitis possibly related to treatment occurred in one patient treated at the 60 mg/m^2 dose level. No moderate or severe (>grade 2) hematological AEs have been observed. The average elimination $t_{1/2}$ was 23 h, and clearance was 49 mL/h/m^2. BB-10901 clearance was non-dose-proportional with greater clearance observed at lower dose levels, perhaps secondary to natural killer cell binding. Two minor responses have

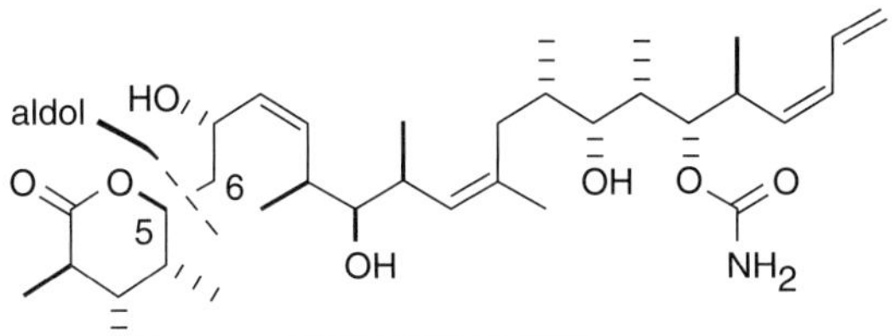

Fig. 11. Discodermolide (XAA296) *(243)*.

been observed (1 neuroendocrine, 1 SCLC patient). The MTD was defined as 60 mg/m^2 weekly ×4 q6w *(242)*. A phase 2 extension of this study is being conducted in patients with relapsed non-pulmonary SCLC. One of the first 10 patients treated had a serious AE of aseptic meningitis, which resolved within a week. No other drug-related severe AEs have occurred. PK analysis revealed an increase in both the terminal $t_{1/2}$ and AUC in week 2 compared with week 1 in two patients. This prolongation of the terminal $t_{1/2}$ after initial dosing is consistent with saturation of CD56+ sites such as natural killer cells. Three patients have shown evidence of clinical activity (one patient each with pre-liminary evidence of a PR, a nonsustained PR, and a minor response).

6.7. Discodermolide

A potent immunosuppressive lactone, discodermolide (XAA296), was originally iso-lated from the marine sponge *Discodermia dissolute* and characterized by researchers at Harbor Branch Oceanographic Institution in 1990 *(244)*. Discodermolide induces tubulin polymerization more potently than does paclitaxel, with a 1:1 stoichiometry with tubulin dimers *(245)*. Discodermolide induces G$_2$/M cell-cycle arrest with an IC$_{50}$ of 3–80 nM, and paclitaxel-resistant cell lines remain sensitive to discodermolide by virtue of its decreased affinity for P-glycoprotein *(246)* (Fig.11). Despite the fact that paclitaxel and discodermolide bind to the same or overlapping sites on β tubulin, these agents exhibit in vitro synergy in multiple cell lines *(247)* as well as enhanced efficacy when combined in xenograft models *(248)*.

Discodermolide is being clinically developed by Novartis Pharma AG, Switzerland. In a phase 1 study, patients with solid tumors were administered IV doses of 0.6–19.2 mg/m^2 at a fixed infusion rate of 0.77 mg/mL/min *(249)*. No drug-related DLTs have occurred. The only grade 4 AE was anemia experienced by 1/26 patients. Grade 3 AEs included one patient each with anemia, vomiting, and fatigue. No neuropathy or neu-tropenia have been observed. One patient with appendix carcinoma had stable disease for four cycles. Pharmacokinetic analyses have shown that C_{max} was achieved at the end of infusion, thereafter declining rapidly in a multiphasic manner. By 24 h postinfusion, blood concentration had decreased to less than 10% of C_{max} but then showed nonlinear PK, suggesting prolonged recycling of drug between tissue and the systemic circulation. C_{max} and AUC ranged from 29 to 482 ng/mL and 103–15211 ng × h/mL, respectively.

7. DISCUSSION

In this chapter, the authors have summarized the clinical data available for anticancer agents that act on microtubules (with the exception of taxane derivatives, discussed else-where in this book). By the number of compounds discussed, it is apparent that there is continuing enthusiasm for the development of agents in this class. Undoubtedly, the source of this enthusiasm is the medical and financial success of the vinca alkaloids and taxanes. The decision to take these agents into the clinic was based on this success and the intent to demonstrate clinical superiority or novelty. All of the agents discussed have at least one of the following properties: novelty of the tubulin binding site, preclinical efficacy in taxane-resistant cancer models, or distinctive pharmacokinetic properties.

It is also important to note that one of the main reasons for choosing these agents for clinical development was their relative lack of cytotoxicity to normal tissue in nonclinical studies. This group of agents had impressive preclinical antitumor activity with minimal degree of toxicity to normal tissues both in vitro and in vivo. It is not clear how this therapeutic index relates to the proximal mechanism of action at the microtubule.

These agents are highly target-specific, as are many of the so-called "molecularly targeted agents" in clinical development. They have been selected for clinical development partly because the only known molecule of relevance with which they interact is tubulin. Yet unlike the kinases, the cancer specificity of these agents is not a function of an alteration in microtubule structure or dynamics in cancer cells. There are no known cancer specific microtubule abnormalities that are either primarily cancer inducing or specific targets of antineoplastic therapy. Whereas it is possible that the specific site of interaction with the microtubule may correlate with the spectrum of activity against human cancers, only the epothilones' clinical development is sufficiently mature to suggest a similar spectrum of activity as the taxanes, which bind an identical tubulin site and similarly alter microtubule dynamics. The clinical data for all the other compounds discussed in this chapter are insufficient at this time to answer this question.

What can we learn from the aggregate clinical data on these agents? There appears to be little correlation between the anticancer activity seen in nonclinical models and that seen in human cancers. Virtually all these compounds exhibited broad-spectrum anticancer activity, yet few have done so in the clinic to date. The one exception as a class is the epothilones, where significant activity was seen in several clinical trials across a diverse spectrum of neoplasms for nearly all agents in the class. The dolastatin data present a mixed picture. Although these agents have shown virtually no anticancer activity in most phase 2 trials, the activity reported against renal cell carcinomas and melanomas merits follow-up. Rhizoxin, though studied less extensively, has also demonstrated mixed efficacy, with responses in head and neck cancer and NSCLC, but no activity against melanoma and breast cancers. The colchicine-like compounds have failed to demonstrate clinically meaningful activity, despite a large body of nonclinical data suggesting otherwise. The data available for the other compounds described in this chapter are insufficient to allow conclusions to be drawn about activity across multiple tumor types.

How will these new agents find a place in the clinical tool chest? The path seems to be set for each class by the first drug within that class to enter the clinic. For example, from the earliest clinical trials it was clear that the epothilones possessed significant antitumor activity, and subsequent trials have shown that the spectrum of activity of these compounds is very similar to that of the taxanes. Therefore, issues concerning ease of delivery, activity against taxane refractory or resistant tumors, and side effect profiles relative to presently approved and investigational taxanes, are likely to influence the uses of the epothilones. Registrational necessities of the companies developing these agents will be as important as the medical rationales for disease-specific development, as it is unlikely that the resources, as well as the clinical trials necessary to distinguish one epothilone from another in the same clinical setting, will be sought before FDA approval for these agents. On the other hand, for classes such as the colchicine-like agents and dolastatins, the clinical data suggest that if the first compound tested within a given class fails to demonstrate meaningful anticancer activity early in clinical development (assuming adequate PK was achieved), it is unlikely that the other members of that class will be successful using the same clinical development paradigm. In addition, based on preclinical data suggesting antiangiogenic activity against tumor endothelial cells, many agents in this class are being developed using clinical trial designs intended to demonstrate antiangiogenic-mediated growth suppression. It is not yet clear whether these agents are uniquely antiangiogenic, or whether the preclinical and clinical development strategies are merely accentuating the antiangiogenic potential of the entire class of tubulin directed agents. When administered using low-dose metronomic schedules, both the taxanes and the epothilones show biological activity against tumors and their associated endothelia, consistent with an antiangiogenic mechanism of action for examples.

The toxicity profiles of these agents are remarkably similar and familiar. Bone marrow suppression manifests most prominently in the clinic as reversible neutropenia. It is both the dose limiting and the most common severe AE seen with many of these agents. A number of phase 1 trials using these agents have been designed with this in mind, and therefore the schedules explored were those thought to maximize drug exposure, while minimizing neutropenia. Gastrointestinal AEs resulting from mucosal toxicity are also prominent. Another common cluster of AEs are peripheral sensory and motor neuropathy, pain syndromes, and neuroconstipation, suggesting that, at least in normal tissues, alteration of the microtubule apparatus by many different mechanisms leads to a common toxicity spectrum. Although there are some nonclinical data to suggest that several of these compounds may possess potent anticancer activity with less neurological injury, these nonclinical observations have yet to be proven in the clinic.

8. CONCLUSIONS

There is a large number of investigational antitubulin agents in early clinical trials. The epothilones are the only class sufficiently advanced in clinical development with promising data to suggest the likelihood of clinical efficacy. For some of the other classes, the dolastatins and colchicine-like compounds in particular, it is unclear that therapeutic efficacy will be established, whereas for others it will take additional clinical testing over the next few years before the potential of the agents can be weighed against the reality of the clinical outcomes.

Virtually all these agents are derivatives of natural products and despite the similarity in their toxicity profiles, the anticancer activity profiles are quite diverse. Until more is known about the cellular consequences of interference with the microtubule by different means, it is unlikely that development of anticancer agents from this class will be different than it has been for the past several decades. There will unavoidably be phase 1 testing of multiple schedules, and phase 2 testing across a wide spectrum of human cancers. However, examination of previous and ongoing efforts to develop this class of agents suggests that acting on the results from early clinical trial data from the first-in-class compound could prevent unnecessarily broad development of other compounds in that class. Unfortunately, owing to factors unrelated to scientific or clinical knowledge, it is likely that multiple agents within a given class will continue to be subject to broad concurrent clinical development strategies. Although inefficient, based on previous experience with successful antimicrotubule agents, it is likely that this process will yield new agents for clinical use that possess therapeutic advantages over the available agents in the clinic today.

REFERENCES

1. Borzilleri RM, Vite GD. Epothilones: new tubulin polymerization agents in preclinical and clinical development. Drugs Future 2002;27:1149–1163.
2. Gerth K, Bedorf N, Hofle G, et al. Epothilons A and B: antifungal and cytotoxic compounds from Sorangium cellulosum (Myxobacteria). Production, physico-chemical and biological properties. J Antibiot (Tokyo) 1996;49:560–563.
3. Höfle G, Bedorf N, Steinmetz H, et al. Epothilone A and B - novel 16-membered macrolides with cytotoxic activity: isolation, crystal structure, and conformation in solution. Angew Chem Int Ed Engl 1996;35:1567–1569.
4. Bollag DM, McQueney PA, Zhu J, et al. Epothilones, a new class of microtubule-stabilizing agents with a taxol-like mechanism of action. Cancer Res 1995;55:2325–2333.
5. Kowalski RJ, Giannakakou P, Hamel E. Activities of the microtubule-stabilizing agents epothilones A and B with purified tubulin and in cells resistant to paclitaxel (Taxol(R)). J Biol Chem 1997;272: 2534–2541.
6. Rubin EH, Rothermel J, Tesfaye F, et al. Phase I dose-finding study of weekly single-agent patupilone in patients with advanced solid tumors. J Clin Oncol 2005;23:9120–9129.
7. Calvert PM, O'Neill V, Twelves C, et al. A phase I clinical and pharmacokinetic study of EPO906 (Epothilone B), given every three weeks, in patients with advanced solid tumors. Proc Am Soc Clin Oncol 2001;20:A429.
8. Melichar B, Tabernero J, Casado E, et al. Phase I dose optimization trial of patupilone in previously treated patients (pts) with advanced colon cancer (ACC). Proc Am Soc Clin Oncol 2005;24:A3688.
9. Thompson JA, Swerdloff J, Escudier B, et al. Phase II trial evaluating the safety and efficacy of EPO906 in patients with advanced renal cancer. Proc Am Soc Clin Oncol 2003;22:A1628 (Poster presentation).
10. Poplin E, Moore M, O'Dwyer P, et al. Safety and efficacy of EPO906 in patients with advanced colorectal cancer: A review of 2 phase II trials. Proc Am Soc Clin Oncol 2003;22:A1135.
11. Gore M, Kaye S, Oza AM, et al. Phase I trial of patupilone plus carboplatin in patients with advanced cancer. Proc Am Soc Clin Oncol 2005;24:A5087.
12. Awada A, Bleiberg H, de Valeriola D, et al. Phase I clinical and pharmacology study of the epothilone analog BMS-247550 given weekly in patients (pts) with advanced solid tumors. Proc Am Soc Clin Oncol 2001;20:A427.
13. Spriggs D, Soignet S, Bienvenu B, et al. Phase I first-in-man study of the epothilone B analog BMS-247550 in patients with advanced cancer. Proc Am Soc Clin Oncol 2001;20:A428.
14. Gadgeel SM, Wozniak A, Boinpally RR, et al. Phase I clinical trial of BMS-247550, a derivative of epothilone B, using accelerated titration 2B design. Clin Cancer Res 2005;11:6233–6239.
15. Mani S, McDaid H, Goel S, et al. Expanded evaluation of BMS-247550 (BMS) using a 1hr infusion at the recommended phase II dose (RPTD). Proc Am Soc Clin Oncol 2003;22:A995.
16. Mani S, McDaid H, Shen HJ, et al. Phase I pharmacokinetic and pharmacodynamic study of an epothilone B analog (BMS-247550) administered as a 1-hour infusion every 3 weeks: an update. Proc Am Soc Clin Oncol 2002;21:A409.
17. Tripathi R, Gadgeel SM, Wozniak AJ, et al. Phase I clinical trial of BMS-247550 (epothilone B derivative) in adult patients with advanced solid tumors. Proc Am Soc Clin Oncol 2002;21:A407.
18. Awada A, Jones S, Piccart M, et al. Final results of the phase I study of the novel epothilone BMS-247550 administered weekly in patients (pts) with advanced solid tumors. Eur J Cancer 2002;38:S41(A122).
19. Hao D, Hammond LA, deBono JS, et al. Continuous weekly administration of the epothilone-B derivative, BMS247,550 (NSC710428): a phase I and pharmacokinetic (PK) study. Proc Am Soc Clin Oncol 2002;21:A411.

20. Burris HA, Awada A, Jones S, et al. Phase I study of the novel epothilone BMS-247550 administered weekly in patients (pts) with advanced malignancies. Proc Am Soc Clin Oncol 2002;21:A412.
21. Abraham J, Agrawal M, Bakke S, et al. Phase I trial and pharmacokinetic study of BMS-247550, an epothilone B analog, administered intravenously on a daily schedule for five days. J Clin Oncol 2003;21:1866–1873.
22. Agrawal M, Kotz H, Abraham J, et al. A phase I clinical trial of BMS 247550 (NSC 71028), an epothilone B derivative, in patients with refractory neoplasms. Proc Am Soc Clin Oncol 2002;21:A410.
23. Peerebom D, Batchelor T, Lesser G, et al. NABTT 2111: A phase I trial of BMS-247550 for patients with recurrent high-grade gliomas. Proc Am Soc Clin Oncol 2005;24:A1563.
24. Thambi P, Edgerly M, Agrawal M, et al. A phase I clinical trial of BMS 247550 (NSC 71028), an epothilone B derivative, given daily for 3 days on a 21 day cycle in patients with refractory neoplasms. Proc Am Soc Clin Oncol 2003;22:A540.
25. Zhuang SH, Agrawal M, Edgerly M, et al. A Phase I clinical trial of ixabepilone (BMS-247550), an epothilone B analog, administered intravenously on a daily schedule for 3 days. Cancer 2005;103: 1932–1938.
26. Widemann BC, Fox E, Goodspeed WJ, et al. Phase I trial of the epothilone B analog BMS-247550 (ixabepilone) in children with reractory solid tumors. Proc Am Soc Clin Oncol 2005;24:A8529.
27. Plummer R, Molife R, Verrill M, et al. Phase I and pharmacokinetic study of BMS-247550 in combination with carboplatin in patients with advanced solid malignancies. Proc Am Soc Clin Oncol 2002;21:A2125 (Poster presentation).
28. Kelly WK, Galsky MD, Small EJ, et al. Multi-institutional trial of the epothilone B analogue BMS-247550 with or without estrumustine phosphate (EMP) in patients with progressive castrate-metastatic prostate cancer (PCMPC): updated results. Proc Am Soc Clin Oncol 2004;23:A4509.
29. Smaletz O, Galsky M, Scher HI, et al. Pilot study of epothilone B analog (BMS-247550) and estramustine phosphate in patients with progressive metastatic prostate cancer following castration. Ann Oncol 2003;14:1518–1524.
30. Anderson S, Dizon D, Sabbatini P, et al. Phase I trial of BMS-247550 and gemcitabine in patients with advanced solid tumor malignancies. Proc Am Soc Clin Oncol 2004;23:A2098.
31. Roché H, Delord JP, Bunnell CA, et al. Phasc II studies of the novel epothilone BMS-247550 in patients (pts) with taxane-naïve or taxane-refractory metastatic breast cancer. Proc Am Soc Clin Oncol 2002;21:A223.
32. Roché H, Cure H, Bunnell C, et al. A phase II study of epothilone analog BMS-247550 in patients with metastatic breast cancer previously treated with an anthracycline. Proc Am Soc Clin Oncol 2003;22:A69 (Poster presentation).
33. Thomas E, Tabernero J, Fornier M, et al. A phase II study of the epothilone B analog BMS-247550 in patients with taxane-resistant metastatic breast cancer. Proc Am Soc Clin Oncol 2003;22:A30 (Poster presentation).
34. Chen T, Molina A, Moore S, et al. Epothilone B analog (BMS-247550) at the recommended phase II dose (RPTD) in patients (pts) with gynecologic (gyn) and breast cancer. Proc Am Soc Clin Oncol 2004;23:A2115.
35. Low JA, Wedam SB, Brufsky A, et al. A phase 2 trial of BMS-247550 (ixabepilone), and epothilone B analog, given daily x 5 in breast cancer. Proc Am Soc Clin Oncol 2004;23:A545.
36. Zhuang SH, Menefee M, Kotz H, et al. A phase II clinical trial of BMS-247550 (ixabepilone), a microtubule-stabilizing agent in renal cell cancer. Proc Am Soc Clin Oncol 2004;23:A4550.
37. Delbaldo C, Lara PN, Vansteenkiste J, et al. Phase II study of the novel epothilone BMS-247550 in patients (pts) with recurrent or metastatic non-small cell lung cancer (NSCLC) who have failed first-line platinum-based chemotherapy. Proc Am Soc Clin Oncol 2002;21:A1211.
38. Vansteenkiste JF, Breton JL, Sandler A, et al. A randomized phase II study of epothilone analog BMS-247550 in patients (pts) with non-small cell lung cancer (NSCLC) who have failed first-line platinum-based chemotherapy. Proc Am Soc Clin Oncol 2003;22:A2519 (Poster presentation).
39. Ajani JA, Shah MA, Bokemeyer C, et al. Phase II study of the novel epothilone BMS-247550 in patients (pts) with metastatic gastric adenocarcinoma previously treated with a taxane. Proc Am Soc Clin Oncol 2002;21:A619.
40. Okuno SH, Geyer SM, Maples WJ, et al. Phase 2 study of epothilone B analog (BMS-247550) in soft tissue sarcomas. Proc Am Soc Clin Oncol 2002;21:A1645 (Poster presentation).
41. Eng C, Kindler HL, Skoog L, et al. The epothilone analogue, BMS-247550, in patients (pts) with advanced colorectal cancer (CRC). Proc Am Soc Clin Oncol 2003;22:A1134.

42. Singh DA, Kindler HL, Eng C, et al. Phase II trial of the epothilone B analog BMS-247550 in patients with hepatobiliary cancer. Proc Am Soc Clin Oncol 2003;22:A1127.
43. Whitehead RP, McCoy SA, Rivkin SE, et al. A phase II trial of epothilone B analogue BMS-247550 (NSC No.710428) in patients with advanced pancreas cancer: a Southwest Oncology Group Study. Proc Am Soc Clin Oncol 2004;23:A4012.
44. Pavlick AC, Millward M, Farrell K, et al. A phase II study of epothilone B analog (EpoB)-BMS 247550 (NSC No. 710428) in stage IV malignant melanoma (MM). Proc Am Soc Clin Oncol 2004;23:A7542.
45. Rosen PJ, Rosen LS, Britten C, et al. KOS-862 (epothilone D): results of a phase I dose-escalating trial in patients with advanced malignancies. Proc Am Soc Clin Oncol 2002;21:A413.
46. Piro LD, Rosen LS, Parson M, et al. KOS-862 (epothilone D): A comparison of two schedules in patients with advanced malignancies. Proc Am Soc Clin Oncol 2003;22:A539 (Poster presentation).
47. Marshall JL, Ramalingam S, Hwang JJ, et al. Phase 1 and pharmacokinetic (PK) study of weekly KOS-862 (Epothilone D) combined with gemcitabine (GEM) in patients (pts) with advanced solid tumors. Proc Am Soc Clin Oncol 2005;24:A2041.
48. Monk JP, Calero-Villalona M, Dupont J, et al. Phase 1 trial of KOS-862 (epothilone D) in combination with carboplating (C) in patients with solid tumors. Proc Am Soc Clin Oncol 2005;24:A2049.
49. Overmoyer B, Waintraub S, Kaufman PA, et al. Phase II trial of KOS-862 (epothilone D) in anthracycline and taxane pretreated metastatic breast cancer. Proc Am Soc Clin Oncol 2005;24:A778.
50. Yee L, Lynch T, Villalona-Calero M, et al. A phase II study of KOS-862 (Epothilone D) as second-line therapy in non-small cell lung cancer. Proc Am Soc Clin Oncol 2005;24:A7127.
51. Smit WM, Sufliarsky J, Spanik S, et al. Phase I/II dose-escalation trial of patupilone every 3 weeks in patients with relapsed/refractory ovarian cancer. Proc Am Soc Clin Oncol 2005;24:A5056.
52. Osterlind K, Sanchez JM, Zatloukal P, et al. Phase I/II dose escalation trial of patupilone every 3 weeks in patients with non-small cell lung cancer. Proc Am Soc Clin Oncol 2005;24:A7110.
53. Aisner J, Gore M, Rubin EH, et al. Two phase IB trials of EPO906 plus carboplatin in patients with advanced malignancies. Proc Am Soc Clin Oncol 2003;22:A574 (Poster presentation).
54. Jordan A, Hadfield JA, Lawrence NJ, et al. Tubulin as a target for anticancer drugs: agents which interact with the mitotic spindle. Med Res Rev 1998;18:259–296.
55. Jordan MA, Walker D, de Arruda M, et al. Suppression of microtubule dynamics by binding of cemadotin to tubulin: possible mechanism for its antitumor action. Biochemistry 1998;37:17,571–17,578.
56. Otani M, Natsume T, Watanabe JI, et al. TZT-1027, an antimicrotubule agent, attacks tumor vasculature and induces tumor cell death. Jpn J Cancer Res 2000;91:837–844.
57. Pettit GR, Singh SB, Hogan F, et al. The absolute configuratiion and synthesis of natural (–)-dolastatin 10. J Am Chem Soc 1989;111:5463–5465.
58. Bai R, Pettit GR, Hamel E. Dolastatin 10, a powerful cytostatic peptide derived from a marine animal. Inhibition of tubulin polymerization mediated through the vinca alkaloid binding domain. Biochem Pharmacol 1990;39:1941–1949.
59. Bai R, Roach MC, Jayaram SK, et al. Differential effects of active isomers, segments, and analogs of dolastatin 10 on ligand interactions with tubulin. Correlation with cytotoxicity. Biochem Pharmacol 1993;45:1503–1515.
60. Bai RL, Pettit GR, Hamel E. Structure-activity studies with chiral isomers and with segments of the antimitotic marine peptide dolastatin 10. Biochem Pharmacol 1990;40:1859–1864.
61. Pettit GR, Srirangam JK, Barkoczy J, et al. Antineoplastic agents 337. Synthesis of dolastatin 10 structural modifications. Anticancer Drug Des 1995;10:529–544.
62. de Arruda M, Cocchiaro CA, Nelson CM, et al. LU103793 (NSC D-669356): a synthetic peptide that interacts with microtubules and inhibits mitosis. Cancer Res 1995;55:3085–3092.
63. Hopper LD, Van Dijk S, Shannon P, et al. Safety and toxicokinetics in a five-day oral toxicity study of a dolastatin-15 analog, ILX651, in beagle dogs. Proc Am Assoc Cancer Res 2003;44:A1749.
64. Li Y, Kobayashi H, Hashimoto Y, et al. Interaction of marine toxin dolastatin 10 with porcine brain tubulin: competitive inhibition of rhizoxin and phomopsin A binding. Chem Biol Interact 1994;93:175–183.
65. Haldar S, Basu A, Croce CM. Serine-70 is one of the critical sites for drug-induced Bcl2 phosphorylation in cancer cells. Cancer Res 1998;58:1609–1615.
66. Kalemkerian GP, Ou X, Adil MR, et al. Activity of dolastatin 10 against small-cell lung cancer *in vitro* and *in vivo*: induction of apoptosis and bcl-2 modification. Cancer Chemother Pharmacol 1999;43:507–515.

67. Natsume T, Watanabe J, Tamaoki S, et al. Characterization of the interaction of TZT-1027, a potent antitumor agent, with tubulin. Jpn J Cancer Res 2000;91:737–747.
68. Madden T, Tran HT, Beck D, et al. Novel marine-derived anticancer agents: a phase I clinical, pharmacological, and pharmacodynamic study of dolastatin 10 (NSC 376128) in patients with advanced solid tumors. Clin Cancer Res 2000;6:1293–1301.
69. Pitot HC, McElroy EA Jr, Reid JM, et al. Phase I trial of dolastatin-10 (NSC 376128) in patients with advanced solid tumors. Clin Cancer Res 1999;5:525–531.
70. Vaishampayan U, Glode M, Du W, et al. Phase II study of dolastatin-10 in patients with hormone-refractory metastatic prostate adenocarcinoma. Clin Cancer Res 2000;6:4205–4208.
71. Saad ED, Kraut EH, Hoff PM, et al. Phase II study of dolastatin-10 as first-line treatment for advanced colorectal cancer. Am J Clin Oncol 2002;25:451–453.
72. Hoffman MA, Blessing JA, Lentz SS. A phase II trial of dolastatin-10 in recurrent platinum-sensitive ovarian carcinoma: a Gynecologic Oncology Group study. Gynecol Oncol 2003;89:95–98.
73. Pitot HC, Frytak S, Croghan GA, et al. Phase II study of dolastatin-10 (dola-10) in patients (pts) with advanced renal cell carcinoma. Proc Am Soc Clin Oncol 2002;21:A2409.
74. Krug LM, Miller VA, Kalemkerian GP, et al. Phase II study of dolastatin-10 in patients with advanced non-small-cell lung cancer. Ann Oncol 2000;11:227–228.
75. Margolin K, Longmate J, Synold TW, et al. Dolastatin-10 in metastatic melanoma: a phase II and pharmokinetic trial of the California Cancer Consortium. Invest New Drugs 2001;19:335–340.
76. Perez EA, Hillman DW, Fishkin PA, et al. Phase II trial of dolastatin-10 in patients with advanced breast cancer. Invest New Drugs 2005;23:257–261.
77. Kindler HL, Tothy PK, Wolff R, et al. Phase II trials of dolastatin-10 in advanced pancreaticobiliary cancers. Invest New Drugs 2005;23:489–493.
78. Yamamoto N, Andoh M, Kawahara M, et al. Phase I study of TZT-1027, an inhibitor of tubulin polymerization, given weekly x 3 as a 1-hour intravenous infusion in patients (pts) with solid tumors. Proc Am Soc Clin Oncol 2002;21:A420.
79. Schoffski P, Thate B, Beutel G, et al. Phase I and pharmacokinetic study of TZT-1027, a novel synthetic dolastatin 10 derivative, administered as a 1-hour intravenous infusion every 3 weeks in patients with advanced refractory cancer. Ann Oncol 2004;15:671–679.
80. de Jonge MJ, van der Gaast A, Planting AS, et al. Phase I and pharmacokinetic study of the dolastatin 10 analogue TZT-1027, given on days 1 and 8 of a 3-week cycle in patients with advanced solid tumors. Clin Cancer Res 2005;11:3806–3813.
81. Horti J, Juhasz E, Bodrogi I. Preliminary results of a phase I trial of TZT-1027, an inhibitor of tubulin polymerization, in patients with advanced non-small cell lung cancer. Proc Am Assoc Cancer Res 2002;43:A2744.
82. Blagden S, Thomas A, De-Bono JS, et al. Phase I study of intravenous TZT-1027 (T) and carboplating (C), administered on day 1 (T and C) and day 8 (T) every three weeks in patients (pts) with advanced solid tumors. Proc Am Soc Clin Oncol 2005;24:A3141.
83. Wolff I, Bruntsch U, Cavalli F, et al. Phase I clinical study of LU 103793 (cemadotin) given on a weekly (wkly) x 4 schedule. Proc Am Soc Clin Oncol 1997;16:A783.
84. Mross K, Berdel WE, Fiebig HH, et al. Clinical and pharmacologic phase I study of Cemadotin-HCl (LU103793), a novel antimitotic peptide, given as 24-hour infusion in patients with advanced cancer. A study of the Arbeitsgemeinschaft Internistische Onkologie (AIO) Phase I Group and Arbeitsgruppe Pharmakologie in der Onkologie und Haematologie (APOH) Group of the German Cancer Society. Ann Oncol 1998;9:1323–1330.
85. Villalona-Cajero M, Von Hoff D, Eckhardt G, et al. Phase I and pharmacokinetic (PK) study of LU 103793, a water soluble analog of dolastatin-15, on a daily x 5 schedule. Proc Am Soc Clin Oncol 1997;16:784.
86. Villalona-Calero MA, Baker SD, Hammond L, et al. Phase I and pharmacokinetic study of the water-soluble dolastatin 15 analog LU103793 in patients with advanced solid malignancies. J Clin Oncol 1998;16:2770–2779.
87. Supko JG, Lynch TJ, Clark JW, et al. A phase I clinical and pharmacokinetic study of the dolastatin analogue cemadotin administered as a 5-day continuous intravenous infusion. Cancer Chemother Pharmacol 2000;46:319–328.
88. Kerbrat P, Dieras V, Pavlidis N, et al. Phase II study of LU 103793 (dolastatin analogue) in patients with metastatic breast cancer. Eur J Cancer 2003;39:317–320.

89. Smyth J, Boneterre ME, Schellens J, et al. Activity of the dolastatin analogue, LU103793, in malignant melanoma. Ann Oncol 2001;12:509–511.

90. Marks RS, Graham DL, Sloan JA, et al. A phase II study of the dolastatin 15 analogue LU 103793 in the treatment of advanced non-small-cell lung cancer. Am J Clin Oncol 2003;26:336–337.

91. Ebbinghaus S, Rubin E, Hersh E, et al. A phase I study of the dolastatin-15 analogue tasidotin (ILX651) administered intravenously daily for 5 consecutive days every 3 weeks in patients with advanced solid tumors. Clin Cancer Res 2005;11:7807–7816.

92. Cunningham C, Appleman LJ, Kirvan-Visovatti M, et al. Phase I and pharmacokinetic study of the dolastatin-15 analogue tasidotin (ILX651) administered intravenously on days 1, 3, and 5 every 3 weeks in patients with advanced solid tumors. Clin Cancer Res 2005;11:7825–7833.

93. Weiss GR, Mita A, Garrison M, et al. Phase I, pharmacokinetic (PK) study of synthadotin (SYN-D; ILX651), a next generation antitubulin, administered IV weekly x3 weeks every 4 weeks (wx3q4w). Proc Am Soc Clin Oncol 2005;24:A3073.

94. Ebbinghaus S, Hersh E, Cunningham CC, et al. Phase II study of synthadotin (SYN-D; ILX651) administered daily for 5 consecutive days once every 3 weeks (qdx5q3w) in patients (pts) with inoperable locally advanced or metastatic melanoma. Proc Am Soc Clin Oncol 2004;24:A7530.

95. McDermott DF, Hersh E, Weber J, et al. ILX651 administered daily for five days every 3 weeks (qdx5dq3w) in patients (pts) with inoperable locally advanced or metastatic melanoma: phase II experience. Proc Am Soc Clin Oncol 2005;24:A7556.

96. Blakey DC, Ashton SE, Westwood FR, et al. ZD6126: a novel small molecule vascular targeting agent. Int J Radiat Oncol Biol Phys 2002;54:1497–1502.

97. Schumacher G, Neuhaus P. The physiological estrogen metabolite 2-methoxyestradiol reduces tumor growth and induces apoptosis in human solid tumors. J Cancer Res Clin Oncol 2001;127:405–410.

98. Sklarin NT, Lathia CD, Benson L, et al. A phase I trial and pharmacokinetic evaluation of CI-980 in patients with advanced solid tumors. Invest New Drugs 1997;15:235–246.

99. Tozer GM, Kanthou C, Parkins CS, et al. The biology of the combretastatins as tumour vascular targeting agents. Int J Exp Pathol 2002;83:21–38.

100. Sackett DL, Varma JK. Molecular mechanism of colchicine action: induced local unfolding of beta-tubulin. Biochemistry 1993;32:13,560–13,565.

101. Davis PD, Dougherty GJ, Blakey DC, et al. ZD6126: a novel vascular-targeting agent that causes selective destruction of tumor vasculature. Cancer Res 2002;62:7247–7253.

102. Micheletti G, Poli M, Borsotti P, et al. Vascular-targeting activity of ZD6126, a novel tubulin-binding agent. Cancer Res 2003;63:1534–1537.

103. Blakey DC, Westwood FR, Walker M, et al. Antitumor activity of the novel vascular targeting agent ZD6126 in a panel of tumor models. Clin Cancer Res 2002;8:1974–1983.

104. Pettit GR, Cragg GM, Singh SB. Antineoplastic agents, 122. Constituents of Combretum caffrum. J Nat Prod 1987;50:386–391.

105. Pettit GR, Temple C Jr, Narayanan VL, et al. Antineoplastic agents 322. synthesis of combretastatin A-4 prodrugs. Anticancer Drug Des 1995;10:299–309.

106. Dark GG, Hill SA, Prise VE, et al. Combretastatin A-4, an agent that displays potent and selective toxicity toward tumor vasculature. Cancer Res 1997;57:1829–1834.

107. Ahmed B, Van Eijk LI, Bouma-Ter Steege JC, et al. Vascular targeting effect of combretastatin A-4 phosphate dominates the inherent angiogenesis inhibitory activity. Int J Cancer 2003;105:20–25.

108. D'Amato RJ, Lin CM, Flynn E, et al. 2-Methoxyestradiol, an endogenous mammalian metabolite, inhibits tubulin polymerization by interacting at the colchicine site. Proc Natl Acad Sci USA 1994;91:3964–3968.

109. Fotsis T, Zhang Y, Pepper MS, et al. The endogenous oestrogen metabolite 2-methoxyoestradiol inhibits angiogenesis and suppresses tumour growth. Nature 1994;368:237–239.

110. de Ines C, Leynadier D, Barasoain I, et al. Inhibition of microtubules and cell cycle arrest by a new 1-deaza-7,8-dihydropteridine antitumor drug, CI 980, and by its chiral isomer, NSC 613863. Cancer Res 1994;54:75–84.

111. Portnow J, Stuart G, Eller S, et al. The intracerebral distribution study of CI-980: A new agent being studied in patients with glioblastoma multiforme. Proc Am Soc Clin Oncol 1999;18:565.

112. Waud WR, Leopold WR, Elliott WL, et al. Antitumor activity of ethyl 5-amino-1,2-dihydro-2-methyl-3-phenyl-pyrido [3,4-b]pyrazin-7-ylcarbamate, 2-hydroxyethanesulfonate, hydrate (NSC 370147) against selected tumor systems in culture and in mice. Cancer Res 1990;50:3239–3244.

113. Radema SA, Beerepoot LV, Witteveen PO, et al. Clinical evaluation of the novel vascular-targeting agent, ZD6126: assessment of toxicity and surrogate markers of vascular damage. Proc Am Soc Clin Oncol 2002;21:A439.

114. Gadgeel SM, LoRusso PM, Wozniak AJ, et al. A dose-escalation study of the novel vascular-targeting agent, ZD6126, in patients with solid tumors. Proc Am Soc Clin Oncol 2002;21:A438.

115. Thorpe PE, Chaplin DJ, Blakey DC. The first international conference on vascular targeting: meeting overview. Cancer Res 2003;63:1144–1147.

116. DelProposto Z, LoRusso P, Latif Z, et al. MRI evaluation of the effects of the vascular-targeting agent ZD6126 on tumor vasculature. Proc Am Soc Clin Oncol 2002;21:A440.

117. Dowlati A, Robertson K, Cooney M, et al. A phase I pharmacokinetic and translational study of the novel vascular targeting agent combretastatin a-4 phosphate on a single-dose intravenous schedule in patients with advanced cancer. Cancer Res 2002;62:3408–3416.

118. Rustin GJ, Galbraith SM, Anderson H, et al. Phase I clinical trial of weekly combretastatin A4 phosphate: clinical and pharmacokinetic results. J Clin Oncol 2003;21:2815–2822.

119. Stevenson JP, Rosen M, Sun W, et al. Phase I trial of the antivascular agent combretastatin A4 phosphate on a 5-day schedule to patients with cancer: magnetic resonance imaging evidence for altered tumor blood flow. J Clin Oncol 2003;21:4428–4438.

120. Bilenker JH, Flaherty KT, Rosen M, et al. Phase I trial of combretastatin a-4 phosphate with carboplatin. Clin Cancer Res 2005;11:1527–1533.

121. Rustin GJ, Nathan PD, Boxall J, et al. A phase Ib trial of combretastatin A-4 phosphate (CA4P) in combination with carboplatin or paclitaxel chemotehrapy in patients with advanced cancer. Proc Am Soc Clin Oncol 2005;24:A3103.

122. Sledge GW Jr, Miller KD, Haney LG, et al. A phase I study of 2-methoxyestradiol (2ME2) in patients (pts) with refractory metastatic breast cancer (MBC). Proc Am Soc Clin Oncol 2002;21:A441.

123. Dahut WL, Lakhani NJ, Gulley JL, et al. Phase I clinical trial of oral 2-methoxyestradiol, an antiangiogenic and apoptotic agent, in patients with solid tumors. Cancer Biol Ther 2006;5:22–27.

124. Miller KD, Murry DJ, Curry E, et al. A phase I study of 2-methoxyestradiol (2ME2) plus docetaxel (D) in patients (pts) with metastatic breast cancer (MBC). Proc Am Soc Clin Oncol 2002;21:A442.

125. Sweeney C, Liu G, Yiannoutsos C, et al. A phase II multicenter, randomized, double-blind, safety trial assessing the pharmacokinetics, pharmacodynamics, and efficacy of oral 2-methoxyestradiol capsules in hormone-refractory prostate cancer. Clin Cancer Res 2005;11:6625–6633.

126. Volker KM, Mercer BG, Treston A, et al. Effect of route of administration, dose, and schedule on the anti-tumor activity of Panzem NCD in a murine orthotopic lung cancer model. Proc Am Assoc Cancer Res 2005;46:A2993.

127. EntreMed I: 2006 EntreMed commences phase 2 studies with panzem NCD. http://www.entremed.com./download/press/GBM_Ph2_FINALOUT.pdf, last date accessed March 6, 2007.

128. Bernstein ML, Baruchel S, Devine S, et al. Phase I and pharmacokinetic study of CI-980 in recurrent pediatric solid tumor cases: a Pediatric Oncology Group study. J Pediatr Hematol Oncol 1999;21:494–500.

129. Fishkin P, Stadler WM, Gibbons J, et al. A university of Chicago phase II consortium study(UCPC) of CI-980 in patients (pts) metastatic renal cell carcinoma (RCC). Proc Am Soc Clin Oncol 1998;17:1275.

130. Gutheil J, Van Echo D, Egorin M, et al. Phase I study of CI-980 in patients with refractory malignancies. Proc Am Assoc Cancer Res 1996;37:1129.

131. Kudelka AP, Hasenburg A, Verschraegen CF, et al. Phase II study of i.v. CI-980 in patients with advanced platinum refractory epithelial ovarian carcinoma. Anticancer Drugs 1998;9:405–409.

132. Kunschner LJ, Fine H, Hess K, et al. CI-980 for the treatment of recurrent or progressive malignant gliomas: national central nervous system consortium phase I-II evaluation of CI-980. Cancer Invest 2002;20:948–954.

133. Mikkelsen T, Phuphanich S, Batchelor T, et al. Phase I/II trial of CI-980 in newly diagnosed malignant glioma - NABTT 9602. Proc Am Soc Clin Oncol 2000;19:A619.

134. Natale R, Waterhouse D, Grove WR, et al. Phase I clinical and pharmacokinetic trial of CI-980, a novel mitotic inhibitor. Proc Am Soc Clin Oncol 1992;11:292.

135. Patel SR, Burgess MA, Papadopolous NE, et al. Phase II study of CI-980 (NSC 635370) in patients with previously treated advanced soft-tissue sarcomas. Invest New Drugs 1998;16:87–92.

136. Pazdur R, Meyers C, Diaz-Canton E, et al. Phase II trial of intravenous CI-980 (NSC 370147) in patients with metastatic colorectal carcinoma. Model for prospective evaluation of neurotoxicity. Am J Clin Oncol 1997;20:573–576.

137. Rowinsky EK, Long GS, Noe DA, et al. Phase I and pharmacological study of CI-980, a novel synthetic antimicrotubule agent. Clin Cancer Res 1997;3:401–407.

138. Rowinsky EK, Noe DA, Grochow LB, et al. Phase I and pharmacological study of CI-980, a synthetic and structurally unique antimicrotubule agent, on a 72-hour continuous infusion schedule in adults with solid tumors. Proc Am Soc Clin Oncol 1995;14:1477.

139. Ryan CW, Shulman KL, Richards JM, et al. CI-980 in advanced melanoma and hormone refractory prostate cancer. Invest New Drugs 2000;18:187–191.

140. Sciortino D, Arrieta R, Masters GA, et al. A phase II trial of CI-980 in advanced non-smallCELL lung cancer (NSCLC). Proc Am Soc Clin Oncol 1998;17:1925.

141. Thomas JP, Moore T, Kraut EH, et al. A phase II study of CI-980 in previously untreated extensive small cell lung cancer: an Ohio State University phase II research consortium study. Cancer Invest 2002;20:192–198.

142. Whitehead RP, Unger JM, Flaherty LE, et al. Phase II trial of CI-980 in patients with disseminated malignant melanoma and no prior chemotherapy. A Southwest Oncology Group study. Invest New Drugs 2001;19:239–243.

143. Yokoi A, Kuromitsu J, Kawai T, et al. Profiling novel sulfonamide antitumor agents with cell-based phenotypic screens and array-based gene expression analysis. Mol Cancer Ther 2002;1:275–286.

144. Patnaik A, Forero L, Goetz A, et al. HMN-214, a novel oral antimicrotubular agent and inhibitor of polo-like- and cyclin-dependent kinases: Clinical, pharmacokinetic (PK) and pharmacodynamic (PD) relationships observed in a phase I trial of a daily x 5 schedule every 28 days. Proc Am Soc Clin Oncol 2003;22:A514 (Poster presentation).

145. Patnaik A, Forero L, Goetz A, et al. HMN-214, a novel oral antimicrotubular agent and inhibitor of polo-like- and cyclin-dependent kinases: Clinical, pharmacokinetic (PK) and pharmacodynamic (PD) relationships observed in a phase I trial of a daily x 5 schedule every 28 days. Proc Am Soc Clin Oncol 2003;22:A514.

146. Owa T, Yoshino H, Okauchi T, et al. Discovery of novel antitumor sulfonamides targeting G1 phase of the cell cycle. J Med Chem 1999;42:3789–3799.

147. Yoshino H, Ueda N, Niijima J, et al. Novel sulfonamides as potential, systemically active antitumor agents. J Med Chem 1992;35:2496–2497.

148. Supuran CT. Indisulam: an anticancer sulfonamide in clinical development. Expert Opin Invest Drugs 2003;12:283–287.

149. Yoshimatsu K, Yamaguchi A, Yoshino H, et al. Mechanism of action of E7010, an orally active sulfonamide antitumor agent: inhibition of mitosis by binding to the colchicine site of tubulin. Cancer Res 1997;57:3208–3213.

150. Shoemaker AR, Oleksijew A, Credo B, et al. Evaluation of the Antimitotic Agent ABT-751 in the ApcMin Model of Intestinal Tumorigenesis. Proc Am Assoc Cancer Res 2003;44:A2738.

151. Yamamoto K, Noda K, Yoshimura A, et al. Phase I study of E7010. Cancer Chemother Pharmacol 1998;42:127–134.

152. Yee KW, Hagey A, Verstovsek S, et al. Phase 1 study of ABT-751, a novel microtubule inhibitor, in patients with refractory hematologic malignancies. Clin Cancer Res 2005;11:6615–6624.

153. Cho SY, Adamson PC, Hagey A, et al. Phase I trial and pharmacokinetic (PK) study of ABT-751, and orally bioavailable tubulin agent, in pediatric patients with refractory solid tumors. Proc Am Soc Clin Oncol 2004;23:A2080.

154. Fox E, Adamson PC, Hagey A, et al. Phase I trial of oral ABT-751 in pediatric patients: preliminary evidence of activity in neuroblastoma (NBL). Proc Am Soc Clin Oncol 2005;24:A8527.

155. Sprague E, Fleming GF, Carr R, et al. Phase I study of 21-day continuous dosing of the oral antimitotic agent ABT-751. Proc Am Soc Clin Oncol 2003;22:A518 (Poster presentation).

156. Hande KR, Meek K, Lockhart AC, et al. Pharmacokinetics and toxicity of ABT-751, a novel orally administered microtubulin inhibitor. Proc Am Soc Clin Oncol 2003;22:A520 (Poster presentation).

157. Kobayashi H, Hande KR, Berlin JD, et al. Phase I results of ABT-751, a novel microtubulin inhibitor, administered daily x 7 every 3 weeks. Proc Am Soc Clin Oncol 2004;23:A2079.

158. Benson AB, Kindler HL, Jodrell D, et al. Phase 2 study of ABT-751 in patients with refratory metastatic colorectal carcinoma (CRC). Proc Am Soc Clin Oncol 2005;24:A3537.

159. Hagey A, Figlin RA, Moldawer N, et al. Preliminary phase 2 results of ABT-751 in subjects with advanced renal cell carcinoma (RCC). Proc Am Soc Clin Oncol 2005;24:A4603.
160. Mauer AM, Szeto L, Belt RJ, et al. Preliminary results of a phase 2 study of ABT-751 in patients (pts) with taxane-refractory non-small cell lung carcinoma (NSCLC). Proc Am Soc Clin Oncol 2005;24: A7137.
161. Washington DK, Storniolo AV, Saleh M, et al. Phase 2 results of ABT-751 in subjects with taxane-refractory breast cancer: interim analysis. Proc Am Soc Clin Oncol 2005;24:A724.
162. Taylor C, Dragovich T, Simpson A, et al. A phase I and pharmacokinetic study of HMN-214 administered orally for 21 consecutive days, repeated every 28 days to patients with advanced solid tumors. Proc Am Soc Clin Oncol 2002;21:A419.
163. Von Hoff DD, Taylor C, Rubin S, et al. A phase I and pharmacokinetic study of HMN-214, a novel oral polo-like kinase inhibitor, in patients with advanced solid tumors. Proc Am Soc Clin Oncol 2005;23:A3034.
164. Towle MJ, Salvato KA, Budrow J, et al. *In vitro* and *in vivo* anticancer activities of synthetic macrocyclic ketone analogues of halichondrin B. Cancer Res 2001;61:1013–1021.
165. Bai RL, Paull KD, Herald CL, et al. Halichondrin B and homohalichondrin B, marine natural products binding in the vinca domain of tubulin. Discovery of tubulin-based mechanism of action by analysis of differential cytotoxicity data. J Biol Chem 1991;266:15,882–15,889.
166. Hirata Y, Uemura D. Halichondrins: antitumor polyether macrolides from a marine sponge. Pure Appl Chem 1986;58:701–710.
167. Pettit GR, Herald CL, Boyd MR, et al. Isolation and structure of the cell growth inhibitory constituents from the western Pacific marine sponge Axinella sp. J Med Chem 1991;34:3339–3340.
168. Aicher TD, Buszek KR, Fang FG, et al. Total synthesis of halichondrin B and norhalichondrin B. J Am Chem Soc 1992;114:3162–3164.
169. Synold TW, Morgan RJ, Newman EM, et al. A phase I pharmacokinetic and target validation study of the novel anti-tubulin agent E7389: a California Cancer Consortium trial. Proc Am Soc Clin Oncol 2005;24:A3036.
170. Rubin E, Rosen L, Rajeev V, et al. Phase I study of E7389 administered by 1 hour infusion every 21 days. Proc Am Soc Clin Oncol 2005;24:A2054.
171. Silberman S, O'Shaughnessy J, Vahdat L, et al. E7389, a novel anti-tubulin is safe and effective in patients with refractory breast cancer. San Antonio Breast Cancer Symposium. 2005; Poster 1063.
172. Loganzo F, Discafani CM, Annable T, et al. HTI-286, a synthetic analogue of the tripeptide hemiasterlin, is a potent antimicrotubule agent that circumvents P-glycoprotein-mediated resistance *in vitro* and *in vivo*. Cancer Res 2003;63:1838–1845.
173. Anderson HJ, Coleman JE, Andersen RJ, et al. Cytotoxic peptides hemiasterlin, hemiasterlin A and hemiasterlin B induce mitotic arrest and abnormal spindle formation. Cancer Chemother Pharmacol 1997;39:223–226.
174. Bai R, Durso NA, Sackett DL, et al. Interactions of the sponge-derived antimitotic tripeptide hemiasterlin with tubulin: comparison with dolastatin 10 and cryptophycin 1. Biochemistry 1999;38: 14,302–14,310.
175. Loganzo F, Annable T, Tan X, et al. Cells resistant to HTI-286 do not over-express P-glycoprotein but have low drug accumulation and a point mutation in a-tubulin. Proc Am Assoc Cancer Res 2003;44:A6535.
176. Poruchynsky MS, Kim JH, Nogales E, et al. Tumor cells resistant to a microtubule-depolymerizing hemiasterlin analog, HTI-286, have mutations in a- or β- tubulin and increased microtubule stability. Proc Am Assoc Cancer Res 2003;44:A2731.
177. Ratain MJ, Undevia S, Janisch L, et al. Phase 1 and pharmacological study of HTI-286, a novel antimicrotubule agent: Correlation of neutropenia with time above a threshold serum concentration. Proc Am Soc Clin Oncol 2003;22:A516 (Poster presentation).
178. Stevenson JP, Sun W, Gallagher M, et al. Phase I trial of the cryptophycin analogue LY355703 administered as an intravenous infusion on a day 1 and 8 schedule every 21 days. Clin Cancer Res 2002;8:2524–2529.
179. Golakoti T, Ogino J, Heltzel CE, et al. Structure determination, conformational analysis, chemical stability studies, and antitumor evaluation of the cryptophycins. Isolation of 18 new analogs from Nostoc sp. strain GSV 224. J Am Chem Soc 1995;117:12,030–12,049.
180. Moore RE. Cyclic peptides and depsipeptides from cyanobacteria: a review. J Ind Microbiol 1996;16: 134–143.

181. Panda D, DeLuca K, Williams D, et al. Antiproliferative mechanism of action of cryptophycin-52: kinetic stabilization of microtubule dynamics by high-affinity binding to microtubule ends. Proc Natl Acad Sci USA 1998;95:9313–9318.

182. Barbier P, Gregoire C, Devred F, et al. *In vitro* effect of cryptophycin 52 on microtubule assembly and tubulin: molecular modeling of the mechanism of action of a new antimitotic drug. Biochemistry 2001;40:13,510–13,519.

183. Panda D, Ananthnarayan V, Larson G, et al. Interaction of the antitumor compound cryptophycin-52 with tubulin. Biochemistry 2000;39:14,121–14,127.

184. Wagner MM, Paul DC, Shih C, et al. *In vitro* pharmacology of cryptophycin 52 (LY355703) in human tumor cell lines. Cancer Chemother Pharmacol 1999;43:115–125.

185. Sessa C, Weigang-Kohler K, Pagani O, et al. Phase I and pharmacological studies of the cryptophycin analogue LY355703 administered on a single intermittent or weekly schedule. Eur J Cancer 2002;38:2388–2396.

186. De Pas TM, Mandala M, Curigliano G, et al. Acute vulvar vestibulitis occurring during chemotherapy with cryptophycin analogue LY355703. Obstet Gynecol 2000;95:1030.

187. Groth G, Schott K, Ohnmacht U, et al. A phase II study of LY355703 (cryptophycine) as first-line therapy for stage IIIb or IV NSCLC subjects: preliminary analysis. Eur J Cancer 2001;37(suppl 6): S48(A167).

188. Edelman MJ, Gandara DR, Hausner P, et al. Phase 2 study of cryptophycin 52 (LY355703) in patients previously treated with platinum based chemotherapy for advanced non-small cell lung cancer. Lung Cancer 2003;39:197–199.

189. D'Agostino G, del Campo J, Mellado B, et al. A multicenter phase II study of the cryptophycin analog LY355703 in patients with platinum-resistant ovarian cancer. Int J Gynecol Cancer 2006;16: 71–76.

190. Iwasaki S, Kobayashi H, Furukawa J, et al. Studies on macrocyclic lactone antibiotics. VII. Structure of a phytotoxin "rhizoxin" produced by Rhizopus chinensis. J Antibiot (Tokyo) 1984;37:354–362.

191. Takahashi M, Iwasaki S, Kobayashi H, et al. Studies on macrocyclic lactone antibiotics. XI. Anti-mitotic and anti-tubulin activity of new antitumor antibiotics, rhizoxin and its homologues. J Antibiot (Tokyo) 1987;40:66–72.

192. Takahashi M, Iwasaki S, Kobayashi H, et al. Rhizoxin binding to tubulin at the maytansine-binding site. Biochim Biophys Acta 1987;926:215–223.

193. Takahashi M, Kobayashi H, Iwasaki S. Rhizoxin resistant mutants with an altered β-tubulin gene in Aspergillus nidulans. Mol Gen Genet 1989;220:53–59.

194. Hendriks HR, Plowman J, Berger DP, et al. Preclinical antitumour activity and animal toxicology studies of rhizoxin, a novel tubulin-interacting agent. Ann Oncol 1992;3:755–763.

195. Onozawa C, Shimamura M, Iwasaki S, et al. Inhibition of angiogenesis by rhizoxin, a microbial metabolite containing two epoxide groups. Jpn J Cancer Res 1997;88:1125–1129.

196. Aoki K, Watanabe K, Sato M, et al. Effects of rhizoxin, a microbial angiogenesis inhibitor, on angio-genic endothelial cell functions. Eur J Pharmacol 2003;459:131–138.

197. Bissett D, Graham MA, Setanoians A, et al. Phase I and pharmacokinetic study of rhizoxin. Cancer Res 1992;52:2894–2898.

198. Graham MA, Bissett D, Setanoians A, et al. Preclinical and phase I studies with rhizoxin to apply a pharmacokinetically guided dose-escalation scheme. J Natl Cancer Inst 1992;84:494–500.

199. Goh BC, Fleming GF, Janisch L, et al. Development of a schedule-dependent population pharmaco-dynamic model for rhizoxin without quantitation of plasma concentrations. Cancer Chemother Pharmacol 2000;45:489–494.

200. Tolcher AW, Aylesworth C, Rizzo J, et al. A phase I study of rhizoxin (NSC 332598) by 72-hour con-tinuous intravenous infusion in patients with advanced solid tumors. Ann Oncol 2000;11:333–338.

201. Hanauske AR, Catimel G, Aamdal S, et al. Phase II clinical trials with rhizoxin in breast cancer and melanoma. The EORTC Early Clinical Trials Group. Br J Cancer 1996;73:397–399.

202. Verweij J, Wanders J, Gil T, et al. Phase II study of rhizoxin in squamous cell head and neck cancer. The EORTC Early Clinical Trials Group. Br J Cancer 1996;73:400–402.

203. Kaplan S, Hanauske AR, Pavlidis N, et al. Single agent activity of rhizoxin in non-small-cell lung cancer: a phase II trial of the EORTC Early Clinical Trials Group. Br J Cancer 1996;73:403–405.

204. McLeod HL, Murray LS, Wanders J, et al. Multicentre phase II pharmacological evaluation of rhi-zoxin. EORTC early clinical studies (ECSG)/pharmacology and molecular mechanisms (PAMM) groups. Br J Cancer 1996;74:1944–1948.

205. Kerr DJ, Rustin GJ, Kaye SB, et al. Phase II trials of rhizoxin in advanced ovarian, colorectal and renal cancer. Br J Cancer 1995;72:1267–1269.

206. Rubenstein SM, Baichwal V, Beckmann H, et al. Hydrophilic, pro-drug analogues of T138067 are efficacious in controlling tumor growth *in vivo* and show a decreased ability to cross the blood brain barrier. J Med Chem 2001;44:3599–3605.

207. Schumaker RD, Mani S, Wright M, et al. Phase I study of T900607-sodium, a novel microtubule inhibitor, in patients with advanced solid tumors. Proc Am Soc Clin Oncol 2001;20:A442 (Poster presentation).

208. Shan B, Medina JC, Santha E, et al. Selective, covalent modification of beta-tubulin residue Cys-239 by T138067, an antitumor agent with *in vivo* efficacy against multidrug-resistant tumors. Proc Natl Acad Sci USA 1999;96:5686–5691.

209. Frankmoelle WP, Medina JC, Shan B, et al. Glutathione S-transferase metabolism of the antineoplastic pentafluorophenylsulfonamide in tissue culture and mice. Drug Metab Dispos 2000;28:951–958.

210. Schwendner SW, Hoffman LA, Thoolen MJ, et al. Efficacy of the novel tubulin binding agent, T900607, against human tumor xenografts in mice. Proc Am Assoc Cancer Res 2000;41:A1919.

211. Schwendner SW, Hoffman LA, Thoolen MJ, et al. Efficacy of combination therapy with the tubulin binding agent, T900607, against MX-1 human mammary tumor xenografts in mice. Proc Am Assoc Cancer Res 2000;41:A1914.

212. Zhang W, Timmermans P, Rosenblum M. *In vitro* and *in vivo* studies of the novel tubulin binding agent T900607: activity against MDR and MRP expressing human tumors. Proc Am Assoc Cancer Res 2000;41:A1706.

213. Budman DR, Berg WB, Spriggs DR, et al. A phase I study of a novel antimicrotubule agent: T138067. Clin Cancer Res 2000;6(suppl):A563.

214. Donehower RC, Schwartz GH, Wolff AC, et al. A phase I pharmacokinetic study of T138067 administered as a weekly 3-hour infusion. Proc Am Soc Clin Oncol 2001;20:A438 (Poster presentation).

215. Venook AP, Rowinsky E, Donehower RC, et al. Safety and pharmacokinetics (PK) of T138067 (T67) administered as a weekly 3-hour infusion in subjects with hepatocellular carcinoma (HCC) in a phase 1 study. Proc Am Soc Clin Oncol 2004;23:A4087.

216. Molpus K, Schwartz G, O'Dwyer P, et al. A phase I study of the anti-microtubule agent T138067-sodium administered daily x 5 every 3 weeks. Proc Am Soc Clin Oncol 2002;21:A415.

217. Leung TW, Feun L, Posey J, et al. A phase II study of T138067-sodium in patients (pts) with unresectable hepatocellular carcinoma (HCC). Proc Am Soc Clin Oncol 2002;21:A572.

218. Jahan TM, Sandler A, Burris H, et al. A phase II study of T138067-sodium in prior taxane-treated patients (pts) with locally advanced or metastatic non-small cell lung cancer (NSCLC). Proc Am Soc Clin Oncol 2002;21:A1282.

219. Kirby S, Gertler SZ, Mason W, et al. Phase 2 study of T138067-sodium in patients with malignant glioma: Trial of the National Cancer Institute of Canada Clinical Trials Group. Neurooncology 2005;7:183–188.

220. Posey J, Johnson P, Mok T, et al. Results of a phase 2/3 open-label, randomized trial of T138067 versus doxorubicin (DOX) in chemotherapy-naive, unresectable hepatocellular carcinoma (HCC). Proc Am Soc Clin Oncol 2005;A4035.

221. Lockhart AC, Mani S, Olsen R, et al. T900607-sodium administered daily x 5 as a 60-minute infusion every 3 weeks: a phase I study of T900607-sodium in patients (pts) with refractory cancer. Proc Am Soc Clin Oncol 2002;21:A417.

222. Stagg RJ, Killham P, Asif-Suleman S, et al. A phase I study of T900607-sodium administered weekly in patients with refractory cancer. Proc Am Soc Clin Oncol 2002;21:A416.

223. Gelmon KA, Belanger K, Soulieres D, et al. A phase I study of T900607 given once every 3 weeks in patients with advanced refractory cancers; National Cancer Institute of Canada Clinical Trials Group (NCIC-CTG) IND 130. Invest New Drugs 2005;23:445–453.

224. Garrett CR, Becerra CR, Chan R, et al. A phase II study of T900607 (T607) in subjects with chemotherapy-naive unresectable hepatocellular carcinoma (HCC). Proc Am Soc Clin Oncol 2004; 23:A4125.

225. Tolcher AW, Ochoa L, Hammond LA, et al. Cantuzumab mertansine, a maytansinoid immunoconjugate directed to the CanAg antigen: a phase I, pharmacokinetic, and biologic correlative study. J Clin Oncol 2003;21:211–222.

226. Blum RH, Kahlert T. Maytansine: a phase I study of an ansa macrolide with antitumor activity. Cancer Treat Rep 1978;62:435–438.
227. Blum RH, Wittenberg BK, Canellos GP, et al. A therapeutic trial of maytansine. Cancer Clin Trials 1978;1:113–117.
228. Cabanillas F, Bodey GP, Burgess MA, et al. Results of a phase II study of maytansine in patients with breast carcinoma and melanoma. Cancer Treat Rep 1979;63:507–509.
229. Cabanillas F, Rodriguez V, Hall SW, et al. Phase I study of maytansine using a 3-day schedule. Cancer Treat Rep 1978;62:425–428.
230. Chabner BA, Levine AS, Johnson BL, et al. Initial clinical trials of maytansine, an antitumor plant alkaloid. Cancer Treat Rep 1978;62:429–433.
231. Eagan RT, Creagan ET, Ingle JN, et al. Phase II evaluation of maytansine in patients with metastatic lung cancer. Cancer Treat Rep 1978;62:1577–1579.
232. Eagan RT, Ingle JN, Rubin J, et al. Early clinical study of an intermittent schedule for maytansine (NSC-153858): brief communication. J Natl Cancer Inst 1978;60:93–96.
233. Issell BF, Crooke ST. Maytansine. Cancer Treat Rev 1978;5:199–207.
234. Chari RVJ, Derr SM, Widdison WC, et al. SB 408075: A tumor-activated prodrug with exceptional activity against colon, pancreatic and lung tumor xenografts. Clin Cancer Res 1999;5(suppl):A462.
235. Liu C, Tadayoni BM, Bourret LA, et al. Eradication of large colon tumor xenografts by targeted delivery of maytansinoids. Proc Natl Acad Sci USA 1996;93:8618–8623.
236. Iwasaki S. Natural organic compounds that affect to microtubule functions. Yakugaku Zasshi. 1998; 118:112–126.
237. Smith SV. Technology evaluation: cantuzumab mertansine, ImmunoGen. Curr Opin Mol Ther 2004; 6:666–674.
238. Helft PR, Schilsky RL, Hoke FJ, et al. A phase I study of cantuzumab mertansine administered as a single intravenous infusion once weekly in patients with advanced solid tumors. Clin Cancer Res 2004;10:4363–4368.
239. Chari RVJ, Steeves RM, Xie Hongsheng, et al. Preclinical development of huN901-DM1: A tumor-activated prodrug directed against small cell lung cancer. Eur J Cancer 2000;36(suppl 6):A118.
240. Smith SV. Technology evaluation: huN901-DM1, ImmunoGen. Curr Opin Mol Ther 2005;7: 394–401.
241. Tolcher A, Forouzesh B, McCreery H, et al. A phase I and pharmacokinetic study of BB10901, a maytansinoid immunoconjugate, in CD56 expressing tumors. Eur J Cancer 2002;38(suppl 7):152(A509).
242. Fossella F, McCann J, Tolcher A, et al. Phase II trial of BB-10901 (huN901-DM1) given weekly for four consecutive weeks every 6 weeks in patients with relapsed SCLC and CD56-positive small cell carcinoma. Proc Am Soc Clin Oncol 2005;24:A7159.
243. Paterson I, Delgado O, Florence GJ, et al. 1,6-asymmetric induction in boron-mediated aldol reactions: application to a practical total synthesis of (+)-discodermolide. Org Lett 2003;5:35–38.
244. Gunasekera SP, Gunasekera M, Longley RE, et al. Discodermolide: a new bioactive polyhydroxylated lactone from the marine sponge Discodermia dissoluta. J Org Chem 1990;55:4912.
245. ter Haar E, Kowalski RJ, Hamel E, et al. Discodermolide, a cytotoxic marine agent that stabilizes microtubules more potently than taxol. Biochemistry 1996;35:243–250.
246. Balachandran R, ter Haar E, Welsh MJ, et al. The potent microtubule-stabilizing agent (+)-discodermolide induces apoptosis in human breast carcinoma cells—preliminary comparisons to paclitaxel. Anticancer Drugs 1998;9:67–76.
247. Martello LA, McDaid HM, Regl DL, et al. Taxol and discodermolide represent a synergistic drug combination in human carcinoma cell lines. Clin Cancer Res 2000;6:1978–1987.
248. Huang GS, McDaid HM, Kotla VR, et al. In vivo evaluation of combination treatment with taxol and discodermolide against ovarian carcinoma xenografts in nude mice. Proc Am Assoc Cancer Res 2003;44:LB–166.
249. Mita A, Lockhart AC, Chen TL, et al. A phase I pharmacokinetic (PK) trial of XAA296A (discodermolide) administered every 3 wks to adult patients with advanced solid malignancies. Proc Am Soc Clin Oncol 2004;23:A2025.

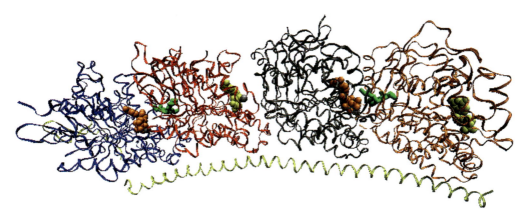

Color Plate 1. Fig. 3 Chapter 2: A ribbon diagram representation of ISA0.pdb, the structure of stathmin–colchicines-tubulin, was rendered with VMD. (*See* complete caption on p. 26.)

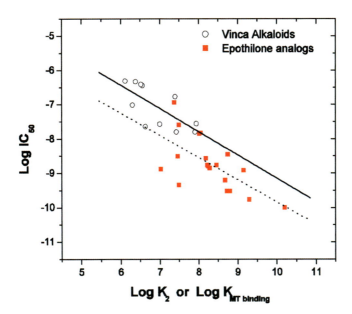

Color Plate 2. Fig. 6 Chapter 2: Log IC50 vs log K for two classes of antimitotic drugs. (*See* complete caption on p. 35.)

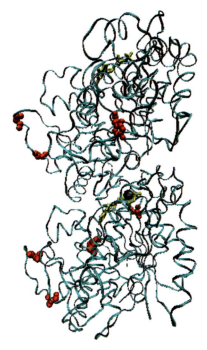

Color Plate 3. Fig. 7 Chapter 2: A ribbon diagram representation of 1JFF.pdb, the structure of tubulin derived from Zn-induced sheets, where the mutations that cause stabilization in cold adapted fish are highlighted in red, was rendered with VMD. (*See* complete caption on p. 40.)

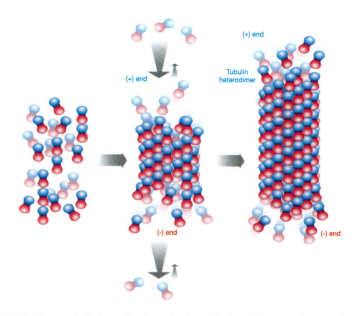

Color Plate 4. Fig. 1 Chapter 3: Polymerization of microtubules. (*See* complete caption on p. 48.)

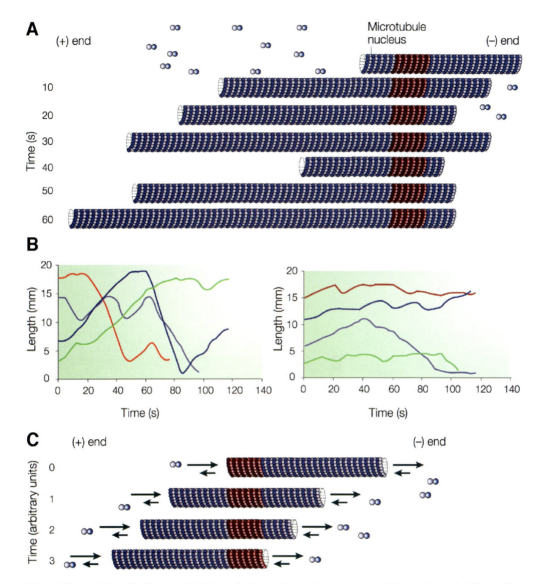

Color Plate 5. Fig. 2 Chapter 3: Microtubules undergo dynamic instability and treadmilling. (*See* complete caption on p. 50.)

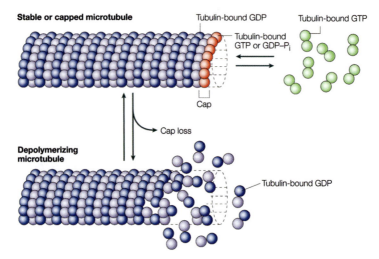

Color Plate 6. Fig. 3 Chapter 3: Dynamic instability and the GTP cap. (*See* complete caption on p. 55.)

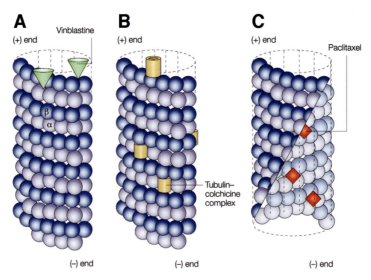

Color Plate 7. Fig. 4 Chapter 3: Antimitotic drugs bind to microtubules at diverse sites that can mimic the binding of endogenous regulators. (*See* complete caption on p. 58.)

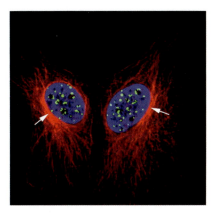

Color Plate 8. Fig. 5 Chapter 3: Microtubules in two human osteosarcoma cells in interphase of the cell cycle. (*See* complete caption on p. 63.)

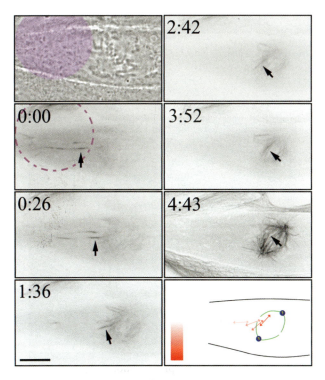

Color Plate 9. Fig. 7 Chapter 3: Caged fluorescence. (*See* complete caption on p. 65.)

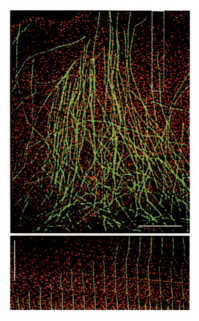

Color Plate 10. Fig. 8 Chapter 3: Speckle microscopy. (*See* complete caption on p. 66.)

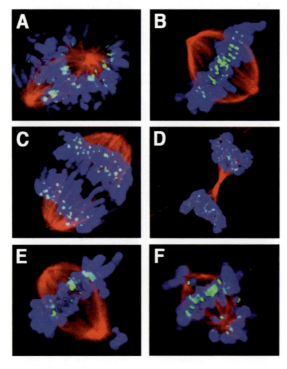

Color Plate 11. Fig. 10 Chapter 3: Organization of the mitotic spindle in human osteosarcoma cells throughout mitosis and in the presence of antimitotic drugs. (*See* complete caption on p. 70.)

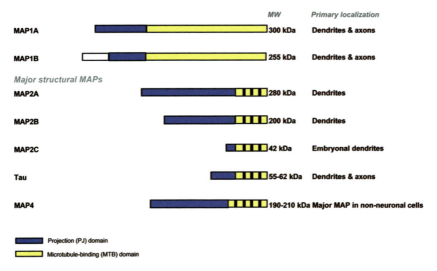

Color Plate 12. Fig. 1 Chapter 4: Schematic diagram of the major MAPs. (*See* complete caption on p. 85.)

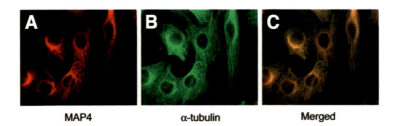

Color Plate 13. Fig. 2 Chapter 4: MAP binding to microtubules. (See complete caption on p. 85.)

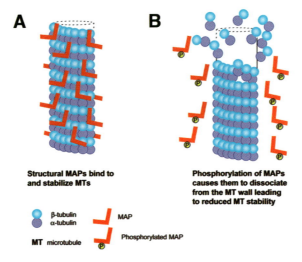

Color Plate 14. Fig. 3 Chapter 4: Schematic model of interactions of MAPs with microtubules. (*See* complete caption on p. 86.)

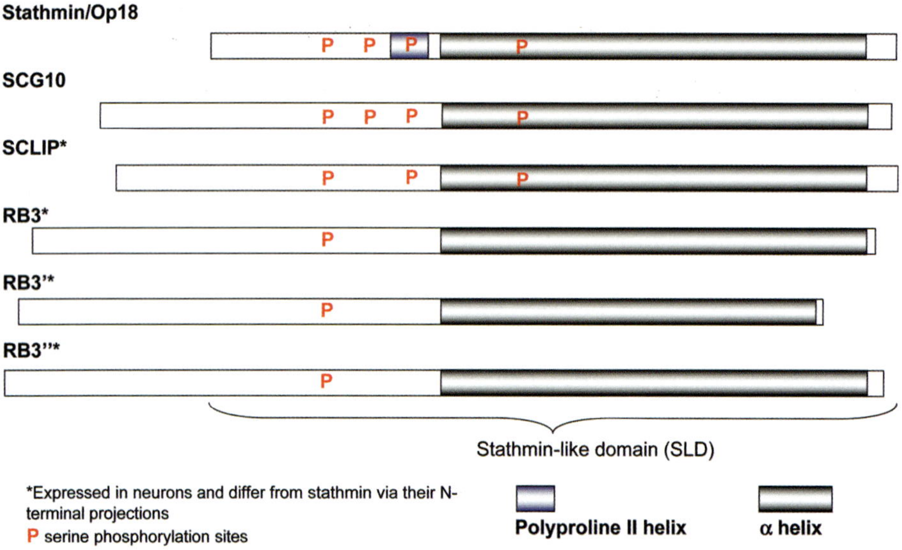

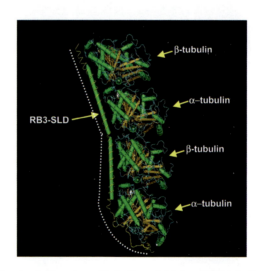

Stathmin/Op18

SCG10

SCLIP*

RB3*

RB3'*

RB3''*

Stathmin-like domain (SLD)

*Expressed in neurons and differ from stathmin via their N-terminal projections
P serine phosphorylation sites

Polyproline II helix   **α helix**

Color Plate 15. Fig. 4 Chapter 4: Schematic diagram of the stathmin-family of proteins. (*See* complete caption on p. 95.)

β-tubulin

α–tubulin

RB3-SLD

β-tubulin

α–tubulin

Color Plate 16. Fig. 5 Chapter 4: Structure of tubulin-colchicine: RB3-tubulin complex. (*See* complete caption on p. 96.)

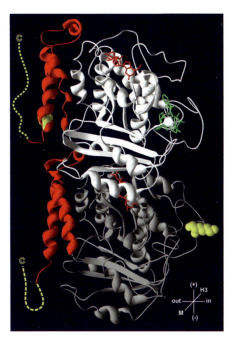

Color Plate 17. Fig. 1 Chapter 8: Model of the tubulin α/β-heterodimer. (*See* complete caption on p. 196.)

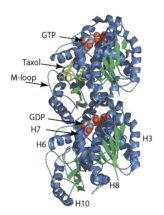

Color Plate 18. Fig. 1 Chapter 9: Ribbon diagram of the tubulin dimer structure. (*See* complete caption on p. 213.)

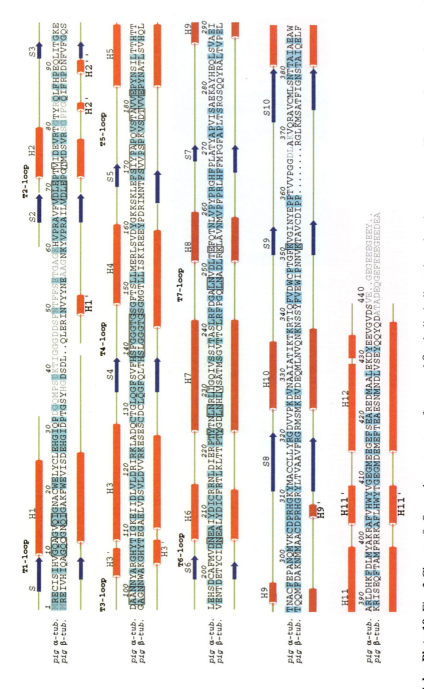

Color Plate 19. Fig. 2 Chapter 9: Secondary structure for α- and β-tubulin indicated on the pig sequences. (*See* complete caption on p. 215.)

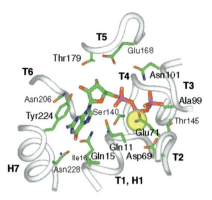

Color Plate 20. Fig. 3 Chapter 9: Topological structures around the nonexchangeable nucleotide site. (*See* complete caption on p. 215.)

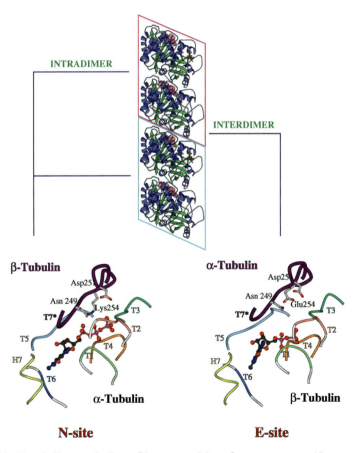

Color Plate 21. Fig. 4 Chapter 9: Protofilament and interface structures. (*See* complete caption on p. 216.)

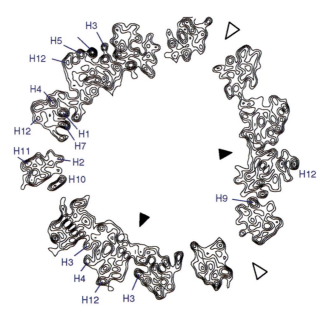

Color Plate 22. Fig. 5 Chapter 9: Cross-section of the three-dimensional microtubule density map. (*See* complete caption on p. 218.)

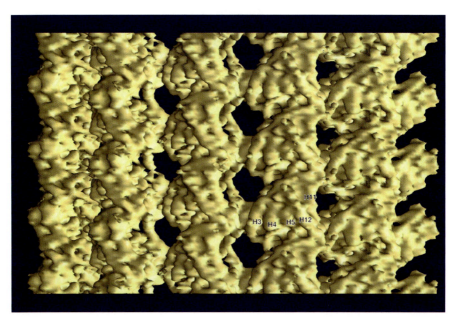

Color Plate 23. Fig. 6 Chapter 9: Surface view of the microtubule density map, seen from the outside with the plus end up. (*See* complete caption on p. 219.)

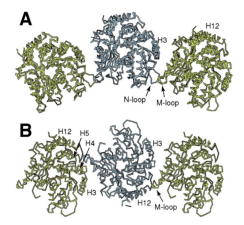

Color Plate 24. Fig. 7 Chapter 9: Protofilament arrangements and contacts in **(A)** microtubules, **(B)** the Zn-sheets. (*See* complete caption on p. 220.)

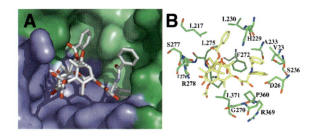

Color Plate 25. Fig. 8 Chapter 9: The Taxol-binding site. (*See* complete caption on p. 221.)

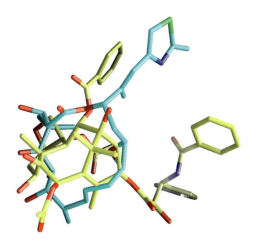

Color Plate 26. Fig. 9 Chapter 9: Superposition of Taxol (yellow) and epothilone-A (blue) in the conformations and orientations in which they bind tubulin. (*See* complete caption on p. 222.)

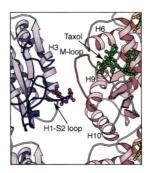

Color Plate 27. Fig. 10 Chapter 9: Detail of the interface between protofilaments, as seen from the inside of the microtubule. (*See* complete caption on p. 223.)

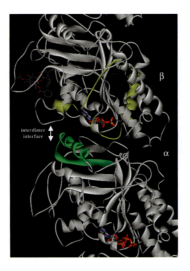

Color Plate 28. Fig. 4 Chapter 10: Model of the HTI-286 and vinblastine photolabeling domains in tubulin. (*See* complete caption on p. 242.)

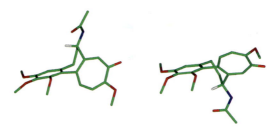

Color Plate 29. Fig. 4 Chapter 11: Structures of (7*S*,a*R*)-(_)-colchicine (left) and (7*R*,a*S*)-(+)-colchicine (right), illustrating the atropisomers of the colchicine ring system. (*See* complete caption on p. 268.)

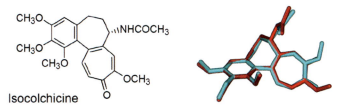

Color Plate 30. Fig. 6 Chapter 11: Isocolchicine vs colchicine. (*See* complete caption on p. 269.)

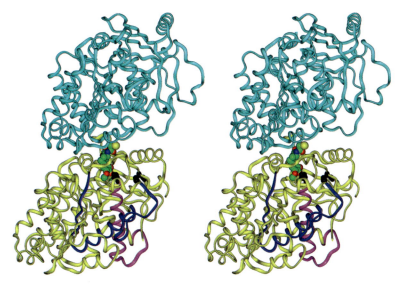

Color Plate 31. Fig. 8 Chapter 11: Stereoview of one of the tubulin-colchicine complexes from the crystal structure of tubulin/ colchicine:stathmin-like domain complex (PDB 1SA0). (*See* complete caption on p. 272.)

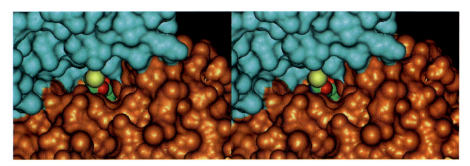

Color Plate 32. Fig. 9 Chapter 11: Stereoview of the colchicine-binding site of tubulin/colchicine: stathmin-likedomain complex (PDB 1SA0). (*See* complete caption on p. 273.)

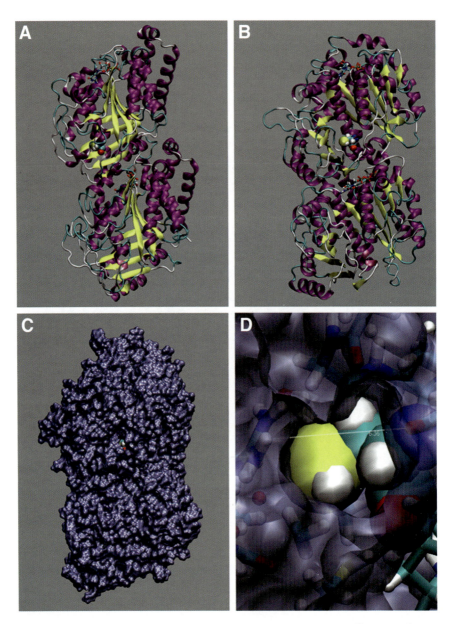

Color Plate 33. Fig. 1 Chapter 12: The position and exposure of cysteine 239 in β-tubulin (βcys239). (*See* complete caption on p. 286.)

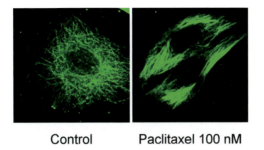

Control   Paclitaxel 100 nM

Color Plate 34. Fig. 3 Chapter 13: Paclitaxel-induced tubulin polymerization *in vitro*. (*See* complete caption on p. 311.)

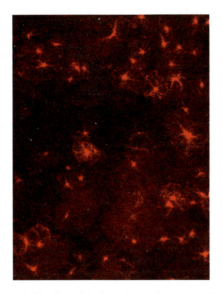

Color Plate 35. Fig. 4 Chapter 13: Paclitaxel-induced mitotic arrest with aster formation *in vivo*. (*See* complete caption on p. 312.)

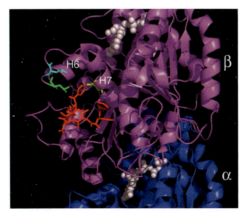

Color Plate 36. Fig. 3 Chapter 14: Structure of an αβ-tubulin heterodimer. (*See* complete caption on p. 347.)

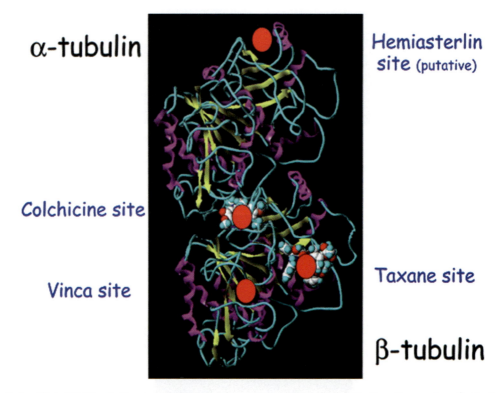

Color Plate 37. Fig. 1 Chapter 15: An overview of some of the ligand-binding sites on the αβ-tubulin dimer. (*See* complete caption on p. 359.)

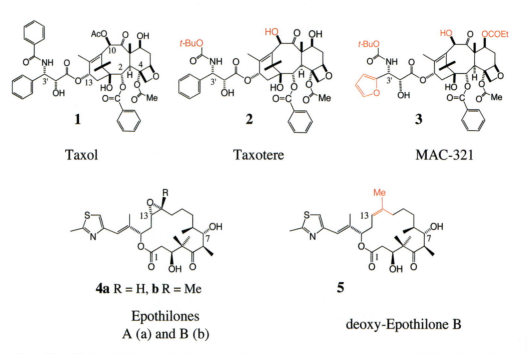

Color Plate 38. Fig. 2 Chapter 15: Structures of taxane-binding site microtubule stabilizing drugs. (*See* complete caption on p. 362.)

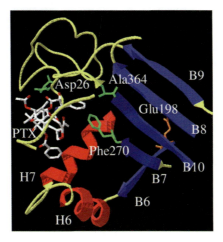

Color Plate 39. Fig. 3 Chapter 15: The Asp26, Phe270, Ala364, and Glu198 residues are mutated following drug selection with Taxol™. (*See* complete caption on p. 367.)

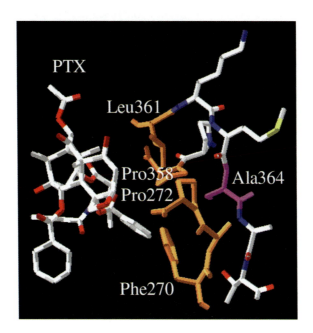

Color Plate 40. Fig. 4 Chapter 15: The Ala364 cluster representing the βAla364Thr mutation seen in the ovarian carcinoma 1A9 cells. (*See* complete caption on p. 368.)

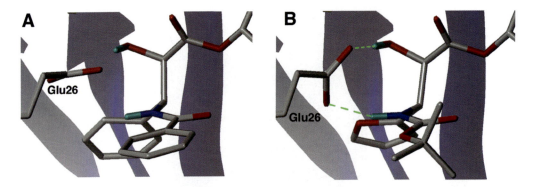

Color Plate 41. Fig. 5 Chapter 15: Binding models of Taxol™ and taxotere in the taxane-binding pocket of the βAsp26Glu mutant seen in the epidermoid carcinoma KB-31 cells. (*See* complete caption on p. 369.)

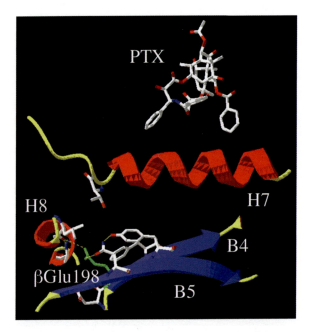

Color Plate 42. Fig. 6 Chapter 15: Representation of the locus of the βGlu198Gly mutation induced by Taxol™-selection in the MDA-MB-231 breast cancer cells. (*See* complete caption on p. 369.)

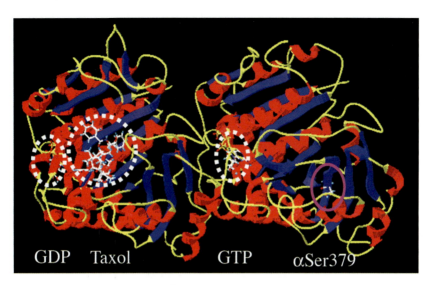

Color Plate 43. Fig. 7 Chapter 15: The αβ-tubulin dimer illustrating the spatial relationships of GDP, GTP, Taxol™, and αSer379 in the context of the Taxol™-induced αSer379Arg mutation in the A549 lung cancer cells. (*See* complete caption on p. 370.)

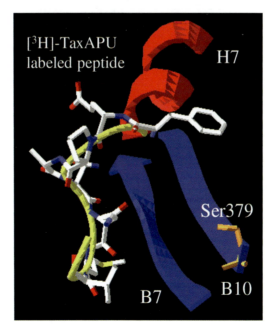

Color Plate 44. Fig. 8 Chapter 15: The αSer379Arg mutation is located on strand B10 in α-tubulin. (*See* complete caption on p. 371.)

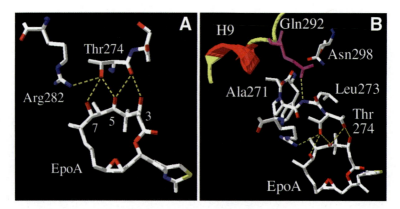

Color Plate 45. Fig. 9 Chapter 15: Environment around the C-3 to C-7 oxygen substituents of EpoA bound to β-tubulin. (*See* complete caption on p. 372.)

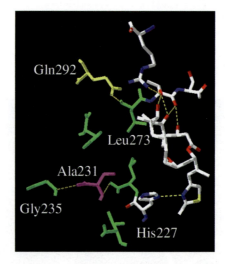

Color Plate 46. Fig. 10 Chapter 15: At lower left, Ala231 (magenta) surrounded by a hydrophobic pocket (green) is subjected to the Ala231Thr mutation, which presumably perturbs the His227 interaction with dEpoB (represented by EpoA at their binding pocket). (*See* complete caption on p. 374.)

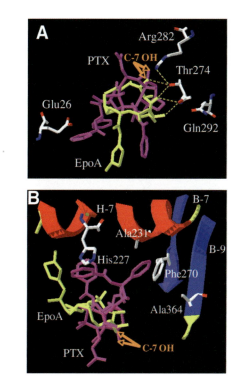

Color Plate 47. Fig. 11 Chapter 15: Taxol™ and EpoA tubulin-bound structures overlap as determined by the corresponding electron crystallographic structures. (*See* complete caption on p. 376.)

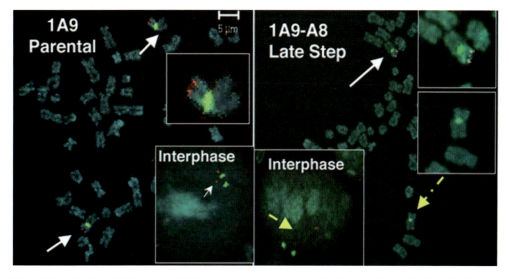

Color Plate 48. Fig. 12 Chapter 15: Fluorescence *in situ* hybridization analysis of 1A9 cells Depicts tubulin LOH in the drug-resistant 1A9-A8 cells. (*See* complete caption on p. 379.)

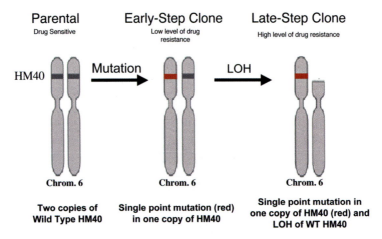

Color Plate 49. Fig. 13 Chapter 15: Schematic representation of Evolution of Microtubule-Targeting Drug-Induced Resistance. (*See* complete caption on p. 379.)

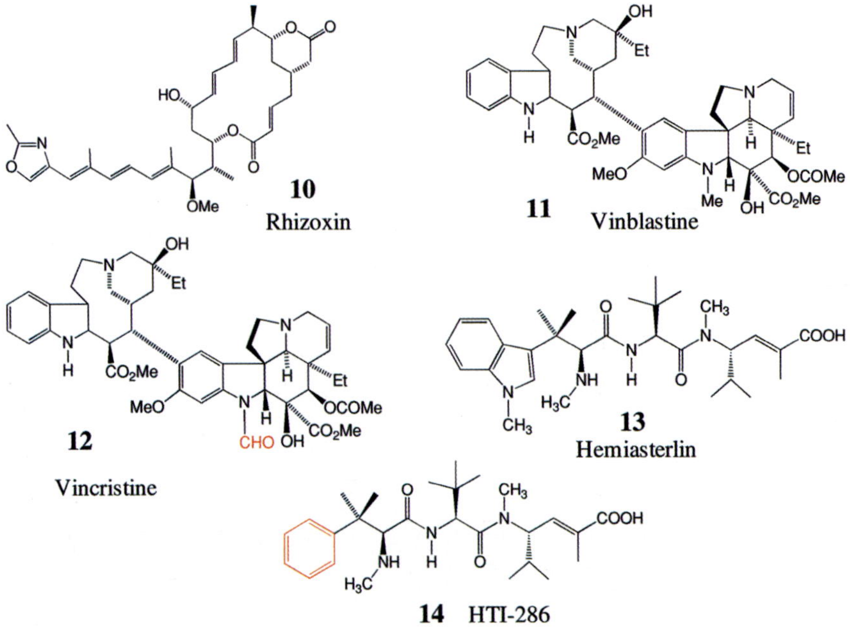

10
Rhizoxin

11
Vinblastine

12
Vincristine

13
Hemiasterlin

14   HTI-286

Color Plate 50. Fig. 15 Chapter 15: Structures of *Vinca*-binding site microtubule-destabilizing drugs. (*See* complete caption on p. 380.)

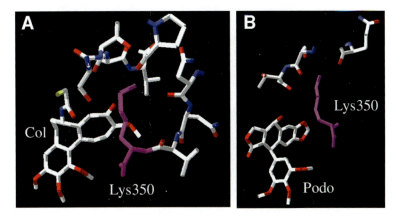

Color Plate 51. Fig. 16 Chapter 15: The colchicine and podophyllotoxin binding sites as determined by X-ray crystallography. (*See* complete caption on p. 381.)

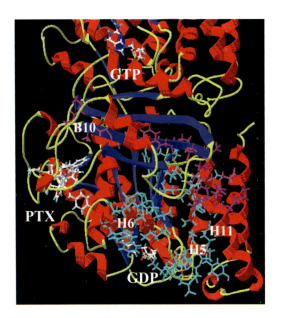

Color Plate 52. Fig. 17 Chapter 15: Photoaffinity labeled peptides and mutations around the *Vinca* site. (*See* complete caption on p. 383.)

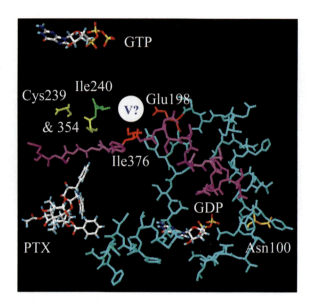

Color Plate 53. Fig. 18 Chapter 15: Closeup of the photoaffinity labeled peptides and mutations around the *Vinca* site. (*See* complete caption on p. 384.)

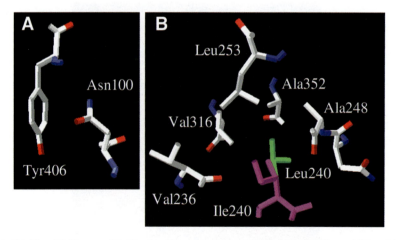

Color Plate 54. Fig. 19 Chapter 15: The local environments of the βAsn100Ile and βLeu240Ile mutations. (*See* complete caption on p. 384.)

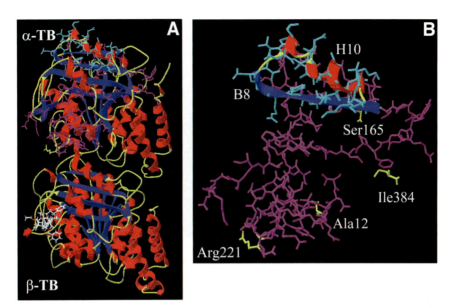

Color Plate 55. Fig. 20 Chapter 15: Photoaffinity labeled peptides and mutations around the α-tubulin hemiasterlin-binding site. (*See* complete caption on p. 386.)

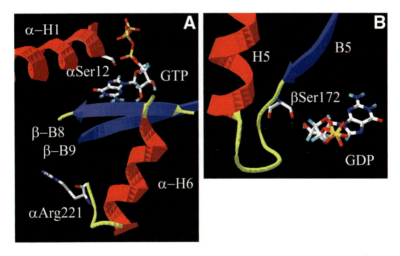

Color Plate 56. Fig. 21 Chapter 15: Local environments of the αAla12Ser, βSer172Ala, and βArg221His mutations. (*See* complete caption on p. 387.)

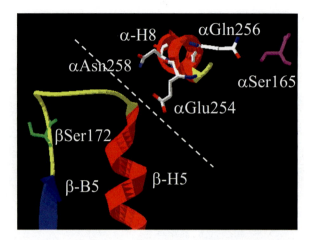

Color Plate 57. Fig. 22 Chapter 15: A potential interaction train for αSer165Pro and βSer172Ala across the αβ-tubulin dimer interface (dotted line); αSer165 (magenta), βSer172 (green). (*See* complete caption on p. 388.)

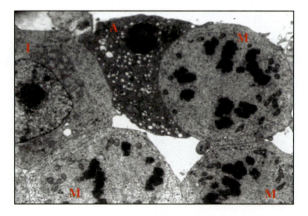

Color Plate 58. Fig. 1 Chapter 18: Cellular morphology of human mitotic and apoptotic cells. (*See* complete caption on p. 480.)

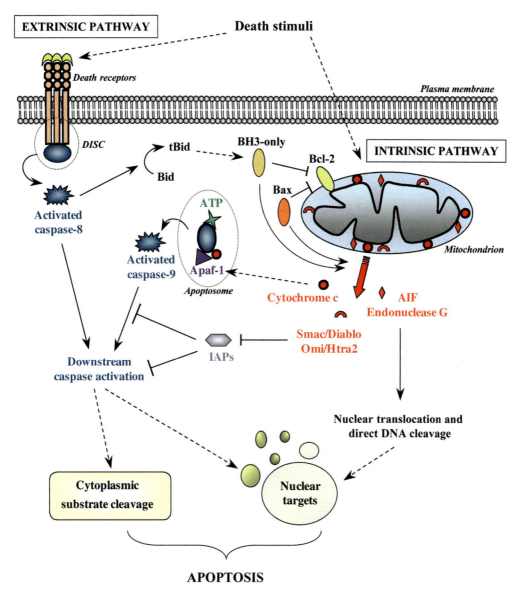

Color Plate 59. Fig. 2 Chapter 18: Two main pathways to apoptosis. (*See* complete caption on p. 483.)

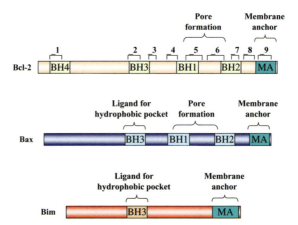

Color Plate 60. Fig. 3 Chapter 18: Linear structure of antiapoptotic Bcl-2, proapoptotic Bax, and proapoptotic Bim. (*See* complete caption on p. 485.)

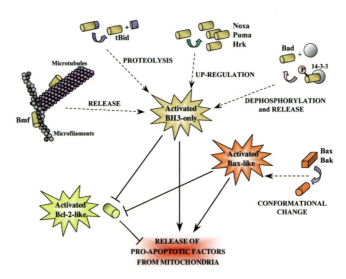

Color Plate 61. Fig. 4 Chapter 18: Potential role of Bcl-2 family proteins in the mitochondrial permeability. (*See* complete caption on p. 486.)

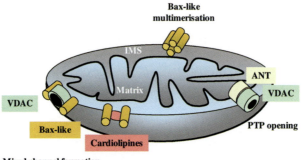

Color Plate 62. Fig. 5 Chapter 18: Models for release of mitochondrial intermembrane space factors. (*See* complete caption on p. 488.)

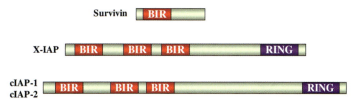

Color Plate 63. Fig. 6 Chapter 18: Schematic representations of mammalian IAPs. (*See* complete caption on p. 490.)

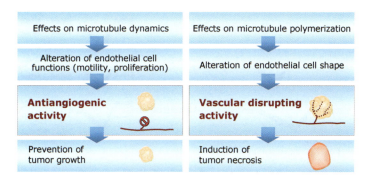

Color Plate 64. Fig. 1 Chapter 19: Tubulin-targeting agents. (*See* complete caption on p. 520.)

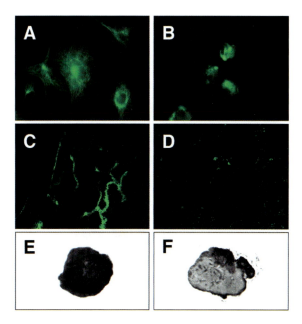

Color Plate 65. Fig. 2 Chapter 19: Effects of vascular disrupting agents. (*See* complete caption on p. 521.)

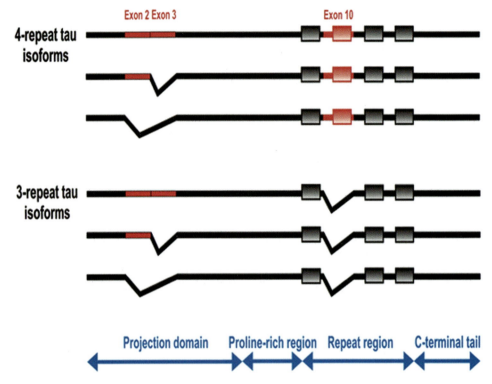

Color Plate 66. Fig. 1 Chapter 21: Domain structure of the 6-tau isoforms expressed in the central nervous system. (*See* complete caption on p. 561.)

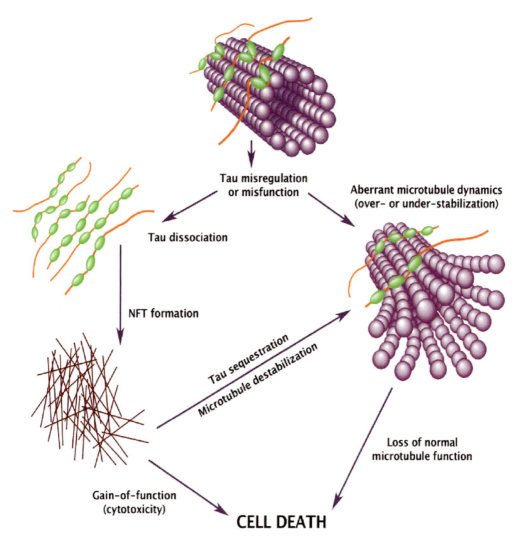

Tau misregulation
or misfunction

Aberrant microtubule dynamics
(over– or under–stabilization)

Tau dissociation

NFT formation

Tau sequestration
Microtubule destabilization

Loss of normal
microtubule function

Gain-of-function
(cytotoxicity)

CELL DEATH

Color Plate 67. Fig. 4 Chapter 21: Three possible pathways of tau-mediated neuronal cell death. (*See* complete caption on p. 567.)

18 Microtubule Damaging Agents and Apoptosis

Manon Carré and Diane Braguer

Contents

Summary

Nowadays, molecules that affect microtubule functions, the so-called Microtubule-Damaging Agents (MDAs), constitute a class of anti-cancer drugs largely used in the clinics. Interest for MDAs is accompanied with advances in the fundamental understanding of their mechanism of action, including tumor cell death induction. MDAs have shown a high ability to induce apoptosis, programmed and tightly regulated cell death that is not or insufficiently activated in cancer cells. Here, the major intracellular signaling cascades responsible for apoptosis are first reviewed, focusing on the mitochondrial pathway. Then, the molecular and cellular mechanisms involved in the pro-apoptotic activity of MDAs are precised. Especially, the modulation of Bcl-2 family members that triggers mitochondrial membrane permeabilization is described, as well as the release of pro-apoptotic factors from the intermembrane space, and the final activation of caspases that leads to the biochemical destruction of the cell. Since MDAs inhibit microtubule functions and generally perturbate cell cycle progression, microtubule-linked proteins and cell-cycle progression regulators are proposed as candidates to control the apoptotic machinery. Other factors, such as MAPKs, stress markers, and survival factors, are also reviewed as modulators of the cell survival/death balance. Lastly, the direct effects of MDAs on mitochondria and the possible involvement of the tubulin/microtubule system in this phenomenon are discussed. Altogether, these data highlight the crucial role played by mitochondria in MDA-induced apoptosis, and propose that mitochondria should be investigated as a target of choice to improve cancer therapy.

Key Words: Apoptosis; mitochondria; microtubule; tumor cell; anti-cancer drug; taxane; *Vinca* alkaloid.

From: *Cancer Drug Discovery and Development: The Role of Microtubules in Cell Biology, Neurobiology, and Oncology* Edited by: Tito Fojo © Humana Press, Totowa, NJ

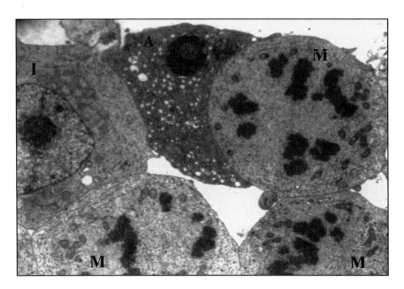

Fig. 1. Cellular morphology of human mitotic and apoptotic cells. Neuroblastoma SK-N-SH cells were incubated with paclitaxel for 24 h and visualized by transmission electron microscopy. One cell remains apparently intact (I), three are arrested in mitosis (M), as indicating by the distribution of chromosomes on the metaphasic plate, and one is apoptotic (A), with a highly condensed DNA and numerous vacuoles. To view this figure in color, see the insert and the companion CD-ROM.

1. INTRODUCTION

The word "apoptosis" comes from the ancient Greek, meaning the falling off of leaves from a tree in autumn. The phenomenon of cellular apoptosis (Fig. 1) was first described by Kerr in 1972 (1). Apoptosis is associated with biochemical and physical changes involving the cytoplasm, the nucleus, and the plasma membrane in an orderly fashion. Apoptotic cells are characterized by loss of cell membrane phospholipid asymmetry, condensation of chromatin, reduction in nuclear size, internucleosomal DNA cleavage (recognized as a DNA ladder on agarose gels), shrinkage of the cell, membrane blebbing, and breakdown of the cell into membrane-bound apoptotic bodies. These bodies are subsequently phagocytized. Normal apoptosis is genetically regulated in *Caenorhabditis elegans* and mammalian homologs of the *C. elegans* death genes have been identified (2,3). Apoptosis is a programmed and highly regulated physiological response in which cell death occurs without collateral damage to surrounding tissues, in contrast with necrosis. For maintaining cell homeostasis, a balance between the increase (by proliferation) and the decrease (by terminal differentiation and cell death) in cell number has to be tightly regulated (4,5). Apoptosis is one of today's more active fields of biomedical research because of the recognition that many deseases involve too much apoptosis (i.e., AIDS and neurodegenerative diseases such as Parkinson's disease and Alzheimer's disease) or too little apoptosis (i.e., autoimmune diseases and cancer) (6,7).

Diverse signals are able to trigger apoptosis including ultraviolet (UV) or γ-irradiation, oxidative damage, cytokines such as tumor necrosis factor (TNF)-α and transforming growth factor-β, growth factor withdrawal as well as chemotherapeutic drugs. Studies performed during a decade demonstrated that all classes of currently available anticancer drugs, including those that target DNA integrity or cytokinesis, induce apoptosis in susceptible cell types, but apoptotic pathways are not clearly understood. The induction of apoptosis is correlated with tumor response and clinical outcome in cancer patients. Resistance to apoptosis causes a decrease in the sensitivity of cancer cells to drugs, resulting in the failure of chemotherapy. The realization that apoptosis is inhibited in cancer cells, coupled with improved understanding of apoptotic pathways, leads to development of new therapies that directly target the apoptotic machinery.

Paclitaxel has been the most studied microtubule-damaging agent (MDA), and many of its actions in signaling and apoptosis are thought to be extended to other MDAs that affect the tubulin-microtubule equilibrium. Among MDAs, taxanes interact with microtubules (polymerized tubulin) and prevent microtubule depolymerization, whereas the *Vinca* alkaloids interact both with dimeric tubulin and microtubules and prevent polymerization. Both classes of MDAs suppress microtubule dynamics. These molecules were initially characterized as mitotic inhibitors, and their anticancer activity was attributed to their capacity to inhibit cell division through microtubule network disruption *(8–10)*. Mitotic arrest induced by MDAs is believed to be a consequence of suppression of mitotic spindle microtubule dynamics, which prevents proper chromosome alignment at the metaphase plate, resulting in a sustained block at the metaphase–anaphase transition and ultimately an apoptotic cell death *(11)*. However, the biochemical events that lead to apoptosis downstream of inhibition of microtubule dynamics are not understood *(12)*.

Although the arrest of the cell cycle at mitosis is often related to MDA-induced apoptosis, substantial evidence indicates that it is not the only cell cycle disturbance that can lead to cell death. Actually, apoptosis can occur either directly after a sustained mitotic arrest or following an aberrant exit from mitosis into a multinucleate interphase *(13)*, or even following a blockage in other cell-cycle phases. Vincristine induces apoptotic cell death in chronic lymphocytic leukemia B-cells that are in G0/G1 phase of the cell cycle indicating that the antimitotic action of *Vinca* alkaloids cannot explain the death of these cells *(14)*. The relationship between the distinct phases of cell-cycle arrest and apoptosis varies according to drug concentration. At low-paclitaxel concentrations, A549 cell death may occur after an aberrant mitosis by a Raf-1-independent pathway, whereas at higher paclitaxel concentrations, death of the same cells may be the result of a terminal mitotic arrest occurring by a Raf-1-dependent pathway *(15)*. Low concentrations of paclitaxel can induce a p53- and/or p21-dependent G1 and G2 arrest before apoptosis *(16)*.

The relationship among alteration of the microtubule network, the slowing or complete block at mitosis, and ensuing apoptosis, may also depend on the cell line. Indeed, Blajeski et al. *(17)* showed that high concentrations of nocodazole or vincristine block seven of ten breast cancer lines in mitosis, whereas they cause a p21-associated G1 and G2 arrest in the other three. Moreover, mitosis and apoptosis may be uncoupled according to the differentiation status of the cell. In colon adenocarcinoma HT29-D4 proliferating cells, paclitaxel induces mitosis and then apoptosis whereas in differentiated HT29-D4 cells, apoptosis is only induced at very high drug concentration without significant change in the microtubule network or mitotic block *(18)*. Most of MDAs promote similar signal transduction events leading to apoptosis although it is conceivable that different drug-binding sites and different ways of inhibiting of cell dynamics may result in specific alterations in certain signaling pathways.

In the Subheading 2, of this chapter, the current state of knowledge of the apoptotic signal pathways in cancer cells is described, especially detailing those involved in MDA-induced apoptosis. Then how MDAs can trigger the cascade of apoptotic events is explained, focusing in the Subheading 4, on their direct action on mitochondria.

2. DESCRIPTION OF APOPTOTIC SIGNALING PATHWAYS

2.1. Two Major Apoptotic Routes

Within the past few years, most researchers who have focused on apoptosis have put mitochondria at the center of the apoptotic pathways *(19)*. These organelles, well-known as the main source of cellular energy, play a critical role in the regulation of apoptosis by acting as a reservoir for proapoptotic proteins such as cytochrome-*c*, Smac/Diablo (second mitochondria-derived activator of caspases/direct inhibitors of apoptosis [IAP]-binding protein), apoptosis-inducing factor (AIF), endonuclease G, and procaspases. The release of these proteins activates appropriate downstream signaling cascades leading to the morphological hallmarks of apoptosis. Cytotoxic agents induce two main apoptotic signaling pathways according to the cascade of events upstream of mitochondria, the extrinsic and the intrinsic pathways *(20,21)* (Fig. 2).

2.1.1. DEATH RECEPTOR PATHWAY

The extrinsic pathway is also called the death receptor pathway. Death receptors identified to date include Fas/CD95, TNFR 1, DR3, DR4, DR5, and DR6 of the TRAIL receptor superfamily *(22)*. Activation of the death receptor, by binding of a specific ligand to its extracellular domain, induces the formation of the death-inducing signaling complex *(23,24)*, which serves for the activation of caspase-8 and the subsequent apoptotic signaling pathway *(25)*. The extrinsic pathway is independent of mitochondria with caspase-8 directly activating downstream caspases such as caspase-3 *(26)*. In this case, apoptosis cannot be inhibited by Bcl-2-like survival proteins *(27)*. However, the extrinsic pathway cross-interacts with the mitochondrial pathway when caspase-8 cleaves Bid, leading to the translocation of the truncated form of Bid to mitochondria where it works to release cytochrome-*c* *(28)*. The death receptor pathway has been proposed to play a role in the development and functioning of the immune system *(29)*, and also in apoptosis induced by some anticancer drugs such as doxorubicin in neuroblastoma *(30)*. However, it does not seem to be activated by MDAs *(31–36)*, even though these drugs can modify the level of expression of death receptors and their ligands *(35,36)*.

2.1.2. INTRINSIC MITOCHONDRIAL APOPTOTIC PATHWAY

By contrast, the intrinsic pathway is mostly triggered by death receptor-independent apoptotic stimuli such as UV and γ-irradiation, chemotherapeutic drugs, viruses, bacteria, the removal of cytokines, neurotrophins, and growth factors, or by anoikis. In the intrinsic pathway, most of the cellular apoptotic signals converge on the mitochondria, triggering the release of apoptogenic proteins through an increase in mitochondrial membrane permeability. This process is under the control of the proapoptotic/antiapoptotic ratio of the Bcl-2 family members. However, it involves a mechanism not well-defined as discussed later. The proteins released into cytosol promote apoptosis either through caspase activation (cytochrome-*c*, Smac/Diablo, Omi/HtrA2), or in a caspase-independent manner (endonuclease G, AIF). Caspase activation leads to the cleavage of a number of nuclear and cytoplasmic substrates, including those responsible for nuclear integrity, cell-cycle progression, and DNA repair.

It has now become evident that mitochondria act as integrators of proapoptotic signals, transducing them to the final execution machinery of apoptosis. The following

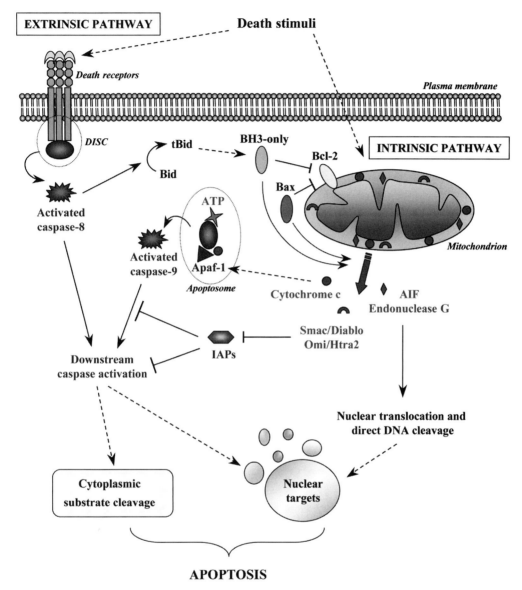

Fig. 2. Two main pathways to apoptosis. Activation of the death receptor pathway, by binding of death-inducing ligands, directly results in the caspase cascade initiation and subsequent substrate cleavage, without involvement of mitochondria. The intrinsic pathway activation leads to the mitochondrial permeabilization and thus to the release of proapoptotic factors. Among them, cytochrome-*c* forms a cytosolic complex (apoptosome) with the apoptosis activating factor-1 (Apaf-1), ATP, and the procaspase-9, which results in the caspase cascade initiation. Smac/Diablo and Omi/Htra2 are also released from mitochondria and enhance the caspase activity by preventing action of the inhibitor of apoptosis proteins (IAPs). The intrinsic pathway may also operate through a caspase-independent mechanism, which involve the release from mitochondria and translocation to the nucleus of AIF or endonuclease G. Mitochondrial integrity is maintained by Bcl-2-like survival factors, whereas BH3-only and Bax-like proteins induce mitochondrial permeabilization upon apoptotic signals (*see* Fig. 3). Finally, caspase-8 can make a link between the two pathways by activation of Bid in a truncated form. To view this figure in color, see the insert and the companion CD-ROM.

Table 1
The Bcl-2 Family Members

Bcl-2 family groups	Antiapoptotic Bcl-2-like	Proapoptotic Bax-like	Proapoptotic BH3-only
	Bcl-2	Bax	Bik/Nbk
	Bcl-xL	Bak	Blk
	Bcl-w	Bok/Mtd	Hrk/DP5
	Mcl-1	Bcl-x	BNIP3
	A1/Bfl-1		BimL/Bod
	NR-13		Bad
	Boo/Diva/Bcl1-L-10		Bid
	Bcl-B		Noxa
			PUMA/Bbc3
			Bmf

Proteins are classified in three groups: one corresponds to survival factors (Bcl-2-like) and the two others to apoptosis-inducers (Bax-like and BH3-only proteins).

Subheading 2.2, will describe the Bcl-2 family proteins involved upstream of mitochondria, then the changes in mitochondrial permeability leading to the release of proapoptotic factors, and finally the subsequent events downstream of mitochondria.

2.2. Bcl-2 Family Proteins

Higher eukaryotes possess up to 30 Bcl-2 family homologs that can be grouped into three categories: one group of Bcl-2-like survival factors and two groups that promote cell death named the Bax-like and the BH3-only families (Table 1) (reviewed in refs. *37,38*). Their structure reveals a structural homology with bacterial pore-forming toxins forming pores, suggesting that members of the Bcl-2 family may form membrane pores *(39)*. Two post-translational modifications, phosphorylation, and proteolytic cleavage have regulatory function on the activity of most Bcl-2 members.

2.2.1. BCL-2-LIKE SURVIVAL FACTORS

Overexpression of Bcl-2-like proteins has been largely shown to suppress apoptosis induced by several factors, including MDAs *(40)*. Members of this subfamily contain 3–4 Bcl-2 homology domains, which are required for their survival functions (Fig. 3). These domains mediate the interaction of Bcl-2-like proteins with other protein partners as proapoptotic members of the Bcl-2 family. The BH1–3 domains form a hydrophobic groove that constitutes the functional part of the protein. The N-terminal BH4 domain stabilizes this structure, but is not present in all Bcl-2-like members (reviewed in ref. *38*). The C-terminal domain helps to insert these proteins at the cytoplasmic face of organelles such as the outer mitochondrial membrane.

Bcl-2-like members maintain mitochondrial integrity in the absence of any apoptotic signal. Through their hydrophobic groove, they can interact with proteins that contain a BH3 region. Moreover, proteins that do not contain a BH3 domain have been found to interact with Bcl-2, either in the hydrophobic groove or in the BH4 domain *(41)*. The antiapoptotic function of the Bcl-2-like protein family is regulated by phosphorylation of residues in the loop domain. Once hyperphosphorylated during mitosis, Bcl-2-like members become ineffective in preventing apoptosis induced by anticancer drugs *(42)*.

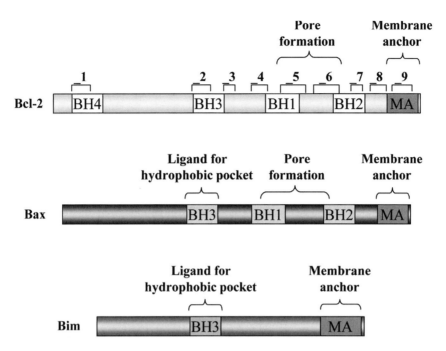

Fig. 3. Linear structure of antiapoptotic Bcl-2, proapoptotic Bax, and proapoptotic Bim. BH: Bcl-2 homology domain. α5/α6 helix are pore-forming units and the C-terminal hydrophobic α-9 anchors the proteins in intracellular membranes. The BH3 domain in the proapoptotic members is a ligand for the hydrophobic pocket formed by the BH1–3 domains of the survival factors. To view this figure in color, see the insert and the companion CD-ROM.

Although the kinase responsible for Bcl-2 phosphorylation has not been clearly identi-fied, good candidates are cyclin-dependent kinases (CDKs) and mitogen-activated pro-tein kinase (MAPKs)—especially [c-Jun N-terminal kinase] JNK—as described in the Subheading 3. Bcl-2-like survival factors can also be converted into proapoptotic pro-teins after proteolytic cleavage by caspases and calpains *(43)*.

2.2.2. Bax-Like Death Factors

Members of the Bax-like family have sequences similar to those found in Bcl-2 (BH1–3 domains). Most of them lack the N-terminal BH4 domain of the Bcl-2-like sur-vival proteins, converting them into proapoptotic factors (Fig. 3). Bcl-X_S is an exeption as it is the only proapoptotic member of this subfamily with a BH4 domain and with-out BH1 and BH2 domains *(38)*. Bax was the first family member discovered and thus the most studied member of this group. It is cytosolic and the mechanism by which it associates with mitochondrial membranes when apoptosis is triggered is not fully understood *(38,44)*. It has been proposed that Bax or Bak undergoes a conformational change and thus translocates to mitochondria where it stably inserts into the outer mem-brane by a transmembrane domain *(45,46)*. The apoptosis-associated speck-like protein (ASC) could function as an adaptator molecule for Bax and, on p53 induction, may participate in its translocation to mitochondria *(47)*. It has become widely accepted that Bax acts on mitochondria to increase permeability and to mediate the release of proapoptotic factors from the intermembrane space into the cell cytosol *(48,49)*. At this point, Bax can be inhibited by an interaction with Bcl-2-like survival factors. When

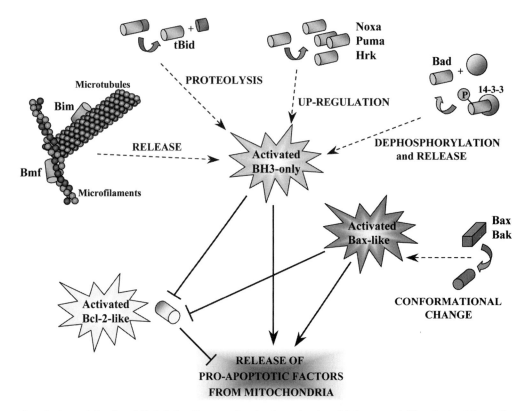

Fig. 4. Potential role of Bcl-2 family proteins in the mitochondrial permeability. In healthy cells, active Bcl-2-like survival proteins localize on mitochondrial membranes, whereas inactive BH3-only and Bax-like proteins are normally cytoplasmic (and probably also loosely bound to mitochondria in a nonrepresented inactive form). In response to apoptotic signals, rates of active BH3-only proteins increase through different mechanisms. Sequestered Bim and Bmf are released from microtubules and microfilaments respectively. Bid is truncated through a caspase 8-mediated proteolysis. Noxa, Puma, and Hrk are upregulated at the transcriptional level. Bad is dephosphorylated and thus released from the 14-3-3 scaffold protein. These activated BH3-only proteins then translocate to mitochondria where they neutralize Bcl-2-like survival factors. In addition, the apoptotic signals trigger a strong change in Bax-like protein conformation. It leads to their stable insertion into the outer mitochondrial membrane that can provoke mitochondrial membrane permeabilization (*see* Fig. 7). Some BH3-only proteins, like tBid, may stimulate the Bax-like protein oligomerization and thus enhance their proapoptotic effects. To view this figure in color, see the insert and the companion CD-ROM.

these factors are absent or inactivated, Bax can oligomerize in the outer membrane to form a pore *(50)*. It is still not certain whether Bax forms channels on its own or works by interacting with pre-existing outer mitochondrial membrane channels (*see* Section 2.3).

2.2.3. BH3-Only Proteins

In contrast with the other proapoptotic proteins (Bax-like), these factors have only the short BH3 domain, hence their name (Fig. 3). In mammalian cells, BH3-only proteins are activated by different mechanisms in response to an apoptotic signal. Although they have a common proteolytic sequence, Bad and the other members of this subfamily have distinct localizations and/or different ways of activation (Fig. 4). On induction of apoptosis, the BH3-only proteins are activated and translocated to mitochondria where they participate in membrane permeabilization. When phosphorylated at serine residues by Akt, Raf-1, or protein kinase A (PKA) *(51–53)*, Bad is sequestered in the cytoplasm by binding to 14-3-3 proteins *(54)*. It is dephosphorylated when apoptosis is triggered, released from 14-3-3, and thus becomes free to interact with Bcl-2-like survival factors. In contrast with Bad, phosphorylation of Bik increases its proapoptotic potency by a mechanism that does not affect its affinity for Bcl-2-like survival proteins *(55)*. Bid is normally activated by a caspase-8-dependent proteolytic cleavage, in response to death-receptor activation. However, it can also be cleaved by other caspases, suggesting that its function may not be limited to the death-receptor pathway and that it could have a role in the intrinsic pathway. The inactive cytosolic form of Bid is cleaved into a truncated fragment (tBid) that translocates to mitochondria. This targeting of tBid is facilitated by N-myristoylation *(56)* and by its high-affinity binding to cardiolipin, a mitochondrial specific lipid.

In contrast with most of the BH3-only proteins that are localized in the cytoplasm until their activation, Bim and Bmf are kept inactive by sequestration by the cytoskeleton. Bim is produced as three major alternative spliced products from the same gene, BimEL, BimL, and BimS. The first two of these, BimEL and BimL, are normally sequestered in the microtubular dynein motor complex by binding to the dynein light chain LC8 *(57)* whereas Bmf is sequestered to the actin cytoskeleton-based myosin-V motor complex by association with the dynein light chain 2 *(58)*. Translocation of Bim or Bmf proteins to mitochondria is triggered by disruption of the cytoskeleton, by a mechanism not yet fully elucidated *(see* Subheading 3.2).

2.2.4. INTERACTIONS AMONG BCL-2 FAMILY MEMBERS

The relative levels of pro and antiapoptotic members of the Bcl-2 family determine the cell's susceptibility to apoptosis. Several members of this family are capable of forming death-promoting or death-inhibiting homo and/or heterodimers *(59,60)*. How these factors interact to regulate the apoptotic process is not fully understood. Activated BH3-only proteins can interact with Bcl-2-like and Bax-like proteins because members from these two subfamilies contain a hydrophobic pocket, the binding site of BH3-only. However, these interactions are restricted by intracellular compartmentalization in the membranes where Bcl-2 is localized. Additional regulation by cellular proteins that are not present under in vitro-binding conditions cannot be excluded. Both types of proapoptotic proteins seem to be required to initiate apoptosis: the BH3-only proteins acting as damage sensors and direct antagonists of the prosurvival proteins, and the Bax-like proteins acting further downstream, probably in mitochondrial disruption *(61–63)*.

Current models emphasize the role of BH3-only proteins (Fig. 4). In response to an apoptotic stress, a particular BH3-only protein is activated. It then interacts with Bcl-2-like survival protein and neutralizes its survival function, probably through alteration of its conformation and enhancement of its association with mitochondrial membranes *(64)*. At the same time, inactive forms of Bax-like factors, soluble or loosely attached to the mitochondrial membrane, undergo a conformational change, eventually assisted by BH3-only proteins *(65)*. Bax-like factors then become active by stably inserting into the mitochondrial outer membranes. They provoke membrane permeabilization to release caspase-activating and other proapoptotic factors. However, the mechanism allowing the release of proteins localized in the intermembrane space to the cytosol remains to be further defined.

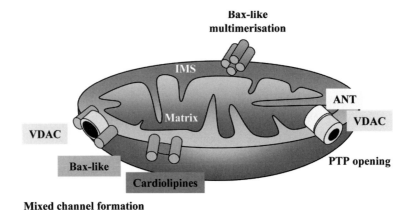

Fig. 5. Models for release of mitochondrial intermembrane space factors. PTP: permeability transition pore; IMS: intermembrane space; ANT: adenine nucleotide translocator; VDAC: voltage-dependent anion channel. Proapoptotic factors translocate to cytosol by PTP opening or by formation of channels made up of Bax-like proteins, alone or in association with other mitochondrial components (proteins or lipids). As described before (*see* Fig. 6), Bcl-2 proteins inhibit the mitochondrial permeabilization, whatever the way used for proapoptotic factor release. To view this figure in color, see the insert and the companion CD-ROM.

2.3. How are Mitochondrial Proapoptotic Factors Released?

A large variety of proteins are released when the outer mitochondrial membrane is perforated, as revealed by proteomic analysis under in vitro tBid-treatment of mitochondria *(66,67)*. Different models have been proposed to explain how mitochondria release apoptogenic factors (Fig. 5). Cytochrome-*c* was the first characterized mitochondrial factor shown to be released from the mitochondrial intermembrane space and much effort has been directed toward elucidating its mechanism of release. A massive cytochrome-*c* release has been largely associated with a mitochondrial membrane potential ($\Delta\Psi$m) dissipation (reviewed in ref. *68*).

The mitochondrion is surrounded by a double membrane that divides it into two compartments: the intermembrane space and the matrix. A transmembrane channel, called the permeability transition pore (PTP), is localized at the contact sites between the outer and the inner membrane. It is a complex of several proteins not yet fully determined. The main proteins are the voltage-dependent anion channel (VDAC, a mitochondrial porin) in the outer membrane and the adenine nucleotide translocator in the inner membrane. The PTP is closed during normal mitochondrial functioning, as a proper $\Delta\Psi$m is essential to maintain oxidative phosphorylation, adenosine 5′-triphosphate (ATP) synthesis, and thus cell survival. PTP opening can operate in two distinct states: a low-conductance state when the pore is partly open, permeable to molecules <300 Da, with a reversible decrease in $\Delta\Psi$m, and the high-conductance state when the pore is permeable to molecules <1500 Da with an irreversible collapse of the $\Delta\Psi$m *(69)*.

Two main models of how PTP opening may be involved in the induction of cytochrome-*c* release have been proposed *(70)*. The first is the PTP-induced mitochondrial swelling model, also called the VDAC opening model. This model requires that the apoptosis inducing agent directly interacts with the PTP, leading to its opening, to a rapid mitochondria depolarization ($\Delta\Psi$m collapse) and to matrix swelling. Opening of the PTP can be reversed by agents as cyclosporin A. Moreover, this mechanism of PTP opening can be preceded by the VDAC closure model. Actually, an early transient hyperpolarization of the mitochondrion has been reported. Closure of VDAC leads to the failure of mitochondria to maintain ATP/ADP exchange with the cytosol and a decrease in F_0F_1-ATP synthetase activity, resulting in hyperpolarization of the mitochondria. When it occurs, this event is followed by the mitochondrial swelling, rupture of the outer membrane, and release of proapoptotic factors into the cytosol (68). The second model is the PTP-nonswelling model that has been observed in isolated mitochondria. Cyclosporin A can also inhibit PTP opening in this model. The release of cytochrome-c is induced by low concentrations of recombinant Bax or truncated Bid without mitochondrial matrix swelling or rupture of the outer membrane. The transient opening of the PTP may explain the absence of swelling.

However, PTP involvement remains controversial and other models for cytochrome-c release exist. A nonspecific rupture of the outer membrane and the formation of conducting channels in the outer membrane have been postulated (71). Contrary to the former models, Bcl-2 familly members, and not the PTP, are the major players. The proapoptotic members insert into the outer membrane where they oligomerize and form protein-permeable channels in an autonomous fashion. To date, formation of a Bax channel (consisting four Bax molecules) and Bax-Bid-lipidic pores have been described (72). Moreover, VDAC can play an essential role in the increase of outer membrane permeability, independently of the occurrence of permeability transition events (73). It is regulated by the Bcl-2 family proteins through a direct interaction. For example, channels formed by a collaboration between Bax and VDAC have been proposed, allowing an increase in the size of the pore for releasing molecules of molecular masses >15,000 Da (reviewed in ref. 48).

Finally, it is still not clearly defined whether the other mitochondrial intermembrane proteins are released simultaneously by a mechanism similar to cytochrome-c release. The action of caspases would be required for release of AIF (74). These data raise the question about the functional hierarchy among caspase activation, AIF release, and mitochondrial permeabilization (75). The involvement of mitochondrial bioenergetics in outer membrane permeabilization is not yet elucidated. Particularly, the role of the transient mitochondrial hyperpolarization before cytochrome-c release is unknown.

2.4. Signaling Downstream of Mitochondria

2.4.1. ROLE OF APOPTOGENIC PROTEINS RELEASED FROM MITOCHONDRIA

Release of cytochrome-c from mitochondria into the cytosol has been implicated as an important step in apoptosis. Once in the cytosol, cytochrome-c binds to the apoptotic protease activating factor (Apaf)-1, triggering formation of the apoptosome (76). This oligomeric complex contains seven molecules each of Apaf-1, cytochrome-c, (d)ATP, and procaspase-9 (Fig. 2). Procaspase-9 is the initiator caspase of the apoptosome, the first caspase activated downstream of the mitochondria.

There have been 14 mammalian caspases identified to date. They are synthesized as precursors that, under apoptotic conditions, undergo a proteolysis-mediated activation by other caspases in a cascade. Functionally, three classes of caspases have been distinguished: the initiator caspases that are characterized by long prodomains containing

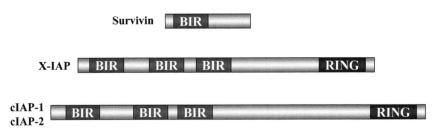

Fig. 6. Schematic representations of mammalian IAPs. These proteins contain one to three N-terminal repeats of an 70-amino acid motif known as a baculovirus IAP repeat (BIR). This motif plays an important role in caspase inhibition by all IAPs. A C-terminal RING finger domain is also required for the antiapoptotic activity of some, but not all, IAPs. To view this figure in color, see the insert and the companion CD-ROM.

either a death effector domain (caspases-8 and caspase-10) or a caspase recruitement domain (caspase-2 and caspase-9), the effector caspases, which contain short prodomains (caspase-3, caspase-6, and caspase-7), and the remaining caspases for which the main role depends on cytokine maturation more than on apoptosis. The majority of procaspases and activated-caspases are present in the cytosol, but some have also been localized in subcellular compartments such as mitochondria and the nucleus, and may be redistributed during the apoptotic process (77–80). Each caspase participates in promotion of apoptosis. Upon activation, the prodomains are cleaved off adjacent to aspartic acid residues, separating the large and small subunits. Initiator caspases cleave and activate effector caspases. The later caspase in the cascade then cleaves cellular substrates, leading to the final apoptotic phenotype. The most studied effector caspase is the cytosolic caspase-3, which is capable of cleaving important cellular substrates such as the DNA repair enzyme poly(ADP-ribose)polymerase (PARP) and the inhibitor of caspase-activated DNAse (reviewed in ref. 81). Mitochondria play a key role in the control of apoptosis because they release cytochrome-c that is a necessary cofactor to initiate the caspase cascade (82,83). However, caspase-2 has also been described as being required for permeabilization of mitochondria (84,85). In this case, mitochondria are amplifiers of caspase activity rather than initiators of caspase activation.

In addition to the control of zymogen-form activation, caspase regulation is achieved by IAPs-mediated inhibition. So far, eight human IAPs have been identified among which the most studied are X-IAP, c-IAP1, c-IAP2, and survivin (Fig. 6).

They are thought to directly inhibit caspase-3, caspase-7, and caspase-9 (86,87). However, their activity is not restricted to caspase inhibition and they have roles in cell-cycle regulation, protein degradation, and caspase-independent signal-transduction cascades (reviewed in ref. 88). Survivin, the smallest member of the mammalian IAP family, is overexpressed in most of human cancers. A direct association between survivin and polymerised tubulin has been demonstrated in vitro (89). When expressed during mitosis, survivin has been shown to localize to various components of the mitotic apparatus such as the centrosomes, microtubules of the metaphase and anaphase spindle, and kinetochores (90). Survivin shows a clear cell-cycle-dependent post-translational regulation. Polyubiquitylation of survivin and proteasome-dependent degradation have been demonstrated in interphase cells and mitotic phosphorylation of survivin by CDK1-cyclin B1 has been associated with increased protein stability at metaphase. It is a particularly interesting protein in MDA-induced apoptosis because it may be considered both as an apoptotic inhibitor and a mitotic regulator (91).

Inhibition of caspases by IAPs is relieved by proteins like Smac/Diablo and Omi/Htra2 that are released by mitochondria during changes in permeability *(92,93)*. The proapoptotic activity of Omi/Htra2 in vivo involves a serine protease activity in addition to its IAP-binding ability *(94)*, and could also induce a caspase-independent apoptosis *(95)*.

Besides cytochrome-*c* and IAP inhibitors, other proapoptotic factors of mitochondrial origin are released to induce apoptosis independently of caspases *(67)*. AIF and endonuclease G are released from mitochondria and translocate to the nucleus where they cleave chromatin DNA into nucleosomal fragments *(96,97)*. Although their proapoptotic activity is independent of caspases, their translocation may depend, in part, on caspase activation *(75)*. Mutations that abolish the AIF-DNA interaction suppress AIF-induced chromatin condensation, yet have no effect on its NADH oxidase activity *(98)*. Thus, the apoptogenic and oxidoreductase functions of AIF can be dissociated *(99)*. Recent studies suggest that AIF could be a major factor involved in caspase-independent neuronal death, emphasizing the central role of mitochondria in the control of physiological and pathological cell demise.

2.4.2. MODULATION BY REACTIVE OXYGEN SPECIES, CA^{2+}, AND HEAT SHOCK PROTEINS

Besides the release of proapoptotic factors from mitochondrial intermembrane space, intracellular "mediators" such as reactive oxygen species (ROS), calcium ion (Ca^{2+}), or heat shock proteins (HSPs) may be involved in apoptotic signaling. Mitochondria are the major site of ROS production, because of the activity of complex I and III in the respiratory chain, and ROS increase may result from an impairment of the mitochondrial respiratory chain. ROS have been proposed as mediators of apoptosis *(100)*. As shown in TNF-α-induced apoptosis, Bcl-2 could act as an antioxidant by blocking ROS-mediated steps in the cascade of apoptotic events *(101)*. An early increase in ROS levels has been found to precede mitochondrial membrane permeabilization, Bax relocalization, and cytochrome-*c* release in apoptosis triggered by several signals *(102)*. In contrast, some data raise the possibility that ROS increase occurs later, in the execution of the death program *(103)*. The specific targets of ROS in the apoptotic process have not been elucidated. Mitochondrial ROS lead to general damaging effects on cellular structures and particularly on mitochondria, including mitochondrial DNA, which in turn, could activate downstream cascades of events *(100,104)*. Alternatively, the increase in ROS production could induce an imbalance of intracellular redox status and a subsequent oxidative stress. The glutathione (GSH) redox cycle is an important antioxidant defense system that also plays a role in the integrity of mitochondrial proteins and lipids involved in the permeabilization of mitochondrial membranes *(105)*.

The role of calcium ions in the regulation of apoptosis has been recently emphasized *(43)*. Bcl-2 family proteins affect Ca^{2+} fluxes from endoplasmic reticulum (ER) and mitochondria, and in turn, cytosolic Ca^{2+}-related signals can trigger apoptosis *(106,107)*. There are many potential targets for Ca^{2+} signaling in apoptosis. Calcineurin is a Ca^{2+}/calmodulin-dependent phosphatase involved in the dephosphorylation of Bad, which results in its translocation to mitochondria *(108)*. Calpain family of Ca^{2+}-activated cysteine proteases crosstalk with caspases for regulation of apoptotic process *(43)*. These findings provide evidence for a link between Ca^{2+} signals and apoptosis, which remains to be evaluated.

HSPs interact with a number of cellular systems and form efficient cytoprotective mechanisms. However, recent data show that HSPs have a role in the regulation of apoptosis *(109)*. Under normal conditions, HSPs play numerous roles in cell functions, modulating protein activity, conformation, degradation, translocation, and folding. HSPs are produced in response to stress conditions, making cells resistant to cell damage *(110)*. HSP90 is involved in cell resistance to apoptosis through the conformational maturation of oncogenic signaling client proteins, including HER-2, Bcr-Abl, HIF-1α, Akt, and Raf-1. HSP90 inhibitors such as 17-AAG, bind to HSP90 and induce the proteasomal degradation of HSP90 client proteins. They have shown promising antitumor activity in preclinical models *(111)*. In tumor cells, HSP90 is present in highly active multichaperone complexes and its binding affinity for 17-AAG is 100 times higher than in normal cells, thus conferring tumor selectivity for HSP90 inhibitors *(112)*.

Interestingly, HSPs have been demonstrated to directly interact with various components of the cell death machinery upstream and downstream of mitochondria (reviewed in ref. *113*). HSP27 inteferes with the process of apoptotic cell death by preventing Bax translocation. Moreover, by sequestering cytochrome-*c* or Apaf-1, HSP27, HSP70, and HSP90 inhibit functional apoptosome assembly and thus prevent activation of the caspase cascade *(114–116)*. In addition, HSP70 directly binds to AIF and, in a manner independent of its chaperone-activity, inhibits AIF-mediated apoptosis *(117)*. HSP27 also represses procaspase-3 activation through a direct association *(114)*. In contrast, HSP60 and HSP10 localize to mitochondria and can participate in apoptosis by accelerating caspase-3 activation *(118)*.

To summarize, this section has reviewed current data on the role of various proteins in apoptosis promotion and amplification. Although many questions remain to be resolved, it clearly appears that mitochondria, the cells' arsenals, orchestrate apoptosis by both caspase-dependent and -independent pathways. They are the convergent point for numerous upstream signals and they trigger downstream signal routes leading to apoptosis. The involvement of mitochondria in apoptosis induced by MDAs will be presented in the Subheading 3. Moreover, as apoptosis and proliferation are intimately coupled *(119)*, the role of proteins controlling both apoptosis and cell-cycle progression will also be evaluated. Furthermore, activation of survival pathways upon treatment with MDAs will also be discussed, as such activation may constitute a cellular response to escape to death signals.

3. APOPTOTIC SIGNALING PATHWAYS INDUCED BY MDAS

Although MDAs differently disturb microtubule network integrity (by inducing polymerization or depolymerization and suppression of dynamics) and induce various signals, most of these agents trigger similar molecular mechanisms to promote apoptosis of tumor cells. Important links between disturbance of microtubule integrity and induction of apoptosis are still lacking, but several intracellular signaling pathways have been elucidated in apoptosis induced by MDAs.

The cytotoxicity of MDAs is not correlated with any single parameter, but rather with a combination of effects on the cell cycle and apoptosis. Modulation of the Bcl-2-family member activity is a common activity for MDAs that induce release of mitochondrial proapoptotic factors and caspase activation. Upstream of mitochondria, microtubule disruption could also be accompanied by the release of microtubule-sequestered factors as described for Bim and survivin. In addition, cell-cycle progression regulators (as CDKs, MAPKs, p53 or p21), cell stress actors (such as ROS and HSPs), and cell survival factors (NFκB, Akt, and P70S6K) have been investigated as signaling pathways promoting or preventing apoptosis induced by these drugs. The cellular response is likely to be amplified by a transcriptional effect of MDAs as they can regulate the expression of genes encoding for proteins involved in the apoptotic machinery.

3.1. Integration of MDA-Triggered Apoptotic Signals by Mitochondria

Mitochondria constitute the heart of the intrinsic signaling pathway activated by MDAs. Release of mitochondrial apoptotic factors, induced by MDAs, is under the control of endogenous effectors, among them the Bcl-2 family members remain the most studied.

3.1.1. MODULATION OF BCL-2 FAMILY MEMBER STATUS

Upregulation of *Bcl-2* and *Bcl-xL* genes is a mechanism by which tumor cells resist to MDAs *(119–122)*. Pretreatment of cells with Bcl-2 antisense RNA mediates an increase (more than twofold) in docetaxel- and paclitaxel-sensitivity *(123)*. Bcl-xL overexpression inhibits progression of molecular events that lead to paclitaxel-induced apoptosis *(40,124)*. Likewise, overexpression of Bcl-2 protects cells against apoptosis, without affecting microtubule network disruption and cell-cycle arrest *(125)*. Thus, inactivation of Bcl-2-like proteins seems to be necessary for MDA-induced apoptosis, downstream of mitotic block. It is largely thought that the loss of function of these anti-apoptotic members results from their hyperphosphorylation following microtubule disruption by paclitaxel, vinorelbine, vincristine, vinblastine, colchicine, or nocodazole *(126–131)*. As DNA damaging agents do not induce its hyperphosphorylation, Bcl-2 has been proposed to be a specific "guardian of microtubule integrity" *(42,132)*. This process could be mediated by different kinases such as PKA *(132)*, p34^{cdc2} kinase, JNK, or Raf-1 kinase as described later. Once hyperphosphorylated following paclitaxel or vinorelbine treatment, Bcl-2 is less likely to heterodimerize with the proapoptotic Bax, leading to an increase of mitochondrial free Bax levels and thus to mitochondrial permeabilization and apoptosis *(130,133,134)*. Inactivation of Bcl-2 by hyperphosphorylation during mitosis is an attractive model to link mitotic block and apoptotic signaling pathways induced by MDAs. However, this possibility remains controversial as hyperphosphorylation of Bcl-2 occurs in normally cycling cells at the G2/M phase under conditions that do not lead to apoptosis, suggesting that hyperphosphorylated Bcl-2 is mainly a marker of mitosis *(135–137)*. Nevertheless, if the paclitaxel concentration is too low to induce a sustained arrest in G2/M and apoptosis, exit from mitosis occurs, because of mitotic slippage, with the rapid dephosphorylation of Bcl-2 by PP1 phosphatase *(138)*. When the MDA concentration is sufficient to maintain a G2/M block, it is the extent, duration, and/or irreversible nature of the mitosis-associated signals that distinguish a preapoptotic cell from one destined to divide *(139)*. In this case, the persistent Bcl-2 hyperphosphorylation could constitute an apoptotic signal.

Even if the role of Bcl-2 hyperphosphorylation *per se* in apoptosis promotion is not well-defined, inactivation of Bcl-2 probably at least increases the susceptibility of cells to death signals induced by MDAs. Interestingly, Bcl-2 may be cleaved by activated caspase 3, turning its function from antiapoptotic to proapoptotic *(140)*. The resulting proapoptotic 22 kDa product is likely to participate in the apoptotic signaling pathways induced by paclitaxel *(141,142)*. In parallel with these post-translational modifications, treatments with MDAs can decrease Bcl-xL and Bcl-2 protein levels *(141,143)*. This could be a consequence of p53 induction (*see* Section 3.3.2) that downregulates Bcl-2 *(144)* rather than a result of proteasome-dependent degradation of phosphorylated forms of Bcl-2 *(145)*.

Like apoptosis initiated by many stimuli, paclitaxel-induced apoptosis can be enhanced by the Bax-like and BH3-only proteins, proapoptotic members of the Bcl-2 family. Bcl-X_S, Bax, or Bad overexpression sensitises cancer cells to paclitaxel and vincristine *(146–148)*. One of the earliest events induced by paclitaxel exposure is the upregulation of Bad, Bax, and/or Bak *(141,149)*. A significant increase in Bax level is also observed after a 24 h-treatment with vinorelbine *(130)*. Moreover, treating cells with these drugs triggers a conformational change in the Bax protein, its subsequent translocation from cytosol to mitochondria, and its complex formation, which disturbs $\Delta\Psi$m *(150,151)*. Translocation of the proapoptotic protein Bim from microtubules to mitochondria following MDA treatment (as described in Section 3.2.1) also participates in the release of mitochondrial intermembrane space factors. Finally, increase in Bax levels has been described to promote paclitaxel intracellular accumulation, amplifying its cytotoxic effect *(152)*.

Thus, by modulating both the activity and the expression levels of Bcl-2 family proteins, the amounts of active proapoptotic members increase at mitochondrial membranes, whereas those of active antiapoptotic members decrease, turning their ratio in favor of mitochondrial permeability and caspase activation.

3.1.2. CASPASE CASCADE ACTIVATION

Caspases are involved in MDA-induced signaling pathways as pretreatment with caspase inhibitors such as Benzyloxycarbonyl-Val-Ala-Asp (OMe) fluoromethylketone (Z-VAD-fmk) prevents, at least in part, apoptosis *(131,153–155)*. Although an increase in expression of death receptor and their ligands is described in some cell types *(35,36)*, the extrinsic signaling pathway, mediating caspase-8 activation, is generally not involved in MDA-induced apoptosis *(31–34)*. In the intrinsic signaling pathway induced by MDAs, mitochondrial membrane permeability that promotes caspase activation is not affected by caspase-8 inhibition *(153,156)*, confirming that this caspase is dispensable upstream of mitochondria. Thus, the caspases, required as mediators of apoptosis for MDAs, are only localized downstream of the mitochondria. Induction of cytochrome-*c* release from mitochondria by MDAs, under the control of Bcl-2 family members, permits the assembly of the apoptosome complex proteins.

Overexpression of Apaf-1, the adaptator molecule of the apoptosome, which interacts with procaspase-9, enhances paclitaxel-induced apoptosis *(157)*. In response to paclitaxel-treatment, release of proapoptotic factors triggers an early increase in caspase-9 activity, and the caspase-9-specific inhibitor Z-Leu-Glu(OMe)-His-Asp(OMe)-FMK.TFA (z-LEHD-fmk) effectively protects cells from MDA-mediated apoptosis *(131,154,158)*. Once activated by paclitaxel, caspase-9 mainly induces caspase-3 activation, leading to cleavage of substrates such as PARP, DFF45 /inhibitor of caspase-activated DNAse, topoisomerase I, or Cdc6 *(159–162)* and thus to the final apoptotic events. Caspase-8 is also a target of caspase-3, and it can be significantly activated downstream of mitochondria during paclitaxel exposure *(31,33,156, 160,163)*. In this case, the caspase-3-specific inhibitor impairs MDA-induced caspase-8 cleavage *(33)*. Caspase-8 activation then leads to Bid cleavage, triggering a signal amplification loop, which could be required for an optimal cytochrome-*c* release *(34,154)*. Finally, paclitaxel-activated caspase-3 could also mediate disruption of cell adhesion through the cleavage of APC protein, β- and γ-catenin, that might contribute to paclitaxel-induced apoptosis *(164)*.

Thus, caspase-3 cleaves different targets, and appears to have a key role in the mitochondrial apoptotic signaling pathway. In support of this, in situations where activation of the mitochondrial apoptosome is disturbed, caspase-3 overexpression restores cancer cell sensitivity to MDAs *(165)*. However, it should be noted that apoptotic cell death can be induced by MDAs such as docetaxel through a caspase-3-independent mechanism *(166)*. It may be considered that another caspase effector could assume the caspase-3 role in apoptotic routes, bypassing the need for activated caspase 3. Actually, paclitaxel- or vinorelbine-induced caspase-9 activation can also lead to an increase in caspase-7 *(130,167,168)* and caspase-2 activity *(131)*. As IAPs inhibit caspase-3, -7, and -9, their inactivation by Second mitochondria-derived activator of caspases/Direct IAP binding protein (Smac/DIABLO) peptides enhances paclitaxel efficacy *in situ (169)*. In like manner, paclitaxel produces higher levels of apoptosis when combined with a recombinant adenovirus encoding Smac/DIABLO *(170)*.

So, the caspase cascade-activated downstream of mitochondria is thought to be involved in apoptosis induced by MDAs. However, paclitaxel has been described to trigger apoptosis in a nonsmall-cell lung cancer cell line through a caspase-independent mechanism, in which the apoptotic machinery described earlier is only coactivated *(171,172)*. Apoptosis following paclitaxel treatment also occurs without involvement of either caspase-9 or caspase-3 in human ovarian and breast carcinoma cells *(173)*. The role of mitochondria-released factors such as AIF or endonuclease G, mediating apoptosis through a caspase-independent pathway, is currently unknown. Finally, paclitaxel can induce a slow nonapoptotic cell death, without activation of caspase-3, caspase-8, or PARP cleavage *(174)*. Thus, novel signaling pathways still remain to be discovered in order to understand how MDAs induce cell death.

3.2. Release of Microtubule-Sequestering Factors

Microtubules serve as a depot for different signaling molecules and thus they are able to affect a more number of biological processes in cells. In particular Bim and survivin are located on microtubules and regulate apoptosis upstream of mitochondria. Their release, by disturbance of microtubule integrity, could affect the activities of these apoptosis regulators. As polymerizing and depolymerizing agents share the common property of suppressing microtubule dynamics, it could be argued that involvement of the microtubule-sequestered factors in apoptosis is likely to be similar among the different MDAs.

3.2.1. BIM

Bim is a proapoptotic factor of the Bcl-2 family that is sequestered by microtubules *(57)*, and therefore, it represents an important link between the microtubule network and the apoptotic machinery (Fig. 4). Bim protein levels increase dramatically after paclitaxel treatment, and gene-silencing experiments show that the transcriptional upregulation of Bim can be a direct cause of apoptosis in cancer cells *(175)*. Bim appears to be required for apoptotic response to MDAs, as lymphocytes $bim^{-/-}$ are refractory to microtubule perturbation *(176)*. Knockdown of Bim by siRNA confers resistance to apoptosis induced by MDAs in breast cancer cells (reviewed in ref. *177*). Upon disruption of microtubule network functions by MDAs, freed Bim translocates to mitochondria where it neutralizes Bcl-2 or activates Bax, and therefore, constitutes an initiating event in apoptotic signaling *(57,178)*. Last, by activating JNK *(see* later), MDAs may induce Bim phosphorylation and thus its translocation from microtubules to mitochondria as described for UV-stimulated apoptosis *(179)*.

3.2.2. Survivin

A pool of the IAP survivin associates with microtubules and participates in mitotic spindle function *(89,90)*. In skin and lung fibroblasts, surviving colocalizes with centrosomes in the cytoplasm during interphase, then moves to centromeres during mitosis, and finally localizes to the midbody spindle microtubules during telophase. In paclitaxel treated cells, survivin is extensively relocalized to α-tubulin in microtubules both during either interphase and mitosis, but the functional consequence of this redistribution remains to be evaluated *(180)*. Expression of the survivin gene correlates with paclitaxel resistance in human ovarian cancer *(181)*. In endothelial cells, survivin induction by vascular endothelial growth factor (VEGF) strongly decreases the sensitivity to paclitaxel and vinblastine *(182)*. Forced expression of survivin in epitheloid carcinoma cells profoundly influences microtubule dynamics with reduction of pole-to-pole distance at metaphase and stabilization of microtubules against nocodazole-induced depolymerization *(183)*.

There is still debate about survivin's function in regulating apoptosis, cell division, or both *(184)*, and the mechanism of action of MDAs on survivin activity is not clearly defined. Data indicate that paclitaxel induces survivin as an early event independent of G2/M arrest *(185)*. Inhibition of PI3K/Akt and MAPK pathways diminishes survivin induction and sensitizes cells to paclitaxel-mediated cell death *(185)*. In addition, a direct interaction between survivin and Smac/Diablo has been shown to reduce the neutralizing effect of Smac/Diablo on other IAPs, leading to decrease in paclitaxel-induced apoptosis *(186)*. Survivin induction appears to be a mechanism by which cancer cells evade apoptosis, and targeting this survival pathway may result in novel approaches for cancer therapeutics. However, further investigation is needed as knocking down survivin expression abrogated the cyclin B1 stabilization and the mitotic arrest induced by paclitaxel *(187)*.

3.3. Involvement of Cell-Cycle Progression Regulators

3.3.1. p34[CDC2] Kinase

One of the main questions concerning MDA-induced apoptosis is how the apoptotic machinery and cell-cycle regulation might be linked. This is why studies on cyclins and CDKs, which regulate cell-cycle progression, are still of major interest.

The kinase activity of p34[cdc2] (or CDK1), is specifically activated at the G2/M transition by induction of cyclin B1-expression. Inhibition of p34[cdc2] kinase activity with olomoucine or with cyclin B1-specific antisense oligonucleotide prevents paclitaxel-induced apoptosis, suggesting that the complex cyclin B1/p34[cdc2] kinase plays an important role in this process *(188)*. Use of a dominant-negative mutant of p34[cdc2] confirms that activation of this kinase plays a critical role in paclitaxel-mediated apoptosis *(189)*. This drug is responsible for an early peak in p34[cdc2] kinase activity *(137,188,190)*. An upregulation of p34[cdc2] is observed during paclitaxel-induced apoptosis, and inhibition of protein synthesis by cycloheximide prevents p34[cdc2] accumulation, G2/M arrest, and apoptosis *(191)*. Synthesis of p34[cdc2] protein may play a role in paclitaxel-induced apoptosis through a positive feedback loop that increases the activity of cyclin B/p34[cdc2] kinase. Furthermore, by inducing microtubule damage, MDAs activate the mitotic spindle checkpoint, leading to the blockade of proteasome-dependent cyclin B1 degradation *(192)*. All these events converge in a sustained activation of cyclin B1/p34^{cdc2} kinase, maintaining cells in mitosis.

The link between p34^{cdc2} kinase activation by MDAs and induction of apoptosis remains unclear. O'Connor et al. *(193)* report that microtubule stabilization by MDAs engenders a survival pathway that depends on elevated activity of p34^{cdc2} kinase. However, several proteins playing crucial roles in mitosis are phosphorylated by cyclin B1/p34^{cdc2} kinase, and it could be envisaged that the persistent activation of some of these proteins ultimately triggers an apoptotic signal *(194)*. Interestingly, an association between p34^{cdc2} and the antiapoptotic protein Bcl-2 has also been shown, and may lead to Bcl-2 inactivation by phosphorylation *(137,195,196)*.

3.3.2. P53 AND/OR P21

Mutations in p53 are present in more than 60% of human cancers *(197)*, but the p53 status of cancer cells is generally considered not to play a role in paclitaxel cytotoxicity *(198–200)*. The IC$_{50}$ of paclitaxel for mutated p53-expressing ovarian cancer cells is not higher than that of wild-type p53-expressing cells *(201)*. Rates of clinical response to treatment with paclitaxel for patients with p53 mutant metastatic nonsmall-cell lung carcinoma have been shown to be as good as those for patients with wild-type p53 *(202)*. Surprisingly, loss of normal p53 function has been described to sensitize ovarian cancer cells to paclitaxel-induced G2/M arrest and apoptosis *(203)*. These data could highlight an advantage of paclitaxel in p53-mutated cancer therapy. However, in another human ovarian cancer cell line, the inactivation of p53 results in a strong decrease in paclitaxel cytotoxicity. Moreover, paclitaxel-, vinorelbine-, or vincristine-treatment results in the upregulation and activation of p53 as well as p21WAF1, a p53-downstream gene *(16,182,204–207)*. Induction of p21WAF1 expression could also be p53-independent, as described for vinorelbine and paclitaxel *(204,207)*. Expression of p21WAF1, in stably transfected anaplastic thyroid cancer cells with p21WAF1 complementary DNA, does not induce apoptosis *per se* but does enhance the cytotoxic effects of paclitaxel, suggesting that p21WAF1 might be required for cell sensitivity to paclitaxel *(208)*. The potential involvement of p53 and/or p21WAF1 in apoptotic signal pathways induced by MDAs is still not resolved *(209)*.

The discrepancy in the results obtained by silencing p53 may result from substitution by a p53-like protein. The variability in cell type sensitivity to p53 induction is another cause of conflicting observations as two groups of cells could be distinguished: those with MDA-inducible p53, and those with p53 that does not respond to these agents as well as cells lacking p53 *(16)*. Last, the main source of contradictory data is likely to involve the drug concentration studied. Actually, low concentrations of paclitaxel, vinorelbine, or vinflunine increase p53 protein levels, whereas high concentrations do not affect or even inhibit these levels *(182,210)*. Thus, apoptosis induced by low concentrations of MDAs possibly involves upregulation of p53, whereas apoptosis induced by high concentrations of the drugs occurs in a p53-independent manner.

Induction of p53 and/or p21WAF1 by MDAs is often associated with mitotic slippage, thus preventing mitotic arrest-mediated apoptosis *(15,16,211)*. Nevertheless, increases in p53/p21WAF1 protein levels block cells in G1 and/or G2 phase(s), promoting apoptosis in a mitotic arrest-independent way *(15,16,210,212)*. Interestingly, like *p21WAF1*, *Bcl-2*, and *Bax* genes are transcriptional targets for p53 *(16,144,213)*. Following p53 induction by MDAs, Bcl-2 is downregulated whereas Bax is upregulated, leading to a Bax/Bcl-2 ratio in favor of apoptosis. BH3-only members levels can also be modulated through a p53-mediated induction, as described for PUMA and Noxa *(214,215)*. These modifications in expression of p53-downstream target genes may result from enhanced p53 nuclear accumulation, as a consequence of microtubule dynamics suppression by paclitaxel and vinflunine *(210,216)*.

Finally, p53 has been described to induce apoptosis through a transcription-independent mechanism *(216a)*, by directly interacting with Bax, Bad or Bak and by directly mediating mitochondrial membrane permeability *(216b,217)*. Thus, accumulating evidence indicates a possible involvement of p53 and/or p21 in apoptotic pathways following disruption of microtubule network function. It should also not be forgotten that, like p53 and p21, MDAs can also induce other proteins controlling cell-cycle progression and apoptosis, as described for the proliferation-associated protein p120 in human cancer cells following exposure to vincristine and paclitaxel *(218)*.

3.4. Activation of MAPKs

Signal transduction pathways induced by MDA treatment include activation of the MAPK superfamily. The extracellular signal-regulated kinases (ERKs) are mainly associated with proliferation and differentiation while the p38 MAPKs and JNKs regulate responses to cellular stresses. Involvement of Raf-1 kinase, a central component of the MAPK-related pathways, has also been evaluated in MDA-induced apoptosis. The distinct role of these kinases after microtubule damage is not clearly defined, and the difficulty is increased by the fact that the signaling pathways mediated by their activation may be redundant.

3.4.1. EXTRACELLULAR SIGNAL-REGULATED KINASES

ERK is generally related to survival signaling, and the exposure to paclitaxel, vinblastine, vincristine, or colchicine can produce a reduction in ERK activity, a key event in the cellular response leading to apoptosis *(219,220)*. Surprisingly, disruption of microtubules by MDAs has also been described to activate ERK in various human cancer cell lines *(221–224)*. In human cervix carcinoma cells, time-dependent activation of ERK coincided with G2/M arrest and the biochemical events of apoptosis *(225)*. Interestingly, Seidman et al. *(226)* described human ovarian carcinoma cells in which low concentrations of paclitaxel (1–100 nM) for short exposure (0.5–6 h) activate ERK in a transient-manner. In contrast, longer exposure (24 h) resulted in abrogation of ERK, and high concentrations (1–10 μM) resulted in a downregulation of ERK activity. Thus, whether ERK inhibition affects MDA-induced apoptosis remains controversial, and the results probably depend on the study conditions. Long-term inhibition of MEK in a clonogenic assay antagonizes paclitaxel toxicity in epitheloid carcinoma cells *(221)*, and activated ERK participates in apoptotic signaling pathways in paclitaxel-treated neuroblastoma cells *(222)*. In contrast, paclitaxel activates the ERK pathway independently of activating the programmed cell death machinery in human esophageal squamous cancer cells *(223)*. Finally, ERK activation can constitute a cell survival signal in response to paclitaxel treatment *(225–227)*. This last notion is supported by the fact that various types of tumors exhibit high levels of ERK activity *(228,229)*, and can resist to antitumor treatments. ERK-mediated resistance to paclitaxel could be because of Bad inactivation through its phosphorylation on Ser-112 *(230)*. In addition, through ERK activation, vinblastine and colchicine upregulate Mcl-1, a viability-promoting member of the Bcl-2 family *(224)*. As described with other stimuli, activated ERK may also phosphorylate Bim on Ser-69, selectively leading to its proteasomal degradation *(230–232)*. For that matter, combinations of paclitaxel with ERK inhibitors are proposed as a possible new approach to chemotherapy *(225,227,233)*.

3.4.2. p38 MAPK

Because the kinase p38 can function as a component of the spindle assembly checkpoint *(234)*, its involvement in the apoptotic response induced by MDAs has been of considerable interest. Disruption of microtubules with nocodazole activates p38 MAPK in epitheloid carcinoma cells and human lymphoid cells *(234,235)*. High concentrations of paclitaxel also rapidly (within 2 h) activate p38 kinase, which then remains active for more than 24 h in human ovarian carcinoma cells *(226)*. In breast cancer cells, paclitaxel treatment stimulates p38 MAPK activity in a concomitant manner with PKA, leading to apoptosis through inhibition of the Na^+/H^+ exchanger *(236)*. In addition, p38 MAPK inhibition partially protects cells from paclitaxel *(226)*. An involvement of p38 MAPK pathways could be necessary for the transition from proliferation state to paclitaxel-induced apoptosis. However, contrary results indicate that paclitaxel, vinblastine, vincristine, or colchicine cause a reduction in basal p38 MAPK activity in human carcinoma cells *(220)* and that paclitaxel fails to induce p38 MAPK activation in leukemia cells *(142,237)*. In addition, even when activated upon paclitaxel treatment, p38 MAPK is described to be not linked to activation of the cell death machinery *(223)*. Thus, the role played by p38 MAPK in apoptotic signal pathways remains unclear and its role seems mainly to depend on the cell type rather than on the MDA.

3.4.3. c-Jun N-Terminal Kinase

Interestingly, an association between JNK and microtubules, through motor proteins, has been reported *(238)*. By interacting with tubulin and/or microtubules, paclitaxel, docetaxel, vinblastine, vincristine, nocodazole, and colchicine cause a rapid increase in JNK activity, leading to apoptosis in a variety of human cancer cells *(125,219,220,239,240)*. Apoptosis induced by paclitaxel is significantly promoted in cells expressing JNK1, whereas it is effectively suppressed in cells expressing the dominant negative JNK1 mutant or JBD, a JNK inhibitor protein *(241)*. The defect in JNK activation in *mekk1*$^{-/-}$ cells abrogates apoptosis induced by vinblastine *(242)*. JNK could be responsible for BCL-2 inactivation as treatment with a JNK-specific antisense oligonucleotide prevents paclitaxel-induced JNK activation, Bcl-2 hyperphosphorylation, and apoptosis *(243)*. A combination of dominant-negative ASK1 and JNK inhibits paclitaxel-induced Bcl-2 hyperphosphorylation, indicating that activation of the ASK1/JNK pathway by paclitaxel could be necessary for this process *(135,244)*. Similarly, vinblastine-induced phosphorylation of Bcl-2 and Bcl-X$_L$ is mediated by JNK *(245)*. However, opposite results have shown that activation of JNK is not required for paclitaxel-induced Bcl-2 hyperphosphorylation *(246)*, and that the loss of JNK activation increases apoptosis in response to paclitaxel *(247)*. Interestingly, JNK activation is described to be transiently required *(241,246)*. The fact that different phases of paclitaxel-induced apoptosis may involve JNK or alternately proceed through JNK-independent signaling pathways could explain the discrepancy in the observations.

3.4.4. RAF-1 KINASE

MDAs specifically induce Raf-1 kinase activation through disruption of microtubules *(248)*. Downregulation of Raf-1 is associated with paclitaxel-acquired resistance in human breast cancer cells *(249)*. Paclitaxel-induced Raf-1 activation coincides with inactivation of Bcl-2 by phosphorylation and depletion of Raf-1 prevents Bcl-2 phosphorylation and apoptosis *(248,250)*. Raf-1 kinase functions as a central component of the MAPK signal transduction pathway, but in leukemia cells, MDAs have been shown to induce Raf-1/Bcl-2 hyperphosphorylation in a MAPK-independent manner *(142)*. These data are in agreement with the description of a cellular fraction of Raf-1 directly associated with Bcl-2 *(251)*. In contrast with these observations, Raf-1 is described to be not involved in Bcl-2 hyperphosphorylation in human leukemia cells *(40)*. Moreover, the association between Bcl-2 and Raf-1 seems to target the kinase to mitochondria, allowing Raf-1 to contribute to cellular survival by phosphorylating Bad *(51)*. Last, Raf-1 is inactivated by MDAs in epidermoid carcinoma cells *(220)* and evidence indicates that paclitaxel-induced apoptosis is mediated by JNK and occurs in parallel with suppression of the Raf-1 kinase activity. Thus, Raf-1 kinase appears to play a central role in some cases of MDA-induced apoptosis, but its impact remains unclear in other cases. Its role is probably cell type-dependent.

3.5. Involvement of Cellular Stress Markers

3.5.1. REACTIVE OXYGEN SPECIES

ROS production is modified by MDAs. The production of peroxide free radicals and superoxide anions is enhanced by paclitaxel, docetaxel, and 2-methoxyestradiol *(252)*. Paclitaxel treatment also rapidly results in an increase in nitric oxide (NO) production, whereas nocodazole treatment induces its decrease *(253)*. Vinblastine treatment has been shown to induce apoptosis with generation of ROS *(254)*. However, it remains unclear whether ROS *per se* participate in the induction of apoptosis by MDAs or only constitute a cellular stress marker during the apoptotic process. Actually, although the antioxidant magnonol is effective in decreasing paclitaxel, docetaxel and 2-methoxyestradiol-induced peroxide production, it does not prevent drug-induced cell-cycle arrest and apoptosis in hepatoma cells *(252)*. Similarly, pretreatment with the antioxidants *N*-acetyl-L-cysteine, ascorbic acid, or vitamin E has no effect on JNK activation by MDAs in human breast cancer cells *(255)*. On the other hand, *N*-acetyl-L-cysteine and glutathione (GSH) abolish vinblastine-induced cell death in human lung cancer cells *(254)*, suggesting that ROS play a role in some types of MDA-induced apoptosis. In agreement, hydrogen peroxide generation has been identified as a determining event for paclitaxel effectiveness *(255a)*. Last, as described for low-level oxidative signals *(235)*, ROS could contribute to the mitotic block induced by MDAs, but this role remains to be evaluated.

3.5.2. HEAT SHOCK PROTEINS

Paclitaxel and vincristine have been shown to selectively increase the amount of membrane and cytoplasmic HSP70 in tumor cells *(256)*. Such accumulation of HSP70 could be mediated by an increase in hydrogen peroxide production *(257)*. HSP70-over-expressing cells show some resistance to paclitaxel *(258)*, indicating that an increase in HSP activity might constitute a survival response to MDAs. Inhibition of HSP90 functions sensitizes tumors to paclitaxel *(259)*, confirming that HSPs are involved in resistance to MDAs. Interestingly, HSP90 binds to tubulin dimers in vitro *(260)*, and HSP70 binds to polymerized tubulin in vitro *(261)*. HSP90 colocalizes with microtubules in fibroblasts, PtK cells, and endothelial cells *(262,263)*. MDAs do not affect the HSP90 protein content in pulmonary artery endothelial cells, but paclitaxel increases the amount of HSP90 binding to NO synthase (NOS) whereas nocodazole decreases the amount of HSP90 binding to NOS *(253)*. Inactivation of HSP90 by geldanamycin prevents the paclitaxel-induced increase in NOS activity. Thus, the changes in NO production induced by MDAs depend, at least in part, on the ability of HSP90 to bind and activate NOS *(253)*. It has been suggested that increased tubulin polymerization by paclitaxel could bring HSP90 and NOS physically closer to each other, thus promoting their association, or the increased tubulin polymerization could alter the conformation of bound HSP90, increasing its affinity for NOS *(264)*, but it is not clear how this might occur. Last, it has become apparent that, independently of its binding to microtubules, paclitaxel can directly bind to HSP90 and HSP70 *(265,266)*, but the clinical value of this putative activity against HSPs remains to be determined.

3.6. Modulation of Cell Survival Factors

3.6.1. NFκB

The transcription factor NFκB upregulates the expression of antiapoptotic genes such as the *IAPs*, *Bcl-X$_L$*, *Bcl-2*, and *c-myc*. Treatment with paclitaxel rapidly activates NFκB in mouse macrophages *(267)*, as does vinblastine in colon carcinoma cells, apparently through the degradation of the NFκB inhibitor, IκB *(268)*. Diverse components that inactivate NFκB increase docetaxel and paclitaxel-induced apoptosis *(269,270)*, strongly suggesting that NFκB may participate in cellular resistance to MDAs. Likewise, overexpression of IκB or blockade of its intracellular degradation makes cells more sensitive to paclitaxel *(269,271)*. Surprisingly, solid tumor cells stably transfected with antisense IκBα-expression vectors, have increased sensitivity to paclitaxel-induced apoptosis *(272)*. Furthermore, stable transfection of a mutant IκBα that was insensitive to degradation resulted in reduced sensitivity of human tumor cells to paclitaxel-induced apoptosis *(273)*. Thus, activation of the NFκB/IκB signaling pathway might play an important role in determining the susceptibility of tumor cells to paclitaxel-induced apoptosis. All of these data highlight the paradoxical role of NFκB in proliferation or apoptosis. Its activation by MDAs may either constitute a cell survival signal to resist to imminent apoptosis or contribute to the mediation of tumor cell death.

3.6.2. Akt

Akt, also called PKB, a serine/threonine kinase known as an important survival factor has been investigated as a possible target for MDA-induced apoptosis. Akt inhibits apoptosis through inactivation, by phosphorylation, of various targets such as Bax and caspase-9 *(52,274)*. By inhibiting the release of cytochrome-*c*, overexpression of Akt has been described as a resistance factor to paclitaxel treatment *(275)*. However, whether suppression of Akt activity affects tumor cell sensitivity to paclitaxel remains to be elucidated, as the opposite results have also been described *(259,276)*.

3.6.3. 70 kDa Ribosomal S6 Kinase

The 70 kDa ribosomal S6 kinase (p70S6K) is involved in cell growth and survival, and is constitutively activated in cancer cells *(277)*. Activation of p70S6K requires a phosphorylation triggered by growth factors. In multiple breast and ovarian cancer cell

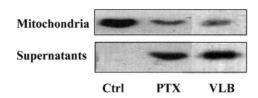

Mitochondria

Supernatants

Ctrl   PTX   VLB

Fig. 7. Release of cytochrome-*c* from isolated mitochondria incubated with MDAs. Isolated mito-chondria from SK-N-SH cells were untreated (Ctrl) or incubated with 50 μ*M* paclitaxel (PTX) or 1.5 μ*M* vinblastine (VLB) for 2 h at 37°C. Mitochondrial pellets and their respective supernatants were then analysed by immunoblotting using cytochrome-*c* antibody. Both the stabilizing and the depolymeris-ing agent affected isolated mitochondria by directly inducing the release of cytochrome-*c*.

lines, paclitaxel is able to induce p70S6K phosphorylation at threonine 421 and serine 424. Despite the phosphorylation of p70S6K, paclitaxel inactivates this kinase in a concentration- and time-dependent manner *(278)*. The kinases PKC and JNK seem to be responsible for this phosphorylation, as their inhibition prevents paclitaxel-induced p70S6K inactivation. Interestingly, the paclitaxel-induced phosphorylation and low activity of p70S6K mainly occurs during mitosis. Finally, Bad has been described as a substrate for this p70S6K *(279)*, indicating that its inactivation may allow the dephos-phorylation-mediated activation of Bad. Thus, the antitumor effects of paclitaxel may involve, at least in part, inhibition of p70S6K.

4. DIRECT EFFECTS OF MDAS ON MITOCHONDRIA

4.1. Action of MDAs on Isolated Mitochondria

In parallel with their effects on diverse cellular proteins leading to the activation of the intrinsic signaling pathway (*see* Section 3), MDAs could also activate the apoptotic path-way through a direct action on mitochondria. Interestingly, incubation of isolated mito-chondria with paclitaxel, vinorelbine, or vinblastine induces a PTP-dependent cytochrome-*c* release *(280–282)* (Fig. 7). This effect is specific to MDAs, as doxorubucin, a DNA-damaging agent is not able to induce release of cytochrome-*c* from isolated mitochondria *(281)*. In addition, other MDAs as docetaxel, nocodazole, vincristine, discodermolide, and CI 980 (personnal data) and drugs such as arsenic trioxide that bind to SH groups of tubulin also induce the release of cytochrome-*c* from isolated mitochondria *(282)*.

Paclitaxel induces a large amplitude swelling of purified mitochondria and a collapse of the mitochondrial $\Delta\Psi m$ *(280,281)*. It provokes a loss of mitochondrial Ca^{2+} through PTP opening in isolated mitochondria *(283)*. The direct effect of MDAs on mitochon-dria is relevant in intact cells and it occurs early in the apoptotic pathway. During the apoptotic process triggered by paclitaxel in human neuroblastoma cells, cytochrome-*c* release occurs before caspase activation *(153)*. Caspase inhibition does not prevent cytochrome-*c* release in neuroblastoma cells nor in gastric cancer cells treated by pacli-taxel and docetaxel respectively *(36,153)*. In mouse cells, interference of paclitaxel with the mitochondrial Ca^{2+} signal cascade is observed upstream of a significant effect on microtubule organization *(283)*. In parallel with its effect on mitochondrial perme-ability, paclitaxel induces changes in the respiration rate of mitochondria. It signifi-cantly increases the cytochrome oxidase-mediated ROS production by purified liver mitochondria *(280)*. The loss of $\Delta\Psi m$ induced by paclitaxel in isolated mitochondria is amplified by stimulation of the respiratory state *(284)*. In agreement with these data, paclitaxel-induced cytochrome-*c* release is inhibited by respiratory chain inhibitors *(282)*. An increase in the respiration rate of mitochondria is also detected following paclitaxel treatment *(153)*. Thus, modulation of the respiration rate and ROS levels as well as cytochrome-*c* release mediated by the decrease in $\Delta\Psi m$, may trigger apoptotic signaling pathways as a result of direct effects of MDAs on mitochondria.

4.2. Putative Targets of MDAs on Mitochondria

It has been shown that paclitaxel binds to mitochondria *(284)*, but the mitochondrial target for paclitaxel and other MDAs remains undefined.

4.2.1. BCL-2

Bcl-2 was identified as a potential mitochondrial target for paclitaxel by screening a library of phage-displayed peptides *(285)*, and a model for the interaction of paclitaxel with the Bcl-2 loop domain has been proposed *(286)*. These data are supported by the observation that is associated with a strong resistance to MDAs in ovarian tumor cells *(284,284a)*. Examination of Bcl-2 by immunohistochemistry in tumors from a small number of ovarian cancer patients with paclitaxel-resistant tumors revealed that Bcl-2 is downregulated in this clinical setting when levels are compared with those found in drug-sensitive tumors *(284)*. Similarly, treatment of prostate cancer cells expressing Bcl-2 with paclitaxel induces apoptosis, whereas treatment of the Bcl-2 negative cells does not induce apoptosis *(287)*. Thus, even if Bcl-2 is an antiapoptotic protein, a basal level may be necessary for the direct binding of paclitaxel to mitochondria. Whether Vinca alkaloids can bind to Bcl-2 remains unknown.

4.2.2. MITOCHONDRIAL TUBULIN

Because MDAs specifically act on mitochondria, another putative target of interest could be tubulin, the primary target of these drugs. It has been found that tubulin is present on mitochondrial membranes, both in mitochondria in purified suspensions and in whole cells *(288)*. In neuroblastoma SK-N-SH cells, mitochondrial tubulin represents $2.2 \pm 0.5\%$ of the total cellular tubulin, and the mitochondrial tubulin is enriched in acetylated and tyrosinated α-tubulin as well as in class III β-tubulin *(288)*. Moreover, by coimmunoprecipitation, a specific association between mitochondrial tubulin and VDAC, the main outer membrane PTP component has been reported (Fig. 8) *(288)*. Such an association strongly suggests that mitochondrial tubulin could play a direct role in the control of apoptosis, and that paclitaxel-induced release of cytochrome-*c* may involve its binding to mitochondrial tubulin. Finally, mitochondrial tubulin has also been found to be associated with Bcl-2 (Fig. 9A) *(284)*, and a direct interaction between Bcl-2 and tubulin purified from lamb brains has been shown (Fig. 9B). Then, a complex between VDAC, tubulin and Bcl-2 could regulate initiation of apoptotic signaling pathways in mitochondria.

Although the direct target of MDAs on mitochondria is not yet well-defined, these data suggest that some of the fastest effects of the MDAs might be because of a direct action on mitochondrial integrity.

5. PERSPECTIVES

In this chapter, the major mitochondrial apoptotic signaling pathway have been focused, and based on the available data, it would be proposed that mitochondria should

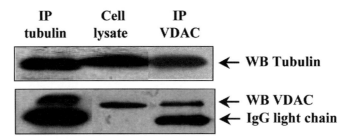

Fig. 8. Association between mitochondrial tubulin and VDAC. Isolated mitochondrial lysates were subjected to immunoprecipation (IP) with antibodies to α-tubulin or to VDAC. The resulting immunoprecipitates were analyzed by Western blotting (WB) with both anti-α-tubulin and antiVDAC antibodies. Immunoglobulin (IgGs) present in the precipitates are represented by the light chain on the figure. The antibody to α-tubulin coimmunoprecipitated tubulin and VDAC, and immunoprecipitation of VDAC brought down tubulin. Thus, mitochondrial tubulin is associated with the mitochondrial outer membrane through its binding to VDAC.

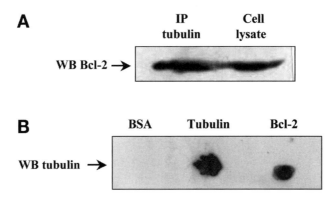

Fig. 9. Tubulin/Bcl-2 interaction. **(A)** Lysates from isolated mitochondria were subjected to immunoprecipitation (IP) with antibodies to α-tubulin. The resulting immunoprecipitates were analyzed by Western blotting (WB) with antiBcl-2 antibodies. As immunoprecipitation of tubulin brought down Bcl-2, these two proteins are likely to be partners at the mitochondrial membrane. **(B)** Dot blot experiments were performed by loading 2 μg of bovine serum albumin, pure tubulin, and Bcl-2 on a nitrocellulose membrane. The membrane was incubated with a suspension of tubulin and analyzed by WB with antiα-tubulin antibodies. The binding of tubulin to Bcl-2 indicates a direct interaction between the two proteins.

be investigated as an additional target for MDAs. Nevertheless, it cannot be excluded that other apoptotic pathways, previously considered as being minor till now, may also be involved. Besides its association with mitochondria, tubulin has been shown to be embedded or anchored in different membranes such as plasma membrane *(289,290)*. In addition, βII-tubulin has been localized to the nucleus *(291)* and γ-tubulin is bound to the Golgi apparatus, allowing possible nucleation of microtubules from this organelle *(292)*. The function(s) of tubulin in these unusual locations is not known, but it could be involved in integrating signals between different subcellular compartments.

Importantly, these subcellular compartments also contain apoptotic signaling proteins that ensure cross-talk *(293)*. In particular, antiapoptotic Bcl-2 members and proapoptotic Bax and Bak reside on the ER membranes *(294)* and might participate in the regulation of apoptosis by modifying ER Ca^{2+} storage (reviewed in refs. *295,296*) as described earlier for mitochondrial Ca^{2+}. There is accumulating evidence that mitochondria and the ER are connected in regulating apoptosis. Overexpression of Bcl-2 leads to the reduction of mitochondrial Ca^{2+} stores and to the enhancement of Ca^{2+} sequestration in ER *(294)*. In addition, ER-Bcl-2 inhibits Bax activation and cytochrome-*c* release from mitochondria thanks to an intermediate that is likely a BH3-only protein *(297)*. Furthermore, ER-targeted but not mitochondria-targeted Bak leads to a progressive depletion of ER Ca^{2+} and induces caspase-12 cleavage. Thus, this organelle might also initiate an alternative pathway of caspase activation and a mitochondria-independent apoptotic process *(294)*. ER contributions to apoptosis are only now becoming apparent, and require to be evaluated following treatment with MDAs.

Last, few studies have addressed the eventual effects of MDAs on tumor cell migration and interaction with host cells or the extracellular matrix *(298–301)*, major events involved in tumor invasion. Angiogenesis is another process largely participating in tumor progression, and its inhibition by drugs that can increase apoptosis of endothelial cells is a current challenge in cancer research. Interestingly, studies have shown evidence for an antiangiogenic effect of MDAs when used at low concentrations *(302–304)*. The signal pathway involved in the inhibition of endothelial cell proliferation by MDAs may differ from those involved in cancer cells *(305,306)* and it remains to be further investigated.

6. CONCLUSIONS

MDAs trigger apoptosis by activating the intrinsic mitochondrial signaling pathway. They modulate both the activity and the expression levels of Bcl-2 family members, turning their ratio in favor of mitochondrial permeability, leading to apoptogenic factor release and caspase activation. In general, MDA concentrations that trigger apoptosis also induce microtubular network disturbance and mitotic block. Thus, microtubule-linked signaling proteins and cell-cycle progression regulators are likely to interfere with the MDA-activated apoptotic machinery. In addition, tubulin associated with cell compartment membranes, as mitochondrial membranes, may be also involved in transduction of MDA-induced signals. Finally, further investigation of the MDA-activated apoptotic pathways will improve the knowledge on their mechanism of action in preventing tumor progression. Identification of key proteins involved in apoptotic pathways may allow the combinaison of new pharmacological modulators with MDAs to amplify their anticancer efficacy.

REFERENCES

1. Kerr JF, Wyllie AH, Currie AR. Apoptosis: a basic biological phenomenon with wide-ranging implications in tissue kinetics. Br J Cancer 1972;26:239–257.
2. Conradt B, Horvitz HR. The C. elegans protein EGL-1 is required for programmed cell death and interacts with the Bcl-2-like protein CED-9. Cell 1998;93:519–529.
3. Stergiou L, Hengartner MO. Death and more: DNA damage response pathways in the nematode C. elegans. Cell Death Differ 2004;11:21–28.
4. Wyllie AH. Apoptosis and the regulation of cell numbers in normal and neoplastic tissues: an overview. Cancer Metastasis Rev 1992;11:95–103.
5. Meier P, Finch A, Evan G. Apoptosis in development. Nature 2000;407:796–801.
6. Thompson CB. Apoptosis in the pathogenesis and treatment of disease. Science 1995;267:1456–1462.
7. Martin SJ, Green DR. Apoptosis and cancer: the failure of controls on cell death and cell survival. Crit Rev Oncol Hematol 1995;18:137–153.

8. Cutts JH, Beer CT, Noble RL. Biological properties of Vincaleukoblastine, an alkaloid in Vinca rosea Linn, with reference to its antitumor action. Cancer Res 1960;20:1023–1031.

9. Jordan MA, Thrower D, Wilson L. Mechanism of inhibition of cell proliferation by Vinca alkaloids. Cancer Res 1991;51:2212–2222.

10. Schiff PB, Fant J, Horwitz SB. Promotion of microtubule assembly in vitro by taxol. Nature 1979; 277:665–667.

11. Jordan MA, Wilson L. Microtubules as a target for anticancer drugs. Nat Rev Cancer 2004;4: 253–265.

12. Honore S, Pasquier E, Braguer D. Understanding microtubule dynamics for improved cancer therapy Cell Mol Life Sci 2005;62:3039–3056.

13. Wang TH, Wang HS, Soong YK. Paclitaxel-induced cell death: where the cell cycle and apoptosis come together. Cancer 2000;88:2619–2628.

14. Vilpo JA, Koski T, Vilpo LM. Selective toxicity of vincristine against chronic lymphocytic leukemia cells in vitro. Eur J Haematol 2000;65:370–378.

15. Torres K, Horwitz SB. Mechanisms of Taxol-induced cell death are concentration dependent. Cancer Res 1998;58:3620–3626.

16. Giannakakou P, Robey R, Fojo T, Blagosklonny MV. Low concentrations of paclitaxel induce cell type-dependent p53, p21 and G1/G2 arrest instead of mitotic arrest: molecular determinants of paclitaxel-induced cytotoxicity. Oncogene 2001;20:3806–3813.

17. Blajeski AL, Phan VA, Kottke TJ, Kaufmann SH. G(1) and G(2) cell-cycle arrest following microtubule depolymerization in human breast cancer cells. J Clin Invest 2002;110:91–99.

18. Carles G, Braguer D, Dumontet C, et al. Differentiation of human colon cancer cells changes the expression of beta-tubulin isotypes and MAPs. Br J Cancer 1999;80:1162–1168.

19. Finkel E. The mitochondrion: is it central to apoptosis? Science 2001;292:624–626.

20. Strasser A. Life and death during lymphocyte development and function: evidence for two distinct killing mechanisms. Curr Opin Immunol 1995;7:228–234.

21. Lawen A. Apoptosis-an introduction. Bioessays 2003;25:888–896.

22. Locksley RM, Killeen N, Lenardo MJ. The TNF and TNF receptor superfamilies: integrating mammalian biology. Cell 2001;104:487–501.

23. Boldin MP, Goncharov TM, Goltsev YV, Wallach D. Involvement of MACH, a novel MORT1/FADD-interacting protease, in Fas/APO-1- and TNF receptor-induced cell death. Cell 1996;85:803–815.

24. Muzio M, Chinnaiyan AM, Kischkel FC, et al. FLICE, a novel FADD-homologous ICE/CED-3-like protease, is recruited to the CD95 (Fas/APO-1) death—inducing signaling complex. Cell 1996;85: 817–827.

25. Medema JP, Scaffidi C, Kischkel FC, et al. FLICE is activated by association with the CD95 death-inducing signaling complex (DISC). Embo J 1997;16:2794–2804.

26. Scaffidi C, Fulda S, Srinivasan A, et al. Two CD95 (APO-1/Fas) signaling pathways. Embo J 1998; 17:1675–1687.

27. Huang DC, Hahne M, Schroeter M, et al. Activation of Fas by FasL induces apoptosis by a mechanism that cannot be blocked by Bcl-2 or Bcl-x(L). Proc Natl Acad Sci USA 1999;96:14,871–14,876.

28. Li H, Zhu H, Xu CJ, Yuan J. Cleavage of BID by caspase 8 mediates the mitochondrial damage in the Fas pathway of apoptosis. Cell 1998;94:491–501.

29. Green DR, Ferguson TA. The role of Fas ligand in immune privilege. Nat Rev Mol Cell Biol 2001;2:917–924.

30. Fulda S, Susin SA, Kroemer G, Debatin KM. Molecular ordering of apoptosis induced by anticancer drugs in neuroblastoma cells. Cancer Res 1998;58:4453–4460.

31. Goncalves A, Braguer D, Carles G, Andre N, Prevot C, Briand C. Caspase-8 activation independent of CD95/CD95-L interaction during paclitaxel-induced apoptosis in human colon cancer cells (HT29-D4). Biochem Pharmacol 2000;60:1579–1584.

32. Ferreira CG, Tolis C, Span SW, et al. Drug-induced apoptosis in lung cnacer cells is not mediated by the Fas/FasL (CD95/APO1) signaling pathway. Clin Cancer Res 2000;6:203–212.

33. Wieder T, Essmann F, Prokop A, et al. Activation of caspase-8 in drug-induced apoptosis of B-lymphoid cells is independent of CD95/Fas receptor-ligand interaction and occurs downstream of caspase-3. Blood 2001;97:1378–1387.

34. von Haefen C, Wieder T, Essmann F, Schulze-Osthoff K, Dorken B, Daniel PT. Paclitaxel-induced apoptosis in BJAB cells proceeds via a death receptor-independent, caspases-3/-8-driven mitochondrial amplification loop. Oncogene 2003;22:2236–2247.

35. Pucci B, Bellincampi L, Tafani M, Masciullo V, Melino G, Giordano A. Paclitaxel induces apoptosis in Saos-2 cells with CD95L upregulation and Bcl-2 phosphorylation. Exp Cell Res 1999;252: 134–143.

36. Kim R, Tanabe K, Emi M, Uchida Y, Toge T. Death receptor-dependent and -independent pathways in anticancer drug-induced apoptosis of breast cancer cells. Oncol Rep 2003;10:1925–1930.

37. Adams JM, Cory S. Life-or-death decisions by the Bcl-2 protein family. Trends Biochem Sci 2001; 26:61–66.

38. Borner C. The Bcl-2 protein family: sensors and checkpoints for life-or-death decisions. Mol Immunol 2003;39:615–647.

39. Suzuki M, Youle RJ, Tjandra N. Structure of Bax: coregulation of dimer formation and intracellular localization. Cell 2000;103:645–654.

40. Ibrado AM, Liu L, Bhalla K. Bcl-xL overexpression inhibits progression of molecular events leading to paclitaxel-induced apoptosis of human acute myeloid leukemia HL-60 cells. Cancer Res 1997;57:1109–1115.

41. Reed JC. Bcl-2 family proteins. Oncogene 1998;17:3225–3236.

42. Haldar S, Basu A, Croce CM. Bcl2 is the guardian of microtubule integrity. Cancer Res 1997;57: 229–233.

43. Orrenius S, Zhivotovsky B, Nicotera P. Regulation of cell death: the calcium-apoptosis link. Nat Rev Mol Cell Biol 2003;4:552–565.

44. Priault M, Cartron PF, Camougrand N, Antonsson B, Vallette FM, Manon S. Investigation of the role of the C-terminus of Bax and of tc-Bid on Bax interaction with yeast mitochondria. Cell Death Differ 2003;10:1068–1077.

45. Griffiths GJ, Corfe BM, Savory P, et al. Cellular damage signals promote sequential changes at the N-terminus and BH-1 domain of the pro-apoptotic protein Bak. Oncogene 2001;20:7668–7676.

46. Cartron PF, Priault M, Oliver L, Meflah K, Manon S, Vallette FM. The N-terminal end of Bax contains a mitochondrial-targeting signal. J Biol Chem 2003;278:11,633–11,641.

47. Ohtsuka T, Ryu H, Minamishima YA, et al. ASC is a Bax adaptor and regulates the p53-Bax mitochondrial apoptosis pathway. Nat Cell Biol 2004;6:121–128.

48. Desagher S, Martinou JC. Mitochondria as the central control point of apoptosis. Trends Cell Biol 2000;10:369–377.

49. Heimlich G, McKinnon AD, Bernardo K, et al. Bax-induced cytochrome c release from mitochondria depends on alpha-helices-5 and -6. Biochem J 2004;378:247–255.

50. Mikhailov V, Mikhailova M, Pulkrabek DJ, Dong Z, Venkatachalam MA, Saikumar P. Bcl-2 prevents Bax oligomerization in the mitochondrial outer membrane. J Biol Chem 2001;276:18,361–18,374.

51. Wang HG, Rapp UR, Reed JC. Bcl-2 targets the protein kinase Raf-1 to mitochondria. Cell 1996;87:629–638.

52. Datta SR, Dudek H, Tao X, et al. Akt phosphorylation of BAD couples survival signals to the cell-intrinsic death machinery. Cell 1997;91:231–241.

53. Harada H, Becknell B, Wilm M, et al. Phosphorylation and inactivation of BAD by mitochondria-anchored protein kinase A. Mol Cell 1999;3:413–422.

54. Zha J, Harada H, Yang E, Jockel J, Korsmeyer SJ. Serine phosphorylation of death agonist BAD in response to survival factor results in binding to 14-3-3 not BCL-X(L). Cell 1996;87:619–628.

55. Verma S, Zhao LJ, Chinnadurai G. Phosphorylation of the pro-apoptotic protein BIK: mapping of phosphorylation sites and effect on apoptosis. J Biol Chem 2001;276:4671–4676.

56. Zha J, Weiler S, Oh KJ, Wei MC, Korsmeyer SJ. Posttranslational N-myristoylation of BID as a molecular switch for targeting mitochondria and apoptosis. Science 2000;290:1761–1765.

57. Puthalakath H, Huang DC, O'Reilly LA, King SM, Strasser A. The proapoptotic activity of the Bcl-2 family member Bim is regulated by interaction with the dynein motor complex. Mol Cell 1999;3:287–296.

58. Puthalakath H, Villunger A, O'Reilly LA, et al. Bmf: a proapoptotic BH3-only protein regulated by interaction with the myosin V actin motor complex, activated by anoikis. Science 2001;293:1829–1832.

59. Oltvai ZN, Milliman CL, Korsmeyer SJ. Bcl-2 heterodimerizes in vivo with a conserved homolog, Bax, that accelerates programmed cell death. Cell 1993;74:609–619.

60. Sattler M, Liang H, Nettesheim D, et al. Structure of Bcl-xL-Bak peptide complex: recognition between regulators of apoptosis. Science 1997;275:983–986.

61. Willis SN, Adams JM. Life in the balance: how BH3-only proteins induce apoptosis. Curr Opin Cell Biol 2005;17:617–625.

62. Zong WX, Lindsten T, Ross AJ, MacGregor GR, Thompson CB. BH3-only proteins that bind pro-survival Bcl-2 family members fail to induce apoptosis in the absence of Bax and Bak. Genes Dev 2001;15:1481–1486.
63. Kaufmann T, Schinzel A, Borner C. Bcl-w(edding) with mitochondria. Trends Cell Biol 2004;14: 8–12.
64. Wilson-Annan J, O'Reilly LA, Crawford SA, et al. Proapoptotic BH3-only proteins trigger membrane integration of prosurvival Bcl-w and neutralize its activity. J Cell Biol 2003;162:877–887.
65. Liu FT, Newland AC, Jia L. Bax conformational change is a crucial step for PUMA-mediated apoptosis in human leukemia. Biochem Biophys Res Commun 2003;310:956–962.
66. Van Loo G, Demol H, van Gurp M, et al. A matrix-assisted laser desorption ionization post-source decay (MALDI-PSD) analysis of proteins released from isolated liver mitochondria treated with recombinant truncated Bid. Cell Death Differ 2002;9:301–308.
67. van Gurp M, Festjens N, van Loo G, Saelens X, Vandenabeele P. Mitochondrial intermembrane proteins in cell death. Biochem Biophys Res Commun 2003;304:487–497.
68. Scarlett JL, Sheard PW, Hughes G, Ledgerwood EC, Ku HH, Murphy MP. Changes in mitochondrial membrane potential during staurosporine-induced apoptosis in Jurkat cells. FEBS Lett 2000;475:267–272.
69. Ichas F, Mazat JP. From calcium signaling to cell death: two conformations for the mitochondrial permeability transition pore. Switching from low- to high-conductance state. Biochim Biophys Acta 1998;1366:33–50.
70. Ly JD, Grubb DR, Lawen A. The mitochondrial membrane potential (deltapsi(m)) in apoptosis; an update. Apoptosis 2003;8:115–128.
71. Waterhouse NJ, Ricci JE, Green DR. And all of a sudden it's over: mitochondrial outer-membrane permeabilization in apoptosis. Biochimie 2002;84:113–121.
72. Degli Esposti M, Dive C. Mitochondrial membrane permeabilisation by Bax/Bak. Biochem Biophys Res Commun 2003;304:455–461.
73. Tsujimoto Y, Shimizu S. The voltage-dependent anion channel: an essential player in apoptosis. Biochimie 2002;84:187–193.
74. Arnoult D, Parone P, Martinou JC, Antonsson B, Estaquier J, Ameisen JC. Mitochondrial release of apoptosis-inducing factor occurs downstream of cytochrome c release in response to several proapoptotic stimuli. J Cell Biol 2002;159:923–929.
75. Penninger JM, Kroemer G. Mitochondria, AIF and caspases—rivaling for cell death execution. Nat Cell Biol 2003;5:97–99.
76. Li P, Nijhawan D, Budihardjo I, et al. Cytochrome c and dATP-dependent formation of Apaf-1/caspase-9 complex initiates an apoptotic protease cascade. Cell 1997;91:479–489.
77. Susin SA, Lorenzo HK, Zamzami N, et al. Mitochondrial release of caspase-2 and -9 during the apoptotic process. J Exp Med 1999;189:381–394.
78. Zhivotovsky B, Samali A, Gahm A, Orrenius S. Caspases: their intracellular localization and translocation during apoptosis. Cell Death Differ 1999;6:644–651.
79. Qin ZH, Wang Y, Kikly KK, et al. Pro-caspase-8 is predominantly localized in mitochondria and released into cytoplasm upon apoptotic stimulation. J Biol Chem 2001;276:8079–8086.
80. Chandra D, Tang DG. Mitochondrially localized active caspase-9 and caspase-3 result mostly from translocation from the cytosol and partly from caspase-mediated activation in the organelle. Lack of evidence for Apaf-1-mediated procaspase-9 activation in the mitochondria. J Biol Chem 2003;278:17,408–17,420.
81. Grutter MG. Caspases: key players in programmed cell death. Curr Opin Struct Biol 2000;10: 649–655.
82. Slee EA, Harte MT, Kluck RM, et al. Ordering the cytochrome c-initiated caspase cascade: hierarchical activation of caspases–2, –3, –6, –7, –8, and –10 in a caspase-9-dependent manner. J Cell Biol 1999;144:281–292.
83. Gottlieb RA. Mitochondria: execution central. FEBS Lett 2000;482:6–12.
84. Lassus P, Opitz-Araya X, Lazebnik Y. Requirement for caspase-2 in stress-induced apoptosis before mitochondrial permeabilization. Science 2002;297:1352–1354.
85. Guo Y, Srinivasula SM, Druilhe A, Fernandes-Alnemri T, Alnemri ES. Caspase-2 induces apoptosis by releasing proapoptotic proteins from mitochondria. J Biol Chem 2002;277:13,430–13,437.
86. Roy N, Deveraux QL, Takahashi R, Salvesen GS, Reed JC. The c-IAP-1 and c-IAP-2 proteins are direct inhibitors of specific caspases. Embo J 1997;16:6914–6925.

87. Liu T, Brouha B, Grossman D. Rapid induction of mitochondrial events and caspase-independent apoptosis in Survivin-targeted melanoma cells. Oncogene 2004;23:39–48.

88. Salvesen GS, Duckett CS. IAP proteins: blocking the road to death's door. Nat Rev Mol Cell Biol 2002;3:401–410.

89. Li F, Ambrosini G, Chu EY, et al. Control of apoptosis and mitotic spindle checkpoint by survivin. Nature 1998;396:580–584.

90. Fortugno P, Wall NR, Giodini A, et al. Survivin exists in immunochemically distinct subcellular pools and is involved in spindle microtubule function. J Cell Sci 2002;115:575–585.

91. Altieri DC. Survivin, versatile modulation of cell division and apoptosis in cancer. Oncogene 2003;22:8581–8589.

92. Shi Y. A conserved tetrapeptide motif: potentiating apoptosis through IAP-binding. Cell Death Differ 2002;9:93–95.

93. Chai J, Du C, Wu JW, Kyin S, Wang X, Shi Y. Structural and biochemical basis of apoptotic activation by Smac/DIABLO. Nature 2000;406:855–862.

94. Verhagen AM, Silke J, Ekert PG, et al. HtrA2 promotes cell death through its serine protease activity and its ability to antagonize inhibitor of apoptosis proteins. J Biol Chem 2002;277:445–454.

95. Suzuki Y, Nakabayashi Y, Takahashi R. Ubiquitin-protein ligase activity of X-linked inhibitor of apoptosis protein promotes proteasomal degradation of caspase-3 and enhances its anti-apoptotic effect in Fas-induced cell death. Proc Natl Acad Sci USA 2001;98:8662–8667.

96. Susin SA, Lorenzo HK, Zamzami N, et al. Molecular characterization of mitochondrial apoptosis-inducing factor. Nature 1999;397:441–446.

97. Widlak P, Li LY, Wang X, Garrard WT. Action of recombinant human apoptotic endonuclease G on naked DNA and chromatin substrates: cooperation with exonuclease and DNase I. J Biol Chem 2001;276:48,404–48,409.

98. Cande C, Cohen I, Daugas E, et al. Apoptosis-inducing factor (AIF): a novel caspase-independent death effector released from mitochondria. Biochimie 2002;84:215–222.

99. Miramar MD, Costantini P, Ravagnan L, et al. NADH oxidase activity of mitochondrial apoptosis-inducing factor. J Biol Chem 2001;276:16,391–16,398.

100. Fleury C, Mignotte B, Vayssiere JL. Mitochondrial reactive oxygen species in cell death signaling. Biochimie 2002;84:131–141.

101. Sidoti-de Fraisse C, Rincheval V, Risler Y, Mignotte B, Vayssiere JL. TNF-alpha activates at least two apoptotic signaling cascades. Oncogene 1998;17:1639–1651.

102. Kirkland RA, Franklin JL. Bax, reactive oxygen, and cytochrome c release in neuronal apoptosis. Antioxid Redox Signal 2003;5:589–596.

103. Cai J, Jones DP. Superoxide in apoptosis. Mitochondrial generation triggered by cytochrome c loss. J Biol Chem 1998;273:11,401–11,404.

104. Lee HC, Wei YH. Mitochondrial biogenesis and mitochondrial DNA maintenance of mammalian cells under oxidative stress. Int J Biochem Cell Biol 2005;37:822–834.

105. Fernandez-Checa JC. Redox regulation and signaling lipids in mitochondrial apoptosis. Biochem Biophys Res Commun 2003;304:471–479.

106. Nutt LK, Chandra J, Pataer A, et al. Bax-mediated $Ca2^{(}$ mobilization promotes cytochrome c release during apoptosis. J Biol Chem 2002;277:20,301–20,308.

107. Zhu L, Ling S, Yu XD, et al. Modulation of mitochondrial $Ca(2^{()}$ homeostasis by Bcl-2. J Biol Chem 1999;274:33,267–33,273.

108. Wang HG, Pathan N, Ethell IM, et al. $Ca2^{(}$-induced apoptosis through calcineurin dephosphorylation of BAD. Science 1999;284:339–343.

109. Sreedhar AS, Csermely P. Heat shock proteins in the regulation of apoptosis: new strategies in tumor therapy; A comprehensive review. Pharmacol Ther 2004;101:227–257.

110. Hartl FU. Molecular chaperones in cellular protein folding. Nature 1996;381:571–579.

111. Neckers L. Hsp90 inhibitors as novel cancer chemotherapeutic agents. Trends Mol Med 2002;8:S55–S61.

112. Kamal A, Thao L, Sensintaffar J, et al. A high-affinity conformation of Hsp90 confers tumour selectivity on Hsp90 inhibitors. Nature 2003;425:407–410.

113. Parcellier A, Gurbuxani S, Schmitt E, Solary E, Garrido C. Heat shock proteins, cellular chaperones that modulate mitochondrial cell death pathways. Biochem Biophys Res Commun 2003;304: 505–512.

114. Pandey P, Saleh A, Nakazawa A, et al. Negative regulation of cytochrome c-mediated oligomerization of Apaf-1 and activation of procaspase-9 by heat shock protein 90. Embo J 2000;19:4310–4322.

115. Bruey JM, Ducasse C, Bonniaud P, et al. Hsp27 negatively regulates cell death by interacting with cytochrome c. Nat Cell Biol 2000;2:645–652.

116. Beere HM, Wolf BB, Cain K, et al. Heat-shock protein 70 inhibits apoptosis by preventing recruitment of procaspase-9 to the Apaf-1 apoptosome. Nat Cell Biol 2000;2:469–475.

117. Ravagnan L, Gurbuxani S, Susin SA, et al. Heat-shock protein 70 antagonizes apoptosis-inducing factor. Nat Cell Biol 2001;3:839–843.

118. Samali A, Cai J, Zhivotovsky B, Jones DP, Orrenius S. Presence of a pre-apoptotic complex of pro-caspase-3, Hsp60 and Hsp10 in the mitochondrial fraction of jurkat cells. Embo J 1999; 18:2040–2048.

119. Vermeulen K, Berneman ZN, Van Bockstaele DR. Cell cycle and apoptosis. Cell Prolif 2003; 36:165–175.

120. Tang C, Willingham MC, Reed JC, et al. High levels of p26BCL-2 oncoprotein retard taxol-induced apoptosis in human pre-B leukemia cells. Leukemia 1994;8:1960–1969.

121. Huang Y, Ray S, Reed JC, et al. Estrogen increases intracellular p26Bcl-2 to p21Bax ratios and inhibits taxol-induced apoptosis of human breast cancer MCF-7 cells. Breast Cancer Res Treat 1997; 42:73–81.

122. Lebedeva I, Rando R, Ojwang J, Cossum P, Stein CA. Bcl-xL in prostate cancer cells: effects of overexpression and down-regulation on chemosensitivity. Cancer Res 2000;60:6052–6060.

123. Kim R, Tanabe K, Uchida Y, Emi M, Toge T. Effect of Bcl-2 antisense oligonucleotide on drug-sensitivity in association with apoptosis in undifferentiated thyroid carcinoma. Int J Mol Med 2003;11:799–804.

124. Basu A, Haldar S. Identification of a novel Bcl-xL phosphorylation site regulating the sensitivity of taxol- or 2-methoxyestradiol-induced apoptosis. FEBS Lett 2003;538:41–47.

125. Gajate C, Barasoain I, Andreu JM, Mollinedo F. Induction of apoptosis in leukemic cells by the reversible microtubule-disrupting agent 2-methoxy-5-(2',3',4'-trimethoxyphenyl)-2,4,6-cycloheptatrien-1 -one: protection by Bcl-2 and Bcl-X(L) and cell cycle arrest. Cancer Res 2000;60:2651–2659.

126. Haldar S, Jena N, Croce CM. Inactivation of Bcl-2 by phosphorylation. Proc Natl Acad Sci USA 1995;92:4507–4511.

127. Poruchynsky MS, Wang EE, Rudin CM, Blagosklonny MV, Fojo T. Bcl-xL is phosphorylated in malignant cells following microtubule disruption. Cancer Res 1998;58:3331–3338.

128. Basu A, Haldar S. Microtubule-damaging drugs triggered bcl2 phosphorylation-requirement of phosphorylation on both serine-70 and serine-87 residues of bcl2 protein. Int J Oncol 1998;13:659–664.

129. Fang G, Chang BS, Kim CN, Perkins C, Thompson CB, Bhalla KN. "Loop" domain is necessary for taxol-induced mobility shift and phosphorylation of Bcl-2 as well as for inhibiting taxol-induced cytosolic accumulation of cytochrome c and apoptosis. Cancer Res 1998;58:3202–3208.

130. Liu XM, Wang LG, Kreis W, Budman DR, Adams LM. Differential effect of vinorelbine versus paclitaxel on ERK2 kinase activity during apoptosis in MCF-7 cells. Br J Cancer 2001;85:1403–1411.

131. Yuan SY, Hsu SL, Tsai KJ, Yang CR. Involvement of mitochondrial pathway in Taxol-induced apoptosis of human T24 bladder cancer cells. Urol Res 2002;30:282–288.

132. Srivastava RK, Srivastava AR, Korsmeyer SJ, Nesterova M, Cho-Chung YS, Longo DL. Involvement of microtubules in the regulation of Bcl2 phosphorylation and apoptosis through cyclic AMP-dependent protein kinase. Mol Cell Biol 1998;18:3509–3517.

133. Shitashige M, Toi M, Yano T, Shibata M, Matsuo Y, Shibasaki F. Dissociation of Bax from a Bcl-2/Bax heterodimer triggered by phosphorylation of serine 70 of Bcl-2. J Biochem (Tokyo) 2001;130: 741–748.

134. Salah-Eldin AE, Inoue S, Tsukamoto S, Aoi H, Tsuda M. An association of Bcl-2 phosphorylation and Bax localization with their functions after hyperthermia and paclitaxel treatment. Int J Cancer 2003;103:53–60.

135. Yamamoto K, Ichijo H, Korsmeyer SJ. BCL-2 is phosphorylated and inactivated by an ASK1/Jun N-terminal protein kinase pathway normally activated at G(2)/M. Mol Cell Biol 1999;19:8469–8478.

136. Scatena CD, Stewart ZA, Mays D, et al. Mitotic phosphorylation of Bcl-2 during normal cell cycle progression and Taxol-induced growth arrest. J Biol Chem 1998;273:30,777–30,784.

137. Ling YH, Tornos C, Perez-Soler R. Phosphorylation of Bcl-2 is a marker of M phase events and not a determinant of apoptosis. J Biol Chem 1998;273:18,984–18,991.

138. Brichese L, Valette A. PP1 phosphatase is involved in Bcl-2 dephosphorylation after prolonged mitotic arrest induced by paclitaxel. Biochem Biophys Res Commun 2002;294:504–508.

139. Fan M, Du L, Stone AA, Gilbert KM, Chambers TC. Modulation of mitogen-activated protein kinases and phosphorylation of Bcl-2 by vinblastine represent persistent forms of normal fluctuations at G2-M1. Cancer Res 2000;60:6403–6407.

140. Cheng EH, Kirsch DG, Clem RJ, et al. Conversion of Bcl-2 to a Bax-like death effector by caspases. Science 1997;278:1966–1968.

141. Tudor G, Aguilera A, Halverson DO, Laing ND, Sausville EA. Susceptibility to drug-induced apoptosis correlates with differential modulation of Bad, Bcl-2 and Bcl-xL protein levels. Cell Death Differ 2000;7:574–586.

142. Blagosklonny MV, Chuman Y, Bergan RC, Fojo T. Mitogen-activated protein kinase pathway is dispensable for microtubule-active drug-induced Raf-1/Bcl-2 phosphorylation and apoptosis in leukemia cells. Leukemia 1999;13:1028–1036.

143. Liu QY, Stein CA. Taxol and estramustine-induced modulation of human prostate cancer cell apoptosis via alteration in bcl-xL and bak expression. Clin Cancer Res 1997;3:2039–2046.

144. Basu A, Haldar S. The relationship between BcI2, Bax and p53: consequences for cell cycle progression and cell death. Mol Hum Reprod 1998;4:1099–1109.

145. Brichese L, Barboule N, Heliez C, Valette A. Bcl-2 phosphorylation and proteasome-dependent degradation induced by paclitaxel treatment: consequences on sensitivity of isolated mitochondria to Bid. Exp Cell Res 2002;278:101–111.

146. Sumantran VN, Ealovega MW, Nunez G, Clarke MF, Wicha MS. Overexpression of Bcl-XS sensitizes MCF-7 cells to chemotherapy-induced apoptosis. Cancer Res 1995;55:2507–2510.

147. Strobel T, Tai YT, Korsmeyer S, Cannistra SA. BAD partly reverses paclitaxel resistance in human ovarian cancer cells. Oncogene 1998;17:2419–2427.

148. Sawa H, Kobayashi T, Mukai K, Zhang W, Shiku H. Bax overexpression enhances cytochrome c release from mitochondria and sensitizes KATOIII gastric cancer cells to chemotherapeutic agent-induced apoptosis. Int J Oncol 2000;16:745–749.

149. Jones NA, Turner J, McIlwrath AJ, Brown R, Dive C. Cisplatin- and paclitaxel-induced apoptosis of ovarian carcinoma cells and the relationship between bax and bak up-regulation and the functional status of p53. Mol Pharmacol 1998;53:819–826.

150. Yamaguchi H, Paranawithana SR, Lee MW, Huang Z, Bhalla KN, Wang HG. Epothilone B analogue (BMS-247550)-mediated cytotoxicity through induction of Bax conformational change in human breast cancer cells. Cancer Res 2002;62:466–471.

151. Makin GW, Corfe BM, Griffiths GJ, Thistlethwaite A, Hickman JA, Dive C. Damage-induced Bax N-terminal change, translocation to mitochondria and formation of Bax dimers/complexes occur regardless of cell fate. EMBO J 2001;20:6306–6315.

152. Strobel T, Kraeft SK, Chen LB, Cannistra SA. BAX expression is associated with enhanced intracellular accumulation of paclitaxel: a novel role for BAX during chemotherapy-induced cell death. Cancer Res 1998;58:4776–4781.

153. Andre N, Carre M, Brasseur G, et al. Paclitaxel targets mitochondria upstream of caspase activation in intact human neuroblastoma cells. FEBS Lett 2002;532:256–260.

154. Perkins CL, Fang G, Kim CN, Bhalla KN. The role of Apaf-1, caspase-9, and bid proteins in etoposide- or paclitaxel-induced mitochondrial events during apoptosis. Cancer Res 2000;60:1645–1653.

155. Suzuki A, Kawabata T, Kato M. Necessity of interleukin-1beta converting enzyme cascade in taxotere-initiated death signaling. Eur J Pharmacol 1998;343:87–92.

156. Pan J, Xu G, Yeung SC. Cytochrome c release is upstream to activation of caspase-9, caspase-8, and caspase-3 in the enhanced apoptosis of anaplastic thyroid cancer cells induced by manumycin and paclitaxel. J Clin Endocrinol Metab 2001;86:4731–4740.

157. Perkins C, Kim CN, Fang G, Bhalla KN. Overexpression of Apaf-1 promotes apoptosis of untreated and paclitaxel- or etoposide-treated HL-60 cells. Cancer Res 1998;58:4561–4566.

158. Uyar D, Takigawa N, Mekhail T, et al. Apoptotic pathways of epothilone BMS 310705. Gynecol Oncol 2003;91:173–178.

159. Au JL, Kumar RR, Li D, Wientjes MG. Kinetics of hallmark biochemical changes in paclitaxel-induced apoptosis. AAPS Pharm Sci 1999;1:E8.

160. Oyaizu H, Adachi Y, Taketani S, Tokunaga R, Fukuhara S, Ikehara S. A crucial role of caspase 3 and caspase 8 in paclitaxel-induced apoptosis. Mol Cell Biol Res Commun 1999;2:36–41.

161. Samejima K, Svingen PA, Basi GS, et al. Caspase-mediated cleavage of DNA topoisomerase I at unconventional sites during apoptosis. J Biol Chem 1999;274:4335–4340.

162. Yim H, Jin YH, Park BD, Choi HJ, Lee SK. Caspase-3-mediated cleavage of Cdc6 induces nuclear localization of p49-truncated Cdc6 and apoptosis. Mol Biol Cell 2003;14:4250–4259.
163. Ferreira CG, Span SW, Peters GJ, Kruyt FA, Giaccone G. Chemotherapy triggers apoptosis in a caspase-8-dependent and mitochondria-controlled manner in the non-small cell lung cancer cell line NCI-H460. Cancer Res 2000;60:7133–7141.
164. Ling Y, Zhong Y, Perez-Soler R. Disruption of cell adhesion and caspase-mediated proteolysis of beta- and gamma-catenins and APC protein in paclitaxel-induced apoptosis. Mol Pharmacol 2001;59:593–603.
165. Friedrich K, Wieder T, Von Haefen C, et al. Overexpression of caspase-3 restores sensitivity for drug-induced apoptosis in breast cancer cell lines with acquired drug resistance. Oncogene 2001;20: 2749–2760.
166. Kolfschoten GM, Hulscher TM, Duyndam MC, Pinedo HM, Boven E. Variation in the kinetics of caspase-3 activation, Bcl-2 phosphorylation and apoptotic morphology in unselected human ovarian cancer cell lines as a response to docetaxel. Biochem Pharmacol 2002;63:733–743.
167. Yang LX, Zhu J, Wang HJ, Holton RA. Enhanced apoptotic effects of novel paclitaxel analogs on NCI/ADR-RES breast cancer cells. Anticancer Res 2003;23:3295–3301.
168. Panvichian R, Orth K, Pilat MJ, et al. Signaling network of paclitaxel-induced apoptosis in the LNCaP prostate cancer cell line. Urology 1999;54:746–752.
169. Arnt CR, Chiorean MV, Heldebrant MP, Gores GJ, Kaufmann SH. Synthetic Smac/DIABLO peptides enhance the effects of chemotherapeutic agents by binding XIAP and cIAP1 in situ. J Biol Chem 2002;277:44,236–44,243.
170. McNeish IA, Bell S, McKay T, Tenev T, Marani M, Lemoine NR. Expression of Smac/DIABLO in ovarian carcinoma cells induces apoptosis via a caspase-9-mediated pathway. Exp Cell Res 2003;286: 186–198.
171. Huisman C, Ferreira CG, Broker LE, et al. Paclitaxel triggers cell death primarily via caspase-independent routes in the non-small cell lung cancer cell line NCI-H460. Clin Cancer Res 2002;8: 596–606.
172. Broker LE, Huisman C, Span SW, Rodriguez JA, Kruyt FA, Giaccone G. Cathepsin B mediates caspase-independent cell death induced by microtubule stabilizing agents in non-small cell lung cancer cells. Cancer Res 2004;64:27–30.
173. Ofir R, Seidman R, Rabinski T, et al. Taxol-induced apoptosis in human SKOV3 ovarian and MCF7 breast carcinoma cells is caspase-3 and caspase-9 independent. Cell Death Differ 2002;9:636–642.
174. Blagosklonny MV, Robey R, Sheikh MS, Fojo T. Paclitaxel-induced FasL-independent apoptosis and slow (non-apoptotic) cell death. Cancer Biol Ther 2002;1:113–117.
175. Sunters A, Fernandez de Mattos S, Stahl M, et al. FoxO3a transcriptional regulation of Bim controls apoptosis in paclitaxel-treated breast cancer cell lines. J Biol Chem 2003;278:49,795–49,805.
176. Bouillet P, Metcalf D, Huang DC, et al. Proapoptotic Bcl-2 relative Bim required for certain apoptotic responses, leukocyte homeostasis, and to preclude autoimmunity. Science 1999;286:1735–1738.
177. Bhalla KN. Microtubule-targeted anticancer agents and apoptosis. Oncogene 2003;22:9075–9086.
178. Li R, Moudgil T, Ross HJ, Hu H-M. Apoptosis of non-small-cell lung cancer cell lines after paclitaxel treatment involves the BH3-only proapoptotic protein Bim. Cell Death Differ 2005;12:292–303.
179. Lei K, Davis RJ. JNK phosphorylation of Bim-related members of the Bcl2 family induces Bax-dependent apoptosis. Proc Natl Acad Sci USA 2003;100:2432–2437.
180. Jiang X, Wilford C, Duensing S, Munger K, Jones G, Jones D. Participation of Survivin in mitotic and apoptotic activities of normal and tumor-derived cells. J Cell Biochem 2001;83:342–354.
181. Zaffaroni N, Pennati M, Colella G, et al. Expression of the anti-apoptotic gene survivin correlates with taxol resistance in human ovarian cancer. Cell Mol Life Sci 2002;59:1406–1412.
182. Tan G, Heqing L, Jiangbo C, et al. Apoptosis induced by low-dose paclitaxel is associated with p53 upregulation in nasopharyngeal carcinoma cells. Int J Cancer 2002;97:168–172.
183. Giodini A, Kallio MJ, Wall NR, et al. Regulation of microtubule stability and mitotic progression by survivin. Cancer Res 2002;62:2462–2467.
184. Li F. Survivin study: what is the next wave? J Cell Physiol 2003;197:8–29.
185. Ling X, Bernacki RJ, Brattain MG, Li F. Induction of survivin expression by taxol (paclitaxel) is an early event which is independent of taxol-mediated G2/M arrest. J Biol Chem 2004;279:15,196–15,203.
186. Song Z, Yao X, Wu M. Direct interaction between survivin and Smac/DIABLO is essential for the anti-apoptotic activity of survivin during taxol-induced apoptosis. J Biol Chem 2003;278:23,130–23,140.

187. Lens SM, Wolthuis RM, Klompmaker R, et al. Survivin is required for a sustained spindle checkpoint arrest in response to lack of tension. EMBO J 2003;22:2934–2947.
188. Shen SC, Huang TS, Jee SH, Kuo ML. Taxol-induced p34cdc2 kinase activation and apoptosis inhibited by 12-O-tetradecanoylphorbol-13-acetate in human breast MCF-7 carcinoma cells. Cell Growth Differ 1998;9:23–29.
189. Makino K, Yu D, Hung MC. Transcriptional upregulation and activation of p55Cdc via p34(cdc2) in Taxol-induced apoptosis. Oncogene 2001;20:2537–2543.
190. Ibrado AM, Kim CN, Bhalla K. Temporal relationship of CDK1 activation and mitotic arrest to cytosolic accumulation of cytochrome C and caspase-3 activity during Taxol-induced apoptosis of human AML HL-60 cells. Leukemia 1998;12:1930–1936.
191. Chadebech P, Truchet I, Brichese L, Valette A. Up-regulation of cdc2 protein during paclitaxel-induced apoptosis. Int J Cancer 2000;87:779–786.
192. Huang TS, Shu CH, Chao Y, Chen SN, Chen LL. Activation of MAD 2 checkprotein and persistence of cyclin B1/CDC 2 activity associate with paclitaxel-induced apoptosis in human nasopharyngeal carcinoma cells. Apoptosis 2000;5:235–241.
193. O'Connor DS, Wall NR, Porter AC, Altieri DC. A p34(cdc2) survival checkpoint in cancer. Cancer Cell 2002;2:43–54.
194. Mollinedo F, Gajate C. Microtubules, microtubule-interfering agents and apoptosis. Apoptosis 2003;8:413–450.
195. Lu QL, Hanby AM, Nasser Hajibagheri MA, et al. Bcl-2 protein localizes to the chromosomes of mitotic nuclei and is correlated with the cell cycle in cultured epithelial cell lines. J Cell Sci 1994;107(pt 2):363–371.
196. Pathan N, Aime-Sempe C, Kitada S, Haldar S, Reed JC. Microtubule-targeting drugs induce Bcl-2 phosphorylation and association with Pin1. Neoplasia 2001;3:70–79.
197. Wang TH, Wang HS. p53, apoptosis and human cancers. J Formos Med Assoc 1996;95:509–522.
198. Woods CM, Zhu J, McQueney PA, Bollag D, Lazarides E. Taxol-induced mitotic block triggers rapid onset of a p53-independent apoptotic pathway. Mol Med 1995;1:506–526.
199. Bacus SS, Gudkov AV, Lowe M, et al. Taxol-induced apoptosis depends on MAP kinase pathways (ERK and p38) and is independent of p53. Oncogene 2001;20:147–155.
200. Fan S, Cherney B, Reinhold W, Rucker K, O'Connor PM. Disruption of p53 function in immortalized human cells does not affect survival or apoptosis after taxol or vincristine treatment. Clin Cancer Res 1998;4:1047–1054.
201. Debernardis D, Sire EG, De Feudis P, et al. p53 status does not affect sensitivity of human ovarian cancer cell lines to paclitaxel. Cancer Res 1997;57:870–874.
202. King TC, Akerley W, Fan AC, et al. p53 mutations do not predict response to paclitaxel in metastatic nonsmall cell lung carcinoma. Cancer 2000;89:769–773.
203. Vikhanskaya F, Vignati S, Beccaglia P, et al. Inactivation of p53 in a human ovarian cancer cell line increases the sensitivity to paclitaxel by inducing G2/M arrest and apoptosis. Exp Cell Res 1998;241:96–101.
204. Liu XM, Jiang JD, Ferrari AC, Budman DR, Wang LG. Unique induction of p21(WAF1/CIP1)expression by vinorelbine in androgen-independent prostate cancer cells. Br J Cancer 2003;89:1566–1573.
205. Vayssade M, Faridoni-Laurens L, Benard J, Ahomadegbe JC. Expression of p53-family members and associated target molecules in breast cancer cell lines in response to vincristine treatment. Biochem Pharmacol 2002;63:1609–1617.
206. Ciciarello M, Mangiacasale R, Casenghi M, et al. p53 displacement from centrosomes and p53-mediated G1 arrest following transient inhibition of the mitotic spindle. J Biol Chem 2001;276:19,205–19,213.
207. Blagosklonny MV, Schulte TW, Nguyen P, Mimnaugh EG, Trepel J, Neckers L. Taxol induction of p21WAF1 and p53 requires c-raf-1. Cancer Res 1995;55:4623–4626.
208. Yang HL, Pan JX, Sun L, Yeung SC. p21 Waf-1 (Cip-1) enhances apoptosis induced by manumycin and paclitaxel in anaplastic thyroid cancer cells. J Clin Endocrinol Metab 2003;88:763–772.
209. Fojo T. p53 as a therapeutic target: unresolved issues on the road to cancer therapy targeting mutant p53. Drug Resist Updat 2002;5:209–216.
210. Pourroy B, Carre M, Honore S, et al. Low concentrations of vinflunine induce apoptosis in human SK-N-SH neuroblastoma cells through a postmitotic G1 arrest and a mitochondrial pathway. Mol Pharmacol 2004;66:580–591.

211. Barboule N, Chadebech P, Baldin V, Vidal S, Valette A. Involvement of p21 in mitotic exit after paclitaxel treatment in MCF-7 breast adenocarcinoma cell line. Oncogene 1997;15:2867–2875.
212. Long BH, Fairchild CR. Paclitaxel inhibits progression of mitotic cells to G1 phase by interference
with spindle formation without affecting other microtubule functions during anaphase and telephase.
Cancer Res 1994;54:4355–4361.
213. Wu Y, Mehew JW, Heckman CA, Arcinas M, Boxer LM. Negative regulation of bcl-2 expression by
p53 in hematopoietic cells. Oncogene 2001;20:240–251.
214. Han J, Flemington C, Houghton AB, et al. Expression of bbc3, a pro-apoptotic BH3-only gene, is regulated by diverse cell death and survival signals. Proc Natl Acad Sci USA 2001;98:11,318–11,323.
215. Nakano K, Vousden KH. PUMA, a novel proapoptotic gene, is induced by p53. Mol Cell
2001;7:683–694.
216. Giannakakou P, Nakano M, Nicolaou KC, et al. Enhanced microtubule-dependent trafficking and p53
nuclear accumulation by suppression of microtubule dynamics. Proc Natl Acad Sci USA 2002;99:
10,855–10,860.
216a. Moll UM, Marchenko N, Zhang XK. p53 and Nur 77/TR3-transcription factors that directly target
mitochondria for cell death induction. Oncogene 2006;25:4725–4743.
216b. Jiang P, Du W, Heese K, Wu M. The Bad guy cooperates with a good cop p53: Bad is transcriptionally
up-regulated by p53 and forms Bad/p53 complex at the mitochondria to induce apoptosis. Mol Cell
Biol 2006;26:9071–9082.
217. Perfettini JL, Kroemer RT, Kroemer G. Fatal liaisons of p53 with Bax and Bak. Nature Cell Biol
2004;6:386–388.
218. Blagosklonny MV, Iglesias A, Zhan Z, Fojo T. Like p53, the proliferation-associated protein p120
accumulates in human cancer cells following exposure to anticancer drugs. Biochem Biophys Res
Commun 1998;244:368–373.
219. Amato SF, Swart JM, Berg M, Wanebo HJ, Mehta SR, Chiles TC. Transient stimulation of the
c-Jun-NH2-terminal kinase/activator protein 1 pathway and inhibition of extracellular signal-regulated
kinase are early effects in paclitaxel-mediated apoptosis in human B lymphoblasts. Cancer Res
1998;58:241–247.
220. Stone AA, Chambers TC. Microtubule inhibitors elicit differential effects on MAP kinase (JNK,
ERK, and p38) signaling pathways in human KB-3 carcinoma cells. Exp Cell Res 2000;254:110–119.
221. Boldt S, Weidle UH, Kolch W. The role of MAPK pathways in the action of chemotherapeutic drugs.
Carcinogenesis 2002;23:1831–1838.
222. Guise S, Braguer D, Carles G, Delacourte A, Briand C. Hyperphosphorylation of tau is mediated by
ERK activation during anticancer drug-induced apoptosis in neuroblastoma cells. J Neurosci Res
2001;63:257–267.
223. Okano J, Rustgi AK. Paclitaxel induces prolonged activation of the Ras/MEK/ERK pathway independently of activating the programmed cell death machinery. J Biol Chem 2001;276:19,555–19,564.
224. Townsend KJ, Trusty JL, Traupman MA, Eastman A, Craig RW. Expression of the antiapoptotic
MCL1 gene product is regulated by a mitogen activated protein kinase-mediated pathway triggered
through microtubule disruption and protein kinase C. Oncogene 1998;17:1223–1234.
225. McDaid HM, Horwitz SB. Selective potentiation of paclitaxel (taxol)-induced cell death by mitogen-
activated protein kinase kinase inhibition in human cancer cell lines. Mol Pharmacol 2001;60:
290–301.
226. Seidman R, Gitelman I, Sagi O, Horwitz SB, Wolfson M. The role of ERK 1/2 and p38 MAP-kinase
pathways in taxol-induced apoptosis in human ovarian carcinoma cells. Exp Cell Res 2001;268:
84–92.
227. MacKeigan JP, Collins TS, Ting JP. MEK inhibition enhances paclitaxel-induced tumor apoptosis.
J Biol Chem 2000;275:38,953–38,956.
228. Schmidt CM, McKillop IH, Cahill PA, Sitzmann JV. Increased MAPK expression and activity in
primary human hepatocellular carcinoma. Biochem Biophys Res Commun 1997;236:54–58.
229. Loda M, Capodieci P, Mishra R, et al. Expression of mitogen-activated protein kinase phosphatase-1
in the early phases of human epithelial carcinogenesis. Am J Pathol 1996;149:1553–1564.
230. Mabuchi S, Ohmichi M, Kimura A, et al. Inhibition of phosphorylation of BAD and Raf-1 by Akt
sensitizes human ovarian cancer cells to paclitaxel. J Biol Chem 2002;277:33,490–33,500.
231. Luciano F, Jacquel A, Colosetti P, et al. Phosphorylation of Bim-EL by Erk1/2 on serine 69 promotes
its degradation via the proteasome pathway and regulates its proapoptotic function. Oncogene 2003;
22:6785–6793.

232. Ley R, Balmanno K, Hadfield K, Weston C, Cook SJ. Activation of the ERK1/2 signaling pathway promotes phosphorylation and proteasome-dependent degradation of the BH3-only protein, Bim. J Biol Chem 2003;278:18,811–18,816.

233. Taxman DJ, MacKeigan JP, Clements C, Bergstralh DT, Ting JP. Transcriptional profiling of targets for combination therapy of lung carcinoma with paclitaxel and mitogen-activated protein/extracellular signal-regulated kinase kinase inhibitor. Cancer Res 2003;63:5095–5104.

234. Takenaka K, Moriguchi T, Nishida E. Activation of the protein kinase p38 in the spindle assembly checkpoint and mitotic arrest. Science 1998;280:599–602.

235. Kurata S. Selective activation of p38 MAPK cascade and mitotic arrest caused by low level oxidative stress. J Biol Chem 2000;275:23,413–23,416.

236. Reshkin SJ, Bellizzi A, Cardone RA, Tommasino M, Casavola V, Paradiso A. Paclitaxel induces apoptosis via protein kinase A- and p38 mitogen-activated protein-dependent inhibition of the Na^+/H^+ exchanger (NHE) NHE isoform 1 in human breast cancer cells. Clin Cancer Res 2003;9:2366–2373.

237. Yu C, Wang S, Dent P, Grant S. Sequence-dependent potentiation of paclitaxel-mediated apoptosis in human leukemia cells by inhibitors of the mitogen-activated protein kinase kinase/mitogen-activated protein kinase pathway. Mol Pharmacol 2001;60:143–154.

238. Nagata K, Puls A, Futter C, et al. The MAP kinase kinase kinase MLK2 co-localizes with activated JNK along microtubules and associates with kinesin superfamily motor KIF3. EMBO J 1998;17: 149–158.

239. Wang TH, Wang HS, Ichijo H, et al. Microtubule-interfering agents activate c-Jun N-terminal kinase/stress-activated protein kinase through both Ras and apoptosis signal-regulating kinase pathways. J Biol Chem 1998;273:4928–4936.

240. Lee LF, Li G, Templeton DJ, Ting JP. Paclitaxel (Taxol)-induced gene expression and cell death are both mediated by the activation of c-Jun NH2-terminal kinase (JNK/SAPK). J Biol Chem 1998;273: 28,253–28,260.

241. Ham YM, Choi JS, Chun KH, Joo SH, Lee SK. The c-Jun N-terminal kinase 1 activity is differentially regulated by specific mechanisms during apoptosis. J Biol Chem 2003;278:50,330–50,337.

242. Kwan R, Burnside J, Kurosaki T, Cheng G. MEKK1 is essential for DT40 cell apoptosis in response to microtubule disruption. Mol Cell Biol 2001;21:7183–7190.

243. Shiah SG, Chuang SE, Kuo ML. Involvement of Asp-Glu-Val-Asp-directed, caspase-mediated mitogen-activated protein kinase kinase 1 Cleavage, c-Jun N-terminal kinase activation, and subsequent Bcl-2 phosphorylation for paclitaxel-induced apoptosis in HL-60 cells. Mol Pharmacol 2001;59: 254–262.

244. Srivastava RK, Mi QS, Hardwick JM, Longo DL. Deletion of the loop region of Bcl-2 completely blocks paclitaxel-induced apoptosis. Proc Natl Acad Sci USA 1999;96:3775–3780.

245. Fan M, Goodwin M, Vu T, Brantley-Finley C, Gaarde WA, Chambers TC. Vinblastine-induced phosphorylation of Bcl-2 and Bcl-XL is mediated by JNK and occurs in parallel with inactivation of the Raf-1/MEK/ERK cascade. J Biol Chem 2000;275:29,980–29,985.

246. Wang TH, Popp DM, Wang HS, et al. Microtubule dysfunction induced by paclitaxel initiates apoptosis through both c-Jun N-terminal kinase (JNK)-dependent and -independent pathways in ovarian cancer cells. J Biol Chem 1999;274:8208–8216.

247. Yujiri T, Fanger GR, Garrington TP, Schlesinger TK, Gibson S, Johnson GL. MEK kinase 1 (MEKK1) transduces c-Jun NH2-terminal kinase activation in response to changes in the microtubule cytoskeleton. J Biol Chem 1999;274:12,605–12,610.

248. Blagosklonny MV, Giannakakou P, el-Deiry WS, et al. Raf-1/bcl-2 phosphorylation: a step from microtubule damage to cell death. Cancer Res 1997;57:130–135.

249. Lee M, Koh WS, Han SS. Down-regulation of Raf-1 kinase is associated with paclitaxel resistance in human breast cancer MCF-7/Adr cells. Cancer Lett 2003;193:57–64.

250. Blagosklonny MV, Schulte T, Nguyen P, Trepel J, Neckers LM. Taxol-induced apoptosis and phosphorylation of Bcl-2 protein involves c-Raf-1 and represents a novel c-Raf-1 signal transduction pathway. Cancer Res 1996;56:1851–1854.

251. Wang HG, Miyashita T, Takayama S, et al. Apoptosis regulation by interaction of Bcl-2 protein and Raf-1 kinase. Oncogene 1994;9:2751–2756.

252. Lin HL, Liu TY, Chau GY, Lui WY, Chi CW. Comparison of 2-methoxyestradiol-induced, docetaxel-induced, and paclitaxel-induced apoptosis in hepatoma cells and its correlation with reactive oxygen species. Cancer 2000;89:983–994.

253. Su Y, Zharikov SI, Block ER. Microtubule-active agents modify nitric oxide production in pulmonary artery endothelial cells. Am J Physiol Lung Cell Mol Physiol 2002;282:L1183–L1189.

254. Simizu S, Takada M, Umezawa K, Imoto M. Requirement of caspase-3(-like) protease-mediated hydrogen peroxide production for apoptosis induced by various anticancer drugs. J Biol Chem 1998;273:26,900–26,907.

255. Shtil AA, Mandlekar S, Yu R, et al. Differential regulation of mitogen-activated protein kinases by microtubule-binding agents in human breast cancer cells. Oncogene 1999;18:377–384.

255a. Alexandre J, Batteux F, Nicco C, Chereau C, Laurent A, Guillevin L, Weill B, Goldwasser F. Accumulation of hydrogen peroxide is an early and crucial step for paclitaxel-induced cancer cell death both in vitro and in vivo. Int J Cancer 2006;119:41–48.

256. Gehrmann M, Pfister K, Hutzler P, Gastpar R, Margulis B, Multhoff G. Effects of antineoplastic agents on cytoplasmic and membrane-bound heat shock protein 70 (Hsp70) levels. Biol Chem 2002; 383:1715–1725.

257. Kemp TJ, Causton HC, Clerk A. Changes in gene expression induced by H_2O_2 in cardiac myocytes. Biochem Biophys Res Commun 2003;307:416–421.

258. Kwak HJ, Jun CD, Pae HO, et al. The role of inducible 70-kDa heat shock protein in cell cycle control, differentiation, and apoptotic cell death of the human myeloid leukemic HL-60 cells. Cell Immunol 1998;187:1–12.

259. Solit DB, Basso AD, Olshen AB, Scher HI, Rosen N. Inhibition of heat shock protein 90 function down-regulates Akt kinase and sensitizes tumors to Taxol. Cancer Res 2003;63:2139–2144.

260. Garnier C, Barbier P, Gilli R, Lopez C, Peyrot V, Briand C. Heat-shock protein 90 (hsp90) binds in vitro to tubulin dimer and inhibits microtubule formation. Biochem Biophys Res Commun 1998;250: 414–419.

261. Sanchez C, Padilla R, Paciucci R, Zabala JC, Avila J. Binding of heat-shock protein 70 (hsp70) to tubulin. Arch Biochem Biophys 1994;310:428–432.

262. Sanchez ER, Redmond T, Scherrer LC, Bresnick EH, Welsh MJ, Pratt WB. Evidence that the 90-kilo-dalton heat shock protein is associated with tubulin-containing complexes in L cell cytosol and in intact PtK cells. Mol Endocrinol 1988;2:756–760.

263. Czar MJ, Welsh MJ, Pratt WB. Immunofluorescence localization of the 90-kDa heat-shock protein to cytoskeleton. Eur J Cell Biol 1996;70:322–330.

264. Skidgel RA. Proliferation of regulatory mechanisms for eNOS: an emerging role for the cytoskeleton. Am J Physiol Lung Cell Mol Physiol 2002;282:L1179–L1182.

265. Byrd CA, Bornmann W, Erdjument-Bromage H, et al. Heat shock protein 90 mediates macrophage activation by Taxol and bacterial lipopolysaccharide. Proc Natl Acad Sci USA 1999;96:5645–5650.

266. Ochel HJ, Gademann G. Heat-shock protein 90: potential involvement in the pathogenesis of malignancy and pharmacological intervention. Onkologie 2002;25:466–473.

267. Hwang S, Ding A. Activation of NF-kappa B in murine macrophages by taxol. Cancer Biochem Biophys 1995;14:265–272.

268. Bourgarel-Rey V, Vallee S, Rimet O, et al. Involvement of nuclear factor kappaB in c-Myc induction by tubulin polymerization inhibitors. Mol Pharmacol 2001;59:1165–1170.

269. Patel NM, Nozaki S, Shortle NH, et al. Paclitaxel sensitivity of breast cancer cells with constitutively active NF-kappaB is enhanced by IkappaBalpha super-repressor and parthenolide. Oncogene 2000; 19:4159–4169.

270. Zhang H, Morisaki T, Nakahara C, et al. PSK-mediated NF-kappaB inhibition augments docetaxel-induced apoptosis in human pancreatic cancer cells NOR-P1. Oncogene 2003;22:2088–2096.

271. Dong QG, Sclabas GM, Fujioka S, et al. The function of multiple IkappaB : NF-kappaB complexes in the resistance of cancer cells to Taxol-induced apoptosis. Oncogene 2002;21:6510–6519.

272. Huang Y, Johnson KR, Norris JS, Fan W. Nuclear factor-kappaB/IkappaB signaling pathway may contribute to the mediation of paclitaxel-induced apoptosis in solid tumor cells. Cancer Res 2000;60: 4426–4432.

273. Huang Y, Fan W. IkappaB kinase activation is involved in regulation of paclitaxel-induced apoptosis in human tumor cell lines. Mol Pharmacol 2002;61:105–113.

274. Cardone MH, Roy N, Stennicke HR, et al. Regulation of cell death protease caspase-9 by phosphorylation. Science 1998;282:1318–1321.

275. Page C, Lin HJ, Jin Y, et al. Overexpression of Akt/AKT can modulate chemotherapy-induced apoptosis. Anticancer Res 2000;20:407–416.

276. Mitsuuchi Y, Johnson SW, Selvakumaran M, Williams SJ, Hamilton TC, Testa JR. The phosphatidyli-nositol 3-kinase/AKT signal transduction pathway plays a critical role in the expression of p21WAF1/CIP1/SDI1 induced by cisplatin and paclitaxel. Cancer Res 2000;60:5390–5394.

277. Kwon HK, Bae GU, Yoon JW, et al. Constitutive activation of p70S6k in cancer cells. Arch Pharm Res 2002;25:685–690.

278. Le XF, Hittelman WN, Liu J, et al. Paclitaxel induces inactivation of p70 S6 kinase and phosphory-lation of Thr421 and Ser424 via multiple signaling pathways in mitosis. Oncogene 2003;22:484–497.

279. Harada H, Andersen JS, Mann M, Terada N, Korsmeyer SJ. p70S6 kinase signals cell survival as well as growth, inactivating the pro-apoptotic molecule BAD. Proc Natl Acad Sci USA 2001;98: 9666–9670.

280. Varbiro G, Veres B, Gallyas F Jr, Sumegi B. Direct effect of Taxol on free radical formation and mito-chondrial permeability transition. Free Radic Biol Med 2001;31:548–558.

281. Andre N, Braguer D, Brasseur G, et al. Paclitaxel induces release of cytochrome c from mitochon-dria isolated from human neuroblastoma cells. Cancer Res 2000;60:5349–5353.

282. Carre M, Carles G, Andre N, et al. Involvement of microtubules and mitochondria in the antagonism of arsenic trioxide on paclitaxel-induced apoptosis. Biochem Pharmacol 2002;63:1831–1842.

283. Kidd JF, Pilkington MF, Schell MJ, et al. Paclitaxel affects cytosolic calcium signals by opening the mitochondrial permeability transition pore. J Biol Chem 2002;277:6504–6510.

284. Ferlini C, Raspaglio G, Mozzetti S, et al. Bcl-2 down-regulation is a novel mechanism of paclitaxel resistance. Mol Pharmacol 2003;64:51–58.

284a. Estève MA, Carré M, Bourgarel-Rey V, Kruczynski A, Raspaglio G, Ferlini C, Braguer D. Bcl-2 down-regulation and tubulin subtype composition are involved in resistance of ovarian cancer cells to vinflunine. Mol Cancer Ther 2006;5:1–10.

285. Rodi DJ, Janes RW, Sanganee HJ, Holton RA, Wallace BA, Makowski L. Screening of a library of phage-displayed peptides identifies human bcl-2 as a taxol-binding protein. J Mol Biol 1999;285: 197–203.

286. Wu JH, Batist G, Zamir LO. A model for the interaction of paclitaxel with the Bcl-2 loop domain: a chemical approach to induce conformation-dependent phosphorylation. Anticancer Drug Des 2000; 15:441–446.

287. Haldar S, Chintapalli J, Croce CM. Taxol induces bcl-2 phosphorylation and death of prostate cancer cells. Cancer Res 1996;56:1253–1255.

288. Carre M, Andre N, Carles G, et al. Tubulin is an inherent component of mitochondrial membranes that interacts with the voltage-dependent anion channel. J Biol Chem 2002;277:33,664–33,669.

289. Zambito AM, Wolff J. Plasma membrane localization of palmitoylated tubulin. Biochem Biophys Res Commun 2001;283:42–47.

290. Palestini P, Pitto M, Tedeschi G, et al. Tubulin anchoring to glycolipid-enriched, detergent-resistant domains of the neuronal plasma membrane. J Biol Chem 2000;275:9978–9985.

291. Xu K, Luduena RF. Characterization of nuclear betaII-tubulin in tumor cells: a possible novel target for taxol. Cell Motil Cytoskeleton 2002;53:39–52.

292. Chabin-Brion K, Marceiller J, Perez F, et al. The Golgi complex is a microtubule-organizing organelle. Mol Biol Cell 2001;12:2047–2060.

293. Ferri KF, Kroemer G. Organelle-specific initiation of cell death pathways. Nat Cell Biol 2001;3: E255–E263.

294. Zong WX, Li C, Hatzivassiliou G, et al. Bax and Bak can localize to the endoplasmic reticulum to initiate apoptosis. J Cell Biol 2003;162:59–69.

295. Scorrano L, Oakes SA, Opferman JT, et al. BAX and BAK regulation of endoplasmic reticulum Ca^{2+}: a control point for apoptosis. Science 2003;300:135–139.

296. Breckenridge DG, Germain M, Mathai JP, Nguyen M, Shore GC. Regulation of apoptosis by endo-plasmic reticulum pathways. Oncogene 2003;22:8608–8618.

297. Thomenius MJ, Wang NS, Reineks EZ, Wang Z, Distelhorst CW. Bcl-2 on the endoplasmic reticu-lum regulates Bax activity by binding to BH3-only proteins. J Biol Chem 2003;278:6243–6250.

298. Kadi A, Pichard V, Lehmann M, et al. Effect of microtubule disruption on cell adhesion and spread-ing. Biochem Biophys Res Commun 1998;246:690–695.

299. Zhou X, Li J, Kucik DF. The microtubule cytoskeleton participates in control of beta2 integrin avidity. J Biol Chem 2001;276:44,762–44,769.

300. Niggli V. Microtubule-disruption-induced and chemotactic-peptide-induced migration of human neutrophils: implications for differential sets of signalling pathways. J Cell Sci 2003;116:813–822.
301. Hu YL, Li S, Miao H, Tsou TC, del Pozo MA, Chien S. Roles of microtubule dynamics and small GTPase Rac in endothelial cell migration and lamellipodium formation under flow. J Vasc Res 2002;39:465–476.
302. Belotti D, Vergani V, Drudis T, et al. The microtubule-affecting drug paclitaxel has antiangiogenic activity. Clin Cancer Res 1996;2:1843–1849.
303. Griggs J, Metcalfe JC, Hesketh R. Targeting tumour vasculature: the development of combretastatin A4. Lancet Oncol 2001;2:82–87.
304. Micheletti G, Poli M, Borsotti P, et al. Vascular-targeting activity of ZD6126, a novel tubulin-binding agent. Cancer Res 2003;63:1534–1537.
305. Pasquier E, Carre M, Pourroy B, et al.Antiangiogenic activity of paclitaxel is associated with its cytostatic effect, mediated by the initiation but not completion of a mitochondrial apoptotic signaling pathway. Mol Cancer Ther 2004;3:1301–1310.
306. Pasquier E, Honore S, Pourroy B, et al. Antiangiogenic concentrations of paclitaxel induce an increase in microtubule dynamics in endothelial cells but not in cancer cells. Cancer Res 2005;65: 2433–2440.

19 Microtubule Targeting Agents and the Tumor Vasculature

Raffaella Giavazzi, Katiuscia Bonezzi, and Giulia Taraboletti

CONTENTS

SUMMARY

As microtubules are important regulators of endothelial cell biology, it is not surprising that tubulin binding compounds have the potential to target the tumor vasculature. Two main uses of microtubule targeting compounds have been proposed. Tubulin binding agents can prevent the formation of new vessels (acting as inhibitors of angiogenesis) or damage the existing tumor vasculature (acting as vascular disrupting agents, VDA). Antiangiogenic and vascular disrupting microtubule targeting agents are hereby reviewed, with particular emphasis on their potentiality and limits in the clinical practice.

Key Words: Angiogenesis inhibitors; vascular disrupting agent (VDA); endothelial cells; tumor vasculature; combination therapy.

1. INTRODUCTION

In the recent years, the tumor vasculature has emerged as a promising target for cancer therapy. The development of a functional blood vessel network is critical for tumor growth, for the progression from a premalignant tumor to an invasive cancer, and for metastasis formation *(1,2)*. The identification of agents affecting the tumor vasculature has become a highly active area of investigation from basic to clinical research. Theoretically, several advantages characterize this therapeutic approach. Damage to tumor vessels would have severe consequences on the hundreds of tumor cells that depend on them for survival, both at the early stage of tumor progression and later on advanced stage tumors *(3)*. This therapy would be effective on all angiogenesis-dependent

From: *Cancer Drug Discovery and Development: The Role of Microtubules in Cell Biology, Neurobiology, and Oncology* Edited by: Tito Fojo © Humana Press, Totowa, NJ

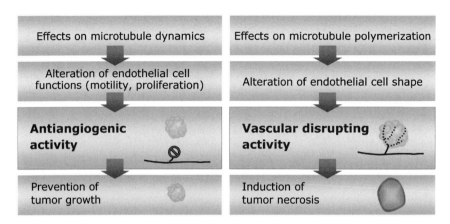

Fig. 1. Tubulin-targeting agents. Tubulin-targeting agents can act as antiangiogenic and VDAs. By altering microtubule dynamics they affect endothelial cell functions essential to the angiogenic process; this results in inhibition of tumor growth. By affecting microtubule organization they cause vessel damage; this induces tumor necrosis. To view this figure in color, see the insert and the companion CD-ROM.

solid tumors, independently on their histological type. The target endothelial cells are adjacent to the bloodstream, therefore easily accessible to the drug. Finally, as the target endothelial cells are a population of "normal", nontransformed cells, the development of genetically driven resistance to therapy is unlikely.

Two approaches can be foreseen:

1. Antiangiogenic therapy, aimed at inhibiting angiogenesis, the formation of new vessels from pre-existing ones;
2. Vascular disrupting therapy, aimed at selectively destroying the already formed tumor vascular bed.

The mechanisms at the base of the two approaches, the effects on the tumors, and the therapeutic applications are very different (Fig. 1).

Antiangiogenic compounds prevent the formation of new vessels *(1)* either directly, by blocking endothelial cell functional response to stimulating angiogenic factors, or indirectly by interfering with the production, availability, or activity of angiogenic factors *(2,4–6)*. The antiangiogenic approach exerts mainly a cytostatic effect, as it leads to prevention of tumor growth and metastasis, without eradicating the existing tumor. Therefore, the assumption is that these agents work best on small tumors, even before the angiogenic switch occurs. This therapeutic approach has been proposed to prevent the growth of the tumor mass and to maintain metastasis in a dormant state *(7)*. The antiangiogenic effect is expected to last as long as the drug is present. Hence a chronic treatment is required, with obvious implications in terms of safety and side effects associated with the clinical use of these regimens.

In contrast, vascular disrupting agents (VDA, previously defined vascular-targeting agents, VTA) exploit the antigenic and functional differences between blood vessels in tumors and in normal tissues *(8–11)*, to cause a selective damage of the vessel bed within the tumor *(12–14)*. Strategies to affect the tumor vasculature are ligand-directed vascular targeting compounds (antibodies and peptides that deliver an effector to the endothelium) and small molecules, which include cytokine-inducer flavonoids flavone-8-acetic acid

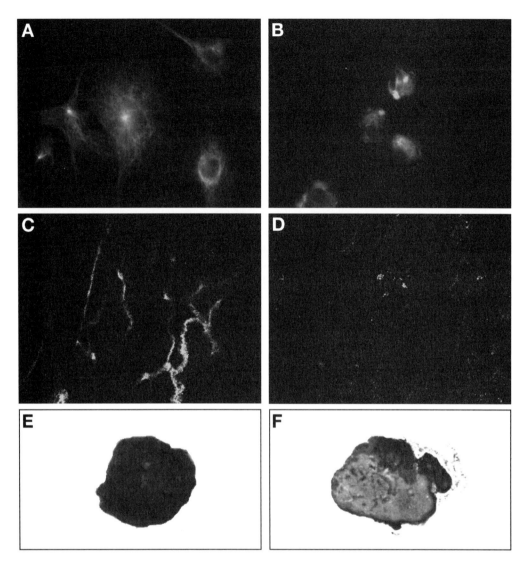

Fig. 2. Effects of vascular disrupting agents. Microtubules destabilizing agents affect cytoskeletal organization and the morphology of endothelial cells that shortly after treatment retract (**A**, vehicle; **B**, VDA). In vivo, microtubules destabilizing agents cause rapid shutdown of newly formed vessel in the Matrigel plug implanted in mice (**C,D**). Confocal microscopy images (×200) show functional perfused vessels, visualized with FITC-conjugated lectin, in vehicle-treated animals (C), and a rapid occlusion of the vessels after a single injection of the VDA (D). In a subcutaneous tumor model, microtubules destabilizing agents cause massive central necrosis of the tumor mass, evident 24 h after a single treatment, with a thin rim of viable tumor cells remaining at the periphery (**E**, vehicle; **F**, VDA). (For details *see* ref. *42*). To view this figure in color, see the insert and the companion CD-ROM.

(FAA) and 5,6-dimethylxant henone-4-acetic acid (DMXAA) and tubulin targeting agents (colchicine-like or *Vinca* alkaloid-like compounds) (reviewed in refs. *12,13,15,16*). This therapy causes a cascade of events, starting with endothelial cell damage and consequent vessel congestion and loss of blood flow, which lead to central necrosis of the tumor tissue (Fig. 2). Vessel occlusion and induction of tumor necrosis occur in the inner part of

Table 1
Tubulin-Targeting Agents Endowed With Antiangiogenic or Vascular Disrupting Activity

Antiangiogenic agents		
Paclitaxel	Taxane	*23,29,34,82*
Docetaxel	Taxane	*28,30,34*
IDN 5390	Taxane	*40,41*
BMS-275183	Taxane	*31*
Epothilone B	Epothilone	*31*
2-methoxyestradiol	Estrogen metabolite	*32,38*
Vinblastine	*Vinca* alcaloid	*36*
Vascular disrupting agents		
Combretastatin A-4 Phospate	Combretastatin A-4	*48,61*
ZD6126	Colchinol	*50–52,83*
AVE8062	Combretastatin A-4 deriv.	*53*
OXI 4503	Combretastatin A-1 deriv.	*54*
ABT-751	Methoxybenzene-sulfonamide	*55*
Vinflunine	*Vinca* alcaloid	*56*
TZT-1027	Dolastatin	*57*

Listed are the principal compounds for which antiangiogenic or vascular disrupting activity has been described. In many cases the distinction is not absolute; this may depend on the drug concentration, tumor type, or treatment modality.

the tumor mass, particularly of large tumors, supposedly because the high interstitial pressure inside the tumor favors vessel collapse *(13)*. However, induction of tumor necrosis by single administration of the compound does not lead to tumor eradication, as a rim of viable tumor cells survives and proliferates in the periphery of the tumor *(13)*. This can be overcome by repeated administrations of the agent, or by combination regimens with radio- or chemotherapy (*see* discussion next).

1.1. Tubulin-Targeting Agents

Tubulin-targeting compounds can act as both inhibitors of angiogenesis and vascular disrupting compounds (Fig. 1 and Table 1). The rationale beyond the use of microtubule-targeting agents to target tumor vessels is based on the crucial role of microtubules in several cell functions, including those that characterize endothelial cells in the developing tumor neovasculature. Therefore, microtubule-targeting agents not only act as antimitotic drugs, but also as vascular affecting compounds, often at doses below those required to affect microtubule polymerization *(17,18)*.

By affecting microtubule dynamics, tubulin-targeting agents inhibit endothelial cell functions crucial to the process of angiogenesis, and in particular cell proliferation and motility. Microtubule stabilizing compounds such as the taxanes have this as their principal effect on endothelial cells (Fig. 1). In contrast, by affecting microtubule organization, tubulin-targeting agents cause endothelial cell shape changes, selective damage of tumor vessels and tumor necrosis, hence acting as VDAs. Microtubule destabilizing compounds, such as those related to colchicine and the *Vinca* alkaloids, have this as their principal effect (Fig. 1). However, in many cases the distinction between antiangiogenic and vascular disrupting activity of the tubulin targeting agents is not absolute, and, depending on the drug concentration and in vivo setting, these agents can act as angio-preventive or VDAs.

2. MICROTUBULE TARGETING AGENTS
AS ANTIANGIOGENIC DRUGS

It has been known for some time that conventional cytotoxic chemotherapeutic drugs are "accidental" inhibitors of angiogenesis. This activity derives from the ability of chemotherapeutic drugs to target not only tumor cells but also normal host cell types involved in angiogenesis, including proliferating endothelial cells in the tumor vasculature, endothelial progenitor cells in the bone marrow or peripheral blood, and bone marrow progenitors of proangiogenic inflammatory cells *(19)*. Although in many cases, the antiangiogenic activity is observed only at near-full cytotoxic concentrations, in some cases a true antiangiogenic activity can be recognized *(2,12,20–23)*. Attempts to improve and exploit this activity have been made, in particular by optimizing the dose and schedule of the treatment. As mentioned earlier, an antiangiogenic treatment is effective as long as the active concentration of the inhibitor (usually far below the maximum tolerated dose [MTD]) is present. For this reason the administration of these chemotherapeutic agents at low doses on a frequent and continuous schedule (metronomic regimen) has been proposed in order to optimally exploit their antiangiogenic activity *(22,24–26)*. An antiangiogenic effect is believed to contribute to the efficacy of dose-dense chemotherapeutic regimens used in the clinic; including those employing the taxanes *(22,27)*. However this conclusion derives from retrospective analysis of clinical precedents and it remains to be validated in randomized prospective clinical trials.

Although preclinical studies have shown antitumor activity with metronomic regimens, in some cases associated with alterations of the tumor vasculature, it remains to be shown if this is the only antitumor mechanism. A direct activity on tumor cells could be partially excluded by showing activity on tumors resistant to that particular drug *(24)*, but other pharmacodynamic mechanisms induced by the prolonged administration of the drug could play a role.

Among the tubulin-targeting agents, the microtubule-stabilizing taxanes paclitaxel and docetaxel, and new derivatives such as IDN 5390 and BMS 275183 have been described to affect angiogenesis at subcytotoxic concentrations *(23,28–35)*. Also microtubule destabilizers, such as vinblastine *(36)* and the soft alkylating agents phenyl-3-(2-chloroethyl)ureas *(37)* are able to exert this effect. Interesting is the antiangiogenic activity of the estrogen metabolite 2-methoxyestradiol (2ME2) for which several potential molecular targets and pathways of activation have been suggested *(38)*. Although 2ME2 has both antitumor and antiangiogenic effects, its promising activity in preclinical studies has led to its clinical development as an orally active, small-molecule inhibitor of angiogenesis (Table 1).

The antiangiogenic activity has been associated with the effect of tubulin-targeting agents on microtubule dynamic instability, and consequently on endothelial cell functions relevant to angiogenesis including cell motility, proliferation, and cord formation. Proliferating endothelial cells, but not other cell types, are particularly sensitive to the inhibitory effects of protracted, ultralow concentrations of paclitaxel, vinblastine, docetaxel, and epothilone B (reviewed in ref. *22*). In addition, the authors earlier studies underlined the potent role of low concentration of taxanes on endothelial cell motility in determining their antiangiogenic activity *(23,39)*. This observation led to the selection of a lead compound, IDN 5390, a seco-derivative with an open C-ring at C-7 and C-8, characterized by a potent antimotility activity on endothelial cells and poor cytotoxicity *(40,41)*.

Tubulin-targeting agents can also act as indirect inhibitors of angiogenesis, by affecting the production of proangiogenic and antiangiogenic factors. Paclitaxel, docetaxel, and vincristine decrease the production of vascular endothelial growth factor (VEGF) by small cell lung carcinoma cells (29,42–44). Paclitaxel, docetaxel, vinblastine, BMS-275183, and epothilone B promote the production of the angiogenesis inhibitor thrombospondin-1 (43,45). Interestingly, it has been shown that taxotere, epothilone B, discodermolide, vincristine, 2ME2, and colchicines reduce the levels and transcriptional activity of HIF-1α (46,47), a major transcription factor associated to hypoxia is that it positively regulates the expression of proangiogenic factors. This effect, which depends on the activity of the compounds on microtubules, indicates an upstream antiangiogenic event, common to different classes of tubulin-targeting agents.

3. MICROTUBULE TARGETING AGENTS AS VASCULAR DISRUPTING DRUGS

The damaging effects on tumor vessels of the first generation microtubules destabilizing agents, such as colchicine and the *Vinca* alkaloids, vincristine and vinblastine, although recognized for many years, have not been successfully exploited, as it occurred at doses close to the MTD. Only recently, microtubules destabilizing agents have been developed with the characteristic of exerting vascular disrupting activity at subtoxic concentrations. In contrast to former compounds, such as colchicine, which bound to tubulin in an almost irreversible fashion, these novel compounds are usually characterized by a short-time retention in the cell and a reversible binding to tubulin, sufficient to cause changes in the shape of the target endothelial cells, but not enough to cause severe toxicity (18).

Among the compounds that bind to the colchicine site, the combretastatins, specifically developed as VDAs are the most studied example of this family of vascular disrupting compounds (48,49). The main compound, combretastatin A-4 disodiumphosphate (CA4-P), is the prodrug of the active drug combretastatin A-4, derived from the African shrub *Combretum caffrum*. Recent additions to this family of compounds include ZD6126, the water-soluble phosphate prodrug of the colchicine derivative *N*-acetylcolchinol (50–52), AVE8062 (formerly AC-7700), also a derivative of Combretastatin A-4 (53), and Oxi 4503, a derivative of Combretastatin A-1 (54). Also ABT-751, a methoxybenzene-sulfonamide that binds to the colchicine site on tubulin and is orally active has shown selective induction of vascular damage in growing tumors (55). All these compounds are in development in clinical trials as VDAs against solid tumors. (Table 1) Among the second-generation compounds that bind to the *Vinca* domain on tubulin, a vascular disrupting activity has been described for vinflunine (56) and dolastatins (TZT-1027) (57).

The activity of these compounds is conceivably based on the difference between endothelial cells in tumor and in normal tissues in terms of dependence on the tubulin cytoskeleton. These compounds, at subtoxic concentrations cause a rapid reduction in tumor of blood flow, followed by vessel shut down and tumor necrosis. This cascade of events is thought to be triggered by the rapid endothelial cell morphological and functional alterations associated with the disruption of the tubulin cytoskeleton, which include activation of the small GTPase RhoA, reorganization of the actin cytoskeleton, phosphorilation of the myosin light chain, alteration of the cell shape, disengagement of cell-cell junctional molecules, disruption of the VE-cadherin/β-catenin/Akt signaling pathway, and increased permeability to macromolecules (Fig. 2) *(52,58–62)*. In some cases, apoptosis has been indicated as the final event caused by these compounds on endothelial cells *(63,64)*. Interestingly, although endothelial cells are considered the main target of VDAs, growing evidence point to a role of other mechanisms in the action of VDAs, including effects on blood cells, platelet activation, coagulation, and active vasoconstriction *(58,65)*.

This cascade of events leads to a rapid drop in tumor blood flow (observed within few minutes of drug administration), followed by a reduction of vascular volume, a loss of vessel integrity, and vessel shut down (in about 20 min) *(58)*. Finally, sustained inhibition of blood flow results, in a few hours, in massive central necrosis of the tumor, with a rim of viable tumor cells remaining at the periphery (Fig. 2), the hallmark of the action of tubulin-binding VDAs *(13,48,50–52,66)*.

The induction of vascular shutdown demonstrated in experimental tumor models as the mechanism of action of these compounds has been confirmed in clinical studies, where positron emission tomography (PET) and magnetic resonance imaging used to measure the effects of VDAs have shown suppressive effects on tumor perfusion and blood flow *(67)*. Phase I studies have found that these compounds are in general tolerated and manageable with no significant hematological/chemical toxicity, although reversible blood pressure changes and cardiac adverse effects have been reported *(13,68)*.

It seems plausible that the antitumor effect of VDA is owing to the stasis of the tumor blood flow, but many questions remain regarding the exact mechanism of action of these compounds. Finally one has to consider, as with any of these VDAs, a possible direct effect on the tumor cell compartment, the magnitude of which might depend on the sensitivity of the tumor itself. An understanding of the mechanism of action at the molecular level is necessary to further clarify the apparent selectivity of these compounds for the tumor vasculature and ultimately to optimize their development.

4. POTENTIALS, LIMITS, AND CHALLENGING ISSUES

Although the theory at the basis of targeting the tumor stroma is intriguing and opens a wide array of possibilities, several challenges have been raised by the first clinical experimentations of these approaches. Some limits, as discussed earlier, are intrinsic with the concept of antiangiogenic and vascular disrupting therapy *per se*. In addition, in the case of tubulin-targeting compounds, issues are raised by the ability of most of these agents to target both endothelial cells and other cells types, including tumor cells. It is therefore difficult to identify and exploit a pure effect on endothelial cells rather than a combination of vessel-targeting effects and antitumor effect. Moreover, as other cell types can be affected, the toxicity of these compounds is always an important issue to consider.

4.1. Exploitment of the Vascular Targeting Effect

In the case of compounds used as "antiangiogenic" agents (Fig. 1), the otpimization of their use as inhibitors of endothelial cells has been mainly addressed through changes in the schedule of administration *(22,24,25)*. A further adjustment that might be needed is the optimization of the formulation of the compounds. For example, it has been shown that clinically achievable concentrations of the formulation vehicle can affect the antiangiogenic activity of paclitaxel and docetaxel *(69)*. The identification of IDN 5390 has indicated that another approach to exploit the antiangiogenic activity of taxanes and other tubulin-targeting agents is the development of less cytotoxic compounds that retain the effect on endothelial cell functions; in this case the effect on cell motility *(40,41)*.

In the case of microtubule-destabilzing VDAs, the efforts to optimize their applicability have been mainly oriented toward the development of compounds active on tumor vessels but possessing low toxicity against the normal vasculature and tissues (Fig. 1). Studies are ongoing to identify specific moieties in tumor blood vessels (on smooth muscle or endothelial cells), or molecular mechanisms at the basis of the selective effects on tumor vasculature. In the long-term, the identification of novel small molecules that exploit pathophysiological differences between tumor and normal vasculature should provide more potent antitumor VDAs.

4.1.1. Drug Resistance

An important issue in the development of antiangiogenic compounds is the potential occurrence of drug resistance. Although some findings are challenging this belief (70,71), the target endothelial cells are considered a genetically stable population, and hence the induction of genetically based mechanisms of drug resistance is not expected. However, these cells are exposed to an environment rich of survival factors, in particular VEGF, which might protect endothelial cells from the antiproliferative, proapoptotic effects of antiangiogenic compounds. Indeed, the antiangiogenic effect of taxanes has been shown to be counteracted by VEGF and potentiated by VEGF inhibitors (24,33,72). The implication of this observation is that the use of tubulin-targeting compounds as antiangiogenic drugs would benefit from a combination with inhibitors of tumor derived survival factors for endothelial cells (22). In the case of vascular disrupting compounds, a synergy has been observed with inhibitors of nitric oxide synthase; whereas nitric oxide exerts a protective effect against the compounds (66,73).

4.1.2. Combination Therapies

The clinical studies completed to date are encouraging. However, used as monotherapy, antiangiogenic or VDAs have so far had limited clinically relevant antitumor effects. Antiangiogenic compounds are believed to be mainly cytostatic, thus at their best they might cause a delay of the tumor growth, whereas VDAs leave a rim of viable tumor cells that proliferate and allow the tumor to resume its growth. As indicated from preclinical studies, better therapeutic results with these agents are likely to be obtained when used in combination with other treatment modalities. Phase I/II clinical studies combining combretastatin-A4 with radiotherapy and chemotherapy or radioimmunotherapy on different tumor types are ongoing.

A combination therapy using these agents can be designed with one of two goals. On the one hand, it may potentiate the activity of the vessel-targeting compound *per se* on the vascular compartment. As discussed earlier, the use of VEGF inhibitors or nitric oxide synthase inhibitors can potentiate the antiproliferative/antiapoptotic effects of the tubulin-targeting agent on endothelial cells. Accordingly, docetaxel was synergistic with a monoclonal antibody against VEGF, whereas 2ME2 (33) and vinblastine plus an antibody against the VEGF receptor-2, induced sustained tumor regression (24). The authors found that the vascular targeting property of paclitaxel is enhanced by SU6668, a receptor tyrosine kinase inhibitor (74). In this study, paclitaxel and SU6668 acted synergistically, causing apoptosis of endothelial cells.

On the other hand, the combination can be aimed to also target the tumor cells thereby obtaining a therapeutic modality that hits both the vascular (antiangiogenic/ vascular disrupting compounds) and the tumor (selective antitumor agent) compartment.

This is the case for VDAs, where approaches to affect the peripheral rim of surviving tumor cells is needed. Accordingly, CA4-P and ZD6126 have been found to enhance the activity of chemotherapeutic agents *(75,76)* and radiotherapy *(77–79)*. Emerging evidence shows the relevance of the mechanism of action of the drugs, schedule and sequence in the outcome of combination treatments *(84)*.

5. CONCLUSIONS

Tubulin targeting agents have demonstrated a clear clinical success. New compounds are under development, with the aim of improving their therapeutic/pharmacological/ toxicological profile and their ability to overcome drug resistance *(18,80,81)*. The finding that these agents are also able to target tumor vessels, introduces a new possible use of these compounds.

Although the potential activity of tubulin-binding agents as "antiangiogenic/vascular targeting" tools has been reviewed, there are several significant challenges that must be overcome to increase the chances of their success in the clinic. Foremost understanding the molecular basis of their "selective" action should improve their correct use and might aid in the development of new agents. The development of combined modality regimens faces the difficulties of determining the optimal dosing, scheduling and sequencing that have thus far been mainly based on empiricism. Because antiangiogenic and vascular disrupting tubulin-binding agents are often not used at their MTD, new criteria to establish the active dose and to evaluate the efficacy of the treatment should help in establishing treatment modalities. New molecular and functional surrogate markers, which include blood levels of angiogenesis-related molecules and circulating endothelial or endothelial progenitor cells are being validated. Functional imaging analysis is being used in clinical trials to detect changes in tumor blood flow, and represents one of the most attractive approaches under investigation.

ACKNOWLEDGMENTS

Part of the work presented here was supported by a grant from the European Union FPS (LSHC-CT-2003-503233).

REFERENCES

1. Carmeliet P, Jain RK. Angiogenesis in cancer and other diseases. Nature 2000;407:249–257.
2. Kerbel R, Folkman J. Clinical translation of angiogenesis inhibitors. Nat Rev Cancer 2002;2: 727–739.
3. Bergers G, Javaherian K, Lo KM, Folkman J, Hanahan D. Effects of angiogenesis inhibitors on multistage carcinogenesis in mice. Science 1999;284:808–812.
4. Giavazzi R, Nicoletti MI. Small molecules in anti-angiogenic therapy. Curr Opin Invest Drugs 2002;3:482–491.
5. Cristofanilli M, Charnsangavej C, Hortobagyi GN. Angiogenesis modulation in cancer research: novel clinical approaches. Nat Rev Drug Discovery 2002;1:415–426.
6. Ferrara N, Hillan KJ, Gerber HP, Novotny W. Discovery and development of bevacizumab, an anti-VEGF antibody for treating cancer. Nat Rev Drug Discovery 2004;3:391–400.
7. Folkman J. Seminars in Medicine of the Beth Israel Hospital, Boston. Clinical applications of research on angiogenesis. N Engl J Med 1995;333:1757–1763.
8. Ruoslahti E. Targeting tumor vasculature with homing peptides from phage display. Semin Cancer Biol 2000;10:435–442.
9. Hoffman JA, Giraudo E, Singh M, et al. Progressive vascular changes in a transgenic mouse model of squamous cell carcinoma. Cancer Cell 2003;4:383–391.

10. Rafii S, Avecilla ST, Jin DK. Tumor vasculature address book: identification of stage-specific tumor vessel zip codes by phage display. Cancer Cell 2003;4:331–333.

11. St Croix B, Rago C, Velculescu V, et al. Genes expressed in human tumor endothelium. Science 2000;289:1197–1202.

12. Chaplin DJ, Dougherty GJ. Tumour vasculature as a target for cancer therapy. Br J Cancer 1999;80(suppl 1)57–64.

13. Thorpe PE. Vascular targeting agents as cancer therapeutics. Clin Cancer Res 2004;10:415–427.

14. Siemann DW, Chaplin DJ, Horsman MR. Vascular-targeting therapies for treatment of malignant disease. Cancer 2004;100:2491–2499.

15. Neri D, Bicknell R. Tumour vascular targeting. Nat Rev Cancer 2005;5:436–446.

16. Taraboletti G, Margosio B. Antiangiogenic and antivascular therapy for cancer. Curr Opin Pharmacol 2001;1:378–384.

17. Jordan MA, Wilson L. Microtubules and actin filaments: dynamic targets for cancer chemotherapy. Curr Opin Cell Biol 1998;10:123–130.

18. Jordan MA, Wilson L. Microtubules as a target for anticancer drugs. Nat Rev Cancer 2004;4:253–265.

19. Ferrara N, Kerbel RS. Angiogenesis as a therapeutic target. Nature 2005;438:967–974.

20. Miller KD, Sweeney CJ, Sledge GW Jr. Redefining the target: chemotherapeutics as antiangiogenics. J Clin Oncol 2001;19:1195–1206.

21. Schirner M. Antiangiogenic chemotherapeutic agents. Cancer Metastasis Rev 2000;19:67–73.

22. Kerbel RS, Kamen BA. The anti-angiogenic basis of metronomic chemotherapy. Nat Rev Cancer 2004;4:423–436.

23. Belotti D, Vergani V, Drudis T, et al. The microtubule-affecting drug paclitaxel has antiangiogenic activity. Clin Cancer Res 1996;2:1843–1849.

24. Klement G, Baruchel S, Rak J, et al. Continuous low-dose therapy with vinblastine and VEGF receptor-2 antibody induces sustained tumor regression without overt toxicity. J Clin Invest 2000;105:R15–R24.

25. Browder T, Butterfield CE, Kraling BM, et al. Antiangiogenic scheduling of chemotherapy improves efficacy against experimental drug-resistant cancer. Cancer Res 2000;60:1878–1886.

26. Hanahan D, Bergers G, Bergsland E. Less is more, regularly: metronomic dosing of cytotoxic drugs can target tumor angiogenesis in mice. J Clin Invest 2000;105:1045–1047.

27. Gasparini G. Metronomic scheduling: the future of chemotherapy? Lancet Oncol 2001;2:733–740.

28. Hotchkiss KA, Ashton AW, Mahmood R, Russell RG, Sparano JA, Schwartz EL. Inhibition of endothelial cell function in vitro and angiogenesis in vivo by docetaxel (Taxotere): association with impaired repositioning of the microtubule organizing center. Mol Cancer Ther 2002;1:1191–1200.

29. Lau DH, Xue L, Young LJ, Burke PA, Cheung AT. Paclitaxel (Taxol): an inhibitor of angiogenesis in a highly vascularized transgenic breast cancer. Cancer Biother Radiopharm 1999;14:31–36.

30. Vacca A, Ribatti D, Iurlaro M, et al. Docetaxel versus paclitaxel for antiangiogenesis. J Hematother Stem Cell Res 2002;11:103–118.

31. Bocci G, Nicolaou KC, Kerbel RS. Protracted low-dose effects on human endothelial cell proliferation and survival in vitro reveal a selective antiangiogenic window for various chemotherapeutic drugs. Cancer Res 2002;62:6938–6943.

32. Klauber N, Parangi S, Flynn E, Hamel E, D'Amato RJ. Inhibition of angiogenesis and breast cancer in mice by the microtubule inhibitors 2-methoxyestradiol and taxol. Cancer Res 1997;57:81–86.

33. Sweeney CJ, Miller KD, Sissons SE, et al. The antiangiogenic property of docetaxel is synergistic with a recombinant humanized monoclonal antibody against vascular endothelial growth factor or 2-methoxyestradiol but antagonized by endothelial growth factors. Cancer Res 2001;61:3369–3372.

34. Grant DS, Williams TL, Zahaczewsky M, Dicker AP. Comparison of antiangiogenic activities using paclitaxel (taxol) and docetaxel (taxotere). Int J Cancer 2003;104:121–129.

35. Petrangolini G, Cassinelli G, Pratesi G, et al. Antitumour and antiangiogenic effects of IDN 5390, a novel C-seco taxane, in a paclitaxel-resistant human ovarian tumour xenograft. Br J Cancer 2004;90: 1464–1468.

36. Vacca A, Iurlaro M, Ribatti D, et al. Antiangiogenesis is produced by nontoxic doses of vinblastine. Blood 1999;94:4143–4155.

37. Petitclerc E, Deschesnes RG, Cote MF, et al. Antiangiogenic and antitumoral activity of phenyl-3-(2-chloroethyl)ureas: a class of soft alkylating agents disrupting microtubules that are unaffected by cell adhesion-mediated drug resistance. Cancer Res 2004;64:4654–4663.

38. Mooberry SL. Mechanism of action of 2-methoxyestradiol: new developments. Drug Resist Updat 2003;6:355–361.

39. Belotti D, Rieppi M, Nicoletti MI, et al. Paclitaxel (Taxol(R)) inhibits motility of paclitaxel-resistant human ovarian carcinoma cells. Clin Cancer Res 1996;2:1725–1730.

40. Taraboletti G, Micheletti G, Rieppi M, et al. Antiangiogenic and Antitumor Activity of IDN 5390, a New Taxane Derivative. Clin Cancer Res 2002;8:1182–1188.

41. Taraboletti G, Micheletti G, Giavazzi R, Riva A. IDN 5390: a new concept in taxane development. Anticancer Drugs 2003;14:255–258.

42. Keyes K, Cox K, Treadway P, et al. An in vitro tumor model: analysis of angiogenic factor expression after chemotherapy. Cancer Res 2002;62:5597–5602.

43. Yoo GH, Piechocki MP, Ensley JF, et al. Docetaxel induced gene expression patterns in head and neck squamous cell carcinoma using cDNA microarray and PowerBlot. Clin Cancer Res 2002;8: 3910–3921.

44. Avramis IA, Kwock R, Avramis VI. Taxotere and vincristine inhibit the secretion of the angiogenesis inducing vascular endothelial growth factor (VEGF) by wild-type and drug-resistant human leukemia T-cell lines. Anticancer Res 2001;21:2281–2286.

45. Bocci G, Francia G, Man S, Lawler J, Kerbel RS. Thrombospondin 1, a mediator of the antiangiogenic effects of low-dose metronomic chemotherapy. Proc Natl Acad Sci USA 2003;100:12,917–12,922.

46. Mabjeesh NJ, Escuin D, LaVallee TM, et al. 2ME2 inhibits tumor growth and angiogenesis by disrupting microtubules and dysregulating HIF. Cancer Cell 2003;3:363–375.

47. Escuin D, Kline ER, Giannakakou P. Both microtubule-stabilizing and microtubule-destabilizing drugs inhibit hypoxia-inducible factor-1alpha accumulation and activity by disrupting microtubule function. Cancer Res 2005;65:9021–9028.

48. Chaplin DJ, Hill SA. The development of combretastatin A4 phosphate as a vascular targeting agent. Int J Radiat Oncol Biol Phys 2002;54:1491–1496.

49. West CM, Price P. Combretastatin A4 phosphate. Anticancer Drugs 2004;15:179–187.

50. Blakey DC, Ashton SE, Westwood FR, Walker M, Ryan AJ. ZD6126: a novel small molecule vascular targeting agent. Int J Radiat Oncol Biol Phys 2002;54:1497–1502.

51. Davis PD, Dougherty GJ, Blakey DC, et al. ZD6126: a novel vascular-targeting agent that causes selective destruction of tumor vasculature. Cancer Res 2002;62:7247–7253.

52. Micheletti G, Poli M, Borsotti P, et al. Vascular-targeting activity of ZD6126, a novel tubulin-binding agent. Cancer Res 2003;63:1534–1537.

53. Nihei Y, Suga Y, Morinaga Y, et al. A novel combretastatin A-4 derivative, AC-7700, shows marked antitumor activity against advanced solid tumors and orthotopically transplanted tumors. Jpn J Cancer Res 1999;90:1016–1025.

54. Hill SA, Toze GM, Pettit GR, Chaplin DJ. Preclinical evaluation of the antitumour activity of the novel vascular targeting agent Oxi 4503. Anticancer Res 2002;22:1453–1458.

55. Segreti JA, Polakowski JS, Koch KA, et al. Tumor selective antivascular effects of the novel antimitotic compound ABT-751: an in vivo rat regional hemodynamic study. Cancer Chemother Pharmacol 2004;54:273–281.

56. Holwell SE, Hill BT, Bibby MC. Anti-vascular effects of vinflunine in the MAC 15A transplantable adenocarcinoma model. Br J Cancer 2001;84:290–295.

57. Otani M, Natsume T, Watanabe JI, et al. TZT-1027, an antimicrotubule agent, attacks tumor vasculature and induces tumor cell death. Jpn J Cancer Res 2000;91:837–844.

58. Tozer GM, Kanthou C, Baguley BC. Disrupting tumour blood vessels. Nat Rev Cancer 2005;5: 423–435.

59. Vincent L, Kermani P, Young LM, et al. Combretastatin A4 phosphate induces rapid regression of tumor neovessels and growth through interference with vascular endothelial-cadherin signaling. J Clin Invest 2005;115:2992–3006.

60. Bayless KJ, Davis GE. Microtubule depolymerization rapidly collapses capillary tube networks in vitro and angiogenic vessels in vivo through the small GTPase Rho. J Biol Chem 2004;279:11,686–11,695. Epub 12003 Dec 11629.

61. Kanthou C, Tozer GM. The tumor vascular targeting agent combretastatin A-4-phosphate induces reorganization of the actin cytoskeleton and early membrane blebbing in human endothelial cells. Blood 2002;99:2060–2069.

62. Krendel M, Zenke FT, Bokoch GM. Nucleotide exchange factor GEF-H1 mediates cross-talk between microtubules and the actin cytoskeleton. Nat Cell Biol 2002;4:294–301.

63. Iyer S, Chaplin DJ, Rosenthal DS, Boulares AH, Li LY, Smulson ME. Induction of apoptosis in proliferating human endothelial cells by the tumor-specific antiangiogenesis agent combretastatin A-4. Cancer Res 1998;58:4510–4514.

64. Sheng Y, Hua J, Pinney KG, et al. Combretastatin family member OXI4503 induces tumor vascular collapse through the induction of endothelial apoptosis. Int J Cancer 2004;111:604–610.

65. Hori K, Saito S. Microvascular mechanisms by which the combretastatin A-4 derivative AC7700 (AVE8062) induces tumour blood flow stasis. Br J Cancer 2003;89:1334–1344.

66. Tozer GM, Prise VE, Wilson J, et al. Mechanisms associated with tumor vascular shut-down induced by combretastatin A-4 phosphate: intravital microscopy and measurement of vascular permeability. Cancer Res 2001;61:6413–6422.

67. Anderson HL, Yap JT, Miller MP, Robbins A, Jones T, Price PM. Assessment of pharmacodynamic vascular response in a phase I trial of combretastatin A4 phosphate. J Clin Oncol 2003;21:2823–2830.

68. Rustin GJ, Galbraith SM, Anderson H, et al. Phase I clinical trial of weekly combretastatin A4 phosphate: clinical and pharmacokinetic results. J Clin Oncol 2003;21:2815–2822.

69. Ng SS, Figg WD, Sparreboom A. Taxane-mediated antiangiogenesis in vitro: influence of formulation vehicles and binding proteins. Cancer Res 2004;64:821–824.

70. Streubel B, Chott A, Huber D, et al. Lymphoma-specific genetic aberrations in microvascular endothelial cells in B-cell lymphomas. N Engl J Med 2004;351:250–259.

71. Hida K, Hida Y, Amin DN, et al. Tumor-associated endothelial cells with cytogenetic abnormalities. Cancer Res 2004;64:8249–8255.

72. Tran J, Master Z, Yu JL, Rak J, Dumont DJ, Kerbel RS. A role for survivin in chemoresistance of endothelial cells mediated by VEGF. Proc Natl Acad Sci USA 2002;99:4349–4354.

73. Davis PD, Tozer GM, Naylor MA, Thomson P, Lewis G, Hill SA. Enhancement of vascular targeting by inhibitors of nitric oxide synthase. Int J Radiat Oncol Biol Phys 2002;54:1532–1536.

74. Naumova E, Ubezio P, Garofalo A, et al. The vascular targeting property of paclitaxel is enhanced by SU6668, a receptor tyrosine kinase inhibitor, causing apoptosis of endothelial cells, inhibition of angiogenesis and potentiation of its antitumor activity. Clin Cancer Res 2006;12:1839–1849.

75. Wildiers H, Ahmed B, Guetens G, et al. Combretastatin A-4 phosphate enhances CPT-11 activity independently of the administration sequence. Eur J Cancer 2004;40:284–290.

76. Siemann DW, Mercer E, Lepler S, Rojiani AM. Vascular targeting agents enhance chemotherapeutic agent activities in solid tumor therapy. Int J Cancer 2002;99:1–6.

77. Pedley RB, Hill SA, Boxer GM, et al. Eradication of colorectal xenografts by combined radioimmunotherapy and combretastatin a-4 3-O-phosphate. Cancer Res 2001;61:4716–4722.

78. Siemann DW, Horsman MR. Enhancement of radiation therapy by vascular targeting agents. Curr Opin Invest Drugs 2002;3:1660–1665.

79. Siemann DW, Rojiani AM. Enhancement of radiation therapy by the novel vascular targeting agent ZD6126. Int J Radiat Oncol Biol Phys 2002;53:164–171.

80. Wood KW, Cornwell WD, Jackson JR. Past and future of the mitotic spindle as an oncology target. Curr Opin Pharmacol 2001;1:370–377.

81. Rowinsky EK. Taxane analogues: distinguishing royal robes from the Emperor's New Clothes. Clin Cancer Res 2002;8:2759–2763.

82. Wang J, Lou P, Lesniewski R, Henkin J. Paclitaxel at ultra low concentrations inhibits angiogenesis without affecting cellular microtubule assembly. Anticancer Drugs 2003;14:13–19.

83. McCarty MF, Takeda A, Stoeltzing O, et al. ZD6126 inhibits orthotopic growth and peritoneal carcinomatosis in a mouse model of human gastric cancer. Br J Cancer 2004;90:705–711.

84. Taraboletti G, Micheletti G, Dossi R, Borsotti P, Martinelli M, Fiordaliso F, Ryan AJ, Giavazzi R. Potential antagonism of tubulin-binding anticancer agents in combination therapies. Clin Cancer Res 2005;11:2720–2726.

20 Neurodegenerative Diseases

Tau Proteins in Neurodegenerative Diseases Other Than AD

André Delacourte, Nicolas Sergeant, and Luc Buée

CONTENTS

SUMMARY

Microtubules are essential rails for the transport of molecules from the neuronal cell body to the synapse through the axon and vice versa. Dynamic stabilization or remodeling of microtubules is essential for the neuronal network and plasticity and thus, any disturbance of this equilibrium can have devastating consequences on brain function. Microtubule-associated Tau proteins are essential molecules regulating the dynamics of the microtubule network. They are also the basic component of neurofibrillary degeneration observed in many neurological disorders, the so-called Tauopathies. Many etiological factors, including mutations, splicing, and phosphorylation, relate Tau proteins to neurodegeneration and the microtubule network is the most possibly essential for the dynamic and progressive propagation of neurofibrillary degeneration in Tauopathies. A better knowledge of the etiological factors responsible

From: *Cancer Drug Discovery and Development: The Role of Microtubules in Cell Biology, Neurobiology, and Oncology* Edited by: Tito Fojo © Humana Press, Totowa, NJ

for the microtubule network collapse in brain diseases is essential for development of future differential diagnosis and therapeutic strategies. They would hopefully find their application against Alzheimer's disease but also in many neurological disorders for which a dysfunction of Tau biology has been identified.

Key Words: Alternative splicing; axons; cell death; mutations; neurons, phosphorylation.

1. TAU PROTEINS

Tau proteins belong to the microtubule-associated proteins (MAP) family. They are found in many animal species. In human, they are found in neurons (for review, *see* refs. *1,2*), although nonneuronal cells usually have trace amounts. For instance, Tau proteins can be expressed in glial cells, mainly in pathological conditions *(3)*, and it is possible to detect Tau messenger RNA (mRNA) and proteins in several peripheral tissues such as heart, kidney, lung, muscle, pancreas, testis as well as in fibroblasts *(4–6)*.

2. GENE

The human *Tau* gene is unique and contains 16 exons located over more than 100 kb on the long arm of chromosome 17 at band position 17q21 *(7,8)* (Fig. 1). The restriction analysis and sequencing of the gene shows that it contains two CpG islands, one associated with the promoter region, the other with exon 9 *(8)*. The CpG island in the putative Tau promoter region resembles a previously described neuron-specific promoters. Two regions homologous to the mouse Alu-like sequence are present. The sequence of the promoter region also reveals a TATA-less sequence that is likely to be related to the presence of multiple initiation sites, typical of housekeeping genes. Three SP1-binding sites that are important in directing transcription initiation in other TATA-less promoters, are also found in the proximity of the first transcription initiation site *(9)* (Fig. 1). The SP1-binding sites are suggested to control neuronal-specific expression of Tau *(10)*. Several single nucleotide polymorphisms (SNPs) are located in the large intron 0. The *A* allele of the haplotype tagging SNP167 abolishes putative-binding sites for LBP-1c/LSF/CP2 (Fig. 1). The *G* allele of htSNP167 promotes a higher expression of luciferase in mouse and human neuroblastoma cells *(11)*. Noteworthy that polymorphism in *LBP-1c/LSF/CP2* gene is a genetic determinant for Alzheimer's disease (AD) *(12,13)*.

3. TAU SPLICING

The Tau primary transcript contains 16 exons (Fig. 1). However, two of them (exons 4A, and 8) are skipped in human brain. They are specific to peripheral Tau proteins. Exon 4A is found in bovine, human, and rodent peripheral tissues with a high degree of homology. Cryptic splicing sites are described in exon 6 that would generate Tau mRNA lacking the remaining 3′-exon cassettes *(14)*. Those are found in muscle and the spinal cord but the presence of the protein remains to be determined. Exon 1 is part of the promoter, and is transcribed but not translated (Fig. 1). Exons 1, 4, 5, 7, 9, 11, 12, and 13 are constitutive exons (Fig. 2A). Exon 14 is part of the 3′-unstranslated region of Tau mRNA *(15–18)*. Exons 2, 3, and 10 are alternatively spliced and are adult brain-specific *(8)*. Exon 3 never appears independently of exon 2 although exon 2 can appear without

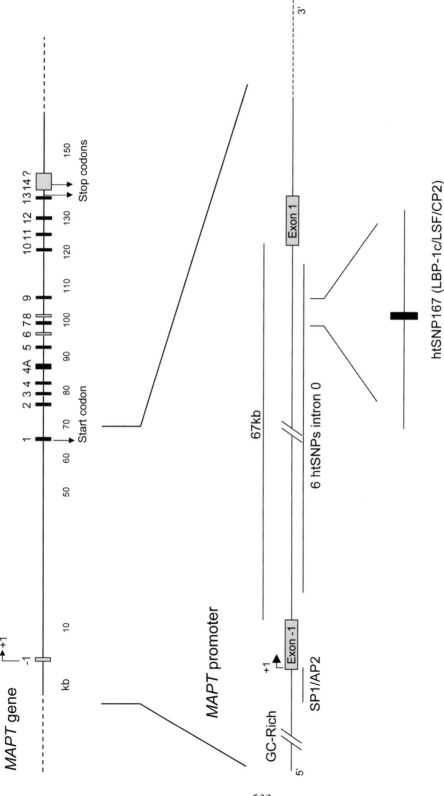

Fig. 1. *Tau* gene structure and promoter. The human *tau* gene, located on chromosome 17 at position q21, spans over 110 kb. It is made up of 16 exons numbered from −1 to 14. The start codon is located in the exon 1 and two stop codons are described; one in the intron between exon 13 and 14 and the second in the exon 14. The 3′ ending of exon 14 is not completely characterized in humans. The initiation of transcription is indicated by +1. The promoter region of tau encompasses the sequence upstream exon −1, the exon −1 and the intron 0. The promoter is GC-rich and contains SP1 and AP2 binding-motifs proximal to exon 1. Intron 0 contains many haplotype tagging single nucleotide polymorphisms among those, htSNP167 modulates tau promoter function depending on the allele.

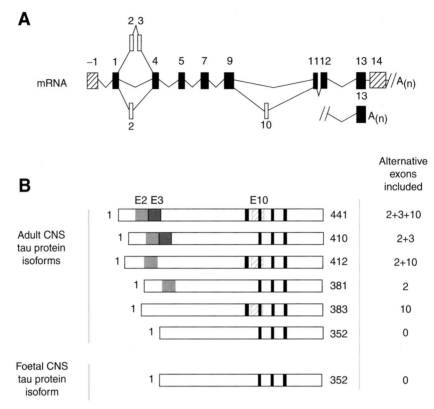

Fig. 2. Tau pre-mRNA alternative splicing in the CNS and protein isoforms translated from alternative tau mRNAs. **(A)** Schematic representation of Tau mRNAs. In the CNS, exons 4A, 6, and 8 are constitutively skipped. A small proportion of tau mRNA might arise from the use of cryptic sites in the exon 6 and would give tau proteins lacking the microtubule-binding domain. This potential use of cryptic splicing site is not detailed in the figure. Exon 1 and 14 correspond to the 5′- and 3′-untranslated region of Tau mRNAs, respectively. Two cleavage-polyadenylation sites are described; one in intron 13/14 and one in exon 14. The polyadenylated Tau mRNA including exon 14 is less represented in human. In the CNS, the alternative splicing of exon 2, 3, and 10 generates six tau proteins. Inclusion of exon 3 occurs only with exon 2, whereas exon 2 can be included alone. **(B)** The six human brain tau isoforms are represented, as they are resolved by polyacrylamide gel electrophoresis. The amino acid sequences corresponding respectively to exons 2, 3, and 10 are detailed. The tau isoform lacking the alternative exons 2, 3, and 10 is the only tau isoform expressed in fetal CNS.

exon 3 *(17)*. Thus, alternative splicing of these three exons allows for six mRNAs ($2^-3^-10^-$; $2^+3^-10^-$; $2^+3^+10^-$; $2^-3^-10^+$; $2^+3^-10^+$; $2^+3^+10^+$) and in the human brain, the Tau primary transcript is consisted of six mRNAs *(15,16,19)* (Fig. 2A).

4. STRUCTURE AND FUNCTIONS

In the human brain, Tau proteins constitute a family of six isoforms (six mRNAs) that range from 352 to 441 amino acids (Fig. 2B). Their molecular weight ranges from 45 to 65 kDa when resolved on sodium dodecyl sulfate-polyacrylamide gel electrophoresis. The Tau isoforms differ from each other by the presence of either three (3R) or four repeat-regions (4R) in the carboxy-terminal (C-terminal) part of the molecule and the absence or presence of one or two inserts (29 or 58 amino acids) in the amino-terminal

(N-terminal) part *(15–18)*. Each of these isoforms is likely to have particular physiolog- ical roles as they are differentially expressed during development. For instance, only one Tau isoform, characterized by 3R and no N-terminal inserts, is present during fetal stages, whereas the six isoforms (with one or two N-terminal inserts and 3 or 4R) are expressed during adulthood *(20,21)*. Thus, Tau isoforms are likely to have specific func- tions related to the absence or presence of regions encoded by the cassette exons 2, 3, and 10. Furthermore, the six Tau isoforms are not be equally expressed in neurons. For example, Tau mRNAs containing exon 10 are not found in granular cells of the dentate gyrus *(15)*. Thus, Tau isoforms are differentially distributed in neuronal subpopulations.

5. THE PROJECTION DOMAIN

The two 29 amino acids sequences encoded by exons 2 and 3 give different lengths to the N-terminal part of Tau proteins. These two additional inserts are highly acidic, and are followed by a basic proline-rich region. The N-terminal part of Tau proteins is referred to as the projection domain as it projects from the microtubule surface where it may interact with other cytoskeletal elements and the plasma membrane *(22,23)* (*see also* Chapter 10).

In mice lacking the *Tau* gene, an increase in MAP1A, which may compensate for the functions of Tau proteins has been observed *(24)*. In contrast to this report, another study has shown that embryonic hippocampal cultures from tau deficient mice show a significant delay in maturation as measured by axonal and neuritic extensions *(25)*. However, in both studies, axonal growth and axonal diameter were particularly affected. This may be related to the particular length of the N-terminal domain (with or without sequences encoded by exons 2 and 3) of Tau proteins in specific axons. In fact, the pro- jection domain of Tau determines spacing between microtubules in an axon and may increase axonal diameter *(26)*. It should be noted that in peripheral neurons, which often have a very long axon with large diameter, an additional N-terminal Tau sequence encoded by exon 4A is present, generating a specific Tau isoform called "big Tau" *(8,27)*. These results strongly suggest that the N-terminal regions of Tau proteins are crucial in the stabilization and organization of certain types of axons.

Tau proteins bind to spectrin and actin filaments *(28–31)*. Through these interactions, Tau proteins may allow microtubules to interconnect with other cytoskeletal compo- nents such as neurofilaments *(8,32)* and may restrict the flexibility of the microtubules *(33)*. There is also evidence that Tau proteins interact with cytoplasmic organelles. Such interactions may allow for binding between microtubules and mitochondria *(34)*. The Tau N-terminal projection domain also permits interactions with the neural plasma membrane *(22)*. Thus, Tau may act as a mediator between microtubules and the plasma membrane. This interaction has been defined as involving a binding between the proline-rich sequence in the N-terminal part of Tau proteins and the SH3 domains of src-family nonreceptor tyrosine kinases such as fyn. Studies have determined that human tau tyr18 and 29 are phosphorylated by the src family tyrosine kinase fyn *(35,36)*. The same proline-rich region of Tau proteins is likely involved in the interac- tion with phospholipase C (PLC)-γ isozymes *(37,38)*. Hwang and colleagues have demonstrated in vitro that Tau proteins complex specifically with the SH3 domain of PLC-γ, and enhance its activity in the presence of unsaturated fatty acids such as arachi- donic acid (AA) (Fig. 3). These results suggest that in cells that express Tau proteins,

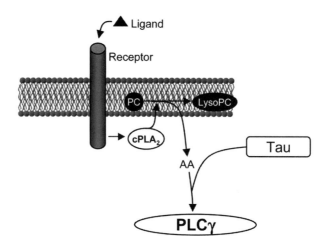

Fig. 3. Tau proteins complexes and activates phospholipase C-γ in concerted action with AA. Receptor-mediated activation of PLC-γ by concerted action of Tau and AA generated by cytosolic phospholipase A2 (cPLA$_2$). PC, phosphatidylcholine, LysoPC, lysophophatidylcholine.

receptors coupled to cytosolic phospholipase A2 may activate PLC-γ indirectly, in the absence of the usual tyrosine phosphorylation, through the hydrolysis of phosphatidylcholine to generate AA *(37,38)*. Altogether, these data indicate that Tau proteins may also play a role in the signal transduction pathway involving PLC-γ (Fig. 3) (for review *see* ref. *39*).

6. THE MICROTUBULE ASSEMBLY DOMAIN

Tau proteins bind microtubules through repetitive regions in their C-terminal part (*see* Chapter 10). These repetitive regions are the repeat domains (R1-R4) encoded by exons 9–12 *(40)* (Fig. 1C). The three (3R) or four copies (4R) are made of a highly conserved 18-amino acid repeat *(16,19,40,41)* separated from each other by less conserved 13- or 14-amino acid interrepeat domains. Tau proteins are known to act as promoter of tubulin polymerization in vitro, and are involved in axonal transport *(42–45)*. They have been shown to increase the rate of microtubule polymerization, and to inhibit the rate of depolymerization *(46)*. The 18-amino acid repeats bind to microtubules through a flexible array of distributed weak sites *(40,47)*. It has been demonstrated that adult Tau isoforms with 4R (R1–4) are more efficient at promoting microtubule assembly than the fetal isoform with 3R (R1, R3, R4) *(20,47,48)*. Interestingly, the most potent part to induce microtubule polymerization is the interregion between repeats 1 and 2 (R1–R2 interregion) and more specifically peptide 274KVQIINKK281 within this sequence. This R1–R2 interregion is unique to 4R Tau (as it occurs between exons 9 and 10), adult-specific and responsible for differences in the binding affinities between 3R and 4R Tau *(49)*. Using a cysteine mutant of the repeat domain of Tau enabling for a nanogold-labeling of the microtubule-binding domain, Tau protein is shown to decorate the inner surface of microtubule close to the taxol-binding site on β-tubulin *(50,51)*.

Recent evidence supports a role for the microtubule-binding domain in the modulation of the phosphorylation state of Tau proteins. A direct and competitive binding has been demonstrated between residues 224–236 (according to the numbering of the longest isoform) and microtubules on one hand, and residues 224–236 and protein phosphatase (PP)2A on the other hand *(52)*. As a consequence, microtubules could inhibit PP2A activity by competing for binding to Tau at the microtubule-binding domains.

7. POST-TRANSLATIONAL MODIFICATIONS

7.1. O-Glycosylation

O-Glycosylation is a dynamic and abundant post-translational modification that is characterized by the addition of a *O*-linked *N*-acetylglucosamine (*O*-GlcNAc) residue on Ser or Thr in the proximity of proresidues *(53)* by an *O*-GlcNAc transferase *(54)*. Although the functional significance of a *O*-GlcNAc modification is not yet understood, it is implicated in transcriptional regulation, protein degradation, cell activation, cell-cycle regulation, and the proper assembly of multimeric protein complexes *(55)*. This modification is often reciprocal to phosphorylation (for review *see* ref. *56*). It occurs in neurofilaments *(57)*, and in MAPs including MAP2 and Tau proteins *(58)*. The number of *O*-GlcNAcylated sites on Tau proteins is lower than the number of phosphorylation sites. Site-specific or stoichiometric changes in *O*-GlcNAcylation may modulate Tau function. In fact, phosphorylation and *O*-GlcNacylation may have opposite effects (*see* later for the role of Tau phosphorylation). For instance, *O*-GlcNacylation of Tau proteins and other MAPs suggest a role for *O*-GlcNac in mediating their interactions with tubulin. *O*-GlcNacylation may also play a role in the subcellular localization and degradation of Tau proteins *(56,58,59)*.

7.2. Phosphorylation

7.2.1. SITES OF PHOSPHORYLATION

There are eighty putative Ser or Thr phosphorylation sites on the longest brain Tau isoform (441 amino acids). Using phosphorylation-dependent monoclonal antibodies against Tau, mass spectrometry and sequencing, at least 35 phosphorylation sites have been described, including Thr39, Ser46Pro, Thr50Pro, Thr69Pro, Ser131, Thr135, Thr153Pro, Thr175Pro, Thr181Pro, Ser195, Ser198, Ser199Pro, Ser202Pro, Thr205Pro, Ser208, Ser210, Thr212Pro, Ser214, Thr217Pro, Thr231Pro, Ser235Pro, Ser237, Ser241, Ser262, Ser285, Ser305, Ser324, Ser352, Ser356, Ser396Pro, Ser400, Thr403, Ser404Pro, Ser409, Ser412, Ser413, Ser416, and Ser422Pro *(60–66)*. All of these sites are localized outside the microtubule-binding domains with the exception of Ser 262 (R1), Ser285 (R1-R2 interrepeat), Ser305 (R2-R3 interrepeat), Ser324 (R3), Ser352 (R4), and Ser356 (R4) *(67,68)*. Both Ser/Thr-Pro and non-Ser/Thr-Pro sites have been identified *(63)*. The different states of Tau phosphorylation result from the activity of specific kinases and phosphatases toward these sites.

7.2.2. KINASES

Most of the kinases involved in Tau phosphorylation are part of the proline-directed protein kinases, which include mitogen-activated protein kinase (MAP) *(69–72)*, Tau-tubulin kinase *(73)* and cyclin-dependent kinases including cdc2 and cdk5 *(74,75)*. Stress-activated protein kinases have been shown to be involved in Tau phosphorylation *(70,76,77)*. Non-Ser/Thr-Prosites can be phosphorylated by many other protein kinases, including microtubule-affinity regulating kinase *(78)*, Ca^{2+}/calmodulin-dependent protein kinase II *(79,80)*, cyclic adenosine monophosphate-dependent kinase (PKA) *(81,82)* and casein kinases I and II *(83,84)*.

Glycogen synthase kinase (GSK)3 is a Tau kinase able to phosphorylate both non Ser/Thr-Pro sites and Ser/Thr-Pro sites (For review *see* ref. *85*). Numerous kinases, proline-directed and nonproline directed, have to be used in tandem in order to observe a complete phosphorylation of recombinant Tau, and may be positively modulated at the substrate level by nonproline-directed protein kinases-catalyzed phosphorylations *(86)*. A recent example is dual-specificity tyrosine phosphorylation-regulated kinase (DYRK) and GSK3β *(87)*.

7.2.3. Phosphatases

Tau proteins from brain tissue or neuroblastoma cells are rapidly dephosphorylated by endogenous phosphatases *(88–91)*. Ser/Thr phosphatase proteins 1, 2A, 2B (calcineurin), and 2C are present in the brain *(92,93)*, and are developmentally regulated *(94)*. Like kinases, phosphatases have many direct or indirect physiological effects, and counterbalance the action of kinases. They are associated directly or indirectly with microtubules *(52,94,95)*. Thus, Tau proteins have been demonstrated to act as a link between PP1 and tubulin, whereas PP2A is directly linked to the microtubules by ionic interactions *(52)*.

Purified phosphatase proteins 1, 2A, and 2B can dephosphorylate Tau proteins in vitro *(96–99)*. For instance, in fetal rat primary cultured neurons, the use of phosphatase 2A inhibitors induces phosphorylation of Tau proteins on some sites, whereas phosphatase 2B inhibitors allow phosphorylation on other sites *(100,101)*, suggesting that phosphatases 2A and 2B are involved in dephosphorylation of different sites on Tau proteins in neurons.

PP5 is a 58-kDa novel phosphoseryl/phosphothreonyl protein phosphatase. It is ubiquitously expressed in all mammalian tissues examined, with a high level in the brain, but little is known about its physiological substrates. This phosphatase dephosphorylated recombinant tau phosphorylated with cyclic adenosine monophosphate-dependent protein kinase and GSK3β. The specific activity of PP5 toward Tau was comparable with those reported with other protein substrates examined to date. Immunostaining demonstrated that PP5 was primarily cytoplasmic in PC12 cells. A small pool of PP5 associated with microtubules. Expression of active PP5 in PC12 cells resulted in reduced phosphorylation of tau, suggesting that PP5 can also dephosphorylate tau in cells *(102)*.

7.2.4. Tau Phosphorylation and Microtubule Assembly

Tau proteins bind microtubules through the microtubule-binding domains. The heptapeptide 224KKVAVVR230 located in the proline-rich region has a high microtubule-binding activity in combination with the repeats regions *(103)*, suggesting intramolecular interactions between the both regions. However, microtubule assembly depends partially upon the phosphorylation state as phosphorylated Tau proteins have less effect than non-phosphorylated Tau proteins on microtubule polymerization *(43,44,98,104–107)*. Phosphorylation of Ser262 dramatically reduces the affinity of Tau for microtubules in vitro *(104)*. Nevertheless, this site alone, which is present in fetal Tau, adult Tau as well as in hyperphosphorylated Tau proteins found in neurofibrillary tangles (NFT), is insufficient to eliminate Tau binding to microtubules *(67)*. Thus, phosphorylation outside the microtubule-binding domains can strongly influence tubulin assembly by modifying the affinity between Tau and microtubules. For instance, phosphorylation at Thr231 was shown to be one of the major phosphorylation sites of which inhibits tau's binding to microtubules *(108,109)*.

7.2.5. Tau Phosphorylation and Cell Sorting

Tau is a phosphoprotein, as was first demonstrated with the monoclonal antibody Tau-1 raised against a dephosphorylated site. As Tau-1 labels preferentially axons, Tau were tagged as "axonal proteins" *(110)*. However, the state of phosphorylation of Tau proteins is likely different according to the cell compartments *(111)*, and Tau-1 immunoreactivity was observed in the somatodendritic compartment of neurons after dephosphorylation *(112)*. In fact, the labeling of cell bodies and dendrites with phosphorylation-independent antibodies such as Alz-50, demonstrates that these proteins are found in all compartments of the nerve cells, and are not exclusively "axonal proteins." However, compared with other MAPs, Tau proteins are preferentially axonal. Both phosphorylation and transcription factors may be involved in Tau trafficking and cell sorting (nuclear, axonal, or somatodendritic) *(23)*.

8. TAU AGGREGATION IN NEUROLOGICAL DISORDERS OTHER THAN AD: TAUOPATHIES

In neurodegenerative disorders, other than AD, referred to as tauopathies, abnormally and hyperphosphorylated tau proteins aggregate in the absence of amyloid deposits. Comparative biochemistry of the Tau aggregates shows that they differ in both Tau isoform phosphorylation and content of, which enables a molecular classification of Tauopathies. Five classes of Tauopathies have been defined depending on the type of Tau aggregates that constitute the "Bar Code" for neurodegenerative disorders (Fig. 4).

8.1. Class 0: Frontal Lobe Dementia Non-Alzheimer Non-Pick

Frontal lobe dementia is a neurological disorder that has been recently characterized, despite the fact that it is the most common presenile dementing disorder in Europe, after AD. Like Pick's disease (PiD), it has a "frontal" pathology; however, whereas PiD is neuropathologically characterized by Pick bodies, frontal lobe dementia has no specific neuropathological hallmarks. Morphological changes consist of neuronal cell loss, spongiosis, and gliosis mainly in the superficial cortical layers of the frontal and temporal cortex. No tau aggregates are observed although a loss of tau-protein expression is observed in this disorder *(113–115)*.

8.2. Class I: A Major Tau Triplet at 60, 64, 69

Class I is characterized by a pathological tau triplet at 60, 64, and 69 kDa and a minor pathological tau at 72/74 kDa (Fig. 2B). It is now well-established that this pathological tau triplet corresponds to the aggregation of the six tau isoforms *(116,117)*. The pathological tau 60 is made up of the shortest tau isoform $(2^-3^-10^-)$. The pathological Tau 64 and 69 are each made up of two tau isoforms. Tau isoforms either with the exon 2 or exon 10 alone make up the pathological Tau 64. The pathological Tau 69 is made of tau isoforms either with exon 2+10 or exon 2+3. The longest tau isoform containing exons 2, 3, and 10 $(2^+3^+10^+)$ constitute the 72/74 kDa pathological component. The prototypical neurological disorder that characterizes this class is AD, but includes nine additional neurological disorders such as hippocampal tauopthy in cerebral aging, amyotrophic lateral sclerosis parkinsonism–dementia complex of Guam, Parkinsonism with dementia of Guadeloupe, Niemann-Pick disease of type C, postencephalitic parkinsonism, Familial

A **Bar code classification of Tauopathies**

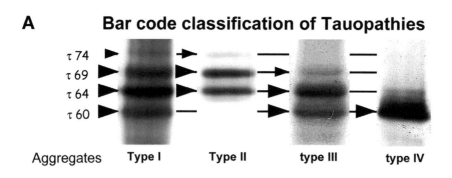

τ 74
τ 69
τ 64
τ 60

Aggregates   Type I   Type II   type III   type IV

B   **Tauopathies: 5 Classes**

Class 0   • Frontal lobe dementia non-AD, non-Pick   **Loss of tau protein expression**

Class 1
• Cerebral aging (indivuduals aged over 75 years)
• Alzheimer 's disease (sporadic and familial)
• ALS/parkinsonism-dementia complex of Guam
• Parkinson with dementia of Guadeloupe
• Niemann-Pick disease type C
• Postencephalitic parkinsonism
• Familial British dementia
• Dementia pugilistica
• Down's syndrome
• FTDP-17

Aggregates types

Type I

Tau triplet:

Tau 60, 64 and 69

Class 2
• Corticobasal degeneration
• Argyrophilic grain dementia
• Progressive supranuclear palsy
• FTDP-17

Type II

Tau doublet:
Tau 64 and 69

Class 3
• Pick's disease
• FTDP-17

Type III

Tau doublet:
Tau 60 and 64

Class 4   • Myotonic dystrophy of type I and II   Type IV   **Tau 60**

Fig. 4. The bar code of Tauopathies and their classification. **(A)** Human brain tissue from patients affected by different Tauopathies are separated by polyacrylamide gel electrophoresis and pathological tau proteins are revealed by Western-blotting using phospho-dependent Tau antibodies (e.g., AD2 monoclonal antibody recognizing the phosphorylated Ser396/404 form of Tau, numbering according to the longest Tau isoform). Four different electrophorectic patterns of pathological Tau proteins are illustrated. Those are made up of pathological Tau bands at 60, 64, 69, and 74 kDa, which correspond to pathological Tau that are found in aggregates. The type I aggregate is characterized by the presence of the four pathological Tau components, whereas the type II and III include 2 major pathological tau components at 64 and 69 kDa, or 60 and 64 kDa, respectively. Finally, the fourth aggregate type is characterized by a strong pathological Tau component at 60 kDa, with the 64 kDa and 69 kDa components often observed depending on the severity of the affected region analyzed. Those four main patterns of pathological Tau thus represent a bar code of tauopathies. **(B)** Neurological disorders for which a pathological Tau pattern has been defined are classified according to the "bar code." Five classes are defined including a unique or multiple neurological disorders. The four aggregate types are detailed as well as the pathological-tau pattern identified. Class 0 includes a unique neurological disorder characterized by the loss of expression of Tau protein and lacks distinctive neuropathological features, that is Frontal dementia of non-AD, non-Pick type.

British dementia, dementia pugislistica, Down's syndrome as well as Frontotemporal dementia (FTD) with parkinsonism linked to chromosome 17 (FTDP-17) *(118–128)*.

8.3. Class II: A Major Tau Doublet at 64 and 69 kDa

The class II profile is characterized essentially by the aggregation of 4R-tau isoforms. This pathological tau profile is observed in corticobasal degeneration (CBD), Argyrophylic grain dementia, progressive supranuclear palsy (PSP), and FTDP-17 *(129–131)*. PSP is a late-onset atypical Parkinsonism disorder described by Steele, Richardson, and Olszewski in 1964 *(132,133)*. Dementia is also a common feature at the end-stage of the disease *(134,135)*. Neuropathologically, PSP is characterized by neuronal loss, gliosis, and NFT formation. NFT were first described in the basal ganglia, brainstem, and cerebellum *(132)*. Subsequently, the degenerating process has been described in the perirhinal, inferior temporal, and prefrontal cortex, with the same features as subcortical NFT *(136,137)*. Glial fibrillary tangles have also been described *(138–141)*. CBD was first described in 1967 and referred to as corticodentatonigral degeneration with neuronal achromasia *(142,143)*. It is a rare, sporadic, and slowly progressive late-onset neurodegenerative disorder, characterized clinically by cognitive disturbances and extrapyramidal motor dysfunction. Moderate dementia emerges sometimes late in the course of the disease *(144)*. There is a clinical and pathological overlap between PSP and CBD *(134,135)*. Neuropathological examination reveals severe glial and neuronal abnormalities. The glial pathology in PSP is characterized by "tufted" plaques whereas that in CBD is consisted of astrocytic plaques and numerous tau-immunoreactive inclusions in the white matter. In CBD, achromatic ballooned neurons are often detected in the cortex, brainstem, and subcortical structures, as are neuritic changes and NFT *(145,146)*. In both PSP and CBD, the pathological tau profile consists essentially of the aggregation of 4R-tau isoforms, although a study of a large series of PSP patients suggests that the pathological Tau profile is heterogeneous and includes variable amounts of 3R-tau isoforms as well. Thus, an increased ratio of 4R/3R pathological tau isoform may better define the class II of Tauopathies.

A third class II tauopathy (argyrophilic grain dementia [AGD]) was described in 1987, when Braak and coworkers *(147)* reported a series of eight patients with a non-Alzheimer, late-onset dementia. Clinically, AGD is characterized by behavioral disturbances such as personality change and emotional imbalance as well as memory and cognitive impairment *(148)*. At the neuropathological level, AGD is characterized by the occurrence of argyrophilic grains (ArG) on light microscopy of the brain tissue, hence the name AGD. ArGs are neuronal inclusions stained by silver dyes *(149)*. The diagnosis of Dementia with ArGs is based on the widespread occurrence of minute, spindle or comma-shaped, argyrophilic, tau-immunoreactive structures distinct from neuropile threads and predominantly located in the hippocampus and related limbic areas *(150)*. The 4R/3R ratio has been shown to be increased in AGD, thus demonstrating that AGD is most likely a 4R tau-pathology *(131,151)*. Although the tau pathology in AGD principally affects the limbic system, the tau pathology has been shown to extend throughout the cerebral cortex, very distant from the limbic temporal region, in brain areas that are considered to be spared in AGD *(152)*. The neocortical extension of tau pathology in AGD is a further feature that it shares with PSP and CBD, and might be a clue to a pathological continuum between PSP, CBD, limbic AGD, and diffuse

AGD *(152)*. Many features are shared by all four repeat (4R) Tauopathies. For example, like in PSP and CBD, the subthalamic nuclei are selectively involved by Tau aggregates in AGD patients *(153)*. A series of 10 polymorphisms within the *tau* gene, including 8 SNP and 1 deletion, have been found to be inherited in complete linkage disequilibrium with each other and with the dinucleotide polymorphism 116507(TG)*n* defining two extended haplotypes (H1 and H2) that cover at least the entire *tau* gene *(154*, for reviews *see* refs. *155,156)*. The H1/H1 haplotype is more frequent in PSP/CBD patients than in controls or other tauopathies *(11,154,157)*. H1/H1 may be more frequent in AGD patients than in controls, though recent studies have failed to establish statistically a significant difference *(158,159)*.

8.4. Class III: A Major Tau Doublet at 60 and 64 kDa

Class III Tauopathes include PiD and autosomal dominant inherited FDTP-17 (*see also* Table 1). PiD is a rare form of neurodegenerative disorder characterized by a progressive dementing process. Early in the clinical course, patients show signs of frontal disinhibition *(160)*. Neuropathologically, PiD is characterized by prominent fronto-temporal lobar atrophy, gliosis, severe neuronal loss, ballooned neurons, and the presence of neuronal inclusions called Pick bodies *(161,162)*. Pick bodies are labeled by Tau antibodies, with a higher density in the hippocampus than in the neocortex *(161–163)*. The laminar distribution of Pick bodies is clearly different from other tauopathies such as PSP and CBD. In the hippocampus, Pick bodies are numerous in granular cell neurons of the dentate gyrus, in CA1, subiculum and entorhinal cortex, whereas in the neocortex, they are mainly found in layers II and VI of the temporal and frontal lobes. Ultrastructurally, Pick bodies consist of accumulation of both random coiled and straight filaments.

Biochemical analysis using a quantitative Western blot approach with phosphorylation-dependent antitau antibodies has revealed that in all cases of PiD studied, a major 60 and 64 kDa pathological Tau doublet is observed in the isocortex, in the limbic areas and in subcortical nuclei *(145,162)*. A faint pathological tau band is observed at 69 kDa *(164)*. The pathological Tau profile of PiD contrasts with that of class II Tauopathies, with the pathological Tau isoforms consisting essentially of the 3R-tau isoforms *(164)*. In addition, aggregated Tau proteins in PiD are not detected by the monoclonal antibody 12E8 raised against the phosphorylated residue ser262/ser356 whereas in other neurodegenerative disorders, this phosphorylation site is detected *(162,165)*. The lack of phosphorylation at ser262 and 356 sites is likely to be related to either a kinase inhibition in neurons that degenerate in PiD or an absence of these kinases within degenerating neurons *(166)*.

Interestingly, many patients with mutations on presenilin 1 gene ([*M146L*] and [*G183V*]) exhibit a clinical feature of fronto-temporal dementia with Pick bodies and a major 60 and 64 kDa tau doublet (167,168, for review, 156).

8.5. Class IV: A Major Tau 60

Class IV is also represented by a single neurogical disorder, myotonic dystrophy (DM) of types I and II. DM is the commonest form of adult-onset muscular dystrophy. It is a multisystemic disease affecting many systems including the central nervous system (CNS) (cognitive and neuropsychiatric impairments), the heart (cardiac conduction defects), genital tract (testicular atrophy), eyes (cataracts), ears (deafness), gastrointestinal tract (smooth muscle), and endocrine system (insulin resistance), thus leading to a wide

Table 1
FTDP-17: Effects of *Tau* Mutations

Mutation	Localization	Aggregates	Isoforms
R5H	Exon 1	Glial	4R
R5L	Exon 1	Neu(DNF)	4R + 1N3R
K257T	Exon 9	Neu(PiD)	3R > 4R
L266V	Exon 9	Neu/Glial	3R + 4R
G272V	Exon 9	Neu(PiD)	3R + 4R
I260V	Exon 9	ND	ND
N279K	Exon 10	Neu/Glial	4R
ΔK280	Exon 10	ND	ND
L284L	Exon 10	Neu	4R?
N296H	Exon 10	Neu/Glial	4R
N296N	Exon 10	Neu/Glial	4R
ΔN296	Exon 10	ND	ND
P301L	Exon 10	Neu	4R
P301S	Exon 10	Neu	4R
G303V	Exon 10	Neu/Glial	4R
S305N	Exon 10	Neu	4R
S305S	Exon 10	Neu	4R
+3, +11, +12, +13, +14, +16	Intron 10	Neu/Glial	4R
+19, +29	Intron 10	ND	3R>>4R7
+33	Intron 10	ND	ND
L315R	Exon 11	ND	ND
K317M	Exon 11	Neu/Glial	4R?
S320F	Exon 11	Neu(PiD)	ND
G335V	Exon 12	Neu(PiD)	3R+4R?
Q336R	Exon 12	Neu(PiD)	3R+4R?
V337M	Exon 12	Neu	3R + 4R
E342V	Exon 12	Neu	4R (0N4R>1N4R>2N4R)
S352V	Exon 12	Neu	ND
K369I	Exon 12	Neu(PiD)/Glial	3R + 4R
G389R	Exon 13	Neu(PiD)	4R > 3R
R406W	Exon 13	Neu	3R + 4R

Neuronal aggregates are indicated by Neu. Neuronal inclusions corresponding to Pick bodies are pre-cised (PiD). The cellular types of aggregates or pathological tau isoform profiles that have not been deter-mined are indicated by ND.

and variable complex panel of symptoms *(169,170)*. Clinically DM includes two entities designated as myotonic dystrophy of type I (DM1) and myotonic dystrophy of type II (DM2), as recommended by the International Myotonic Dystrophy Consortium (2000) *(171)* as well as at the European NeuroMuscular Centre DM2/PROMM Workshop (2003) *(172)*. Although DM2 is also referred to as proximal myotonic myopathy, because of subtle clinical differences *(173)*, DM2 now refers to the gene locus and mutation *(174)*.

DM1, the most common form of DM is an inherited autosomal dominant disorder caused by a single gene mutation consisting of expansion of a CTG trinucleotide motif in the 3′-untranslated region of the DM protein kinase gene (*dmpk*), located on chromo-some 19q *(175)*. The mutation that causes DM2 corresponds also to an expansion of a

CCTG tetranucleotide in the first intron (untranslated sequence) of the *ZNF9* gene, located on chromosome 3q, that encodes a nuclear protein *(174,176)*. Both mutations are very unstable. The length changes from one generation to another as well as in somatic cells of an individual. Thus, DM1 and DM2 mutations share many pathogenic similarities (for review *see* ref. *177*).

Cognitive impairment, including memory, visuo-spatial recall, verbal scale, with cortical atrophy of the frontal and the temporal lobe and white matter lesions are often described in both DM1 and DM2 *(178,179)*. Neuropathological lesions such as NFT, have been observed in adult DM1 individuals more than 50 yr of age *(180,181)*.

The pathological tau profile of DM1 is characterized by a strong pathological tau band at 60 kDa and to a lesser extent, a pathological tau component at 64 and 69kDa. This typical pathological tau profile is reflected by a reduced number of tau isoform expression in the brain of individuals with DM1, both at the protein and mRNA levels *(182)*. In addition, tau protein expression is also demonstrated to be altered in transgenic with human DM1 locus *(183)*. The analysis of multiple brain regions of one genetically confirmed DM2 patient aged of 71 yr, showed some neurofibrillary degenerating processes. Using specific immunological probes against amino acid sequences corresponding to exon 2 and exon 3 corresponding, the neurofibrillary lesions were shown to be devoid of tau isoforms with N-terminal inserts *(184)*. An altered splicing of tau characterized by a reduced expression of tau isoforms containing the N-terminal inserts characterizes both DM1 and DM2. Overall, it demonstrates that the CNS is affected and that DM are real tauopathies. The direct relationship between the altered splicing of Tau and neurofibrillary degeneration in DM remains to be established. Indeed, such an altered splicing of Tau is commonly observed in FTDP-17 and considered as reminiscent to neurofibrillary degeneration and Tauopathies.

9. TAU MUTATIONS AND FAMILIAL FRONTOTEMPORAL DEMENTIA AND CHROMOSOME 17-LINKED PATHOLOGIES

Historically, FTD were often classified as a form of PiD, even when Pick cells or Pick bodies were not found *(185)*. However, this denomination may involve different subgroups of pathologies, and the Lund and Manchester groups published in 1994 a consensus on Clinical and Neuropathological Criteria for Frontotemporal Dementia (1994) *(186)*. This publication clarified the position of PiD within FTD, and several of the reported cases of familial PiD were probably cases of familial FTD. Indeed, it is difficult to ascertain families, which have the classic pathological features of PiD from the literature *(187)*, because they often have unusual clinical features.

In 1994, Wilhelmsen and colleagues *(188)* described an autosomal dominantly–inherited disease related to familial FTD, characterized by adult-onset behavioral disturbances, frontal lobe dementia, Parkinsonism, and amyotrophy. They demonstrated a genetic linkage between this pathology, designated disinhibition-dementia-parkinsonism-amyotrophy complex, and chromosome 17q21–22 *(188)*. Since then, several families sharing strong clinical and pathological features and for which there is a linkage with chromosome 17q22–22 have been described *(189–192)*. They have been included in a group of pathologies referred to as FTDP-17 *(193)*.

Although clinical heterogeneity exists between and within the families with FTDP-17, usual symptoms include behavioral changes, loss of frontal executive functions, language deficit, and hyperorality. Parkinsonism and amyotrophy are described in some families, but are not consistent features. Neuropathologically, brains of FTD patients exhibit an atrophy of frontal and temporal lobes, severe neuronal cell loss, gray and white matter gliosis, and a superficial laminar spongiosis. One of the main important characteristics is the filamentous pathology affecting the neuronal cells, or both neuronal and glial cells in some cases. The absence of amyloid aggregates is usually established *(193,194)*.

FTDP-17 has been related to mutations on the *Tau* gene *(195–197)*. Tau mutations always segregate with the pathology and are not found in the control subjects, suggesting their pathogenical role. To date, several mutations have been described in the *Tau* gene among the different families with cases diagnosed as FTDP-17 (Table 1). Twenty missense mutations in coding regions R5H *(198)*, R5L *(199)*, K257T *(200,201)*, I260V *(202)*, L266V *(203)*, N279K, G272V *(195,204)*, N279K *(205–208)*, N296H *(209,210)*, G303V *(211)*, P301L *(190,212,213)*, P301S *(214,215)*, S305N *(206,216)*, L315R *(217)*, K317M *(218)*, S320F *(217)*, G335V *(219)*, Q336R *(220)*, V337M *(221)*, E342V, S352V, K369I *(222)*, G389R *(223)*, R406W *(195)*, three silent mutations L284L *(224)*, N296N *(225)*, S305S *(226)*, two single amino acid deletions ΔK280 and ΔN296 *(210,224,227)*, and nine intronic mutations in the splicing region following exon 10 at position +3 *(204)*, +11, +12, +13 *(195)*, +14 *(195)*, +16 *(228)*, +19, +29, and +33 have been reported (Table 1).

9.1. Mutations Affecting Tau Splicing

Depending on their functional effects, Tau protein mutations may be divided into two groups: mutations affecting the alternative splicing of exon 10, and leading to changes in the proportion of 4R- and 3R-Tau isoforms, and mutations modifying Tau interactions with microtubules. In patients with FTDP-17 mutations affecting splicing include intronic mutations in the splicing region following exon 10 (+3, +13, +14, +16) and some missense mutations. Intronic mutations disturb a stem loop structure in the 5′-splice site of exon 10 that stabilizes this region of the pre-mRNA *(195,204,229,230)*. Sequence analysis of this splicing region in different animals indicates that the lack of the stem loop structure is associated with an increase in Tau mRNAs containing exon 10 *(229)*. Indeed, without this stem loop, access of U1snRNP to this site may be facilitated, increasing the formation of exon 10+ Tau mRNAs and thus the 4R-Tau isoform *(224,229,230)*. Interestingly, in these families, abnormally phosphorylated 4R-Tau isoforms aggregate into filaments and display a Tau-electrophoretic profile similar to the major Tau doublet at 64 and 69 kDa found in PSP and CBD *(129,161,204,231)*. Some missense mutations (N279K and S305N) also modify the splicing of exon 10 *(123,224)*. For instance, the change in nucleotide for N279K and S305N mutations also creates an exon-splicing enhancer sequence *(224)*. The silent mutation L284L increases the formation of Tau mRNAs containing exon 10, presumably by destroying an exon splicing silencing element *(224)*. Families with one of these three missense mutations display the same electrophoretic Tau pattern than those having intronic mutations, namely a Tau doublet at 64 and 69 kDa *(224,232,233)*. Finally, the twisted ribbon filaments described in neurons and glial cells are a common neuropathological feature in all of the neurodegenerative disorders belonging to this first group.

9.2. Tau Missence Mutations and Tau Aggregation

The second group of Tau mutations found in FTDP-17 affects tau interaction with microtubules and includes several missense mutations as well as a deletion mutation. The effects of mutations G272V, P301L, P301S, V337M, G389R, and R406W in an in vitro system of microtubule assembly were reported by Goedert and colleagues *(221)*. These authors showed that mutated Tau isoforms bind microtubules to a lesser extent than wild-type isoforms. They suggest that the mutated isoforms may induce microtubule disassembly *(221,234)*. These data were confirmed by a number of studies *(224,232,235)*. When missense mutations are located in Tau regions common to all isoforms, outside exon 10 (G272V, V337M, G389R, R406W), the six Tau isoforms do not bind properly to microtubules. These proteins aggregate into *paired helical filaments* and straight filaments similar to those described in AD, and are present in neuronal cells. Their biochemical characterization shows a Tau-electrophoretic profile similar to the AD Tau-triplet. Conversely, when missense mutations are located in exon 10 (P301L, P301S), 4R-Tau isoforms are affected, and these do not bind to microtubules, but instead aggregate into twisted ribbon filaments. This type of filamentous inclusions is described in both neurons and glial cells. The biochemical characterization shows a Tau electrophoretic profile similar to the major Tau doublet encountered in PSP and CBD.

The ΔK280 mutation is particular, as it modifies the ratio 4R-Tau/3R-Tau ratio and could also affect the interaction between Tau and microtubules. This mutation may decrease the formation of Tau mRNAs containing exon 10 and thus enhance the formation of 3R-Tau isoforms. Moreover, this deletion mutation is also responsible for a considerably reduced ability of Tau to promote microtubule assembly, and is stronger than the effect of the P301L mutation. No data are currently available on the biochemistry of Tau aggregates in this family *(224,227)*.

Mutations of the *Tau* gene and their involvement in FTDP-17 emphasize the fact that abnormal Tau proteins may play a central role in the etiopathogenesis of neurodegenerative disorders, without any implication of the amyloid cascade. The functional effects of the mutations suggest that a reduced ability of Tau to interact with microtubules may be upstream of hyperphosphorylation and aggregation. These mutations may also lead to an increase in free cytoplasmic Tau (especially the 4R-Tau isoforms), and therefore facilitating their aggregation into filaments *(235)*. Finally, some mutations may have a direct effect on Tau fibrillogenesis *(236)*.

9.3. Tau Missence Mutations and In Vivo Microtubule Stabilization

Regarding the functional effects of Tau mutants, it was shown that mutations G272V, ΔK280, P301L, P301S, and V337M cause a decreased ability of Tau to promote microtubule assembly in vitro *(221)*. However, the magnitude of the observed effects and the reported rank order of potency of individual mutations have been variable. Overexpression of mutant Tau in transfected cells has given inconsistent results as far as effects on microtubule binding and stability are concerned *(206,227,235)*. Such studies are confounded by the problem that high expression of mutant Tau may override any effects that are present at more physiological levels.

For in vivo investigation, recombinant tau 3R, 4R, or mutated tau are injected in Xenopus oocytes and maturation of oocytes is used as an indicator of microtubule function *(237)*. Normal oocyte maturation, as visualized by the appearance of a white spot indicates that oocyte meiosis driven by microtubules is normal. Wild-type 4R-tau inhibits maturation of oocytes in a concentration-dependent manner, whereas 3R-tau has no effect. These data suggest that there is a direct interaction of 4R-Tau and microtubules that interferes with oocyte maturation. Whatever the concentration of the 4R-mutant Tau (G272V, P301L, P301S, and V337M) injected in oocyte, it fails to affect oocyte maturation although few changes in the organization of meiotic spindles are observed. In contrast to wild-type 4R-tau, these mutations are likely to reduce the interaction of tau with microtubules. Two additional mutations (R406W and S305N) have been analyzed and found to perturb oocyte maturation. Differential phosphorylation cannot explain the data obtained for the first group of Tau mutants as wild-type 3R- and 4R-tau are found to be phosphorylated at the same extent as the G272V, P301L, P301S, and V337M 4R Tau mutants. Therefore, these mutations strongly reduce the ability of Tau to interact with microtubules *(237)*.

Regarding the R406W mutation, there are some controversial effects: state of phosphorylation and reduced microtubule binding. At low concentrations, R406W Tau mutant strongly interferes with oocyte maturation even when no effect is observed with wild-type 4R-tau. Conversely, at high concentrations similar to those of injected wild-type 4R-tau that block maturation, oocyte maturation is never completely abolished. Altogether, these data suggest that at low concentrations the R406W mutant Tau is less phosphorylated and thus show a better microtubule binding whereas at high concentrations, the phosphorylation state is not sufficient to thwart the reduction of microtubule binding of this Tau mutant *(237)*.

The most pronounced inhibitory effect on oocyte maturation is observed with the S305N Tau, even following microinjections of small amounts of Tau protein. This effect is independent of phosphorylation. It was shown that S305N Tau has a slightly increased ability to promote microtubule assembly in vitro when compared with wild-type protein *(206)*. Altogether, these data demonstrate that in vivo Tau missense mutations either strongly reduce interactions with microtubules or increase these interactions. The observed phenotype is dependent on the combined effect of Tau phosphorylation and concentration. It also demonstrates that the *Xenopus* oocyte is an interesting heterologous model system for following perturbations in microtubule function.

10. FAMILIAL VS SPORADIC TAUOPATHIES

The mapping of the spatiotemporal distribution of tau pathology in the different brain areas is important to understand how the disease spreads in the brain. Although *tau* gene mutations affect simultaneoulsy different brain areas, many "sporadic" tauopathies affect first a precise vulnerable area: the entorhinal and hippocampal area in AD, the brain stem in PSP and CBD.

Indeed, there is a precise biochemical pathway of tau pathology in aging and in AD. The progression of tau pathology is sequential, invariable, hierarchical, and predictable. 10 stages (S1–S10) were defined, corresponding to 10 brain areas sequentially affected *(238)*. The extension of the tauopathy fits well with the evolution of cognitive deficits, from memory disorders when the hippocampal formation is affected (stages 1–3) to language impairment (temporal areas affected at stages 4–6) and then to apraxia, agnosia when frontal and parietal areas are involved (stage 7–10).

It is interesting to note that the pathway of tau pathology in PSP and in CBD is quite different from the one in AD, and roughly the opposite, emerging from subcortical nuclei toward the neocortex, and especially the frontal motor cortex. In most sporadic neurodegenerative disorders, the spreading of tau pathology follows specific neuronal connections, like a precise neuronal chain reaction. This pathway starts in the vulnerable brain area characteristic of the disease to specific neocortical brain areas. At that level of tau extension, a dementing process is occuring. These observations in the human brain demonstrate the progressive collapse of neuronal populations, along cortico-cortical of subcortico-cortical connections.

11. CONCLUSIONS

Aggregation of tau proteins in filamentous inclusions is a common feature of more than 20 neurodegenerative disorders. The laminar and regional distribution of NFT or other inclusions are different among dementing conditions. Likewise, a "bar code" of pathological tau proteins electrophoretic profiles permits a classification of disorders sharing similar biochemical signatures. Many parameters can explain this classification such as the selective aggregation of specific sets of tau isoforms, the differential vulnerability of subneuronal population, in addition to possibly variable sets of enzymes (e.g., kinases, phosphatases). Not withstanding the regional or laminar distribution or the electrophoretic pattern of pathological tau proteins, their aggregation is always correlated to dementia when association neocortical areas are involved. The recent discovery of mutations on the *tau* gene, resulting in an abnormal aggregation of tau isoforms into filamentous inclusions in FTDP-17, demonstrates that abnormal tau metabolism is sufficient to induce nerve cell degeneration, in relationship with the property of mutant tau to regulate microtubule dynamics. The role of *tau* haplotypes and mutations in neurodegeneration and tau aggregation is likely to be central. Is PiD really a sporadic disorder? Are PSP and CBD related to specific *tau* H1 haplotype? How PS1 mutations related to Tau pathology? Numerous questions have recently emerged and should allow for a better understanding of tau pathology.

Altogether, these data indicate that numerous mechanisms including cell vulnerability, regulation of many enzymes, and tau mutations could interact to disturb tau metabolism, and result in the disorganization of the cystoskeleton, including the microtubule network, commonly observed in all of these neurodegenerative illnesses. But the key event is always the disorganization of the cytoskeleton leading to nerve cell degeneration.

ACKNOWLEDGMENTS

This work was supported by the Institut de la Santé et de la Recherche Médicale, the Center National de la Recherche Scientifique, the European Community (APOPIS LSHM-CT-2003-503330), AFM, FEDER, GIS-Longévité, Conseil Régional Nord Pas-de-Calais (Neuronox).

REFERENCES

1. Schoenfeld TA, Obar RA. Diverse distribution and function of fibrous microtubule-associated proteins in the nervous system. Int Rev Cytol 1994;151:67–137.
2. Tucker RP. The roles of microtubule-associated proteins in brain morphogenesis: a review. Brain Res Brain Res Rev 1990;15:101–120.
3. Chin SS, Goldman JE. Glial inclusions in CNS degenerative diseases. J Neuropathol Exp Neurol 1996;55:499–508.
4. Gu Y, Oyama F, Ihara Y. Tau is widely expressed in rat tissues. J Neurochem 1996;67:1235–1244.
5. Ingelson M, Vanmechelen E, Lannfelt L. Microtubule-associated protein tau in human fibroblasts with the Swedish Alzheimer mutation. Neurosci Lett 1996;220:9–12.

6. Vanier MT, Neuville P, Michalik L, Launay JF. Expression of specific tau exons in normal and tumoral pancreatic acinar cells. J Cell Sci 1998;111(pt 10):1419–1432.

7. Neve RL, Harris P, Kosik KS, Kurnit DM, Donlon TA. Identification of cDNA clones for the human microtubule-associated protein tau and chromosomal localization of the genes for tau and micro-tubule-associated protein 2. Brain Res 1986;387:271–280.

8. Andreadis A, Broderick JA, Kosik KS. Relative exon affinities and suboptimal splice site signals lead to non-equivalence of two cassette exons. Nucleic Acids Res 1995;23:3585–3593.

9. Andreadis A, Wagner BK, Broderick JA, Kosik KS. A tau promoter region without neuronal speci-ficity. J Neurochem 1996;66:2257–2263.

10. Heicklen-Klein A, Ginzburg I. Tau promoter confers neuronal specificity and binds Sp1 and AP-2. J Neurochem 2000;75:1408–1418.

11. Rademakers RS, Melquist M, Cruts J, et al. High-density SNP haplotyping suggests altered regulation of tau gene expression in progressive supranuclear palsy. Hum Mol Genet 2005;14(21):3281–3292.

12. Lambert JCL, Goumidi FW, Vrieze B, et al. The transcriptional factor LBP-1c/CP2/LSF gene on chro-mosome 12 is a genetic determinant of Alzheimer's disease. Hum Mol Genet 2000;9(15):2275–2280.

13. Bertram LM, Parkinson MB, McQueen K, et al. Further evidence for LBP-1c/CP2/LSF association in Alzheimer's disease families. J Med Genet 2005;42(11):857–862.

14. Wei ML, Andreadis A. Splicing of a regulated exon reveals additional complexity in the axonal microtubule-associated protein tau. J Neurochem 1998;70:1346–1356.

15. Goedert M, Spillantini MG, Jakes R, Rutherford D, Crowther RA. Multiple isoforms of human microtubule-associated protein tau: sequences and localization in neurofibrillary tangles of Alzheimer's disease. Neuron 1989a;3:519–526.

16. Goedert M, Spillantini MG, Potier MC, Ulrich J, Crowther RA. Cloning and sequencing of the cDNA encoding an isoform of microtubule-associated protein tau containing four tandem repeats: differen-tial expression of tau protein mRNAs in human brain. EMBO J 1989b;8:393–399.

17. Andreadis A, Brown WM, Kosik KS. Structure and novel exons of the human tau gene. Biochemistry 1992;31:10,626–10,633.

18. Ikeda S, Tokuda T, Yanagisawa N, Kametani F, Ohshima T, Allsop D. Variability of beta-amyloid pro-tein deposited lesions in Down's syndrome brains. Tohoku J Exp Med 1994;174:189–198.

19. Himmler A. Structure of the bovine tau gene: alternatively spliced transcripts generate a protein fam-ily. Mol Cell Biol 1989;9:1389–1396.

20. Goedert M, Jakes R. Expression of separate isoforms of human tau protein: correlation with the tau pattern in brain and effects on tubulin polymerization. EMBO J 1990;9:4225–4230.

21. Kosik KS, Orecchio LD, Bakalis S, Neve RL. Developmentally regulated expression of specific tau sequences. Neuron 1989;2:1389–1397.

22. Brandt R, Leger J, Lee G. Interaction of tau with the neural plasma membrane mediated by tau's amino-terminal projection domain. J Cell Biol 1995;131:1327–1340.

23. Hirokawa N, Shiomura Y, Okabe S. Tau proteins: the molecular structure and mode of binding on microtubules. J Cell Biol 1988;107:1449–1459.

24. Harada A, Oguchi K, Okabe S, et al. Altered microtubule organization in small-calibre axons of mice lacking tau protein. Nature 1994;369:488–491.

25. Dawson HN, Ferreira A, Eyster MV, Ghoshal N, Binder LI, Vitek MP. Inhibition of neuronal matu-ration in primary hippocampal neurons from tau deficient mice. J Cell Sci 2001;114:1179–1187.

26. Chen J, Kanai Y, Cowan NJ, Hirokawa N. Projection domains of MAP2 and tau determine spacings between microtubules in dendrites and axons. Nature 1992;360:674–677.

27. Georgieff IS, Liem RK, Couchie D, Mavilia C, Nunez J, Shelanski ML. Expression of high molecu-lar weight tau in the central and peripheral nervous systems. J Cell Sci 1993;105(pt 3):729–737.

28. Carlier MF, Simon C, Cassoly R, Pradel LA. Interaction between microtubule-associated protein tau and spectrin. Biochimie 1984;66:305–311.

29. Correas I, Padilla R, Avila J. The tubulin-binding sequence of brain microtubule-associated proteins, tau and MAP-2, is also involved in actin binding. Biochem J 1990;269:61–64.

30. Henriquez JP, Cross D, Vial C, Maccioni RB. Subpopulations of tau interact with microtubules and actin filaments in various cell types. Cell Biochem Funct 1995;13:239–250.

31. Selden SC, Pollard TD. Phosphorylation of microtubule-associated proteins regulates their interac-tion with actin filaments. J Biol Chem 1983;258:7064–7071.

32. Miyata Y, Hoshi M, Nishida E, Minami Y, Sakai H. Binding of microtubule-associated protein 2 and tau to the intermediate filament reassembled from neurofilament 70-kDa subunit protein. Its regula-tion by calmodulin. J Biol Chem 1986;261:13,026–13,030.

33. Matus A. Microtubule-associated proteins. Curr Opin Cell Biol 1990;2:10–14.

34. Jung D, Filliol D, Miehe M, Rendon A. Interaction of brain mitochondria with microtubules reconstituted from brain tubulin and MAP2 or TAU. Cell Motil Cytoskeleton 1993;24:245–255.

35. Lee G, Thangavel R, Sharma VM, et al. Phosphorylation of tau by fyn: implications for Alzheimer's disease. J Neurosci 2004;24:2304–2312.

36. Williamson R, Scales T, Clark BR, et al. Rapid tyrosine phosphorylation of neuronal proteins including tau and focal adhesion kinase in response to amyloid-beta peptide exposure: involvement of Src family protein kinases. J Neurosci 2002;22:10–20.

37. Jenkins SM, Johnson GV. Tau complexes with phospholipase C-gamma in situ. Neuroreport 1998;9: 67–71.

38. Hwang SC, Jhon DY, Bae YS, Kim JH, Rhee SG. Activation of phospholipase C-gamma by the concerted action of tau proteins and arachidonic acid. J Biol Chem 1996;271:18,342–18,349.

39. Rhee SG. Regulation of phosphoinositide-specific phospholipase C. Annu Rev Biochem 2001;70: 281–312.

40. Lee G, Neve RL, Kosik KS. The microtubule binding domain of tau protein. Neuron 1989;2: 1615–1624.

41. Lee G, Cowan N, Kirschner M. The primary structure and heterogeneity of tau protein from mouse brain. Science 1988;239:285–288.

42. Brandt R, Lee G. The balance between tau protein's microtubule growth and nucleation activities: implications for the formation of axonal microtubules. J Neurochem 1993;61:997–1005.

43. Cleveland DW, Hwo SY, Kirschner MW. Purification of tau, a microtubule-associated protein that induces assembly of microtubules from purified tubulin. J Mol Biol 1977a;116:207–225.

44. Cleveland DW, Hwo SY, Kirschner MW. Physical and chemical properties of purified tau factor and the role of tau in microtubule assembly. J Mol Biol 1977b;116:227–247.

45. Weingarten MD, Lockwood AH, Hwo SY, Kirschner MW. A protein factor essential for microtubule assembly. Proc Natl Acad Sci USA 1975;72:1858–1862.

46. Drechsel DN, Hyman AA, Cobb MH, Kirschner MW. Modulation of the dynamic instability of tubulin assembly by the microtubule-associated protein tau. Mol Biol Cell 1992;3:1141–1154.

47. Butner KA, Kirschner MW. Tau protein binds to microtubules through a flexible array of distributed weak sites. J Cell Biol 1991;115:717–730.

48. Gustke N, Trinczek B, Biernat J, Mandelkow EM, Mandelkow E. Domains of tau protein and interactions with microtubules. Biochemistry 1994;33:9511–9522.

49. Goode BL, Feinstein SC. Identification of a novel microtubule binding and assembly domain in the developmentally regulated inter-repeat region of tau. J Cell Biol 1994;124:769–782.

50. Kar S, Fan J, Smith MJ, Goedert M, Amos LA. Repeat motifs of tau bind to the insides of microtubules in the absence of taxol. EMBO J 2003a;22:70–77.

51. Kar S, Florence GJ, Paterson I, Amos LA. Discodermolide interferes with the binding of tau protein to microtubules. FEBS Lett 2003b;539:34–36.

52. Sontag E, Nunbhakdi-Craig V, Lee G, et al. Molecular interactions among protein phosphatase 2A, tau, and microtubules. Implications for the regulation of tau phosphorylation and the development of tauopathies. J Biol Chem 1999;274:25,490–25,498.

53. Haltiwanger RS, Busby S, Grove K, et al. O-glycosylation of nuclear and cytoplasmic proteins: regulation analogous to phosphorylation? Biochem Biophys Res Commun 1997;231:237–242.

54. Kreppel LK, Blomberg MA, Hart GW. Dynamic glycosylation of nuclear and cytosolic proteins. Cloning and characterization of a unique O-GlcNAc transferase with multiple tetratricopeptide repeats. J Biol Chem 1997;272:9308–9315.

55. Hart GW, Kreppel LK, Comer FI, et al. O-GlcNAcylation of key nuclear and cytoskeletal proteins: reciprocity with O-phosphorylation and putative roles in protein multimerization. Glycobiology 1996;6:711–716.

56. Kamemura K, Hart GW. Dynamic interplay between O-glycosylation and O-phosphorylation of nucleocytoplasmic proteins: a new paradigm for metabolic control of signal transduction and transcription. Prog Nucleic Acid Res Mol Biol 2003;73:107–136.

57. Dong DL, Xu ZS, Hart GW, Cleveland DW. Cytoplasmic O-GlcNAc modification of the head domain and the KSP repeat motif of the neurofilament protein neurofilament-H. J Biol Chem 1996;271: 20,845–20,852.

58. Lefebvre T, Ferreira S, Dupont-Wallois L, et al. Evidence of a balance between phosphorylation and O-GlcNAc glycosylation of Tau proteins—a role in nuclear localization. Biochim Biophys Acta 2003a;1619:167–176.

59. Lefebvre T, Caillet-Boudin ML, Buee L, Delacourte A, Michalski JC. O-GlcNAc glycosylation and neurological disorders. Adv Exp Med Biol 2003b;535:189–202.

60. Hasegawa M, Morishima-Kawashima M, Takio K, Suzuki M, Titani K, Ihara Y. Protein sequence and mass spectrometric analyses of tau in the Alzheimer's disease brain. J Biol Chem 1992;267: 17,047–17,054.

61. Hasegawa M. Phosphorylation in tau protein. Seikagaku 1993;65:469–473.

62. Lovestone S, Reynolds CH. The phosphorylation of tau: a critical stage in neurodevelopment and neurodegenerative processes. Neuroscience 1997;78:309–324.

63. Morishima-Kawashima M, Hasegawa M, Takio K, et al. Proline-directed and non-proline-directed phosphorylation of PHF-tau. J Biol Chem 1995;270:823–829.

64. Buee L, Bussiere T, Buee-Scherrer V, Delacourte A, Hof PR. Tau protein isoforms, phosphorylation and role in neurodegenerative disorders. Brain Res Brain Res Rev 2000;33:95–130.

65. Connell JW, Gibb GM, Betts JC, et al. Effects of FTDP-17 mutations on the in vitro phosphorylation of tau by glycogen synthase kinase 3β identified by mass spectometry demonstrate certain mutations exert long-range conformational changes. *FEBS* Lett 2001;493(1):40–44.

66. Yoshimura Y, Ichinose I, Yamauchi T. Phosphorylation of tau protein to sites found in Alzheimer's disease brain is catalyzed by Ca2+/calmodulin–dependent protein kinase II as demonstrated tandem mass spectometry. Neurosci Lett. 2003;353(3):185–188.

67. Seubert P, Mawal-Dewan M, Barbour R, et al. Detection of phosphorylated Ser262 in fetal tau, adult tau, and paired helical filament tau. J Biol Chem 1995;270:18,917–18,922.

68. Roder HM, Fracasso RP, Hoffman FJ, Witowsky JA, Davis G, Pellegrino CB. Phosphorylation-dependent monoclonal Tau antibodies do not reliably report phosphorylation by extracellular signal-regulated kinase 2 at specific sites. J Biol Chem 1997;272:4509–4515.

69. Drewes G, Lichtenberg-Kraag B, Doring F, et al. Mitogen activated protein (MAP) kinase transforms tau protein into an Alzheimer-like state. EMBO J 1992;11:2131–2138.

70. Goedert M, Hasegawa M, Jakes R, Lawler S, Cuenda A, Cohen P. Phosphorylation of microtubule-associated protein tau by stress-activated protein kinases. FEBS Lett 1997;409:57–62.

71. Reynolds CH, Utton MA, Gibb GM, Yates A, Anderton BH. Stress-activated protein kinase/c-jun N-terminal kinase phosphorylates tau protein. J Neurochem 1997;68:1736–1744.

72. Vulliet R, Halloran SM, Braun RK, Smith AJ, Lee G. Proline-directed phosphorylation of human Tau protein. J Biol Chem 1992;267:22,570–22,574.

73. Takahashi M, Tomizawa K, Ishiguro K, Takamatsu M, Fujita SC, Imahori K. Involvement of tau protein kinase I in paired helical filament-like phosphorylation of the juvenile tau in rat brain. J Neurochem 1995;64:1759–1768.

74. Baumann K, Mandelkow EM, Biernat J, Piwnica-Worms H, Mandelkow E. Abnormal Alzheimer-like phosphorylation of tau-protein by cyclin-dependent kinases cdk2 and cdk5. FEBS Lett 1993;336:417–424.

75. Liu WK, Williams RT, Hall FL, Dickson DW, Yen SH. Detection of a Cdc2-related kinase associated with Alzheimer paired helical filaments. Am J Pathol 1995;146:228–238.

76. Jenkins SM, Zinnerman M, Garner C, Johnson GV. Modulation of tau phosphorylation and intracellular localization by cellular stress. Biochem J 2000;345(pt 2):263–270.

77. Buee-Scherrer V, Goedert M. Phosphorylation of microtubule-associated protein tau by stress-activated protein kinases in intact cells. FEBS Lett 2002;515:151–154.

78. Drewes G, Ebneth A, Preuss U, Mandelkow EM, Mandelkow E. MARK, a novel family of protein kinases that phosphorylate microtubule-associated proteins and trigger microtubule disruption. Cell 1997;89:297–308.

79. Baudier J, Cole RD. Interactions between the microtubule-associated tau proteins and S100b regulate tau phosphorylation by the Ca2+/calmodulin-dependent protein kinase II. J Biol Chem 1988;263: 5876–5883.

80. Steiner B, Mandelkow EM, Biernat J, et al. Phosphorylation of microtubule-associated protein tau: identification of the site for Ca2(+)-calmodulin dependent kinase and relationship with tau phosphorylation in Alzheimer tangles. EMBO J 1990;9:3539–3544.

81. Jicha GA, Weaver C, Lane E, et al. cAMP-dependent protein kinase phosphorylations on tau in Alzheimer's disease. J Neurosci 1999;19:7486–7494.

82. Litersky JM, Johnson GV. Phosphorylation by cAMP-dependent protein kinase inhibits the degradation of tau by calpain. J Biol Chem 1992;267:1563–1568.

83. Pierre M, Nunez J. Multisite phosphorylation of tau proteins from rat brain. Biochem Biophys Res Commun 1983;115:212–219.

84. Greenwood JA, Scott CW, Spreen RC, Caputo CB, Johnson GV. Casein kinase II preferentially phosphorylates human tau isoforms containing an amino-terminal insert. Identification of threonine 39 as the primary phosphate acceptor. J Biol Chem 1994;269:4373–4380.

85. Planel E, Sun X, Takashima A. Role of GSK-3beta in Alzheimer's Disease Pathology. Drug Dev Res 2002;56:491–510.

86. Singh TJ, Zaidi T, Grundke-Iqbal I, Iqbal K. Non-proline-dependent protein kinases phosphorylate several sites found in tau from Alzheimer disease brain. Mol Cell Biochem 1996;154:143–151.

87. Woods YL, Cohen P, Becker W, et al. The kinase DYRK phosphorylates protein-synthesis initiation factor eIF2Bepsilon at Ser539 and the microtubule-associated protein tau at Thr212: potential role for DYRK as a glycogen synthase kinase 3-priming kinase. Biochem J 2001;355:609–615.

88. Buee-Scherrer V, Condamines O, Mourton-Gilles C, et al. AD2, a phosphorylation-dependent monoclonal antibody directed against tau proteins found in Alzheimer's disease. Brain Res Mol Brain Res 1996;39:79–88.

89. Sergeant N, Bussiere T, Vermersch P, Lejeune JP, Delacourte A. Isoelectric point differentiates PHF-tau from biopsy-derived human brain tau proteins. Neuroreport 1995;6:2217–2220.

90. Matsuo ES, Shin RW, Billingsley ML, et al. Biopsy-derived adult human brain tau is phosphorylated at many of the same sites as Alzheimer's disease paired helical filament tau. Neuron 1994;13: 989–1002.

91. Soulie C, Lepagnol J, Delacourte A, Caillet-Boudin ML. Dephosphorylation studies of SKNSH-SY 5Y cell Tau proteins by endogenous phosphatase activity. Neurosci Lett 1996;206:189–192.

92. Cohen P, Cohen PT. Protein phosphatases come of age. J Biol Chem 1989;264:21,435–21,438.

93. Ingebritsen TS, Cohen P. Protein phosphatases: properties and role in cellular regulation. Science 1983;221:331–338.

94. Dudek SM, Johnson GV. Postnatal changes in serine/threonine protein phosphatases and their association with the microtubules. Brain Res Dev Brain Res 1995;90:54–61.

95. Liao H, Li Y, Brautigan DL, Gundersen GG. Protein phosphatase 1 is targeted to microtubules by the microtubule-associated protein Tau. J Biol Chem 1998;273:21,901–21,908.

96. Goedert M, Jakes R, Qi Z, Wang JH, Cohen P. Protein phosphatase 2A is the major enzyme in brain that dephosphorylates tau protein phosphorylated by proline-directed protein kinases or cyclic AMP-dependent protein kinase. J Neurochem 1995;65:2804–2807.

97. Goto S, Yamamoto H, Fukunaga K, Iwasa T, Matsukado Y, Miyamoto E. Dephosphorylation of microtubule-associated protein 2, tau factor, and tubulin by calcineurin. J Neurochem 1985;45:276–283.

98. Yamamoto H, Saitoh Y, Fukunaga K, Nishimura H, Miyamoto E. Dephosphorylation of microtubule proteins by brain protein phosphatases 1 and 2A, and its effect on microtubule assembly. J Neurochem 1988;50:1614–1623.

99. Yamamoto H, Saitoh Y, Yasugawa S, Miyamoto E. Dephosphorylation of tau factor by protein phosphatase 2A in synaptosomal cytosol fractions, and inhibition by aluminum. J Neurochem 1990;55:683–690.

100. Ono T, Yamamoto H, Tashima K, et al. Dephosphorylation of abnormal sites of tau factor by protein phosphatases and its implication for Alzheimer's disease. Neurochem Int 1995;26:205–215.

101. Saito T, Ishiguro K, Uchida T, Miyamoto E, Kishimoto T, Hisanaga S. In situ dephosphorylation of tau by protein phosphatase 2A and 2B in fetal rat primary cultured neurons. FEBS Lett 1995;376:238–242.

102. Gong CX, Liu F, Wu G, et al. Dephosphorylation of microtubule-associated protein tau by protein phosphatase 5. J Neurochem 2004;88:298–310.

103. Goode BL, Denis PE, Panda D, et al. Functional interactions between the proline-rich and repeat regions of tau enhance microtubule binding and assembly. Mol Biol Cell 1997;8:353–365.

104. Biernat J, Gustke N, Drewes G, Mandelkow EM, Mandelkow E. Phosphorylation of Ser262 strongly reduces binding of tau to microtubules: distinction between PHF-like immunoreactivity and microtubule binding. Neuron 1993;11:153–163.

105. Bramblett GT, Goedert M, Jakes R, Merrick SE, Trojanowski JQ, Lee VM. Abnormal tau phosphorylation at Ser396 in Alzheimer's disease recapitulates development and contributes to reduced microtubule binding. Neuron 1993;10:1089–1099.

106. Drubin DG, Kirschner MW. Tau protein function in living cells. J Cell Biol 1986;103:2739–2746.

107. Lindwall G, Cole RD. Phosphorylation affects the ability of tau protein to promote microtubule assembly. J Biol Chem 1984;259:5301–5305.

108. Sengupta A, Kabat J, Novak M, Wu Q, Grundke-Iqbal I, Iqbal K. Phosphorylation of tau at both Thr 231 and ser 262 is required for maximal inhibition of its binding to microtubules. Arch Biochem Biophys 1998;357(2):299–309.

109. Hamdane M, Sambo Av, Delobel P, et al. Mitoctic-like tau phosphorylation by p25-C dk5 kinase complex. J Biol Chem 2003;278(36):34,026–34,034.

110. Binder LI, Frankfurter A, Rebhun LI. The distribution of tau in the mammalian central nervous system. J Cell Biol 1985;101:1371–1378.

111. Riederer BM, Binder LI. Differential distribution of tau proteins in developing cat cerebellum. Brain Res Bull 1994;33:155–161.

112. Papasozomenos SC. The heat shock-induced hyperphosphorylation of tau is estrogen-independent and prevented by androgens: implications for Alzheimer disease. Proc Natl Acad Sci USA 1997;94: 6612–6617.

113. Delacourte A, Buee L. Normal and pathological Tau proteins as factors for microtubule assembly. Int Rev Cytol 1997;171:167–224.

114. Zhukareva V, Vogelsberg-Ragaglia V, Van Deerlin VM, et al. Loss of brain tau defines novel sporadic and familial tauopathies with frontotemporal dementia. Ann Neurol 2001;49:165–175.

115. Zhukareva V, Sundarraj S, Mann D, et al. Selective reduction of soluble tau proteins in sporadic and familial frontotemporal dementias: an international follow-up study. Acta Neuropathol (Berl) 2003; 105:469–476.

116. Goedert M, Spillantini MG, Cairns NJ, Crowther RA. Tau proteins of Alzheimer paired helical filaments: abnormal phosphorylation of all six brain isoforms. Neuron 1992;8:159–168.

117. Sergeant N, David JP, Goedert M, et al. Two-dimensional characterization of paired helical filament-tau from Alzheimer's disease: demonstration of an additional 74-kDa component and age-related biochemical modifications. J Neurochem 1997a;69:834–844.

118. Hof PR, Bouras C, Buee L, Delacourte A, Perl DP, Morrison JH. Differential distribution of neurofibrillary tangles in the cerebral cortex of dementia pugilistica and Alzheimer's disease cases. Acta Neuropathol (Berl) 1992a;85:23–30.

119. Hof PR, Nimchinsky EA, Buee-Scherrer V, et al. Amyotrophic lateral sclerosis/parkinsonism-dementia complex of Guam: quantitative neuropathology, immunohistochemical analysis of neuronal vulnerability, and comparison with related neurodegenerative disorders. Acta Neuropathol (Berl) 1994a;88:397–404.

120. Buee-Scherrer V, Buee L, Hof PR, et al. Neurofibrillary degeneration in amyotrophic lateral sclerosis/parkinsonism-dementia complex of Guam. Immunochemical characterization of tau proteins. Am J Pathol 1995;146:924–932.

121. Buee-Scherrer V, Buee L, Leveugle B, et al. Pathological tau proteins in postencephalitic parkinsonism: comparison with Alzheimer's disease and other neurodegenerative disorders. Ann Neurol 1997;42:356–359.

122. Delacourte A, Buee L. Tau pathology: a marker of neurodegenerative disorders. Curr Opin Neurol 2000;13:371–376.

123. Caparros-Lefebvre D, Sergeant N, Lees A, et al. Guadeloupean parkinsonism: a cluster of progressive supranuclear palsy-like tauopathy. Brain 2002;125:801–811.

124. Delacourte A, Sergeant N, Wattez A, et al. Tau aggregation in the hippocampal formation: an ageing or a pathological process? Exp Gerontol 2002;37:1291–1296.

125. Love S, Bridges LR, Case CP. Neurofibrillary tangles in Niemann-Pick disease type C. Brain 1995;118(pt 1):119–129.

126. Spillantini MG, Tolnay M, Love S, Goedert M. Microtubule-associated protein tau, heparan sulphate and alpha-synuclein in several neurodegenerative diseases with dementia. Acta Neuropathol (Berl) 1999;97:585–594.

127. Revesz T, Holton JL, Doshi B, Anderton BH, Scaravilli F, Plant GT. Cytoskeletal pathology in familial cerebral amyloid angiopathy (British type) with non-neuritic amyloid plaque formation. Acta Neuropathol (Berl) 1999;97:170–176.

128. Holton JL, Ghiso J, Lashley T, et al. Regional distribution of amyloid-Bri deposition and its association with neurofibrillary degeneration in familial British dementia. Am J Pathol 2001;158:515–526.

129. Sergeant N, Wattez A, Delacourte A. Neurofibrillary degeneration in progressive supranuclear palsy and corticobasal degeneration: tau pathologies with exclusively exon 10 isoforms. J Neurochem 1999;72:1243–1249.

130. Flament S, Delacourte A, Verny M, Hauw JJ, Javoy-Agid F. Abnormal Tau proteins in progressive supranuclear palsy. Similarities and differences with the neurofibrillary degeneration of the Alzheimer type. Acta Neuropathol (Berl) 1991;81:591–596.

131. Tolnay M, Sergeant N, Ghestem A, et al. Argyrophilic grain disease and Alzheimer's disease are distinguished by their different distribution of tau protein isoforms. Acta Neuropathol (Berl) 2002;104:425–434.

132. Steele JC, Richardson JC, Olszewski J. Progressive Supranuclear Palsy. A Heterogeneous Degeneration Involving the Brain Stem, Basal Ganglia and Cerebellum with Vertical Gaze and Pseudobulbar Palsy, Nuchal Dystonia and Dementia. Arch Neurol 1964;10:333–359.

133. Richardson JC, Steele J, Olszewski J. Supranuclear Ophthalmoplegia, Pseudobulbar Palsy, Nuchal Dystonia and Dementia. A Clinical Report on Eight Cases of Heterogenous System Degeneration. Trans Am Neurol Assoc 1963;88:25–29.

134. Litvan I, Baker M, Hutton M. Tau genotype: no effect on onset, symptom severity, or survival in progressive supranuclear palsy. Neurology 2001;57:138–140.

135. Litvan I, Hutton M. Clinical and genetic aspects of progressive supranuclear palsy. J Geriatr Psychiatry Neurol 1998;11:107–114.

136. Hauw JJ, Verny M, Delaere P, Cervera P, He Y, Duyckaerts C. Constant neurofibrillary changes in the neocortex in progressive supranuclear palsy. Basic differences with Alzheimer's disease and aging. Neurosci Lett 1990;119:182–186.

137. Hof PR, Delacourte A, Bouras C. Distribution of cortical neurofibrillary tangles in progressive supranuclear palsy: a quantitative analysis of six cases. Acta Neuropathol (Berl) 1992b;84:45–51.

138. Bergeron C, Pollanen MS, Weyer L, Lang AE. Cortical degeneration in progressive supranuclear palsy. A comparison with cortical-basal ganglionic degeneration. J Neuropathol Exp Neurol 1997;56: 726–734.

139. Verny M, Duyckaerts C, Agid Y, Hauw JJ. The significance of cortical pathology in progressive supranuclear palsy. Clinico-pathological data in 10 cases. Brain 1996;119(pt 4):1123–1136.

140. Komori T. Tau-positive glial inclusions in progressive supranuclear palsy, corticobasal degeneration and Pick's disease. Brain Pathol 1999;9:663–679.

141. Komori T, Arai N, Oda M, et al. Astrocytic plaques and tufts of abnormal fibers do not coexist in corticobasal degeneration and progressive supranuclear palsy. Acta Neuropathol (Berl) 1998;96: 401–408.

142. Rebeiz JJ, Kolodny EH, Richardson EP Jr. Corticodentatonigral degeneration with neuronal achromasia: a progressive disorder of late adult life. Trans Am Neurol Assoc 1967;92:23–26.

143. Rebeiz JJ, Kolodny EH, Richardson EP Jr. Corticodentatonigral degeneration with neuronal achromasia. Arch Neurol 1968;18:20–33.

144. Rinne JO, Lee MS, Thompson PD, Marsden CD. Corticobasal degeneration. A clinical study of 36 cases. Brain 1994;117(pt 5):1183–1196.

145. Buée-Scherrer V, Hof PR, Buee L, et al. Hyperphosphorylated tau proteins differentiate corticobasal degeneration and Pick's disease. Acta Neuropathol (Berl) 1996;91(4):351–359.

146. Wakabayashi K, Takahashi H. Pathological heterogeneity in progressive supranuclear palsy and corticobasal degeneration. Neuropathology 2004;24(1):79–86.

147. Braak H, Braak E. Argyrophilic grains: characteristic pathology of cerebral cortex in cases of adult onset dementia without Alzheimer changes. Neurosci Lett 1987;76:124–127.

148. Braak H, Braak E. Argyrophilic grain disease: frequency of occurrence in different age categories and neuropathological diagnostic criteria. J Neural Transm 1998;105:801–819.

149. Braak H, Braak E. Cortical and subcortical argyrophilic grains characterize a disease associated with adult onset dementia. Neuropathol Appl Neurobiol 1989;15:13–26.

150. Tolnay M, Spillantini MG, Goedert M, Ulrich J, Langui D, Probst A. Argyrophilic grain disease: widespread hyperphosphorylation of tau protein in limbic neurons. Acta Neuropathol (Berl) 1997;93: 477–484.

151. Togo T, Sahara N, Yen SH, et al. Argyrophilic grain disease is a sporadic 4-repeat tauopathy. J Neuropathol Exp Neurol 2002;61:547–556.

152. Maurage CA, Sergeant N, Schraen-Maschke S, et al. Diffuse form of argyrophilic grain disease: a new variant of four-repeat tauopathy different from limbic argyrophilic grain disease. Acta Neuropathol (Berl) 2003;106(6):575–583.

153. Mattila P, Togo T, Dickson D. The subthalamic nucleus has neurofibrillary tangles in argyrophilic grain disease and advanced Alzheimer's disease. Neurosci Lett 2002;320:81–85.

154. Baker M, Litvan I, Houlden H, et al. Association of an extended haplotype in the tau gene with progressive supranuclear palsy. Hum Mol Genet 1999;8:711–715.

155. Schraen-Maschke S, Dhaenens CM, Delacourte A, Sablonniere B. Microtubule-associated protein tau gene: a risk factor in human neurodegenerative diseases. Neurobiol Dis 2004;15(3):449–460.

156. Dermaut B, Kumar-Singh S, Rademakers R, Theuns J, Cruts M, Van Broeckhoven C. Tau is central in the genetic Alzheimer frontotemporal dementia spectrum. Trends Genet 2005;21(12): 664–672.

157. Hutton M. Molecular genetics of chromosome 17 tauopathies. Ann N Y Acad Sci 2000;920: 63–73.

158. Ishizawa T, Ko LW, Cookson N, Davias P, Espinoza M, Dickson DW. Selective neurofibrillary degeneration of the hippocampal CA2 sector is associated with four-repeat tauopathies. J Neuropathol Exp Neurol 2002;61:1040–1047.

159. Miserez AR, Clavaguera F, Monsch AU, Probst A, Tolnay M. Argyrophilic grain disease: molecular genetic difference to other four-repeat tauopathies. Acta Neuropathol (Berl) 2003;106(4):363–366.

160. Brion S, Plas J, Jeanneau A. Pick's disease. Anatomo-clinical point of view. Rev Neurol (Paris) 1991;147:693–704.

161. Buee Scherrer V, Hof PR, Buee L, et al. Hyperphosphorylated tau proteins differentiate corticobasal degeneration and Pick's disease. Acta Neuropathol (Berl) 1996;91:351–359.

162. Delacourte A, Robitaille Y, Sergeant N, et al. Specific pathological Tau protein variants characterize Pick's disease. J Neuropathol Exp Neurol 1996;55:159–168.

163. Hof PR, Bouras C, Perl DP, Morrison JH. Quantitative neuropathologic analysis of Pick's disease cases: cortical distribution of Pick bodies and coexistence with Alzheimer's disease. Acta Neuropathol (Berl) 1994b;87:115–124.

164. Sergeant N, David JP, Lefranc D, Vermersch P, Wattez A, Delacourte A. Different distribution of phosphorylated tau protein isoforms in Alzheimer's and Pick's diseases. FEBS Lett 1997b;412: 578–582.

165. Probst A, Tolnay M, Langui D, Goedert M, Spillantini MG. Pick's disease: hyperphosphorylated tau protein segregates to the somatoaxonal compartment. Acta Neuropathol (Berl) 1996;92:588–596.

166. Mailliot C, Sergeant N, Bussiere T, Caillet-Boudin ML, Delacourte A, Buee L. Phosphorylation of specific sets of tau isoforms reflects different neurofibrillary degeneration processes. FEBS Lett 1998;433:201–204.

167. Dermaut B, Kumar-Singh S, Engelborghs S, et al. A novel presenilin 1 mutation associated with Pick's disease but not beta–amyloid plaques. Ann Neurol 2004;55(5):617–626.

168. Halliday GM, Song YJ, Lepar G, et al. Pick bodies in a family with presenilin-1 Alzheimer's disease. Ann Neurol 2005;57(1):139–143.

169. Reardon W, Harper PS. Advances in myotonic dystrophy: a clinical and genetic perspective. Curr Opin Neurol Neurosurg 1992;5:605–609.

170. Meola G. Clinical and genetic heterogeneity in myotonic dystrophies. Muscle Nerve 2000;23:1789–1799.

171. New nomenclature and DNA testing guidelines for myotonic dystrophy type 1 (DM1). The International Myotonic Dystrophy Consortium (IDMC). Neurology 2000;4:1218–1221.

172. Udd B, Meola G, Krahe R, et al. Report of the 115th ENMC workshop: DM2/PROMM and other myotonic dystrophies. 3rd Workshop, 14-16 February 2003, Naarden, The Netherlands. Neuromuscul Disord 2003;13:589–596.

173. Ricker K, Koch MC, Lehmann-Horn F, et al. Proximal myotonic myopathy: a new dominant disorder with myotonia, muscle weakness, and cataracts. Neurology 1994;44:1448–1452.

174. Liquori CL, Ricker K, Moseley ML, et al. Myotonic dystrophy type 2 caused by a CCTG expansion in intron 1 of ZNF9. Science 2001;293:864–867.

175. Brook JD, McCurrach ME, Harley HG, et al. Molecular basis of myotonic dystrophy: expansion of a trinucleotide (CTG) repeat at the 3′ end of a transcript encoding a protein kinase family member. Cell 1992;68:799–808.

176. Ranum LP, Rasmussen PF, Benzow KA, Koob MD, Day JW. Genetic mapping of a second myotonic dystrophy locus. Nat Genet 1998;19:196–198.

177. Mankodi A, Thornton CA. Myotonic syndromes. Curr Opin Neurol 2002;15:545–552.

178. Meola G, Sansone V, Perani D, et al. Reduced cerebral blood flow and impaired visual-spatial function in proximal myotonic myopathy. Neurology 1999;53:1042–1050.

179. Meola G, Sansone V, Perani D, et al. Executive dysfunction and avoidant personality trait in myotonic dystrophy type 1 (DM-1) and in proximal myotonic myopathy (PROMM/DM-2). Neuromuscul Disord 2003;13:813–821.

180. Vermersch P, Sergeant N, Ruchoux MM, et al. Specific tau variants in the brains of patients with myotonic dystrophy. Neurology 1996;47:711–717.

181. Yoshimura N, Otake M, Igarashi K, Matsunaga M, Takebe K, Kudo H. Topography of Alzheimer's neurofibrillary change distribution in myotonic dystrophy. Clin Neuropathol 1990;9:234–239.

182. Sergeant N, Sablonniere B, Schraen-Maschke S, et al. Dysregulation of human brain microtubule-associated tau mRNA maturation in myotonic dystrophy type 1. Hum Mol Genet 2001;10:2143–2155.

183. Seznec H, Agbulut O, Sergeant N, et al. Mice transgenic for the human myotonic dystrophy region with expanded CTG repeats display muscular and brain abnormalities. Hum Mol Genet 2001;10:2717–2726.

184. Maurage CA, Udd B, Ruchoux MM, et al. Similar brain tau pathology in DM2/PROMM and DM1/Steinert disease. Neurology 2005;5(10):1636–1638.

185. Constantinidis J, Richard J, Tissot R. Pick's disease. Histological and clinical correlations. Eur Neurol 1974;11:208–217.

186. Clinical and neuropathological criteria for frontotemporal dementia. The Lund and Manchester Groups. J Neurol Neurosurg Psychiatry 1994;57:416–418.

187. Brown J. Pick's disease. Baillieres Clin Neurol 1992;1:535–557.

188. Wilhelmsen KC, Lynch T, Pavlou E, Higgins M, Nygaard TG. Localization of disinhibition-dementia-parkinsonism-amyotrophy complex to 17q21-22. Am J Hum Genet 1994;55:1159–1165.

189. Bird TD, Wijsman EM, Nochlin D, et al. Chromosome 17 and hereditary dementia: linkage studies in three non-Alzheimer families and kindreds with late-onset FAD. Neurology 1997;48:949–954.

190. Heutink P, Stevens M, Rizzu P, et al. Hereditary frontotemporal dementia is linked to chromosome 17q21-q22: a genetic and clinicopathological study of three Dutch families. Ann Neurol 1997;41:150–159.

191. Murrell JR, Koller D, Foroud T, et al. Familial multiple-system tauopathy with presenile dementia is localized to chromosome 17. Am J Hum Genet 1997;61:1131–1138.

192. Wijker M, Wszolek ZK, Wolters EC, et al. Localization of the gene for rapidly progressive autosomal dominant parkinsonism and dementia with pallido-ponto-nigral degeneration to chromosome 17q21. Hum Mol Genet 1996;5:151–154.

193. Foster NL, Wilhelmsen K, Sima AA, Jones MZ, D'Amato CJ, Gilman S. Frontotemporal dementia and parkinsonism linked to chromosome 17: a consensus conference. Conference Participants. Ann Neurol 1997;41:706–715.

194. Spillantini MG, Bird TD, Ghetti B. Frontotemporal dementia and Parkinsonism linked to chromosome 17: a new group of tauopathies. Brain Pathol 1998a;8:387–402.

195. Hutton M, Lendon CL, Rizzu P, et al. Association of missense and 5′-splice-site mutations in tau with the inherited dementia FTDP-17. Nature 1998;393:702–705.

196. Poorkaj P, Bird TD, Wijsman E, et al. Tau is a candidate gene for chromosome 17 frontotemporal dementia. Ann Neurol 1998;43:815–825.

197. Spillantini MG, Goedert M. Tau gene mutations and tau pathology in frontotemporal dementia and parkinsonism linked to chromosome 17. Adv Exp Med Biol 2001;487:21–37.

198. Hayashi S, Toyoshima Y, Hasegawa M, et al. Late-onset frontotemporal dementia with a novel exon 1 (Arg5His) tau gene mutation. Ann Neurol 2002;51:525–530.

199. Poorkaj P, Muma NA, Zhukareva V, et al. An R5L tau mutation in a subject with a progressive supranuclear palsy phenotype. Ann Neurol 2002;52:511–516.

200. Pickering-Brown S, Baker M, Yen SH, et al. Pick's disease is associated with mutations in the tau gene. Ann Neurol 2000;48:859–867.

201. Rizzini C, Goedert M, Hodges JR, et al. Tau gene mutation K257T causes a tauopathy similar to Pick's disease. J Neuropathol Exp Neurol 2000;59:990–1001.

202. Grover A, England E, Baker M, et al. A novel tau mutation in exon 9 (I260V) causes a four-repeat tauopathy. Exp Neurol 2003;184:131–140.

203. Kobayashi T, Ota S, Tanaka K, et al. A novel L266V mutation of the tau gene causes frontotemporal dementia with a unique tau pathology. Ann Neurol 2003;53:133–137.

204. Spillantini MG, Murrell JR, Goedert M, Farlow MR, Klug A, Ghetti B. Mutation in the tau gene in familial multiple system tauopathy with presenile dementia. Proc Natl Acad Sci USA 1998b;95:7737–7741.

205. Delisle MB, Murrell JR, Richardson R, et al. A mutation at codon 279 (N279K) in exon 10 of the Tau gene causes a tauopathy with dementia and supranuclear palsy. Acta Neuropathol (Berl) 1999;98:62–77.

206. Hasegawa M, Smith MJ, Iijima M, Tabira T, Goedert M. FTDP-17 mutations N279K and S305N in tau produce increased splicing of exon 10. FEBS Lett 1999;443:93–96.

207. Yasuda M, Kawamata T, Komure O, et al. A mutation in the microtubule-associated protein tau in pallido-nigro-luysian degeneration. Neurology 1999;53:864–868.

208. Wszolek ZK, Tsuboi Y, Uitti RJ, Reed L. Two brothers with frontotemporal dementia and parkinson-ism with an N279K mutation of the tau gene. Neurology 2000;55:1939.

209. Iseki E, Matsumura T, Marui W, et al. Familial frontotemporal dementia and parkinsonism with a novel N296H mutation in exon 10 of the tau gene and a widespread tau accumulation in the glial cells. Acta Neuropathol (Berl) 2001;102:285–292.

210. Yoshida H, Crowther RA, Goedert M. Functional effects of tau gene mutations deltaN296 and N296H. J Neurochem 2002;80:548–551.

211. Ros R, Thobois S, Streichenberger N, et al. A new mutation of the tau gene, G303V, in early-onset familial progressive supranuclear palsy. Arch Neurol 2005;62(9):1444–1450.

212. Dumanchin C, Camuzat A, Campion D, et al. Segregation of a missense mutation in the microtubule-associated protein tau gene with familial frontotemporal dementia and parkinsonism. Hum Mol Genet 1998;7:1825–1829.

213. Poorkaj P, Grossman M, Steinbart E, et al. Frequency of tau gene mutations in familial and sporadic cases of non-Alzheimer dementia. Arch Neurol 2001;58:383–387.

214. Bugiani O, Murrell JR, Giaccone G, et al. Frontotemporal dementia and corticobasal degeneration in a family with a P301S mutation in tau. J Neuropathol Exp Neurol 1999;58:667–677.

215. Sperfeld AD, Collatz MB, Baier H, et al. FTDP-17: an early-onset phenotype with parkinsonism and epileptic seizures caused by a novel mutation. Ann Neurol 1999;46:708–715.

216. Iijima M, Tabira T, Poorkaj P, et al. A distinct familial presenile dementia with a novel missense muta-tion in the tau gene. Neuroreport 1999;10:497–501.

217. Rosso SM, Kaat LD, Baks T, et al. Frontotemporal dementia in The Netherlands: patient characteris-tics and prevalence estimates from a population-based study. Brain 2003;126:2016–2022.

218. Zarranz JJ, Ferrer I, Lezcano E, et al. A novel mutation (K317M) in the MAPT gene causes FTDP and motor neuron disease. Neurology 2005;64(9):1578–1585.

219. Neumann M, Diekmann S, Bertsch U, Vanmassenhove B, Bogerts B, Kretzschmar HA. Novel G335V mutation in the tau gene associated with early onset familial frontotemporal dementia. Neurogenetics 2005;6(2):91–95.

220. Pickering-Brown SM, Baker M, Nonaka T, et al. Frontotemporal dementia with Pick-type histology associated with Q336R mutation in the tau gene. Brain 2004;127(pt 6):1415–1426.

221. Hasegawa M, Smith MJ, Goedert M. Tau proteins with FTDP-17 mutations have a reduced ability to promote microtubule assembly. FEBS Lett 1998;437:207–210.

222. Neumann M, Schulz-Schaeffer W, Crowther RA, et al. Pick's disease associated with the novel Tau gene mutation K369I. Ann Neurol 2001;50:503–513.

223. Murrell JR, Spillantini MG, Zolo P, et al. Tau gene mutation G389R causes a tauopathy with abun-dant pick body-like inclusions and axonal deposits. J Neuropathol Exp Neurol 1999;58:1207–1226.

224. D'Souza I, Poorkaj P, Hong M, et al. Missense and silent tau gene mutations cause frontotemporal dementia with parkinsonism-chromosome 17 type, by affecting multiple alternative RNA splicing regulatory elements. Proc Natl Acad Sci USA 1999;96:5598–5603.

225. Spillantini MG, Yoshida H, Rizzini C, et al. A novel tau mutation (N296N) in familial dementia with swollen achromatic neurons and corticobasal inclusion bodies. Ann Neurol 2000;48:939–943.

226. Stanford PM, Halliday GM, Brooks WS, et al. Progressive supranuclear palsy pathology caused by a novel silent mutation in exon 10 of the tau gene: expansion of the disease phenotype caused by tau gene mutations. Brain 2000;123(pt 5):880–893.

227. Rizzu P, Van Swieten JC, Joosse M, et al. High prevalence of mutations in the microtubule-associated protein tau in a population study of frontotemporal dementia in the Netherlands. Am J Hum Genet 1999;64:414–421.

228. Goedert M, Spillantini MG, Crowther RA, et al. Tau gene mutation in familial progressive subcorti-cal gliosis. Nat Med 1999a;5:454–457.

229. Grover A, Houlden H, Baker M, et al. 5' splice site mutations in tau associated with the inherited dementia FTDP-17 affect a stem-loop structure that regulates alternative splicing of exon 10. J Biol Chem 1999;274:15,134–15,143.

230. Varani L, Hasegawa M, Spillantini MG, et al. Structure of tau exon 10 splicing regulatory element RNA and destabilization by mutations of frontotemporal dementia and parkinsonism linked to chromosome 17. Proc Natl Acad Sci USA 1999;96:8229–8234.

231. Spillantini MG, Crowther RA, Kamphorst W, Heutink P, van Swieten JC. Tau pathology in two Dutch families with mutations in the microtubule-binding region of tau. Am J Pathol 1998c;153:1359–1363.

232. Hong M, Zhukareva V, Vogelsberg-Ragaglia V, et al. Mutation-specific functional impairments in distinct tau isoforms of hereditary FTDP-17. Science 1998;282:1914–1917.

233. Reed LA, Schmidt ML, Wszolek ZK, et al. The neuropathology of a chromosome 17-linked autosomal dominant parkinsonism and dementia (pallido-ponto-nigral degeneration). J Neuropathol Exp Neurol 1998;57:588–601.

234. Goedert M, Jakes R, Crowther RA. Effects of frontotemporal dementia FTDP-17 mutations on heparin-induced assembly of tau filaments. FEBS Lett 1999b;450:306–311.

235. Yen SH, Hutton M, DeTure M, Ko LW, Nacharaju P. Fibrillogenesis of tau: insights from tau missense mutations in FTDP-17. Brain Pathol 1999;9:695–705.

236. Chiti F, Stefani M, Taddei N, Ramponi G, Dobson CM. Rationalization of the effects of mutations on peptide and protein aggregation rates. Nature 2003;424:805–808.

237. Delobel P, Flament S, Hamdane M, et al. Functional characterization of FTDP-17 tau gene mutations through their effects on Xenopus oocyte maturation. J Biol Chem 2002;277:9199–9205.

238. Delacourte A, David JP, Sergeant N, et al. The biochemical pathway of neurofibrillary degeneration in aging and Alzheimer's disease. Neurology 1999;52:1158–1165.

21 Structure, Function, and Regulation of the Microtubule Associated Protein Tau

Janis Bunker and Stuart C. Feinstein

CONTENTS

SUMMARY

The neural protein tau was first identified and purified in 1975 as a protein that copurifies with tubulin and assembles tubulin subunits into microtubules *(1)*. Although tau has been studied intensively for almost 30 yr, many aspects of tau structure, function, and regulation remain unclear. Whereas tau is widely appreciated to be important in normal neurodevelopment (reviewed in refs. *2,3*), it is most widely known because it is the major component of the intraneuronal "neurofibrillary tangles" associated with Alzheimer's disease pathology *(4–7)*. In addition to Alzheimer's disease, these insoluble and abnormal tau tangles are also associated with numerous other neurodegenerative disorders known collectively as "tauopathies." This chapter will discuss historical and current research regarding normal tau structure, function, and regulation as well as possible molecular mechanisms of tau-mediated neuronal cell death in disease.

Key Words: Microtubule dynamics; neurodegeneration; Alzheimer's disease; FTOP-17; dementia; tauopathy.

1. IN VIVO/CELLULAR STUDIES OF NORMAL TAU FUNCTION

Tau is expressed primarily in the cell body and axons of neurons, although it is also expressed at lower levels in glial cells (reviewed in ref. *8*). The initial expression and action of tau coincides temporally with the initial extension of neuronal processes during development *(9,10)*, and its expression is maintained and tightly regulated throughout adulthood. A variety of necessity and sufficiency studies have been performed in order to elucidate the functional roles of tau in neurons. Antisense experiments demonstrated

From: *Cancer Drug Discovery and Development: The Role of Microtubules in Cell Biology, Neurobiology, and Oncology* Edited by: Tito Fojo © Humana Press, Totowa, NJ

that normal levels of tau expression are necessary for the establishment of neuronal cell polarity, as well as for neuronal process extension and stability (10–13). Furthermore, expression of tau in non-neuronal cells demonstrated that tau is sufficient to increase microtubule density, stability, and bundling; in cases of very high levels of expression, tau actually induced process outgrowth in some non-neuronal cells (14–19). Taken together, these studies led to the conclusion that the assembly and stabilization of microtubules by tau is required for the establishment of neuronal cell polarity, as well as outgrowth and maintenance of neuronal processes.

Several later studies brought this conclusion into question. The first question arose when the production of a tau knockout mouse failed to produce a strong phenotype (20). These mice were immunohistologically normal and exhibited normal axonal outgrowth. The only significant deficit was a slight decrease in microtubule stability in small caliber axons. However, another microtubule-associated protein, MAP1A, was shown to be upregulated in these mice; the authors suggested that the increased level of MAP1A might compensate for the loss of tau. Subsequently, another study found that inactivation of tau by microinjection of function-blocking antibodies did not prevent normal axonal outgrowth in cultured sympathetic neurons (21). Shortly thereafter, a second tau knockout mouse was generated that exhibited a stronger phenotype including delayed axonal extension, but again, neuronal cell polarity was established in the absence of tau (22). To address directly the possibility that the limited phenotypes of tau knockout mice were owing to functional redundancy, double-knockout mice were produced in which both tau and MAP1B (known to be highly expressed in neurons) were disrupted (23). These mice displayed a severe phenotype, having a life-span of only 4 wk (23). Thus, MAP1B and tau appear to be able to substitute for one another in mice, preventing a clear conclusion regarding the functional roles of tau based on mouse genetic analysis. The relevance of such functional redundancy to normal human development remains uncertain. Taken together, the weight of the evidence continues to support important roles for tau in neural development, with the additional note that at least in mice, other MAPs can be redundant with tau function.

2. THE TAU: MICROTUBULE INTERACTION

2.1. Breaking It Down—Tau Structure and Function

Tau is a member of a family of closely related microtubule-associated proteins, which also includes MAP2 and MAP4 (24,25) The domain structure of tau consists of an N-terminal "projection domain," a proline-rich basic region, a conserved microtubule-binding region consisting of a series of evenly spaced imperfect repeats and a C-terminal tail. The repeat region contains the microtubule-binding domain (Fig. 1). Whereas only a single tau isoform is expressed in fetal brain, alternative RNA splicing of tau RNA leads to the synthesis of six different tau isoforms in the adult central nervous system (26–28). These six isoforms arise from the exclusion/inclusion of exons 2 and 3 encoded sequences in the N-terminal portion of the protein, and the exclusion/ inclusion of exon 10 encoded sequences in the repeat region (Fig. 1). Notably, the exclusion of exon 10 encoded sequences results in a protein containing only three imperfect repeats (3-repeat tau), whereas its inclusion generates an isoform with 4 imperfect repeats (4-repeat tau).

The fact that the region of tau responsible for promoting microtubule assembly and stability consists of a series of evenly spaced repeats led to a straight-forward "linear" model for tau action in which each 18 amino acid long repeat serves as an independent

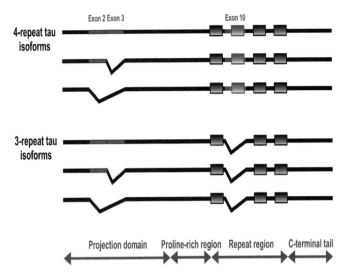

Fig. 1. Domain structure of the 6-tau isoforms expressed in the central nervous system. The domains of tau consist of a projection domain, a proline-rich region, a repeat region, and a C-terminal tail as indicated by blue arrows at the bottom of the figure. Alternative splicing of exons 2, 3, and 10 (shown in red) results in the expression of 6 different isoforms of tau in the adult central nervous system. Exon 10 encodes the first inter-repeat and the second imperfect repeat and the exclusion of this exon results in 3-repeat tau. Shaded boxes indicate repeats. To view this figure in color, see the insert and the companion CD-ROM.

tubulin-binding domain separated from one another by 13–14 amino acid long spacers or linkers that allowed for the proper positioning of the repeats on the tubulin lattice *(26,29,30)*. By binding in this manner, tau was proposed to assemble and stabilize microtubules by crosslinking the tubulin subunits together. A number of structure–function studies were performed to test this model and gain an understanding of how the domains of tau function to bind, assemble, and stabilize microtubules. These studies generally involved one of two in vitro assays. The first was a microtubule-binding assay in which taxol-stabilized microtubules were incubated with either full-length tau or a fragment of tau. After reaching equilibrium, the microtubules were pelleted by centrifugation and the proportion of tau that cosedimented with the microtubules and was quantified in order to determine the binding affinity of the tau molecule being assayed. The second assay was a microtubule assembly assay in which the ability of a given tau molecule to promote microtubule assembly was quantified using spectrophotometric light scattering to monitor the level of microtubule polymer present in a reaction as a function of time.

Both binding and assembly assays demonstrated that the 6-tau isoforms fell into two functional groups—4-repeat tau and 3-repeat tau. In general, 4-repeat tau exhibited two to threefold greater microtubule binding and assembly activities than 3-repeat tau; by comparison, the presence or absence of sequences encoded by exons 2 and 3 had little effect *(31,32)*. The fact that an additional repeat enhanced microtubule binding and assembly activities supported the view that individual repeats serve as independent tubulin-binding domains.

Several early tau fragment and truncation studies also supported this model. For example, each repeat was shown to contribute to tau's microtubule-binding affinity

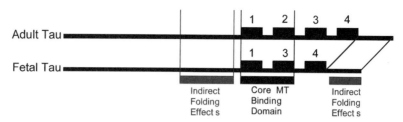

Fig. 2. "Core microtubule-binding domain" model for Tau's interaction with microtubules. Based on truncation and deletion analysis, Goode et al., 2000 *(32)* proposed that tau interacts with microtubules through a core microtubule-binding domain made up of the first two imperfect repeats and the inter-repeat between them. The regions flanking the repeat region, i.e., the proline rich region on the amino side and the C-terminus on the carboxyl side, influence the nature of the microtubule-binding event indirectly, likely through folding effects. That the core microtubule-binding domain differs significantly between 3-repeat (fetal) and 4-repeat (adult) tau leads to the possibility of isoform specific mechanisms of action (*see* also Fig. 3), which is predicted by the developmentally regulated expression of tau isoforms and demanded by the FTDP-17 RNA splicing mutations.

(30,30a,33,34). However, other work indicated that the actual mechanism of tau binding to microtubules was significantly more complex. First, Ennulat et al. *(35)* found that a single repeat can assemble microtubules, which was not easily consistent with the simple crosslinking model. Similarly, synthetic peptides corresponding to individual repeats stabilize microtubule dynamics *(36)*. Additionally, Goode and Feinstein, 1994 *(33)* showed that the inter-repeat between repeats 1 and 2 (which is specific to 4-repeat tau) possessed very potent microtubule binding and assembly capability; indeed, it was several times more active than any of the repeats. Finally, more detailed truncation studies showed that the first two repeats (in both 3-repeat and 4-repeat tau) and their intervening inter-repeat make a much stronger contribution to microtubule-binding affinity than the remaining repeats and inter-repeats, thus defining a core microtubule binding domain *(32,33)* (Fig. 2).

The regions that flank the repeat region (the proline-rich region on the amino side and the C-terminal tail), whereas unable to bind microtubules on their own *(30,32)*, can contribute significantly to the binding activity of the repeat region *(30,32,34,37)*. Thus, these regions regulate the microtubule-binding activity of the repeat region, perhaps through protein-folding effects. In contrast, the N-terminus does not contribute to the binding and assembly of microtubules *(30,34)*, but rather appears to affect microtubule spacing plasma membrane attachment and/or tau phosphorylation *(38,39a)*.

Finally, the C-terminal tail of tau affects microtubule binding in an especially interesting manner, that is, it plays a much greater role in the binding of 3-repeat tau than 4-repeat tau *(32)*. This indicates that 3-repeat and 4-repeat tau may bind to microtubules with different conformations. The possibility of isoform specific structure-function features was first suggested by earlier competition studies in which a small peptide corresponding to the 4-repeat specific inter-repeat between repeats 1 and 2 was able to compete for binding with 4-repeat tau, but not with 3-repeat tau *(33)*.

Taken together, these studies have led to a refined model in which 3-repeat and 4-repeat tau adopt isoform-specific folded conformations on binding to microtubules *(32)*. Because structure determines function, it follows that these different tau conformations could confer different functional effects on the region of microtubule to which they are bound (*see* Fig. 3 and next subheading for more details).

3R tau   4R tau

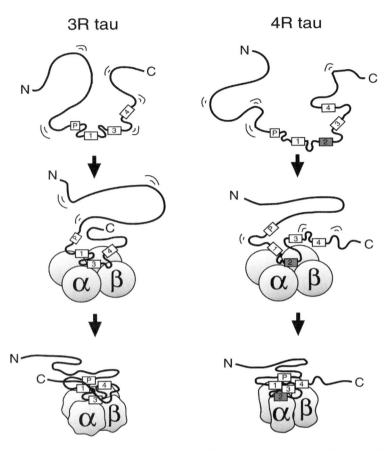

Fig. 3. Isoform-specific tau:microtubule interaction. This model, proposed in Goode et al., 2000 *(32)*, integrates an induced fit perspective along with isoform-specific action by 3-repeat and 4-repeat tau. Tau has little higher ordered structure when it is in solution, consistent with biophysical studies *(123)*. On interaction with microtubules, each tau isoform adopts a more complex and isoform-specific folded structure involving intramolecular interactions between the different core microtubule-binding domains and flanking regions, which in turn differentially influences microtubule structure and function, for example, the regulation of microtubule dynamics. Each sphere represents a α or β-tubulin monomer, as labeled. The "P" box corresponds to the proline-rich region that can regulate microtubule-binding activity indirectly.

2.2. The Tubulin-Binding Site(s) for Tau

Considerable effort has also been expended seeking to define the site(s) on tubulin that interact with tau. Although early biochemical approaches concluded that the C-terminal ends of both α- and β-tubulin are important for the tau–tubulin interaction, there remain a number of discrepancies regarding the exact locations and number of tau-binding sites on the tubulin subunits *(40–45)*. More recently, Chan et al. used a chemical crosslinking study to identify two distinct sites on each tubulin subunit, one located within the 12 C-terminal amino acids on each α- and β-tubulin polypeptide and a second "internal" site located within the C-terminal one-third of each monomer but not within the last 12 amino acids *(46)*. Their data further indicated that tau crosslinks to the C-terminal sites of α- and β-tubulin through repeat 1 and/or the adjacent inter-repeat. The crosslinking to the internal site is most likely mediated by the second repeat, i.e., repeat 2 in 4-repeat tau and repeat 3 in 3-repeat tau.

On a larger scale, biophysical approaches have also been used to address the nature of the tau–microtubule interaction. Al-Bassam et al. *(47)* compared tau decorated and control microtubules using cryo-EM and helical image analysis. Their data suggested that tau forms an ordered structure along protofilament ridges on the microtubule exterior (i.e., the highest point on each protofilament relative to the center of the microtubule, as opposed to the "valleys" between adjacent protofilaments), interacting with the carboxyl end of each tubulin subunit and perhaps bridging the tubulin interfaces to stabilize longitudinal interactions (i.e., the interactions between sequential tubulin subunits along each protofilament). In contrast, Makrides et al. *(48)* used atomic force microscopy to visualize the tau–microtubule interaction. At low molar ratios of tau:tubulin, tau oligomers were observed in ring-like structures encircling the outer surface of microtubules, perhaps suggesting stabilization of lateral protofilament–protofilament interactions. At saturating tau:tubulin ratios, tau covered the entire surface of microtubules in a regular structural pattern. Finally, Kar et al. *(49)* used three-dimensional electron cryomicroscopy and found that tau binds to the inner surface of microtubules, close to the taxol-binding site on β-tubulin. This result suggested that tau might stabilize microtubules in a similar way to taxol, although with a lower affinity allowing for reversible assembly. Indeed, tau and taxol do stabilize microtubule dynamics in quite similar manners *(50)*.

Thus, there is uncertainty concerning the number and location of tau-binding site(s) on microtubules. At least a partial resolution to this situation may have been revealed by a recent equilibrium-competition binding study, suggesting that tau can bind to two distinct sites on microtubules, one that displays reversible-binding kinetics and another with irreversible-binding kinetics *(51)*. Whether these two kinetically defined sites correspond to two distinct physical sites or two forms of a single site is yet to be determined.

3. MICROTUBULES, TAU, AND THE REGULATION OF MICROTUBULE DYNAMICS

Microtubules are involved in many essential cellular processes, including cell division, protein trafficking, and the generation and maintenance of cell shape. The highly elongated morphology of neuronal cells magnifies the importance of microtubules, which are essential for the establishment and maintenance of cell morphology, as well as for axonal transport, which moves cargo up and down the long axonal process.

Microtubules display two types of dynamic behavior: treadmilling and dynamic instability (*see* Chapter 10). Treadmilling (or flux) occurs as a result of a net addition of tubulin subunits at the plus end and a net loss of subunits at the minus end *(52–54)*. Dynamic instability is characterized by transitions between slow growth and rapid shortening at microtubule ends, and is most prominent at the plus ends of microtubules *(55–59)*.

Microtubule dynamic behavior can vary between cell types according to the role of microtubules in that cell, and can even vary from one subcellular location to another. For instance, in neurons, axonal microtubules are relatively stable and this stability is thought to be important for their function *(60)*. On the other hand, growth cones possess both stable and highly dynamic microtubules, and it has been shown that rapid microtubule dynamics are required for axon elongation *(61)*. Pharmacological studies support the idea that proper regulation of microtubule dynamics is important for cell survival. For example, studies with the microtubule-stabilizing drug taxol have shown that overstabilization of microtubules leads to cell arrest, which can then lead to apoptosis *(62,63)*. At the other extreme, elegant somatic cell genetics studies using taxol-resistant cell lines demonstrate that understabilization of microtubules can also lead to cell arrest *(64)*. These results suggest that there is a window of acceptable microtubule dynamics necessary for proper microtubule function and cell viability.

The ability of tau to stabilize microtubules has been well characterized in vitro. During conditions of net microtubule assembly in vitro, tau increases the rate of tubulin polymerization (the microtubule growth rate), decreases the rate of depolymerization (the microtubule shortening rate), and inhibits the transition from growth phase to shortening phase (the catastrophe frequency), resulting in a kinetic stabilization of microtubules *(65,66)*. Under steady-state conditions in vitro, in which there is no net polymer gain and which are likely to be more relevant to conditions within mature neurons, a similar stabilization is seen, except that tau decreases the growth rate as well as the shortening rate *(36)*. This stabilization is detectable even at very low molar ratios of tau to tubulin, far lower than those that produce an effect on microtubule assembly rates *(36)*. In terms of structure and function in vitro dynamics studies using various tau fragments agree relatively well with binding and assembly studies, indicating that the repeat region and the regions flanking the repeat region are important in stabilizing microtubules *(36,66)*. Also consistent with results from microtubule assembly assays, small peptides containing a single repeat or inter-repeat region of tau are able to regulate microtubule dynamics *(36)*.

Only a few studies have directly addressed the effects of tau on microtubule dynamics in living cells. In one study, heat-stable MAPs (consisting mainly of MAP2 and tau) were injected into non-neuronal cells and were found to stabilize microtubules in a similar manner to that seen for tau under steady state conditions in vitro *(67)*. In more recent work by Bunker et al. *(68)*, physiologically relevant concentrations of both 3-repeat and 4-repeat tau were microinjected into non-neuronal cells and the effects on the dynamic instability behavior of individual microtubules was measured by time-lapse microscopy. Both isoforms suppressed microtubule dynamics, though to different extents. Specifically, 4-repeat tau reduced the rate and extent of both growing and shortening events. In contrast, 3-repeat tau stabilized most dynamic parameters about threefold less potently than 4-repeat tau and had only a minimal ability to suppress shortening events. That 4-repeat tau has a much stronger effect on shortening events than does 3-repeat tau in cells is consistent with in vitro work by Panda et al. *(69)* and Levy et al. *(69a)*.

Taken together, the in vitro and in vivo data demonstrate clearly that tau is a potent regulator of microtubule dynamics. Further, 3-repeat and 4-repeat tau exhibit both quantitative and qualitative mechanistic differences. All other factors being equal, these differences suggest that fetal neurons expressing only 3-repeat tau should be considerably more dynamic than mature neurons expressing equal amounts of 3-repeat and 4-repeat tau. This is consistent with the notion that fetal neurons require a more plastic cytoskeleton to successfully navigate to their target than do mature, synapsed neurons. These data also have implications for tau-mediated neuronal cell death (*see* Section 4).

3.1. Regulation of Tau

Since the proper regulation of microtubule function is critical for cell survival and tau is a major regulator of microtubule function, it follows that the regulation of tau activity is likely to be equally important. There are two major mechanisms regulating tau activity: alternative RNA splicing and phosphorylation. As noted earlier, alternative RNA splicing results in the synthesis of 6-tau isoforms in the adult central nervous system that can be grouped into two categories, 3-repeat tau and 4-repeat tau (Fig. 1). Also as noted earlier, tau isoform expression is developmentally regulated, with fetal brain expressing only 3-repeat tau, whereas adult brain expresses approximately equal amounts of 3-repeat and 4-repeat tau *(28,70,71)*.

Tau is also a highly phosphorylated protein. At least 30 Ser–Thr phosphorylation sites have been identified in tau using mass spectrometry and monoclonal antibodies *(72, reviewed in ref. 2)*. The majority of these sites are proline directed (i.e., the +1 residue is a proline) and they are concentrated primarily in the regions of tau that flank the repeat region. However, there are also several KXGS sites located in the repeat region. Additionally, recent work has demonstrated that tau can be tyrosine phosphorylated as well *(73,74)*.

In the central nervous system, tau exists as a highly heterogeneous mixture of partially phosphorylated molecules (reviewed in ref. *72*). However, this phosphorylation is developmentally regulated; fetal tau has a greater average molar ratio of phosphates per tau molecule (~7) than normal adult tau (~2–3) *(75,76)*. As phosphorylation at many of these sites has been demonstrated to decrease the ability of tau to bind and assemble microtubules *(77–79)*, this developmental decrease in phosphorylation would be predicted to increase tau potency later in development. Thus, as with alternative splicing, microtubule stability would increase as development progressed. Interestingly, tau phosphorylation levels are increased in Alzheimer's disease and related dementias *(2)*. Although many kinases have been shown to phosphorylate tau in vitro, there are four kinases for which strong evidence exists for an in vivo role: the proline directed kinases GSK3ββ and CDK5, the nonproline directed kinase MARK and the nonreceptor tyrosine kinase fyn. All of these kinases coimmunoprecipitate with tau from cells, phosphorylate tau in cells, and/or have been shown to alter phosphorylation levels of tau in mouse models *(73,74,80–84)*.

4. TAU DYSFUNCTION IN NEURODEGENERATIVE DISEASE

Intraneuronal neurofibrillary tangles (NFTs) are one of the two hallmark pathological features of Alzheimer's disease, the second being extracellular amyloid plaques composed of amyloid β (Aβ). In 1986, tau was found to be the major component of NFTs found in the brains of patients with Alzheimer's disease *(4–7)*. Subsequently, "tau tangles" have been correlated with numerous additional neurodegenerative diseases, now referred to as "tauopathies," including fronto-temporal dementias, Pick's disease, and supra-nuclear palsy *(85,86)*. However, until recently, it remained unclear whether tau played a causal role in disease or whether it was simply a downstream consequence of neuronal cell death. The answer to this fundamentally important question came in 1998, when a group of disorders known as fronto-temporal dementia with Parkinsonism linked to chromosome 17 (FTDP-17) were genetically linked to multiple mutations in the tau gene. These mutations, all of which exhibited dominant phenotypes,

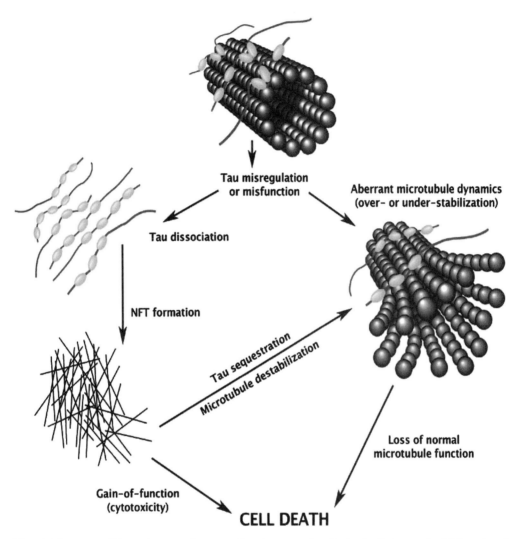

Fig. 4. Three possible pathways of tau-mediated neuronal cell death. Pathway 1—This gain-of-function model suggests that some genetic or environmental factor(s) cause tau to dissociate from microtubules and aggregate into abnormal, cytotoxic neurofibrillary tangles. Pathway 2—This dosage-effect/loss-of-function model suggests that some genetic or environmental factor(s) cause tau to improperly regulate microtubule dynamics, leading to under- or overstabilized microtubules. As a result, these improperly regulated microtubules cannot perform their normal essential cellular functions, leading to cell death. Pathway 3—This model incorporates elements of pathways 1 and 2. In this model, some genetic or environmental factor(s) cause tau to dissociate from microtubules and aggregate into abnormal neurofibrillary tangles. The sequestration of tau in these aggregates results in less tau to regulate microtubule dynamics, leading to overactive microtubules that cannot perform their essential functions, leading to cell death. To view this figure in color, see the insert and the companion CD-ROM.

demonstrated unequivocally that errors in tau action can cause neuronal cell death and dementia *(87–89)*. Interestingly, FTDP-17 patients do not exhibit Aβ plaques. These findings led to a model in which tau dysfunction is a downstream consequence of Aβ action in Alzheimer's disease, whereas tau dysfunction is the initiating event in FTDP-17, causing neuronal cell death without Aβ pathology. In both cases, errors in tau action are in the pathway leading to neuronal cell death. This model was supported by subsequent work using cultured cells from a tau knockout mouse that demonstrated that Aβ mediated toxicity depends on the presence of tau *(90)*.

Although the genetic data clearly demonstrate that errors in tau action can cause neurodegeneration, the underlying molecular mechanisms of tau dysfunction remain to be determined. There are a number of hypotheses that have been put forth, some of which are schematized in Figure 4.

1. Defects in tau cause NFTs to form, which are cytotoxic and directly cause neuronal cell death;
2. Defects in tau cause NFTs to form, sequestering tau so that it is no longer able to perform its normal function of stabilizing microtubules, leading to disruption of microtubule function and subsequent cell death;
3. Defects in tau directly decrease the ability of tau to regulate microtubule dynamics, which leads to disruption of microtubule function and cell death.

In this last model, NFT formation occurs after the cell has died (or when it is committed and well on the way to death) and is therefore not a part of the mechanism(s) causing cell death. The first of these models is a gain-of-function model, whereas the second and third are loss-of-function models. Each will be discussed further later.

4.1. Two Classes of FTDP-17 Tau Mutations

Upon the discovery of tau FTDP-17 mutations, researchers immediately began to try to determine the molecular consequences of the mutations in order to gain insight into the mechanism(s) of disease. There has now been at least 33 point mutations found in more than 80 families (reviewed in ref. *86*). As noted earlier, it is especially important to emphasize the fact that all of the mutations display genetic dominance; therefore, any proposed mechanism must account for the fact that only one mutant allele is required to cause disease. Also of interest is that whereas some FTDP-17 mutations result in changes in the amino acid sequence of tau, others are silent or intronic. Thus the mutations are divided into two categories: those that act at the protein level and those that act at the RNA level.

The mutations that act at the protein level consist of single amino acid substitution or deletion mutations. The fact that the vast majority of these mutations map to the microtubule-binding region of tau, or the flanking regions that regulate microtubule binding, immediately raises the notion of loss-of-function. Indeed, in vitro studies have found that these mutations generally lead to a decreased interaction with microtubules, as measured by a reduction in the ability of the mutant tau to bind and assemble microtubules *(87,91,92)*. A similar loss of microtubule association has been reported in transfected cells *(93,94)*. These findings suggest that the mutations could result in the misregulation of microtubule dynamics owing to a haplo-insufficiency mechanism in which both alleles must produce functional tau for proper action. However, it is also possible that the decreased tau–microtubule interaction leads to increased levels of non-microtubule bound tau in the cytosol which could be available for filament formation. In fact, some studies have found that some missense mutations (though not all) increase the propensity of tau to form filaments in vitro *(95,96)*. These NFTs could then contribute to cell death either by sequestering tau or by a direct toxic effect.

The mutations that act at the RNA level (or splicing mutations) are mainly silent or intronic mutations that alter the efficiency of exon 10 splicing *(88,97–101)*. The consequence of these mutations is that the normal adult pattern of 50% 4-repeat tau and 50% 3-repeat tau is altered, increasing the amount of 4-repeat tau relative to 3-repeat tau. Theoretically, keeping in mind that only one allele is mutated, the 4-repeat tau to 3-repeat tau molar ratio should shift to 3:1, and a shift of approximately this magnitude has been confirmed at the protein level in brain extracts from FTDP-17 patients *(87)*. The fact that neurodegeneration results from simply altering the ratio of otherwise wild-type proteins indicates that there must be important functional differences between the two isoforms, such as in their respective abilities to bind microtubules and regulate their actions *(see* Section 3). Altering the isoform ratio could therefore result in changes in microtubule behavior. Alternatively, perhaps an overabundance of one isoform could increase the levels of that isoform in the cytosol *(102)*, possibly resulting in increased incorporation of the overabundant isoform into NFTs. Indeed, it has been demonstrated that the NFTs formed in patients with splicing mutations do consist mainly of the overabundant isoform *(88,103,104)*.

4.2. Tau Transgenic Models

Tau transgenic studies in several species have also been used to gain insight into normal and pathological tau action. Transgenic mice overexpressing 4-repeat wild-type tau in neurons, recapitulating the overabundance of 4-repeat tau seen in splicing mutations, exhibit defects in neuronal function and axonal degeneration without the presence of NFTs *(105–107)*. However, the interpretation of these results is complicated by the fact that adult rodents normally express only 4-repeat tau rather than the human pattern of equal amounts of 3-repeat and 4-repeat tau. Similar studies have been performed using flies, worms, and lampreys as model systems *(108–110)*. In all these systems, tau overexpression leads to neurodegeneration, however, NFTs were detected only in lamprey neurons. These findings demonstrate that tau-mediated neuronal cell death can occur in the absence of NFTs, and favor the loss-of-function hypotheses that neurodegeneration results from the disruption of normal microtubule function rather than from cytotoxic NFTs. However, the possibility remains that accumulation of excess soluble tau in the cytoplasm is toxic. Furthermore, expression of high levels of exogenous tau on top of endogenous tau may be detrimental to cells through a mechanism independent of the actual human disease mechanism.

In addition to wild-type tau, FTDP-17 mutant tau has also been transgenically expressed in mice, *Drosophila,* and *Caenorhabditis elegans (109–114)*. The overexpression of FTDP-17 mutant tau consistently produces more severe phenotypes than the overexpression of wild-type tau. As FTDP-17 tau is expressed on top of endogenous tau in these animal models, it would be expected that microtubules should be overstabilized in all cases, and more so by wild-type than FTDP-17 tau. Thus, these results are not easily reconciled with the model that neuronal cell death results from misregulation of microtubule dynamics. However, interpretation of data from transgenic mouse models can be complicated by ill-defined compensatory mechanisms, varying levels of expression, and divergent genetic backgrounds. Nonetheless, the simplest interpretation of these data is that the mutated forms of tau are more toxic than wild-type tau, supporting the hypothesis that a toxic effect of tau causes neurodegeneration. Yet, NFTs are still not formed in most of the model systems, supporting the idea that NFTs themselves may not be a part of the pathway to cell death and disease, but rather be a byproduct of the disease process.

4.3. Loss-of-Microtubule-Function Models Vs Filament-Formation Models of Neurodegeneration

It is useful to assess how the observations and experimental data that have been accumulated regarding Alzheimer's disease and FTDP-17 fit with the models of how tau might cause neurodegeneration (Fig. 4 and Section 3.1). In these models there are two major cellular events that can precede neuronal cell death: filament formation and/or microtubule misregulation. Filament formation is central to the first two models presented, in which tau filaments are either toxic to the cell or serve to sequester tau and thereby prevent it from performing its normal regulatory function on microtubule behavior. Microtubule misregulation is central to the second and third models, in which sequestered tau fails to properly regulate microtubules, or mutations in tau directly interfere with the ability of tau to regulate microtubules. Importantly, these two ideas are not necessarily mutually exclusive, as seen in the sequestration model.

Models to which filament formation is central are supported by the fact that several mutations increase the propensity of tau to form fibers in in vitro reactions (95,96) and the fact that the number of filaments observed in the brain correlates with the severity of dementia (115). However, as just described, numerous animal models have shown that tau-mediated neurodegeneration can occur in the absence of NFTs, indicating that they are not necessary for cell death (107–111). This finding may be reconciled with the toxicity and sequestration models if there is a "pretangle" tau oligomer or misfolded form of tau that cannot bind microtubules and/or has a toxic effect before tangles are fully formed. This would be consistent with popular models for other disease causing proteins, including Aβ (116).

The models in which loss of microtubule function contributes to neurodegeneration, are supported by several lines of evidence. First, there is significant microtubule loss in Alzheimer's disease brains, both in number and length (117). Second, overexpression of tau in astrocytes decreases the number of stable microtubules in the cells, leading to disruption of the cytoskeleton, abnormal trafficking, and eventually cell death (118). Third, overexpression of tau in neurons has been shown to disrupt normal axonal transport (119,120) and it has been shown that disruption of microtubule-based axonal transport in mice is sufficient to cause age-dependent neurodegeneration (121). Moreover, the microtubule dynamics stabilizing drug paclitaxel has been shown to restore relatively normal axonal transport and proper motor function to mice exhibiting impaired motor function in a mouse tauopathy model (122). These in vivo findings, along with the finding that tau mutations decrease the affinity of tau for microtubules in vitro, support the idea that loss of microtubule function plays a key role in neurodegeneration. The strongest evidence against these models is that FTDP-17 tau is a more potent inducer of neurodegeneration than wild-type tau when expressed on top of endogenous wild-type tau in animal models (110–114), suggesting that the mechanism is not a loss-of-function mechanism. Taken together, the data do not yet clearly support one model over the others. In the final analysis, further work will be necessary to elucidate the mechanism(s) underlying tau-induced neuronal cell death in dementia (123,124).

ACKNOWLEDGMENTS

We are extremely grateful to Les Wilson and Mary Ann Jordan for many years of collaboration and innumerable valuable discussions. We also thank the members of the Wilson and Jordan labs for generously sharing their talents, reagents, and equipment with us over the years, especially Herb Miller, Kathy Kamath, and Dulal Panda, as well as the members of the Feinstein lab for too many things to list. We are also grateful to Dmitri Leonoudakis, Carol Vandenberg, and Kathy Foltz for comments on the manuscript. Finally, we also thank Maura Jess and Michelle Massie for assistance with figures. This work was supported by NIH grant NS35010 and NSF ITR grant 0331697 to SCF.

REFERENCES

1. Weingarten MD, Lockwood AH, Hwo SY, Kirschner MW. A protein factor essential for microtubule assembly. Proc Natl Acad Sci USA 1975;72:1858–1862.
2. Buee L, Bussiere T, Buee-Scherrer V, Delacourte A, Hof PR. Tau protein isoforms, phosphorylation and role in neurodegenerative disorders. Brain Res Brain Res Rev 2000;33:95–130.
3. Paglini G, Peris L, Mascotti F, Quiroga S, Caceres A. Tau protein function in axonal formation. Neurochem Res 2000;25:37–42.
4. Delacourte A, Defossez A. Alzheimer's disease: Tau proteins, the promoting factors of microtubule assembly, are major components of paired helical filaments. J Neurol Sci 1986;76:173–186.
5. Grundke-Iqbal I, Iqbal K, Quinlan M, Tung YC, Zaidi MS, Wisniewski HM. Microtubule-associated protein tau. A component of Alzheimer paired helical filaments. J Biol Chem 1986;261:6084–6089.
6. Kosik KS, Joachim CL, Selkoe DJ. Microtubule-associated protein tau (tau) is a major antigenic component of paired helical filaments in Alzheimer disease. Proc Natl Acad Sci USA 1986;83: 4044–4048.
7. Wood JG, Mirra SS, Pollock NJ, Binder LI. Neurofibrillary tangles of Alzheimer disease share antigenic determinants with the axonal microtubule-associated protein tau (tau). Proc Natl Acad Sci USA 1986;83:4040–4043.
8. Schoenfeld TA, Obar RA. Diverse distribution and function of fibrous microtubule-associated proteins in the nervous system. Int Rev Cytol 1994;151:67–137.
9. Drubin DG, Feinstein SC, Shooter EM, Kirschner MW. Nerve growth factor-induced neurite outgrowth in PC12 cells involves the coordinate induction of microtubule assembly and assembly-promoting factors. J Cell Biol 1985;101:1799–1807.
10. Caceres A, Kosik KS. Inhibition of neurite polarity by tau antisense oligonucleotides in primary cerebellar neurons. Nature 1990;343:461–463.
11. Caceres A, Potrebic S, Kosik KS. The effect of tau antisense oligonucleotides on neurite formation of cultured cerebellar macroneurons. J Neurosci 1991;11:1515–1523.
12. Hanemaaijer R, Ginzburg I. Involvement of mature tau isoforms in the stabilization of neurites in PC12 cells. J Neurosci Res 1991;30:163–171.
13. Esmaeli-Azad B, McCarty JH, Feinstein SC. Sense and antisense transfection analysis of tau function: tau influences net microtubule assembly, neurite outgrowth and neuritic stability. J Cell Sci 1994;107(pt 4):869–879.
14. Drubin DG, Kirschner MW. Tau protein function in living cells. J Cell Biol 1986;103:2739–2746.
15. Kanai Y, Takemura R, Oshima T, et al. Expression of multiple tau isoforms and microtubule bundle formation in fibroblasts transfected with a single tau cDNA. J Cell Biol 1989;109:1173–1184.
16. Baas PW, Pienkowski TP, Kosik KS. Processes induced by tau expression in Sf9 cells have an axon-like microtubule organization. J Cell Biol 1991;115:1333–1344.
17. Knops J, Kosik KS, Lee G, Pardee JD, Cohen-Gould L, McConlogue L. Overexpression of tau in a nonneuronal cell induces long cellular processes. J Cell Biol 1991;114:725–733.
18. Lee G, Rook SL. Expression of tau protein in non-neuronal cells: microtubule binding and stabilization. J Cell Sci 1992;102(pt 2):227–237.
19. Takemura R, Okabe S, Umeyama T, Kanai Y, Cowan NJ, Hirokawa N. Increased microtubule stability and α tubulin acetylation in cells transfected with microtubule-associated proteins MAP1B, MAP2 or tau. J Cell Sci 1992;103(pt 4):953–964.

20. Harada A, Oguchi K, Okabe S, et al. Altered microtubule organization in small-calibre axons of mice lacking tau protein. Nature 1994;369:488–491.
21. Tint I, Slaughter T, Fischer I, Black MM. Acute inactivation of tau has no effect on dynamics of microtubules in growing axons of cultured sympathetic neurons. J Neurosci 1998;18:8660–8673.
22. Dawson HN, Ferreira A, Eyster MV, Ghoshal N, Binder LI, Vitek MP. Inhibition of neuronal maturation in primary hippocampal neurons from tau deficient mice. J Cell Sci 2001;114:1179–1187.
23. Takei Y, Teng J, Harada A, Hirokawa N. Defects in axonal elongation and neuronal migration in mice with disrupted tau and map1b genes. J Cell Biol 2000;150:989–1000.
24. Lewis SA, Wang DH, Cowan NJ. Microtubule-associated protein MAP2 shares a microtubule binding motif with tau protein. Science 1988;242:936–939.
25. Aizawa H, Kawasaki H, Murofushi H, et al. A common amino acid sequence in 190-kDa microtubule-associated protein and tau for the promotion of microtubule assembly. J Biol Chem 1989;264:5885–5890.
26. Lee G, Cowan N, Kirschner M. The primary structure and heterogeneity of tau protein from mouse brain. Science 1988;239:285–288.
27. Goedert M, Spillantini MG, Jakes R, Rutherford D, Crowther RA. Multiple isoforms of human microtubule-associated protein tau: sequences and localization in neurofibrillary tangles of Alzheimer's disease. Neuron 1989;3:519–526.
28. Himmler A. Structure of the bovine tau gene: alternatively spliced transcripts generate a protein family. Mol Cell Biol 1989;9:1389–1396.
29. Lee G, Neve RL, Kosik KS. The microtubule binding domain of tau protein. Neuron 1989;2:1615–1624.
30. Butner KA, Kirschner MW. Tau protein binds to microtubules through a flexible array of distributed weak sites. J Cell Biol 115, 1991;115:717–730.
30a. Trinczek B, Biernat J, Baumann K, Mandelkow EM, Mandelkow E. Domains of tau protein, differential phosphorylation and dynamic instability of microtubules. Mol Bio Cell 1995;6:1887–1902.
31. Goedert M, Jakes R. Expression of separate isoforms of human tau protein: correlation with the tau pattern in brain and effects on tubulin polymerization. EMBO J 1990;9:4225–4230.
32. Goode BL, Chau M, Denis PE, Feinstein SC. Structural and functional differences between 3-repeat and 4-repeat tau isoforms. Implications for normal tau function and the onset of neurodegenetative disease. J Biol Chem 2000;275:38,182–38,189.
33. Goode BL, Feinstein SC. Identification of a novel microtubule binding and assembly domain in the developmentally regulated inter-repeat region of tau. J Cell Biol 1994;124:769–782.
34. Gustke N, Trinczek B, Biernat J, Mandelkow EM, Mandelkow E. Domains of tau protein and interactions with microtubules. Biochemistry 1994;33:9511–9522.
35. Ennulat DJ, Liem RK, Hashim GA, Shelanski ML. Two separate 18-amino acid domains of tau promote the polymerization of tubulin. J Biol Chem 1989;264:5327–5330.
36. Panda D, Goode BL, Feinstein SC, Wilson L. Kinetic stabilization of microtubule dynamics at steady state by tau and microtubule-binding domains of tau. Biochemistry 1995;34:11,117–11,127.
37. Goode BL, Denis PE, Panda D, et al. Functional interactions between the proline-rich and repeat regions of tau enhance microtubule binding and assembly. Mol Biol Cell 1997;8:353–365.
38. Chen J, Kanai Y, Cowan NJ, Hirokawa N. Projection domains of MAP2 and tau determine spacings between microtubules in dendrites and axons. Nature 1992;360:674–677.
39. Brandt R, Leger J, Lee G. Interaction of tau with the neural plasma membrane mediated by tau's amino-terminal projection domain. J Cell Biol 1995;131:1327–1340.
39a. Sengupta A, Novak M, Grundke-Iqbal K. Regulation of phosphorylation of tau by cyclin-dependent kinase 5 and glycogen synthase kinase-3 at substrate level. FEBS Lett 2006;58:5925–5933.
40. Serrano L, Montejo de Garcini E, Hernandez MA, Avila J. Localization of the tubulin binding site for tau protein. Eur J Biochem 1985;153:595–600.
41. Littauer UZ, Giveon D, Thierauf M, Ginzburg I, Ponstingl H. Common and distinct tubulin binding sites for microtubule-associated proteins. Proc Natl Acad Sci USA 1986;83:7162–7166.
42. Maccioni RB, Rivas CI, Vera JC. Differential interaction of synthetic peptides from the carboxyl-terminal regulatory domain of tubulin with microtubule-associated proteins. EMBO J 1988;7:1957–1963.
43. Vera JC, Rivas CI, Maccioni RB. Antibodies to synthetic peptides from the tubulin regulatory domain interact with tubulin and microtubules. Proc Natl Acad Sci USA 1988;85:6763–6767.

44. Hagiwara H, Yorifuji H, Sato-Yoshitake R, Hirokawa N. Competition between motor molecules (kinesin and cytoplasmic dynein) and fibrous microtubule-associated proteins in binding to microtubules. J Biol Chem 1994;269:3581–3589.

45. Saoudi Y, Paintrand I, Multigner L, Job D. Stabilization and bundling of subtilisin-treated microtubules induced by microtubule associated proteins. J Cell Sci 1995;108(pt 1):357–367.

46. Chau MF, Radeke MJ, de Ines C, Barasoain I, Kohlstaedt LA, Feinstein SC. The microtubule-associated protein tau cross-links to two distinct sites on each α and β tubulin monomer via separate domains. Biochemistry 1998;37:17,692–17,703.

47. Al-Bassam J, Ozer RS, Safer D, et al. MAP2 and tau bind longitudinally along the outer ridges of microtubule protofilaments. J Cell Biol 2002;157:1187–1196.

48. Makrides V, Shen TE, Bhatia R, et al. Microtubule-dependent oligomerization of tau. Implications for physiological tau function and tauopathies. J Biol Chem 2003;278:33,298–33,304.

49. Kar S, Fan J, Smith MJ, Goedert M, Amos LA. Repeat motifs of tau bind to the insides of microtubules in the absence of taxol. EMBO J 2003;22:70–77.

50. Derry WB, Wilson L, Jordan MA. Substoichiometric binding of taxol suppresses microtubule dynamics. Biochemistry 1995;34:2203–2211.

51. Makrides V, Massie MR, Feinstein SC, Lew J. Evidence for two distinct binding sites for tau on microtubules. Proc Natl Acad Sci USA 2004;101:6746–6751.

52. Margolis RL, Wilson L. Opposite end assembly and disassembly of microtubules at steady state in vitro. Cell 1978;13:1–8.

53. Rodionov VI, Borisy GG. Microtubule treadmilling in vivo. Science 1997;275:215–218.

54. Margolis RL, Wilson L. Microtubule treadmilling: what goes around comes around. Bioassays 1998;20:830–836.

55. Horio T, Hotani H. Visualization of the dynamic instability of individual microtubules by dark-field microscopy. Nature 1986;321:605–607.

56. Kirschner MW, Mitchison T. Microtubule dynamics. Nature 1986;324:621.

57. Hotani H, Horio T. Dynamics of microtubules visualized by darkfield microscopy: treadmilling and dynamic instability. Cell Motility Cytoskeleton 1988;10:229–236.

58. Sammak PJ, Borisy GG. Direct observation of microtubule dynamics in living cells. Nature 1988;332:724–726.

59. Walker RA, O'Brien ET, Pryer NK, et al. Dynamic instability of individual microtubules analyzed by video light microscopy: rate constants and transition frequencies. J Cell Biol 1988;107:1437–1448.

60. Okabe S, Hirokawa N. Rapid turnover of microtubule-associated protein MAP2 in the axon revealed by microinjection of biotinylated MAP2 into cultured neurons. Proc Natl Acad Sci USA 1989;86:4127–4131.

61. Tanaka E, Ho T, Kirschner MW. The role of microtubule dynamics in growth cone motility and axonal growth. J Cell Biol 1995;128:139–155.

62. Jordan MA, Wendell K, Gardiner S, Derry WB, Copp H, Wilson L. Mitotic block induced in HeLa cells by low concentrations of paclitaxel (Taxol) results in abnormal mitotic exit and apoptotic cell death. Cancer Res 1996;56:816–825.

63. Yvon AM, Wadsworth P, Jordan MA. Taxol suppresses dynamics of individual microtubules in living human tumor cells. Mol Biol Cell 1999;10:947–959.

64. Goncalves A, Braguer D, Kamath K, et al. Resistance to Taxol in lung cancer cells associated with increased microtubule dynamics. Proc Natl Acad Sci USA 2001;98:11,737–11,742.

65. Drechsel DN, Hyman AA, Cobb MH, Kirschner MW. Modulation of the dynamic instability of tubulin assembly by the microtubule-associated protein tau. Mol Biol Cell 1992;3:1141–1154.

66. Trinczek B, Biernat J, Baumann K, Mandelkow EM, Mandelkow E. Domains of tau protein, differential phosphorylation, and dynamic instability of microtubules. Mol Biol Cell 1995;6:1887–1902.

67. Dhamodharan R, Wadsworth P. Modulation of microtubule dynamic instability in vivo by brain microtubule associated proteins. J Cell Sci 1995;108(pt 4):1679–1689.

68. Bunker JM, Wilson L, Jordan MA, Feinstein SC. Modulation of microtubule dynamics by tau in living cells: implications for development and neurodegeneration. Mol Biol Cell 2004;15:2720–2728.

69. Panda D, Samuel J, Massie M, Feinstein SC, Wilson L. Differential regulation of microtubule dynamics by 3-repeat and 4-repeat tau: Implications for the onset of neurodegenerative disease. Proc Natl Acad Sci USA 2003; in press.

69a. Levy S, Le Boeuf A, Massie MR, Jordan MA, Wilson L, Feinstein SC. Three- and four- repeat tau regulate the dynamic instability of two distinct microtubule subpopulations in qualitatively different manners–implications for neurodegeneration. J Biol Chem 2005;280:13520–13528.

70. Himmler A, Drechsel D, Kirschner MW, Martin DW Jr. Tau consists of a set of proteins with repeated C-terminal microtubule-binding domains and variable N-terminal domains. Mol Cell Biol 1989;9:1381–1388.

71. Kosik KS, Orecchio LD, Bakalis S, Neve RL. Developmentally regulated expression of specific tau sequences. Neuron 1989;2:1389–1397.

72. Lovestone S, Reynolds CH. The phosphorylation of tau: a critical stage in neurodevelopment and neurodegenerative processes. Neuroscience 1997;78:309–324.

73. Lee G, Newman ST, Gard DL, Band H, Panchamoorthy G. Tau interacts with src-family non-receptor tyrosine kinases. J Cell Sci 1998;111(pt 21):3167–3177.

74. Lee G, Thangavel R, Sharma VM, et al. Phosphorylation of tau by fyn: implications for Alzheimer's disease. J Neurosci 2004;24:2304–2312.

75. Kenessey A, Yen SH. The extent of phosphorylation of fetal tau is comparable to that of PHF-tau from Alzheimer paired helical filaments. Brain Res 1993;629:40–46.

76. Kopke E, Tung YC, Shaikh S, Alonso AC, Iqbal K, Grundke-Iqbal I. Microtubule-associated protein tau. Abnormal phosphorylation of a non-paired helical filament pool in Alzheimer disease. J Biol Chem 1993;268:24,374–24,384.

77. Lindwall G, Cole RD. Phosphorylation affects the ability of tau protein to promote microtubule assembly. J Biol Chem 1984;259:5301–5305.

78. Gustke N, Steiner B, Mandelkow EM, et al. The Alzheimer-like phosphorylation of tau protein reduces microtubule binding and involves Ser-Pro and Thr-Pro motifs. FEBS Lett 1992;307:199–205.

79. Bramblett GT, Goedert M, Jakes R, et al. Abnormal tau phosphorylation at Ser396 in Alzheimer's disease recapitulates development and contributes to reduced microtubule binding. Neuron 1993;10:1089–1099.

80. Wagner U, Utton M, Gallo JM, Miller CC. Cellular phosphorylation of tau by GSK-3β influences tau binding to microtubules and microtubule organisation. J Cell Sci 1996;109(pt 6):1537–1543.

81. Michel G, Mercken M, Murayama M, et al. Characterization of tau phosphorylation in glycogen synthase kinase-3beta and cyclin dependent kinase-5 activator (p23) transfected cells. Biochem Biophys Acta 1998;1380:177–182.

82. Ahlijanian MK, Barrezueta NX, Williams RD, et al. Hyperphosphorylated tau and neurofilament and cytoskeletal disruptions in mice overexpressing human p25, an activator of cdk5. Proc Natl Acad Sci USA 2000;97:2910–2915.

83. Spittaels K, Van den Haute C, Van Dorpe J, et al. Glycogen synthase kinase-3beta phosphorylates protein tau and rescues the axonopathy in the central nervous system of human four-repeat tau transgenic mice. J Biol Chem 2000;275:41,340–41,349.

84. Biernat J, Wu YZ, Timm T, et al. Protein kinase MARK/PAR-1 is required for neurite outgrowth and establishment of neuronal polarity. Mol Biol Cell 2002;13:4013–4028.

85. Lee VM, Goedert M, Trojanowski JQ. Neurodegenerative tauopathies. Annu Rev Neurosci 2001;1121–1159 (Review).

86. Goedert M, Jakes R. Mutations causing neurodegenerative tauopathies. Biochem Biophys Acta 2005;1739(2–3):240–250.

87. Hong M, Zhukareva V, Vogelsberg-Ragaglia V, et al. Mutation-specific functional impairments in distinct tau isoforms of hereditary FTDP-17. Science 1998;282:1914–1917.

88. Hutton M, Lendon CL, Rizzu P, et al. Association of missense and 5'-splice-site mutations in tau with the inherited dementia FTDP-17. Nature 1998;393:702–705.

89. Spillantini MG, Crowther RA, Kamphorst W, Heutink P, van Swieten JC. Tau pathology in two Dutch families with mutations in the microtubule-binding region of tau. Am J Pathol 1998;153:1359–1363.

90. Rapoport M, Dawson HN, Binder LI, Vitek MP, Ferreira A. Tau is essential to β-amyloid-induced neurotoxicity. Proc Natl Acad Sci USA 2002;99:6364–6369.

91. Hasegawa M, Smith MJ, Goedert M. Tau proteins with FTDP-17 mutations have a reduced ability to promote microtubule assembly. FEBS Lett 1998;437:207–210.

92. DeTure M, Ko LW, Yen S, et al. Missense tau mutations identified in FTDP-17 have a small effect on tau-microtubule interactions. Brain Res 2000;853:5–14.

93. Frappier T, Liang NS, Brown K, et al. Abnormal microtubule packing in processes of SF9 cells expressing the FTDP-17 V337M tau mutation. FEBS Lett 1999;455:262–266.

94. Nagiec EW, Sampson KE, Abraham I. Mutated tau binds less avidly to microtubules than wild-type tau in living cells. J Neurosci Res 2001;63:268–275.
95. Arrasate M, Perez M, Armas-Portela R, Avila J. Polymerization of tau peptides into fibrillar structures. The effect of FTDP-17 mutations. FEBS Lett 1999;446:199–202.
96. Nacharaju P, Lewis J, Easson C, et al. Accelerated filament formation from tau protein with specific FTDP-17 missense mutations. FEBS Lett 1999;447:195–199.
97. D'Souza I, Poorkaj P, Hong M, et al. Missense and silent tau gene mutations cause frontotemporal dementia with parkinsonism-chromosome 17 type, by affecting multiple alternative RNA splicing regulatory elements. Proc Natl Acad Sci USA 1999;96:5598–5603.
98. Grover A, Houlden H, Baker M, et al. 5′ splice site mutations in tau associated with the inherited dementia FTDP-17 affect a stem-loop structure that regulates alternative splicing of exon 10. J Biol Chem 1999;274:15,134–15,143.
99. Hasegawa M, Smith MJ, Iijima M, Tabira T, Goedert M. FTDP-17 mutations N279K and S305N in tau produce increased splicing of exon 10. FEBS Lett 1999;443:93–96.
100. Varani L, Hasegawa M, Spillantini MG, et al. Structure of tau exon 10 splicing regulatory element RNA and destabilization by mutations of frontotemporal dementia and parkinsonism linked to chromosome 17. Proc Natl Acad Sci USA 1999;96:8229–8234.
101. Jiang Z, Cote J, Kwon JM, Goate AM, Wu JY. Aberrant splicing of tau pre-mRNA caused by intronic mutations associated with the inherited dementia frontotemporal dementia with parkinsonism linked to chromosome 17. Mol Cell Biol 2000;20:4036–4048.
102. Lu M, Kosik KS. Competition for microtubule-binding with dual expression of tau missense and splice isoforms. Mol Biol Cell 2001;12:171–184.
103. Clark LN, Poorkaj P, Wszolek Z, et al. Pathogenic implications of mutations in the tau gene in pallido-ponto-nigral degeneration and related neurodegenerative disorders linked to chromosome 17. Proc Natl Acad Sci USA 1998;95:13,103–13,107.
104. Spillantini MG, Murrell JR, Goedert M, Farlow MR, Klug A, Ghetti B. Mutation in the tau gene in familial multiple system tauopathy with presenile dementia. Proc Natl Acad Sci USA 1998;95: 7737–7741.
105. Spittaels K, Van den Haute C, Van Dorpe J, et al. Prominent axonopathy in the brain and spinal cord of transgenic mice overexpressing four-repeat human tau protein. Am J Pathol 1999;155:2153–2165.
106. Probst A, Gotz J, Wiederhold KH, et al. Axonopathy and amyotrophy in mice transgenic for human four-repeat tau protein. Acta Neuropathol (Berl) 2000;99:469–481.
107. Santa Cruz K, Lewis J, Spires T, et al. Tau suppression in a neurodegenerative mouse model improves memory function. Science. 2005;309:476–481.
108. Hall GF, Chu B, Lee G, Yao J. Human tau filaments induce microtubule and synapse loss in an in vivo model of neurofibrillary degenerative disease. J Cell Sci 2000;113(pt 8):1373–1387.
109. Wittmann CW, Wszolek MF, Shulman JM, et al. Tauopathy in Drosophila: neurodegeneration without neurofibrillary tangles. Science 2001;293:711–714.
110. Kraemer BC, Zhang B, Leverenz JB, Thomas JH, Trojanowski JQ, Schellenberg GD. Neurodegeneration and defective neurotransmission in a Caenorhabditis elegans model of tauopathy. Proc Natl Acad Sci USA 2003;100:9980–9985.
111. Santacruz K, Lewis J, Spires T, et al. Tau suppression in a neurodegenerative mouse model improves memory function. Science 2005;309:476–481.
112. Lewis J, McGowan E, Rockwood J, et al. Neurofibrillary tangles, amyotrophy and progressive motor disturbance in mice expressing mutant (P301L) tau protein. Nat Genet 2000;25:402–405.
113. Gotz J, Chen F, Barmettler R, Nitsch RM. Tau filament formation in transgenic mice expressing P301L tau. J Biol Chem 2001;276:529–534.
114. Tanemura K, Murayama M, Akagi T, et al. Neurodegeneration with tau accumulation in a transgenic mouse expressing V337M human tau. J Neurosci 2002;22:133–141.
115. Arriagada PV, Growdon JH, Hedley-Whyte ET, Hyman BT. Neurofibrillary tangles but not senile plaques parallel duration and severity of Alzheimer's disease. Neurology 1992;42:631–639.
116. Walsh DM, Klyubin I, Fadeeva JV, et al. Naturally secreted oligomers of amyloid beta protein potently inhibit hippocampal long-term potentiation in vivo. Nature 2002;416:535–539.
117. Cash AD, Aliev G, Siedlak SL, et al. Microtubule reduction in Alzheimer's disease and aging is independent of tau filament formation. Am J Pathol 2003;162:1623–1627.
118. Yoshiyama Y, Zhang B, Bruce J, Trojanowski JQ, Lee VM. Reduction of detyrosinated microtubules and golgi fragmentation are linked to tau-induced degeneration in astrocytes. J Neurosci 2003;23: 10,662–10,671.

119. Trinczek B, Ebneth A, Mandelkow EM, Mandelkow E. Tau regulates the attachment/detachment but not the speed of motors in microtubule-dependent transport of single vesicles and organelles. J Cell Sci 1999;112(pt 14):2355–2367.
120. Stamer K, Vogel R, Thies E, Mandelkow E, Mandelkow EM. Tau blocks traffic of organelles, neurofilaments, and APP vesicles in neurons and enhances oxidative stress. J Cell Biol 2002;156:1051–1063.
121. LaMonte BH, Wallace KE, Holloway BA, et al. Disruption of dynein/dynactin inhibits axonal transport in motor neurons causing late-onset progressive degeneration. Neuron 2002;34:715–727.
122. Zhang B, Maiti A, Shively S, et al. Microtubule-binding drugs oofset tau sequestration by stabilizing microtubules and reversing fast axonal transport deficits in a tauopathy model. Proc Nat Acad Sci USA 2005;102:227–231.
123. Schweers O, Schonbrunn-Hanebeck E, Marx A, Mandelkow E. Structural studies of tau protein and Alzheimer paired helical filaments show no evidence for β-structure.J Biol Chem 1994;269: 24,290–24,297.
124. Drewes G, Ebneth A, Preuss U, Mandelkow EM, Mandelkow E. MARK, a novel family of protein kinases that phosphorylate microtubule-associated proteins and trigger microtubule disruption. Cell 1997;89:297–308.

INDEX